ORGANIC REACTION MECHANISMS · 1967

ORGANIC REACTION MECHANISMS · 1967

An annual survey covering the literature dated December 1966 *through November* 1967

B. CAPON University of Leicester

M. J. PERKINS King's College, University of London

C. W. REES University of Leicester

INTERSCIENCE PUBLISHERS a division of
John Wiley & Sons London · New York · Sydney

Library of Congress Catalog Card Number 66-23143
SBN 470 13477 1

Printed in Great Britain by
Spottiswoode, Ballantyne & Co. Ltd., London and Colchester

Preface

This book is a survey of the work on organic reaction mechanisms published in 1967. For convenience, the literature dated from December 1966 to November 1967, inclusive, was actually covered. The principal aim has been to scan all the chemical literature and to summarize the progress of work on organic reaction mechanism generally and fairly uniformly, and not just on selected topics. Therefore, certain of the sections are somewhat fragmentary and all are concise. Of the 3000 or so papers which have been reported, those which seemed at the time to be the more significant are normally described and discussed, and the remainder are listed.

Our other major aim, second only to comprehensive coverage, has been early publication since we felt that the immediate value of such a survey as this, that of "current awareness", would diminish rapidly with time. In this we have been fortunate to have the expert cooperation of the English office of John Wiley and Sons.

If this book proves to be generally useful, we will continue these annual surveys, and then hope that the series will have some lasting value; some form of cumulative reporting or indexing may even be desirable.

It is not easy to deal rigidly and comprehensively with so ubiquitous and fundamental a subject as reaction mechanism. Any subdivision is a necessary encumbrance and our system, exemplified by the chapter headings, has been supplemented by cross-references and by the form of the subject index. We should welcome suggestions for improvements in future volumes.

March 1968

B.C.
M.J.P.
C.W.R.

Contents

CHAPTER 1

Classical and Non-classical Carbonium Ions[1, 2]

Bicyclic Systems

It has been suggested that torsional strain is a factor contributing to the faster solvolysis of tertiary *exo*-2-norbornyl derivatives than of their *endo*-isomers and to the greater tendency of tertiary 2-norbornyl cations to react with nucleophiles from the *exo*- than from the *endo*-direction.[3] In the initial state for the *exo*-isomer (**1**), the $C_{(2)}$–X and $C_{(1)}$–H bonds are partially eclipsed and the resulting

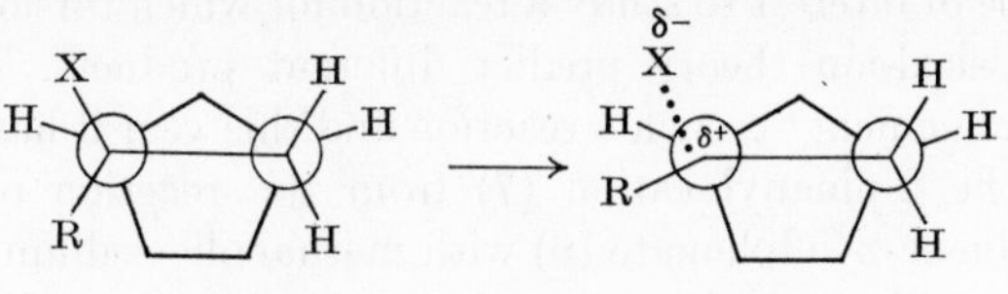

(**1**) (**2**)

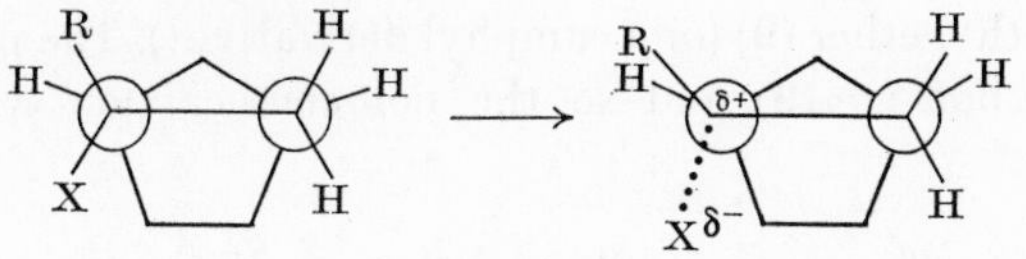

(**3**)

torsional strain will therefore be reduced in the transition state (**2**), when the $C_{(2)}$–X bond is partly broken. Also the *endo*-R group will have moved away from the *endo*-6-hydrogen atom and this movement relieves the partial eclipsing of the $C_{(2)}$–R bond with the $C_{(1)}$–$C_{(6)}$-bond. On the other hand, as the $C_{(2)}$–X bond of the *endo*-isomer (**3**) starts to break the dihedral angle between the $C_{(2)}$–R and the $C_{(1)}$–H bond is decreased and there should be an increase in torsional strain. Also the *endo*-C–X bond moves inward and there may be an increase in non-bonding interactions between X and the *endo*-6-hydrogen atom, as envisaged by Brown.[4] It was suggested that these effects together should be great enough to explain the *exo*:*endo*-rate ratio of *ca.* 500 found with tertiary norbornyl derivatives, and by the principle of microscopic reversibility they should also explain the high *exo*:*endo*-product ratios. Whether or not they will explain these ratios for secondary norbornyl derivatives was, however, not discussed.

1 D. Bethell and V. Gold, "Carbonium Ions—An Introduction", Academic Press, London, 1967.
2 H. C. Brown, *Chem. Eng. News*, 13th February, 1967, **45**, No. 7, p. 87.
3 P. von R. Schleyer, *J. Am. Chem. Soc.*, **89**, 701 (1967).
4 See *Organic Reaction Mechanisms*, **1965**, 10.

Torsional effects were also invoked[5] to explain the preferred *exo*- over *endo*-migration of groups from position 3 to position 2 in 2-norbornyl cations.[6] It was suggested that in the transition state (**4**) for *exo*-migration the arrangement around bonds $C_{(1)}$–$C_{(2)}$ and $C_{(3)}$–$C_{(4)}$ are almost ideally skewed, but in that for *endo*-migration (**5**) they are almost exactly eclipsed, and that this could lead to a difference in their energies of up to 6 kcal mole^{-1}.

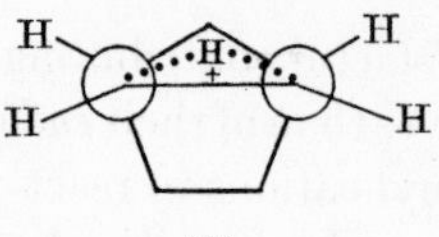

(**4**)

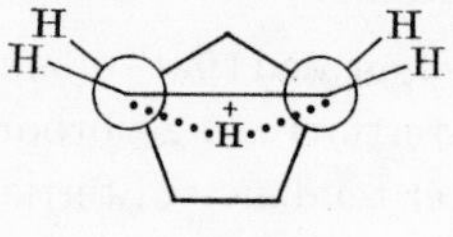

(**5**)

It would be of interest to study a reaction for which torsional-strain theory and non-classical-ion theory predict different products. The 2-pinanyl → bornyl rearrangement[7] is such a reaction and this year it has been studied by generating the 2-pinanyl cation (**7**) from the reaction of *cis*(and *trans*)-myrtanyl toluene-*p*-sulphonate (**6**) with methanolic sodium methoxide. This could rearrange through either non-classical ion (**10**) or classical ion (**8**). The former would be expected to yield bornyl methyl ether (**11**) and the latter isobornyl methyl ether (**9**) (or a camphyl derivative?). The product contained 2% of (**11**) but no (**9**) and so the non-classical ion was the preferred intermediate.[8]

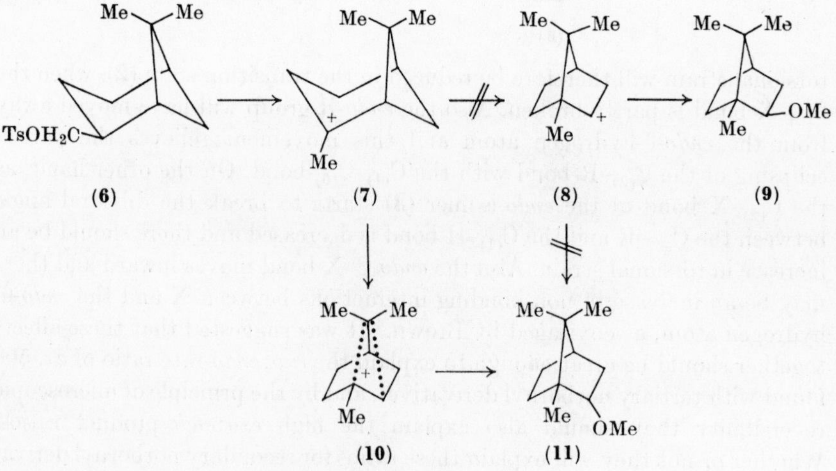

[5] P. von R. Schleyer, *J. Am. Chem. Soc.*, **89**, 699 (1967).

[6] See *Organic Reaction Mechanisms*, **1965**, 23; **1966**, 7—9.

[7] See J. A. Berson in "Molecular Rearrangements", P. de Mayo, ed., Interscience, New York, N.Y., 1963, p. 185; W. Hückel and D. Holzwarth, *Ann. Chem.*, **697**, 69 (1966); H. Schmidt, M. Mühlstädt, and P. Son, *Chem. Ber.*, **99**, 2736 (1966).

[8] J. R. Salmon and D. Whittaker, *Chem. Commun.*, **1967**, 491.

Further discussions of torsional effects are given in references 9 and 16.

Steric hindrance to ionization, another factor which may be important in determining the relative rates of solvolysis of *exo*- and *endo*-norbornyl derivatives,[10] has also received considerable attention this year. Two difficulties in accepting this as an important factor have been that no *bona-fide* examples of a magnitude comparable to that necessary to explain the *exo*:*endo*-norbornyl rate ratio are known, and that substitution at the *endo*-6-position, which at first sight should strongly increase any steric hindrance to ionization, has a relatively small effect on the solvolysis of *endo*-norbornyl derivatives (cf. **12**, **13**, and **14**).[11] In answer to the second point it has been noted that substitution

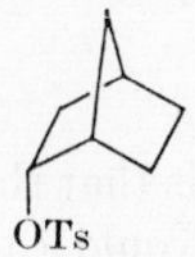

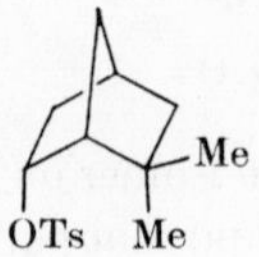

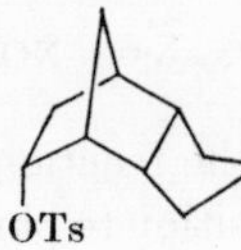

	(**12**)	(**13**)	(**14**)
Rel. rates of acetolysis at 25°	1.00	0.054	0.10
$\nu_{C=O}$ for ketone (cm^{-1})	1751	1746	1743
Calc. rel. rate	1	420	1000

at the *endo*-6-position should also increase the initial-state energy of an *endo*-norbornyl derivative and this may largely compensate for any increase in the transition-state energy. In support of this view, calculation of the relative rates of solvolysis of compounds (**13**) and (**14**) by means of the Foote–Schleyer equation (based on $\nu_{C=O}$ for corresponding ketones), and with the assumption of no non-bonding interaction between the leaving group and norbornyl skeleton in the transition state, leads to much larger values than were observed. The relatively small rate decreases found therefore represent substantial effects when compared with the calculated rates.[12]

Several striking *bona-fide* examples of steric hindrance to ionization have also been reported. Brown and his co-workers have studied *inter alia* the hydrolysis of compounds (**15**)—(**18**) in 80% aqueous acetone. The rate decrease found with the *endo*-compound (**15**) [compared with (**16**)][13] is

9 C. F. Wilcox and R. G. Jesaitis, *Tetrahedron Letters*, **1967**, 2567; *Chem. Commun.*, **1967**, 1046.
10 See *Organic Reaction Mechanisms*, **1965**, 10.
11 Cf. S. Winstein, *J. Am. Chem. Soc.*, **87**, 381 (1965).
12 H. C. Brown, I. Rothberg, P. von R. Schleyer, M. M. Donaldson, and J. J. Harper, *Proc. Natl. Acad. Sci. U.S.*, **56**, 1653 (1966).
13 H. C. Brown and W. J. Hammar, *J. Am. Chem. Soc.*, **89**, 6378 (1967).

only 17-fold but for (**17**) [compared with (**18**)] it is 4300-fold.[14] Both systems yield unrearranged products and the product of substitution from (**17**) and (**18**) is the all-*exo*-alcohol. It was thought unlikely that the high *exo*:*endo* rate ratio with (**18**) and (**17**) would be the result of participa-

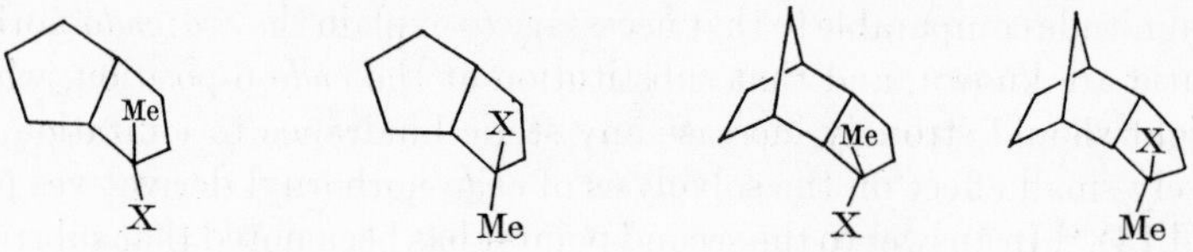

(**15**) (**16**) (**17**) (**18**)

X = *p*-$NO_2 \cdot C_6H_4 \cdot CO \cdot O$

tion in the reaction of the *exo*-isomer on the grounds that the 4,5-bond is in a poor position to participate, and that participation would lead to a very highly strained tetracyclic cation. Also, as mentioned above, no rearranged product is formed. It was considered that torsional effects would not be important (see ref. 16 below, however) and the simplest explanation is therefore that hydrolysis of the *endo*-isomer is slow as a result of steric hindrance to ionization. The formation of the *exo*-alcohol as the sole product of substitution is particularly striking and shows that attack on the intermediate carbonium ion from the *endo*-direction is highly hindered. Similar effects are also important in several other reactions of derivatives of this system (e.g., additions to the corresponding olefin and ketone) which proceed with predominant attack from the *exo*-face.[15] Brown and his co-workers suggest that steric hindrance to ionization and attack from the *endo*-direction in norbornyl derivatives is intermediate between that found in the 2-methylbicyclo[3.3.0]oct-2-yl and 8-methyl-*endo*-5,6-trimethylene-8-norbornyl systems just discussed, and that it is "necessary to recognize steric effects as a factor" in determining the *exo*:*endo*-rate ratios.

The acetolysis of the secondary compounds (**19**) and (**20**) yielded an *exo*:*endo*-rate ratio of only 6. Interpretation is complicated, however, by the much greater initial-state energy of the *endo*-isomer and, when this is taken into account, the Foote–Schleyer equation predicts that it should react about 100 times faster than its *exo*-isomer. Hence steric strain in the transition state or some other factor must be important here also.[12]

The S_N2 reaction of the *endo*-isomer (**19**) with potassium iodide in acetone is also slow with a rate 857 times less than that of cyclopentyl toluene-*p*-sulphonate and 106 times less than that of its 9-*endo*-isomer. Important factors

[14] H. C. Brown, I. Rothberg, and D. L. Vander Jagt, *J. Am. Chem. Soc.*, **89**, 6380 (1967).
[15] H. C. Brown, W. J. Hammar, J. H. Kawakami, I. Rothberg, and D. L. Vander Jagt, *J. Am. Chem. Soc.*, **89**, 6382 (1967).

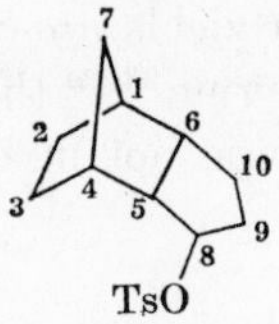

(19)

(20)

in causing this slow rate may be steric hindrance to departure of the toluene-*p*-sulphonyloxy-group and torsional strain. If $C_{(9)}$ of (**19**) is flexed downwards in the initial state, there will be a decrease in the dihedral angle between the *exo*-hydrogen at $C_{(8)}$ and the *exo*-hydrogens at $C_{(5)}$ and $C_{(9)}$ on going to the transition state, and hence an increase in torsional strain. Presumably this factor would also be important in the solvolysis of (**19**) and in that of (**17**) as well.[16]

Other examples of steric hindrance to ionization have been reported in references 17—19.

In contrast to the above behaviour, *endo*-3-chlorobicyclo[3.2.1]octane (**22**) is solvolysed 263 times faster than its *exo*-isomer (**21**) in 80% aqueous alcohol at 130°, probably as a result of steric acceleration.[20] The *endo*:*exo* ratio for the acetolysis of the corresponding toluene-*p*-sulphonates at 25° is only 10.[21]

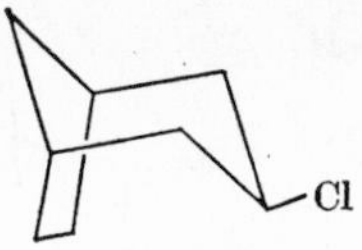

(21)

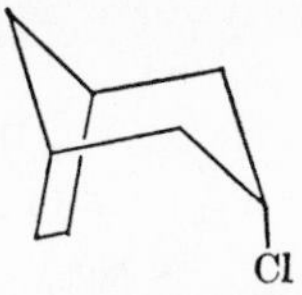

(22)

Several examples have been reported of electrophilic additions to unsaturated norbornyl systems which, although yielding *exo*-products exclusively, cannot have proceeded *via* non-classical ions. Thus 1-[$^2H_3$]methyl-2-methylenenorbornane (**23**), on careful treatment with HCl in ether at 0°, yields 1,2-dimethyl-*exo*-norbornyl chloride in which the deuterium label is only about 50% scrambled.[22] This reaction cannot then have proceeded wholly *via* a non-classical ion as shown in equation (1). 7,7-Dimethyl-2-

[16] I. Rothberg and R. V. Russo, *Chem. Commun.*, **1967**, 998; see also I. Rothberg and R. V. Russo, *J. Org. Chem.*, **32**, 2003 (1967).
[17] J. P. Schaefer and C. A. Flegal, *J. Am. Chem. Soc.*, **89**, 5729 (1967).
[18] R. Baker and J. Hudec, *Chem. Commun.*, **1967**, 929.
[19] A. S. Kende and T. L. Bogard, *Tetrahedron Letters*, **1967**, 3386.
[20] C. A. Grob and A. Weiss, *Helv. Chim. Acta*, **49**, 2605 (1966).
[21] C. W. Jefford, J. Gunsher, and B. Waegell, *Tetrahedron Letters*, **1965**, 3405.
[22] H. C. Brown and K.-T. Liu, *J. Am. Chem. Soc.*, **89**, 466 (1967).

methylenenorbornane (α-fenchene) (**24**) also yields *exo*-chloride (**25**) exclusively and this reaction is an exception to the rule[23, 24] that *endo*-approach is favoured in reactions of norbornyl derivatives, not involving non-classical ions, when C_7 bears a *syn*-alkyl group.

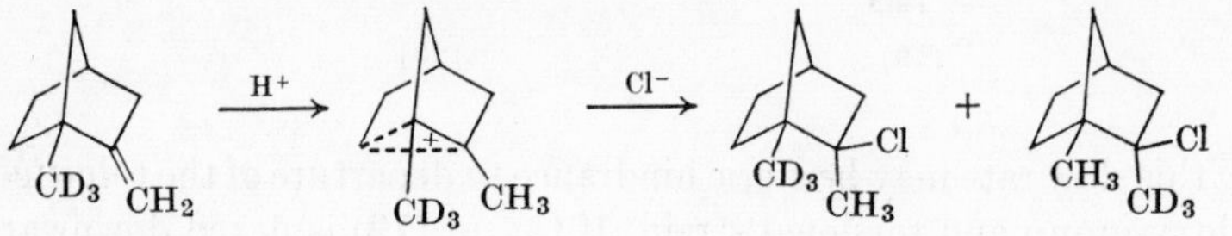

... (1)

(**23**)

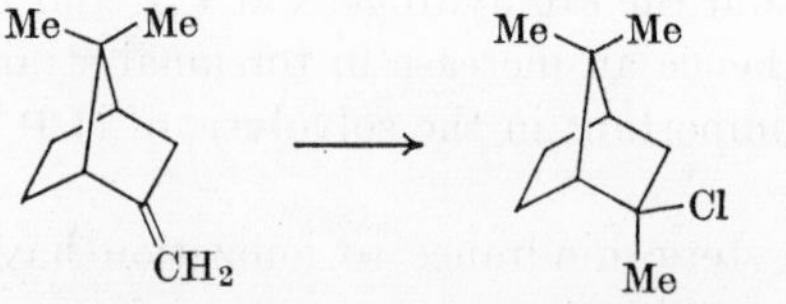

(**24**) (**25**)

Similar behaviour was found with a secondary system. Thus addition of DCl to norbornene at –78° yields 60% of unrearranged 2-*exo*-norbornyl chloride (eqn. 2), and addition to 7,7-dimethylnorbornene yields no *endo*-chloride (eqn. 3).[25]

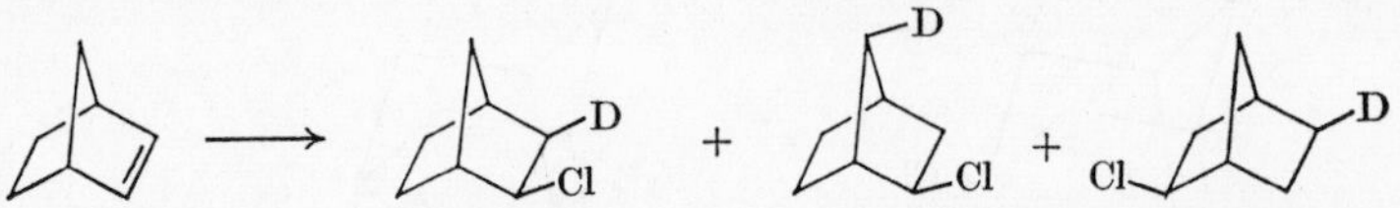

... (2)

~ 60% ~35% ~5%

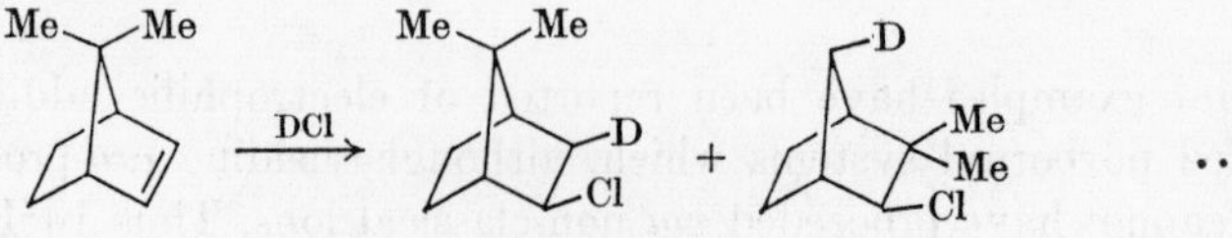

... (3)

90% 10%

Acetolysis of 1-methyl-2-*exo*-norbornyl toluene-*p*-sulphonate occurs 68 times faster than that of 2-*exo*-norbornyl toluene-*p*-sulphonate, indicating that participation by the 1,6-bond is occurring and that the transition state is bridged, although the intermediate 2-methylnorbornyl ion is probably not (see eqn. 4).[26] It follows then that if the transition states for electrophilic

[23] J. A. Berson in "Molecular Rearrangements", Part 1, P. de Mayo, ed., Interscience Publishers, Inc., New York, N.Y., 1963, p. 133.
[24] See *Organic Reaction Mechanisms*, **1965**, 11—12.
[25] H. C. Brown and K.-T. Liu, *J. Am. Chem. Soc.*, **89**, 3900 (1967).
[26] P. von R. Schleyer, *J. Am. Chem. Soc.*, **89**, 3901 (1967).

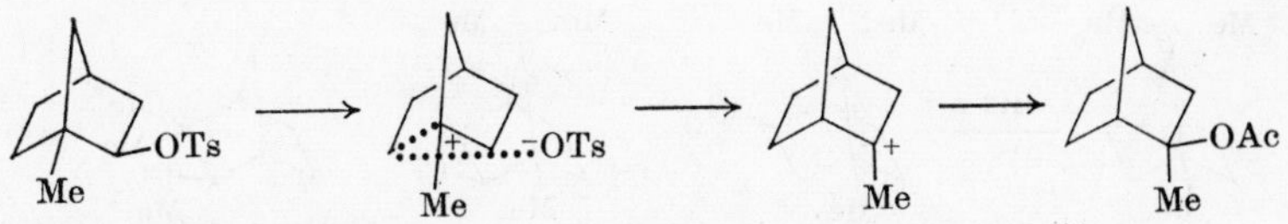
... (4)

addition to 1-methylnorbornene (**26**) were bridged there should be a marked preference for addition of the electrophile to $C_{(3)}$. It has been shown, however, that with three quite different reagents this is not so, despite there being some Wagner–Meerwein rearrangement (see Table 1).[26, 27] Nevertheless exclusive

Table 1. Products of additions to 1-methylnorbornene.[26]

Conditions	**28**	**29**	**27**
HCOOH, 70°, X = OOCH	~ 50%		~ 50%
HCl–H_2O, reflux, X = Cl	43%	6%	51%
HCl–Et_2O, X = Cl		54%	46%

exo-addition was found even when 7,7-dimethyl substituents were present (cf. eqn. 5).[27] It is difficult to avoid the conclusion that these reactions involve classical ions which are trapped from the *exo*-direction exclusively, as a result of torsional and other effects.

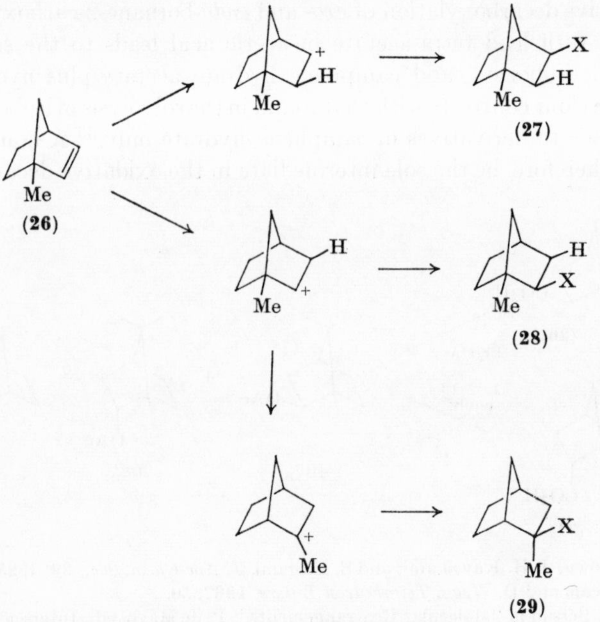

[27] H. C. Brown and K.-T. Liu, *J. Am. Chem. Soc.*, **89**, 3898 (1967).

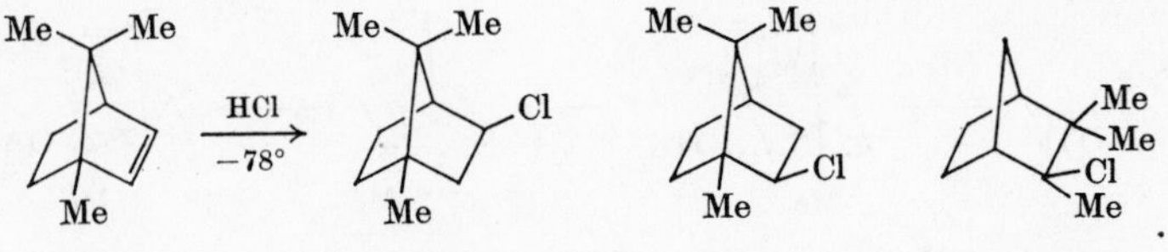

. . . (5)

CH_2Cl_2	4.5 min	44%	8%	48%
Et_2O		46%	30%	24%

Similar behaviour is found in the oxymercuration–demercuration (see p. 142) of norbornenes (see eqns. 6 and 7).[28]

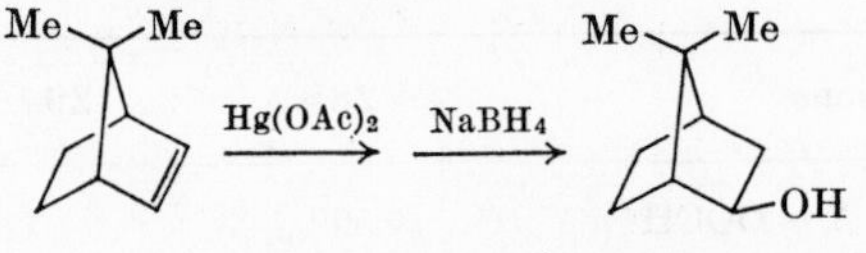

. . . (6)

Yield 84%; > 99.8% *exo*

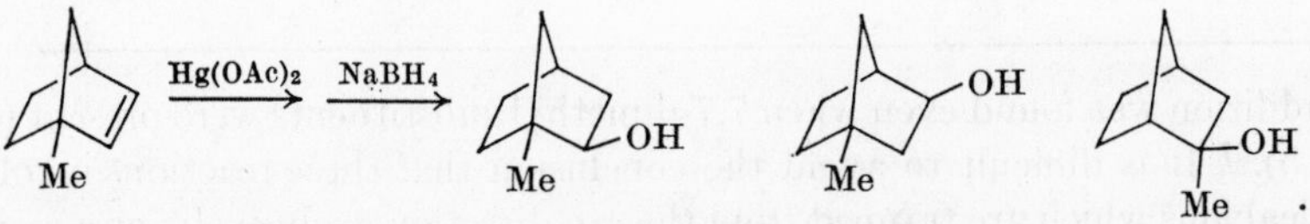

. . . (7)

45% 48% 4%

Oxidative decarboxylation of *exo*- and *endo*-bornane-2-carboxylic acid (**30**) and (**31**) with lead tetra-acetate in acetic acid leads to the same mixture of bornyl, isobornyl, and camphene hydrate acetate plus hydrocarbons.[29] This behaviour contrasts with that found in the solvolysis of isobornyl chloride which leads to derivatives of camphene hydrate only.[30] A non-classical ion cannot, therefore, be the sole intermediate in the oxidative decarboxylations.

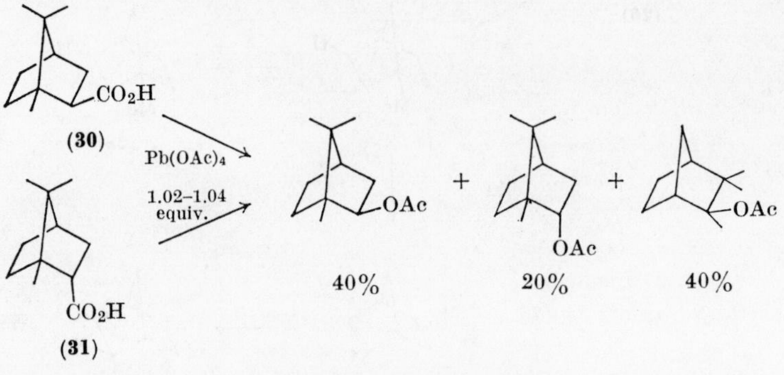

(**30**) (**31**)

40% 20% 40%

[28] H. C. Brown, J. H. Kawakami, and S. Ikegami, *J. Am. Chem. Soc.*, **89**, 1525 (1967).
[29] G. E. Gream and D. Wege, *Tetrahedron Letters*, **1967**, 503.
[30] See J. A. Berson in "Molecular Rearrangements," P. de Mayo, ed., Interscience, New York, N.Y., 1963, p. 121.

In opposition to the large body of evidence being built up by Brown and Schleyer and their co-workers to support the view that carbon bridging is not an important factor in reactions of 2-*exo*-norbornyl derivatives, much significant work suggesting that it is continues to be reported. Thus the secondary deuterium isotope effects for the acetolysis (and solvolysis in 80% aqueous ethanol) of [6-*exo*-^{2}H]- and [6-*endo*-^{2}H]-2-*exo*-norbornyl *p*-bromobenzenesulphonate are $k_H/k_D = 1.09$ and 1.11, respectively, whereas for the corresponding 2-*endo*-isomers they are 0.98 and 0.99. The measurements of these effects is complicated by the concurrent rearrangement of the *p*-bromobenzenesulphonates by ion-pair return which proceeds with an unknown amount of 6 → 2 hydride shift. It was estimated, however, that if complete scrambling of deuterium over positions 1, 2, and 6 occurred with every ionization, irrespective of whether it lead to solvolysis or ion-pair return, the magnitude of the isotope effect for solvolysis of [2-*endo*-^{2}H]-2-*exo*-norbornyl *p*-bromobenzenesulphonate, 1.20, would not account for the effects found with the [6-^{2}H]-isomers. Isotope effects of this magnitude suggest strongly that rehybridization of $C_{(6)}$ is occurring in the transition state and that this involves bridging.[31, 32]

The volumes of activation for the solvolysis of *exo*-2-norbornyl, *endo*-2-norbornyl, and cyclopentyl *p*-bromobenzenesulphonate in 94% aqueous acetone are –14.3, –17.8, and –17.7 cm^3 mole^{-1}, respectively. The less negative value for the *exo*-norbornyl compound was interpreted as resulting from participation by the 1,6-electrons, so that the positive charge is more diffuse in the transition state and electrostriction of the solvent is reduced.[33]

The Wagner–Meerwein rearrangement of norbornyl derivatives results in substituents at positions 5 and 6 which are *exo*- becoming *endo*- and *vice versa* (see **32** → **33**). In an attempt to find if this change had started to occur in the transition state for reactions of 2-*exo*-norbornyl derivatives Corey and Glass studied the acetolysis of 4,5-*exo*-trimethylene-2-*exo*-norbornyl toluene-*p*-sulphonate (**34**). It was estimated that the strain energy of the 4,5-*exo*-trimethylenenorbornyl system is about 6—7 kcal mole^{-1} less than that of the 4,5-*endo*-trimethylene one. Hence if there were bridging in the transition state for acetolysis of 2-*exo*-norbornyl toluene-*p*-sulphonate the rate should be decreased by fusion of a 4,5-*exo*-trimethylene bridge. This expectation was realized, as shown by the relative rates given with formulae (**34**) to (**37**), and so bridging was considered to be important.[34] Although 4,5-*exo*-trimethylene-2-*exo*-norbornyl toluene-*p*-sulphonate reacts only 4 times faster than its 2-*endo*-isomer the products from both contained more than 90% of 2-*exo*-acetate.

[31] B. L. Murr, A. Nickon, T. D. Swartz, and N. H. Werstiuk, *J. Am. Chem. Soc.*, **89**, 1730 (1967).
[32] J. M. Jerkunica, S. Borčić, and D. E. Sunko, *J. Am. Chem. Soc.*, **89**, 1732 (1967).
[33] W. J. le Noble, B. L. Yates, and A. W. Scaplehorn, *J. Am. Chem. Soc.*, **89**, 3751 (1967).
[34] E. J. Corey and R. S. Glass, *J. Am. Chem. Soc.*, **89**, 2600 (1967).

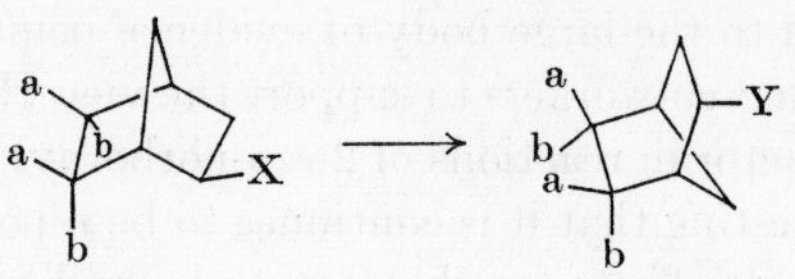

(32) (33)

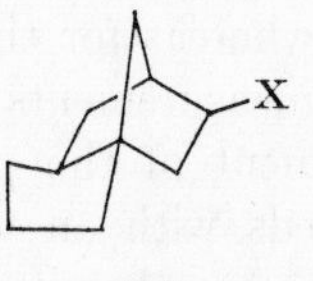
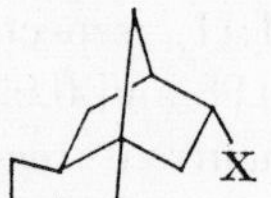
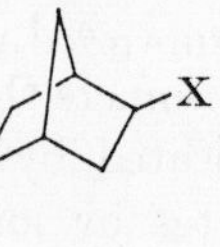
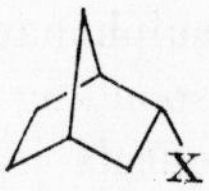

	(34)	(35)	(36)	(37)
Rel. rate of acetolysis at 25°	3.4	0.4	280	1.0

High *exo*:*endo*-rate ratios in reactions of norbornyl derivatives are usually associated with high *exo*:*endo*-product ratios, and these two types of behaviour probably have a common origin, since capture of a carbonium ion by solvent is approximately the microscopic reverse of the ionization process.[35] However, although the *exo*-toluene-*p*-sulphonate (**38**) is acetolysed only 1.7 times faster than its *endo*-isomer (**39**), the product from both, and from their isomers (**41**) and (**42**), is wholly *exo*-acetate (**40**).[36, 37] In our opinion there are three possible explanations for this behaviour.

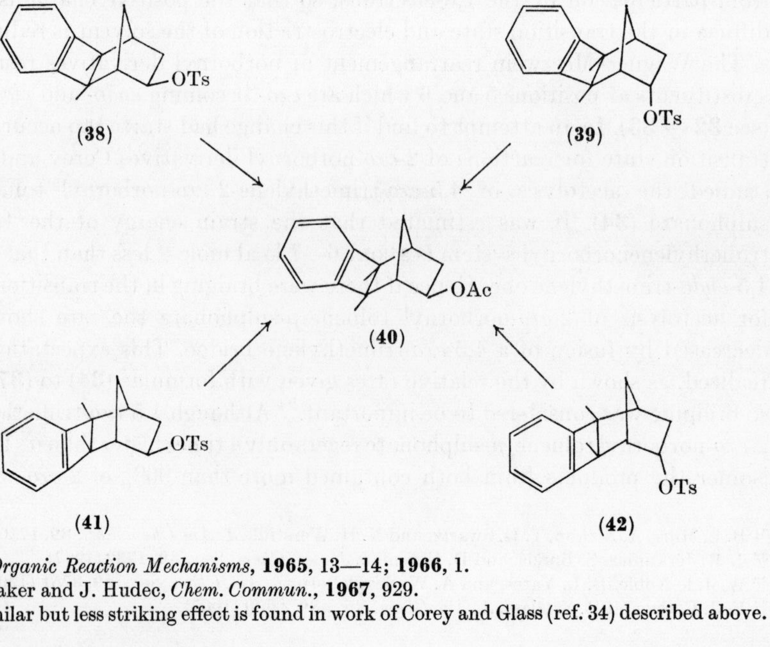

(38) (39) (40) (41) (42)

[35] See *Organic Reaction Mechanisms*, **1965**, 13—14; **1966**, 1.
[36] R. Baker and J. Hudec, *Chem. Commun.*, **1967**, 929.
[37] A similar but less striking effect is found in work of Corey and Glass (ref. 34) described above.

(i) The difference in energy between the transition states for ionization of the *exo*- and *endo*-isomers (**38** and **39**) is much less than the difference in energy between the transition states for the capture of the ionic intermediates by solvent in the *exo*- and the *endo*-direction. The most obvious dissimilarity in these two types of transition state is in the dipole of the breaking and forming bond.[38] So, if there were an unfavourable dipole effect in the transition state for ionization, this would probably not be so in the transition state for solvent capture.

(ii) The *endo*-isomer (**39**) has abnormally high initial-state energy. This is difficult to accept since it reacts 11 times more slowly than 2-*endo*-norbornyl *p*-bromobenzenesulphonate.

(iii) The *exo*-isomer (**38**) has an abnormally low initial-state energy. It is difficult to see any reason why this should be so.

None of these explanations is wholly convincing but in our opinion (i) is the most likely; further work on this system is awaited with interest. The rate of acetolysis of (**38**) is 650 times less than that of norbornyl toluene-*p*-sulphonate and it was suggested that the latter reacted with, and the former without, participation of the 1,6-bond. A factor of 10 was assigned to the rate-decreasing inductive effect of the benzene ring, leaving a factor of 65 for participation. The low rate of acetolysis of (**42**) was ascribed to steric hindrance to ionization. Compound (**41**) undergoes extensive ion-pair return to yield (**38**).

Details have been published of the extensive investigation of the solvolysis of monomethyl-*exo*-2-norbornyl *p*-bromobenzenesulphonates by Berson and his co-workers.[39–44] The 3-*endo*- (**43**), 5-*endo*- (**44**), and 3-*exo*-methyl (**45**) compounds would yield the Wagner–Meerwein-related pairs of classical ions (**46**, **47** and **48**) (or the equivalent non-classical ions) which would be interconverted into one another by $6 \rightarrow 2$ hydride shifts. The same products, derived from these six classical ions, were obtained from all three *p*-bromobenzenesulphonates, but in different proportions, indicating that, although equilibration by the $6 \rightarrow 2$ hydride shifts was substantial, it was not complete. The rates of the $6 \rightarrow 2$ hydride shifts are competitive, therefore, with the rate of capture by solvent and, consistently with this, they are less extensive in the more nucleophilic solvent ethanol than in acetic acid.[42]

[38] See also C. F. Wilcox and R. G. Jesaitis, *Tetrahedron Letters*, **1967**, 2570.

[39] J. A. Berson, J. H. Hammons, A. W. McRowe, R. G. Bergman, A. Remanick, and D. Houston, *J. Am. Chem. Soc.*, **89**, 2561 (1967).

[40] J. A. Berson, A. W. McRowe, R. G. Bergman, and D. Houston, *J. Am. Chem. Soc.*, **89**, 2563 (1967).

[41] J. A. Berson and R. G. Bergman, *J. Am. Chem. Soc.*, **89**, 2569 (1967).

[42] J. A. Berson, A. W. McRowe, and R. G. Bergman, *J. Am. Chem. Soc.*, **89**, 2573 (1967).

[43] J. A. Berson, R. G. Bergman, J. H. Hammons, and A. W. McRowe, *J. Am. Chem. Soc.*, **89**, 2581, 5314 (1967).

[44] J. A. Berson, J. H. Hammons, A. W. McRowe, R. G. Bergman, A. Remanick, and D. Houston, *J. Am. Chem. Soc.*, **89**, 2590 (1967).

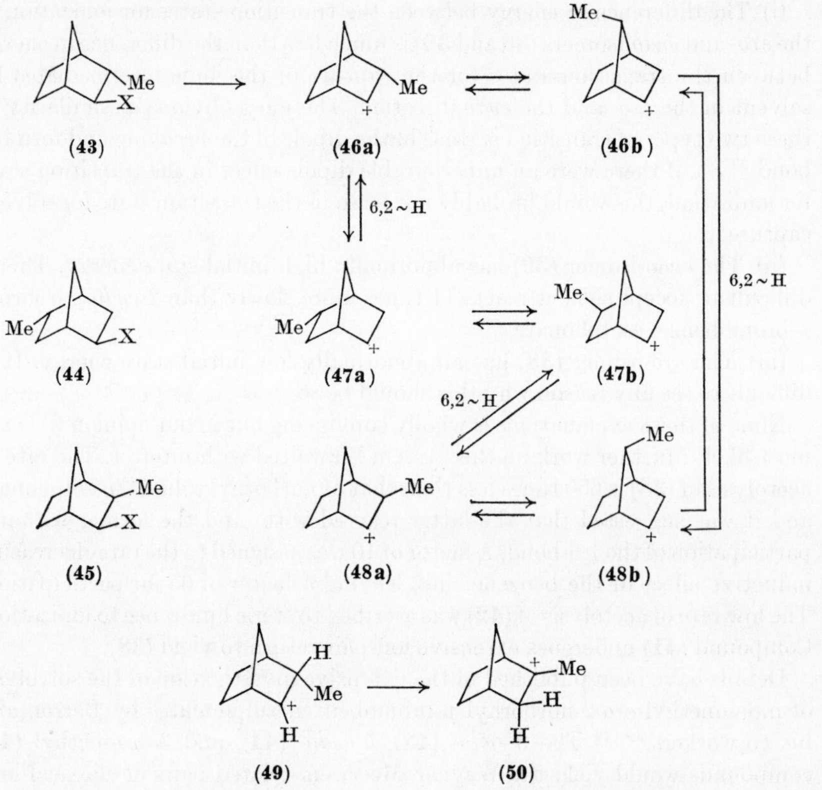

The relative rates of capture by solvent of ions (**46a**) and (**46b**) (and of **47a** and **47b**) are almost identical, which is reasonable as in these the methyl group is remote from the cationic centre. Cation (**48a**), however, is captured 4—8 times less rapidly than (**48b**), as would be expected from the steric effect of the 3-methyl group. Also, as reported last year,[45] 6-*endo*-methyl-2-*exo*-norbornyl *p*-bromobenzenesulphonate yields about 8 times more of the 6-*exo*- than that of the 6-*endo*-methyl compound.[42]

The product from (**43**), (**44**), and (**45**) also contains some *endo*-2-methyl compound formed by the *exo*-tertiary-secondary 3 → 2 hydride shift (**49** → **50**) but none of the products expected from secondary-secondary 3 → 2 hydride shifts, and it was estimated that these are at least 122 times slower than solvent capture in acetic acid at 100°.[43] This is consistent with the much slower 3 → 2 than 6 → 2 hydride shift deduced from the NMR spectrum of the 2-norbornyl cation in SbF_5–SO_2ClF–SO_2,[46] but inconsistent with the 7—10% of 3 → 2 hydride shift reported to occur in the acetolysis of 2-nor-

[45] See *Organic Reaction Mechanisms*, **1966**, 4.
[46] See *Organic Reaction Mechanisms*, **1965**, 23—25; **1966**, 12.

bornyl *p*-bromobenzenesulphonate at 25—45°, on the basis of tritium-labelling experiments.[47]

Methyl substitution also causes an increase in the rate of the 6 → 2 hydride shifts as estimated by the amount occurring in the solvolyses of 6-*endo*-methyl-2-*exo*-norbornyl and [2-*endo*-^{3}H]-2-*exo*-norbornyl *p*-bromobenzenesulphonates.[43] The increase is only by a factor of 5 to 15, but this was considered to be consistent with edge-protonated cyclopropane intermediates[48] in which much of the positive charge is localized on the migrating hydrogen.

Details were also published of the investigation which demonstrated that 2-*endo*-methyl-2-*exo*-norbornyl acetate formed in the acetolysis of 3-*exo*-methyl-2-*endo*-norbornyl *o*-bromobenzenesulphonate results, not from an *endo*-2 → 3 hydride shift, but from a more circuitous route.[44, 49] This was considered to be strong evidence for the 3-methyl-2-norbornyl cation's having a non-classical structure, and the explanation of the preference for *exo*-3 → 2 migrations based on torsional effects (see pp. 1 and 320) was rejected.[50]

A 2 → 3-*endo*-hydride migration occurs when 3-*endo*-phenylbornane-2,3-*exo*-*cis*-diol (**51**) is treated with 1:100 $HClO_4$–AcOH for four hours at room temperature (eqn. 8).[51] The reaction is intramolecular since 2-*endo*-^{2}H-(**51**) yield 3-*endo*-^{2}H-(**52**). It is not clear if the whole process is concerted or if ion (**53**) is an intermediate. If the latter were correct, the question would arise whether hydride migration is faster than solvent capture, or whether the ion reacts rapidly and reversibly with solvent, with hydride migration occurring occasionally but irreversibly. A similar hydride migration does not occur with the analogous phenylnorbornanediol in sulphuric acid.[52]

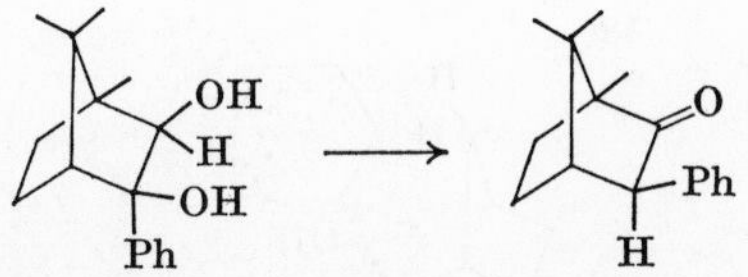

(**51**) (**52**) . . . (8)

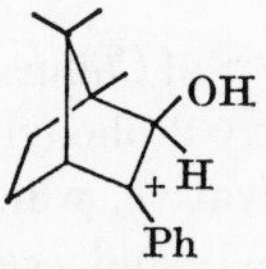

(**53**)

[47] See *Organic Reaction Mechanisms*, **1966**, 6.
[48] See *Organic Reaction Mechanisms*, **1965**, 22.
[49] See *Organic Reaction Mechanisms*, **1965**, 22—23.
[50] See footnote 33g of ref. 44.
[51] A. W. Bushell and P. Wilder, *J. Am. Chem. Soc.*, **89**, 5721 (1967).
[52] Cf. *Organic Reaction Mechanisms*, **1966**, 7—9.

Even *exo*-2 → 3-migration of a phenyl group in the norbornyl system is difficult. Thus 7,2 Wagner–Meerwein rearrangement occurs in the deamination of (**54**) in preference to phenyl migration despite the fact that this should be favoured by the presence of the 3-*endo*-hydroxyl group. It was suggested that in the transition state (**55**) for phenyl migration there was an unfavourable steric interaction between the *ortho*-hydrogen of the phenyl and the 7-*syn*-hydrogen of the norbornane.[53]

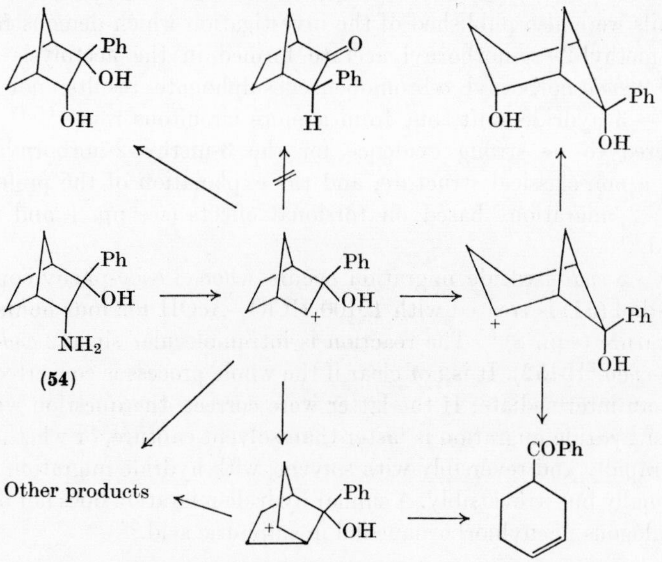

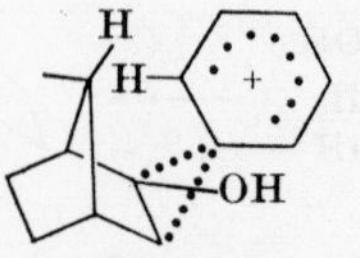

(**55**)

Details have been published[54, 55] of Collins and Benjamin's investigation of the solvolysis of 2-*exo*-hydroxy-2-phenyl-3-*exo*-norbornyl toluene-*p*-sulphonate and its 5,6-*exo*-$^2H_2$-derivative, which demonstrated the occurrence of several stereospecific *exo*-6 → 1- and *endo*-6 → 2-deuteride shifts. These were thought to occur *via* edge-protonated cyclopropanes.

The retarding inductive effect of a β-acetoxy-group on the rate of acetolysis

[53] C. J. Collins, V. F. Raaen, B. M. Benjamin, and I. T. Glover, *J. Am. Chem. Soc.*, **89**, 3940, 5314 (1967).

[54] C. J. Collins and B. M. Benjamin, *J. Am. Chem. Soc.*, **89**, 1652 (1967).

[55] See *Organic Reaction Mechanisms*, **1966**, 7—8.

of norbornyl toluene-*p*-sulphonate, estimated by using the Taft equation, was found to be about 10^3-fold at 25° and so the much greater effects found with compounds (**56**) and (**57**) were thought to have some other origin. This was identified as a strong dipolar repulsion between the lactone dipole and the developing dipole of the C–OTs bond in the transition state, possibly enhanced by the absence of solvent in the cavity of the substrate. The reactions proceed with extensive rearrangement and were formulated as shown in equation (9). The rates were correlated with the carbonyl stretching frequencies of the corresponding ketones by using the Foote–Schleyer equation, and the increases in the frequencies were also ascribed to dipolar repulsions.[56]

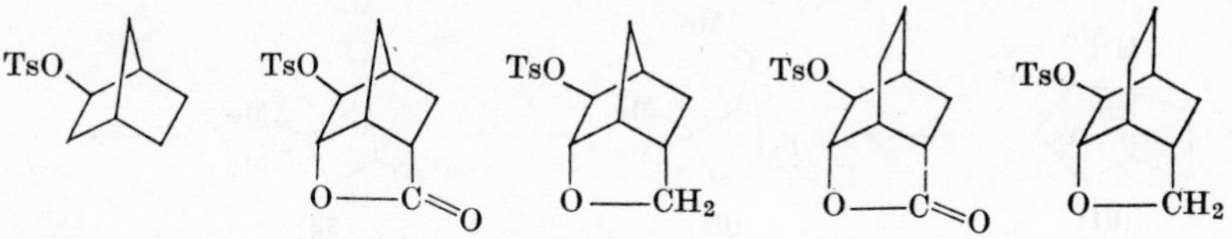

		(**56**)		(**57**)	
Rel. rate of acetolysis at 25°	1	6.5×10^{-8}	1.46×10^{-4}	2.2×10^{-7}	4.7×10^{-4}
$\nu_{C=O}$ for 2-ketone (cm^{-1})	1751	1800	1763	1775	

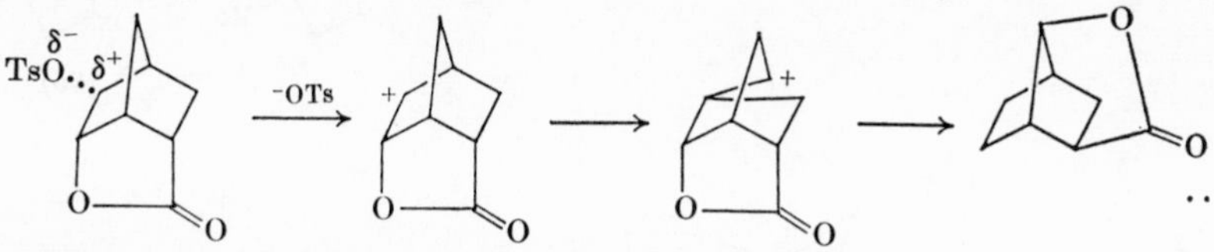

. . . (9)

Acetolysis[57, 58] and formolysis[58] of the specifically deuterated 7-norbornyl toluene-*p*-sulphonate (**58**) proceed with predominant (~ 90%) retention of configuration. A possible reaction intermediate is the non-classical ion (**59**), previously invoked by Winstein and his co-workers to explain their observation that the small amount of bicyclo[3.2.0]hept-2-yl acetate formed in the acetolysis of non-deuterated (**58**) and of *trans*-2-bicyclo[3.2.0]hept-2-yl

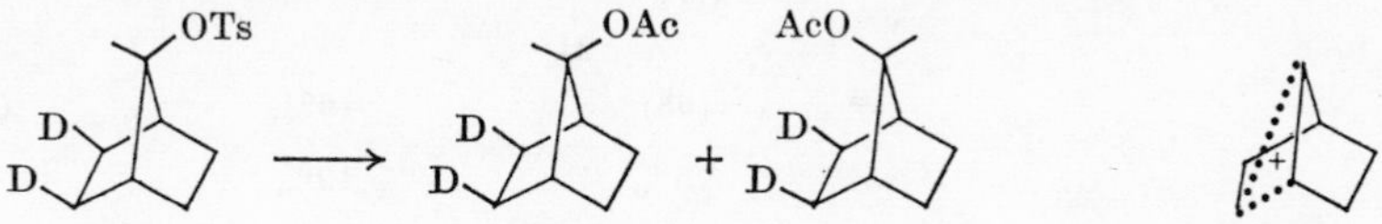

(**58**) 90% 10% (**59**)

[56] R. M. Moriarty, C. R. Romain, and T. O. Lovett, *J. Am. Chem. Soc.*, **89**, 3927 (1967).
[57] P. G. Gassman and J. M. Hornback, *J. Am. Chem. Soc.*, **89**, 2487 (1967).
[58] F. B. Miles, *J. Am. Chem. Soc.*, **89**, 2488 (1967).

toluene-*p*-sulphonate is 100% *trans*. Alternatively the reaction may involve front-side collapse of a classical-ion ion-pair.[58] An ion similar to (**59**) (namely, **64**) may also be involved in the acetolyses of the *cis*- and *trans*-bicyclo[3.2.0]-

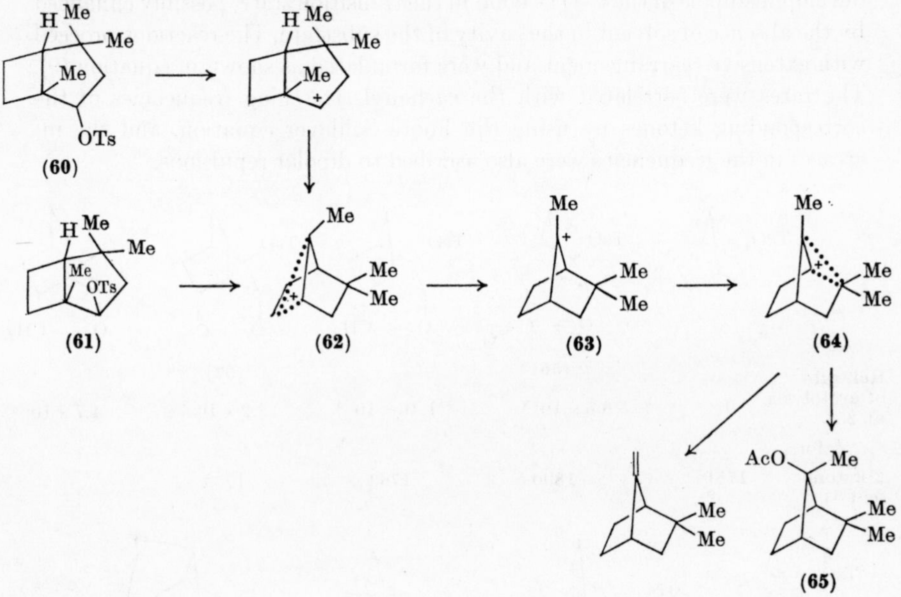

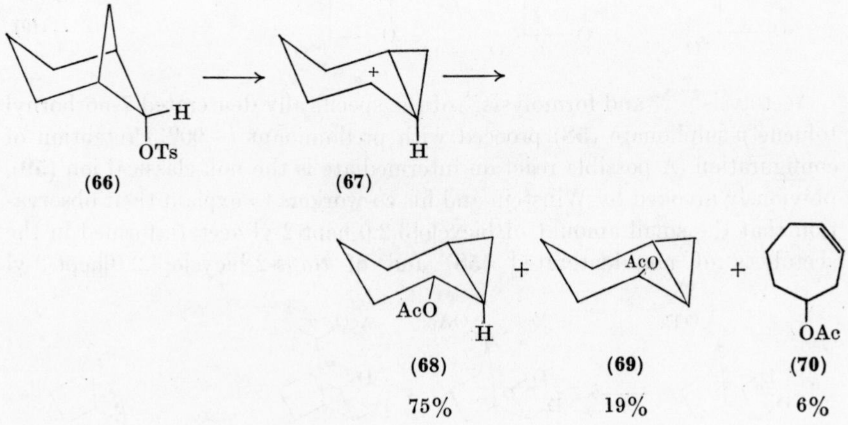

hept-2-yl toluene-*p*-sulphonates (**60**) and (**61**) which yield the 7-norbornyl acetate (**65**) stereospecifically. This is formed with *retention* of configuration at the migration origin and hence initially formed ion (or transition state) (**62**) would have to rearrange to (**64**). Alternatively the intermediate ion, which is

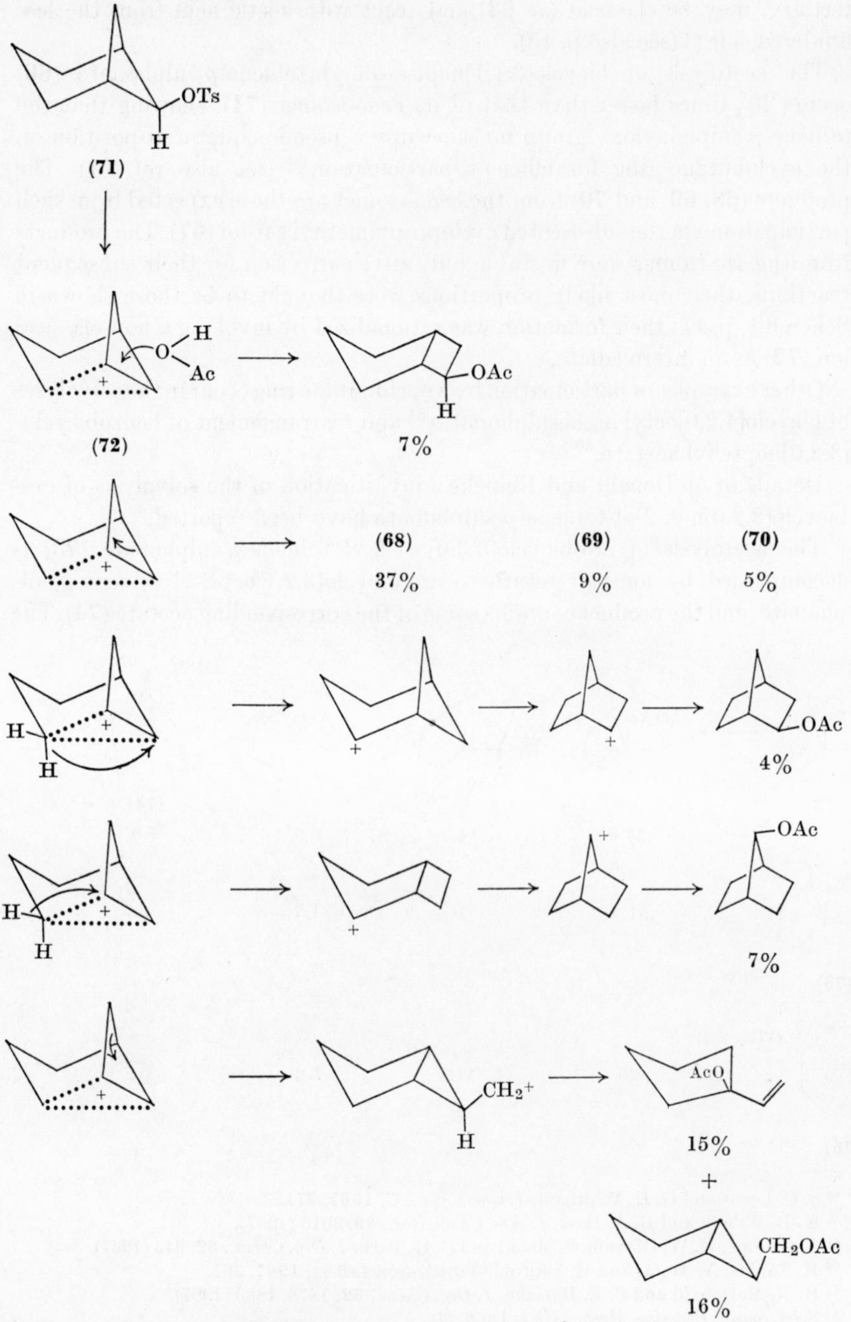
OTs
H
(71)
H
O
Ac
(72)
OAc
H
7%
(68)
37%
(69)
9%
(70)
5%
H
H
OAc
4%
H
H
OAc
7%
CH2+
H
AcO
15%
+
CH2OAc
16%

Scheme 1

tertiary, may be classical (as **63**) and react with acetic acid from the less hindered side[59] (see also p. 15).

The acetolysis of bicyclo[3.1.1]hept-*exo*-6-yl toluene-*p*-sulphonate (**66**) occurs 10^6 times faster than that of its *endo*-isomer (**71**), showing that the toluene-*p*-sulphonyloxy-group must occupy a pseudo-equatorial position on the cyclobutane ring for efficient participation[60] (see also ref. 61). The products (**68**, **69**, and **70**) from the *endo*-isomer are those expected from such participation via the substituted cyclopropylmethyl cation (**67**). The products from the *exo*-isomer were unstable but, after correction for their subsequent reactions, their most likely proportions were thought to be those shown in Scheme 1, p. 17; their formation was rationalized by invoking a non-classical ion (**72**) as an intermediate.

Other examples of participation by a cyclobutane ring occur in the solvolyses of bicyclo[4.2.0]octyl arenesulphonates[61] and rearrangement of benzobicyclo-[3.2.0]heptenyl acetate.[62]

Details of McDonald and Reineke's investigation of the solvolysis of *exo*-bicyclo[2.2.0]hex-2-yl toluene-*p*-sulphonate have been reported.[63, 64]

The acetolysis of *exo*-bicyclo[3.3.0]oct-2-yl toluene-*p*-sulphonate (**73**) is accompanied by ion-pair return to *exo*-bicyclo[3.2.1]oct-8-yl toluene-*p*-sulphonate, and the product contains some of the corresponding acetate (**74**). The

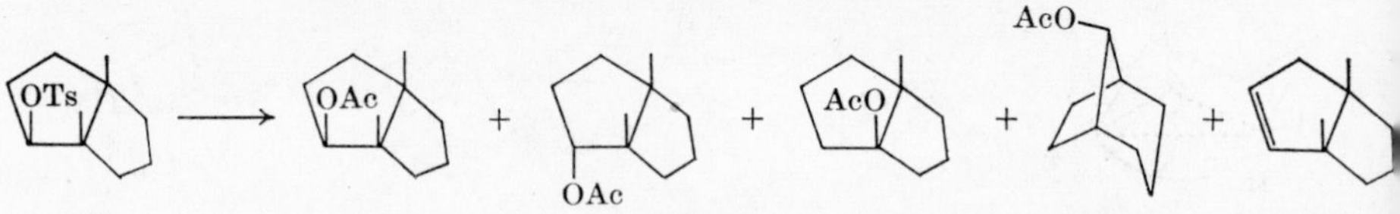

(**73**) (**74**)

(**73**)	24.6	14.7	0.9	7.9	52
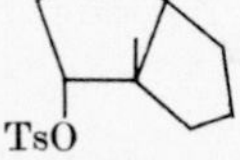 (**75**)	51.3	0	1.7	0	47
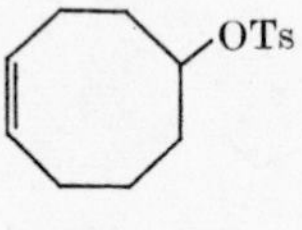 (**76**)	26	41.3	1.2	5	26

[59] S. C. Lewis and G. H. Whitham, *J. Chem. Soc., C*, **1967**, 274.
[60] K. B. Wiberg and B. A. Hess, *J. Am. Chem. Soc.*, **89**, 3015 (1967).
[61] A. C. Cope, R. W. Gleason, S. Moon, and C. H. Park, *J. Org. Chem.*, **32**, 942 (1967).
[62] H. Tanida, Y. Hata, and H. Ishitobi, *Tetrahedron Letters*, **1967**, 361.
[63] R. N. McDonald and C. E. Reineke, *J. Org. Chem.*, **32**, 1878, 1889 (1967).
[64] See *Organic Reaction Mechanisms*, **1965**, 30.

rate is, however, only 1.3 times greater than that for the *endo*-isomer (**75**) and so there is no anchimeric assistance. *cis*-Cyclo-oct-4-enyl toluene-*p*-sulphonate (**76**) also undergoes ion-pair return to *exo*-bicyclo[3.2.1]oct-8-yl toluene-*p*-sulphonate and yields a similar mixture of acetates. The proportions differ from those from the *exo*-compound (**73**), which differ from those from the *endo*-isomer (**74**). It therefore seems likely that product formation is controlled partly by the stereochemistry of ion-pair intermediates.[65]

In contrast to 1-adamantylmethyl toluene-*p*-sulphonate (**78**),[66] 3-noradamantylmethyl toluene-*p*-sulphonate (**77**) is acetolysed 17,000 times faster than neopentyl toluene-*p*-sulphonate.[67] Clearly there is considerable release of ring-strain energy in the transition state similar to that released in solvolyses of 1-methylcyclobutylmethyl arenesulphonates (**79**) (see also p. 76).

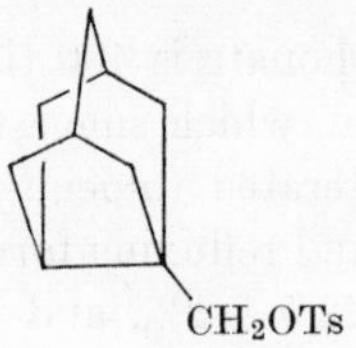

(**77**)

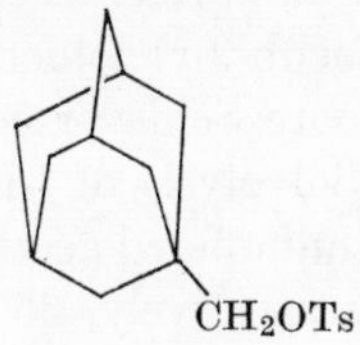

(**78**)

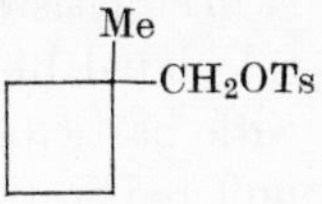

(**79**)

	(77)	(78)	(79)
Rel. rates HOAc 25° ($Me_3C—CH_2OTs$ = 1·0)	17,000	3.9	24,600

Strain energies for a series of bicyclic hydrocarbons and the corresponding bridgehead cations have been calculated by the method of Wiberg. It was concluded that the 5000-fold faster solvolysis of 1-adamantyl compounds than of the corresponding bicyclo[2.2.2]oct-1-yl ones resulted from the rates of the latter being depressed owing to an unfavourable non-bonding interaction between the bridgehead (1 and 4) positions in the bicyclo[2.2.2]oct-1-yl ion and the transition state for its formation. Predictions were made of the relative rate constants for the solvolysis of homoadamant-1-yl, bicyclo-[3.3.1]-1-nonyl, and bicyclo[3.2.1]oct-1-yl derivatives.[68]

The rate of acetolysis of the noradamantyl toluene-*p*-sulphonate (**80**) is 3×10^8 less than that of adamantyl toluene-*p*-sulphonate.[67]

The solvolysis of 1-chlorobicyclo[1.1.1]pentane (a bridgehead chloride) in 80% aqueous ethanol is three times faster than that of *tert*-butyl chloride,

65 W. D. Closson and G. T. Kwiatkowski, *Tetrahedron Letters*, **1966**, 6435.
66 See *Organic Reaction Mechanisms*, **1966**, 67—68.
67 P. von R. Schleyer and E. Wiskott, *Tetrahedron Letters*, **1967**, 2845; see also A. Nickon, G. D. Pandit, and R. O. Williams, *ibid.*, **1967**, 2851; B. R. Vogt and J. R. E. Hoover, *ibid.*, **1967**, 2841.
68 G. J. Gleicher and P. von R. Schleyer, *J. Am. Chem. Soc.*, **89**, 582 (1967); see also R. C. Fort and P. von R. Schleyer, *Advan. Alicyclic Chem.*, **1**, 283 (1966).

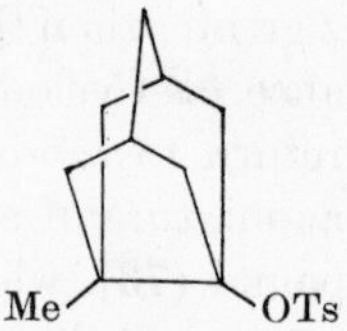

(80)

and the product is 3-methylenecyclobutanol. The high reactivity may result from release of steric strain in the transition state or from stabilization of the transition state through a cross ring interaction between orbitals at $C_{(1)}$ and $C_{(3)}$. 2-Chlorobicyclo[1.1.1]pentane was also reported to be quite reactive.[69]

The kinetics of solvolysis of a bicyclo[3.2.1]octyl bridgehead toluene-*p*-sulphonate are reported in reference 70.

Acetolysis of homocub-9-yl toluene-*p*-sulphonate is 400 times faster than calculated by the Foote–Schleyer correlation, which suggests that it reacts with participation. Solvolysis of the 9-deuterated toluene-*p*-sulphonate in buffered acetic acid, unbuffered acetic acid, and refluxing formic acid yielded products in which, respectively, $36 \pm 4\%$, $22.5 \pm 2\%$, and $10 \pm 2\%$ of the deuterium remained at $C_{(9)}$, indicating that the homocub-9-yl cation must undergo a degenerate rearrangement (cf. eqn. 10) similar to that undergone by bullvalene.[71] This conclusion is also supported by the observation of

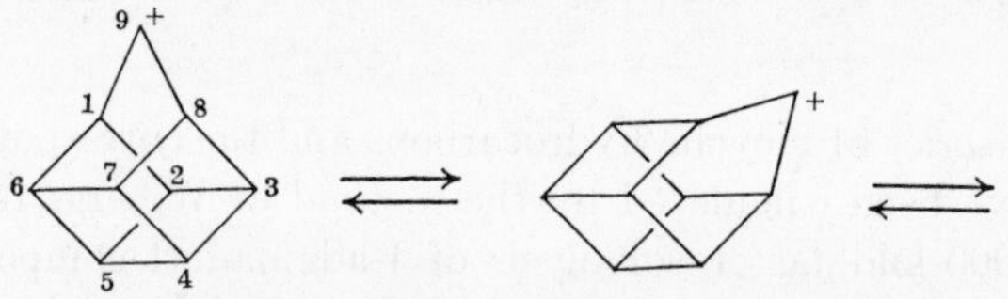

etc. ... (10)

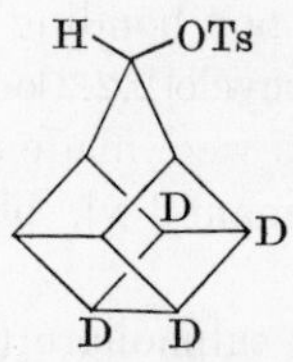

(81)

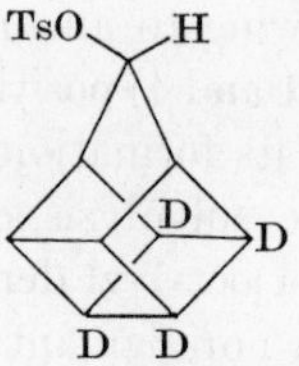

(82)

deuterium scrambling in the product of acetolysis of the tetradeuterated compound (**81**). There was approximately twice as much deuterium at position 9 in the product from (**81**) ($26.1 \pm 3\%$) as in a product from an equimolar mixture of (**81**) and (**82**) (12.3%), which shows that under these condi-

[69] K. B. Wiberg and V. Z. Williams, *J. Am. Chem. Soc.*, **89**, 3373 (1967).

[70] J. MacMillan and R. J. Pryce, *J. Chem. Soc., C*, **1967**, 550.

[71] P. von R. Schleyer, J. J. Harper, G. L. Dunn, V. J. DiPasquo, and J. R. E. Hoover, *J. Am. Chem. Soc.*, **89**, 698, 2242 (1967).

tions the rearrangement is highly stereoselective although the results for formolysis mentioned above indicate some "leakage". When (**81**) was recovered from a partially acetolysed reaction mixture it also showed deuterium scrambling and this had occurred to a greater extent than with the acetates. Extensive rearrangement *via* internal return had thus taken place.[72] Similar rearrangement takes place in the acetolysis of *syn*- and *anti*-homocub-1,3-ylene di(toluene-*p*-sulphonates). Rate enhancements of 10^3–10^4 were calculated by use of the Foote–Schleyer correlation.[73]

Other reactions of bicyclic and polycyclic systems which have received attention include: spontaneous and acid-catalysed solvolyses of esters of camphene hydrate and methylcamphenilol,[74] solvolyses of *exo*- and *endo*-3-bornyl toluene-*p*-sulphonate,[75] acetolysis of 1,3,3-trimethyl-*endo*-2-norbornyl (*endo*-fenchyl) toluene-*p*-sulphonate,[76] conversion of 2-phenylborneol into 4-phenylisobornyl chloride,[77] rearrangement of 2-methylfenchol and 2-methylborneol,[78] rearrangement of camphene hydrochloride into isobornyl chloride,[79] isomerization of methylnorbornanes,[80] sulphonation of camphene,[81] Lewis-acid-catalysed camphene–formaldehyde reaction,[82] acetolysis of tricyclo-[3.3.0.0^{2,6}]oct-3-yl toluene-*p*-sulphonate,[83] solvolyses of allylic bicyclic *p*-bromobenzenesulphonates,[84] of *cis*- and *trans*-2-methylbicyclo[3.2.1]oct-2-yl, 2-methylbicyclo[2.2.2]oct-2-yl, and 1-methylbicyclo[2.2.2]oct-2-yl toluene-*p*-sulphonate,[85] deamination of 2-aminomethylbicyclo[2.2.2]octane,[86] reduction of bicyclo[2.2.2]octenyl toluene-*p*-sulphonates,[87] and rearrangement of the bicyclo[4.2.1]nonatrienyl into the *cis*-bicyclo[4.3.0]nonatrienyl system.[88]

Phenonium Ions

A striking example of phenyl participation occurs in the trifluoroacetolysis of phenethyl toluene-*p*-sulphonate.[89] The ethanolysis, acetolysis, and formolysis

72 J. C. Barborak and R. Pettit, *J. Am. Chem. Soc.*, **89**, 3080 (1967).
73 W. L. Dilling and C. E. Reineke, *Tetrahedron Letters*, **1967**, 2547.
74 C. A. Bunton, C. O'Connor, and D. Whittaker, *J. Org. Chem.*, **32**, 2812 (1967).
75 W. Z. Antkowiak, *Bull. Acad. Polon. Sci., Ser. Sci. Chim.*, **14**, 431 (1966).
76 W. Hückel and E. N. Gabali, *Chem. Ber.*, **100**, 2766 (1967); see also W. Hückel, *J. Prakt. Chem.*, **32**, 320 (1966); *Ann. Acad. Sci. Fennicae*, Ser. A, Nos. 133, 134 (1966).
77 D. Bernstein, *Tetrahedron Letters*, **1967**, 2281.
78 W. Treibs, *Tetrahedron Letters*, **1967**, 4703.
79 P. Schreiber, *Oesterr. Chem. Ztg.*, **66**, 217 (1966).
80 N. A. Belikova, A. A. Bobleva, and A. F. Plate, *Zh. Org. Khim.*, **2**, 2031 (1966); *Chem. Abs.*, **66**, 85247 (1967).
81 J. Wolinsky, D. R. Dimmel, and T. W. Gibson, *J. Org. Chem.*, **32**, 2087 (1967).
82 A. T. Blomquist and R. J. Himics, *Tetrahedron Letters*, **1967**, 3947.
83 J. Meinwald and B. E. Kaplan, *J. Am. Chem. Soc.*, **89**, 2611 (1967).
84 L. E. Gruenewald and D. C. Johnson, *J. Org. Chem.*, **32**, 318 (1967).
85 W. Kraus, *Ann. Chem.*, **708**, 127 (1967).
86 M. Hartmann, *Z. Chem.*, **7**, 305 (1967).
87 R. A. Appleton, J. C. Fairlie, and R. McCrindle, *Chem. Commun.*, **1967**, 690.
88 A. S. Kende and T. L. Bogard, *Tetrahedron Letters*, **1967**, 3383.
89 J. E. Nordlander and W. G. Deadman, *Tetrahedron Letters*, **1967**, 4409.

of this substrate are slower or only slightly faster than that of ethyl toluene-*p*-sulphonate, but the trifluoroacetolysis is over 3000 times faster! There is thus substantial anchimeric assistance by the phenyl group and presumably formation of a phenonium ion (**83**). Consistent with this interpretation there was complete scrambling of the label in the phenethyl trifluoroacetate obtained from [α,α-$^2H_2$]phenethyl toluene-*p*-sulphonate. There was also a small amount of scrambling in unchanged toluene-*p*-sulphonate isolated after one half-life, indicating the occurrence of some ion-pair return.

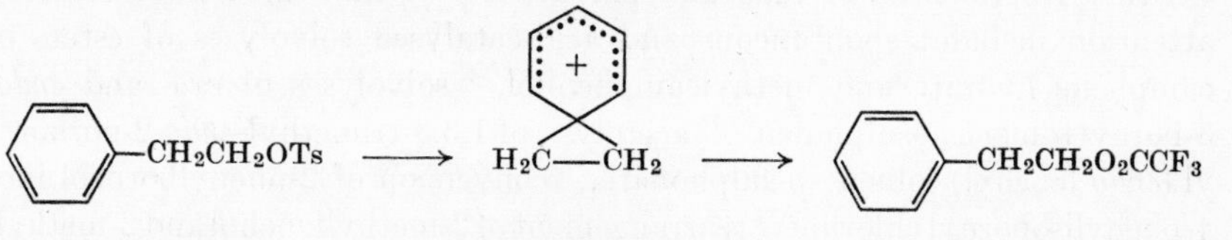

(83)

Reagent: $CF_3CO_2H-CF_3CO_2Na-(CF_3CO)_2O$ (1%).

Contrary to earlier reports, deamination of phenethylamine in acetic acid yields α-methylbenzyl acetate as well as phenethyl acetate (as shown). The

$$PhCH_2CH_2NH_2 \rightarrow \underset{18\%}{PhCH(OAc)CH_3} + \underset{82\%}{PhCH_2CH_2OAc}$$

α-methylbenzyl acetate from [α-^{14}C]phenethylamine shows only 0.23% of scrambling of the label compared with 27% found for the phenethyl acetate. This result shows that a reaction sequence as in Scheme 2 cannot be followed,

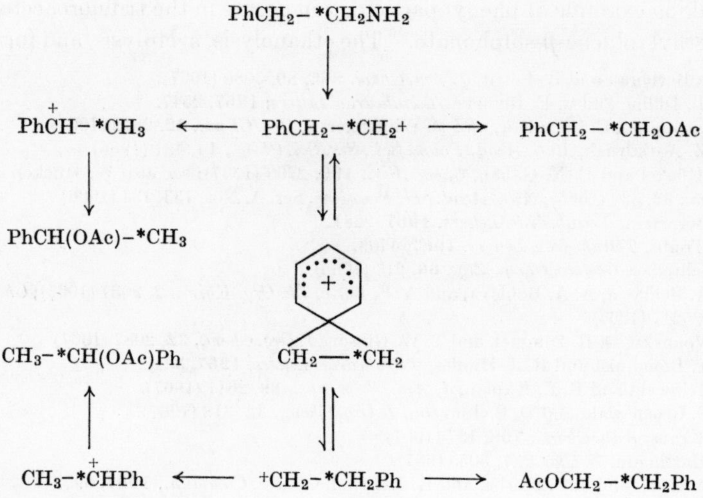

Scheme 2

and that the intermediate phenonium ion (intermediate or transition state) cannot revert to the phenethyl cation if the latter is also the source of the α-methylbenzyl acetate. So either the phenonium ion must be formed irreversibly or nearly all the α-methylbenzyl acetate must be formed by a hydride shift concerted with departure of the nitrogen. The author[90] preferred the former explanation and formulated the phenonium ion as an intermediate formed without the intervention of an unbridged ion.

The solvolyses of several 2-*p*-methoxyphenylalkyl toluene-*p*-sulphonates have been compared with those of the corresponding 2-phenyl compounds (Table 2). The 2-*p*-methoxyphenyl compounds react 20—200 times faster

Table 2. Relative rates of solvolysis of some 2-*p*-methoxyphenylalkyl and 2-phenylalkyl toluene-*p*-sulphonates.

Toluene-*p*-sulphonate	$k_{MeOC_6H_4}/k_{C_6H_5}$ Acetolysis	Formolysis
2-Arylethyl	26	76
2-Aryl-2-methyl-1-propyl	174	74
2-Aryl-1-methylethyl	32	37
2-Aryl-1-methylpropyl (*threo*)	78	22
2-Aryl-1,2-dimethylpropyl	46	

than the corresponding phenyl compounds. This relatively small rate enhancement and the failure of the rates of formolysis of some 2-arylethyl toluene-*p*-sulphonates to be correlated by the σ^+ constants were taken to indicate that the transition state is π-complex rather than σ-complex in character and that equilibrating π-complexed ions were intermediates rather than a σ-complex or bridged phenonium ion.[91]

The diastereoisomeric 1-methyl-2-([2.2]paracyclophanyl)ethyl toluene-*p*-sulphonates (**84**) are solvolysed with a high degree of retention of configuration. It was suggested that this resulted from the intervention of phenonium ions formed by attack of the π-electrons on the external face of the paracyclophane nucleus (see **85a** and **85b**).[92a] An interesting example of Ar_1–6 participation leading to the spirobisdienone (**86**) has been reported.[92b]

An unsuccessful attempt to observe Ar_1–4 participation has been reported.[93, 94]

[90] J. L. Coke, *J. Am. Chem. Soc.*, **89**, 135 (1967).
[91] H. C. Brown, R. Bernheimer, C. J. Kim, and S. E. Scheppele, *J. Am. Chem. Soc.*, **89**, 370 (1967).
[92a] D. J. Cram and F. L. Harris, *J. Am. Chem. Soc.*, **89**, 4642 (1967).
[92b] R. S. Atkinson and A. S. Dreiding, *Helv. Chim. Acta*, **50**, 23 (1967).
[93] P. T. Lansbury and N. T. Boggs, *Chem. Commun.*, **1967**, 1007.
[94] P. T. Lansbury and E. J. Nienhouse, *Chem. Commun.*, **1967**, 1008.

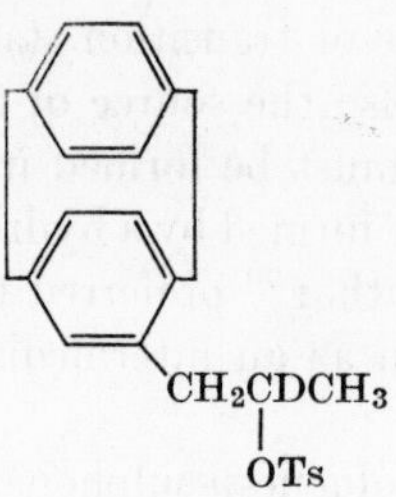

(84)

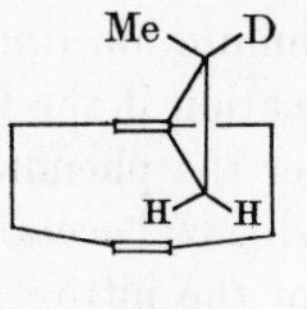

(85a)

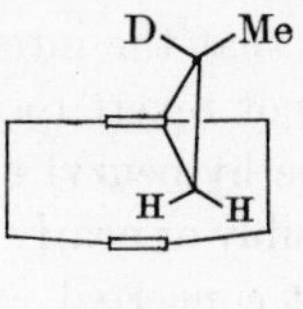

(85b)

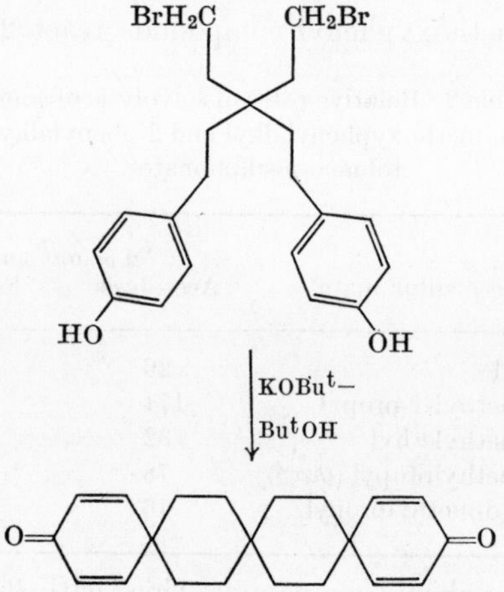

(86)

Although the original reports of the NMR spectra of phenonium ions[95] are incorrect (see below), Olah and his co-workers have now reported convincing spectra for the *p*-methoxy-, 2,4,6-trimethyl-, and pentamethyl-phenonium ions formed by direct participation by the aryl groups (e.g., eqn. 11).[96, 97] The spectrum of the *p*-methoxyphenonium ion has signals at $\delta = 3.47$ (cyclopropyl), 4.25 (methoxyl), and 8.12 and 7.47 ppm (AB quartet). The AB quartet is characteristic of a phenonium ion and excludes a phenyl group which would be present in rapidly equilibrating classical ions. The signal of the methoxyl group is more deshielded than that of the *p*-methoxybenzenonium ion, which indicates that charge is delocalized into the cyclopropane ring (see also p. 42). The *p*-methoxyphenonium ion is formed only from 4-methoxyphenethyl chloride and not from 4-methoxyphenethyl alcohol which yields the dication (**87**). It seems likely that the phenonium ion is formed

[95] See *Organic Reaction Mechanisms*, **1965**, 34.

[96] G. A. Olah, E. Namanworth, M. B. Comisarow, and B. Ramsey, *J. Am. Chem. Soc.*, **89**, 711 (1967).

[97] G. A. Olah, M. B. Comisarow, E. Namanworth, and B. Ramsey, *J. Am. Chem. Soc.*, **89**, 5259 (1967).

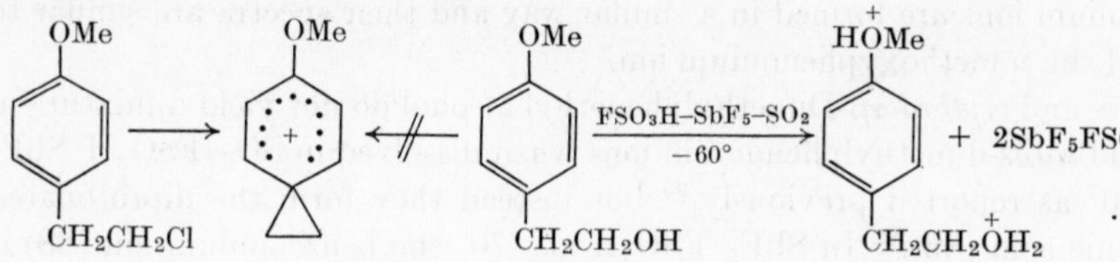

(87) . . . (11)

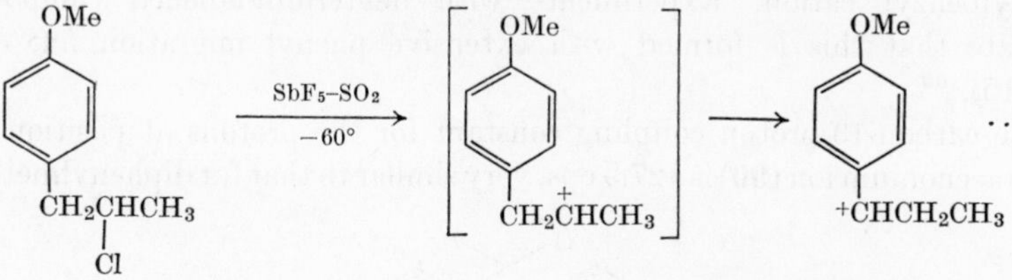

. . . (12)

by direct participation since if the 4-methoxyphenethyl cation were an intermediate it would probably rearrange to the 4-methoxystyryl ion. This is in fact what happens with secondary halides that form much more stable open-chain ions (cf. eqn. 12). The 2,4,6-trimethyl- and pentamethyl-

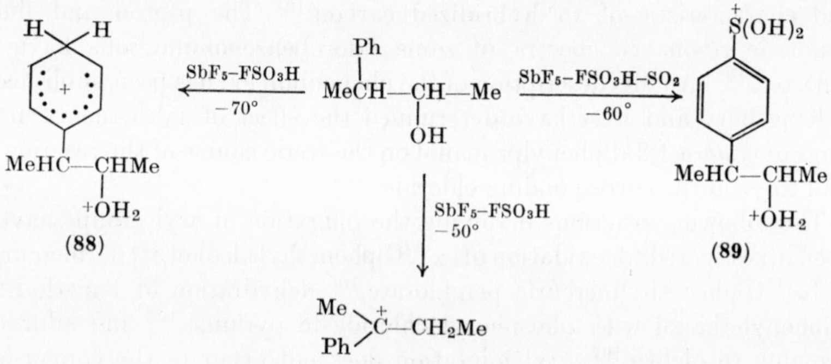

(88) (89)

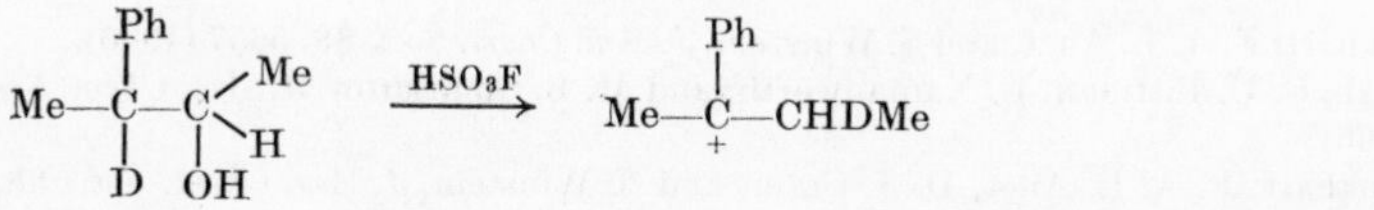

. . . (13)

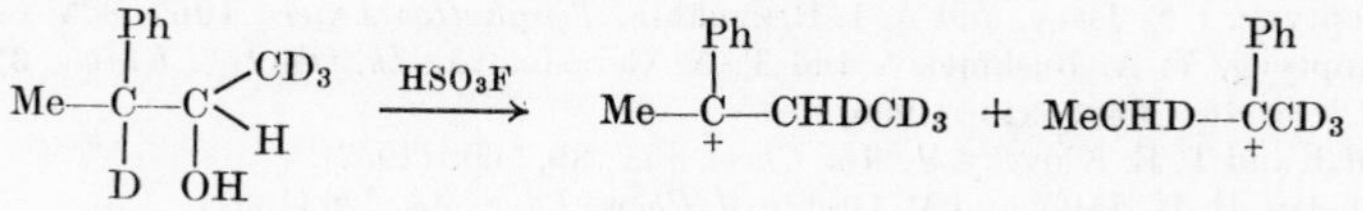

. . . (14)

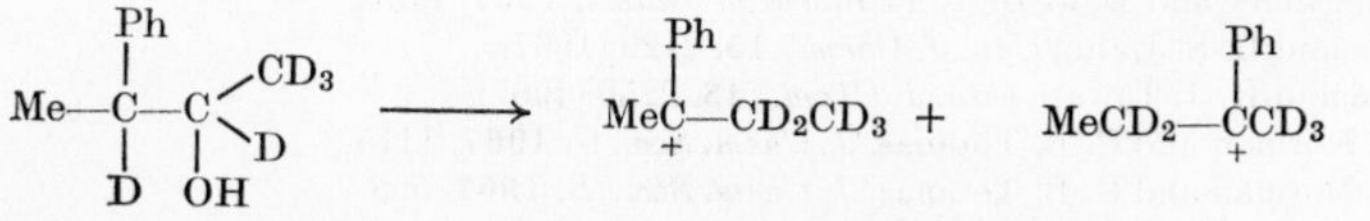

. . . (15)

phenonium ions are formed in a similar way and their spectra are similar to that of the *p*-methoxyphenonium ion.

threo- and *erythro*-$\alpha\beta$-Dimethylphenethyl alcohol do not yield a mixture of *cis*- and *trans*-dimethylphenonium ions when dissolved in SO_2–FSO_3H–SbF_5 at –60° as reported previously,[95] but instead they form the diprotonated sulphinic acid (**89**).[98] In SbF_5–FSO_3H at –70° the benzenonium ion (**88**) is formed[99,100] and, on warming to –50°, this is converted into the α-ethyl-α-methylbenzyl cation. Experiments with deuterium-labelled compounds indicate that this is formed with extensive phenyl migration (cf. eqns. 13—15).[100]

The carbon-13-proton coupling constant for the protons at position 9 of anthracenonium ion (**90**) is 127.5 cps, very similar to that for diphenylmethane

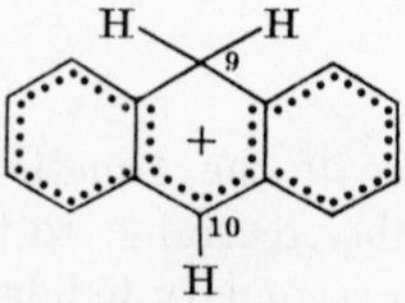

(90)

and characteristic of sp^3-hybridized carbon.[101] The proton and fluorine magnetic resonance spectra of some fluorobenzenonium ions have been reported.[102a] An MO description of the phenonium ion has been published.[102b]

Kingsbury and Best have determined the effect of substituents in both rings of *erythro*-1,2-diphenylpropanol on the steric course of the reaction with HCl to yield the corresponding chlorides.[103]

The following reactions involving the migration of aryl groups have also been investigated: deoxidation of [α-^{14}C]phenethyl alcohol,[104] dermercuration of [α-^{14}C]phenethylmercuric perchlorate,[105] dehydration of 1-methyl-2,2,2-triphenylethanol with phosphoryl chloride in pyridine,[106] and addition of bromine to olefins.[107] Aryl migration does not occur in the conversion of

[98] M. Brookhart, F. A. L. Anet, and S. Winstein, *J. Am. Chem. Soc.*, **88**, 5657 (1966).

[99] G. A. Olah, C. U. Pittman, E. Namanworth, and M. B. Comisarow, *J. Am. Chem. Soc.*, **88**, 5571 (1966).

[100] M. Brookhart, F. A. L. Anet, D. J. Cram, and S. Winstein, *J. Am. Chem. Soc.*, **88**, 5659 (1966).

[101] V. A. Koptyug, I. S. Isaev, and A. I. Rezvukhin, *Tetrahedron Letters*, **1967**, 823; see also V. A. Koptyug, V. A. Bushmelev, and T. N. Gerasimova, *Zh. Obschch. Khim.*, **37**, 140 (1967); *Chem. Abs.*, **66**, 94813e (1966).

[102a] G. A. Olah and T. E. Kiovsky, *J. Am. Chem. Soc.*, **89**, 5692 (1967).

[102b] N. F. Phelan, H. H. Jaffé, and M. Orchin, *J. Chem. Educ.*, **44**, 626 (1967).

[103] C. A. Kingsbury and D. C. Best, *Tetrahedron Letters*, **1967**, 1499.

[104] C. C. Lee and B.-S. Hahn, *Can. J. Chem.*, **45**, 2129 (1967).

[105] C. C. Lee and R. J. Tewari, *Can. J. Chem.*, **45**, 2256 (1967).

[106] R. O. C. Norman and C. B. Thomas, *J. Chem. Soc., C*, **1967**, 1115.

[107] R. O. C. Norman and C. B. Thomas, *J. Chem. Soc., B*, **1967**, 598.

4-nitrophenethyl alcohol into trimethyl-(4-nitrophenethyl)ammonium iodide.[108]

Participation by Double[109] and Triple Bonds

The acetolysis of 2-(norbornen-*syn*-7-yl)ethyl *p*-bromobenzenesulphonate (**91**) yields 22% of tetracyclo[4.3.0.0^{2,4}0^{3,7}]nonane (**100**) ("deltacyclane"), 42% of *exo*-2-brendyl acetate (**98**), and 36% of *exo*-4-brexyl acetate (**99**).[110] Partially solvolysed reaction mixtures contain, in addition to the starting *p*-bromobenzenesulphonate, *exo*-2-brendyl *p*-bromobenzenesulphonate (**97**) but no *exo*-4-brexyl *p*-bromobenzenesulphonate (**96**); the latter product would,

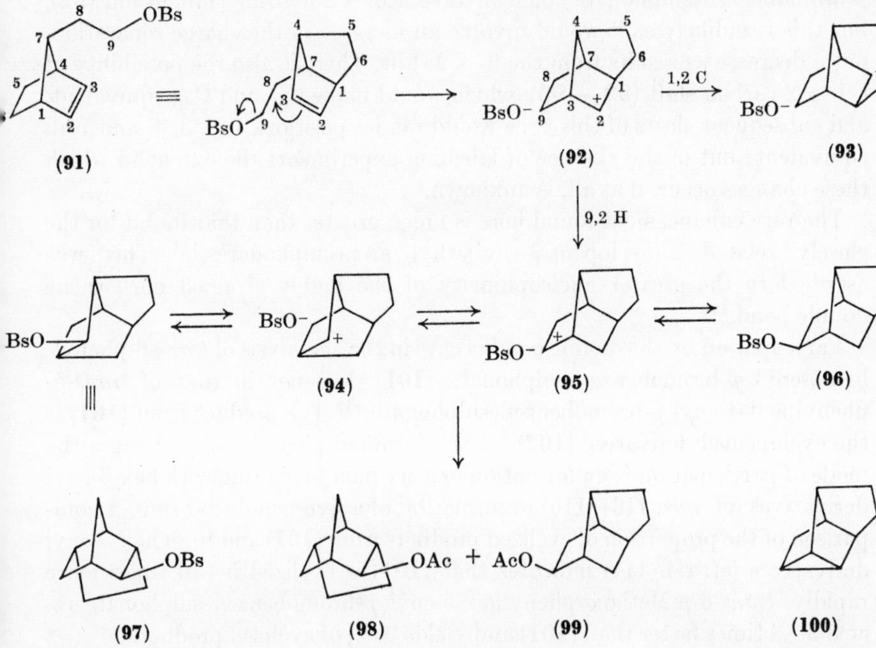

however, have undergone rapid acetolysis under the conditions used, and hence if formed would not have been detected. Some support for its being formed was obtained by showing that (**91**) yielded both it and the brendyl *p*-bromobenzenesulphonate under non-solvolytic conditions in carbon tetrachloride solution. The brendyl and brexyl *p*-bromobenzenesulphonates yield the same products as the 2-(norbornen-*syn*-7-yl)ethyl *p*-bromobenzenesulphonate on acetolysis, and the ratio of brendyl to brexyl acetate is the same (1.1—1.2:1) from all three reactions.

108 E. M. Hodnett and W. J. Dunn, *J. Org. Chem.*, **32**, 3230 (1967).
109 M. Hanack and H.-J. Schneider, *Fortschr. Chem. Forsch.*, **8**, 554 (1967).
110 R. S. Bly, R. K. Bly, A. O. Bedenbaugh, and O. R. Vail, *J. Am. Chem. Soc.*, **89**, 880 (1967).

The rearrangement of (**91**) to the brendyl *p*-bromobenzenesulphonate in acetic acid probably proceeds by ion-pair return rather than through free ions since the proportion occurring was unaffected by the addition of sodium *p*-bromobenzenesulphonate or lithium perchlorate.[110]

Acetolysis of 2-(norbornen-*syn*-7-yl)ethyl *p*-bromobenzenesulphonate is 190,000 times faster than that of its *anti*-isomer and 140,000 times faster than that of 2-(7-norbornyl)ethyl *p*-bromobenzenesulphonate. Clearly there is participation by the double bond in the rate-determining step, and the reaction can be formulated most simply as proceeding through ions (**92**), (**94**), and (**95**) or their non-classical equivalents. The rearrangement to the brendyl *p*-bromobenzenesulphonate could involve a $6 \rightarrow 2$ hydride shift in ion (**92**), but this is unlikely as it would involve an increase in the charge separation, not a decrease as results from the $9 \rightarrow 2$ shift. There is also the possibility of a $1 \rightarrow 2$-carbon shift (**92** → **93**), which would make $C_{(8)}$ and $C_{(9)}$ equivalent, and subsequent shifts of this type would render positions 1, 2, 3, 4, and 7 all equivalent; but in the absence of labelling experiments the extent to which these changes occur, if at all, is unknown.

The rate-enhancement found here is much greater than that found for the closely related 2-(cyclopent-3-enyl)ethyl arenesulphonates.[111] This was ascribed to the greater nucleophilicity of the highly strained norbornene double bond.[110]

Participation by the double bond occurs in the acetolysis of *trans*-6-phenylhex-5-enyl *p*-bromobenzenesulphonate (**101**), but not in that of *trans*-5-phenylpent-4-enyl *p*-bromobenzenesulphonate.[112] The product from (**101**) is the cyclopentyl derivative (**102**), so the terminal phenyl group changes the mode of participation from formation of a six-membered ring with hex-5-enyl derivatives (cf. refs. 113—115) to formation of a five-membered ring. A comparison of the proportion of cyclized products from (**101**) and from hex-5-enyl derivatives (cf. ref. 114) indicates that (**101**) is cyclized 5—10 times more rapidly. *trans*-6-*p*-Methoxyphenylhex-5-enyl *p*-bromobenzenesulphonate reacts 2—3 times faster than (**101**) and yields 90% of cyclized product.

The solvolyses of hex-5-enyl derivatives have also been investigated.[113–115]

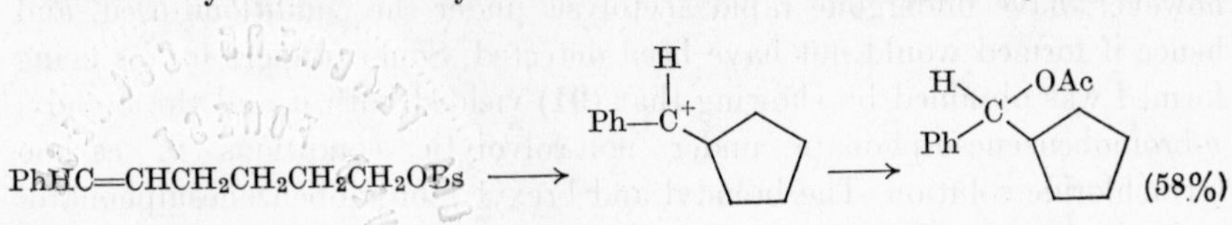

(**101**) (**102**)

[111] See *Organic Reaction Mechanisms*, **1965**, 37.
[112] W. D. Closson and S. A. Roman, *Tetrahedron Letters*, **1966**, 6015.
[113] W. S. Trahanovsky and M. P. Doyle, *J. Am. Chem. Soc.*, **89**, 4867 (1967).
[114] W. S. Trahanovsky, M. P. Doyle, and P. D. Bartlett, *J. Org. Chem.*, **32**, 150 (1967).
[115] W. S. Trahanovsky and M. P. Doyle, *Chem. Commun.*, **1967**, 1021.

Solvolysis of (**103**) in acetone–water (20:80) proceeds with participation by the double bond to yield (**105**)—(**108**). It is possible to write all of these as formed through the cyclopropylmethyl ions (**104**), although Hanack and his

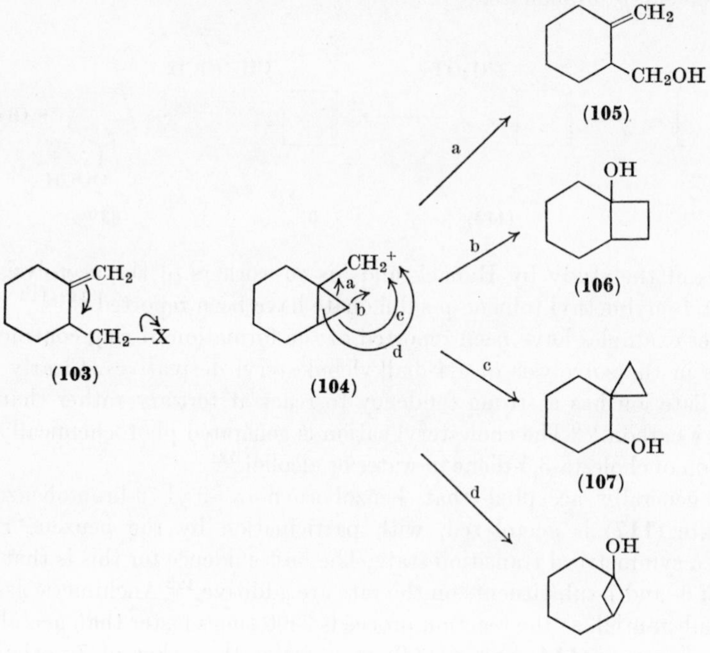

(**108**)

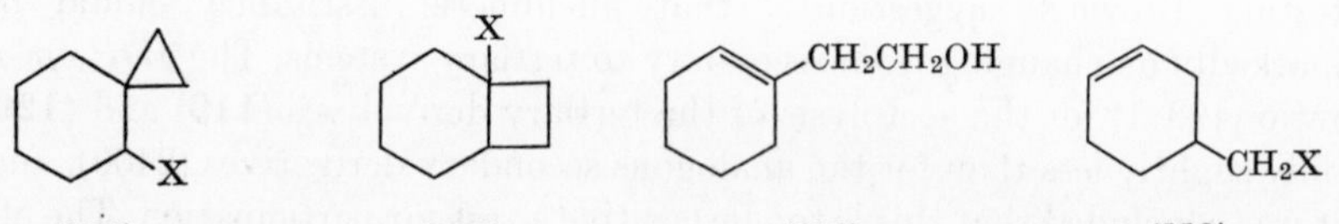

(**109**) (**110**) (**111**) (**112**)

X = p-$NO_2 \cdot C_6H_4 \cdot CO_2$—

co-workers[116] preferred to regard (**108**) as resulting from the methylene-cyclohept-3-yl cation. The hydrolyses of (**109**) and (**110**) also yield (**106**) and (**107**), and some (**111**) as well. In the above-named solvent mixture the rate of solvolysis of (**103**) was only twice as fast as that of (**112**), which reacts without participation, but in 30% acetone–water it was 100 times faster, indicating substantial anchimeric assistance. The *m* constant of the Grunwald equation was 1, similar to that for *tert*-butyl chloride.[116]

[116] H. Schneider-Bernlöhr, H.-J. Schneider, and M. Hanack *Tetrahedron Letters* **1967**, 1425; see also M. Hanack, H.-J. Schneider, and H. Schneider-Bernlöhr, *Tetrahedron*, **23**, 2195 (1967).

Solvolysis of cyclobut-2-enylmethyl toluene-*p*-sulphonate (**113**) does not proceed with homoallylic participation by the double bond but with ring expansion. The rate is slightly less than that of the solvolysis of cyclobutylmethyl toluene-*p*-sulphonate.[117]

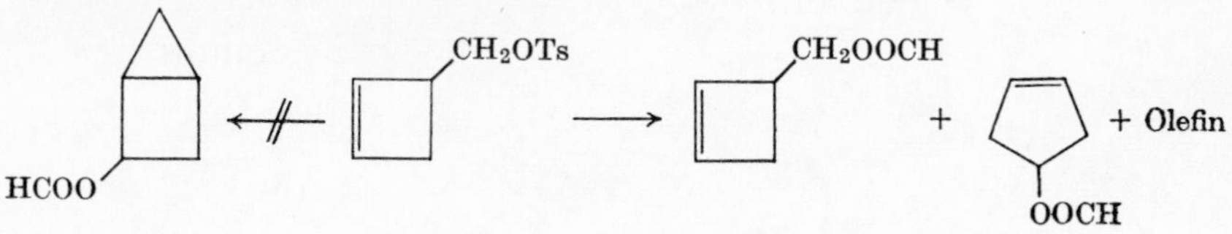

(**113**) 3% 63%

Details of the study by Hanack and his co-workers of the acetolysis of cyclo-oct-4-enylmethyl toluene-*p*-sulphonate have been reported.[118,119]

Further examples have been reported of the formation of ring-contracted products in the solvolysis of 4,4-dialkylcholesteryl derivatives. Clearly the intermediate ion has a strong tendency to react at tertiary rather than at secondary carbon.[120] The cholesteryl cation is generated photochemically by irradiation of cholesta-3,5-diene in water or alcohol.[121]

It is generally accepted that benzobornen-*anti*-9-yl *p*-bromobenzenesulphonate (**117**) is acetolysed, with participation by the benzene ring, through a symmetrical transition state. The best evidence for this is that the effects of 6- and 7-substituents on the rate are additive.[122] Anchimeric assistance is substantial, as the reaction proceeds 2400 times faster than acetolysis of the *syn*-isomer (**114**) and 80,000 times faster than that of 7-norbornyl *p*-bromobenzenesulphonate at 50°. This is therefore an excellent system for testing Brown's suggestion[123] that anchimeric assistance should drop markedly on changing from secondary to tertiary systems. The *anti*:*syn*-rate ratio (493:1) for the acetolysis of the tertiary derivatives (**119**) and (**120**) is only slightly less than for the analogous secondary derivatives (2400), and so it was concluded that this is too insensitive a test for participation. The effect of introducing an α-methyl substituent into isopropyl derivatives ($Me_2HC– \rightarrow Me_3C–$) is to increase the rate of solvolysis by a factor of 55,000 and into α-methylbenzyl derivatives by a factor of 1800. The transition state in which charge is delocalized into the phenyl group makes, therefore, a

[117] M. Hanack and K. Riedlinger, *Chem. Ber.*, **100**, 2107 (1967).
[118] M. Hanack, W. Kraus, W. Rothenwöhrer, W. Kaiser, and G. Wentrup, *Ann. Chem.*, **703**, 44 (1967); K. H. Baggaley, J. R. Dixon, J. M. Evans, and S. H. Graham, *Tetrahedron*, **23**, 299 (1967).
[119] See *Organic Reaction Mechanisms*, **1966**, 27—28.
[120] G. Just, N. D. Hall, and K. St. C. Richardson, *Can. J. Chem.*, **45**, 2521 (1967).
[121] G. Bauslaugh, G. Just, and E. Lee-Ruff, *Can. J. Chem.*, **44**, 2837 (1966).
[122] See *Organic Reaction Mechanisms*, **1966**, 27.
[123] H. C. Brown, *Chem. Brit.*, **2**, 202 (1966).

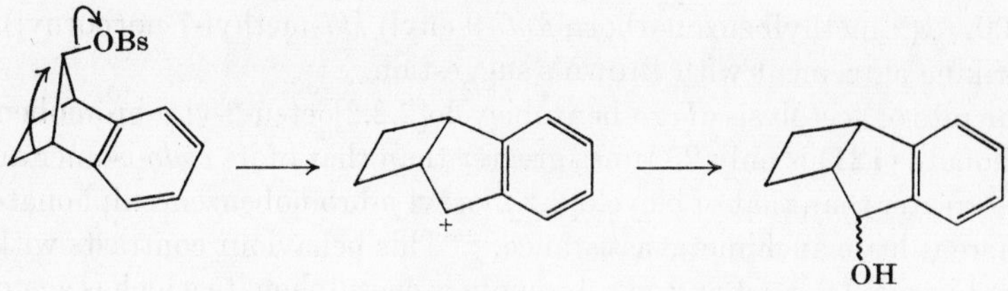

(114) **(115)** **(116)**

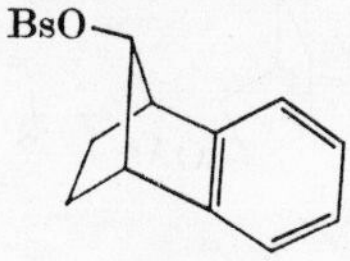

(117)

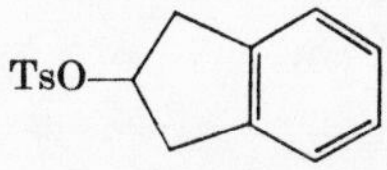

(118)

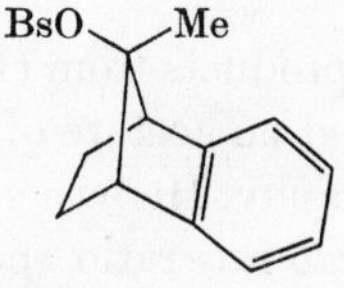

(119)

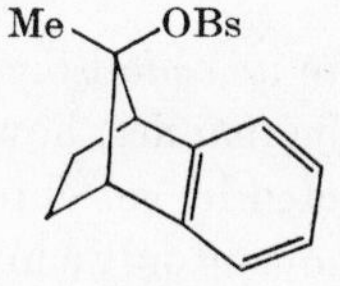

(120)

smaller electronic demand on the α-methyl substituent. The ratios of the rates of acetolysis of the *anti* and *syn* tertiary *p*-bromobenzenesulphonates (**119**) and (**120**) relative to those of the analogous secondary ones (18,000:1 and 86,000:1, respectively) are thus consistent with greater charge delocalization into the benzene ring in the transition state for the *anti*-isomers, but the difference was again considered to be too small to be a reliable test for participation.[124] These arguments depend on the (unstated) premise that the *syn*-isomers react without participation. This is not so, however. Benzonorbornen-*syn*-9-yl *p*-bromobenzenesulphonate is acetolysed 30 times faster than 7-norbornyl *p*-bromobenzenesulphonate despite the presence of the electron-withdrawing benzene ring, and the hydrolysis yields rearranged products (**116**). The driving force for this participation is formation of the benzylic ion (**115**), and consistently with this the effect of substituents on the rate is larger than with inden-2-yl *p*-bromobenzenesulphonate (**118**). The electronic demands made on an α-methyl group in the transition state for solvolysis of the *syn*-isomers may therefore also be reduced. A more valid comparison could possibly be made between the benzonorbornen-*anti*-9-yl and the 7-norbornyl compound, although this is not ideal since the latter also reacted with participation (see p. 15). The figures are *k*(benzonorborn-*anti*-9-enyl)/*k*(7-norbornyl) =

[124] H. Tanida, Y. Hata, S. Ikegami, and H. Ishitobi, *J. Am. Chem. Soc.*, **89**, 2928 (1967).

80,000, k(9-methylbenzonorborn-*anti*-9-enyl)/k(7-methyl-7-norbornyl) = 28, in striking agreement with Brown's suggestion.

The rate of acetolysis of *exo*-benzobicyclo[2.2.2]octen-2-yl *p*-bromobenzenesulphonate (**121**) is only 28 times greater than that of its *endo*-isomer and 2.6 times greater than that of bicyclo[2.2.2]octyl *p*-bromobenzenesulphonate, and so there is little anchimeric assistance.[125] This behaviour contrasts with that of *exo*-bicyclo[2.2.2]octen-2-yl *p*-bromobenzenesulphonate which is acetolysed

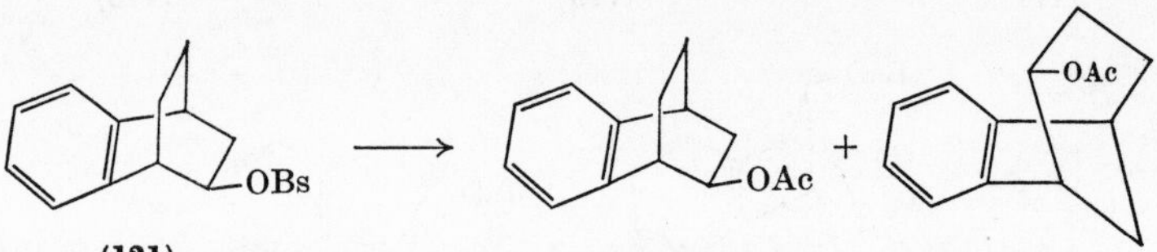

(**121**)

83% +no epimer

17% +no epimer

52 times faster than its *endo*-isomer. The products from (**121**) are formed in a highly stereospecific manner, however, and no acetate of inverted configuration could be detected by gas chromatography. Here then is another reaction which, although showing only a low *exo*:*endo*-rate ratio, shows a high *exo*:*endo*-product ratio. It is difficult to ascribe this to a high initial-state energy for the *endo*-isomer or a low initial-state energy for the *exo*-isomer since if anything the reverse would be expected, but clearly this could be determined. The most likely explanation seems to us to reside in the fact that microscopic reverse of ionization is, in this instance, not a good model for capture of the ionic intermediate by solvent. Again the difference in the dipoles of the breaking and the forming bonds and the role of ion pairs in the ionization process may be important factors (see also p. 11).

The solvolyses of *exo*- and *endo*-benzobicyclo[2.2.2]octadienyl *p*-bromobenzenesulphonates were also investigated. The 49-fold greater rate of acetolysis of bicyclo[2.2.2]octyl *p*-bromobenzenesulphonate than of cyclohexyl *p*-bromobenzenesulphonate was commented on but it was not possible to rationalize its origin.[125] In our opinion it arises most probably from release of eclipsing strain (steric and torsional?) in the transition state.

The norbornadienyl cation undergoes a fascinating degenerate rearrangement[126] whose occurrence was first suggested by a broadening of the signals of protons bound to the bridgehead, $C_{(5)}$ and $C_{(6)}$, but not of those bound to $C_{(1)}$ or $C_{(2)}$, at +77° in FSO_3H. The nature of this rearrangement was made apparent by the NMR spectrum of the 5-$^2H_1$-ion which showed that the label

125 H. Tanida, K. Tori, and K. Kitahonoki, *J. Am. Chem. Soc.*, **89**, 3213 (1967); see also refs. 93 and 94.

126 M. Brookhart, R. K. Lustgarten, and S. Winstein, *J. Am. Chem. Soc.*, **89**, 6350, 6352, 6354 (1967).

was scrambled over positions 1, 7, 4, 5, and 6 at –47° with rate constant 3×10^{-4} sec^{-1} but not over positions 2 and 3; the label appeared first at the bridgehead positions (1 and 4) and then at the bridge position (7). Similar

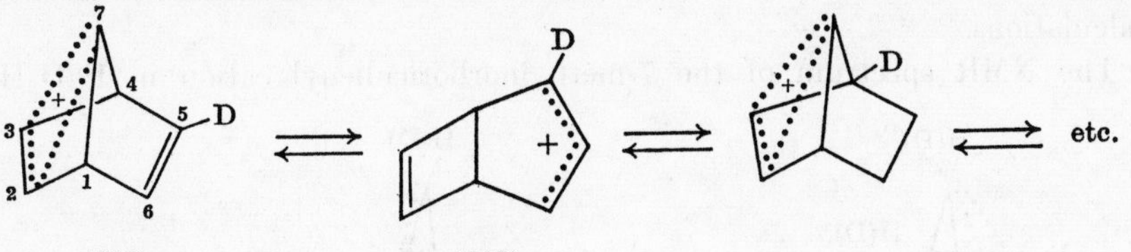

(122) **(123)**

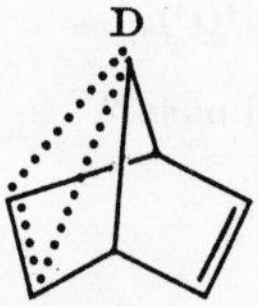

(124)

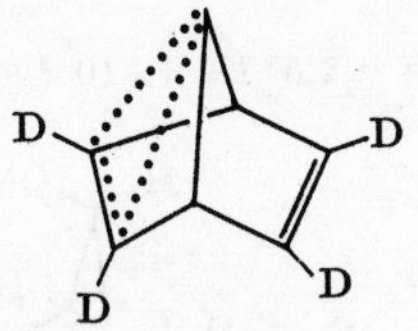

(125)

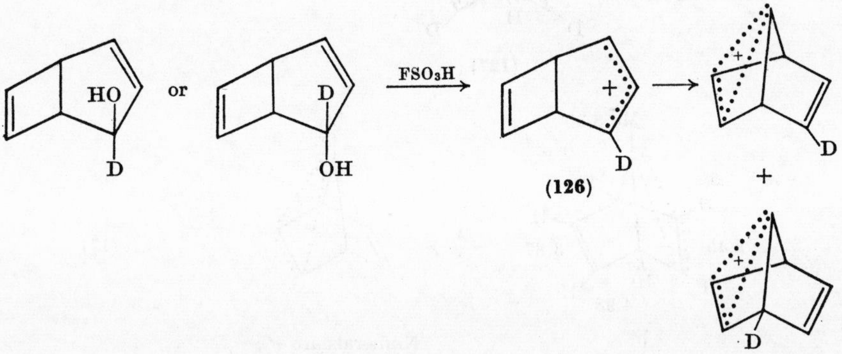

(126)

rearrangements at similar rates were also demonstrated for ions (**124**) and (**125**). Ring contraction to the bicyclo[3.2.0]heptadienyl cation (**123**) followed by ring expansion was proposed as the most reasonable pathway. Consistently with this interpretation the specifically deuterated ion (**126**) (not observed) generated from the deuterated *cis*- and *trans*-alcohols yields equal amounts of 1- and 5-deuterated norbornadienyl cation. It is interesting that neither ring-contraction nor expansion occurs in the solvolysis reactions where the life-time of the carbonium ions is much shorter (see, however, ref. 132, p. 36).

The norbornadienyl cation also undergoes a slower rearrangement which results in $C_{(2)}$ and $C_{(3)}$ becoming equivalent with $C_{(5)}$ and $C_{(6)}$. This was observed by the appearance of signals from protons bound to $C_{(2)}$ and $C_{(3)}$ in the NMR spectrum of the ion derived from (**127**) (see eqn. 16). A possible pathway is bridge flipping, and the value of $\Delta G^\ddagger$ for the rearrangement (+19.6 kcal $mole^{-1}$) therefore sets a lower limit on the energy barrier for this, which is much greater than that (8 kcal $mole^{-1}$) from extended Hückel calculations.

The NMR spectrum of the 7-methylnorbornadienyl cation in FSO_3H

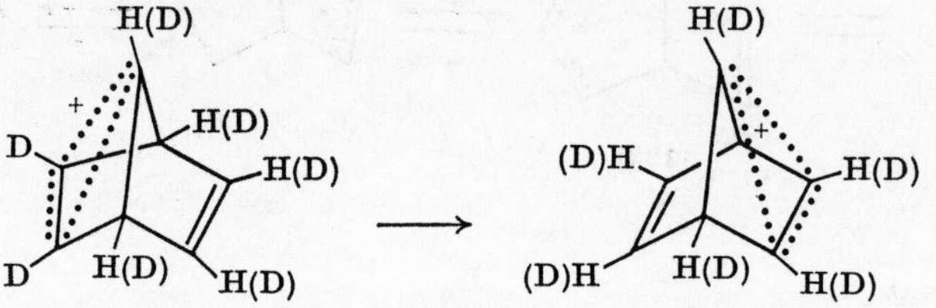

...(16)

At − 2.5° $k = 8 \times 10^{-4}$ sec^{-1}, $\Delta G^\ddagger = 19.6$ kcal $mole^{-1}$

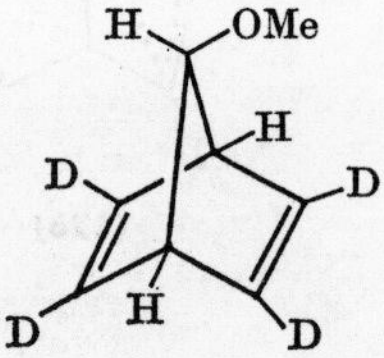

(**127**)

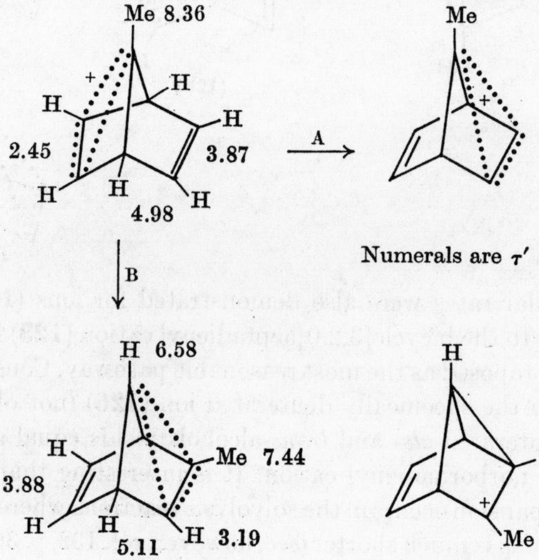

...(17)

Numerals are τ'

(**128**)

For A, k at $-14° = 189$ sec^{-1}, $\Delta G^\ddagger = 12.4$ kcal $mole^{-1}$
For B, k at $-17° = 1.4 \times 10^{-4}$ sec^{-1}, $\Delta G^\ddagger = 18.9$

indicates that this is an unsymmetrical species at –45°, but as the temperature is raised the vinyl signals broaden, and they coalesce at –14°. At the same time the bridgehead signals sharpen, and so only the vinyl protons are being averaged and this must therefore be the result of bridge flipping (eqn. 17). Above –5° rearrangement to the 2-methylnorbornadienyl ion occurs. The chemical shifts of the methyl groups in the 7- and 2-methyl ions ($\tau' = 8.36$ and 7.44 ppm, respectively) supports the view[127] that more of the positive charge resides at $C_{(2)}$ and $C_{(3)}$ than at $C_{(7)}$. The spectrum of the 2-methyl ion is additional evidence against the formulation of norbornadienyl ions as rapidly equilibrating tricyclic ions.[128] If this were so, the 2-methyl ion would be expected to be mainly the tertiary ion (**128**) and the observed spectrum would be derived from this. The chemical shifts of the ring protons of cyclopropylmethyl cations are usually similar (difference < 1 ppm) but here they are quite different (τ' for $H_{(3)} = 3.19$ and for $H_{(7)} = 6.58$ ppm), which argues against this formulation.

The large reduction ($\delta\Delta G^{\ddagger} > 7.2$ kcal mole^{-1}) in the energy barrier for ring flipping caused by the introduction of a 7-methyl group is expected, since much greater stabilization of the symmetrical norbornadienyl ion than of the unsymmetrical ion should result. The introduction of substituents still better able to accommodate positive charge should reduce the energy barrier further and might even cause the symmetrical ion to become the more stable. The first part of this expectation, at least, is borne out when methoxyl and phenyl substituents are introduced, since all four vinyl protons are equivalent in the spectra of the 7-methyl and norbornadien-7-yl cations down to –100°, indicating that, if the ions are unsymmetrical, the barrier to bridge flipping is <7.6 kcal mole^{-1}. It was thought that the 7-phenyl ion is probably unsymmetrical since the phenyl signal was relatively narrow (half-band width 9 cps), whereas in an α-phenyl carbonium ion (as the symmetrical ion would be) the difference in chemical shifts of *ortho*- and *meta*-protons are usually large (> 0.5 ppm). The bridgehead protons of the 7-methoxy-ion are not equivalent at –65° as one is *syn*- and the other *anti*- to the methoxyl-methyl group. $\Delta G^{\ddagger}$ for the *syn–anti*-isomerization is 11.7 kcal mole^{-1}, slightly less than that for the methoxybenzenium ion (13.0 kcal mole^{-1}). Possibly a higher barrier would be expected in a symmetrical methoxynorbornadienyl ion than in a methoxybenzenium ion and so this may be evidence in favour of the unsymmetrical structure.[126] The signal from the OH group in the NMR spectrum of protonated 7-norbornenone is at 1.55 ppm higher field than that from the OH group of protonated 7-norbornanone, consistently with delocalization of the π-electrons of the double bond of the former.[129]

[127] See *Organic Reaction Mechanisms*, **1965**, 12—13.
[128] See *Organic Reaction Mechanisms*, **1966**, 26.
[129] M. Brookhart, G. C. Levy, and S. Winstein, *J. Am. Chem. Soc.*, **89**, 1737 (1967).

The unsubstituted and the 7-phenyl- and 7-methoxy-norbornadienyl ions (but not the 7-methyl ion) undergo a relatively slow rearrangement to the corresponding tropylium ions, rates decreasing in the order 7-Ph > 7-H > 7-MeO.[126]

Deamination of norbornen-*anti*-7-ylamine and benzonorbornen-*anti*-9-ylamine have been investigated.[130]

The structure of the norbornen-7-yl cation has been discussed in terms of a general theory of bicycloaromaticity.[131]

Although, as mentioned above, the norbornadienyl and bicyclo[3.2.0]-heptadienyl cations are apparently not interconverted in solvolysis reactions, the closely related substituted bicyclo[3.2.0]heptenyl cation (**131**) is converted almost completely into the norbornen-7-yl cation (**132**) in acetic acid. Thus acetolysis of both *cis*- and *trans*-toluene-*p*-sulphonates (**129**) and (**130**) yields mainly products with the norbornene skeleton[132] (see also p. 33).

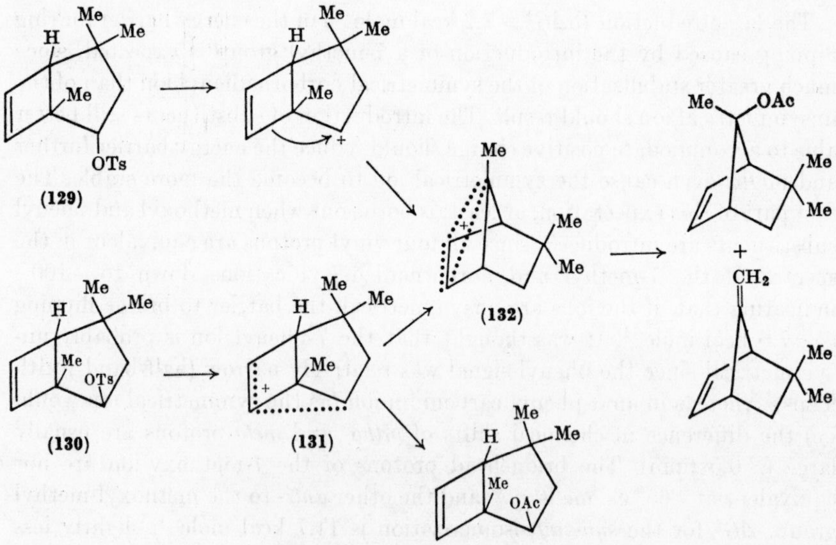

Other investigations of participation by double bonds include studies of solvolyses of 3-indolylethyl derivatives,[133a] 3-(cyclohex-2-enyl)propyl and *cis*- and *trans*-2-(2-vinylcyclopentyl)ethyl *p*-nitrobenzoate,[133b] cyclo-oct-4-enyl toluene-*p*-sulphonate (see p. 19),[134] cyclohept-3-enyldimethylsul-

[130] H. Tanida, T. Tsuji, and T. Irie, *J. Org. Chem.*, **31**, 3941 (1966).
[131] M. J. Goldstein, *J. Am. Chem. Soc.*, **89**, 6357 (1967).
[132] S. C. Lewis and G. H. Whitham, *J. Chem. Soc., C*, **1967**, 274.
[133a] M. Julia, H. Sliwa, and P. Caubère, *Bull. Soc. Chim. France*, **1966**, 3359.
[133b] P. D. Bartlett, E. M. Nicholson, and R. Owyang, *Tetrahedron*, Suppl. 8, Part 2, 399 (1966).
[134] W. D. Closson and G. T. Kwiatkowski, *Tetrahedron Letters*, **1966**, 6435.

phonium ion,[135] and benzonorbornyl halides;[136] also the cyclization of neryl phosphate;[137] the solvolysis of 3-vinylcyclopentyl bromide proceeds without participation by the double bond.[138]

Participation by an allenic double bond occurs in the acetolysis of (**133**) which yields mainly the products shown. The rate is 19 times greater than

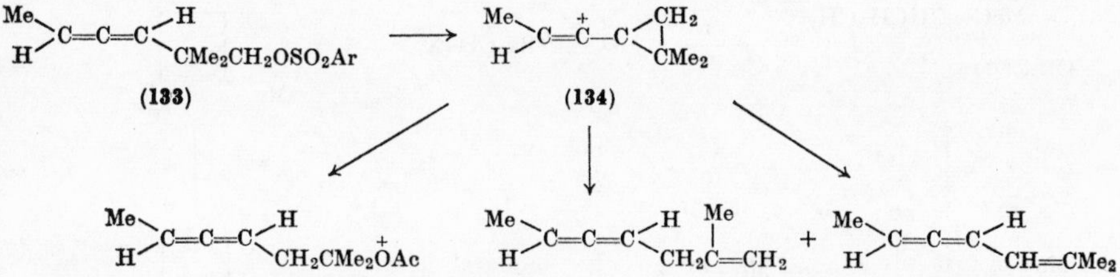

that of 1,2-dimethylbut-3-enyl *p*-bromobenzenesulphonate and 5600 times that of 2,2-dimethylpentyl *p*-bromobenzenesulphonate. The allenic compound lacking the terminal methyl group reacts 3—4 times more slowly. Optically active (**133**) yields optically active products which are formed presumably *via* the vinylic cyclopropylmethyl cation (**134**).[139, 140]

At least one example of apparent participation by a triple bond probably involves an initial addition of solvent followed by participation by the resulting double bond. Addition of formic acid to the triple bond of (**138**) occurs at about the same rate as the solvolysis of (**135**), and it was shown by NMR spectroscopic analysis that (**135**) yields (**136**) in formic acid containing *p*-bromobenzenesulphonic acid, and that this is converted into (**137**) at least 10 times faster than (**135**) in buffered formic acid.[141]

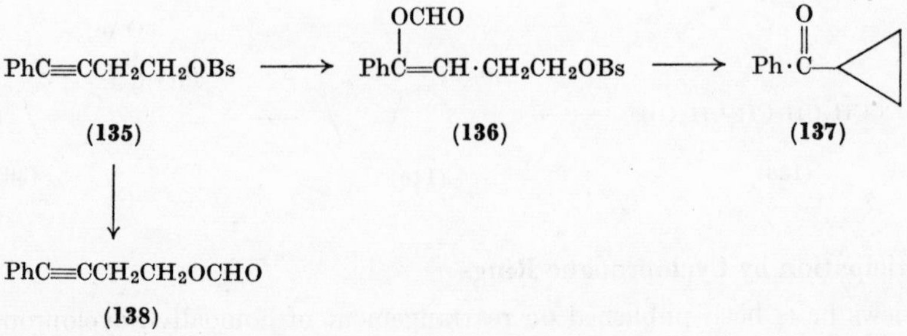

135 C. Chuit and H. Felkin, *Compt. Rend., Ser. C.*, **264**, 1412 (1967).
136 J. W. Wilt, G. Gutman, W. J. Ranus, and A. R. Zigman, *J. Org. Chem.*, **32**, 893 (1967).
137 W. Rittersdorf and F. Cramer, *Tetrahedron*, **23**, 3015 (1967); P. Valenzuela and O. Cori, *Tetrahedron Letters*, **1967**, 3089.
138 J. P. Schaefer and J. Higgins, *J. Org. Chem.*, **32**, 553 (1967).
139 T. L. Jacobs and R. Macomber, *Tetrahedron Letters*, **1967**, 4877.
140 See *Organic Reaction Mechanisms*, **1966**, 31.
141 H. R. Ward and P. D. Sherman, *J. Am. Chem. Soc.*, **89**, 1962 (1967).

Solvolyses of the corresponding alkylacetylenic arenesulphonates do not, however, involve prior addition. The major products of participation are the corresponding cyclobutanones, but in the presence of mercuric acetate, which catalyses additions to triple bonds, alkyl cyclopropyl ketones are formed (see Scheme 3).[142]

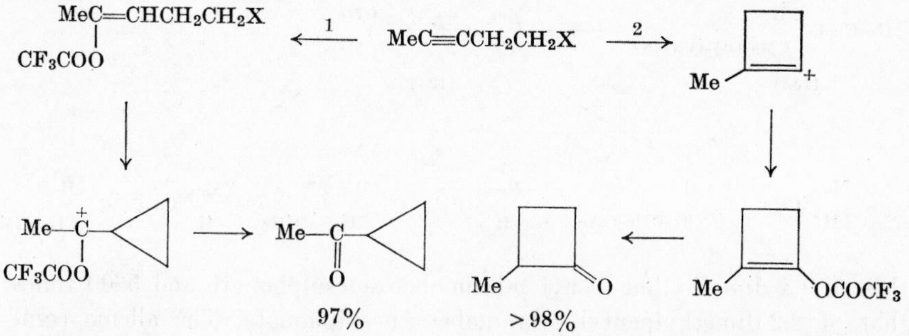

$X = m\text{-}NO_2 \cdot C_6H_4 \cdot SO_3$

Reagents: 1, $CF_3CO_2H–CF_3CO_2Na–Hg(OAc)_2$.
2, $CF_3CO_2H–CF_3CO_2Na$.

Scheme 3

Participation by a more remote triple bond occurs in the acetolysis of (**139**). Unlike the reaction reported last year,[140] in which participation by a terminal triple bond led to formation of a six-membered ring, (**139**) yields a five-membered ring. The stabilizing effect of the phenyl group on the intermediate ion (**140**) and on the transition state for its formation is therefore important.[143]

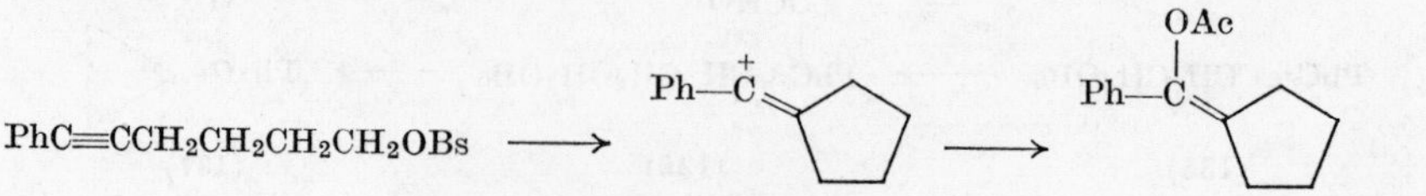

(**139**) (**140**) (36%)

Participation by Cyclopropane Rings

Reviews have been published on rearrangement of homoallyl, cyclopropylmethyl, and cyclobutyl compounds[144] and on conjugation by cyclopropane rings.[145]

142 M. Hanack, I. Herterich, and V. Vött, *Tetrahedron Letters*, **1967**, 3871.
143 W. D. Closson and S. A. Roman, *Tetrahedron Letters*, **1966**, 6015.
144 M. Hanack, and H.-J. Schneider, *Fortschr. Chem. Forsch.*, 8, 554 (1967).
145 J.-P. Pete, *Bull. Soc. Chim. France*, **1967**, 357; see also W. G. Dauben and G. H. Berezin, *J. Am. Chem. Soc.*, **89**, 3449 (1967).

The hydrolysis of (**141**) in 60% aqueous dioxan at 100° is 10^3 times faster than that of norbornen-*anti*-7-yl *p*-nitrobenzoate and hence would be about 10^{14} times faster than that of 7-norbornyl *p*-nitrobenzoate (by alkyl–oxygen fission). The hydrolysis product from (**141**) is a 1:4 mixture of alcohols (**142**)

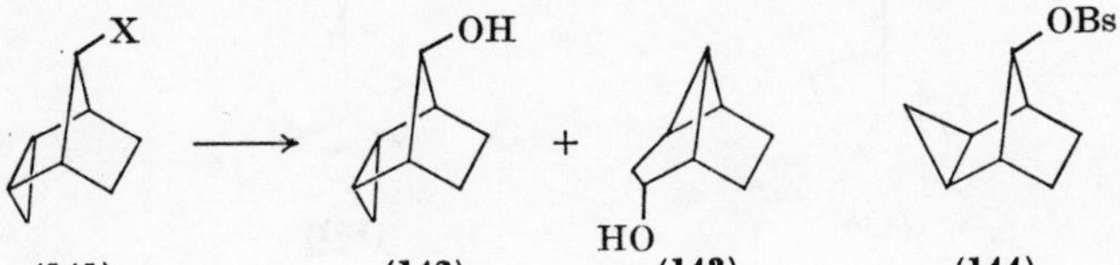

(**141**) (**142**) (**143**) (**144**)

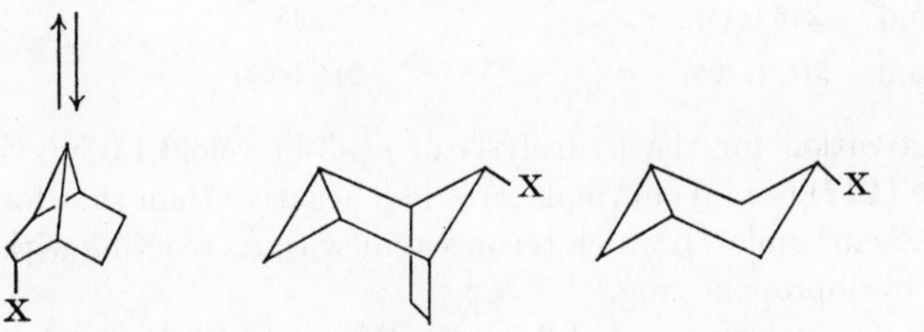

X = *p*-$NO_2 \cdot C_6H_4 \cdot CO \cdot O$

(**145**) (**146**) (**147**)

and (**143**), and some rearranged *p*-nitrobenzoate (**145**) is also formed.[146, 147] These results contrast with the inertness of (**144**)[148] and emphasize the importance of edge-participation by the cyclopropane ring. Presumably the ring opens upwards by a disrotatory process and this would lead to severe steric interactions between the *endo*-hydrogen atoms at positions 2 and 4 in (**144**). The similarity of (**146**) (≡ **141**) to the *cis*-bicyclo[3.1.0]-system (**147**) should also be noted.

Similar considerations hold for the decarbonylation of ketone (**148**) which is much faster than that of the analogous ketone (**149**) with the *exo*-cyclopropane ring. As the cyclopropane ring of (**148**) opens the orbitals on $C_{(2)}$ and $C_{(4)}$ can overlap with the developing *p*-orbitals on $C_{(1)}$ and $C_{(5)}$ which result from the concerted loss of carbon monoxide.[149]

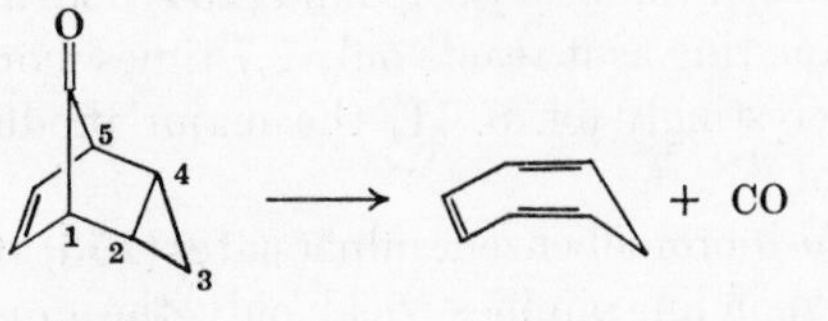

(**148**)

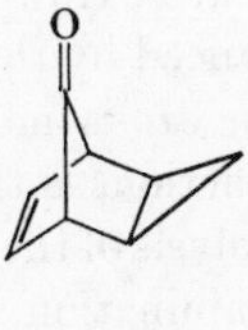

(**149**)

146 H. Tanida, T. Tsuji, and T. Irie, *J. Am. Chem. Soc.*, **89**, 1953 (1967).
147 M. A. Battiste, C. L. Deyrup, R. E. Pincock, and J. Haywood-Farmer, *J. Am. Chem. Soc.*, **89**, 1954 (1967).
148 See *Organic Reaction Mechanisms*, **1966**, 35.
149 B. Halton, M. A. Battiste, R. Rehberg, C. L. Deyrup, and M. E. Brennan, *J. Am. Chem. Soc.*, **89**, 5964 (1967).

It is also interesting that the ultraviolet spectra suggest that there is interaction between the orbitals of the $C_{(2)}$–$C_{(4)}$ bond and those of the carbonyl group in (**150**) but not in (**151**).[150]

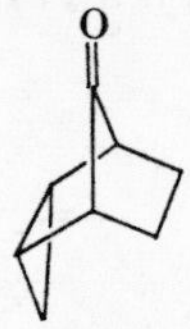

(**150**)

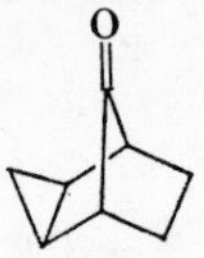

(**151**)

	(150)	(151)
λ_{max} (mμ)	276 (ϵ44)	293
λ_{max} (mμ)	215 (ϵ495)	215 (ϵ63)

The volume of activation for the hydrolysis of *cis*-3-bicyclo[3.1.0]hexyl toluene-*p*-sulphonate (**147**) (–14.0 cm^3 $mole^{-1}$) is less negative than that for its *trans*-isomer (–17.2 cm^3 $mole^{-1}$), which is consistent with its reacting with participation by the cyclopropane ring.[151]

Ultraviolet spectra of solutions of 1,3,5-triarylbicyclo[3.1.0]hexan-3-ol and Lewis acids which may contain trishomocyclopropenyl cations have been reported.[152]

Participation by the cyclopropyl ring does not occur in the acetolysis of either the *exo*- or the *endo*-isomers of (**152**) and (**153**) which proceed at rates similar to those of the analogous norbornyl and norbornenyl toluene-*p*-

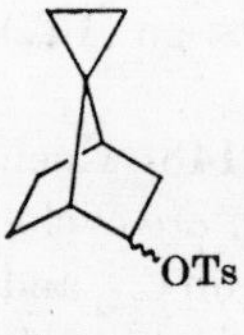

(**152**)

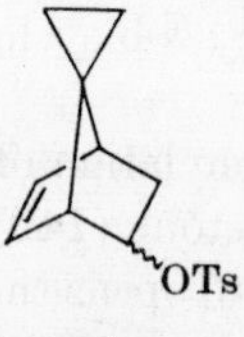

(**153**)

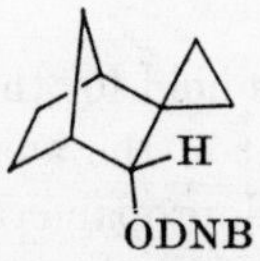

(**154**)

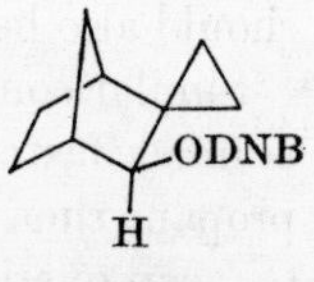

(**155**)

sulphonates. However, hydrolysis of the *endo*-compound (**154**) does appear to be enhanced by the cyclopropane ring as it reacts only 2.7 times more slowly than the *exo*-isomer (**155**). Interestingly (cf. p. 11) the major product from both reactions is the *exo*-alcohol.[153]

Acetolysis of the *exo*- and *endo*-*p*-bromobenzenesulphonates (**156**) and (**157**) and deamination of the corresponding amines yield only the *exo*-acetate. Labelled *endo*-*p*-bromobenzenesulphonate (**159**) yielded an equimolar

[150] R. E. Pincock and J. Haywood-Farmer, *Tetrahedron Letters*, **1967**, 4759.
[151] W. J. le Noble, B. L. Yates, and A. W. Scaplehorn, *J. Am. Chem. Soc.*, **89**, 3751 (1967).
[152] W. Broser and D. Rahn, *Chem. Ber.*, **100**, 3472 (1967).
[153] C. F. Wilcox, R. G. Jesaitis, *Tetrahedron Letters*, **1967**, 2567; see also C. F. Wilcox and R. G. Jesaitis, *Chem. Commun.*, **1967**, 1046.

mixture of acetates (**160**) and (**161**), suggesting that the ion (**158**) (or the analogous classical ions) is an intermediate. The NMR spectrum from *exo*- and *endo*-alcohols in FSO_3H at –55° to –10° is also consistent with presence

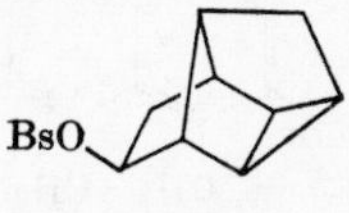

(**156**)

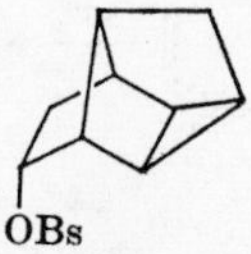

(**157**)

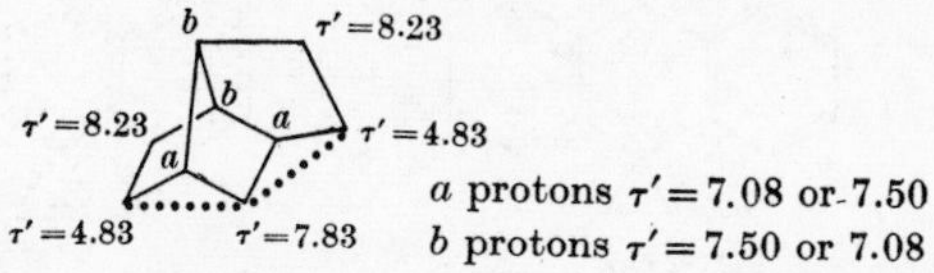

(**158**)

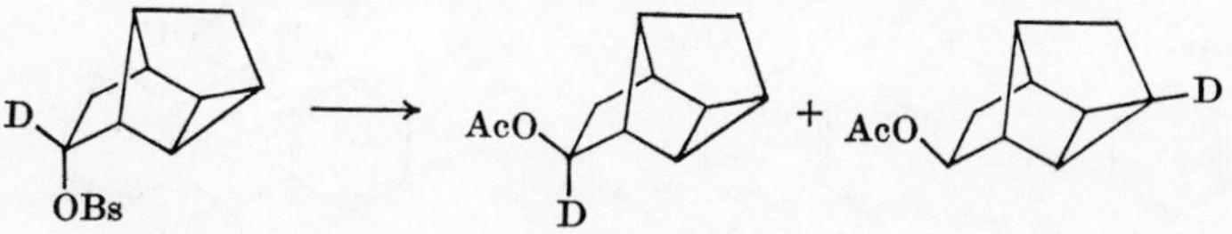

(**159**) (**160**) (**161**)

of this ion, and the signals were assigned as shown. *exo*-*p*-Bromobenzenesulphonate (**156**) was acetolysed about 60 times faster than its *endo*-isomer (**157**), but both rates were considerably larger ($10^{5.9}$ and $10^{3.7}$, respectively) than those calculated from the Foote–Schleyer correlation. The reason for the high rate of the *endo*-isomer is uncertain.[154]

The acetolysis of nortricyclomethyl *p*-bromobenzenesulphonate has been investigated further.[155, 156]

Participation by an even more remote cyclopropyl ring occurs in the acetolysis of (**162**), which yields homoadamant-3-yl acetate (**163**). The anchimeric assistance provided here is probably small since, although (**162**) reacts about 4500 times faster than cyclohexyl toluene-*p*-sulphonate, it reacts only about 5 times faster than (**164**) (cf. ref. 17).[157]

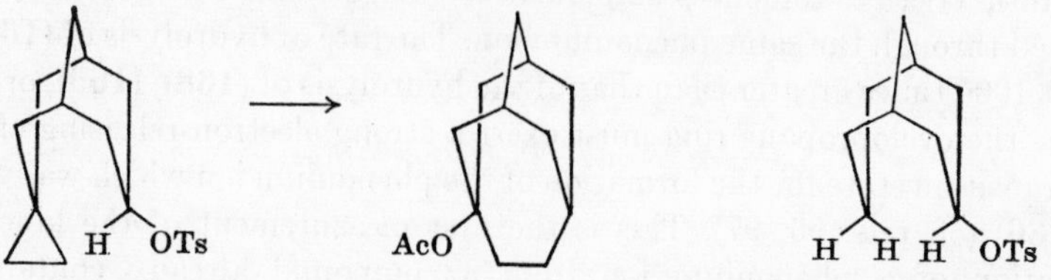

(**162**) (**163**) (**164**)

Two interesting demonstrations of electronic effects exerted by remote cyclopropane rings have been reported by Winstein and his co-workers. The

[154] P. K. Freeman and D. M. Balls, *Tetrahedron Letters*, **1967**, 437.
[155] R. R. Sauers, J. A. Beisler, and H. Feilich, *J. Org. Chem.*, **32**, 569 (1967).
[156] See *Organic Reaction Mechanisms*, **1965**, 16—17.
[157] M. A. Eakin, J. Martin, and W. Parker, *Chem. Commun.*, **1967**, 955.

hydrolysis of the spiroanthryl acetate (**165**) yielded the same products (**167**) and (**170**), in similar proportions, as were obtained by hydrolysis of

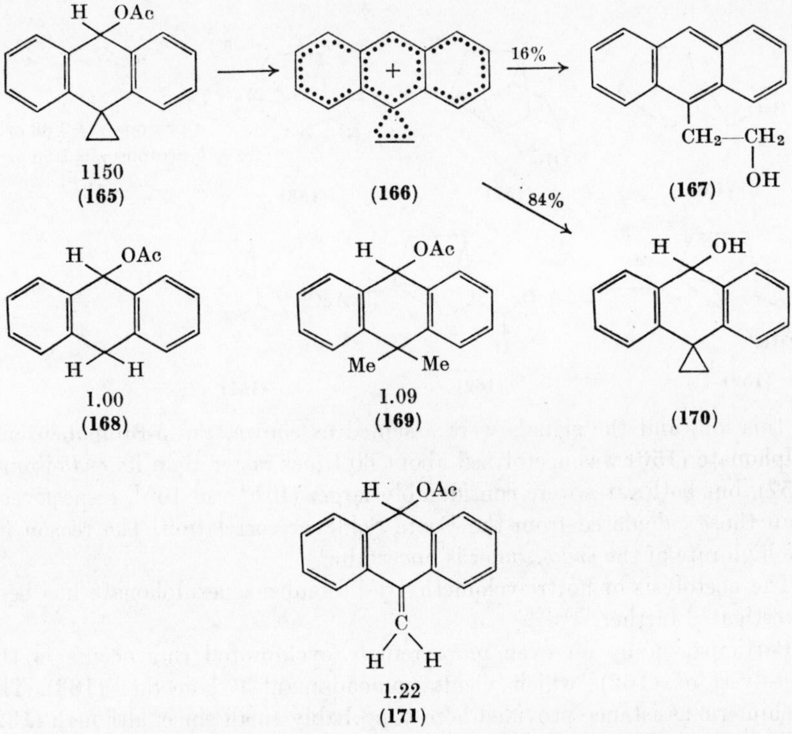

Numerals are relative rates at 25°

2-(9-anthryl)ethyl toluene-*p*-sulphonate,[158] suggesting that both reactions proceed through the same phenonium ion. The rate of hydrolysis of (**165**) was about 1000 times greater than that of the hydrolysis of (**168**), (**169**), or (**171**). Hence the cyclopropane ring must exert a strong electron-releasing effect in the transition state for the formation of the phenonium ion which was written as (**166**) (cf. refs. 96, 97). This is the first measurement of the kinetics of formation of a phenonium ion from a compound already containing a cyclopropane ring.[159]

Hydrolysis of *cis*-acetate (**172**) is about 2×10^2 times faster than that of *trans*-acetate (**174**), and the product from both reactions is mainly the analogous *cis*-alcohol (**176**). The dibenzohomotropylium ion (**173**) was considered to be an intermediate, and there appears therefore to be a stereo-

158 See *Organic Reaction Mechanisms*, **1965**, 34—35.
159 R. Leute and S. Winstein, *Tetrahedron Letters*, **1967**, 2475.

electronic factor which favours its formation from, and reaction to give, *cis*-isomers (**172**) and (**176**), respectively. Interpretation is complicated, however, because compound (**175**) which lacks a cyclopropane ring reacts only slightly more slowly than the *cis*-isomer (**172**).[160]

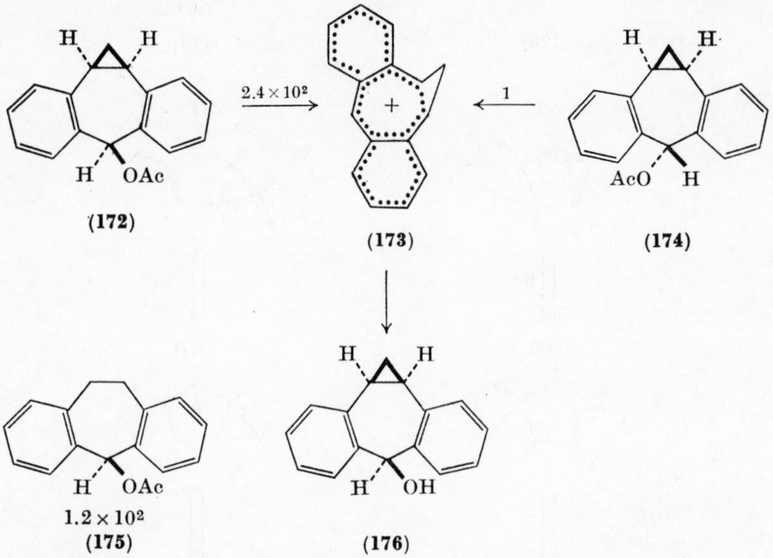

Numerals are relative rates of hydrolysis in 80% aqueous acetone at 25°

Calculations using the ASMO-SCF method predict that the bisected form of the cyclopropylmethyl cation is more stable than the non-bisected form by 0.813 ev. Stabilization arises mainly from a π-type interaction between vacant p-A.O. of the sp^2-carbon and the π-like orbitals of the ring carbons.[161]

The interesting possibility has been explored that a cation such as (**177**) might undergo a five-fold degenerate rearrangement with migration of the cyclopropyl ring around the perimeter of the cyclopentane.[162] Ketone (**178**),

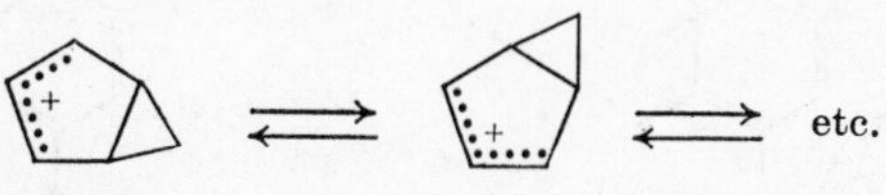

(**177**)

containing six atoms of deuterium and with the protium distribution shown, on treatment with 97% sulphuric acid at 22.6° for 30 min. yielded a ketone (**179**) with the protium distribution shown. It was suggested that (**178**)

[160] R. F. Childs and S. Winstein, *J. Am. Chem. Soc.*, **89**, 6348 (1967).
[161] T. Yonezawa, H. Nakatsuji, and H. Kato, *Bull. Chem. Soc. Japan*, **39**, 2788 (1966).
[162] D. W. Swatton and H. Hart, *J. Am. Chem. Soc.*, **89**, 5075 (1967).

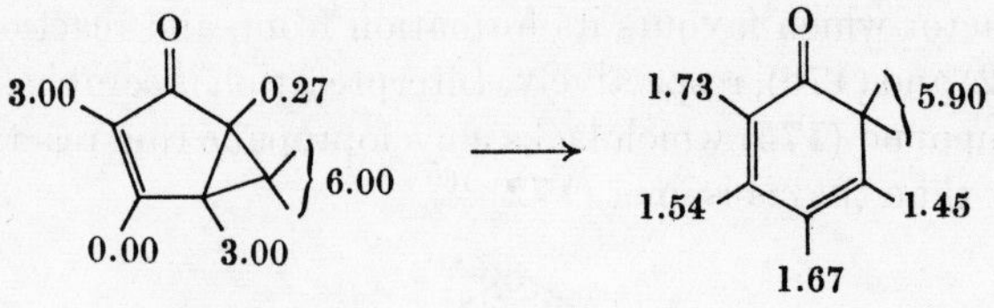

(178) **(179)**

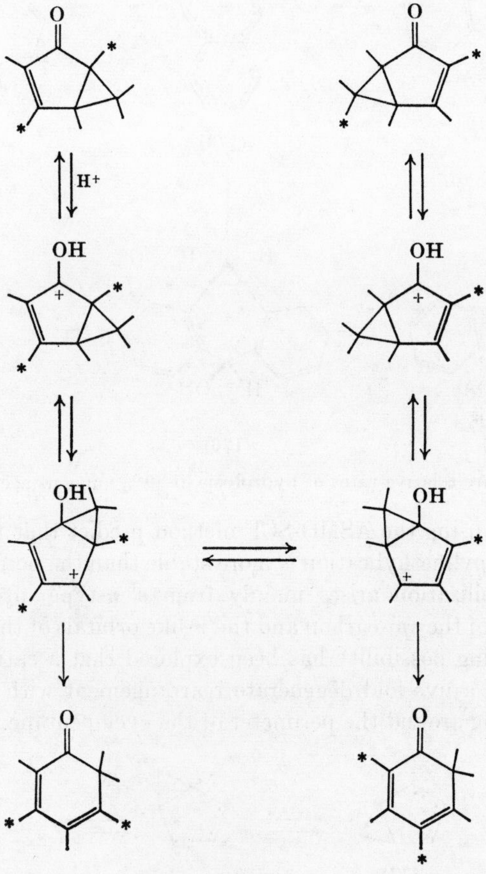

Scheme 4

underwent the rearrangement shown in Scheme 4, involving migration of the cyclopropyl ring around four sides of the cyclopentane. This would lead to the observed protium distribution and should result in scrambling of the label in unused (**178**), which was also observed.

Deamination of methylenecyclopropylmethylamine in aqueous acetic acid yielded the products shown in equation 18. Intermediate ion (**180**) can therefore be written formally as rearranging by pathways *c*, *b*, and/or *a*, but not by pathway *d*.[163]

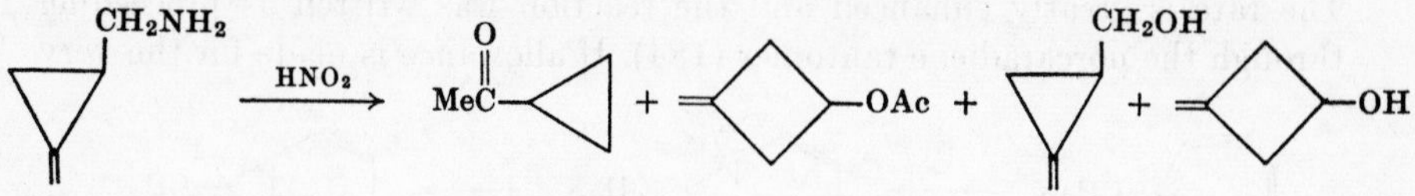

... (18)

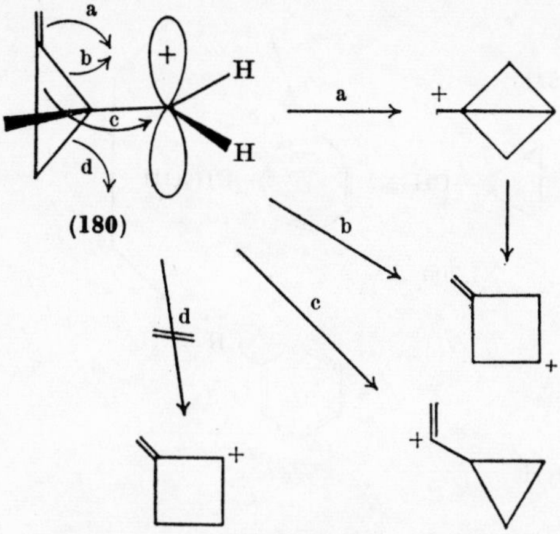

(**180**)

The cyclopropylmethyl cation (**182**) which is an intermediate in the solvolysis of (**181**) undergoes an interesting degenerate rearrangement which results in the product from (**181**), deuterated at position 2, having the label scrambled over positions 2, 8, and 9. The rate enhancement was estimated[164] to be 10^5—10^6.

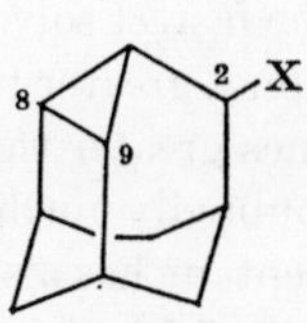

(**181**)

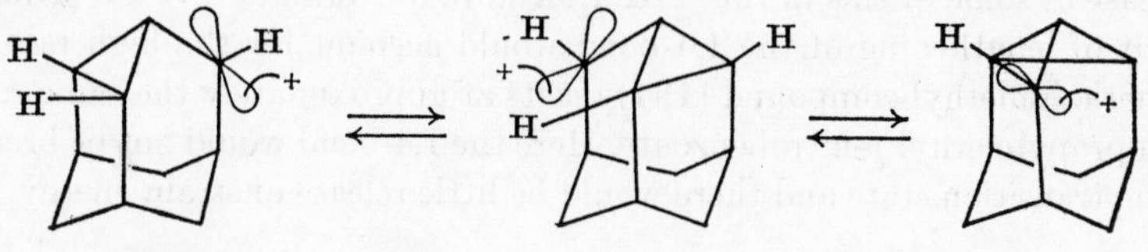

(**182**)

[163] A. Nishimura, H. Kato, and M. Ohta, *J. Am. Chem. Soc.*, **89**, 5083 (1967).
[164] J. E. Baldwin and W. D. Foglesong, *J. Am. Chem. Soc.*, **89**, 6372 (1967).

An interesting example of participation by a cyclopropyl ring formed by valence tautomerization occurs in the solvolysis of cycloheptatrienylmethyl 3,5-dinitrobenzoate (**183**) which yields unrearranged alcohol and styrene. The rate is greatly enhanced and the reaction was written as proceeding through the norcaradiene tautomer (**184**). If allowance is made for the very

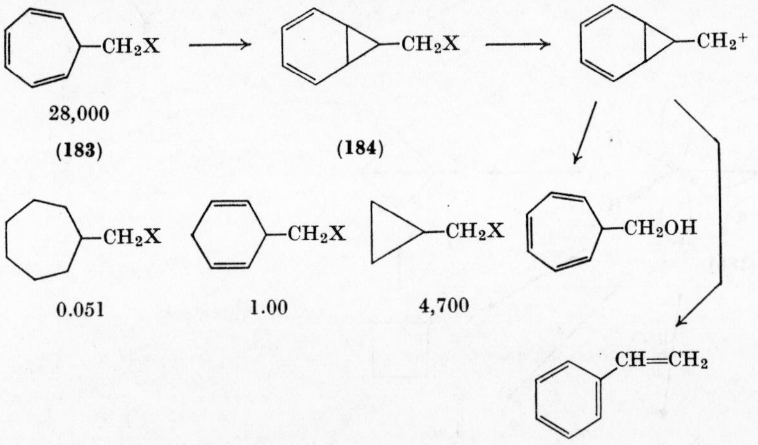

Numerals are relative rates of solvolysis

small concentration of this which must be present (≪ 1%), the rate constant for its solvolysis must be much more than 100 times greater than the measured rate constant.[165, 166]

All the products of hydrolysis of bicyclo[2.1.0]hex-1-ylmethyl *p*-nitrobenzoate (**185**) in 60% aqueous acetone result from breaking of the 1,4-bond (eqn. 19). This was claimed as the fastest solvolysis of any cyclopropylmethyl derivative with a rate 400,000 times greater than that for cyclopropylmethyl *p*-nitrobenzoate and *ca.* 100 times greater than that for bicyclo[1.1.0]but-1-ylmethyl *p*-nitrobenzoate. Undoubtedly much of the driving force comes from release of steric strain. Bicyclopentane has a strain energy of 53.6 kcal $mole^{-1}$ and cyclopentane a strain energy of 6.5 kcal $mole^{-1}$, so that there are probably *ca.* 47 kcal $mole^{-1}$ of strain energy released on going from (**185**) to the products. Release of some of this in the transition state for formation of ion (**186**) as a result of lengthening of the 1,4-bond would account for the high rate. The isomeric 5-methyl compound (**187**) reacts at approximately the same rate as cyclopropylmethyl *p*-nitrobenzoate. Here the 1,4-bond would not be breaking in the transition state and there would be little release of strain energy.[167]

165 G. D. Sargent, N. Lowry, and S. D. Reich, *J. Am. Chem. Soc.*, **89**, 5985 (1967).
166 See also *Organic Reaction Mechanisms,* **1966**, 37.
167 W. G. Dauben and J. R. Wiseman, *J. Am. Chem. Soc.*, **89**, 3545 (1967).

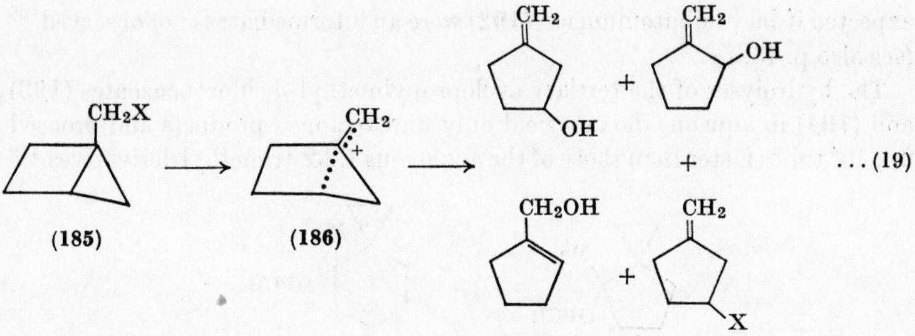

(185) **(186)**

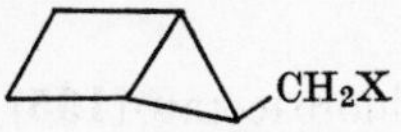

(187)

$X = p\text{-}NO_2 \cdot C_6H_4 \cdot CO \cdot O$

Ion (**189**), with a structure similar to (**186**), has been postulated as intervening in the hydrolysis of the 4-chlorobicyclo[2.2.0]hexane-1-carboxylate ion (**188**). The rate of this reaction is unenhanced, however.

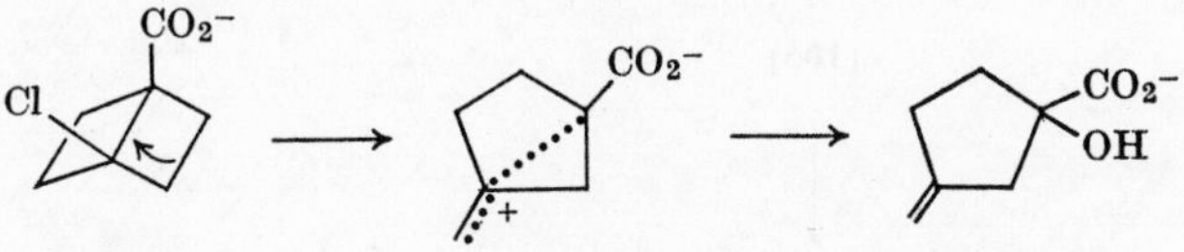

(188) **(189)**

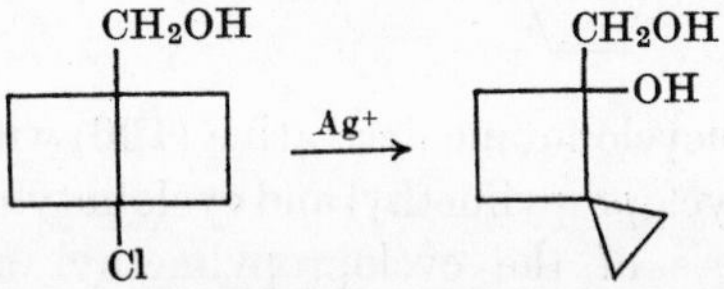

(190) **(191)**

The silver-ion assisted solvolysis of (**190**) proceeds with collapse of the other side of the cyclobutane ring to yield (**191**).[168]

Addition of HCl to bicyclo[3.1.0]hex-2-ene yields, *inter alia*, *cis*- and *trans*-2-chlorobicyclo[3.1.0]hexane (eqn. 20), and the stereospecificity

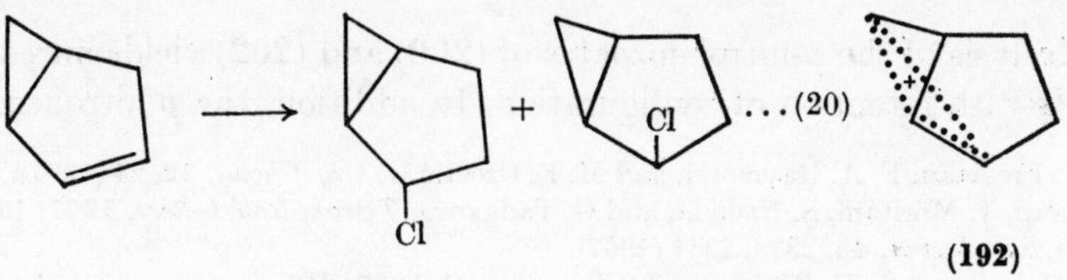

(192)

[168] K. V. Scherer and K. Katsumoto, *Tetrahedron Letters*, **1967**, 3079.

expected if bicyclobutonium ion (**192**) were an intermediate is not observed[169] (see also p. 138).

The hydrolyses of the tertiary cyclopropylmethyl *p*-chlorobenzoates (**193**) and (**194**) in aqueous dioxan yield only unrearranged products and proceed 2×10^4 times faster than those of the analogous 1,2,2-trimethyl derivatives.[170]

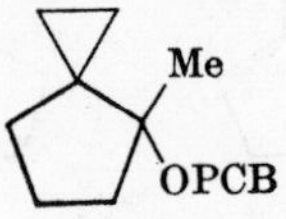

(193)

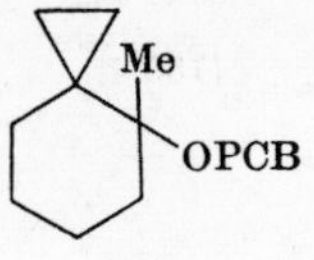

(194)

Hydrolysis of *endo*-bicyclo[3.1.0]hex-2-en-6-ylmethyl trichloroacetate (**195**) yielded *cis*- and *trans*-4-vinylcyclopent-2-enols (90%). No participation by the double bond to generate the nortricyclyl system occurred.[171]

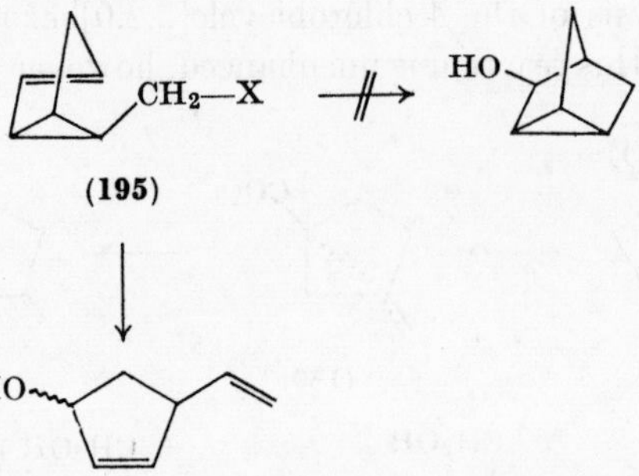

Treatment of the bicyclobutane derivative (**196**) with methanol yields the same mixture of the cyclopropylmethyl and cyclobutylmethyl ethers (**197** and **198**) as the solvolysis of the cyclopropylmethyl *p*-nitrobenzoate (**199**), suggesting that both reactions proceed through the same intermediate ion.[172]

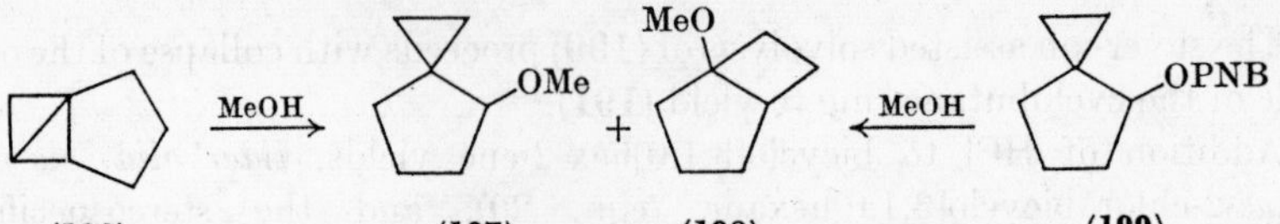

(196) **(197)** **(198)** **(199)**

Hydrolysis of the *p*-nitrobenzoates of (**200**) and (**202**) yields unrearranged alcohols with retention of configuration. In addition, the *p*-nitrobenzoate of

169 P. K. Freeman, F. A. Raymond, and M. F. Grostic, *J. Org. Chem.*, **32**, 24 (1967).
170 T. Tsuji, I. Moritani, S. Nishida, and G. Tadakoro, *Tetrahedron Letters*, **1967**, 1207; *Bull. Chem. Soc. Japan*, **40**, 2338, 2344 (1967).
171 J. T. Lumb and G. H. Whitham, *J. Chem. Soc., C*, **1967**, 216.
172 W. G. Dauben and C. D. Poulter, *Tetrahedron Letters*, **1967**, 3021.

(**202**) yields a ring-expanded alcohol which is the *cis,cis,cis*-isomer (**203**), exclusively, and both alcohols on treatment with dilute perchloric acid

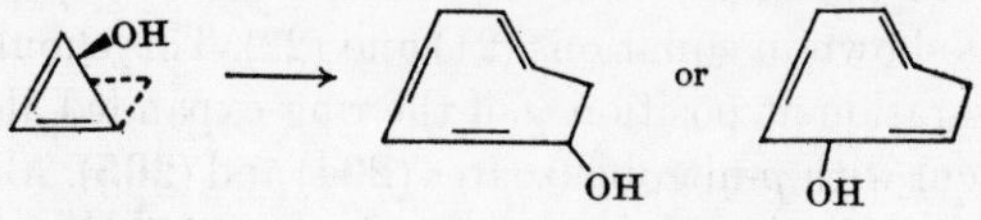

(**200**) (**201**)

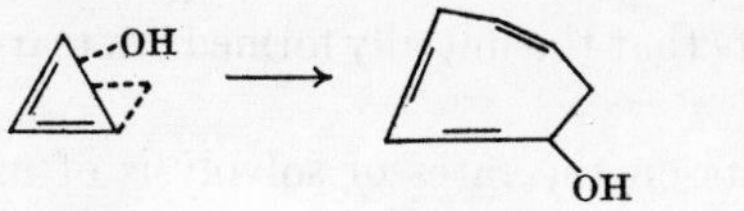

(**202**) (**203**)

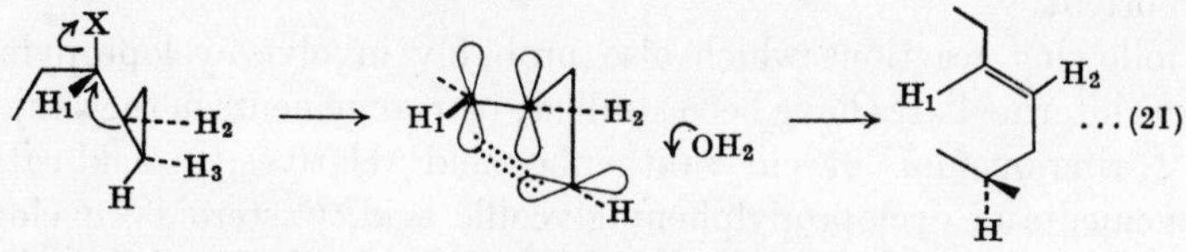

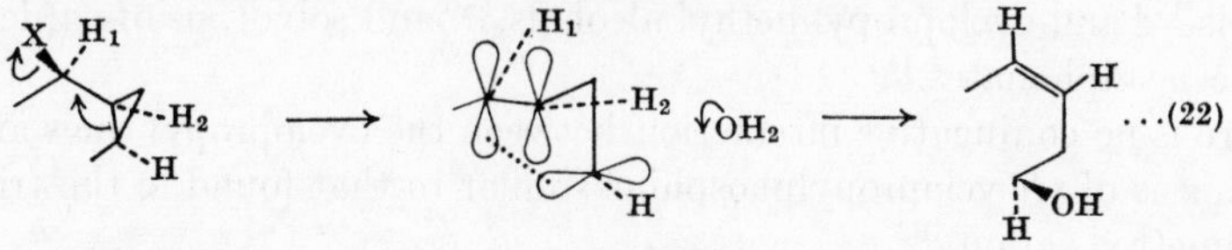

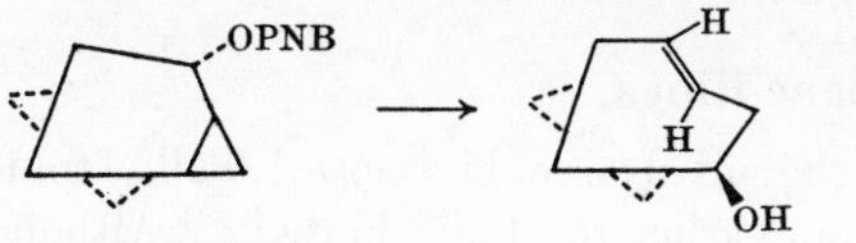

(**204**)

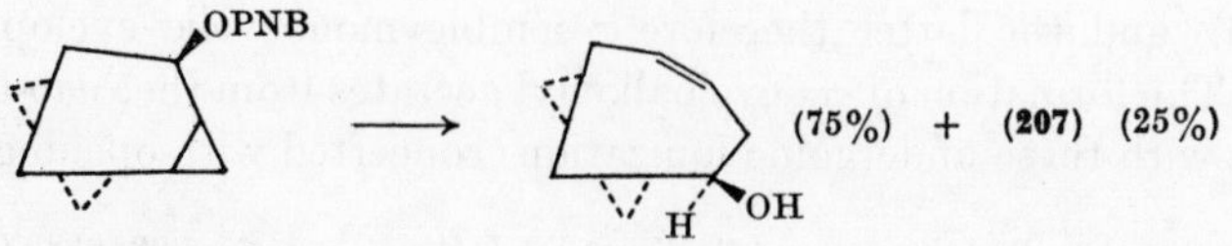

(**205**)

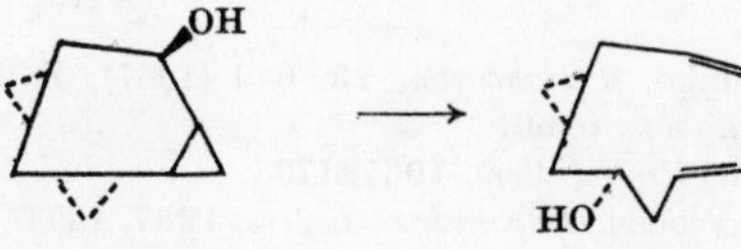

(**206**) (**207**)

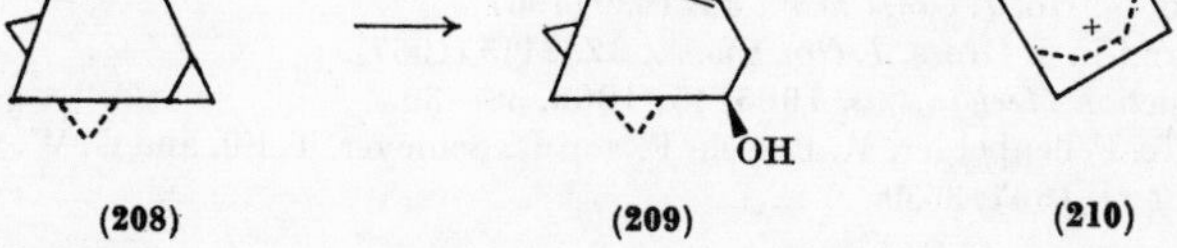

(**208**) (**209**) (**210**)

undergo ring expansion stereospecifically; (**200**) yields *cis,cis,trans*-isomer (**201**), and (**202**) yields the *cis,cis,cis*-isomer (**203**).[173] These results are as expected if reaction involved participation by the electrons in the β,γ-bond of the cyclopropane ring as shown in equations (21) and (22). This should also lead to inversion of configuration at position γ of the ring-expanded alcohols as was demonstrated to occur with *p*-nitrobenzoates (**204**) and (**205**). Alcohols (**206**) and (**208**) yield some bis-ring-expanded products (**207**) and (**209**), not all of which were formed by subsequent ring-expansion of the mono-ring-expanded products. This suggests that the initially formed ion rearranges to ion (**210**).[174]

Secondary hydrogen isotope effects on the rates of solvolysis of methyl-substituted cyclopropylmethyl and cyclobutyl methanesulphonates have been reported.[175]

The following reactions which also probably involve cyclopropylmethyl cations as intermediates have been studied: rearrangements in the thujopsene series,[176] ring-opening of chrysanthemic acid relatives,[177] acid-catalysed rearrangements of cyclopropylphenylglycollic acid,[178] steroidal cyclopropyl alcohols,[179] and cyclopropylmethyl alcohols,[180] and solvolysis of cardenolide toluene-*p*-sulphonates.[181]

There is no conjugative interaction between the cyclopropyl rings and the phosphorus of tricyclopropylphosphine similar to that found in the tricyclopropylmethyl cation.[182]

Cationic Opening of Cyclopropane Rings

The reactivity pattern shown by the solvolysis of bicyclo[*n*,1,0]alkyl toluene-*p*-sulphonates is more complex than earlier results[183] hinted.[184] Although the *endo*-isomer is the more reactive when $n = 3$ or 4, this is not so when $n = 5$ or 6 (cf. Table 3), and the latter therefore resemble monocyclic cyclopropane derivatives. The formation of *cis*-cycloalkenyl acetates from the *endo*-isomers is consistent with these undergoing ionization, concerted with opening of the

173 M. Găsić, D. Whalen, B. Johnson, and S. Winstein, *J. Am. Chem. Soc.*, **89**, 6382 (1967).

174 D. Whalen, M. Găsić, B. Johnson, H. Jones, and S. Winstein, *J. Am. Chem. Soc.*, **89**, 6384 (1967).

175 M. Nikoletić, S. Borčić, and D. E. Sunko, *Tetrahedron*, **23**, 649 (1967); Z. Majerski, M. Nikoletić, S. Borčić, and D. E. Sunko, *ibid.*, p. 661.

176 W. G. Dauben and L. E. Friedrich, *Tetrahedron Letters*, **1967**, 1735.

177 L. Crombie, R. P. Houghton, and D. K. Woods, *Tetrahedron Letters*, **1967**, 4553.

178 L. L. Darko and J. G. Cannon, *J. Org. Chem.*, **32**, 2353 (1967).

179 H. Laurent, H. Müller, and R. Weichert, *Chem. Ber.*, **99**, 3836 (1966).

180 M. Julia, G. Mouzin, and C. Descoins, *Compt. Rend., Ser. C.*, **264**, 330 (1967).

181 M. E. Wolff and W. Ho, *J. Org. Chem.*, **32**, 1839 (1967).

182 D. D. Denney and F. J. Gross, *J. Org. Chem.*, **32**, 2445 (1967).

183 See *Organic Reaction Mechanisms*, **1965**, 45; **1966**, 38—39.

184 U. Schöllkopf, K. Fellenberger, M. Patsch, P. von R. Schleyer, T. Su, and G. W. van Dine, *Tetrahedron Letters*, **1967**, 3639.

Table 3. Acetolyses of bicyclo[n,1,0]alkyl toluene-p-sulphonates.

	endo-Series		*exo*-Series	
n	$k_{rel.}$ (100°)[a]	Product	$k_{rel.}$ (100°)[a]	Product
3	25,000	*cis*-Cyclohex-2-enyl acetate	< 0.01	
4	62[b]	*cis*-Cyclohept-2-enyl acetate	1.7	*exo*-Norcaryl acetate and 1,3-diacetoxy-cycloheptane
5	3.1	*cis*-Cyclo-oct-2-enyl acetate	2,500	*cis*-Cyclo-oct-2-enyl acetate and 1,3-diacetoxycyclo-octane
6	3.5	*cis*-Cyclonon-2-enyl acetate	10,000	*cis*-Cyclonon-2-enyl acetate

[a] Relative to that for cyclopropyl toluene-p-sulphonate, $k = 3.89 \times 10^{-8}$ sec^{-1}.
[b] The much greater value reported last year (see ref. 183) is incorrect.

cyclopropane ring by a disrotatory process, to yield *cis*-allylic cations (cf. eqn. 23). The decrease in rate with increase in n was considered to be related to the decrease in stability of the cyclic allyl cations (**211**) although in our opinion greater ring strain in the initial state when n is small may also be an important factor.

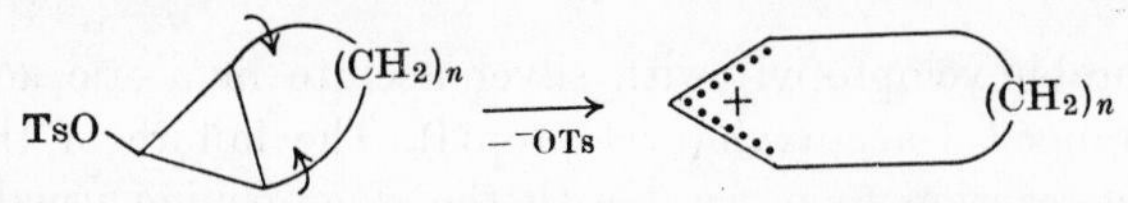

...(23)

(**211**)

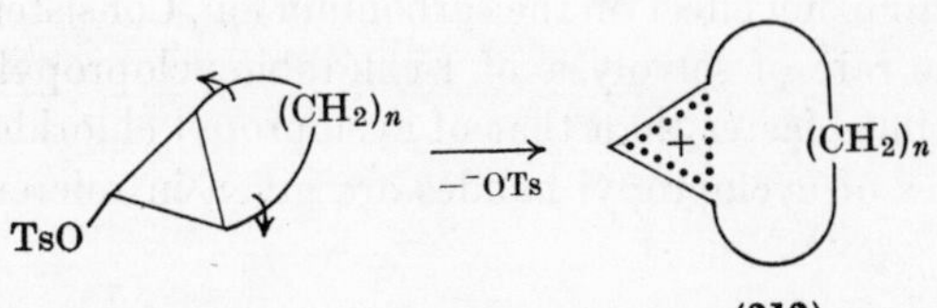

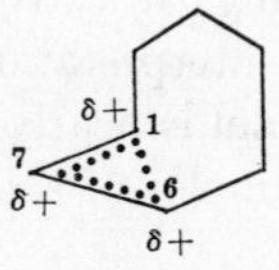

(**212**) (**213**)

In the *exo*-series the rate increases as n increases. The compound with $n = 3$ is very unreactive, and clearly ionization without ring opening or with ring opening in the "wrong" direction is very unfavourable. As n increases the stability of the intermediate ion (and transition state) must also increase and it was suggested that this be formulated as (**212**), intermediate between an allyl and cyclopropyl cation. Consistently with this, *exo*-norcaryl toluene-p-sulphonate ($n = 4$) yields some *exo*-norcaryl acetate. The other product is

1,3-diacetoxycycloheptane, formed presumably by addition of acetic acid to *trans*-cycloheptenyl acetate which would result from attack on (**213**) at positions 1 and 6. The products obtained from the higher members of this series could also result from the *trans*-acetate by isomerization or addition of acetic acid.[184] *exo*-8-Bromobicyclo[5.1.0]octane is also easily hydrolysed to *trans*-cyclo-oct-2-en-1-ol.[185]

Several examples of the rearrangement of *endo*-bicyclic cyclopropyl derivatives to the analogous allylic compounds, occurring under conditions where the *exo*-isomers are unreactive, have been reported[186–190] (cf. eqn. 24).[186]

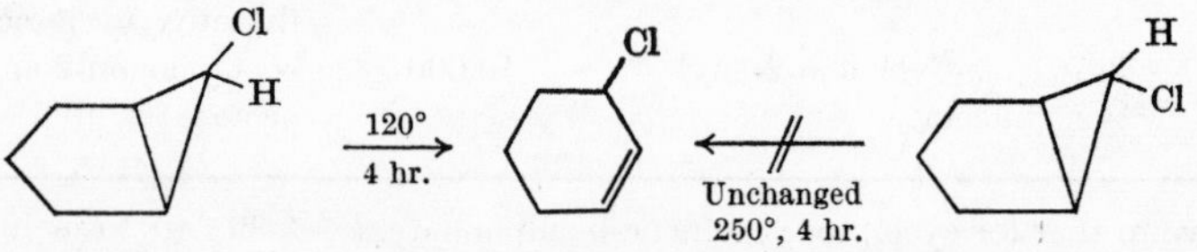

... (24)

Decomposition of the *cis*- and *trans*-cyclopropanediazonium ions (**215**) in methanol yield, *inter alia*, *trans*-cinnamyl methyl ether (**218**), but only a trace of its *cis*-isomer. The diazonium ions were not interconverted (*via* **217**) to the extent of more than 25% since only 0.25 atom of deuterium per molecule was incorporated when the solvent was CH_3OD. Since concerted ring-openings would be expected to be stereospecific (*cis*-**215** → *cis*-**218**; *trans*-**215** → *trans*-**218**) it was concluded that the decompositions yielded the same cyclopropyl cation (**216**).[191]

Treatment of 1-chlorobi(cyclopropyl) with silver acetate in acetic acid yields 42% of unrearranged 1-acetoxybi(cyclopropyl). The failure of this large proportion of the reaction to proceed with the ring-opening usually found with cyclopropyl halides presumably results because the other cyclopropane ring has a strong stabilizing effect on the carbonium ion. Consistently with this interpretation, the rate of solvolysis of 1-chlorobi(cyclopropyl) in 50% ethanol is nearly 10^6 times faster than that of cyclopropyl chloride.[192] Other examples of solvolyses of cyclopropyl halides are given in references 193—195.

185 G. H. Whitham and M. Wright, *Chem. Commun.*, **1967**, 294.
186 M. S. Baird and C. B. Reese, *Tetrahedron Letters*, **1967**, 1379.
187 L. Ghosez, P. Laroche, and G. Slinckx, *Tetrahedron Letters*, **1967**, 2767.
188 L. Ghosez, G. Slinckx, M. Glineur, P. Hoet, and P. Laroche, *Tetrahedron Letters*, **1967**, 2773.
189 T. Ando, H. Yamanaka, and W. Funasaka, *Tetrahedron Letters*, **1967**, 2587.
190 C. W. Jefford and R. T. Medary, *Tetrahedron*, **23**, 4123 (1967).
191 W. Kirmse and H. Schütte, *J. Am. Chem. Soc.*, **89**, 1284 (1967).
192 J. A. Landgrebe and L. W. Becker, *J. Am. Chem. Soc.*, **89**, 2505 (1967).
193 W. E. Parham and R. J. Sperley, *J. Org. Chem.*, **32**, 924 (1967).
194 G. C. Robinson, *J. Org. Chem.*, **32**, 3218 (1967).
195 A. J. Birch, G. M. Iskander, B. I. Magboul, and F. Stansfield, *J. Chem. Soc., C*, **1967**, 358.

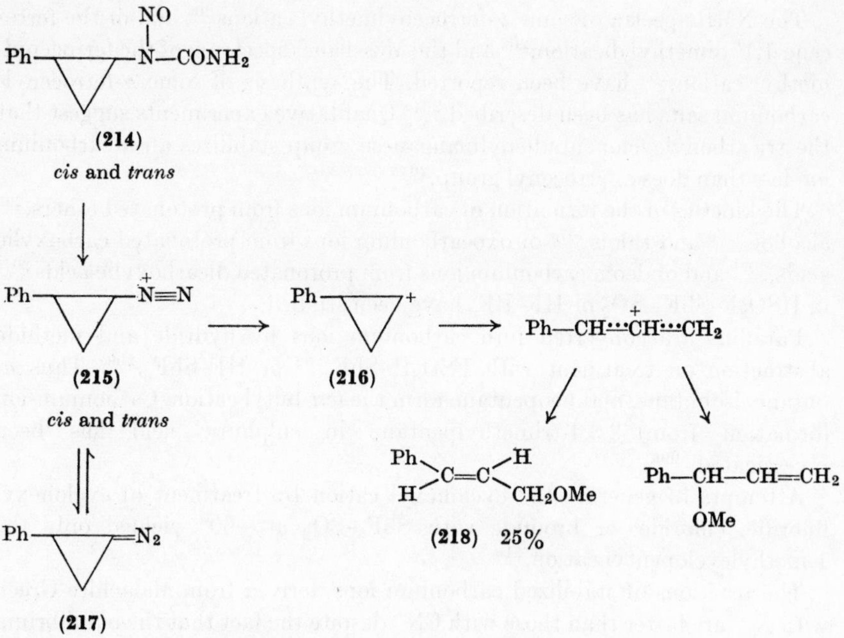

Other Stable Carbonium Ions and Their Reactions[196]

Details have been published of Traylor and Ware's investigation on ferrocenylmethyl cations, and it is concluded that they are not stabilized by direct rear-side participation by the iron orbitals as previously suggested.[197–199] The stereospecificity of formation and reactions of the ions are considered to be controlled by a stereoelectronic factor similar to that which controls the steric course of $E2$ eliminations (see eqn. 25). It was also suggested that the high rate of solvolysis of 2-ferrocenylethyl toluene-*p*-sulphonate[200] resulted from participation of carbon rather than of iron.

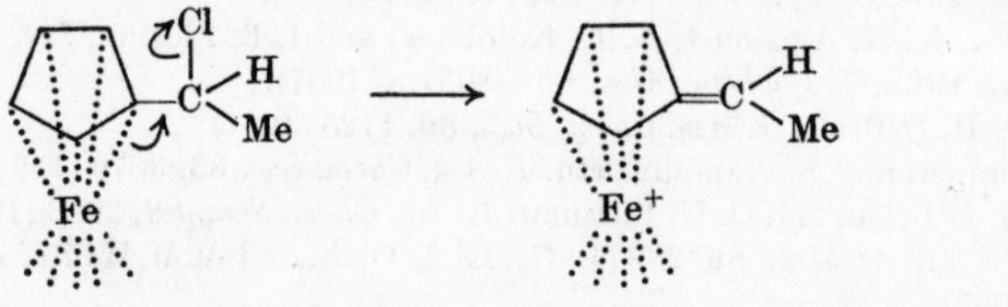

... (25)

[196] G. A. Olah, *Chem. Eng. News*, 27th March, 1967, **45**, No. 14, p. 76.

[197] T. G. Traylor and J. C. Ware, *J. Am. Chem. Soc.*, **89**, 2304 (1967); W. Hanstein and T. G. Traylor, *Tetrahedron Letters*, **1967**, 4451; J. A. Mangravite and T. G. Traylor, *ibid.*, p. 4457, 4461.

[198] See *Organic Reaction Mechanisms*, **1965**, 47—48; **1966**, 41.

[199] See M. Cais, *Organometal. Chem. Rev.*, **1**, 435 (1966).

[200] D. S. Trifan and R. Bacskai, *Tetrahedron Letters*, **1960**, No. 13, p. 1.

The NMR spectra of some α-ferrocenylmethyl cations[201] and of the ferrocene-1,1′-dimethyl dication,[202] and the Mössbauer spectrum of the ferrocenylmethyl cation,[203] have been reported. The synthesis of some α-ferrocenylcarbonium salts has been described.[204a] Qualitative experiments suggest that the tricarbonylcyclopentadienylmanganese group stabilizes an α-carbonium ion less than does a ferrocenyl group.[204b]

The kinetics of the formation of carbonium ions from protonated ethers,[205] alcohols,[206] and thiols[207a] of oxocarbonium ions from protonated carboxylic acids,[207b] and of deoxocarbonium ions from protonated dicarboxylic acids[207c] in $HSO_3F–SbF_5–SO_2$ or $HF–BF_3$ have been studied.

Paraffins are converted into carbonium ions by hydride and methide abstraction on treatment with $FSO_3H–SbF_5$[208] or $HF–SbF_5$.[209a] Thus *n*-butane, isobutane, and neopentane form the *tert*-butyl cation. Carbonium-ion formation from 2,3,4-trimethylpentane in sulphuric acid has been investigated.[209b]

Attempts to generate the cyclohexyl cation by treatment of cyclohexyl fluoride, chloride, or bromide with $SbF_5–SO_2$ at –60° yielded only the 1-methylcyclopentyl cation.[210]

The reactions of stabilized carbonium ions derived from Malachite Green with N_3^- are faster than those with CN^- despite the fact that the equilibrium constant for formation of covalent product is greater with CN^-. The reactions are faster in DMSO and DMF than in methanol and water, but this is not just an initial-state effect since the rates of the reverse reactions are also increased. It was suggested that solvent reorganization plays an essential part in the activation process.[211]

Deuteration of triphenylmethanol in the *o*-, *m*-, and *p*-positions causes a small increase in the degree of dissociation to the triphenylmethyl cation in

201 W. M. Horspool and R. G. Sutherland, *Chem. Commun.*, **1967**, 786.
202 C. U. Pittman, *Tetrahedron Letters*, **1967**, 3619.
203 J. J. Dannenberg and J. H. Richards, *Tetrahedron Letters*, **1967**, 4747.
204a A. N. Nesmeyanov, V. A. Sazonova, G. I. Zudkova, and L. S. Isaeva, *Izv. Akad. Nauk SSSR, Ser. Khim.*, **1966**, 2017; *Chem. Abs.*, **66**, 76133d (1967).
204b A. N. Nesmeyanov, K. N. Anisimov, N. E. Kolobova, and I. B. Zlotina, *Izv. Akad. Nauk SSSR, Ser. Khim.*, **1966**, 729; *Chem. Abs.*, **66**, 38031m (1967).
205 G. A. Olah and D. H. O'Brien, *J. Am. Chem. Soc.*, **89**, 1725 (1967).
206 G. A. Olah, J. Sommer, and E. Namanworth, *J. Am. Chem. Soc.* **89**, 3576 (1967).
207a G. A. Olah, D. H. O'Brien, and C. U. Pittman, *J. Am. Chem. Soc.*, **89**, 2996 (1967).
207b H. Hogeveen, *Rec. Trav. Chim.*, **86**, 289 (1967); G. A. Olah, and A. M. White, *J. Am. Chem. Soc.*, **89**, 3591 (1967).
207c G. A. Olah and A. M. White, *J. Am. Chem. Soc.*, **89**, 4752 (1967).
208 G. A. Olah and J. Lukas, *J. Am. Chem. Soc.*, **89**, 2227, 4739 (1967).
209a A. F. Bickel, C. J. Gaasbeek, H. Hogeveen, J. M. Oelderik, and J. C. Platteeuw, *Chem. Commun.*, **1967**, 634; H. Hogeveen and A. F. Bickel, *ibid.*, p. 635.
209b G. M. Kramer, *J. Org. Chem.*, **32**, 920 (1967).
210 G. A. Olah, J. M. Bollinger, C. A. Cupas, and J. Lukas, *J. Am. Chem. Soc.*, **89**, 2692 (1967).
211 C. D. Ritchie, G. A. Skinner, and V. G. Badding, *J. Am. Chem. Soc.*, **89**, 2063 (1967).

$H_2SO_4(k_H/k_D \approx 1.01$ per deuterium).[212] This cannot be a steric isotope effect[213] (see also ref. 38).

The rearrangements of several trialkylcyclopentenyl cations (e.g., **219**) in H_2SO_4 have been investigated by following the change in the NMR spectrum. The mechanism shown in equation (26) was proposed and shown to be consistent with deuterium-exchange experiments in D_2SO_4.[214]

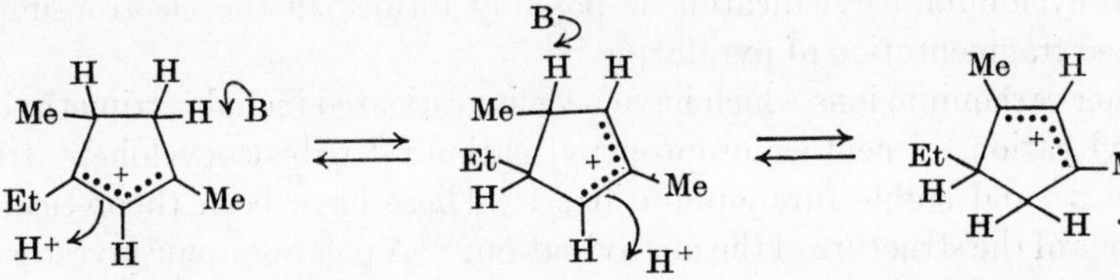

...(26)

(219)

The tris(pentafluorophenyl)methyl cation ($pK_{R^+} - 17.5$) is much less stable than the triphenylmethyl cation ($pK_{R^+} - 6.63$). Its NMR spectrum and those of the mono- and bis-pentafluorophenylmethyl cation have been reported.[215]

A detailed analysis of the NMR spectra of arylmethyl cations has been reported. The spectrum of the 1,3-dimethyl-1,3-diphenylbutyl cation is anomalous in that no phenyl resonance appears at the low-field position usually occupied by signals from the *ortho*-protons; moreover, there is no clear separation between the phenyl conjugated with the carbonium-ion centre and that attached to the quaternary centre. It was suggested that there is an interaction between the two phenyl rings analogous to that in a charge-transfer complex.[216]

The NMR spectra of the following carbonium ions have also been reported: alkyldicarbonium ions,[217] dicarbonium ions in the biphenyl series,[218] 1,3,5-tri(cycloheptatrienylium)benzene trication,[219] bisdioxolenium ions, [220] alkyl- and arylalkyl-fluorocarbonium ions,[221] 1-hydroxyhomotropylium ions,[222]

212 A. J. Kresge and R. J. Preto, *J. Am. Chem. Soc.*, **89**, 5510 (1967).
213 Cf. K. T. Leffek, *Can. J. Chem.*, **45**, 2115 (1967).
214 T. S. Sorensen, *J. Am. Chem. Soc.*, **89**, 3782, 3794 (1967).
215 R. Filler, C.-S. Wang, M. A. McKinney, and F. N. Miller, *J. Am. Chem. Soc.*, **89**, 1026 (1967); G. A. Olah and M. B. Comisarow, *ibid.*, p. 1027.
216 D. G. Farnum, *J. Am. Chem. Soc.*, **89**, 2970 (1967).
217 J. M. Bollinger, C. A. Cupas, K. J. Friday, M. L. Woolfe, and G. A. Olah, *J. Am. Chem. Soc.*, **89**, 156 (1967).
218 H. Hart, T. Sulzberg, R. H. Schwendeman, and R. H. Young, *Tetrahedron Letters*, **1967**, 1337; H. Hart, C.-Y. Wu, R. H. Schwendeman, and R. H. Young, *ibid.*, p. 1343.
219 R. W. Murray and M. L. Kaplan, *Tetrahedron Letters*, **1967**, 1307.
220 H. Hart and D. A. Tomalia, *Tetrahedron Letters*, **1967**, 1347.
221 G. A. Olah, R. C. Chambers, and M. B. Comisarow, *J. Am. Chem. Soc.*, **89**, 1268 (1967).
222 M. Brookhart, M. Ogliaruso, and S. Winstein, *J. Am. Chem. Soc.*, **89**, 1965 (1967).

alkenyloxocarbonium ions,[223] alkoxycarbonium ions,[224] benzyl cations,[225] cyclopropenyl cation,[226] triphenylcyclopropenyl cation.[227]

X-ray crystal-structure determinations have been reported for the *syn*-tetra-*p*-methoxyphenylethylene dication[228] and the bis-4,4′-(diphenylmethyl)biphenyl dication.[229]

Triplet states of some substituted cyclopentadienyl cations have been detected by ESR spectroscopy.[230]

The cyclobutadienyl dication is possibly formed in the electron-impact-induced fragmentation of pyridazine.[231]

Other carbonium ions which have been investigated include: trimethylthiomethyl cation,[232] pentachloropropenyl cation,[233] tribenzocycloheptatrienyl cation,[234] and stable furanonium ions.[235] There have been theoretical discussions of the structure of the methyl cation.[236] A polarographic investigation of organic cations has been reported.[237]

223 G. A. Olah and M. B. Comisarow, *J. Am. Chem. Soc.*, **89**, 2694 (1967).
224 G. A. Olah and J. M. Bollinger, *J. Am. Chem. Soc.*, **89**, 2993 (1967).
225 J. M. Bollinger, M. B. Comisarow, C. A. Cupas, and G. A. Olah, *J. Am. Chem. Soc.*, **89**, 5687 (1967).
226 R. Breslow, J. T. Groves, and G. Ryan, *J. Am. Chem. Soc.*, **89**, 5048 (1967); D. G. Farnum, G. Mehta, and R. S. Silberman, *ibid.*, p. 5048.
227 D. G. Farnum and C. F. Wilcox, *J. Am. Chem. Soc.*, **89**, 5379 (1967); B. Föhlisch and P. Bürgle, *Ann. Chem.*, **701**, 58 (1967); W. Broser and M. Brockt, *Tetrahedron Letters*, **1967**, 3117.
228 N. C. Baenziger, R. E. Buckles, and T. D. Simpson, *J. Am. Chem. Soc.*, **87**, 3405 (1967).
229 J. S. McKechnie and I. C. Paul, *J. Am. Chem. Soc.*, **89**, 5482 (1967).
230 R. Breslow, H. W. Chang, R. Hill, and E. Wasserman, *J. Am. Chem. Soc.*, **89**, 1112 (1967); W. Broser, H. Kurreck, and P. Siegle, *Chem. Ber.*, **100**, 788 (1967).
231 M. H. Benn, T. S. Sorensen, and A. M. Hogg, *Chem. Commun.*, **1967**, 574.
232 W. P. Tucker and G. L. Roof, *Tetrahedron Letters*, **1967**, 2747.
233 K. Kirchoff, F. Boberg, D. Friedemann, and G. R. Schultze, *Tetrahedron Letters*, **1967**, 3861.
234 W. Tochterman and K. Stecher, *Tetrahedron Letters*, **1967**, 3847.
235 U. E. Wiersum and H. Wynberg, *Tetrahedron Letters*, **1967**, 2951.
236 R. E. Davis and A. Ohno, *Tetrahedron*, **23**, 1015 (1967); R. E. Davis, D. Grossee and A. Ohno, *ibid.*, p. 1029.
237 M. I. James and P. H. Plesch, *Chem. Commun.*, **1967**, 508.

Nucleophilic Aliphatic Substitution

Ion-pair Phenomena and Borderline Mechanisms

The suggestion reported last year[1] that the oxygen atoms of a benzhydryl benzoate ion pair may not be equivalent is supported by an investigation of the decomposition of the labelled N-benzhydryl-N-nitrosobenzamide (1) to yield benzhydryl benzoate in which 60% of the label is retained in the carbonyl

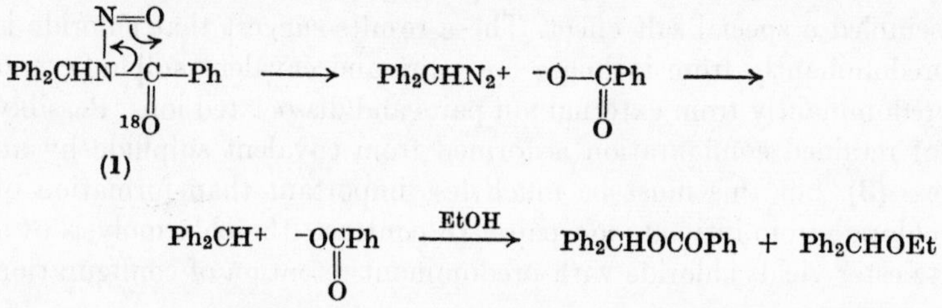

$$\underset{(1)}{Ph_2CHN\overset{\overset{\displaystyle N=O}{\displaystyle |\curvearrowleft}}{\underset{\underset{\displaystyle 18O}{\displaystyle \|}}{C}}-Ph} \longrightarrow Ph_2CHN_2^+ \quad {}^-O-\underset{\underset{\displaystyle O}{\displaystyle \|}}{C}Ph \longrightarrow$$

$$\underset{(2)}{Ph_2CH^+ \quad {}^-O\underset{\underset{\displaystyle O}{\displaystyle \|}}{C}Ph} \xrightarrow{EtOH} Ph_2CHOCOPh + Ph_2CHOEt \qquad \ldots(1)$$

group. If the reaction proceeds as shown in equation 1, more equilibration of the oxygen atoms would be expected in ion pair (2) than in that intervening in solvolysis, since the ions are initially further apart. If then these ion pairs are similar, at least 20% of ion-pair return occurring in solvolysis cannot involve equilibration of the oxygen atoms. It was also shown that the ratio of ester to ether formed on decomposition of (1) in ethanol at 25° was different from the ratio formed in the reaction of diphenyldiazomethane and benzoic acid under the same conditions. The ion pairs intervening in the two reactions must therefore be different. Possibly they are *cis–trans*-isomers, or that intervening in the diazodiphenylmethane reaction is a hydrogen-bonded complex or that intervening in the deamination is vibrationally excited.[2]

Chlorinolysis of carbon–sulphur bonds in acetic acid probably proceeds through a sulphonium ion pair (eqn. 2). Lithium perchlorate increases the rate of reaction of benzyl phenyl sulphide in the manner of a special salt effect,

$$PhSCH_2Ph + Cl_2 \underset{AcOH}{\overset{Fast}{\rightleftharpoons}} \underset{\underset{\displaystyle Cl^-}{\displaystyle |}}{Ph^+S}-CH_2Ph \xrightarrow{Slow} PhCH_2Cl + PhCH_2OAc \qquad \ldots(2)$$

[1] See *Organic Reaction Mechanisms*, **1966**, 47—48.
[2] E. H. White and C. A. Elliger, *J. Am. Chem. Soc.*, **89**, 165 (1967).

and it changes the product ratio [$PhCH_2Cl$]:[$PhCH_2OAc$] from 1.9 when [$LiClO_4$] = 0 to 0.14 when [$LiClO_4$] = 0.06M. Presumably the sulphonium–chloride ion pairs which collapse to benzyl chloride are converted into sulphonium perchlorate ion pairs which collapse to benzyl acetate. Acetolysis of benzyl derivatives does not normally show special salt effects and, as these effects are usually found only with substrates that yield especially stable carbonium ions, it was suggested that the benzyl cation in these reactions is stabilized by having a favourable solvent structure transferred to it from the initial sulphonium ion. Both products from optically active α-ethylbenzyl phenyl sulphide were formed with inversion of configuration, but the optical purity of the chloride (90.5%) was greater than that of the acetate (29%). The optical purity of the chloride decreased linearly with lithium perchlorate concentration (normal salt effect) but that of the acetate decreased in a fashion which resembled a special salt effect. These results suggest that chloride is formed predominantly from intimate ion pairs and covalent sulphide, and acetate predominantly from external ion pairs and dissociated ions. Possibly chloride of retained configuration is formed from covalent sulphide by an S_Ni process (**3**), but this must be much less important than formation of inverted chloride from intimate ion pairs. In contrast, the chlorinolysis of a sulphenate ester yields chloride with predominant retention of configuration (see **4**).[3]

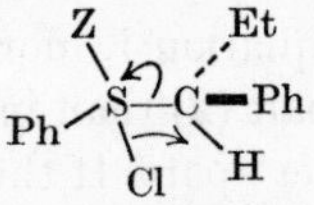

(**3**)

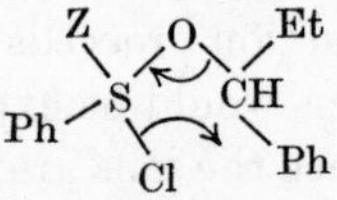

(**4**)

The S_N2 decomposition of trimethylsulphonium thiocyanate and bromide in 88% methanol–water and in dimethylacetamide has been investigated.[4]

The degree of inversion of configuration found in the S_N1 reactions of α-ethyl-α-methylbenzyl hydrogen phthalate and *p*-nitrobenzoate increases with the nucleophile in the order AcOH < MeOH < N_3^-, suggesting that the reactions involve stereospecifically formed ion pairs which are trapped before racemization at relative rates N_3^- > MeOH > AcOH.[5]

The secondary deuterium isotope effect for the solvolysis of α-[$^2H_3$]methylbenzyl chloride in alcohol–water mixtures passes through a maximum ($k_H/k_D = 1.31$) at about 5% water (by volume). It was suggested that there are competing S_N1 and S_N2 processes, with the latter important only at low

[3] H. Kwart, R. W. Body, and D. M. Hoffman, *Chem. Commun.*, **1967**, 765; H. Kwart and P. R. Strilko, *ibid.*, p. 767.

[4] Y. C. Mac, W. A. Millen, A. J. Parker, and D. W. Watts, *J. Chem. Soc., B*, **1967**, 525.

[5] L. H. Sommer and F. A. Carey, *J. Org. Chem.*, **32**, 800, 2473 (1967).

concentrations of water (< 5%) at which the solvent is unable to promote ionization of the C–Cl bond.[6]

Solvolysis of α-methylbenzyl chloride in mixtures of phenol and other solvents, which proceeds with net retention of configuration, has been investigated further.[7] Solvolysis in aqueous acrylonitrile also proceeds with slight

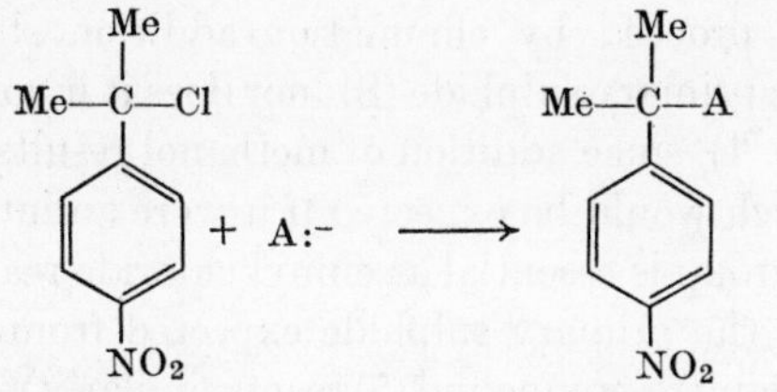

. . . (3)

(5) (6)

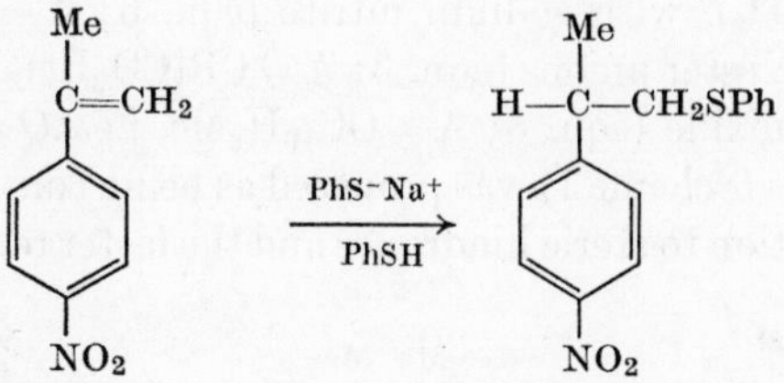

(7) (8)

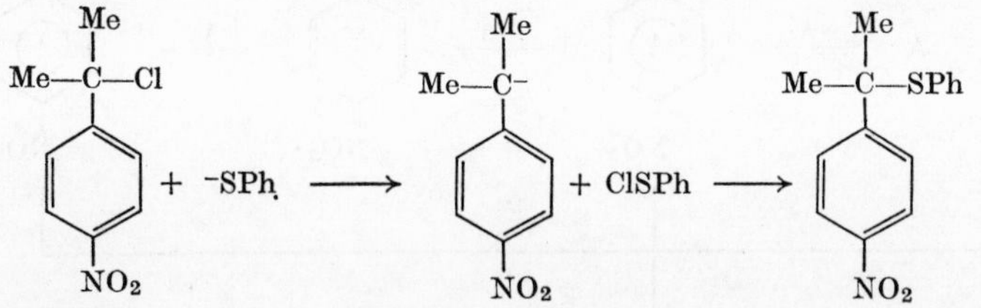

. . . (4)

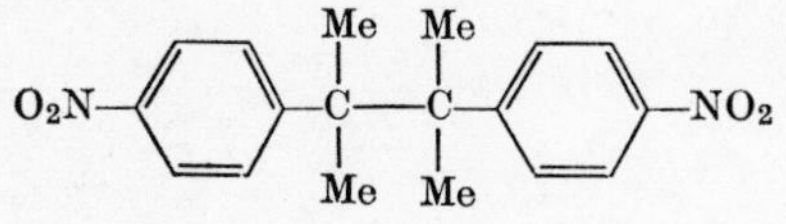

(9)

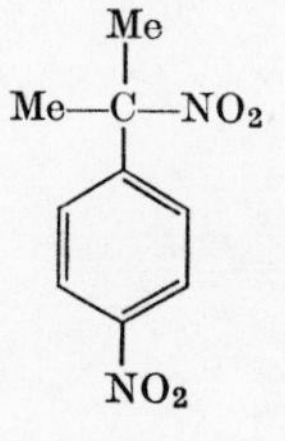

(10)

[6] A. Guinot and G. Lamaty, *Chem. Commun.*, **1967**, 960.
[7] K. Okamoto, M. Hayashi, K. Komatsu, and H. Shingu, *Bull. Chem. Soc. Japan*, **40**, 624 (1967); K. Okamoto, K. Komatsu, and H. Shingu, *ibid.*, p. 1677; see *Organic Reaction Mechanisms*, **1966**, 47.

retention, and some of the acrylonitrile is polymerized, possibly as a result of initiation by the ion-pair intermediate of the solvolysis.[8]

Several apparent S_N2 displacements at tertiary carbon have been reported by Kornblum and his co-workers[9, 10] and shown to be radical-anion processes. Thus αα-dimethyl-4-nitrobenzyl chloride (*p*-nitrocumyl chloride) (**5**) reacts with sodium thiophenoxide in DMF at 0° to yield 95% of the tertiary sulphide (**6**) (A = SPh). Reaction does not proceed by elimination–addition since olefin (**7**) adds thiophenol to give the primary sulphide (**8**); nor does it involve nucleophilic attack on chlorine (eqn. 4), since addition of methanol results in no scavenging of the carbanion, which would be expected if it were an intermediate. The presence of the nitro-group is essential as cumyl chloride reacts much more slowly, to yield mainly the primary sulphide expected from an elimination–addition sequence. The *p*-nitro-compound (**5**) reacts similarly with the lithium salt of 2-nitropropane, producing mainly the more highly branched derivative (eqn. 3; A = CMe_2NO_2), with sodium nitrite (eqn. 3; A = NO_2, i.e., *N*-alkylation), with malonic ester anions [eqn. 3; A = $CR(CO_2Et)_2$], and with sodium 1-methyl-2-naphthoxide (eqn. 3; A = $OC_{10}H_6Me$, i.e., *O*-alkylation). The radical-anion process (Scheme 1) was proposed as being consistent with the insensitivity of the reaction to steric hindrance and the faster reaction

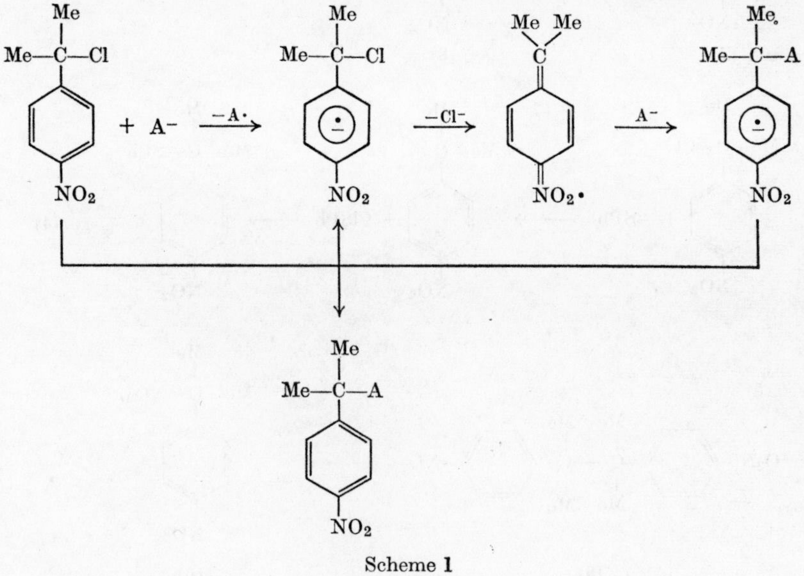

Scheme **1**

[8] K. Okamoto, K. Komatsu, and H. Shingu, *Bull. Chem. Soc. Japan*, **39**, 2785 (1966).

[9] N. Kornblum, T. M. Davies, G. W. Earl, N. L. Holy, R. C. Kerber, M. T. Musser, and D. H. Snow, *J. Am. Chem. Soc.*, **89**, 725 (1967).

[10] N. Kornblum, T. M. Davies, G. W. Earl, G. S. Greene, N. L. Holy, R. C. Kerber, J. W. Manthey, M. T. Musser, and D. H. Snow, *J. Am. Chem. Soc.*, **89**, 5714 (1967).

of $Me(CH_2)_3C^-(CO_2Et)_2$ than of $HC^-(CO_2Et)_2$; alkyl-substituted malonic esters should form less stable carbanions and more stable radicals, and thus alkylation should facilitate the reaction. This mechanism is also consistent with the observations that the reactions are light-catalysed, are retarded by *p*-dinitrobenzene (or oxygen), and afford the dimer (**9**), isolated in 1—5% yields.[9] The aliphatic nitro-group of α,*p*-dinitrocumene (**10**) is also displaced by a wide range of nucleophiles and a similar mechanism is postulated[10] (see also p. 93).

The effect of pyridine on the methanolysis and radiochlorine exchange of triphenylmethyl chloride in benzene has been re-investigated. It appears that a complex is formed without covalent bonding, but with a marked effect on the kinetics.[11] The reaction of triphenylmethyl halides with hexanol in benzene has also been investigated.[12]

The role of ion-pairs in the solvolysis of 9-fluorenyl and diphenylmethyl toluene-*p*-sulphonate has been discussed.[13]

Ion pairing has been investigated by ESR spectroscopy,[14] conductometrically,[15] and by ultrasonic relaxation.[16]

Solvent Effects[17a]

There have been reviews on solvent effects,[17b] ionic reactions in acetonitrile,[18] solvolysis in water,[19] and reactions in dimethyl sulphoxide.[20]

Plots of $\Delta V^‡$ for the solvolysis of benzyl chloride in aqueous alcohols against mole fraction of alcohol show minima for methyl, isopropyl, and *tert*-butyl alcohol as well as for ethyl alcohol.[21] These are mainly an initial-state effect,

11 K. T. Leffek and R. G. Waterfield, *Can. J. Chem.*, **45**, 1497 (1967).
12 I. A. Schneider, N. Hurduc, and M. Popa, *An. Stiint. Univ. "Al. I. Cuza"*, *Iasi*, Sect Ic, **12**, 1 (1966); *Chem. Abs.*, **67**, 32097m (1967).
13 G. W. Cowell and A. Ledwith, *J. Chem. Soc., B*, **1967**, 695; G. W. Cowell, A. Ledwith, and D. G. Morris, *ibid.*, p. 700.
14 M. P. Khakhar, B. S. Prabhananda, and M. R. Das, *J. Am. Chem. Soc.*, **89**, 3100 (1967); T. E. Hogen-Esch and J. Smid, *ibid.*, p. 2764; L. L. Chan and J. Smid, *ibid.*, p. 4547; T. Shimomura, J. Smid, and M. Szwarc, *ibid.*, p. 5743; K. Nakamura, *Bull. Chem. Soc. Japan*, **40**, 1, 1019 (1967); K. Nakamura and Y. Deguchi, *ibid.*, p. 705; M. Iwaizumi, M. Suzuki, T. Isobe, and H. Azumi, *ibid.*, p. 1325; N. Hirota, *J. Am. Chem. Soc.*, **89**, 32 (1967); A. Crowley, N. Hirota, and R. Kreilick, *J. Chem. Phys.*, **46**, 4815 (1967); A. M. Hermann, A. Rembaum, and W. R. Carper, *J. Phys. Chem.*, **71**, 2661 (1967).
15 K. K. Brandes, R. Suhrmann, and R. J. Gerdes, *J. Org. Chem.*, **32**, 741 (1967); R. V. Slates and M. Szwarc, *J. Am. Chem. Soc.*, **89**, 6043 (1967).
16 M. J. Blandamer, D. E. Clarke, T. A. Claxton, M. F. Fox, N. J. Hidden, J. Oakes, M. C. R. Symons, G. S. P. Verma, and M. J. Wootten, *Chem. Commun.*, **1967**, 273.
17a A. J. Parker, *Adv. Phys. Org. Chem.*, **5**, 173 (1967).
17b E. Tommila, *Ann. Acad. Scient. Fennicae*, Ser. A, No. 139 (1967).
18 J. F. Coetzee, *Progr. Phys. Org. Chem.*, **4**, 45 (1967).
19 R. E. Robertson, *Progr. Phys. Org. Chem.*, **4**, 281 (1967).
20 E. Tommila, *Suomen Kemistilehti*, *A*, **40**, 3 (1967); R. Payne, *J. Am. Chem. Soc.*, **89**, 489 (1967).
21 H. S. Golinkin, I. Lee, and J. B. Hyne, *J. Am. Chem. Soc.*, **89**, 1307 (1967); see *Organic Reaction Mechanisms*, **1966**, 50.

as the plots of partial molar volume of benzyl chloride against mole fraction of alcohol show corresponding maxima. Other investigations of *tert*-butyl alcohol–water mixtures are described in references 22 and 23.

The rate of solvolysis of methyl iodide in aqueous dioxan decreases with decreasing water content of the solvent but in aqueous dimethyl sulphoxide it increases. This difference is quantitative rather than qualitative, however, for the activity coefficient of methyl iodide (β_{MeI}) decreases with decreasing water content in both solvent systems. Allowance was made for this by calculating $k/\beta_{MeI}[H_2O]$ which increased with decreasing water content of both solvent systems.[24]

Solvent activity coefficients for transfer of anions and cations from methyl alcohol to water and a range of dipolar solvents have been estimated on the basis of the "tetraphenylarsonium tetraphenylboride assumption".[25]

The thermodynamic properties of, and the kinetics of solvolyses in, 1-methylpyrrolidone have been investigated extensively.[26]

The plots of $\log k$ against acid concentration for the hydrolyses of *n*-propyl chloride, bromide, and iodide in aqueous perchloric acid are straight lines of negative slope, but those for the corresponding isopropyl halides are curves with the rate decreasing at low acid concentration but increasing at high acid concentrations. The increase was attributed to acid-catalysis.[27]

The kinetics of solvolysis of isopropyl chloride in sulphuric acid–water mixtures have been measured.[28]

The *Y*-value of tetrahexylammonium benzoate, determined from the rate of solvolysis of *tert*-butyl chloride in it, is –0.39, intermediate between those of ethanol and water.[29]

High pressures cause an abnormal enhancement of the rate of reaction of triethylamine and ethyl iodide in benzene and nitrobenzene at 25°, possibly as a result of freezing of the solvent.[30] The kinetics of the reaction of pyridine and methyl iodide in benzene–nitrobenzene mixtures have been measured.[31]

The effect of solvent on the rates of solvolysis of *tert*-butyl chloride,[32]

22 M. J. Blandamer, M. C. R. Symons, and M. J. Wootten, *Trans. Faraday Soc.*, **63**, 2337 (1967).
23 J. Burgess, *Chem. Commun.*, **1967**, 1134.
24 P. O. I. Virtanen, *Suomen Kemistilehti*, *B*, **40**, 163 (1967).
25 R. Alexander and A. J. Parker, *J. Am. Chem. Soc.*, **89**, 5549 (1967); R. Alexander, E. C. F. Ko, Y. C. Mac, and A. J. Parker, *ibid.*, p. 3703; W. A. Millen and D. W. Watts, *ibid.*, p. 6051.
26 P. O. I. Virtanen, *Suomen Kemistilehti*, *B*, **39**, 257 (1966); *B*, **40**, 1, 241, 313 (1967); P. O. I. Virtanen and J. Korpela, *ibid.*, *B*, **40**, 99, 316 (1967).
27 J. Koskikallio, *Acta Chem. Scand.*, **21**, 397 (1967); see also *Organic Reaction Mechanisms*, **1966**, 52—53.
28 I. Yrjänä and J. Koskikallio, *Suomen Kemistilehti*, *B* **40**, 190 (1967).
29 C. G. Swain, A. Ohno, D. K. Roe, R. Brown, and T. Maugh, *J. Am. Chem. Soc.*, **89**, 2648 (1967).
30 Y. Kondo, H. Tojima, and N. Tokura, *Bull. Chem. Soc. Japan*, **40**, 1408 (1967).
31 K. Kalliorinne and E. Tommila, *Suomen Kemistilehti*, *B*, **40**, 209 (1967); Y. Kondo and N. Tokura, *Bull. Chem. Soc. Japan*, **40**, 1433, 1438 (1967).
32 P. O. I. Virtanen, *Suomen Kemistilehti*, *B*, **40**, 178 (1967).

9-fluorenyl toluene-*p*-sulphonate,[33] and chloromethyl methyl ether,[34] and on the rate of reaction of *p*-substituted benzyl chlorides with azide ion,[35] of butyl-lithium with butyl bromide,[36] and of alkyl halides with piperidine and morpholine[37] have also been determined.

Isotope Effects

Solvolyses of the chlorides (**11**) and (**12**) which were thought to proceed by limiting S_N1 mechanisms do not show steric secondary isotope effects predicted by calculations using Bartell's model[38] (see also references 212, 213 of Chapter 1).

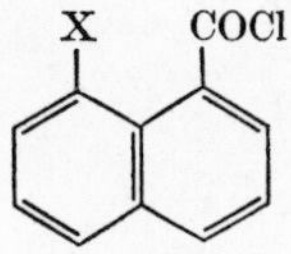

(**11**)

$X = CH_3$ or CD_3

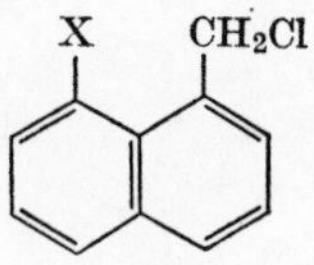

(**12**)

$X = CH_3$ or CD_3

Isotope effects for the reaction of methyl iodide with hydroxide[39] and cyanide[40] ion have been calculated. The secondary deuterium isotope effect for the iodide-exchange reaction of methyl iodide has been determined.[41]

Solvent isotope effects on the hydrolysis of alkyl halides and arenesulphonates are discussed in reference 19; k_{H_2O}/k_{D_2O} for the hydrolysis of dimethylphenylmethyl chloride in aqueous tetrahydrofuran is 1.54.[42] Free energies of transfer of alkali-metal chlorides from H_2O to D_2O have been measured.[43]

Neighbouring-group Participation

Winstein and Allred have published details of their extensive investigation of

33 G. W. Cowell, T. D. George, A. Ledwith, and D. G. Morris, *J. Chem. Soc., B*, **1966**, 1169.
34 T. C. Jones and E. R. Thornton, *J. Am. Chem. Soc.*, **89**, 4863 (1967); see also E. R. Thornton, *ibid.*, p. 2915.
35 U. Miotti, *Gazz. Chim. Ital.*, **97**, 254 (1967).
36 A. I. Shatenshtein, E. A. Kovrizhnykh, and V. M. Basmanova, *Kinet. Katal.*, **7**, 953 (1966); *Chem. Abs.* **66**, 54752c (1967).
37 B. Bariou and M. Kerfanto, *Compt. Rend., Ser. C*, 264, 1134 (1967).
38 G. J. Karabatsos, G. C. Sonnichsen, C. G. Papioannou, S. E. Scheppele, and R. L. Shone, *J. Am. Chem. Soc.*, **89**, 463 (1967).
39 A. V. Willi *Z. Naturforsch.*, a, **21**, 1385 (1966).
40 A. V. Willi, *Z. Naturforsch.*, a, **21**, 1377 (1966).
41 S. Seltzer and A. A. Zavitsas, *Can. J. Chem.*, **45**, 2023 (1967).
42 H. Hübner, K. Kellner, P. Krumbiegel, and M. Mühlstädt, *Abhandl. Deut. Akad. Wiss. Berlin, Kl. Chem., Geol. Biol.*, **1964**, 679 (1963); *Chem. Abs.*, **66**, 37031z (1967).
43 P. Salomaa and V. Aalto, *Acta Chem. Scand.*, **20**, 2035 (1966).

MeO–5 and MeO–6 participation.[44–47] Acetolysis of 4-methoxypentyl or 4-methoxy-1-methylbutyl *p*-bromobenzenesulphonate (**13** and **14**) yields the same product consisting mainly of 4-methoxypentyl acetate (40%) and 4-methoxy-1-methylbutyl acetate (60%) with some tetrahydro-2-methyl-

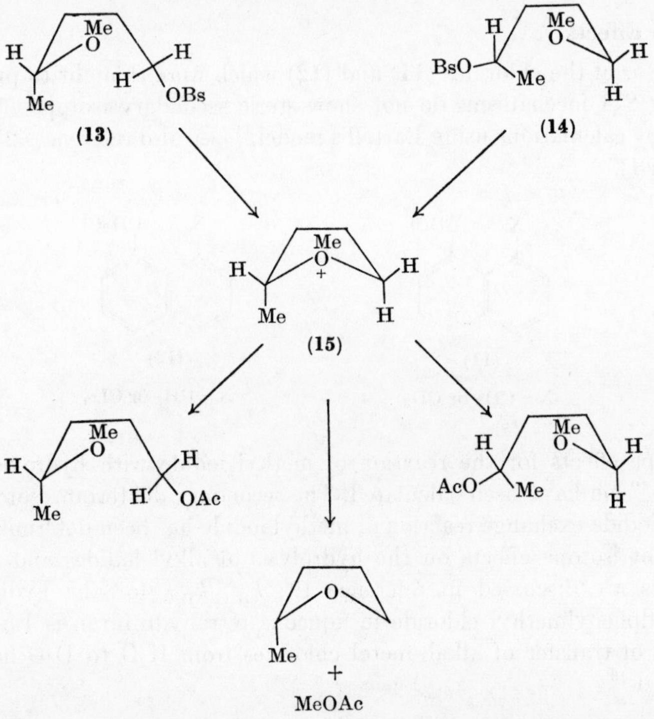

furan (< 2%). The products from the ethanolyses are similar, with both *p*-bromobenzenesulphonates yielding 5-ethoxy-2-methoxypentane (60%) and 4-ethoxy-1-methoxypentane (40%) with some tetrahydro-2-methylfuran (1.7%). This suggests that the reactions involve a common oxonium ion intermediate (**15**).[44] It is interesting that this shows such little preference for attack at the secondary carbon in acetic acid since the ratio of the rates of acetolysis of ethyl and isopropyl toluene-*p*-sulphonate is 1:50, and this suggests to us that the transition state for ring opening of the oxonium ion may have much more S_N2 character than that for the solvolysis of the toluene-*p*-sulphonate. That would be reasonable as the cyclizations of both *p*-bromobenzenesulphonates are concerted processes since they react much faster than

[44] E. L. Allred and S. Winstein, *J. Am. Chem. Soc.*, **89**, 3991 (1967).
[45] E. L. Allred and S. Winstein, *J. Am. Chem. Soc.*, **89**, 3998 (1967).
[46] E. L. Allred and S. Winstein, *J. Am. Chem. Soc.*, **89**, 4008 (1967).
[47] E. L. Allred and S. Winstein, *J. Am. Chem. Soc.*, **89**, 4012 (1967).

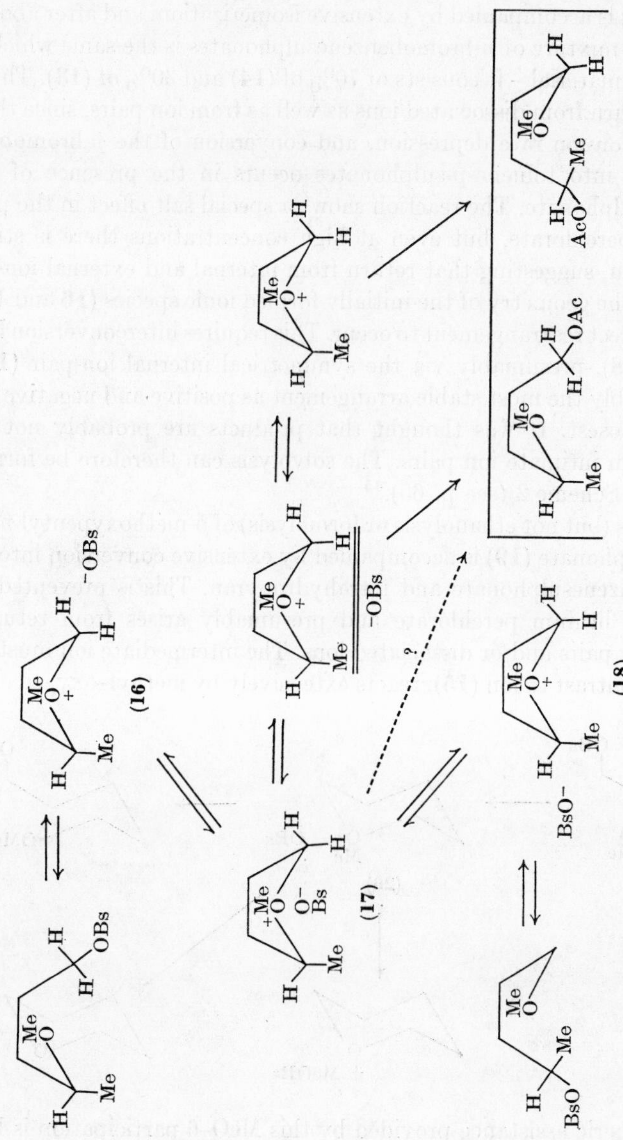

Scheme 2

analogous compounds without methoxyl groups. Therefore, as ring-opening of the cyclic oxonium ions resembles the microscopic reverse of the cyclization it would be expected to be concerted as well.

Acetolysis is accompanied by extensive isomerization, and after about 70% reaction the mixture of *p*-bromobenzenesulphonates is the same whichever is the starting material – it consists of 70% of (**14**) and 30% of (**13**). This is the result of return from dissociated ions as well as from ion pairs, since there is a small common-ion rate depression, and conversion of the *p*-bromobenzenesulphonates into toluene-*p*-sulphonates occurs in the presence of lithium toluene-*p*-sulphonate. The reaction shows a special salt effect in the presence of lithium perchlorate, but even at high concentrations there is still some isomerization, suggesting that return from internal and external ion-pairs is occurring. The geometry of the initially formed ionic species (**16** and **18**) does not allow direct rearrangement to occur. This requires interconversion between (**16**) and (**18**), presumably via the symmetrical internal ion-pair (**17**), this being probably the most stable arrangement as positive and negative charges are there closest. It was thought that products are probably not formed directly from intimate ion pairs. The solvolysis can therefore be formulated as shown in Scheme 2 (see p. 65).[45]

Acetolysis (but not ethanolysis or formolysis) of 5-methoxypentyl *p*-bromobenzene sulphonate (**19**) is accompanied by extensive conversion into methyl *p*-bromobenzenesulphonate and tetrahydropyran. This is prevented by the addition of lithium perchlorate and presumably arises from return from external ion pairs and/or dissociated ions. The intermediate ion must be (**20**) which, in contrast to ion (**15**), reacts extensively by methyl–oxygen cleavage.

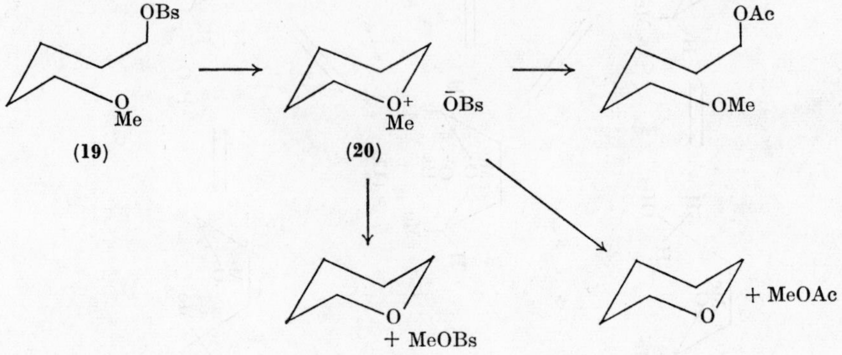

The anchimeric assistance provided by this MeO–6 participation is less than that provided by analogous MeO–5 participation, as 5-methoxypentyl *p*-bromobenzenesulphonate is acetolysed 14 times slower than 4-methoxybutyl *p*-bromobenzenesulphonate. As is reasonable, this is almost all the result of a more negative entropy of activation ($\delta\Delta S^{\ddagger} = 4.7$ eu) for formation of the six-membered ring.[47]

The steric course of the cyclization of 4-methoxy-1-methylbutyl *p*-bromobenzenesulphonate on treatment with lithium chloride in pyridine is inversion of configuration.[48]

Details have been published of Hughes and Speakman's demonstration of MeO–5 participation in the reaction of 2,3,5-tri-*O*-benzyl-4-*O*-toluene-*p*-sulphonyl-D-ribose dimethyl acetal with tetrabutylammonium benzoate in 1-methylpyrrolidone.[49] EtS–5 participation occurs when a solution of 5-*O*-toluene-*p*-sulphonyl-L-arabinose diethyl thioacetal (**21**) in aqueous acetone is heated in the presence of barium carbonate as an acid acceptor.

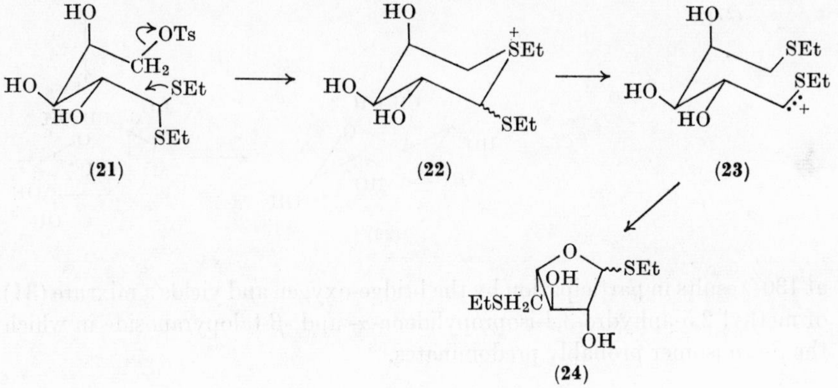

The initially formed sulphonium ion (**22**) presumably undergoes ring opening to form ion (**23**) which recyclizes to products (**24**).[50]

Further interesting examples of neighbouring-group participation by the ring-oxygen of sugar derivatives[51] have also been reported. Three of the products (**27**, **28**, and **29**) of the hydrolysis of methyl 4-*O*-(*p*-nitrobenzenesulphonyl)-α-D-glucopyranoside (**25**) may be rationalized as being formed via a bicyclic oxonium ion (**26**), while the fourth product, methyl α-D-galactopyranoside, probably results from a direct displacement or from an intermediate carbonium ion.[52]

Treatment of 1,6-anhydro-3,4-*O*-isopropylidene-2-*O*-methanesulphonyl-β-D-galactopyranose (**30**) with methanol containing potassium fluoride dihydrate

48 E. R. Novak and D. S. Tarbell, *J. Am. Chem. Soc.*, **89**, 73, 3086 (1967).
49 N. A. Hughes and P. R. H. Speakman, *J. Chem. Soc., C*, **1967**, 1182; see *Organic Reaction Mechanisms*, **1965**, 55—56.
50 N. A. Hughes and R. Robson, *J. Chem. Soc., C*, **1966**, 2366.
51 See *Organic Reaction Mechanisms*, **1966**, 63.
52 P. W. Austin, J. G. Buchanan, and D. G. Large, *Chem. Commun.*, **1967**, 418.

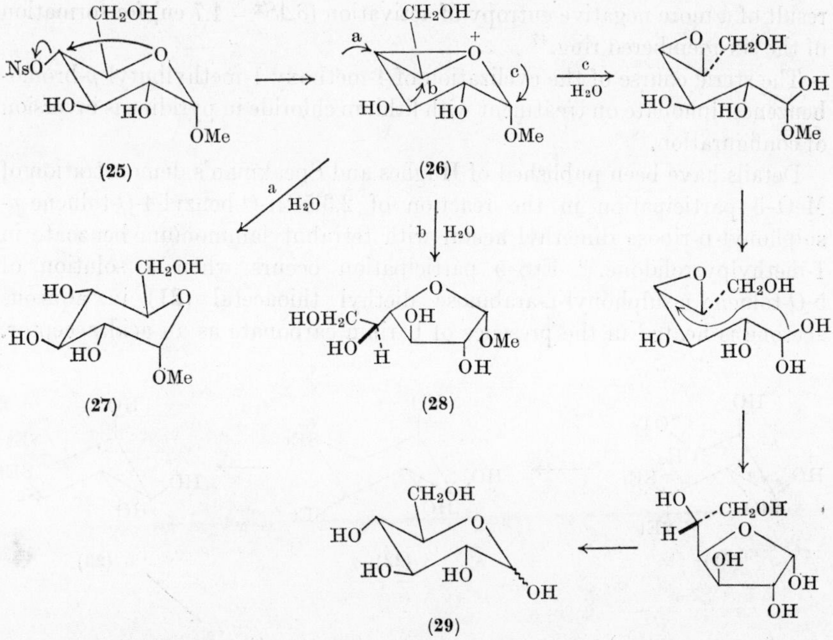

(25) (26) (27) (28) (29)

at 130° results in participation by the bridge-oxygen and yields a mixture (**31**) of methyl 2,5-anhydro-3,4-isopropylidene-α- and -β-talopyranoside in which the β-*exo*-isomer probably predominates.[53]

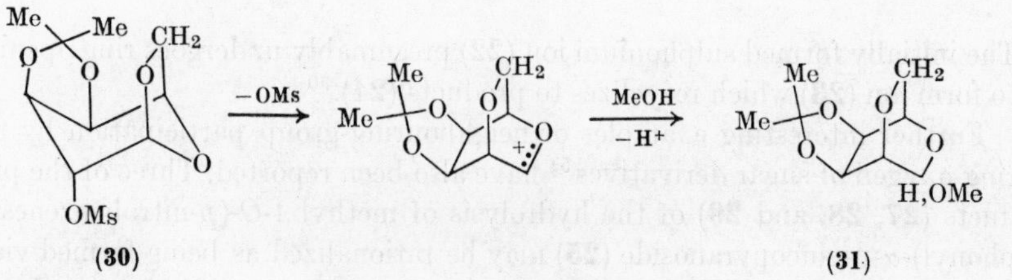

(30) (31)

Other examples of participation by the ring-oxygen of sugar derivatives are described in references 54 and 55.

Treatment of the complex sulphonium salt (**34**) (formed by transannular participation **32** → **33** → **34**) yields the *cis*-glycol (**35**), possibly by transannular S–5 participation as shown.[56]

[53] N. A. Hughes, *Chem. Commun.*, **1967**, 1072.
[54] C. L. Stevens, R. P. Glinski, G. E. Gutowski, and J. P. Dickerson, *Tetrahedron Letters*, **1967**, 649.
[55] L. N. Owen, *Chem. Commun.*, **1967**, 526.
[56] P. Wilder and L. A. Feliu-Otero, *J. Org. Chem.*, **31**, 4264 (1966).

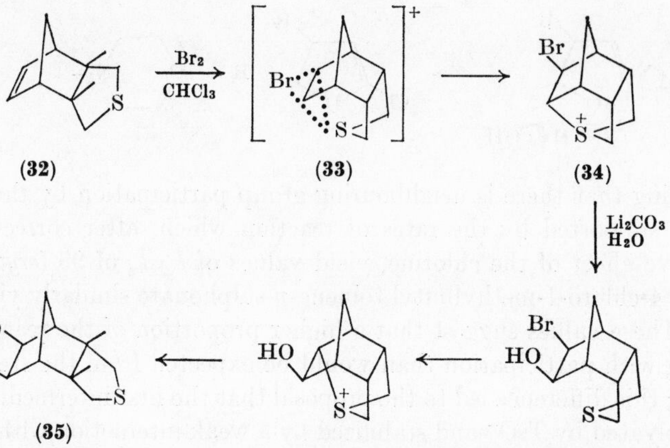

The relative rates of solvolysis of cyclohexyl bromide and *cis*- and *trans*-2-hydroxycyclohexyl bromide in dilute aqueous perchloric acid are 1:2.12:0.64, i.e., they are very similar, despite the strong electron-withdrawing inductive effect of the hydroxyl group. This suggests that the *trans*-hydroxy-compound reacts with neighbouring hydroxyl group participation whereas the *cis*-compound reacts with participation by neighbouring hydrogen. The latter process would yield cyclohexanone, the observed product.[57]

The volume of activation for cyclization of 4-chlorobutanol has been measured.[58]

An example of O^-–5 participation in substitution at vinylic carbon has been reported (see eqn. 5) (p. 70).[59]

Other examples of neighbouring-group participation by ionized[60–67] and un-ionized[60–62] hydroxyl groups have been reported.

The trifluoroacetolyses of *erythro*- and *threo*-4-chloro-1-methylpentyl toluene-*p*-sulphonate occur with about 90% retention of configuration,

57 G. Bodennec, H. Bodot, and A. Nattaghe, *Bull. Soc. Chim. France*, **1967**, 876; see also H. Bodot and É. Laurent-Dieuzeide, *ibid.*, **1966**, 3908; P. Crouzet, É. Laurent-Dieuzeide, and J. Wylde, *ibid.*, **1967**, 4047.

58 A. H. Ewald and D. J. Ottley, *Australian J. Chem.*, **20**, 1335 (1967).

59 R. Dowbenko, *Chem. Ind.* (*London*), **1966**, 2097.

60 J. Myszkowski, A. Z. Zielinski, and R. Lesmian, *Chem. Stosow., Ser. A*, **10**, 439 (1966); *Chem. Abs.* **67**, 11123b (1967).

61 J. Myszkowski, A. Z. Zielinski, and J. Raszkiewicz, *Chem. Stosow., Ser. A*, **10**, 325 (1966); *Chem. Abs.*, **66**, 55289u (1967).

62 J. Myszkowski, A. Z. Zielinski, and J. Habera, *Roczniki Chem.*, **40**, 1343 (1966).

63 E. N. Barantsevich and T. I. Temnikova, *Reakts. Sposobnost Org. Soedin. Tartu. Gos. Univ.*, **3**, 81 (1966); *Chem. Abs.*, **66**, 115200v (1967).

64 E. N. Barantsevich and T. I. Temnikova, *Reakts. Sposobnost Org. Soedin. Tartu. Gos. Univ.*, **3**, 197 (1966); *Chem. Abs.* **67**, 11072j (1967).

65 G. Bodennec, H. Bodot, and A. Nattaghe, *Bull. Soc. Chim. France*, **1967**, 876.

66 W. Reeve and M. Nees, *J. Am. Chem. Soc.*, **89**, 647 (1967).

67 G. Werner and K.-H. Schmidt, *Tetrahedron Letters*, **1967**, 1283.

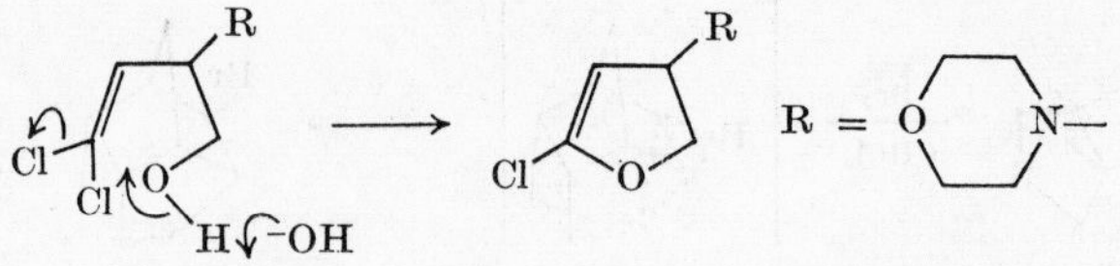

. . . (5)

suggesting that there is neighbouring-group participation by the chlorine.[68] This is supported by the rates of reaction which, after correction for the inductive effect of the chlorine, yield values of k_{Δ}/k_s of 99 (*erythro*) and 65 (*threo*); 4-chloro-1-methylbutyl toluene-*p*-sulphonate similarly yields a value of 33. These values suggest that a higher proportion of the reactions is proceeding with participation than would be expected from the stereochemical results; this difference led to the proposal that the first intermediate is an ion (**36**), solvated by TsO^- and stabilized by a weak interaction with Cl; this ion could still react with inversion of configuration, and could collapse to the cyclic

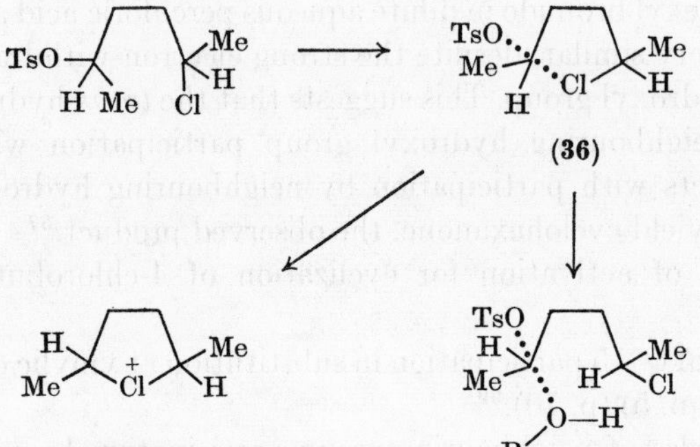

(36)

(37)

chloronium ion (**37**). Qualitatively similar, but much less striking, results were obtained for the formolyses and acetolyses of these substrates. There was also some evidence for F–5 and Cl–6 participation since the trifluoroacetolyses of 4-[68, 69] and 5-chloro-1-methylbutyl[68] toluene-*p*-sulphonate yielded values of k_{Δ}/k_s of 2.4 and 1.1, respectively.

The NMR spectra of several tetramethylethylenehalonium ions in SbF_5–FSO_3H–SO_2 and SbF_5–SO_2 ions have been reported.[70]

Several examples of participation by enolate anions have been reported (cf. eqns. 6—8).[71, 72] Participation by the double bond of an enol ether occurs

[68] P. E. Peterson, R. J. Bopp, D. M. Chevli, E. L. Curran, D. E. Dillard, and R. J. Kamat, *J. Am. Chem. Soc.*, **89**, 5902 (1967).
[69] P. E. Peterson and R. J. Bopp, *J. Am. Chem. Soc.*, **89**, 1283 (1967).
[70] G. A. Olah and J. M. Bollinger, *J. Am. Chem. Soc.*, **89**, 4744 (1967).
[71] A. R. Davies and G. H. R. Summers, *J. Chem. Soc., C*, **1967**, 909.
[72] F. Nerdel, D. Frank, and K. Rehse, *Chem. Ber.*, **100**, 2978 (1967); see also F. Nerdel, D. Frank, K. Gerner, and W. Metasch, *Tetrahedron Letters*, **1967**, 4499.

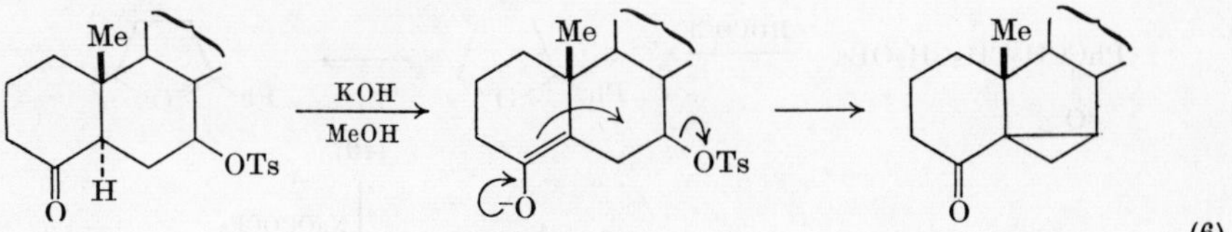

. . . (6)

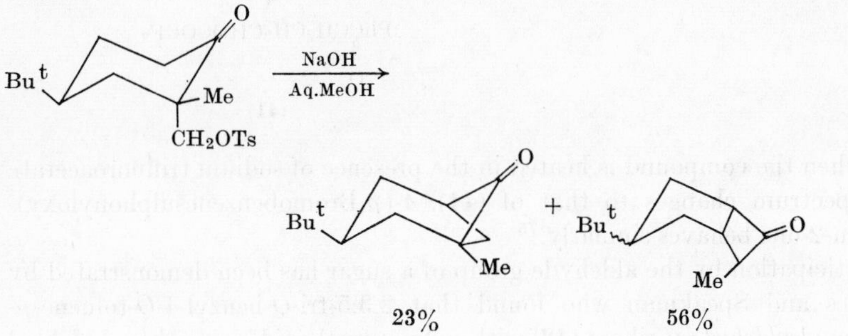

. . . (7)

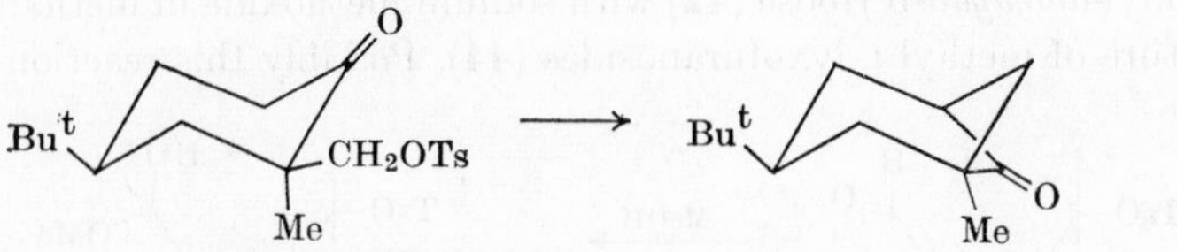

. . . (8)

when the iodide (**38**) is treated with silver nitrate in aqueous ethanol. The product is the ketal (**39**).[73]

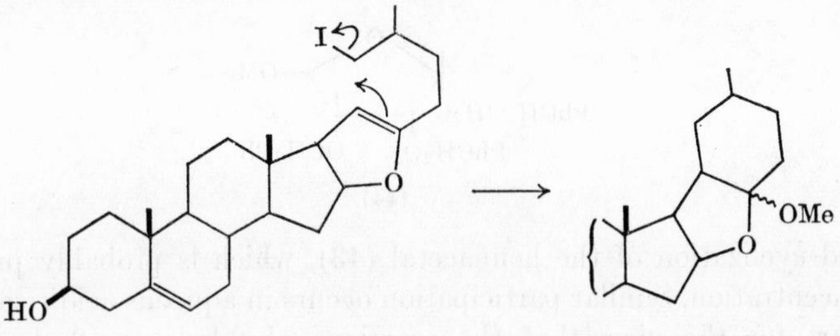

(**38**) (**39**)

Participation by the double bond of an enamine possibly occurs in the reaction of 2-chlorocyclohexanone with piperidine, to yield 6,6-dipiperidinobicyclo[3.1.0]hexane.[74]

The formation of an ion (**40**) from 4-(*p*-bromobenzenesulphonyloxy)butyrophenone in formic and in trifluoroacetic acid has been demonstrated by NMR spectroscopy. The spectrum of (**40**) in trifluoroacetic acid [k(formation) = 1.2×10^{-4} sec^{-1} at $-3°$] is unchanged at room temperature for at least 30 days,

[73] F. C. Uhle, *J. Org. Chem.*, **31**, 4193 (1966); F. C. Uhle, *ibid.*, **32**, 792 (1967).
[74] J. Szmuszkovicz, E. Cerda, M. F. Grostic, and J. F. Zieserl, *Tetrahedron Letters*, **1967**, 3969.

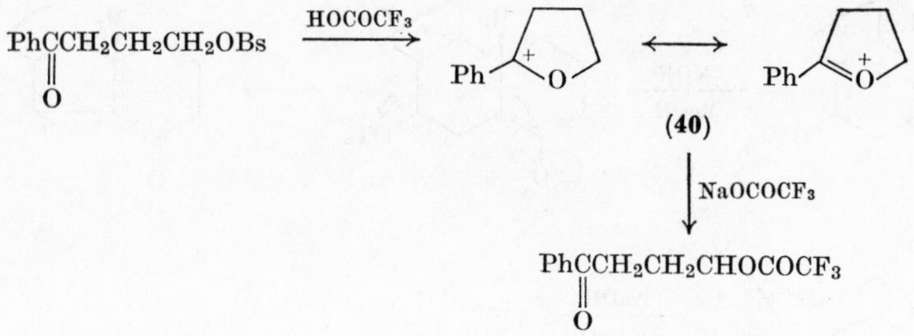

(40)

(41)

but when the compound is heated in the presence of sodium trifluoroacetate the spectrum changes to that of (**41**). 4-(*p*-Bromobenzenesulphonyloxy)-pentan-2-one behaves similarly.[75]

Participation by the aldehyde group of a sugar has been demonstrated by Hughes and Speakman who found that 2,3,5-tri-*O*-benzyl-4-*O*-toluene-*p*-sulphonyl-*aldehydo*-D-ribose (**42**) with sodium methoxide in methanol yielded a mixture of methyl L-lyxofuranosides (**44**). Possibly this reaction is a base-

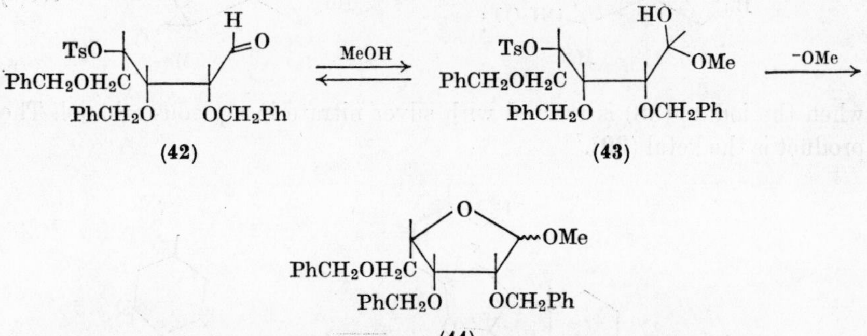

(42) (43)

(44)

catalysed cyclization of the hemiacetal (**43**), which is probably present in high concentration. Similar participation occurs in aqueous acidic solution.[76]

Support for the view that the reactions of 2-halogeno-2-phenylacetophenones with sodium methoxide in methanol proceed through methoxy-epoxides (eqn. 9) has been obtained by isolating these intermediates after reactions of compounds (**45**).[77] Similar participation to form a four-membered ring occurs with (**46**) and sodium methoxide.[78] Neighbouring-group participation does not occur in reactions of α-halogenomethyl ketones with strong nucleophiles (see p. 97).

[75] H. R. Ward and P. D. Sherman, *J. Am. Chem. Soc.*, **89**, 4222 (1967).
[76] N. A. Hughes and P. R. H. Speakman, *J. Chem. Soc., C*, **1967**, 1186.
[77] V. S. Karavan and T. I. Temnikova, *Zh. Organ. Chem.*, **2**, 1410, 1417 (1966).
[78] F. Nerdel and H. Kressin, *Ann. Chem.*, **707**, 1 (1967); see *Organic Reaction Mechanisms*, **1966**, 54; P. Weyerstahl, *Angew. Chem. Intern. Ed. Engl.*, **6**, 1002 (1967).

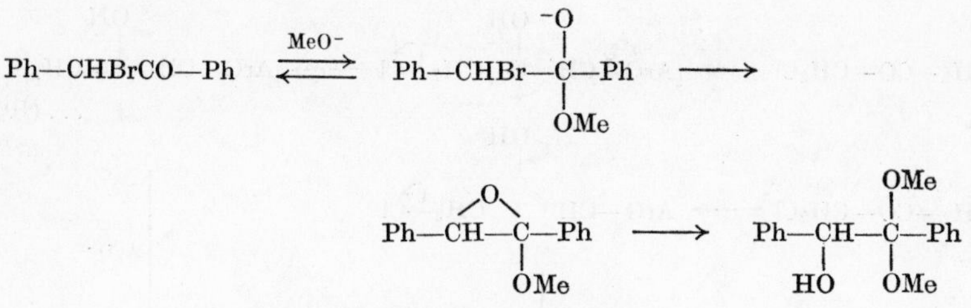

. . . (9)

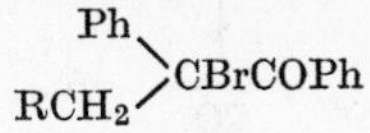

(45)

$R = C_6H_5$ or p-BrC_6H_4

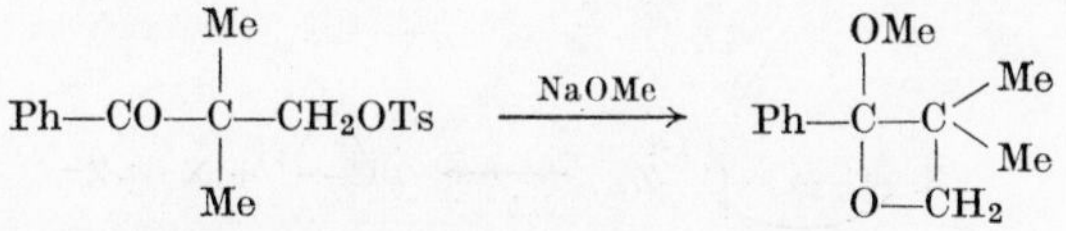

(46)

Acetolysis of a chloroketone (**47**) yields the rearranged product (**48**). 1-Chloro-3-(phenylthio)acetone reacts similarly but the chloro-ketone (**49**) is unreactive. The mechanism shown in equation (10) was proposed,[79] but in our opinion an attractive alternative is that shown in equation (11) (p. 74).

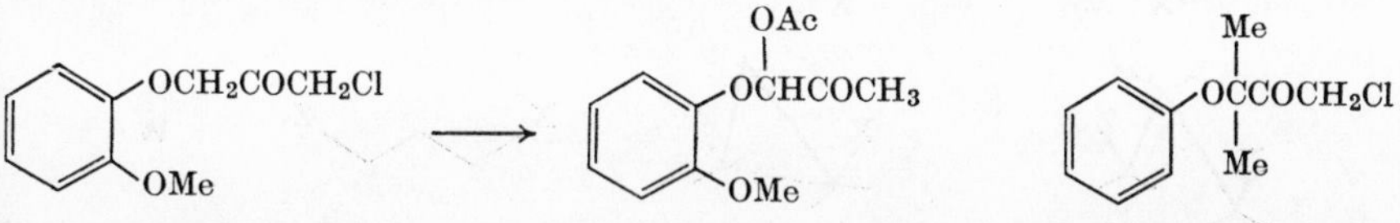

(47) **(48)** **(49)**

The stereochemical course of 1,3-eliminations (i.e., **50** → **51**) has been discussed by Nickon and Werstiuk[80–82] and the arrangements of eliminating groups shown in Scheme 3 were considered. The *apo*-S form refers only to the staggered precursor which must twist to one of the other transition states for ring closure. The preferred steric course for formation of nortricyclene from *exo*-norbornyl toluene-*p*-sulphonate at high concentrations of potassium *tert*-butoxide in *tert*-butyl alcohol was shown to be *exo*-S rather than W by

[79] V. Rosnati, G. Pagani, F. Sannicolò, *Tetrahedron Letters*, **1967**, 1241, 4545; see also V. Rosnati, D. Misiti, and F. De Marchi, *Gazz. Chem. Ital.*, **96**, 497 (1966).
[80] A. Nickon and N. H. Werstiuk, *J. Am. Chem. Soc.*, **89**, 3914 (1967).
[81] A. Nickon and N. H. Werstiuk, *J. Am. Chem. Soc.*, **89**, 3915 (1967).
[82] A. Nickon and N. H. Werstiuk, *J. Am. Chem. Soc.*, **89**, 3917 (1967).

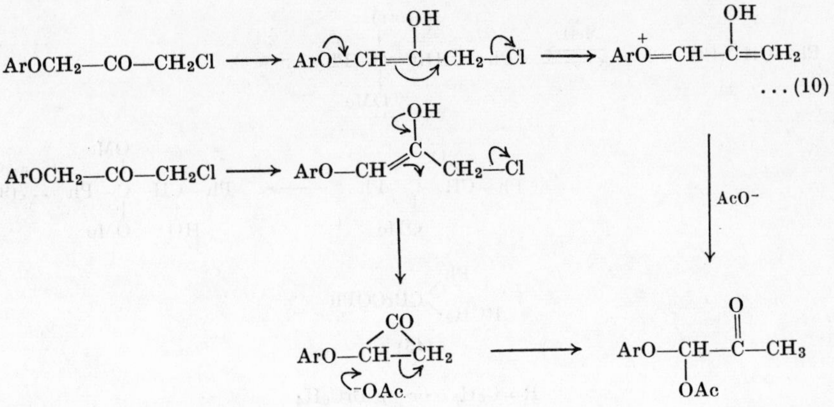

... (10)

... (11)

(50) **(51)**

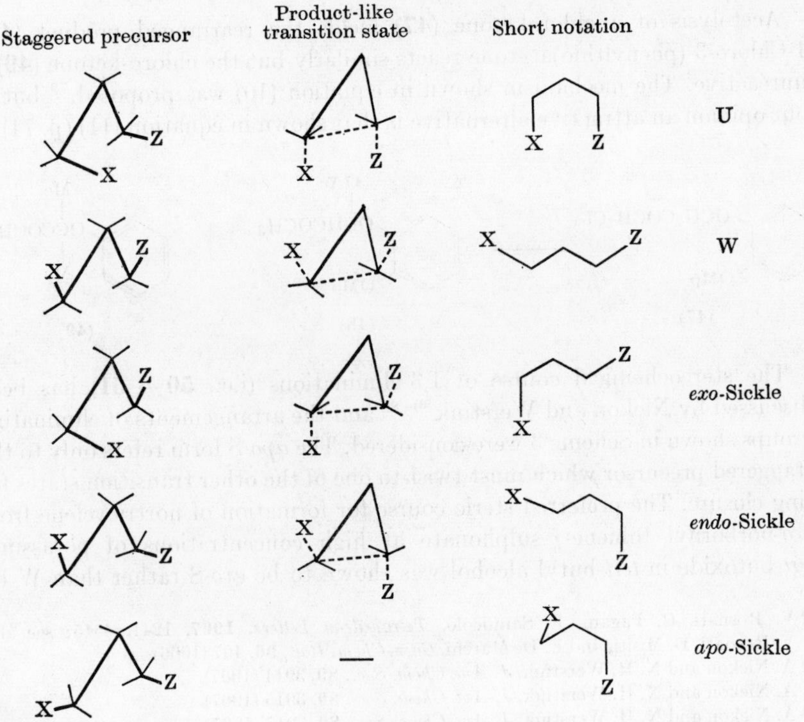

Scheme 3

studying the reactions of *exo*- and *endo*-6-deuterated derivatives. Interpretation is complicated by a concurrent unimolecular reaction which yields nortricyclene through a non-classical ion (or equilibrating classical ions) in which the distinction between 6-*exo*- and 6-*endo*-hydrogens is lost, but the greater loss of deuterium from the 6-*endo*- than from the 6-*exo*-deuterated compound at high *tert*-butoxide concentrations indicates a preference for the *exo*-S geometry.[80] The ratio (P) of the rate of abstraction of a 6-*endo*-hydrogen to that of abstraction of a 6-*exo*-hydrogen was estimated to be 1.5 at 60°.[81] Similar experiments with 6-deuterated *endo*-norbornyl toluene-*p*-sulphonates showed the U- was preferred to the *endo*-S-geometry with $P = 1.7$ at 130°.[82] In view of the normal strong preference for the abstraction of *exo*-hydrogens from norbornane systems, these results may mean that the stereoelectronic factors which favour *exo*-S- and U-transition states are very large indeed.

The cyclization of ω-bromoalkyl *p*-tolyl sulphones (eqn. 12) is 100 times

$$ArSO_2CH_2(CH_2)_{n-2}CH_2Br \xrightarrow[HOBu^t]{KOBu^t} ArSO_2\underset{\diagdown\;(CH_2)_{n-2}\;\diagup}{CH{-}{-}CH_2} \qquad \ldots(12)$$

faster when $n = 3$ than when $n = 5$ and was not observed when $n = 4$ or 6. It was suggested that formation of the three-membered ring is favoured by orbital interaction between the *p*-tolyl sulphone group and the developing cyclopropane ring in the transition state.[83]

To test whether the cyclization of *exo*-chlorosulphone (**52**a) to the cyclopropane (**53**) with potassium *tert*-butoxide in dimethyl sulphoxide (DMSO) proceeded by isomerization to the *endo*-isomer,[84] the reaction of the deuterated compound (**52**b) was studied. When the solvent was 3:1 *tert*-butyl alcohol–DMSO, the product (**53**) contained no deuterium, but with 1:9 *tert*-butyl alcohol–DMSO it contained 25—30% of deuterium. The latter result was thought to arise, not from a front-side displacement by the carbanion, but from an *exo*-*endo*-isomerization in which there was incomplete separation of the carbanion–Bu^tOD pair. Attempts were also made to cyclize the *syn*-chloro-*endo*-sulphone (**54**), which should be epimerized at the chlorine-bearing carbon much less readily, but this was unreactive with sodium ethoxide in refluxing ethanol–dioxan and with potassium *tert*-butoxide in *tert*-butyl alcohol. This was taken as confirmatory evidence that reaction of (**52**a) involved isomerization and not a front-side displacement.[85] Ring opening of sulphone (**55**) by methoxide and thiophenoxide occurs by attack at position 2 and breaking of the 2,8-bond with inversion of configuration.[86]

[83] A. C. Knipe and C. J. M. Stirling, *J. Chem. Soc., B*, **1967**, 808.
[84] See *Organic Reaction Mechanisms*, **1966**, 66—67.
[85] S. J. Cristol and B. B. Jarvis, *J. Am. Chem. Soc.*, **89**, 401 (1967).
[86] S. J. Cristol and B. B. Jarvis, *J. Am. Chem. Soc.*, **89**, 5885 (1967).

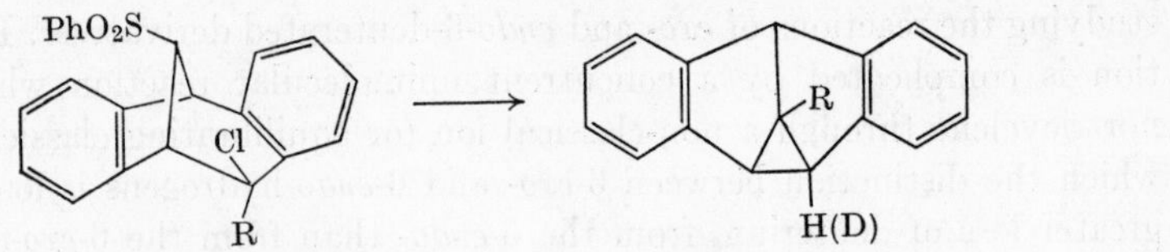

(**52a**) R=H
(**52b**) R=D

(**53**) R=SO₂Ph

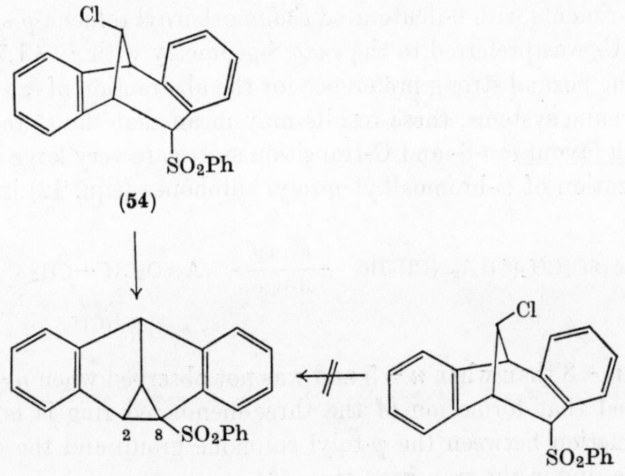

(**54**)

(**55**)

Arguments have been presented that the migrations of alkyl groups which occur in the solvolyses of neopentyl halides and arenesulphonates are concerted with the breaking of the carbon–halogen or carbon–oxygen bonds (cf. p. 15 and ref. 87). Thus the rates of acetolyses and the partial rate factor for migration of the benzyl group ($10^6 k_p^{Bz}$) of compounds p-$XC_6H_4CH_2C(Me)_2CH_2OTs$ are decreased as X becomes more strongly electron-withdrawing. This is consistent with both concerted and stepwise processes, but the observation that the

$$Me_3CCH_2OTs \longrightarrow Me_3CCH_2^+ \ {}^-OTs \longrightarrow Me_2\overset{+}{C}CH_2Me \ {}^-OTs \qquad \ldots (13)$$

$$Me_3CCH_2^+ \ {}^-OTs \xrightarrow{SOH} Me_3CCH_2OS \qquad Me_2\overset{+}{C}CH_2Me \ {}^-OTs \xrightarrow{SOH} Me_2C(OS)CH_2Me + \text{Olefins}$$

migration tendency[88] of the benzyl group was more sensitive to substituent effects in formic than in acetic acid was taken to indicate that this group was

[87] *Organic Reaction Mechanisms*, **1966**, 67—68.
[88] The ratio of one-third of the rate of solvolysis of neopentyl toluene-p-sulphonate to k_p^{Bz} under the same conditions.

participating in the rate-determining step.[89, 90] On the other hand, it has been shown that neopentyl toluene-*p*-sulphonate yields ethyl neopentyl ether and neopentyl alcohol (2—10%) in a wide range of ethanol–water mixtures. Unless these are formed by an S_N2 mechanism, the reaction must involve two carbonium ions and be non-concerted (eqn. 13). Interestingly, a small amount (< 0.3%) of dimethylcyclopropane was also formed.[91]

Details of Austin, Buchanan, and Saunders's investigation of the solvolysis of methyl 3-*O*-(*p*-nitrobenzenesulphonyl)-α-D-glucoside have been published.[92]

Other examples of participation by neighbouring carbon are described in references 93 and 94.

Several examples of transannular hydride shifts have been reported.[95]

The kinetics of cyclization of a large number of β-aminoalkyl sulphates have been reported. One β-methyl group increases the rate by a factor 6.45 and two such groups by a factor 40.5. The relative rates of cyclization of the ω-aminoalkyl sulphates, $H_2N(CH_2)_{n-1}\cdot OSO_3H$ are 445:1:0.08 for $n = 5$, 3, and 4, respectively.[96]

It is reported that the rate of reaction of the chloropiperidine (**56**) with fourteen nucleophiles is independent of the nature of nucleophile and more than 10^4 times faster than the reaction of cyclohexyl chloride with ^{-}OH. Retention of configuration was observed and neighbouring-group participation as shown in equation (14) was postulated. Intermediate (**57**) was prepared by another route.[97]

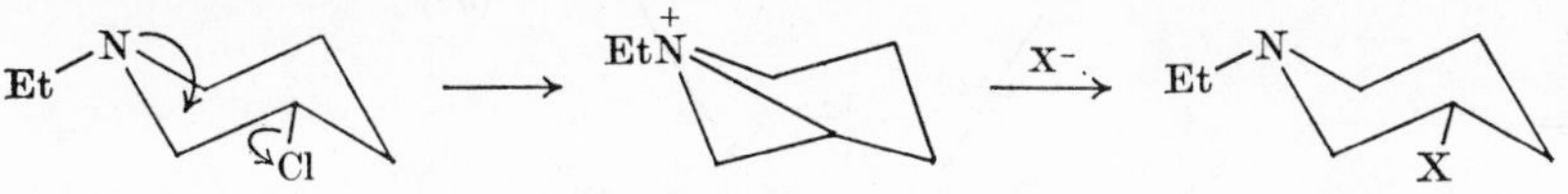

(**56**) (**57**) . . . (14)

Other examples of neighbouring-group participation by amino-groups, including transannular participation (e.g. reaction 15),[98] have been reported.[99]

[89] J. R. Owen and W. H. Saunders, *J. Am. Chem. Soc.*, **88**, 5809 (1966).
[90] R. L. Heidke and W. H. Saunders, *J. Am. Chem. Soc.*, **88**, 5816 (1966).
[91] G. M. Fraser and H. M. R. Hoffmann, *Chem. Commun.*, **1967**, 561.
[92] P. W. Austin, J. G. Buchanan, and R. M. Saunders, *J. Chem. Soc., C*, **1967**, 372; see *Organic Reaction Mechanisms*, **1965**, 61.
[93] A. P. Krapcho and J. E. McCullough, *J. Org. Chem.*, **32**, 2453 (1967).
[94] C. W. Shoppee, R. E. Lack, S. C. Sharma, and L. R. Smith, *J. Chem. Soc., C*, **1967**, 1155.
[95] A. C. Cope, J. M. McIntosh, and M. A. McKervey, *J. Am. Chem. Soc.*, **89**, 4020 (1967); T. S. Cantrell, *J. Org. Chem.*, **32**, 1669 (1967); L. H. Schwartz, M. Feil, A. J. Kascheres, K. Kaufmann, and A. M. Levine, *Tetrehadron Letters*, **1967**, 3785; R. A. Appleton, J. R. Dixon, J. M. Evans, and S. H. Graham, *Tetrahedron*, **23**, 805 (1967).
[96] C. S. Dewey and R. A. Bafford, *J. Org. Chem.*, **32**, 3108 (1967).
[97] C. F. Hammer and S. R. Heller, *Chem. Commun.*, **1966**, 919.
[98] A. J. Sisti and D. L. Lohner, *J. Org. Chem.*, **32**, 2026 (1967).
[99] S. C. Chan and F. Leh, *Australian J. Chem.*, **19**, 2271 (1966); K. Lányi, E. Institoris, M. Lovász, and Z. Szabó, *Chem. Ber.*, **100**, 3045 (1967); V. R. Gaertner, *J. Org. Chem.*, **32**, 2973

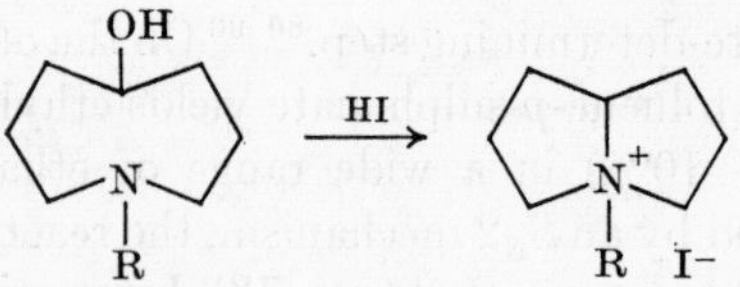

. . . (15)

Amide–3 participation occurs when α-chloro-α,α-diphenylacetamide is dissolved in aliphatic primary and secondary amines (see eqn. 16),[100] and

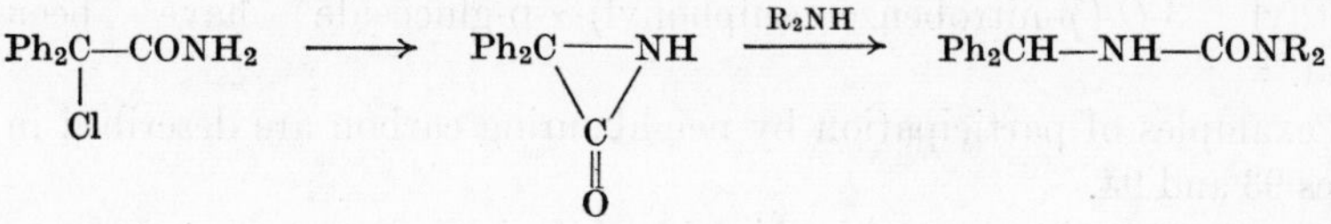

. . . (16)

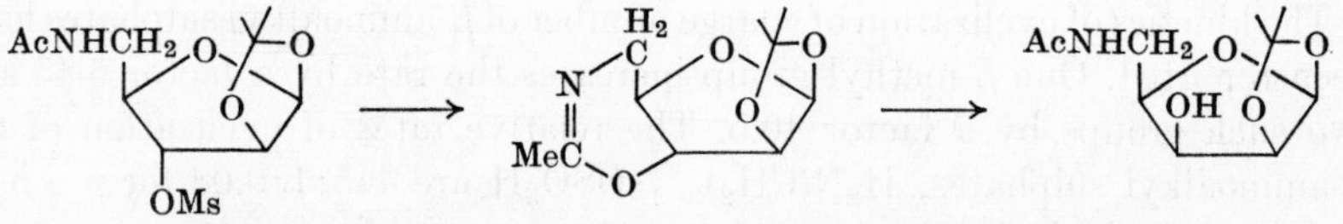

(58)

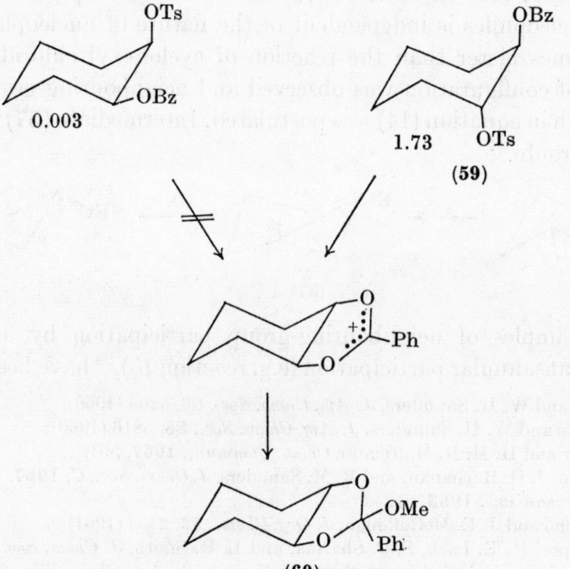

(59)

(60)

Numerals are 10^4k for acetolysis at 100°.

(1967): V. R. Gaertner, *Tetrahedron Letters*, **1967**, 343; B. J. Gaj and D. R. Moore, *ibid.*, p. 2155; U. Burckhardt, C. A. Grob, and H. R. Kiefer, *Helv. Chim. Acta*, **50**, 234 (1967); A. T. Bottini, C. A. Grob, E. Schumacher, and J. Zergenyi, *Helv. Chim. Acta*, **49**, 2516 (1966); E. Schmitz, R. Ohme, and S. Schramm, *Chem. Ber.*, **100**, 2600 (1967); C. R. Dick, *J. Org. Chem.*, **32**, 72 (1967).

[100] S. Sarel, A. Taube, and E. Breuer, *Chem. Ind.* (*London*), **1967**, 1095.

amide–6 participation occurs when the compound (**58**) is treated with sodium benzoate in DMF.[101] Several other examples of participation by amide[102, 103] and thioamide[104] groups have also been published.

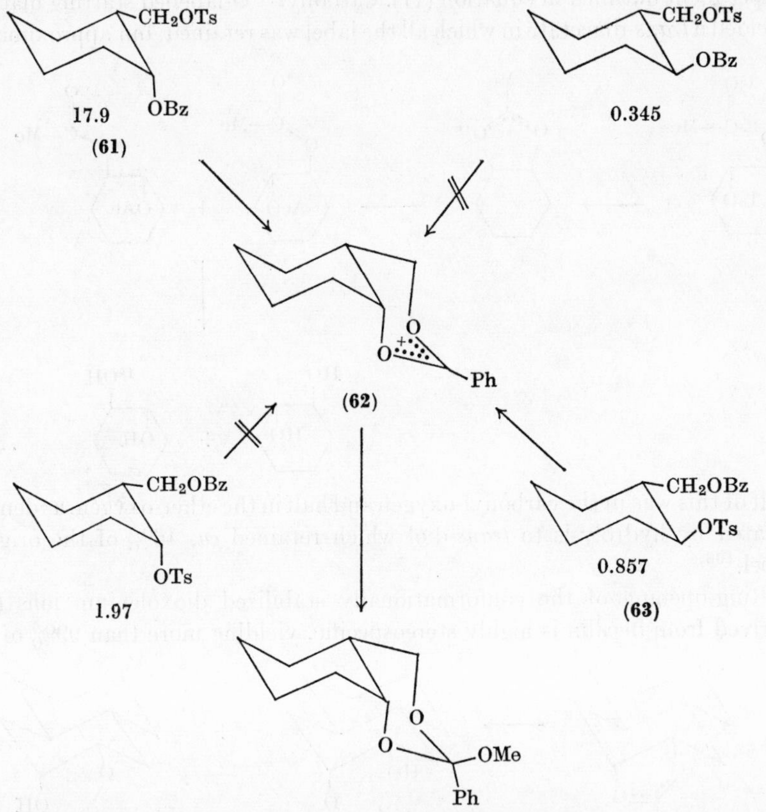

Numerals are 10^4k for acetolysis at 100°.

Neighbouring-group participation by the benzoyloxy-group occurs in the methanolysis of (**59**) (BzO–5), (**61**) (BzO–6), and (**63**) (BzO–6) since the corresponding methyl orthoesters (**60**) and (**62**) are formed. The rates of acetolysis

[101] S. Hanessian, *J. Org. Chem.*, **32**, 163 (1967).

[102] J. W. Huffman, T. Kamiya, and C. B. S. Rao, *J. Org. Chem.*, **32**, 700 (1967); see *Organic Reaction Mechanisms*, **1966**, 55.

[103] M. S. Manhas and S. J. Jeng, *J. Org. Chem.*, **32**, 1246 (1967).

[104] T. Nakai, Y. Ueno, and M. Okawara, *Tetrahedron Letters*, **1967**, 3831; T. Nakai and M. Okawara, *ibid.*, p. 3835; W. Reeve and M. Nees, *J. Am. Chem. Soc.*, **89**, 647 (1967); E. Cherbuliez, B. Baehler, O. Espejo, E. Frankenfeld, and J. Rabinowitz, *Helv. Chem. Acta*, **49**, 2608 (1966).

of (**59**) and (**61**) indicate substantial anchimeric assistance, but that for (**63**) indicates only a moderate rate enhancement.[105]

The intervention of a dioxolenium ion in the acetolysis of *trans*-2-acetoxycyclohexyl toluene-*p*-sulphonate has been demonstrated by the ^{18}O-labelling experiment outlined in equation (17). Carbonyl-^{18}O-labelled starting material yielded a *trans*-diacetate in which all the label was retained, but approximately

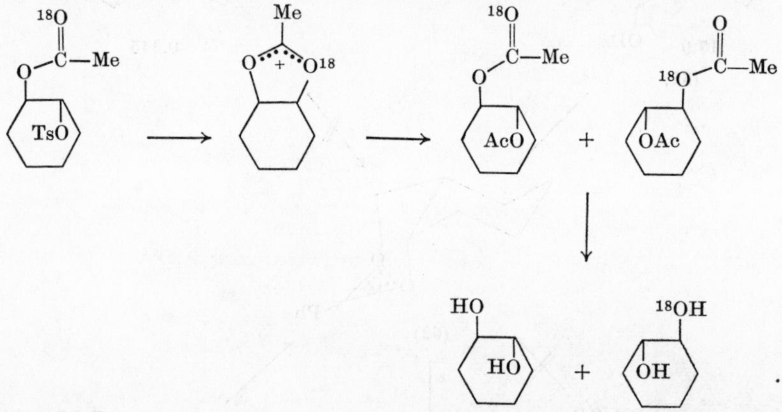

...(17)

half of this was in the carbonyl-oxygen and half in the ether-oxygen, as demonstrated by hydrolysis to *trans*-diol which retained *ca.* 46% of the original label.[106]

Ring-opening of the conformationally stabilized dioxolenium ions (**64**) derived from decalin is highly stereospecific, yielding more than 99% of the

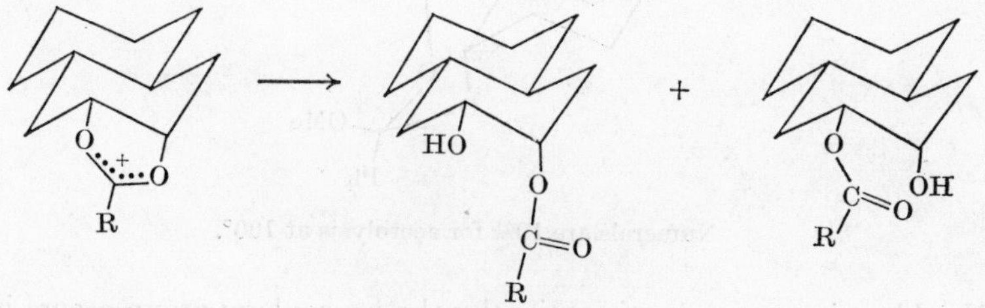

(**64**) (**65**)

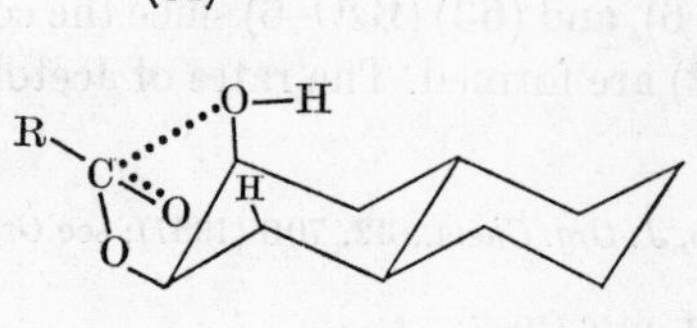

(**66**)

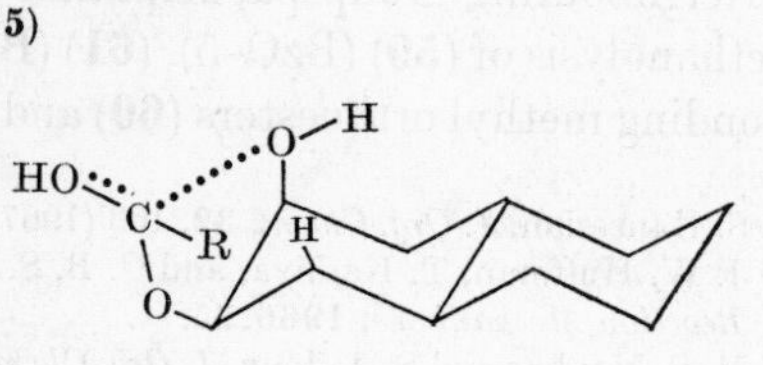

(**67**)

[105] G. Schneider and L. K. Láng, *Chem. Commun.*, **1967**, 13; Ö. K. J. Kovács, G. Schneider, L. K. Láng, and J. Apjok, *Tetrahedron*, **23**, 4181 (1967).

[106] K. B. Gash and G. U. Yuen, *J. Org. Chem.*, **31**, 4234 (1966).

axial ester (**65**). It was suggested that this was the result of an unfavourable non-bonding interaction in the transition state for formation of the equatorial ester between the incipient carbonyl-oxygen or the R-group and the nearest *cis*-axial hydrogen (see **66** and **67**). Consistently with this, when R = H the ration of axial to equatorial ester was only 3:2.[107]

NMR spectra of dioxolenium ions are reported in reference 108, and other examples of neighbouring-group participation by an acetoxy-group in references 109 and 110.

Participation by carboxylate,[111] azide,[112] and oxime groups has also been investigated.[113]

Deaminations and Related Reactions

The proton and fluorine NMR spectra of the 2,2,2-trifluoroethanediazonium ion, $F_3CCH_2N_2^+$, have been reported.[114] The ion was generated by adding a solution of trifluoromethyldiazomethane in [^{2}H]chloroform to FSO_3H at –78°. The ^{1}H and ^{19}F spectra, run at –60° on the FSO_3H solution, both showed the presence of the diazonium ion and the final product, 2,2,2-trifluoroethyl fluorosulphate. The proton spectrum of the ion was a quartet $\delta = 6.3$ ppm, and the fluorine spectrum a triplet at +64.58 ppm; J_{HF} was 6.1 cps in both spectra. On warming to –20°, nitrogen was evolved, to yield 2,2,2-trifluoroethyl fluorosulphate whose proton spectrum was a quartet at $\delta = 5.00$ ppm.

Butan-2-ol formed on aqueous deamination of [1,1-$^2H_2$]isobutylamine had deuterium distribution (**70**) (92.3%) and (**71**) (7.7%) which can be explained by reaction via a species (**68**). No butan-2-ol with deuterium distribution expected from the intervention of an edge-protonated cyclopropane (e.g., **69**) was detected. Clearly (**68**) rearranges to *sec*-butyl cations much faster than it is converted into (**69**).[115] Deamination of [3,3-$^2H_2$]butylamine yields butan-2-ol with deuterium distribution (**72a**) and (**72b**), which also indicates the non-intervention of a protonated cyclopropane.[116] These results contrast with those obtained on deamination of deuterium-labelled *n*-propylamine which indicated the intervention of a protonated cyclopropane intermediate.[117]

107 J. F. King and A. D. Allbutt, *Tetrahedron Letters*, **1967**, 49.
108 C. Pedersen, *Tetrahedron Letters*, **1967**, 511.
109 K. H. Dudley and H. W. Miller, *J. Org. Chem.*, **32**, 2341 (1967).
110 S. J. Angyal and T. S. Stewart, *Australian J. Chem.*, **21**, 2117 (1967); S. J. Angyal, V. J. Bender, P. T. Gilham, R. M. Hoskinson, and M. E. Pitman, *ibid.*, p. 2109.
111 R. B. Sandin, W. J. Rebel, and S. Levine, *J. Org. Chem.*, **31**, 3879 (1966).
112 G. Swift and D. Swern, *J. Org. Chem.*, **32**, 511 (1967); *ibid.*, **31**, 4226 (1966).
113 F. L. Scott and R. J. MacConaill, *Tetrahedron Letters*, **1967**, 3685.
114 J. R. Mohrig and K. Keegstra, *J. Am. Chem. Soc.*, **89**, 5492 (1967).
115 G. J. Karabatsos, N. Hsi, and S. Meyerson, *J. Am. Chem. Soc.*, **88**, 5649 (1966).
116 G. J. Karabatsos, R. A. Mount, D. O. Rickter, and S. Meyerson, *J. Am. Chem. Soc.*, **88**, 5651 (1966).
117 See *Organic Reaction Mechanisms*, **1965**, 63—64.

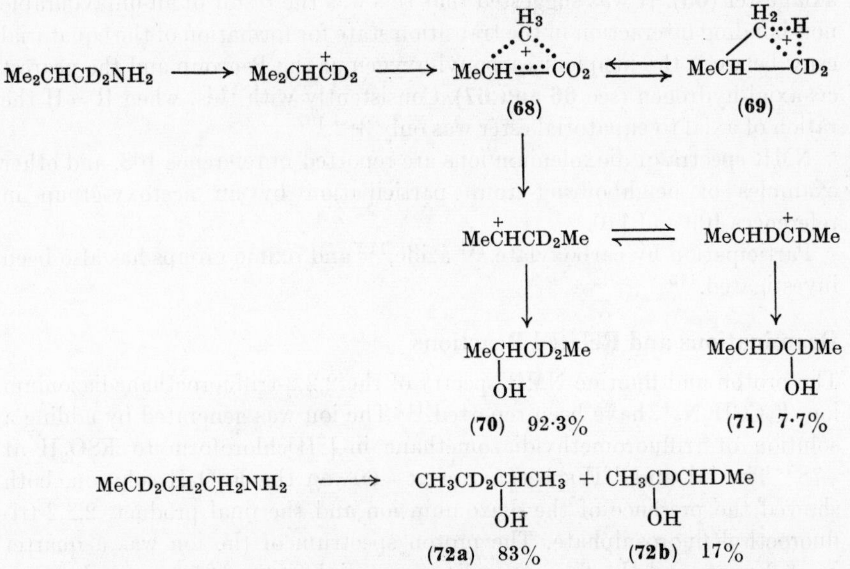

They also show that the *sec*-butyl cation–*sec*-butyl cation interconversion is only slightly slower than capture by solvent.

Evidence was obtained in favour of a protonated cyclopropane intermediate in the rearrangement of ^{13}C- and ^{2}H-labelled *n*-propyl bromide on treatment with $AlBr_3$ (cf. eqns. 18 and 19 after 80% conversion into isopropyl bromide).

$$\underset{100\%}{CH_3CH_2{}^{13}CH_2Br} \longrightarrow \underset{85{\cdot}7\%}{CH_3CH_2{}^{13}CH_2Br} + \underset{3{\cdot}7\%}{CH_3{}^{13}CH_2CH_2Br} + \underset{10{\cdot}6\%}{{}^{13}CH_3CH_2CH_2Br} \quad \ldots (18)$$

$$CH_3CH_2CD_2Br \longrightarrow \underset{79{\cdot}8\%}{C_2H_5\text{–}CD_2Br} + \underset{5{\cdot}0\%}{C_2H_4D\text{–}CHDBr} + \underset{15{\cdot}2\%}{C_2H_3D_2\text{–}CH_2Br} \quad \ldots (19)$$

There was no evidence for the intervention of a protonated cyclopropane in the formolysis of *n*-propyl toluene-*p*-sulphonate.[118]

Diazotization of [*N*,*N*-$^{2}H_2$]isobutylamine (**73a**) in aprotic media (e.g., benzene) and decomposition of *N*-isobutyl-*N*-nitrosoacetamide (**73b**) under similar conditions, but in the presence of one equivalent each of D_2O and [^{2}H]hexanol, yield hydrocarbons into which deuterium has been incorporated (30—40% of $^{2}H_1$, ~ 10% of $^{2}H_2$). It was suggested that the reactions involved a covalent diazonium acetate which underwent an elimination–addition

[118] G. J. Karabatsos, J. L. Fry, and S. Meyerson, *Tetrahedron Letters*, **1967**, 3735.

sequence (Scheme 4). In polar media (e.g., D_2O–DOAc) much less deuterium was incorporated, owing probably to an increase in the proportion of free diazonium ion formed at the expense of the covalent diazonium acetate.[119]

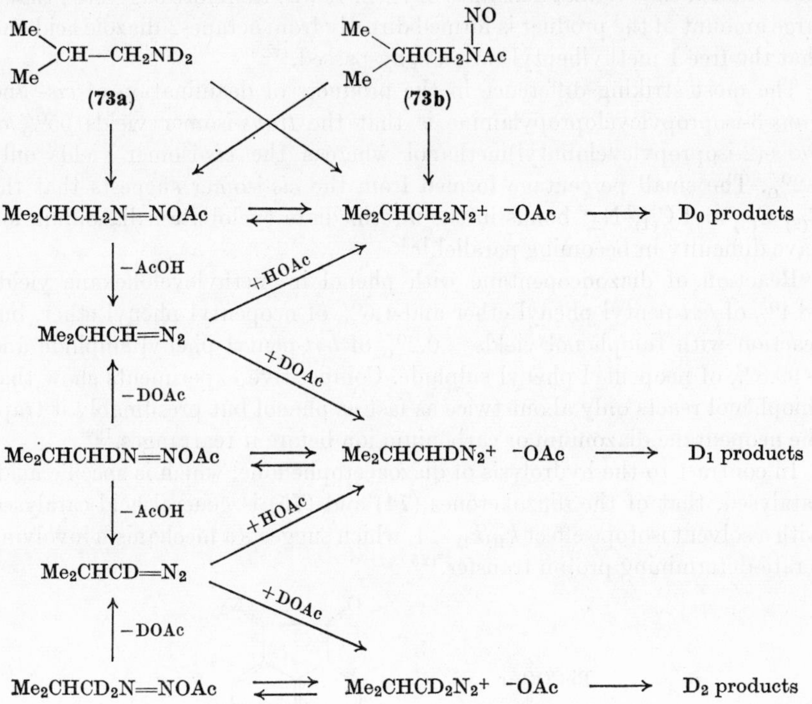

Scheme 4

Cyclopropanes are among the hydrocarbon products of the aprotic diazotization of aliphatic amines.[120] The possibility that these are formed by way of carbenes has now been eliminated by studying the deamination of [1,1-$^2H_2$]isobutylamine. Since it is known that aprotic diazotization involves hydrogen-exchange (see above) the deuterium content of the methylcyclopropane in the product was compared with that of but-1-ene, formation of which only via a cation was thought to be possible. Under a variety of reaction conditions the deuterium contents of the methylcyclopropane and but-1-ene were found to be identical. Appreciable formation of the methylcyclopropane via a carbene is therefore unlikely as this should result in additional loss of deuterium.[121]

[119] J. H. Bayless and L. Friedman, *J. Am. Chem. Soc.*, **89**, 147 (1967).
[120] See *Organic Reaction Mechanisms*, **1966**, 70.
[121] A. T. Jurewicz and L. Friedman, *J. Am. Chem. Soc.*, **89**, 149 (1967).

The decomposition of optically active octane-2-diazotate by aqueous base yields octan-2-ol with 16% net inversion, and in the presence of ether with 40% net inversion. Decomposition in $H_2{}^{18}O$ showed that 40% of the octan-2-ol had retained the original diazotate oxygen. It was therefore suggested that a large amount of the product is formed directly from octane-2-diazoic acid and that the free 1-methylheptyl cation is by-passed.[122]

The most striking difference in the products of deamination of *cis*- and *trans*-3-isopropylcyclopropylamine is that the *trans*-isomer yields 55% of *trans*-(2-isopropylcyclobutyl)methanol whereas the *cis*-isomer yields only 2.2%. The small percentage formed from the *cis*-isomer suggests that the $C_{(2)}$–$C_{(3)}$ and $C_{(1)}$–$N_2{}^+$ bonds in the intermediate cyclobutanediazonium ion have difficulty in becoming parallel.[123]

Reaction of diazoneopentane with phenol in methylcyclohexane yields 98.4% of *tert*-pentyl phenyl ether and 1.6% of neopentyl phenyl ether, but reaction with thiophenol yields < 0.2% of *tert*-pentyl phenyl sulphide and > 99.8% of neopentyl phenyl sulphide. Competitive experiments show that thiophenol reacts only about twice as fast as phenol but presumably it traps the neopentane diazonium or carbonium ion before it rearranges.[124]

In contrast to the hydrolysis of diazoacetophenone, which is specific acid-catalysed, that of the diazoketones (**74**) and (**75**) is general acid-catalysed with a solvent isotope effect $k_H/k_D < 1$, which suggests a mechanism involving a rate-determining proton transfer.[125]

PhCOCAr
‖
N_2

(**74**)

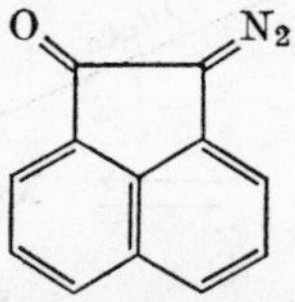

(**75**)

The decomposition of α-diazosulphones, RSO_2CHN_2, in 1,2-dichloroethane in the presence of 70% aqueous perchloric acid yields covalent perchlorates, $RSO_2CH_2OClO_3$.[126]

Nitrosative cleavage of tertiary amines,[127] alkaline decomposition of *N*-butyl-*N*-nitrosourea,[128] decomposition of 1-diazobutane in methanol,[129]

122 R. A. Moss and S. M. Lane, *J. Am. Chem. Soc.*, **89**, 5655 (1967).
123 I. Lillien and R. A. Doughty, *Tetrahedron Letters*, **1967**, 3953.
124 W. Kirmse and K. Horn, *Tetrahedron Letters*, **1967**, 1827.
125 W. Jugelt and D. Schmidt, *Tetrahedron Letters*, **1967**, 985; see also W. Jugelt, *Z. Chem.*, **5**, 455 (1965); W. Jugelt and L. Berseck, *ibid.*, **6**, 420 (1966); L. L. Leveson and C. W. Thomas, *J. Chem. Soc., B*, **1967**, 680; W. Alber and K. Schwabe, *Z. Physik. Chem. (Leipzig)*, **233**, 123 (1966).
126 J. B. F. N. Engberts and B. Zwanenburg, *Tetrahedron Letters*, **1967**, 831.
127 P. A. S. Smith and R. N. Loeppky, *J. Am. Chem. Soc.*, **89**, 1147 (1967).
128 W. Kirmse and G. Wächterhäuser, *Ann. Chem.*, **707**, 44 (1967).
129 W. Kirmse and H. A. Rinkler, *Ann. Chem.*, **707**, 57 (1967).

acid-catalysed decomposition of vinyltriazines,[130] and deamination of the following compounds have also been investigated: 4-aminovaleric acid,[131] dextrorotatory (*S*)-α-methylphenethylamine,[132] [^{14}C]aminoethylcyclopentane,[133] 5-amino-steroids,[134] menthyl- and carvomenthyl-amine,[135] and octylamine.[136]

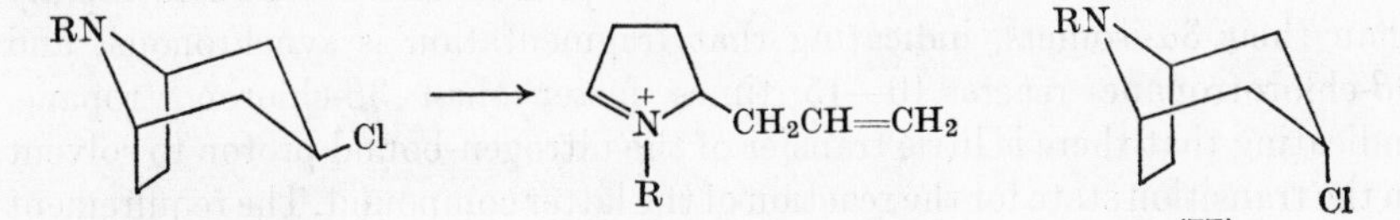

(**76a**) R = Me
(**76b**) R = H
(**77**)

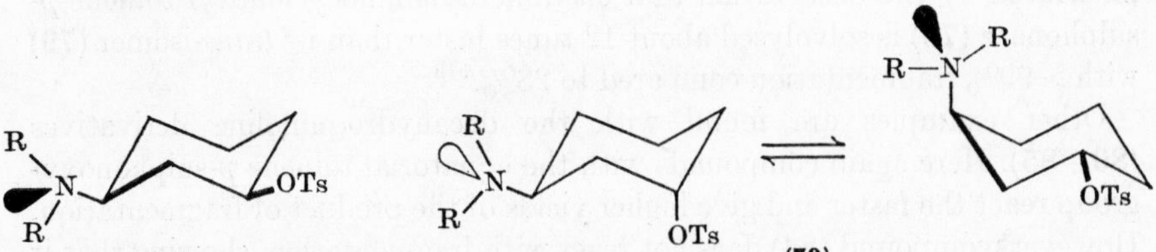

(**78**) (**79**)

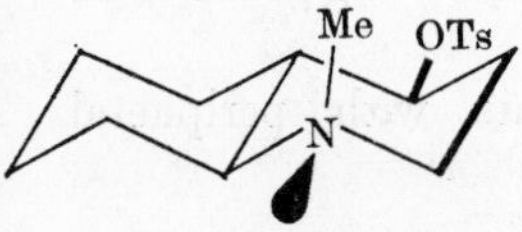

(**80**)
$1{\cdot}32 \times 10^{-2}$ (50°)
100%

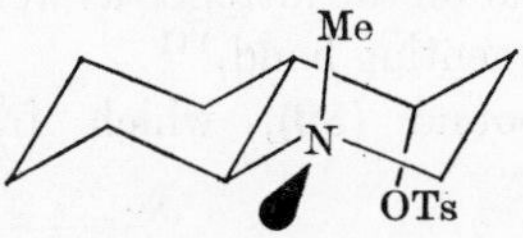

(**81**)
$3{\cdot}03 \times 10^{-4}$ (50°)
77%

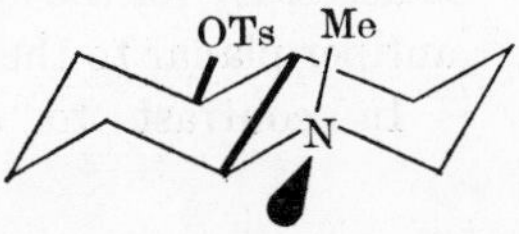

(**82**)
$1{\cdot}24 \times 10^{-4}$ (50°)
100%

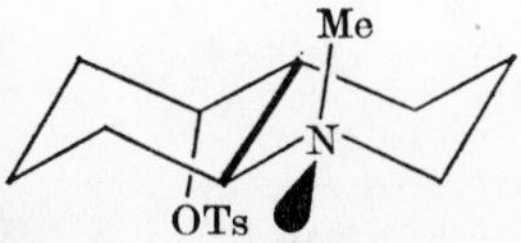

(**83**)
$1{\cdot}75 \times 10^{-5}$ (50°)
0%

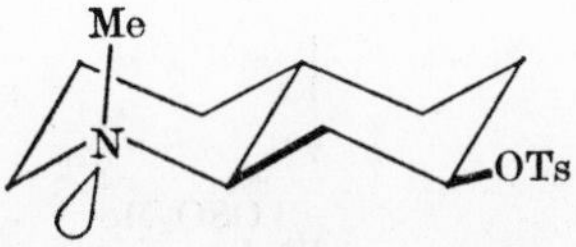

(**84**)
$1{\cdot}95 \times 10^{-5}$ (70°)
0%

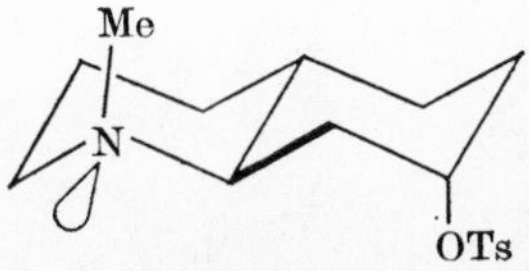

(**85**)
$1{\cdot}24 \times 10^{-5}$ (70°)
0%

Rate constants (sec^{-1}) for solvolysis in 80% ethanol.
Percentage fragmentation; thickened bonds denote fragmentation sites.

130 W. M. Jones and F. W. Miller, *J. Am. Chem. Soc.*, **89**, 1960 (1967).
131 H. Moll, *Chimia*, **20**, 426 (1966).
132 A. Laurent, É. Laurent-Dieuzeide, and P. Mison, *Bull. Soc. Chem. France*, **1967**, 1995.
133 T. N. Shatkina, E. V. Leont'eva, and O. A. Reutov, *Dokl. Akad. Nauk SSSR*, **173**, 113 (1967); *Chem. Abs.*, **67**, 53744k (1967).
134 G. Snatzke and A. Veithen, *Ann. Chem.*, **703**, 159 (1967).
135 H. Feltkamp, F. Koch, and T. N. Thanh, *Ann. Chem.*, **707**, 95 (1967).
136 J. Bakke, *Acta Chem. Scand.*, **21**, 1007 (1967).

Fragmentation Reactions[137] (see formulae on p. 85)

3β-Chlorotropane (**76a**) and 3β-chloronortropane (**76b**) are solvolysed in aqueous ethanol with fragmentation, but their 3α-isomers (**77**) do not undergo this reaction, as expected from the known[138] stereoelectronic requirements for fragmentation reactions.[139] The 3β-compounds also react more rapidly than their 3α-isomers, indicating that fragmentation is synchronous; and 3β-chlorotropane reacts 10—15 times faster than 3β-chloronortropane, indicating that there is little transfer of the nitrogen-bound proton to solvent in the transition state for the reaction of the latter compound. The requirement that the leaving group and the 2,3-bond should be *anti* to one another is also illustrated by the observation that *cis*-dimethylaminocyclohexyl toluene-*p*-sulphonate (**78**) is solvolysed about 12 times faster than its *trans*-isomer (**79**) with > 99% fragmentation compared to 28%.[140]

Other examples are found with the decahydroquinoline derivatives (**80—85**). Here again compounds with the equatorial toluene-*p*-sulphonoxy-group react the faster and give higher yields of the product of fragmentation. However, compound (**84**) does not react with fragmentation, showing that it is necessary for the lone pair on the nitrogen as well as the leaving group to be antiperiplanar to the fragmenting bond.[141]

In contrast to compound (**80**), which fragments with peripheral

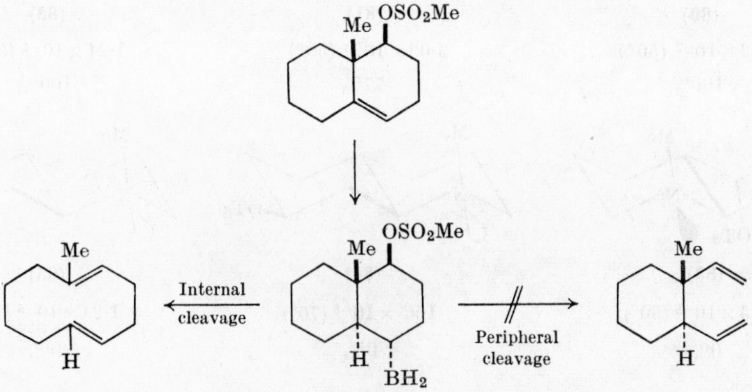

(**86**)

137 C. A. Grob and P. W. Schiess, *Angew. Chem. Intern. Ed. Engl.*, **6**, 1 (1967).
138 See *Organic Reaction Mechanisms*, **1965**, 66.
139 A. T. Bottini, C. A. Grob, E. Schumacher, and J. Zergenyi, *Helv. Chim. Acta*, **49**, 2516 (1966).
140 U. Burckhardt, C. A. Grob, and H. R. Kiefer, *Helv. Chim. Acta*, **50**, 231 (1967).
141 C. A. Grob, H. R. Kiefer, H. J. Lutz, and H. J. Wilkens, *Helv. Chim. Acta*, **50**, 416 (1967).

cleavage, an organoborane (**86**) reacts with predominantly internal cleavage.[142]

Other fragmentation reactions are described in references 143—148.

Displacement Reactions at Elements Other than Carbon[149–151]

Sommer and his co-workers[152–159a] have continued their studies on the steric course of displacement reactions at silicon. The following generalizations were made: (1) Good leaving groups, X, whose conjugate acids have pK_a less than ~ 6 usually undergo nucleophilic displacement from silicon with inversion of configuration regardless of the nature of the solvent, and provided only that the attacking reagent furnishes an entering group Y that is more basic than X. (2) For poor leaving groups, whose conjugate acids have pK_a greater than ~ 6, the stereochemical path may be inversion or retention of configuration; relatively non-polar solvents favour retention. These rules have now been shown to apply for compounds (α-$C_{10}H_7$)PhMeSiX,[152] neopentyl-PhMeSiX,[153] ($PhCH_2$)PhMeSiX,[153] EtPhMeSiX,[153] and Ph_3SiSiXPhMe.[154] An S_N2-Si mechanism was proposed for the inversion reactions, and an S_Ni-Si mechanism for the retention reactions.

The dependence of steric course on the leaving group was investigated with optically active αNpPhMeSiX where X is chlorine, fluorine, methoxyl, and hydrogen. When X = Cl the steric course of the reaction with alkyl-lithiums is always inversion, but when X = H it is always retention, except with benzhydryl-lithium. When X = F or MeO the steric course is inversion with simple alkyls, but retention with allyl-lithium, benzyl-lithium, α-methylbenzyl-lithium, or benzhydryl-lithium. For each alkyl-lithium there is a cross-over value for the pK_a of the conjugate acid of the leaving group at

142 J. A. Marshall and G. L. Bundy, *Chem. Commun.*, **1967**, 854; cf. *Organic Reaction Mechanisms*, **1966**, 74—75.

143 R. A. Olofson and D. M. Zimmerman, *J. Am. Chem. Soc.*, **89**, 5057 (1967).

144 K. Manninen and H. Krieger, *Tetrahedron Letters*, **1967**, 2071.

145 E. V. Crabtree and E. J. Poziomek, *J. Org. Chem.*, **32**, 1231 (1967).

146 W. Pritzkow and W. Rösler, *Ann. Chem.*, **703**, 66 (1967).

147 A. Eschenmoser, D. Felix, and G. Ohloff, *Helv. Chem. Acta*, **50**, 708 (1967); J. Schreiber, D. Felix, A. Eschenmoser, M. Winter, F. Gautschi, K. H. Schulte-Elte, E. Sundt, G. Ohloff, J. Kalvoda, H. Kaufmann, P. Wieland, and G. Anner, *ibid.*, p. 2101; P. Wieland, H. Kaufmann, and A. Eschenmoser, *ibid.*, p. 2108.

148 J. P. Ward, *Tetrahedron Letters*, **1967**, 4031.

149 A. J. Kirby and S. G. Warren, "The Organic Chemistry of Phosphorus", Elsevier, London, 1967.

150 N. Kharasch, B. S. Thyagarajan, and A. I. Khodair, "Mechanisms of Reactions of Sulfur Compounds", Intra-Science Research Foundation, Santa Monica, California, Vol. 1, 1966.

151 E. J. Behman and J. O. Edwards, *Progr. Phys. Org. Chem.*, **4**, 93 (1967); "Nucleophilic Displacements on Peroxide Oxygen and Related Reactions".

152 L. H. Sommer, G. A. Parker, N. C. Lloyd, C. L. Frye, and K. W. Michael, *J. Am. Chem. Soc.*, **89**, 857 (1967).

153 L. H. Sommer, K. W. Michael, and W. D. Korte, *J. Am. Chem. Soc.*, **89**, 868 (1967).

154 L. H. Sommer and K. T. Rosborough, *J. Am. Chem. Soc.*, **89**, 1756 (1967).

which the steric course changes from inversion to retention. With simple alkyl-lithiums this is *ca.* 4 (pK_a of HF), with *p*-$MeOC_6H_4CH_2Li$ it is *ca.* 1.6 (pK_a of MeOH), and with benzyl and allyl-lithium it is *ca.* 40 (pK_a of H_2). The mechanism of the reactions proceeding with retention of configuration is probably S_Ni-Si with either a four- or a six-membered cyclic transition state (**87** or **88**) and with stereochemistry about the silicon either trigonal-bipyramidal or tetragonal-pyramidal.[155, 156]

The steric course of reactions with nitrogen-containing leaving groups was also studied. The reaction of (α-$C_{10}H_7$)PhMeSiNH_2 with benzoic acid in

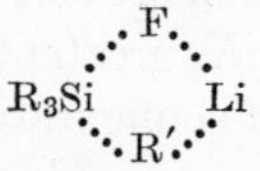

(**87**)

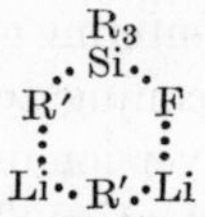

(**88**)

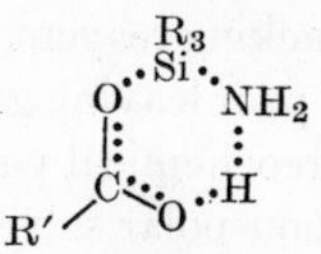

(**89**)

pentane proceeded with 79% retention of configuration and a cyclic transition state (**89**) was proposed. When the steric requirements of either reactant was changed by using mesitoic acid or $R_3SiNHBu^i$ the steric course changed to 70—80% inversion. Presumably the cyclic transition state does not readily accommodate these reactants.[157] The steric course of displacement reactions with sulphur-containing leaving groups was also investigated.[158]

The reaction of silicon hydrides with hydroxylic compounds catalysed by 10% palladium–charcoal or Raney nickel proceed with retention of configuration.[159a] The rearrangement of α-silylcarbinols, (α-$C_{10}H_7$)PhMeSiC(OH)R^1R^2, into silyl ethers, (α-$C_{10}H_7$)PhMeSiOCHR^1R^2 proceeds with retention of configuration at the silicon and probably involves a three-centred cyclic transition state.[159b]

An X-ray crystal structural analysis of dimethylsilylamine at −120° shows it to consist of ten-membered rings, with each silicon penta-coordinated in a trigonal bipyramid having two long apical Si–N bonds.[160]

The solvent isotope effect for the methanolysis of *p*-chlorophenoxytriphenylsilane are $k_H/k_D = 1.25 \pm 0.06$ (acetate catalysis) and 1.35 ± 0.06 (methoxide ion catalysis). The plots of the free energy of activation for the methanolysis of Ph_3COAr against the free energy of ionization of the corresponding phenol,

155 L. H. Sommer, W. D. Korte, and P. G. Rodewald, *J. Am. Chem. Soc.*, **89**, 862 (1967)
156 L. H. Sommer and W. D. Korte, *J. Am. Chem. Soc.*, **89**, 5802 (1967).
157 L. H. Sommer and J. D. Citron, *J. Am. Chem. Soc.*, **89**, 5797 (1967).
158 L. H. Sommer and J. McLick, *J. Am. Chem. Soc.*, **89**, 5806 (1967).
159a L. H. Sommer and J. E. Lyons, *J. Am. Chem. Soc.*, **89**, 1521 (1967).
159b A. G. Brook, C. M. Warner, and W. W. Limburg, *Can. J. Chem.*, **45**, 1231 (1967); A. G. Brook, G. E. LeGrow, and D. M. MacRae, *ibid.*, p. 239.
160 R. Rudman, W. C. Hamilton, S. Novick, and T. D. Goldfarb, *J. Am. Chem. Soc.*, **89**, 5157 (1967).

ArOH, are curves. This may be the result of a change from a stepwise to a concerted mechanism or from variation in Si–O d_π–p_π bonding in the activated complex.[161]

The solvolysis of acetoxyaryldimethylsilanes in propan-1-ol,[162] the alkaline hydrolysis of arylsilanes,[163] the reaction of phenylsilane with nucleophiles,[164] and the reduction of boron trichloride by optically active methyl-1-naphthylphenylsilane (90% retention)[165] have also been investigated.

Unlike most reactions of phosphonium salts with hydroxyl ions which are second-order in the latter, those of phosphonium salts in which the leaving group is 4-nitrobenzyl are of the first order. This can be explained by the usual mechanism (eqn. 20) if k_2[OH] and $k_3 \gg k_1$. The reaction of the five-membered cyclic phosphonium ion (**90**) is about 1700 times faster than that of the analogous six-membered one, probably as a result of the easy formation of a bipyramidal quinquecovalent intermediate caused by a small C–P–C angle (95° ± 3°) in the phospholan ring.[166]

$$R_3\overset{+}{P}{-}CH_2Ar \underset{k_{-1}}{\overset{k_1}{\rightleftharpoons}} R_3P\begin{smallmatrix}OH\\ CH_2Ar\end{smallmatrix} \underset{k_{-2}}{\overset{k_2\ {}^-OH}{\rightleftharpoons}} R_3P\begin{smallmatrix}O^-\\ CH_2Ar\end{smallmatrix} \xrightarrow{k_3} R_3PO + {}^-CH_2Ar \quad \ldots(20)$$

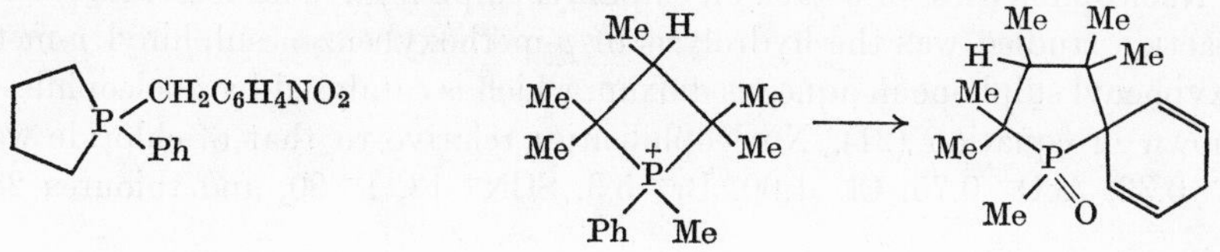

(**90**) (**91**) (**92**)

The interesting observation has been made that the reaction of phosphonium salt (**91**) with hydroxide ion does not result in expulsion of the phenyl anion but instead yields (**92**). Presumably the constraint of the four-membered ring prevents the phenyl group from taking up an apical position in the quinquecovalent intermediate, which it is necessary for it to do before it can depart as an anion. Hence an alternative pathway involving ring expansion is

161 R. L. Schowen and K. S. Latham, *J. Am. Chem. Soc.*, **89**, 4677 (1967); cf. *Organic Reaction Mechanisms*, **1966**, 76.

162 G. Schott and V. Bondybey, *Chem. Ber.*, **100**, 1773 (1967); cf. *Organic Reaction Mechanisms*, **1966**, 76.

163 G. Schott, P. Hansen, S. Kuhla, and P. Zwierz, *Z. Anorg. Allgem. Chem.*, **351**, 37 (1967); J. Hetflejs, F. Mares, and V. Chvalovsky, *Intern. Sym. Org. Chem. Soc. Commun. Prague*, **1965**, p. 282.

164 V. O. Reikhsfel'd and I. E. Saratov, *Zh. Obshch. Khim.*, **37**, 402 (1967); *Chem. Abs.* **67**, 43201s.

165 C. J. Attridge, R. N. Haszeldine, and M. J. Newlands, *Chem. Commun.*, **1966**, 911.

166 G. Aksnes and L. J. Brudvik, *Acta Chem. Scand.*, **21**, 745 (1967).

followed, and there could be considerable driving force for this as a result of the loss of ring strain of the four-membered ring.[167]

Replacement of one of the phenyl groups of the benzyltriphenylphosphonium ion by a ferrocenyl group causes a 60-fold rate decrease in the reaction with HO^-, but replacement of the others has only a slight additional effect. Possibly the phosphonium ion is stabilized by overlap of the non-bonding electrons of the ferrocene with the 3*d*-orbitals of the phosphorus.[168]

The hydrolysis of bisphosphonium salts has been investigated.[169]

The alkaline hydrolysis of *O*-ethyl ethylphosphonochloridothionate **(93)** proceeds with at least 97% inversion of configuration.[170]

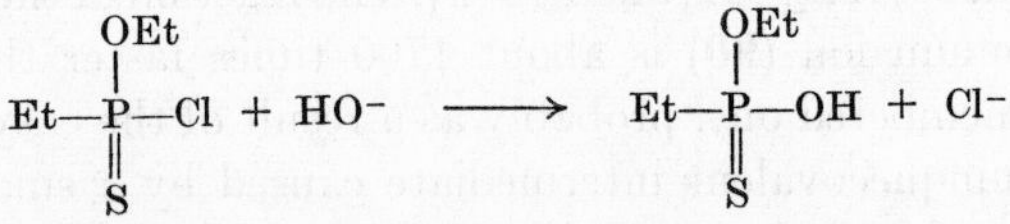

(93)

Other investigations of nucleophilic displacements at phosphorus are described in references 171—174. The Michaelis–Arbuzov reaction between five- and six-membered cyclic phosphites and ethyl iodide has been studied.[175]

Nucleophilicities for attack on sulphinyl-sulphur have been measured. The reaction studied was the hydrolysis of *p*-methoxybenzenesulphinyl *p*-methoxyphenyl sulphone in aqueous dioxan, which is catalysed by nucleophiles as shown in equation (21). Nucleophilicities relative to that of chloride were F^- 0.33, AcO^- 0.75, Cl^- 1.00, Br^- 5.3, SCN^- 13, I^- 90, and thiourea 280.

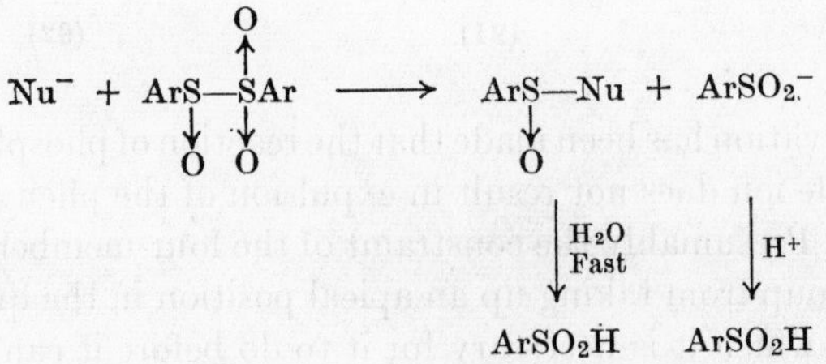

. . . (21)

[167] S. E. Fishwick, J. Flint, W. Hawes, and S. Trippett, *Chem. Commun.*, **1967**, 1113; see also S. E. Cremer, and R. J. Chorvat, *Tetrahedron Letters*, **1968**, 413.
[168] A. W. Smalley, C. E. Sullivan, and W. E. McEwen, *Chem. Commun.*, **1967**, 5.
[169] J. J. Brophy and M. J. Gallagher, *Chem. Commun.*, **1967**, 344.
[170] M. Mikolajczyk, *Tetrahedron*, **23**, 1543 (1967); see also M. Mikolajczyk, *Wiedomosci Chem.*, **21**, 67, 205 (1967); *Chem. Abs.*, **67**, 32080a, 53172x (1967); J. Michalski, *Colloq. Nat. Centre Nat. Rech. Sci.*, **1965**, 203.
[171] W. S. Wadsworth, *J. Org. Chem.*, **32**, 1603 (1967).
[172] J. N. Seiber and H. Tolkmith, *Tetrahedron Letters*, **1967**, 3333.
[173] J. Epstein, P. L. Cannon, H. O. Michel, B. E. Hackley, and W. A. Mosher, *J. Am. Chem. Soc.*, **89**, 2937 (1967).
[174] A. G. Brook, D. M. MacRae, and W. W. Limburg, *J. Am. Chem. Soc.*, **89**, 5493 (1967).
[175] G. Aksnes and R. Eriksen, *Acta Chem. Scand.*, **20**, 2463 (1966).

Interestingly, these values are similar to those for attack at sp^3-hybridized carbon and quite different from those for attack at peroxide-oxygen.[176] Substitution at sulphur and oxygen has been discussed in terms of the theory of hard and soft acids and bases.[177]

Aryl thiosulphinates react with aromatic sulphinic acids in the aqueous acetic acid containing sulphuric acid. The reaction is strongly catalysed by dialkyl and diaryl sulphides when the stoichiometry is: $2ArSO_2H + PhSOSPh \rightarrow 2ArSO_2SPh + H_2O$ and the kinetics are of the first order in both thiolsulphinate and sulphide, and of zero order in sulphinic acid. The mechanism shown in equations (22—26) was proposed. In the absence of sulphides the

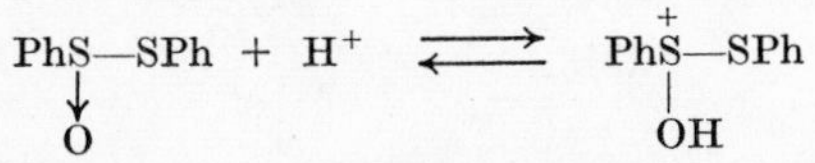

$$PhS(\rightarrow O)\text{—}SPh + H^+ \rightleftharpoons Ph\overset{+}{S}(OH)\text{—}SPh \quad \ldots(22)$$

$$R_2S + Ph\overset{+}{S}(OH)\text{—}SPh \xrightarrow[\text{determining}]{\text{Rate-}} R_2\overset{+}{S}\text{—}SPh + PhSOH \quad \ldots(23)$$

$$R_2\overset{+}{S}\text{—}SPh + ArSO_2H \xrightarrow{\text{Fast}} Ar\text{—}S(\rightarrow O)_2\text{—}SPh + R_2S + H^+ \quad \ldots(24)$$

$$PhSOH + R_2S + H^+ \rightleftharpoons R_2\overset{+}{S}\text{—}SPh + H_2O \quad \ldots(25)$$

$$PhSOH + ArSO_2H \longrightarrow ArS(\rightarrow O)_2\text{—}SPh + H_2O \quad \ldots(26)$$

reaction is of the first order in thiolsulphinate and sulphinic acid. The reaction was considered to be general acid-catalysed as it was faster in AcOH than in AcOD ($k_H k_D = 1.27$) and the mechanism shown in equations (27) and (28) (p. 92) with transition state (**94**) was proposed.[178]

The rate of racemization of optically active diaryl sulphoxides in acetic anhydride is twice the rate of oxygen exchange which suggests a bimolecular mechanism proceding with inversion.[179] Racemization of sulphoxides in acetic anhydride is strongly catalysed by Lewis acids.[180] Oxygen exchange by sulphoxides in alkaline media have also been investigated.[181]

176 J. L. Kice and G. Guaraldi, *Tetrahedron Letters*, **1966**, 6135.
177 B. Saville, *Angew. Chem. Intern. Ed. Engl.*, **6**, 938 (1967); see also R. G. Pearson, *Chem. Brit.*, **3**, 103 (1967); R. F. Hudson, *Structure and Bonding*, **1**, 221 (1966); R. G. Pearson and J. Songstad, *J. Am. Chem. Soc.*, **89**, 1827 (1967); *J. Org. Chem.*, **32**, 2899 (1967).
178 J. L. Kice, C. G. Venier, and L. Heasley, *J. Am. Chem. Soc.*, **89**, 3557 (1967).
179 S. Oae and M. Kise, *Tetrahedron Letters*, **1967**, 1409.
180 E. Jonsson, *Tetrahedron Letters*, **1967**, 3675.
181 S. Oae, M. Kise, N. Furukawa, and Y. H. Khim, *Tetrahedron Letters*, **1967**, 1415.

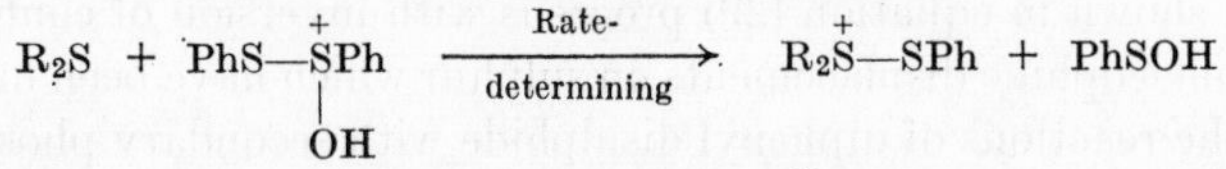

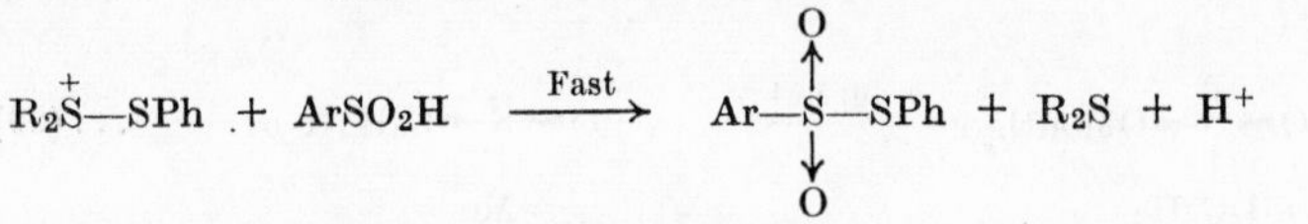

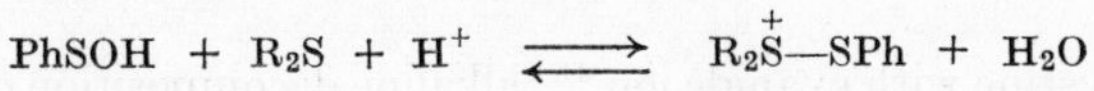

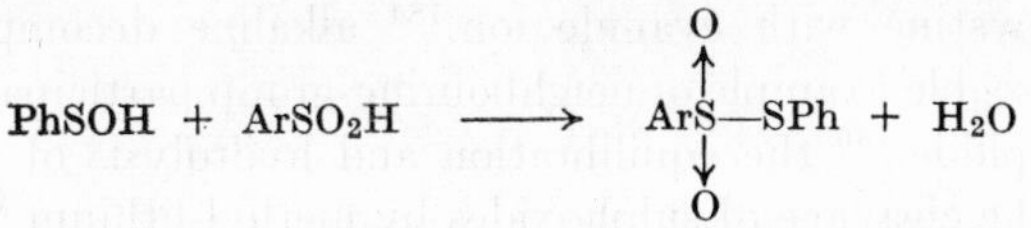

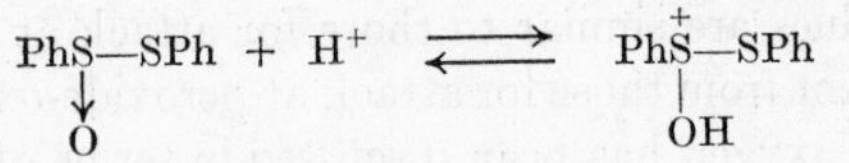

$$PhS—SPh + H^+ \rightleftharpoons PhS^+—SPh \qquad \ldots (27)$$
$$\quad |\qquad\qquad\qquad\qquad\quad |$$
$$\quad O\qquad\qquad\qquad\qquad\quad OH$$

$$\qquad\qquad\qquad\qquad\qquad\qquad\qquad\qquad O$$
$$\qquad\qquad\qquad\qquad\qquad\qquad\qquad\qquad \uparrow$$
$$B + ArSO_2H + PhS^+—SPh \longrightarrow BH^+ + ArS—SPh + PhSOH \quad \ldots (28)$$
$$\qquad\qquad\qquad\quad |\qquad\qquad\qquad\qquad\qquad\qquad\qquad \downarrow$$
$$\qquad\qquad\qquad\quad OH\qquad\qquad\qquad\qquad\qquad\qquad O$$

$$\qquad\qquad\qquad\qquad\qquad\qquad Ar$$
$$\qquad\qquad\qquad\qquad\delta+ \quad |\delta- \quad \delta+$$
$$\qquad\qquad\qquad B\cdots H\cdots O—S\cdots S\cdots SPh$$
$$\qquad\qquad\qquad\qquad\qquad\qquad\quad |\quad |\quad |$$
$$\qquad\qquad\qquad\qquad\qquad\qquad\quad O\ Ph\ OH$$

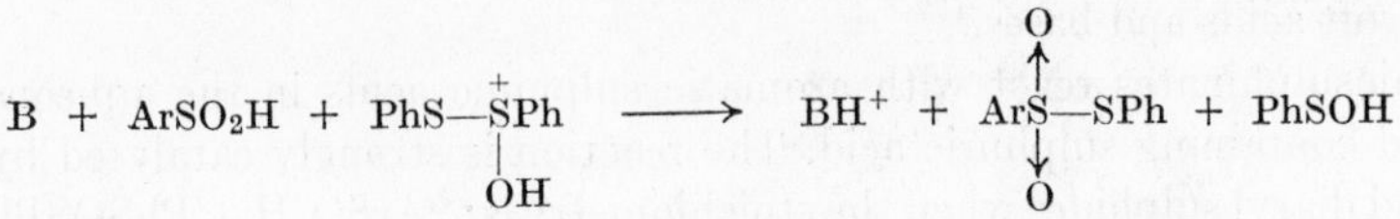

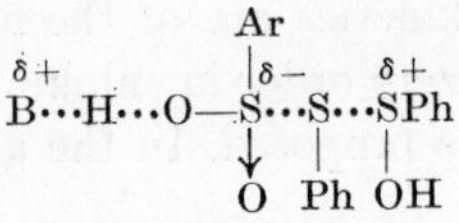

$$(94)$$

The reaction shown in equation (29) proceeds with inversion of configuration.[182] Other nucleophilic displacements on sulphur which have been investigated include the reactions of diphenyl disulphide with secondary phosphine

$$(-)\text{-}O\!\!\blacktriangleright\!\!\overset{..}{S}\!\!\blacktriangleleft\!\!OMenthyl \xrightarrow{\text{MeMgI}} (+)\text{-}O\!\!\blacktriangleright\!\!\overset{..}{S}\!\!\blacktriangleleft\!\!C_6H_4\text{-}I\text{-}p \qquad \ldots (29)$$
$$\qquad\quad |\qquad\qquad\qquad\qquad\qquad\qquad\quad |$$
$$\qquad p\text{-}I\text{-}C_6H_4\qquad\qquad\qquad\qquad\qquad\quad Me$$

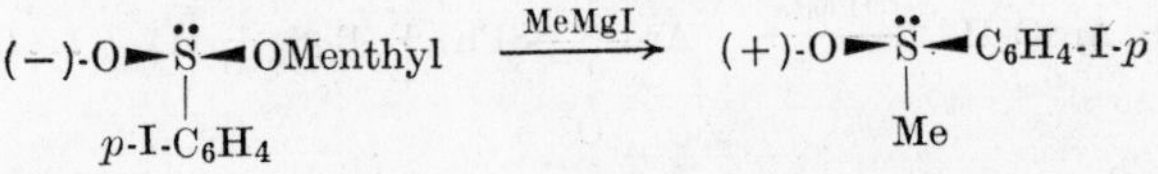

oxides[183] and of cystine with cyanide ion,[184] alkaline decomposition of cystine oxides[183] and of cystine with cyanide ion,[184] alkaline decomposition of disulphides,[185] a possible example of neighbouring-group participation in the reaction of a disulphide,[186] the equilibration and hydrolysis of biotin sulphoxides,[187a] and the cleavage of sulphoxides by methyl-lithium.[187b]

The major products of the ionic decomposition of *p*-nitroperbenzoates (95)

$$\qquad R$$
$$\qquad |$$
$$Me—C—O\diagdown$$
$$\qquad |\qquad\quad O—COAr \longrightarrow Me—\overset{+}{C}\cdots OR + {}^-OCOAr$$
$$\qquad Me\qquad\qquad\qquad\qquad\qquad\qquad\quad |$$
$$\qquad\qquad\qquad\qquad\qquad\qquad\qquad\qquad Me$$

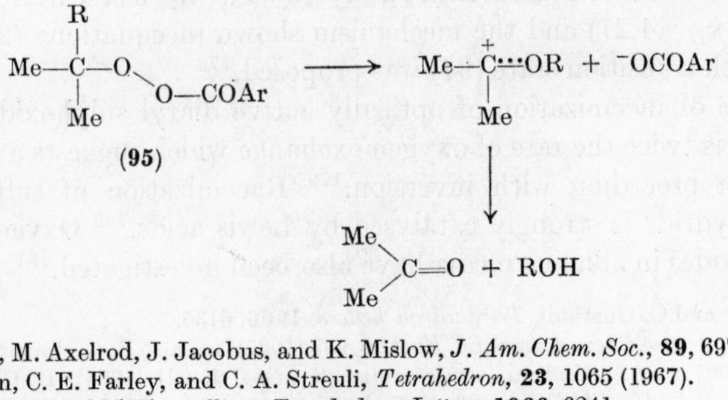

$$(95)$$
$$\qquad\qquad\qquad\qquad\qquad\qquad\qquad\qquad\quad\downarrow$$

$$\qquad\qquad\qquad\qquad\qquad\qquad Me\diagdown$$
$$\qquad\qquad\qquad\qquad\qquad\qquad\qquad\quad C{=}O + ROH$$
$$\qquad\qquad\qquad\qquad\qquad\qquad Me\diagup$$

182 P. Bickart, M. Axelrod, J. Jacobus, and K. Mislow, *J. Am. Chem. Soc.*, **89**, 697 (1967).
183 M. Grayson, C. E. Farley, and C. A. Streuli, *Tetrahedron*, **23**, 1065 (1967).
184 G. H. Wiegand and M. Tremelling, *Tetrahedron Letters*, **1966**, 6241.
185 J. P. Danehy and W. E. Hunter, *J. Org. Chem.*, **32**, 2047 (1967).
186 M. Bellas, D. L. Tuleen, and L. Field, *J. Org. Chem.*, **32**, 2591 (1967).
187a H. Ruis, D. B. McCormick, and L. D. Wright, *J. Org. Chem.*, **32**, 2010 (1967).
187b J. Jacobus and K. Mislow, *J. Am. Chem. Soc.*, **89**, 5228 (1967).

in methanol are acetone, ROH (or olefin), and *p*-nitrobenzoic acid. Thus R migrates rather than Me when it is Et, Pr^i, Bu^t, $PhCH_2$, *m*- or *p*-MeC_6H_4, $PhCH_2CH_2$, or 4-camphanyl. On the basis that the energy for heterolysis of an oxygen–oxygen bond should be 22 kcal $mole^{-1}$ greater than that for a carbon–oxygen bond, and that the energy of activation for the acetolysis of neopentyl toluene-*p*-sulphonate is 30 kcal $mole^{-1}$, it was estimated that the energy of activation for the unassisted heterolysis of the O–O bond should be about 50 kcal $mole^{-1}$. The measured value is 27 kcal $mole^{-1}$ and hence the driving force provided by the migrating R group is about 23 kcal $mole^{-1}$.[188]

The following reactions have also been investigated: nucleophilic displacement of the halogen of α-halogeno-ketones and other positive halogen compounds,[189] reactions of triarylphosphines[190] and amine anions[191] with azides, and nucleophilic displacements from boron,[192] tin,[193] and mercury.[194]

Neighbouring-group participation by an ionized hydroxyl group in a displacement reaction from platinum(II) has been demonstrated (eqns. 30—31).[195]

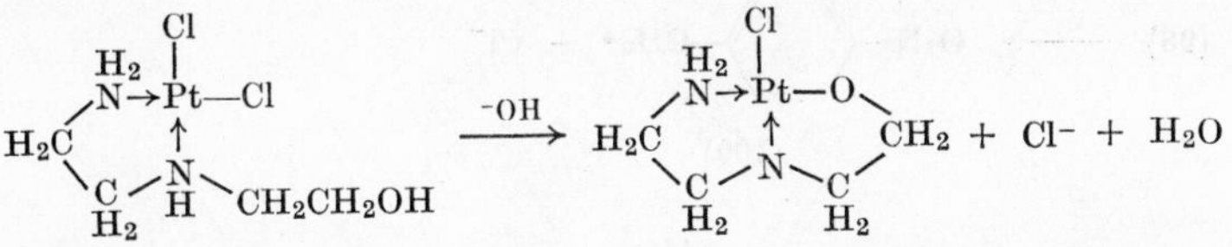

...(30)

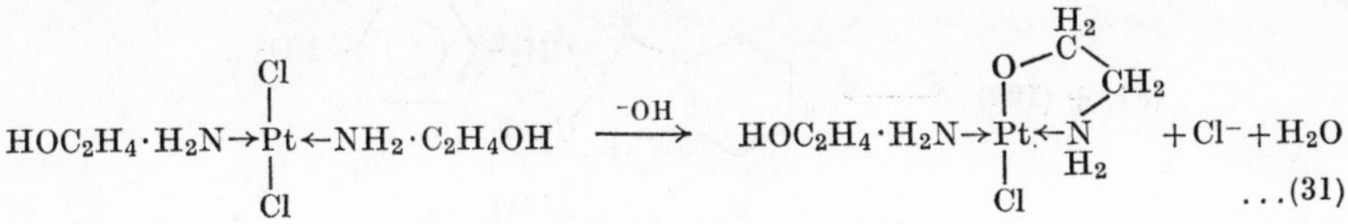

...(31)

Ambident Nucleophiles

Kornblum and his co-workers have now shown that carbon alkylation of an enolate anion, as well as that of the nitroalkane anion,[196] can proceed via a radical anion.[197] The relative rates of *C*-alkylation of ion (**97**) by benzyl, 3-nitrobenzyl, and 4-nitrobenzyl iodides are 1:2:2, by the corresponding

[188] E. Hedaya and S. Winstein, *J. Am. Chem. Soc.*, **89**, 1661, 5314 (1967).
[189] I. J. Borowitz, M. Anschel, and S. Firstenberg, *J. Org. Chem.*, **32**, 1723 (1967); I. J. Borowitz, K. C. Kirby, and R. Virkhaus, *ibid.*, **31**, 4031 (1966); P. A. Chopard, *Chimia*, **20**, 420 (1966).
[190] J. E. Leffler and R. D. Temple, *J. Am. Chem. Soc.*, **89**, 5235 (1967).
[191] W. Fischer and J.-P. Anselme, *J. Am. Chem. Soc.*, **89**, 5284 (1967).
[192] R. E. Davis and R. D. Kenson, *J. Am. Chem. Soc.*, **89**, 1384 (1967).
[193] H. M. J. C. Creemers, F. Verbeek, and J. G. Noltes, *J. Organometal. Chem.*, **8**, 469 (1967).
[194] D. Seyferth, M. E. Gordon, J. Y.-P. Mui, J. M. Burlitch, *J. Am. Chem. Soc.*, **89**, 959 (1967).
[195] K. H. Stephen and F. Basolo, *J. Inorg. Nucl. Chem.*, **29**, 775 (1967).
[196] See *Organic Reaction Mechanisms*, **1965**, 72.
[197] N. Kornblum, R. E. Michel, and R. C. Kerber, *J. Am. Chem. Soc.*, **88**, 5660 (1966); N. Korblum, *Trans. N.Y. Acad. Sci.*, **29**, 1 (1966).

bromides 1:4:4, and by the corresponding chlorides 1:3:900. The rates of *O*-alkylation are changed relatively slightly on going from benzyl to 3-nitrobenzyl to 4-nitrobenzyl whichever the halogen. It was suggested that all these reactions are S_N2 processes except *C*-alkylation of 4-nitrobenzyl chloride, which was postulated to involve radical anions. Supporting evidence was

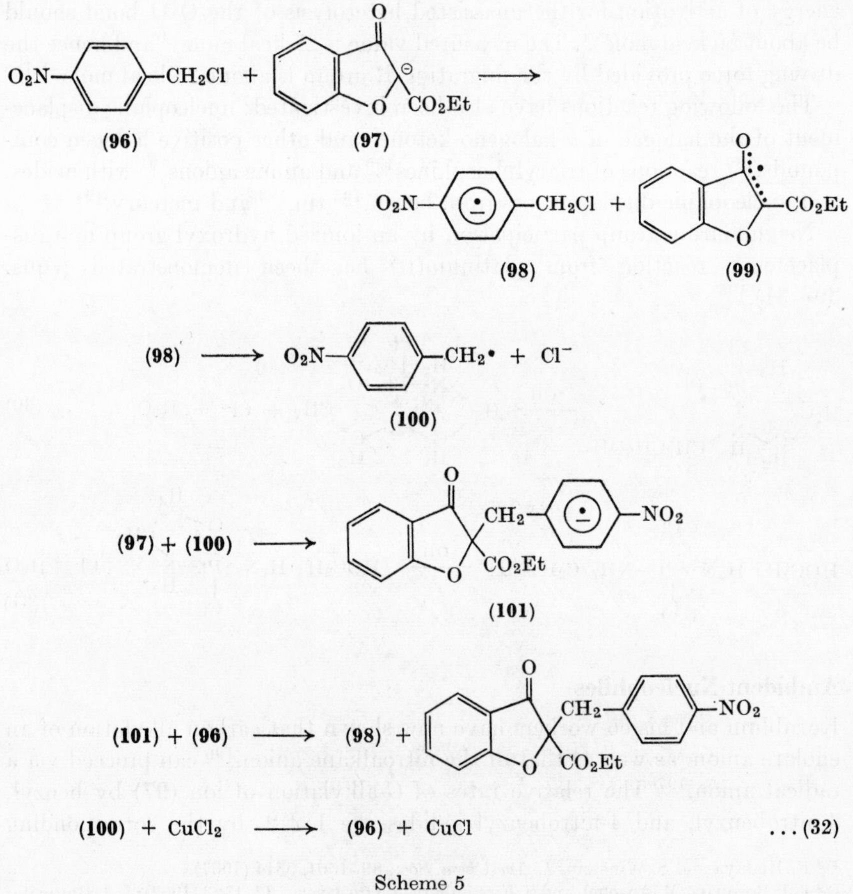

Scheme 5

obtained by showing that the rate of *C*-alkylation of 4-nitrobenzyl chloride is decreased by electron acceptors (e.g., *p*-dinitrobenzene), which results in an increase in the *O*:*C* alkylation ratio, whereas this ratio for the other halides is unchanged.[197] Even more strikingly the rate of *C*-alkylation of 4-nitrobenzyl chloride is drastically reduced by low concentrations of cupric chloride. A concentration of 6×10^{-6} M inhibits the radical-anion process completely but does not affect the S_N2 process. The mechanism shown in Scheme 5 was proposed, with cupric chloride intercepting the chain-carrying radical (**100**) (eqn. 32)[198] (see also p. 60).

Support for the suggestion by Kornblum and his co-workers that *C*-alkylation of the 1-methyl-1-nitroethyl anion by 4-nitrobenzyl chloride is a radical process has been obtained by showing that it is inhibited by oxygen.[199]

The sodium salt of *anti*-benzaldoxime is alkylated mainly on nitrogen, but that of *syn*-benzaldoxime is alkylated on oxygen. This difference possibly arises from the steric effect of the benzene ring.[200]

Reactions of ambident nucleophiles have been discussed in terms of the theory of hard and soft acids and bases.[201]

The chemistry of enolate anions has been reviewed.[202] The ratio of *C*- to *S*-alkylation on reaction of the anion of monothiomalonic ester, $EtO_2CCH_2C(S)OEt$, with ethyl iodide in ethanol is 1:1.7.[203]

Other reactions of ambident ions which have been investigated include: alkylation of alkali-metal and silver salts of 2-hydroxypyrimidines,[204] enolate anions[205,206] (see also p. 327), phosphoramidates,[207] trinitromethane anion,[208] silver cyanate,[209] 1-arylpyrazoles,[210] pyridazines,[211] cinnolines,[212] sulphinate anions,[213] and salts of indole,[214] and the reaction of triethyloxonium fluoroborate with the sodium salt of 2-pyridone.[215]

Other Reactions

Several nucleophilic displacement reactions at vinylic carbon[216] have been investigated. The orientation of displacement reactions in cyclobutenes

198 N. Kornblum, R. E. Michel, and R. C. Kerber, *J. Am. Chem. Soc.*, **88**, 5662 (1966).
199 G. A. Russell and W. C. Danen, *J. Am. Chem. Soc.*, **88**, 5663 (1966); see also G. A. Russell and A. G. Bemis, *ibid.*, p. 5491.
200 E. Buehler, *J. Org. Chem.*, **32**, 261 (1967).
201 R. F. Hudson, "Structure and Bonding", **1**, 221 (1966); R. G. Pearson and J. Songstad. *J. Org. Chem.*, **32**, 2899 (1967).
202 H. O. House, *Record Chem. Progr.*, **28**, 98 (1967).
203 G. Barnikow and G. Strickmann, *Chem. Ber.*, **100**, 1428 (1967).
204 G. C. Hopkins, J. P. Jonak, H. Tieckelmann, and H. J. Minnemeyer, *J. Org. Chem.*, **31**, 3969 (1966).
205 H. O. House and C. J. Blankley, *J. Org. Chem.*, **32**, 1741 (1967).
206 F. H. Bottom and F. J. McQuillin, *Tetrahedron Letters*, **1967**, 1975.
207 J. I. G. Cadogan, R. K. Mackie, and J. A. Maynard, *J. Chem. Soc., C*, **1967**, 1356.
208 V. A. Tartakovskii, L. A. Nikonova, S. S. Novikov, *Izv. Akad. Nauk SSSR, Ser. Khim.*, **1966**, 1290; *Chem. Abs.*, **65**, 16808 (1966).
209 A. Holm and C. Wentrup, *Acta Chem. Scand.*, **20**, 2123 (1966).
210 P. Bouchet, J. Elguevo, and R. Jacquier, *Tetrahedron Letters*, **1966**, 6409.
211 H. Lund and P. Lunde, *Acta Chem. Scand.*, **21**, 1067 (1967).
212 D. E. Ames, G. V. Boyd, R. F. Chapman, A. W. Ellis, A. C. Lovesey, and D. Waite, *J. Chem. Soc., B*, **1967**, 748.
213 K. Schank, *Ann. Chem.*, **702**, 75 (1967).
214 B. Cardillo, G. Casnati, and A. Pochini, *Chim. Ind.*, (*Milan*), **49**, 172 (1967).
215 N. Kornblum and G. P. Coffey, *J. Org. Chem.*, **31**, 3449 (1966).
216 See M. I. Rybinskaya, *Zh. Vses. Khim. Obshchest*, **12**, 11 (1967); *Chem. Abs.*, **66**, 94413z (1967).

appears to be controlled by the ability of the α-substituents to stabilize the intermediate carbanion (see eqn. 33), and, when the α-substituents are the same, by the ability of the β-substituents to do so (see eqn. 34; cf. p. 107).[217]

. . . (33)

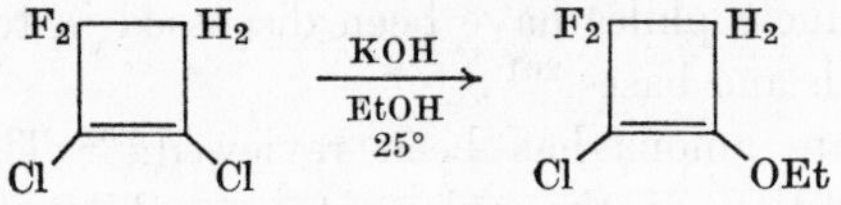

. . . (34)

The reactions of *cis*- and *trans*-β-bromostyrene with lithium diphenylarsenide in tetrahydrofuran proceed with inversion of configuration.[218] The kinetics of chloride exchange and isomerization of *cis*- and *trans*-2-chloro-1-*p*-methoxyphenyl-1-phenylethylene have been analysed in terms of a reaction scheme involving formation of a carbanion intermediate which breaks down to yield *cis*- and *trans*-isomers.[219]

Methyl-blocking and deuterium-labelling (analysis by proton and deuterium magnetic resonance) experiments, and measurements of deuterium isotope effects, have confirmed that the reactions of 1-chloro-cyclopentene, -cyclohexene, and -cycloheptene proceed via a cycloalkyne rather than via a cycloallene.[220, 221]

The proportions of the reactions of several (arylsulphonyl)chloroethylenes with alkoxide ions which proceed by elimination–addition and direct substitution have been determined.[222]

Other investigations of substitution reactions at vinylic carbon are described in references 223—225.

Allenyl halides (**103**) undergo hydrogen exchange 10—20 times faster than halide displacement in aqueous-ethanolic base; they yield the same products as propargyl halides (**102**), and it was suggested that allene-carbene (**104**) was a common intermediate.[226]

The solvolysis of phenacyl bromide in 80% ethanol is twice as fast as that

217 J. D. Park, G. Groppelli, and J. H. Adams, *Tetrahedron Letters*, **1967**, 103.
218 A. M. Aguiar and T. G. Archibald, *J. Org. Chem.*, **32**, 2627 (1967).
219 P. Beltrame, I. R. Bellobono, and A. Feré, *J. Chem. Soc., B*, **1966**, 1165.
220 L. K. Montgomery and L. E. Applegate, *J. Am. Chem. Soc.*, **89**, 2952 (1967).
221 L. K. Montgomery, A. O. Clouse, A. M. Crelier, and L. E. Applegate, *J. Am. Chem. Soc.*, **89**, 3453 (1967).
222 L. Di Nunno, G. Modena, and G. Scorrano, *J. Chem. Soc., B*, **1966**, 1186.
223 J. D. Park, R. Sullivan, and R. J. McMurtry, *Tetrahedron Letters*, **1967**, 173.
224 W. E. Truce, J. E. Parr, and M. L. Gorbarty, *Chem. Ind. (London)*, **1967**, 660.
225 G. Boularand and R. Vessière, *Bull. Soc. Chim. France*, **1967**, 1706.
226 V. J. Shiner and J. S. Humphrey, *J. Am. Chem. Soc.*, **89**, 622 (1967).

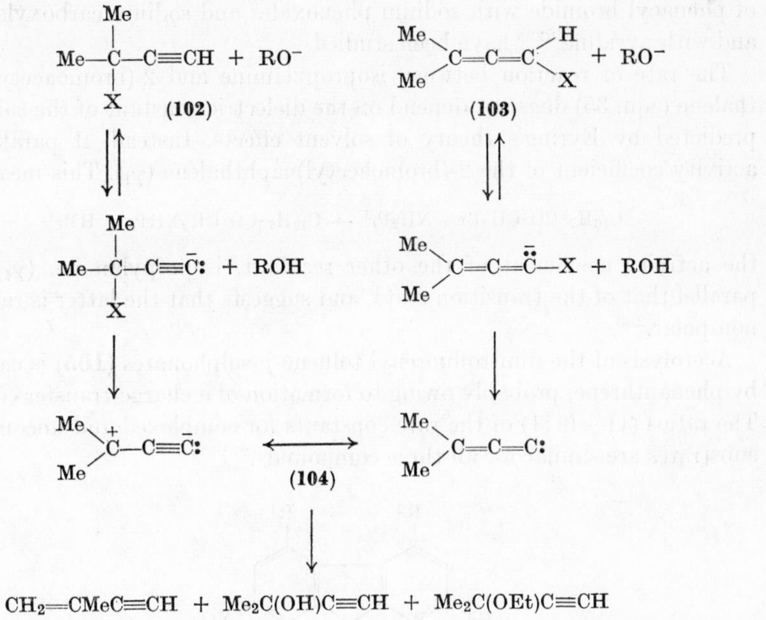

$$\underset{\underset{\underset{X}{|}}{\overset{\overset{Me}{|}}{Me-C}}-C\equiv CH}{}\ +\ RO^- \qquad \underset{Me}{\overset{Me}{>}}C=C=C\overset{H}{\underset{X}{<}}\ +\ RO^-$$

(102)   (103)

$$\underset{\underset{X}{|}}{\overset{\overset{Me}{|}}{Me-C}}-C\equiv\overset{..}{C}:\ +\ ROH \qquad \underset{Me}{\overset{Me}{>}}C=C=\overset{..}{\underset{..}{C}}-X\ +\ ROH$$

$$\underset{Me}{\overset{Me}{>}}\overset{+}{C}-C\equiv\overset{..}{C}: \quad\longleftrightarrow\quad \underset{Me}{\overset{Me}{>}}C=C=\overset{..}{C}:$$

(104)

$$CH_2=CMeC\equiv CH\ +\ Me_2C(OH)C\equiv CH\ +\ Me_2C(OEt)C\equiv CH$$

of ethyl bromide and slightly faster than that of isopropyl bromide, and the product is a 7:3 mixture of phenacyl ether and phenacyl alcohol. The possibility was explored[227] that this relatively high reactivity resulted from neighbouring-group participation, but it was concluded that this is unlikely. Thus participation by the phenyl group was excluded on the grounds that no rearranged product was formed, and by the carbonyl-oxygen on the grounds that $\Delta S^{\ddagger}$ (26.7 eu) was too small in comparison with that found for other ω-chloro ketones known to react in this way. A mechanism involving addition of water or ethanol to the carbonyl group followed by participation by the resulting hydroxyl group was rejected on the basis of the ρ-value (+0.35). The ρ-constant for the addition step would probably be greater than +1.5, so that the ρ-constant for the intramolecular displacement would have to be more negative than −1.2, and this was thought to be an unreasonable value. The Grunwald–Winstein m-value was small (0.20) and it was concluded that reaction involved highly nucleophilic displacement of halide by the solvent molecules.[227]

The ρ-value for the reaction of substituted phenacyl bromides with aniline in methanol is −1.97; the mechanism was thought[228] to be S_N2. The reactions

[227] D. J. Pasto, K. Garves, and M. P. Serve, *J. Org. Chem.*, **32**, 774 (1967); see also D. J. Pasto and K. Garves, *ibid.*, p. 778.
[228] R. K. Mohanty, G. Behera, and M. K. Rout, *Indian J. Chem.*, **5**, 259 (1967).

of phenacyl bromide with sodium phenoxides and sodium carboxylates,[229a] and with pyridine,[229b] have been studied.

The rate of reaction between isopropylamine and 2-(bromoacetyl)naphthalene (eqn. 35) does not depend on the dielectric constant of the solvent as predicted by Eyring's theory of solvent effects. Instead, it parallels the activity coefficient of the 2-(bromoacetyl)naphthalene (γ_I). This means that

$$C_{10}H_7 \cdot CO \cdot CH_2Br + NH_2Pr^i \rightarrow C_{10}H_7 \cdot CO \cdot CH_2NHPr^i + HBr \qquad \ldots(35)$$

the activity coefficient of the other reactant, isopropylamine, (γ_{II}) must parallel that of the transition state, and suggests that the latter is relatively non-polar.[230]

Acetolysis of the dinitrofluorenyl toluene-*p*-sulphonates (**105**) is catalysed by phenanthrene, probably owing to formation of a charge-transfer complex. The ratios (11—14:1) of the rate constants for complexed and uncomplexed substrates are similar for all three compounds.[231]

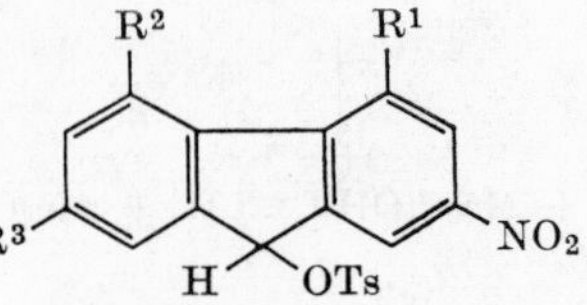

(**105**)

$R^1 = NO_2$; $R^2 = R^3 = H$
$R^1 = R^3 = H$; $R^2 = NO_2$
$R^1 = R^2 = H$; $R^3 = NO_2$

A detailed investigation of the formolysis of androsterone 3α-toluene-*p*-sulphonate has been reported.[232] The major product is 5α-androst-2-en-17-one, but the 3β-, 2β-, 3α-, and a trace of 3α-formate are also formed. 2α-Toluene-*p*-sulphonate is formed in the solvolysis mixture but the amount produced is only sufficient to account for 20% of the 2-formates in the product. Experiments with 2α- and 2β-tritiated toluene-*p*-sulphonates showed that the 2β-formate is formed with migration of the 2β-hydrogen and it was suggested that the toluene-*p*-sulphonate ion stayed on the α-face of the ion and controlled the steric course of its reaction with formic acid.[232]

The methanolysis of methyl perchlorate in benzene is probably bimolecular as the rate is increased markedly by sodium methoxide. The order in methanol

229a K. Okamoto, H. Kushiro, I. Nitta, and H. Shingu, *Bull. Chem. Soc. Japan*, **40**, 1900 (1967).
229b L. M. Litvinenko and L. A. Perel'man, *Zh. Org. Khim.*, **3**, 936 (1967); *Chem. Abs.*, **67**, 53263c (1967).
230 P. J. Taylor, *J. Chem. Soc.*, *B*, **1967**, 904.
231 A. K. Colter, F. F. Guzik, and S. H. Hui, *J. Am. Chem. Soc.*, **88**, 5754 (1966).
232 J. Ramseyer and H. Hirschmann, *J. Org. Chem.*, **32**, 1850 (1967); see also J. Ramseyer, J. S. Williams, and H. Hirschmann, *Steroids*, **9**, 347 (1967).

(0.07—0.4M) is about 2.3, but the rate is decreased by the addition of phenol. Possibly hydrogen-bonded methanol polymers[233] are the nucleophilic species, phenol reducing their nucleophilicity by acting as a hydrogen-bonding donor. Tetra-*n*-butylammonium perchlorate enhances the rate of this reaction about as much as it does the methanolysis of triphenylmethyl chloride. Possibly the perchlorate enhances the nucleophilicity of the methanol by hydrogen bonding to it (i.e., acts as a general-base catalyst).[234a] The methanolysis of methyl perchlorate in methanol–cyclohexane mixtures is known to be general-base catalysed.[234b]

The rate of formation of silver iodide from methyl iodide and silver perchlorate in nitromethane and nitrobenzene tends to a limiting value with increasing methyl iodide concentration at constant silver perchlorate concentration. Possibly a complex is formed between the methyl iodide and silver perchlorate. The reaction products were not determined.[235]

Support for Darwish and Tourigny's inversion mechanism[236] for the racemization of trialkylsulphonium salts has been obtained by showing that adamantanylethylmethylsulphonium perchlorate is racemized more rapidly than *tert*-butylethylmethylsulphonium perchlorate. If dissociation–recombination were an important pathway the adamantanyl compound should react more slowly.[237]

The plots of log (k/T) against $1/T$ for the acetolysis of *cis*- and *trans*-*tert*-butyl-cyclohexyl toluene-*p*-sulphonates are parallel to one another, but not to that for the acetolysis of cyclohexyl toluene-*p*-sulphonate. Above 102.5° cyclohexyl toluene-*p*-sulphonate reacts faster than *cis*-*tert*-butylcyclohexyl toluene-*p*-sulphonate. These results provide additional evidence that the kinetic method for establishing the position of conformational equilibria is of dubious validity.[238]

The rate constants (10^5k, l. mole^{-1} sec^{-1}) for the reaction of methyl iodide with NH_3, $MeNH_2$, Me_2NH, and Me_3N in aqueous solution at 20° are 4.90, 113, 492, and 1410, respectively, and therefore do not parallel the order of basicities, which is $Me_2NH > MeNH_2 > Me_3N > NH_3$. Qualitatively the same order is found for reaction in benzene solution. Amines with other alkyl groups sometimes show decreasing rates on going from primary to secondary to tertiary, probably as a result of steric effects.[239]

233 A. N. Fletcher and C. A. Heller, *J. Phys. Chem.*, **71**, 3742 (1967).
234a D. N. Kevill and H. S. Posselt, *Chem. Commun.*, **1967**, 438.
234b J. Koskikallio, *Suomen Kemistilehti*, *B*, **40**, 131, 199 (1967).
235 D. N. Kevill and V. V. Likhite, *Chem. Commun.*, **1967**, 247.
236 See *Organic Reaction Mechanisms*, **1966**, 85.
237 R. Scartazzini and K. Mislow, *Tetrahedron Letters*, **1967**, 2719.
238 J. L. Mateos, C. Perez, and H. Kwart, *Chem. Commun.*, **1967**, 125; see *Organic Reaction Mechanisms*, **1966**, 354; F. Shah-Malak and J. H. P. Utley, *Chem. Commun.*, **1967**, 69.
239 K. Okamoto, S. Fukui, and H. Shingu, *Bull. Chem. Soc. Japan*, **40**, 1920 (1967); K. Okamoto, S. Fukui, I. Nitta, and H. Shingu, *ibid.*, pp. 2350, 2354.

Bromide ion rapidly displaces SO_2Br from $PhCH_2SO_2Br$ in methylene dichloride solution:

$$PhCH_2SO_2Br + Br^- \rightarrow PhCH_2Br + SO_2 + Br^-$$

Chloride ion similarly displaces SO_2Cl from $PhCH_2SO_2Cl$ but more slowly. The carbon-13 isotope effect, $k_{12}/k_{13} = 1.035$, was considered to indicate a bimolecular mechanism. Compounds with leaving groups $CF_3SO_2^-$, $PhCOCH_2SO_2^-$, $PhSO_2^-$, and $PhCH_2SO_2^-$ did not react, and it was suggested that the $-SO_2Br$ and $-SO_2Cl$ groups underwent fragmentation concerted with nucleophilic attack by Br^- and Cl^- on the benzylic carbon.[240]

There have been more investigations of the steric course of the alkylation of cyclic amines.[241]

The volume of activation of the acid-catalysed hydrolysis of diethyl ether is -10 cm^3 mole^{-1}.[242] The cleavage of the ether linkage has been reviewed.[243]

The ring opening of epoxides,[244] episulphides,[245] and oxetanes[246] has been studied.

There have been theoretical discussions of nucleophilicities[247–251] and

240 J. F. King and D. J. H. Smith, *J. Am. Chem. Soc.*, **89**, 4803 (1967); see *Organic Reaction Mechanisms*, **1965**, p. 54, ref. 16.

241 D. R. Brown, J. McKenna, J. M. McKenna, J. M. Stuart, and B. G. Hutley, *Chem. Commun.*, **1967**, 380; D. R. Brown, R. Lygo, J. McKenna, J. M. McKenna, and B. G. Hutley, *J. Chem. Soc., B*, **1967**, 1184; D. R. Brown, J. McKenna, and J. M. McKenna, *ibid.*, p. 1195; A. T. Bottini and M. K. O'Rell, *Tetrahedron Letters*, **1967**, 423, 429; M. Havel, J. Krupička, J. Sicher, and M. Svoboda, *ibid.*, p. 4009.

242 L. Pyy and J. Koskikallio, *Suomen Kemistilehti, B*, **40**, 134 (1967); see also W. J. Le Noble, *Progr. Phys. Org. Chem.*, **5**, 207 (1967).

243 E. Staude and F. Patai, in "The Chemistry of the Ether Linkage", S. Patai, ed., Interscience, London, 1967, p. 21.

244 W. Reeve and M. Nees, *J. Am. Chem. Soc.*, **89**, 647 (1967); Y. Tanaka, *J. Org. Chem.*, **32**, 2405 (1967); J. Fajkoš, J. Joska, and F. Šorm, *Coll. Czech. Chem. Commun.*, **31**, 4610 (1966); S. J. Angyal and T. S. Stewart, *Australian J. Chem.*, **20**, 2117 (1967); S. Sekiguchi and S. Ishii, *Kogyo Kogaku Zasshi*, **70**, 46 (1967); *Chem. Abs.*, **67**, 32105n (1967); L. A. Paquette, A. A. Youssef, and M. L. Wise, *J. Am. Chem. Soc.*, **89**, 5246 (1967); R. M. Gerkin and B. Rickborn, *ibid.*, p. 5850; S. O. Chan and E. J. Wells, *Can. J. Chem.*, **45**, 2123 (1967); N. S. Isaacs and K. Neelakantan, *ibid.*, p. 1597; M. F. Sorokin and L. G. Shode, *Zh. Organ. Khim.*, **2**, 1469 (1966); *Chem. Abs.*, **66**, 54855 (1967); V. M. Kozlov and N. N. Lebedev, *Tr. Mosk. Khim.-Tekhnol. Inst.*, No. 48, 68 (1965); *Chem. Abs,.* **65**, 18448 (1966); V. M. Kozlov and N. N. Lebedev, *Tr. Mosk. Khim.-Tekhnol. Inst.*, No. 48, 73 (1965); *Chem. Abs.*, **65**, 18446 (1966); N. N. Lebedev and Y. I. Baranov, *Kinetika i Kataliz*, **7**, 619 (1966); *Chem. Abs.*, **65**, 16812 (1966); T. Shono, I. Nishiguchi, A. Oku, and R. Oda, *Tetrahedron Letters*, **1967**, 517; D. J. Goldsmith, B. C. Clark, and R. C. Joines, *ibid.*, p. 1211; D. J. Goldsmith and B. C. Clark, *ibid.*, p. 1215.

245 A. Oddon and J. Wylde, *Bull. Soc. Chim. France*, **1967**, 1603, 1607.

246 P. O. I. Virtanen, *Suomen Kemistilehti, B*, **40**, 185, 193 (1967).

247 R. Scheffold, *Helv. Chim. Acta*, **50**, 1419 (1967).

248 R. V. Vizgert and I. M. Ozdrovskaya, *Reakts. Sposobnost. Org. Soedin. Tartu Gos. Univ.*, **3** 16 (1966).

249 B. Saville, *Angew. Chem. Intern. Ed. Engl.*, **6**, 928 (1967).

250 R. G. Pearson and J. Songstad, *J. Am. Chem. Soc.*, **89**, 1827 (1967).

251 R. G. Pearson and J. Songstad, *J. Org. Chem.*, **32**, 2899 (1967); cf. W. S. Trahanovsky and M. P. Doyle, *Chem. Commun.*, **1967**, 1021.

reactivity in S_N2 displacements.[252, 253] MO calculations on the transition state for the S_N2 mechanisms have been reported.[254]

The non-stationary state kinetic equations for the S_N1 and S_N2 mechanism have been solved and discussed.[255]

Other reactions which have received attention include: hydrolysis of α-halogeno-*sec*-alkyl ester,[256] α-bromo-acids,[257] and 1-(*p*-alkylphenyl)ethyl chlorides;[258] formolysis of benzyl fluorides;[259, 260] acetolysis of polymethylcyclopentyl toluene-*p*-sulphonate[261] and *myo*-inositol benzyl ethers;[262] solvolyses of benzhydryl chlorides,[263] and of cyclo-octyl and *trans*-2-hydroxycyclo-octyl bromides and toluene-*p*-sulphonates;[264] solvolysis[265] and chloride exchange[266] of hydrazidic halides; bromide exchange of bromoacetic acid;[267] iodide exchange of methylene di-iodide;[268] S_N2 and $E2$ reactions of 1,2-dichloroethane,[269] 2-substituted ethyl chlorides,[270] and ω-chloroalkyl chlorides;[271] and the reactions of 2-halogenotropones with alkali,[272] of acetobromoglucose with methyl β-D-glucoside,[273] and of *p*-substituted phenethyl bromides with sodium thiosulphate.[274]

Also, the effect of remote substituents on the rates of solvolysis of steroidal toluene-*p*-sulphonates has been investigated.[275]

252 R. F. Hudson and G. Klopman, *Tetrahedron Letters*, **1967**, 1103.
253 K. Okamoto, I. Nitta, T. Imoto, H. Shingu, *Bull. Chem. Soc. Japan*, **40**, 1905 (1967).
254 W. Drenth, *Rec. Trav. Chim.*, **86**, 318 (1967).
255 K. Frei and H. H. Günthard, *Helv. Chim. Acta*, **50**, 1294 (1967).
256 E. K. Euranto, *Acta Chem. Scand.*, **21**, 721 (1967).
257 J. Leska, *Acta Fac. Rerum Nat. Univ. Comenianae Chim.*, **9**, 535, 547, 567 (1966); *Chem. Abs.*, **67**, 10882m, 10883n 10884p, (1967).
258 R. Anantaraman and M. R. Nair, *Indian J. Chem.*, **5**, 77, 163 (1967).
259 J.-J. Delpuech and C. Beguin, *Bull. Soc. Chim. France*, **1967**, 791.
260 C. Beguin and A. Meary-Tertian, *Bull. Soc. Chim. France*, **1967**, 795.
261 A. P. Krapcho and D. E. Horn, *Tetrahedron Letters*, **1966**, 6107.
262 S. J. Angyal, M. H. Randall, and M. E. Tate, *J. Chem. Soc., C*, **1967**, 919.
263 S. Nishida, *J. Org. Chem.*, **32**, 2692, 2695, 2697 (1967).
264 D. D. Roberts and J. G. Traynham, *J. Org. Chem.*, **32**, 3177 (1967).
265 R. N. Butler and F. L. Scott, *J. Chem. Soc., C*, **1967**, 239.
266 J. S. Clovis, A. Eckell, R. Huisgen, and R. Sustmann, *Chem. Ber.*, **100**, 60 (1967).
267 P. Beronius, *Radiochim. Acta*, **8**, 57 (1967).
268 H. Elias, *Radiochim. Acta*, **6**, 167 (1966).
269 K. Okamoto, H. Matsuda, H. Kawasaki, and H. Shingu, *Bull. Chem. Soc. Japan*, **40**, 1917 (1967).
270 K. Okamoto, T. Kita, K. Araki, and H. Shingu, *Bull. Chem. Soc. Japan*, **40**, 1913 (1967).
271 K. Okamoto, T. Kita, and H. Shingu, *Bull. Chem. Soc. Japan*, **40**, 1908 (1967).
272 E. J. Forbes, D. C. Warrell, and W. J. Fry, *J. Chem. Soc., C*, **1967**, 1693.
273 A. M. Bills and J. W. Green, *J. Chem. Soc., B*, **1967**, 716.
274 S. Oae, K. Akagi, and Y. Yano, *Mem. Fac. Eng., Osaka City Univ.*, **7**, 81 (1965); *Chem. Abs.*, **66**, 94612p (1967).
275 R. Baker and J. Hudec, *Chem. Commun.*, **1967**, 479.

CHAPTER 3

Electrophilic Aliphatic Substitution[1]

A detailed analysis of the kinetics of racemization and deuterium exchange of compound (**1**) in MeOD–MeOK has been reported.[2] Under these conditions

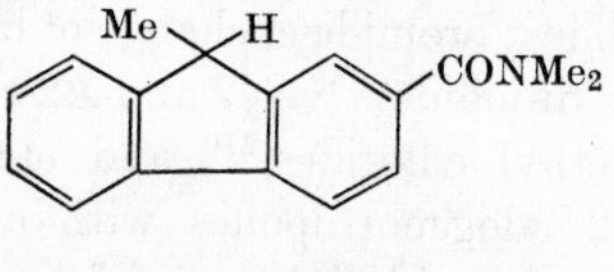

(**1**)

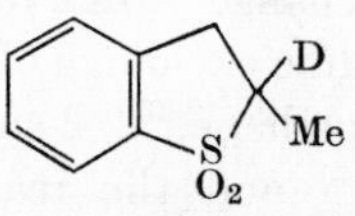

(**2**)

$k_e/k_\alpha = 0.92$. A value close to 1.0 such as this could result from a single racemization mechanism or a combination of inversion and retention mechanisms. The possible processes are:

$$\mathrm{H}_+ \xrightarrow{k_1} \mathrm{D}_+ \qquad \mathrm{H}_- \xrightarrow{k_2} \mathrm{D}_+$$

$$\mathrm{H}_+ \xrightarrow{k_2} \mathrm{D}_- \qquad \mathrm{H}_- \xrightarrow{k_3} \mathrm{H}_+$$

$$\mathrm{H}_+ \xrightarrow{k_3} \mathrm{H}_- \qquad \mathrm{D}_+ \xrightarrow{k_4} \mathrm{D}_-$$

$$\mathrm{H}_- \xrightarrow{k_1} \mathrm{D}_- \qquad \mathrm{D}_- \xrightarrow{k_4} \mathrm{D}_+$$

These lead to an expression for optical rotation (α) at time t:

$$\alpha = \left[\frac{\alpha_0}{(\phi - 1) + \psi}\right]\{(\phi - 1)\exp[-\theta(1-\psi)t] + \psi\exp(-\theta\phi t)\}$$

where $\phi = k_4/(k_2 + k_3)$ (the reciprocal of the kinetic isotope effect k_H/k_D for racemization), $\psi = (k_2 - k_1)/2(k_2 + k_3)$, and $\theta = 2(k_2 + k_3)$ (the rate constant for racemization at zero time). The isotope effect, $\phi = 0.15$, was measured independently and the value of ψ calculated from this; and the time at which rotation first became zero was itself zero within experimental error, so that $k_1 = k_2 = 3.02 \times 10^{-3}$ l. mole^{-1} sec^{-1} at 24.9°. The preferred steric course is therefore racemization, but there is also a small amount of isoinversion (inversion without exchange) with $k_3 = 0.26 \times 10^{-3}$ l. mole^{-1} sec^{-1}. These

[1] D. C. Ayres, "Carbanions in Synthesis", Oldbourne, London, 1966; R. E. Dessy and W. Kitching, *Advan. Organometal. Chem.*, **4**, 267 (1966).

[2] See *Organic Reaction Mechanisms*, **1965**, 81.

values were confirmed by re-resolving partially racemized (**1**) and determining the isotopic composition of the enantiomers.

Similar experiments were performed with *tert*-butyl alcohol and potassium phenoxide.[3] Here $k_1 > k_2$ and there is a small isoinversion component leading to an overall value of $k_e/k_\alpha = 1.0$.

The stereochemical routes followed by reaction of the sulphone (**2**) with potassium *tert*-butoxide in *tert*-butyl alcohol were also determined by re-resolution experiments. Here $k_e/k_\alpha = 0.66$ but this results from a combination of inversion without exchange (isoinversion) (rel. rate 1), net inversion with exchange (rel. rate 3), and racemization (rel. rate 9). The mechanism of Scheme 1 was proposed. Isoinversion probably results from a "conducted tour mechanism" in which the original deuterium migrates from one side of the carbanion to the other via hydrogen-bonding sites of the oxygens of the

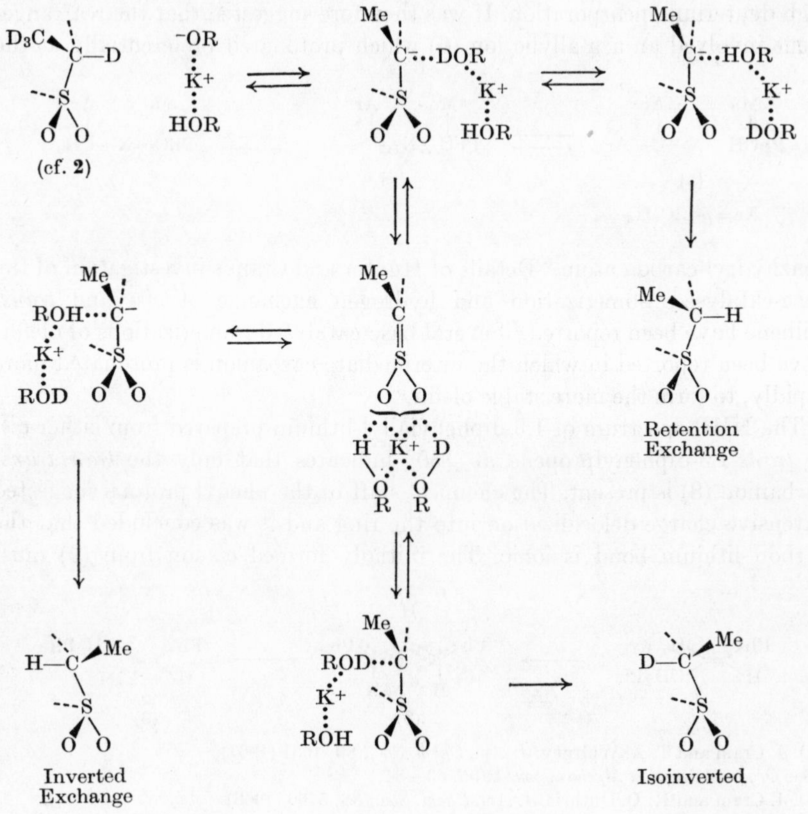

Scheme 1

[3] W. T. Ford, E. W. Graham, and D. J. Cram, *J. Am. Chem. Soc.*, **89**, 689, 690, 4661 (1967).

sulphone group. It was thought that the cyclic sulphonyl ion derived from (**2**) was probably planar.

Conditions have been found (ammonium salt in water at 155°) under which base-catalysed decarboxylations of (**3**) and (**4**) do not proceed with racemization, but probably with net inversion of configuration.[4, 5]

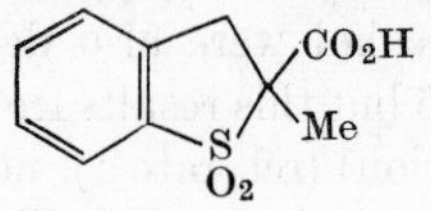

(**3**)

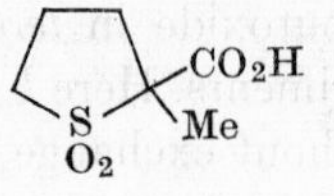

(**4**)

The base-catalysed methylene–azomethine rearrangement (**5**) → (**7**) has been investigated. At low conversions in ethylene [O-$^2H_2$]glycol there was no incorporation of deuterium or loss of optical activity with (**5**), but (**7**) showed high deuterium incorporation. It was therefore suggested that the rearrangement involved an aza-allylic ion (**6**) which protonated preferentially at the

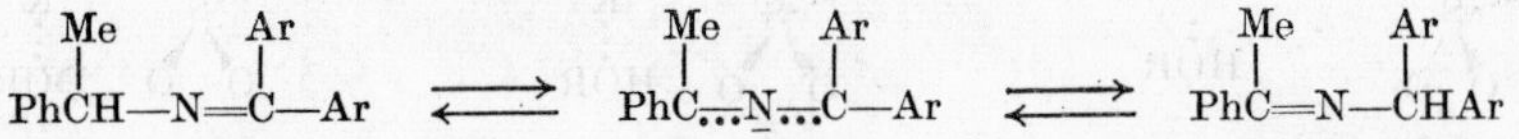

(**5**) (**6**) (**7**)

Ar = p-ClC_6H_4

benzhydryl-carbon atom.[6] Details of Hunter and Cram's investigation of the base-catalysed isomerization and hydrogen exchange of *cis*- and *trans*-stilbene have been reported.[7] Several base-catalysed isomerizations of olefins have been reported in which the intermediate carbanion is protonated more rapidly, to form the more stable olefin.[8]

The NMR spectrum of 1,3-diphenylallyl-lithium prepared from either *cis*- or *trans*-1,3-diphenylpropene at –30° indicates that only the *trans,trans*-carbanion (**8**) is present. The chemical shift of the phenyl protons suggested extensive charge delocalization into the ring and it was concluded that the carbon–lithium bond is ionic. The initially formed *cis*-ion from (**9**) must

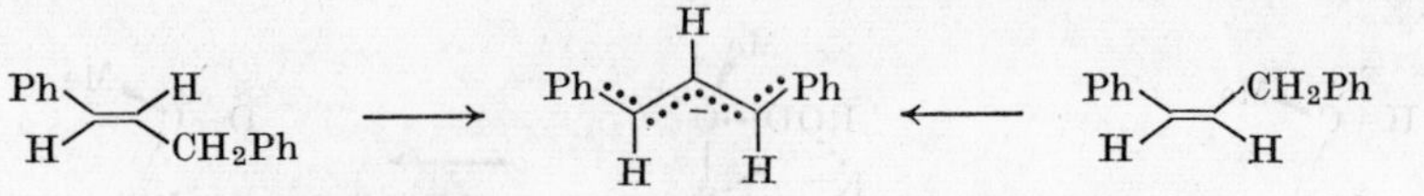

(**8**) (**9**)

[4] D. J. Cram and T. A. Whitney, *J. Am. Chem. Soc.*, **89**, 4651 (1967).
[5] See *Organic Reaction Mechanisms*, **1965**, 83—84.
[6] D. J. Cram and R. D. Guthrie, *J. Am. Chem. Soc.*, **88**, 5760 (1966).
[7] D. H. Hunter and D. J. Cram, *J. Am. Chem. Soc.*, **88**, 5765 (1966); see *Organic Reaction Mechanisms*, **1965**, 83.
[8] S. W. Ela and D. J. Cram, *J. Am. Chem. Soc.*, **88**, 5777, 5791 (1966).

isomerize under the reaction conditions and the barrier for this must be less than the 27 kcal mole^{-1} derived from Hückel calculations.[9]

The reactions of *exo*- and *endo*-2-norbornyl-lithium with bromine in pentane at −70° proceed with inversion of configuration, but those with methyl chloroformate proceed with retention. Carbanions were thought not to be involved, and it was concluded "that inversion and retention pathways for concerted electrophilic aliphatic substitution are inherently about equally available and that changes in reagents and perhaps solvents tip the scales in favour of one or the other."[10]

Inversion of configuration in an electrophilic substitution at saturated carbon also occurs in the reaction of *exo*-norborn-5-ene-2-boronic acid, with mercuric chloride to yield nortricyclylmercuric chloride, and of *exo*-bicyclo-[2.2.2]oct-2-ene-5-boronic acid to yield tricyclo[2.2.2.0^{2,6}]oct-1-ylmercuric chloride.[11]

Inversion barriers of carbanions have been discussed in relation to those of other trivalent species,[12] and interconversion of the enantiomers of the 2,2-dimethyl-1-(phenylsulphonyl)cyclopropyl anion has been studied by NMR spectroscopy.[13]

Molecular-orbital calculations of the preferred conformation of α-sulphinyl carbanions have been described.[14] The deuterons of the methylene group of $PhSO\cdot CD_2\cdot CO_2H$ undergo base-catalysed exchange at different rates.[15]

The UV and NMR spectra of ion (**10**) (a bishomocyclopentadienide ion), previously postulated[16] to account for the rapid base-catalysed exchange of (**11**), support the delocalized structure shown.[17,18] The UV spectrum shows broad overlapping bands at 320 and 380 mμ (ε plateau ~ 10^3), and the signals

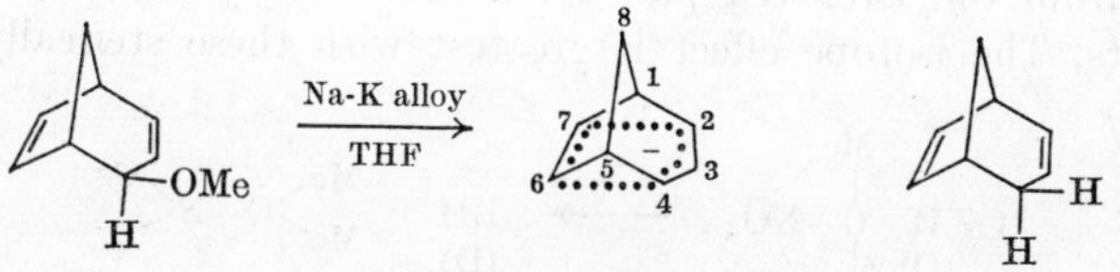

(**10**) (**11**)

[9] H. H. Freedman, V. R. Sandel, and B. P. Thill, *J. Am. Chem. Soc.*, **89**, 1762 (1967).
[10] D. E. Applequist and G. N. Chmurny, *J. Am. Chem. Soc.*, **89**, 875 (1967).
[11] D. S. Matteson and M. L. Talbot, *J. Am. Chem. Soc.*, **89**, 1119, 1123 (1967); cf. *Organic Reaction Mechanisms*, **1965**, 88.
[12] G. W. Koeppl, D. S. Sagatys, G. S. Krishnamurthy, and S. I. Miller, *J. Am. Chem. Soc.*, **89**, 3396 (1967); F. A. L. Anet, R. D. Trepka, and D. J. Cram, *ibid.*, p. 357.
[13] A. Ratajczak, F. A. L. Anet, and D. J. Cram, *J. Am. Chem. Soc.*, **89**, 2072 (1967).
[14] S. Wolfe, A. Rauk, and I. G. Csizmadia, *J. Am. Chem. Soc.*, **89**, 5710 (1967).
[15] E. Bullock, J. M. W. Scott, and P. D. Golding, *Chem. Commun.*, **1967**, 168; cf. *Organic Reaction Mechanisms*, **1966**, 92.
[16] See *Organic Reaction Mechanisms*, **1965**, 85.
[17] J. M. Brown, *Chem. Commun.*, **1967**, 638.
[18] S. Winstein, M. Ogliaruso, M. Sakai, and J. M. Nicholson, *J. Am. Chem. Soc.*, **89**, 3656 (1967); see also S. Winstein in "Aromaticity", Chem. Soc. Special Publication No. 21, p. 34.

in the NMR spectrum for the $C_{(6)}$, $C_{(7)}$, and $C_{(8)}$ protons show upfield shifts [compared to signals from (**11**)] of *ca.* 2 and 1 ppm, respectively. It was calculated[17] from the latter that 30—40% of a fully developed six-π-electron ring current was operative. When the ion was quenched in CH_3OD, the product (**11**) was formed with *exo*- and *endo*-deuterium at $C_{(4)}$ in approximately equal amounts.

Base-catalysed hydrogen exchange of tricyclo[4.3.1.0]deca-2,4,7-triene (**12**) occurs stereospecifically with exchange of $H_{(9')}$ at least 10^4 times faster than that of H_9. It was proposed that the reaction proceeds through the aromatic ten-π-electron aromatic anion (**13**) which is formed and protonated stereospecifically as a result of favourable orbital overlap between the $C_{(9)}$–$H_{(9')}$ and $C_{(1)}$–$C_{(6)}$ bonds.[19]

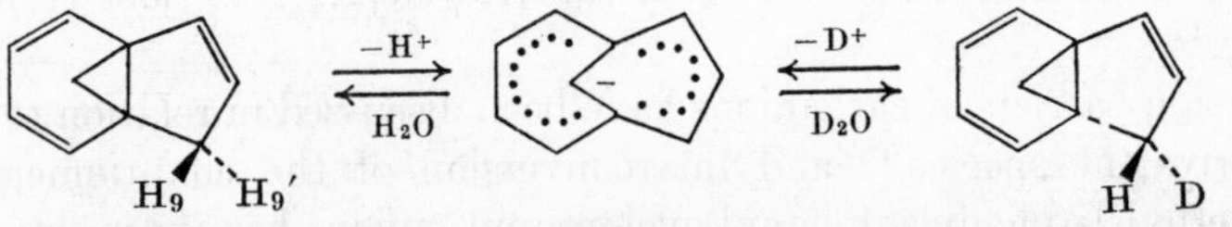

(**12**) (**13**)

Protonation of the monohomocyclo-octatetraene dianion has been investigated.[20]

Extremely large primary isotope effects have been observed in the iodination of 2-nitropropane, catalysed by pyridine and substituted pyridines in *tert*-butyl alcohol–water mixtures.[21] The rate-determining step for these reactions is transfer of a proton or a deuteron from the nitropropane to the base (eqn. 1), and 2,6-dimethyl- and 2-*tert*-butyl-pyridine are poorer catalysts than is predicted from the Brönsted plot for a series of pyridines lacking these substituents. The isotope effect is greatest with these sterically hindered

$$\mathrm{B} + \underset{(\mathrm{D})}{\mathrm{H}}\text{—}\mathrm{C(Me)_2}\text{—}\mathrm{NO_2} \longrightarrow \underset{(\mathrm{D})}{\overset{+}{\mathrm{B}}\mathrm{H}} + \mathrm{Me_2C{=}\overset{+}{N}(O^-)O^-} \quad \ldots(1)$$

pyridines and an extremely high value, $k_H/k_D = 24$, is found with 2,6-dimethylpyridine. Since the maximum isotope effect on classical theory, i.e., that for total loss of zero-point energy in the transition state, is $k_H/k_D = 18$, it was suggested that values larger than this result from tunnelling. Steric hindrance and tunnelling may be connected as much of the potential energy of the transition state of a sterically hindered reaction is non-bonding compression energy, so that any movement away from the maximum in either direction

[19] P. Radlick and W. Rosen, *J. Am. Chem. Soc.*, **89**, 5308 (1967).
[20] M. Ogliaruso and S. Winstein, *J. Am. Chem. Soc.*, **89**, 5290 (1967).
[21] E. S. Lewis and L. H. Funderburk, *J. Am. Chem. Soc.*, **89**, 2322 (1967); cf. *Organic Reaction Mechanisms*, **1966**, 93.

results in a sharp reduction; i.e., the energy barrier is high and thin, which should be ideal for tunnelling. It is also possible that steric interactions cause the hydrogen bond in the transition state to be stretched, so that bending vibrations become less stiff and there is a substantial loss of bonding zero-point energy.[21]

Introduction of a *cis*-3-aryl substituent into *cis*-1-*tert*-butyl-4-nitrocyclohexane (cf. **14**) had only a slight effect on the rate of proton abstraction, but introduction of a 3-*p*-chlorophenyl or a 3-*o*-tolyl group into *trans*-1-*tert*-butyl-4-nitrocyclohexane (cf. **15**) causes, respectively, a 22-fold and a 44-fold rate decrease. It was suggested that there was a bending away of equatorial aryl and nitro-groups in (**15**) which causes the axial hydrogen atom to be screened more effectively by the aryl group.[22]

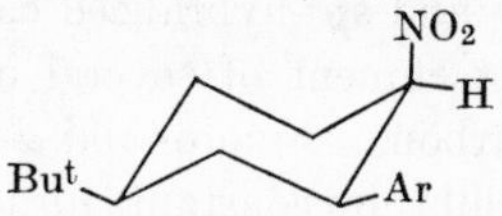

(**14**)

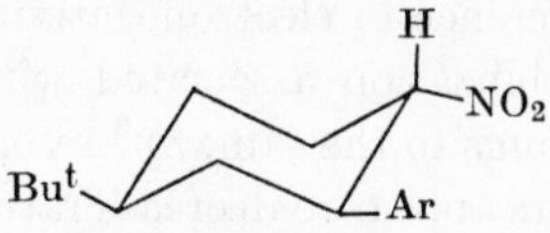

(**15**)

Proton abstraction from nitroethane by hydrazine, methylhydrazine, hydroxylamine, and methoxyamine have been investigated (see p. 335).[23]

The hydrogen isotope effect for the exchange of α-hydrogen of toluene is much greater with lithium cyclohexylamide in cyclohexylamine ($k_H/k_D \approx 10$) than with potassium *tert*-butoxide in DMSO ($k_D/k_T \approx 1$). This unexpected result has now been rationalized as follows. The force constants for the H...A and H...B bonds in the transition state for transfer of a proton between toluene and cyclohexylamide, A...H...B, are approximately equal since the pK_a's of toluene and cyclohexylamine are similar (40 and 35, respectively). Hence the isotope effect should be near to its maximum value. In the transition state for proton transfer between toluene and *tert*-butoxide the force constants of the H...A and H...B bonds should be quite different, the proton should be close to the *tert*-butoxide, and so the isotope effect should be small.[24]

Hydrogen exchange of 1*H*-undecafluorobicyclo[2.2.1]heptane with sodium methoxide in methanol is five times faster than that of tris(trifluoromethyl)-methane, indicating that hyperconjugation as symbolized by (**16**) cannot be an important factor in the carbanion-stabilizing effect of β-fluorine.[25, 26] The

[22] F. G. Bordwell and M. M. Vestling, *J. Am. Chem. Soc.*, **89**, 3906 (1967).
[23] M. J. Gregory and T. C. Bruice, *J. Am. Chem. Soc.*, **89**, 2327 (1967).
[24] J. R. Jones, *Chem. Commun.*, **1967**, 710.
[25] A. Streitwieser and D. Holtz, *J. Am. Chem. Soc.*, **89**, 692 (1967).
[26] See also S. F. Campbell, J. M. Leach, R. Stephens, and J. C. Tatlow, *Tetrahedron Letters*, **1967**, 4269; E. P. Mochalina, B. L. Dyatkin, I. V. Galakhov, and I. L. Knunyants, *Dokl. Akad. Nauk SSSR*, **169**, 1346 (1966); *Chem. Abs.*, **66**, 45901y (1967).

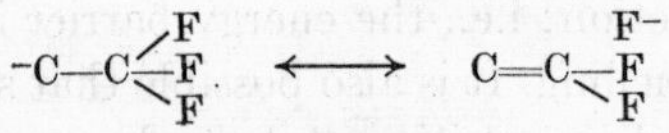

(16)

rate of tritium exchange of 9-trifluoromethyl[9-$^3H_1$] fluorene has approximately the value expected from a linear free-energy plot of the rates of less reactive 9-substituted fluorenes against the pK_a's of the corresponding acetic acids. This result also indicates that the large activating effect of β-fluorine does not result from a special effect such as hyperconjugation.[27]

In contrast to the above results, methoxide-catalysed hydrogen exchange of methoxyacetate, fluoroacetate, dimethoxyacetate, and difluoroacetate is much ($10^{2.6}$—10^{12}) slower than predicted by the Taft equation. This reduced carbanion-stabilizing effect of α-methoxy- and α-fluoro-groups was ascribed to the difference in electronegativity of sp^3- and sp^2-hybridized carbon and to the stabilization associated with the attachment of several oxygen or fluorine atoms to the same sp^3-hybridized carbon. α-Fluoro- and α-methoxy-substituents appear to decrease rate and equilibrium constants for the formation of sp^2-hybridized carbanions or increase them much less than expected from inductive effects.[28]

The kinetics of hydrogen exchange of triphenylmethane and diphenylmethane catalysed by lithium *N*-methylanilide suggests that most of the catalyst is trimeric and that monomer is the catalytically active species.[29]

The second-order rate constant for proton exchange between fluorene and fluorenyl-lithium (0.3M), as measured by a double-resonance NMR method, is approximately 0.5 l. mole^{-1} sec^{-1} in DMSO at 38°.[30] This value is much less than that for proton exchange between the potassium salt of the 4,5-methylenephenanthryl anion (5×10^{-4}M) and 9-methylfluorene (10^{-2}M) which is 1.4×10^3 l. mole^{-1} sec^{-1} in DMSO at 25° as measured by a stopped-flow technique. Possibly the slowness of the reaction of the fluorenyl-lithium is the result of ion pairing.[31]

The second-order rate constants for proton transfer from triphenylmethane to the dimsyl and n-propoxide ions are 8×10^3 and 6×10^3 l. mole^{-1} sec^{-1}, respectively. The latter value is 10^{14} times greater than that for proton transfer from triphenylmethane to methoxide ion in methanol. Proton transfer from triphenylmethane to *tert*-butoxide in dimethyl sulphoxide is much slower than to n-propoxide, possibly as a result of steric hindrance.[32]

Nortricyclone is cleaved by a 10:3 mixture of potassium *tert*-butoxide and

27 A. Streitwieser, A. P. Marchand, and A. H. Pudjaatmaka, *J. Am. Chem. Soc.*, **89**, 693 (1967).
28 J. Hine, L. G. Mahone, and C. L. Liotta, *J. Am. Chem. Soc.*, **89**, 5911 (1967).
29 H. F. Ebel and G. Ritterbusch, *Ann. Chem.*, **704**, 15 (1967).
30 J. I. Brauman, D. F. McMillen, and Y. Kanazawa, *J. Am. Chem. Soc.*, **89**, 1728 (1967).
31 C. D. Ritchie and R. E. Uschold, *J. Am. Chem. Soc.*, **89**, 1730 (1967).
32 C. D. Ritchie and R. E. Uschold, *J. Am. Chem. Soc.*, **89**, 2960 (1967); cf. ref. 211 of Chapter 1; see also C. D. Ritchie and R. E. Uschold, *J. Am. Chem. Soc.*, **89**, 1721, 2752 (1967).

water in aprotic solvents (e.g., ether, see eqn. 2). When water is absent or equimolar amounts of potassium *tert*-butoxide and water are used, there is no cleavage and the reaction probably requires HO^-, acting as a nucleophile,

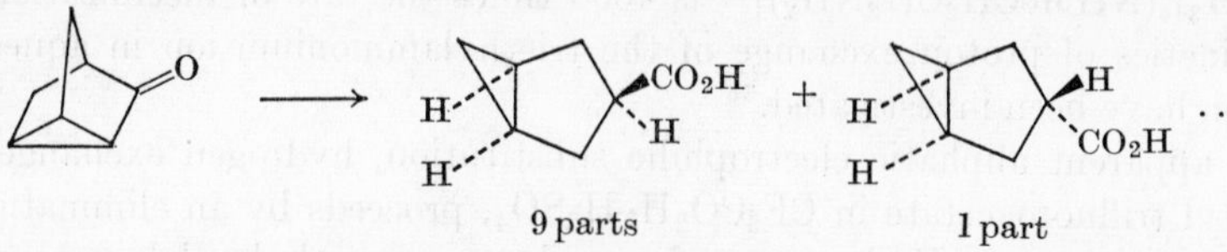

. . . (2)

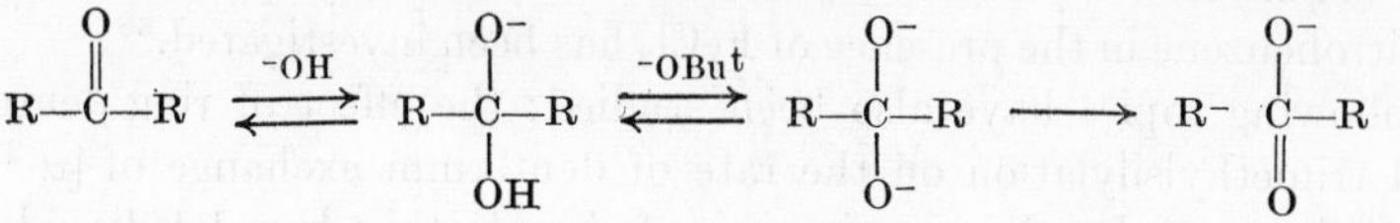

. . . (3)

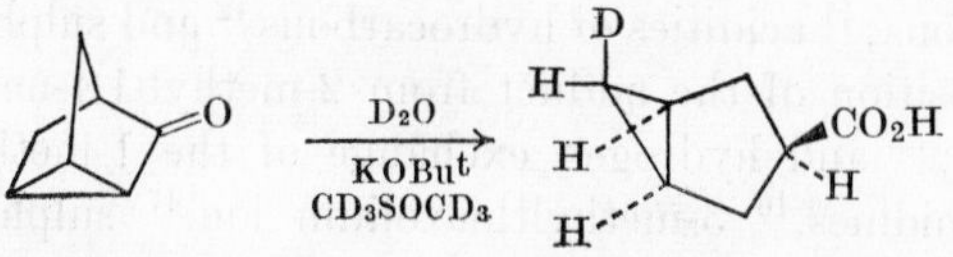

. . . (4)

and Bu^tO^-, acting as a base (eqn. 3). The reaction proceeds with retention of configuration at the cyclopropyl ring (eqn. 4) and thus provides another example of the configurational stability of cyclopropyl carbanions. The reaction was shown to be a general one for the cleavage of non-enolizable ketones.[33]

The cyclopropylmethyl and cyclobutylmethyl carbanions have been generated by fragmentation of the corresponding benzoyl di-imides in sodium

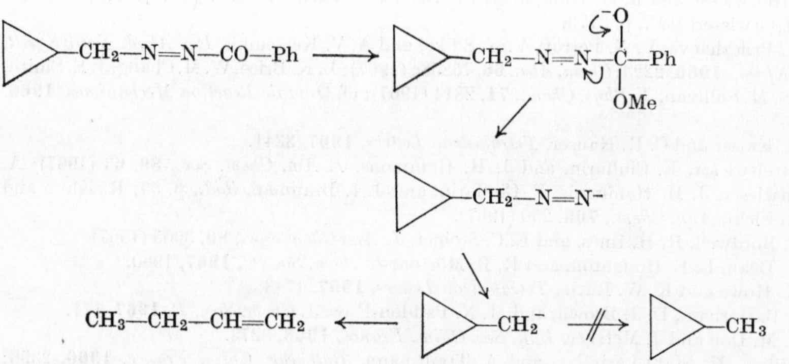

. . . (5)

methoxide–methanol (cf. eqn. 5). The cyclopropylmethyl ion rearranges to the but-3-enyl ion faster than it is protonated, but the cyclobutylmethyl ion is protonated faster than it rearranges. It was estimated that the cyclopropylmethyl ion rearranges at least 10^4 times faster than the cyclobutylmethyl ion

[33] P. G. Gassman, J. T. Lumb, and F. V. Zalar, *J. Am. Chem. Soc.*, **89**, 946 (1967).

owing, presumably, to the greater release of ring-strain energy in the transition state.[34]

The rate of deuteration of the asymmetric nitrogen of $[Co(NH_3)_4(NHMeCH_2CH_2NH_2)]^{3+}$ is 4000 times the rate of racemization.[35] The kinetics of proton exchange of the triethylammonium ion in aqueous solution have been investigated.[36]

The apparent aliphatic electrophilic substitution, hydrogen exchange of *sec*-butyl trifluoroacetate in CF_3CO_2H–H_2SO_4, proceeds by an elimination–addition sequence.[37] Hydrogen exchange between methylcyclohexane and DCl in nitrobenzene in the presence of $FeCl_3$ has been investigated.[38]

The following topics have also been studied: the effect of ring deuteration and trimethylsilylation on the rate of deuterium exchange of [α-^{2}H]-toluene;[39] base-catalysed isomerization of dihydrothiophen 1,1-dioxides[40] and sulphone dianions;[41] acidities of hydrocarbons[42] and sulphones;[43] base-catalysed decomposition of the adduct from 2-methyl-1,4-naphthoquinone and diazomethane;[44] and hydrogen exchange of the 1-methylpyridinium ion,[45] methylpyrimidines,[46] 3-methylthiazolium ion,[47] sulphoxides,[48] and substituted fluorenes.[49]

34 R. W. Hoffmann and K. R. Eicken, *Chem. Ber.*, **100**, 1465 (1967).

35 D. A. Buckingham, L. G. Marzilli, and A. M. Sargeson, *J. Am. Chem. Soc.*, **89**, 825 (1967); cf. *Organic Reaction Mechanisms*, **1966**, 92.

36 E. K. Ralph and E. Grunwald, *J. Am. Chem. Soc.*, **89**, 2963 (1967).

37 V. N. Setkina, E. V. Bykova, A. G. Ginzburg, and D. N. Kursanov, *Dokl. Akad. Nauk SSSR*, **170**, 1344 (1966).

38 V. N. Setkina, I. G. Malakova, and D. N. Kursanov, *Izv. Akad. Nauk SSSR, Ser. Khim.*, **1966**, 1348; *Chem. Abs.*, **66**, 85180u (1967).

39 A. Streitwieser and J. S. Humphrey, *J. Am. Chem. Soc.*, **89**, 3767 (1967); F. Mares and A. Streitwieser, *ibid.*, p. 3770.

40 E. N. Prilezhaeva, V. N. Petrov, V. A. Sil'ke, and A. V. Kessenikh, *Izv. Akad. Nauk SSSR, Ser. Khim.*, **1966**, 2223; *Chem. Abs.*, **66**, 75858g (1967); L. K. Brice, W. M. Chang, J. E. Smith, and S. M. Sullivan, *J. Phys. Chem.*, **71**, 2814 (1967); cf. *Organic Reaction Mechanisms*, **1966**, 94.

41 E. M. Kaiser and C. R. Hauser, *Tetrahedron Letters*, **1967**, 3341.

42 A. Streitwieser, E. Ciuffarin, and J. H. Hammons, *J. Am. Chem. Soc.*, **89**, 63 (1967); A. Streitwieser, J. H. Hammons, E. Ciuffarin, and J. I. Brauman, *ibid.*, p. 59; R. Khun and D. Rewicki, *Ann. Chem.*, **706**, 250 (1967).

43 F. G. Bordwell, R. H. Imes, and E. C. Steiner, *J. Am. Chem. Soc.*, **89**, 3905 (1967).

44 F. M. Dean, L. E. Houghton, and R. B. Morton, *J. Chem. Soc., C*, **1967**, 1980.

45 R. K. Howe and K. W. Ratts, *Tetrahedron Letters*, **1967**, 4743.

46 T. J. Batterham, D. J. Brown, and M. N. Paddon-Row, *J. Chem. Soc., B*, **1967**, 171.

47 H. J. M. Dou and J. Metzger, *Bull. Soc. Chim. France*, **1966**, 3273.

48 J. Jullien, H. Stahl-Larivière, and A. Trautmann, *Bull. Soc. Chim. France*, **1966**, 2398; R. Stewart and J. R. Jones, *J. Am. Chem. Soc.*, **89**, 5069 (1967); cf. E. C. Steiner and J. D. Starkey, *ibid.*, p. 2751.

49 K. Bowden and A. F. Cockerill, *Chem. Commun.*, **1967**, 989; V. A. Bessonov, E. A. Yakovleva, and A. I. Shatenshtein, *Zh. Obshch. Khim.*, **36**, 1362 (1966); *Chem. Abs.*, **66**, 10301v (1967); A. I. Shatenshtein, V. A. Bessonov, and E. A. Yakovleva, *Zh. Obshch. Khim.*, **36**, 2040 (1966); *Chem. Abs.*, **66**, 94430c (1966); V. A. Bessonov, E. A. Yakovleva, F. S. Yakushin, and A. I. Shatenshtein, *Zh. Obshch. Khim.*, **37**, 101 (1967); *Chem. Abs.*, **66**, 115040t (1967).

The reactions of the cyclo-octatetraenyl dianion[50] and the 2-thiopentalenyl anion[51] have been investigated.

Electrophilic substitution in organomercury compounds,[52] and the mechanism of electrophilic substitution reaction of metal alkyls,[53] have been reviewed.

Entropies of activation of A-S_E2 reactions have been discussed.[54]

The claim[55] that dibenzylmercury reacts with HCl by an S_E1 mechanism is incorrect. The reaction is an oxygen-promoted radical process.[56]

An S_E1 mechanism has been proposed for the reaction of *trans*-2-chlorovinylmercuric chloride with iodine in DMSO as the rate is independent of the concentration of the latter. This reaction and the reactions in carbon tetrachloride and benzene, thought to be S_E2 processes, proceed with net retention of configuration.[57, 58]

The rate law for the decomposition of 4- and 2-pyridiniomethylmercuric chloride in dilute aqueous hydrochloric acid has terms which are first- and second-order in chloride ion. A one and two chloride-ion-catalysed S_E1 mechanism (eqn. 6) was proposed. At high concentrations of Cl^- the rate law shows a term which is third-order in chloride, possibly owing to nucelophilic attack on mercury.[59]

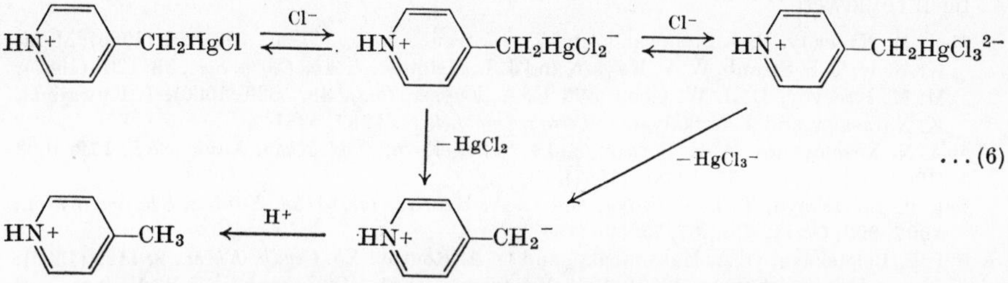

. . . (6)

[50] D. A. Bak and K. Conrow, *J. Org. Chem.*, **31**, 3958 (1966); T. S. Cantrell and H. Shechter, *J. Am. Chem. Soc.*, **89**, 5868, 5877 (1967).

[51] T. S. Cantrell and B. L. Harrison, *Tetrahedron Letters*, **1967**, 4477.

[52] O. A. Reutov, *Fortshr. Chem. Forsch.*, **8**, 61 (1967); *Russian Chem. Rev.*, **36**, 163 (1967); *Usp. Khim.*, **36**, 414 (1967); *Rev. Roum. Chim.*, **12**, 313 (1967); *Chem. Abs.*, **67**, 90174v (1967); *Omagiu. Raluca. Ripan*, **1966**, 481; *Chem. Abs.*, **67**, 53253z (1967).

[53] M. H. Abraham and J. A. Hill, *J. Organometal. Chem.*, **7**, 11 (1967).

[54] M. A. Matesich, *J. Org. Chem.*, **32**, 1258 (1967).

[55] I. P. Beletskaya, L. A. Fedorov, and O. A. Reutov, *Dokl. Akad. Nauk SSSR*, **163**, 1381 (1967).

[56] B. F. Hegarty, W. Kitching, and P. R. Wells, *J. Am. Chem. Soc.*, **89**, 4816 (1967).

[57] I. P. Beletskaya, V. I. Karpov, and O. A. Reutov, *Izv. Akad. Nauk SSSR, Ser. Khim.*, **1966**, 1135; *Chem. Abs.*, **65**, 16808 (1966).

[58] I. P. Beletskaya, V. I. Karpov, and O. A. Reutov, *Izv. Akad. Nauk SSSR, Ser. Khim.*, **1966**, 963.

[59] J. R. Coad and M. D. Johnson, *J. Chem. Soc., B*, **1967**, 633.

Electrophilic substitution reactions of allyl-,[60] vinyl-,[61] benzyl-,[62] and alkyl-mercury compounds[63] have also been studied.

The halogenodemetallation of alkyltin compounds by S_E and radical mechanisms has been investigated.[64a] The dependence on solvent of the rates of iododemetallation of tetra-alkyl-tin and -lead compounds are similar and so the mechanisms are probably also similar. In acetone and acetonitrile the rate is proportional to the concentration of I^- and it was suggested that nucleophilic attack on the metal as well as electrophilic attack on the alkyl group is important.[64b]

The reactions of *cis*- and *trans*-1-(trimethylstannyl)-2-methylcyclopropane with iodine in methanol and acetic acid and with bromine in acetic acid and chlorobenzene proceed with retention of configuration.[65]

The NMR spectra of methylmercury compounds in pyridine show no evidence for methyl-group exchange.[66] Methyl-group exchange of dimethylcadmium has also been investigated.[67]

The following other reactions have also been investigated: reaction of diethylzinc with phenylmercuric chloride,[68] hydrostannolysis of tin–nitrogen and tin–oxygen bonds;[69] *trans*-alkylation of alkylgermanium derivatives;[70] protolysis of dialkylzincs,[71] tetraethyl-lead,[72] and tri-*n*-octylaluminium.[73]

The composition and mechanisms of reactions of Grignard reagents have been reviewed.[74a]

[60] M. M. Kreevoy, P. J. Steinwand, and T. S. Straub, *J. Org. Chem.*, **31**, 4291 (1966); M. M. Kreevoy, T. S. Straub, W. V. Kayser, and J. L. Melquist, *J. Am. Chem. Soc.*, **89**, 1201 (1967); M. M. Kreevoy, D. J. W. Goon, and R. A. Kayser, *ibid.*, **88**, 5529 (1966); I. Kuwajima, K. Narasaka, and T. Mukaiyama, *Tetrahedron Letters*, **1967**, 4281.

[61] A. N. Nesmeyanov, A. E. Borisov, and I. S. Savel'eva, *Dokl. Akad. Nauk SSSR*, **172**, 1093 (1967); *Chem. Abs.*, **67**, 10887s (1967).

[62] I. P. Beletskaya, T. P. Fetisova, and O. A. Reutov, *Izv. Akad. Nauk SSSR, Ser. Khim.*, **1967**, 990; *Chem. Abs.*, **67**, 73059u (1967).

[63] I. P. Beletskaya, O. A. Maksimenko, and O. A. Reutov, *Zh. Organ. Khim.*, **2**, 1129 (1966); *Chem. Abs.*, **66**, 45928n (1967); I. P. Beletskaya, O. A. Maksimenko, V. B. Vol'eva, and O. A. Reutov, *Zh. Organ. Khim.*, **2**, 1132 (1966); *Chem. Abs.*, **66**, 94611n (1967); O. A. Maksimenko, I. P. Beletskaya, and O. A. Reutov, *Zh. Organ. Khim.*, **2**, 1137 (1966); *Izv. Akad. Nauk SSSR, Ser. Khim.*, **1966**, 662.

[64a] S. Boué, M. Gielen, and J. Nasielski, *J. Organometal. Chem.*, **9**, 443, 461, 481 (1967).

[64b] M. Gielen and J. Nasielski, *J. Organometal. Chem.*, **7**, 273 (1967).

[65] P. Baekelmans, M. Gielen, and J. Nasielski, *Tetrahedron Letters*, **1967**, 1149; see also K. Sisido, S. Kozima, and K. Takizawa, *ibid.*, p. 33.

[66] D. N. Ford, P. R. Wells, and P. C. Lauterbur, *Chem. Commun.*, **1967**, 616.

[67] N. S. Ham, E. A. Jeffery, T. Mole, J. K. Saunders, and S. N. Stuart, *J. Organometal. Chem.*, **8**, P7 (1967).

[68] M. H. Abraham and P. H. Rolfe, *J. Organometal. Chem.*, **8**, 395 (1967).

[69] H. M. J. C. Creemers, F. Verbeek, and J. G. Noltes, *J. Organometal. Chem.*, **8**, 469 (1967).

[70] F. Rijkens, E. J. Bulten, W. Drenth, and G. J. M. van der Kerk, *Rec. Trav. Chim.*, **85**, 1223 (1966).

[71] M. H. Abraham and J. A. Hill, *J. Organometal. Chem.*, **7**, 23 (1967).

[72] H. Horn and F. Huber, *Monatsh. Chem.*, **98**, 771 (1967).

[73] E. Kohn and J. M. Gill, *J. Organometal. Chem.*, **7**, 359 (1967).

[74a] E. C. Ashby, *Quart. Rev. (London)*, **21**, 259 (1967).

The NMR spectra of the Grignard reagents prepared from *exo*- and *endo*-5-chloronorborn-2-ene and from 3-chloronortricyclene in diethyl ether are identical and show no signals characteristic of vinyl-protons. A nortricyclyl structure is therefore most likely.[74b]

The kinetics of the reaction of butylmagnesium halides with methyl acetate,[75] of *o*-methoxyarylmagnesium bromides with di-*tert*-butyl ketone,[76] and of ethylmagnesium bromide with 2-ethyl-1,3-dioxolane,[77] have been studied.

The reaction of Grignard reagents with epoxynitriles[78] and with propargyl chloride,[79] and the hydrolysis of but-2-enyl-magnesium, -zinc, and -cadmium bromides[80] have been investigated.

An *X*-ray structure determination of $[EtMgBr,Et_3N]_2$ has been reported.[81]

It has been proposed that the predominant process in reactions of organolithium reagents is charge transfer from the associated reagent to the substrate.[82]

Protonation of pentadienyl-lithiums,[83] lithium-exchange reactions,[84] and the structure of 9-ethyl-9,10-dihydro-9-anthryl-lithium[85] have also been investigated.

[74b] D. O. Cowan, N. G. Krieghoff, J. E. Nordlander, and J. D. Roberts, *J. Org. Chem.*, **32**, 2639 (1967).
[75] T. Holm, *Acta Chem. Scand.*, **20**, 2821 (1966).
[76] M. Oki, *Tetrahedron Letters*, **1967**, 1785.
[77] P. Vink, C. Blomberg, A. D. Vreugdenhil, and F. Bickelhaupt, *Tetrahedron Letters*, **1966**, 6419.
[78] J. Cantacuzène, D. Ricard, and M. Thézé, *Tetrahedron Letters*, **1967**, 1365.
[79] T. L. Jacobs and P. Prempree, *J. Am. Chem. Soc.*, **89**, 6177 (1967).
[80] C. Agami, M. Andrac-Taussig, and C. Prévost, *Bull. Soc. Chim. France*, **1966**, 2596.
[81] J. Toney and G. D. Stucky, *Chem. Commun.*, **1967**, 1168.
[82] C. G. Screttas and J. F. Eastham, *J. Am. Chem. Soc.*, **88**, 5668 (1966).
[83] R. B. Bates, D. W. Gosselink, and J. A. Kaczynski, *Tetrahedron Letters*, **1967**, 199, 205.
[84] L. M. Seitz and T. L. Brown, *J. Am. Chem. Soc.*, **89**, 1602, 1607 (1967).
[85] D. Nicholls and M. Szwarc, *J. Am. Chem. Soc.*, **88**, 5757 (1966).

CHAPTER 4

Elimination Reactions[1]

Sicher and his co-workers have continued their work on eliminations from medium- and large-ring compounds and have published details of the work reported last year[2] which led to the suggestion that the *E*2 eliminations of cycloalkyltrimethylammonium ions with potassium *tert*-butoxide in *tert*-butyl alcohol yielding *cis*-olefins proceed by an *anti*-mechanism and that those yielding *trans*-olefins proceed by a *syn*-mechanism.[3] This work has also been extended to other base–solvent systems.[4] With potassium methoxide in methanol and potassium ethoxide in ethanol the plots of the rates of formation of *trans*-olefins against ring size are also characteristic of a *syn*-mechanism,[2] but the corresponding plots for the formation of *cis*-olefins are slightly different from those found with *tert*-butoxide in *tert*-butyl alcohol.[2] Nevertheless, the latter were still considered to indicate an *anti*-mechanism, with the differences possibly resulting from shifts in the relative timings of the bond changes. The proportion of *trans*-olefin formed from all the cycloalkylammonium salts studied, and from some cycloalkylsulphonium salts as well, increased as the solvent–base system was varied in the series $HOCH_2CH_2OH$–$HOCH_2CH_2OK$, MeOH–MeOK, EtOH–EtOK, Pr^iOH–Pr^iOK, Bu^tOH–Bu^tOK, suggesting that this change favours the *syn*-mechanism. It was considered that it would also result in an increase in the proton-affinity of the base and a decrease in the solvating power of the solvent, and so enhance the *E*1c*B* character of the transition state.[5] These results are therefore in accord with the view that the transition states of *E*2 *syn*-eliminations have *E*1c*B*-like character.[6,7]

A *syn*-mechanism was also demonstrated for the formation of *trans*-olefin from the acyclic quaternary ammonium salts (**1**) (*threo*) and (**2**) (*erythro*) on

1 The following topics have been reviewed: (*a*) eliminations in cyclic *cis-trans*-isomers, W. Hückel and M. Hanack, *Angew. Chem. Intern. Ed. Engl.*, **6**, 534 (1967); (*b*) β-eliminations from β-substituted propionic acids, P. F. Butskus and G. I. Denis, *Usp. Khim.*, **35**, 1999 (1966); *Russian Chem. Rev.*, **35**, 839 (1967); (*c*) gas-phase elimination reactions, A. Maccoll and P. J. Thomas, *Progr. Reaction Kinetics*, **4**, 119 (1967); (*d*) the Hofmann rule, L. D. Freedman, *J. Chem. Educ.*, **43**, 662 (1966); (*e*) carbanion mechanism of olefin-forming eliminations, D. J. McLennan, *Quart. Rev.* (*London*), **21**, 490 (1967).

2 See *Organic Reaction Mechanisms*, **1965**, 103—105.

3 J. Sicher and J. Závada, *Coll. Czech. Chem. Commun.*, **32**, 2122 (1967).

4 J. Závada and J. Sicher, *Coll. Czech. Chem. Commun.*, **32**, 3701 (1967).

5 See, however, reference 19.

6 C. K. Ingold, *Proc. Chem. Soc.*, **1962**, 265.

7 See *Organic Reaction Mechanisms*, **1965**, 91.

reaction with MeOH–MeOK, Bu^tOH–Bu^tOK and Bu^tOK–Me_2SO.[8] The *threo*-isomer (**1**) would yield *trans*-olefin by a *syn*-mechanism with loss of deuterium and by an *anti*-mechanism with retention of deuterium (see eqn. 1),

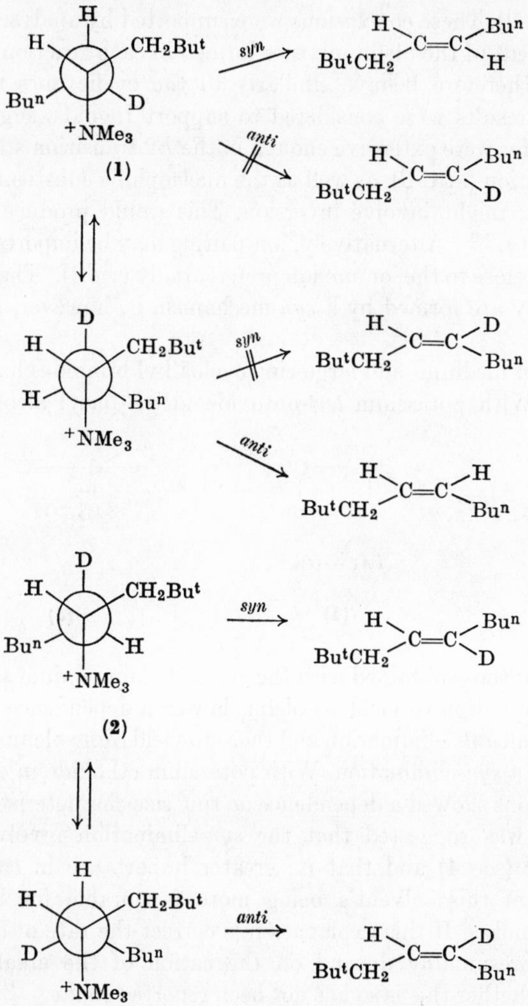

...(1)

...(2)

and consistently with the former it was found that formation of *trans*-olefin showed an isotope effect k_H/k_D in the region 2.3—4.2. Similarly the formation of *cis*-olefin by a *syn*-mechanism would result in retention of deuterium and by an *anti*-mechanism would result in loss of deuterium, and the observation

[8] M. Pánková, J. Sicher, and J. Závada, *Chem. Commun.*, **1967**, 394.

of an isotope effect here ($k_H/k_D = 3.1$—4.7) also favours the latter. Consistently with these results, formation of both *cis*- and *trans*-olefins from the *erythro*-isomer showed virtually no isotope effect ($k_H/k_D = 0.9$—1.2), and hence both were probably formed without loss of deuterium by *anti*- and *syn*-mechanisms, respectively (see eqn. 2). These conclusions were supported by analysis of the total deuterium content of the olefin mixture formed in each reaction. These acyclic compounds therefore behave similarly to the cyclic ones studied previously, and the results were considered to support Ingold's suggestion that "if proton transfer were extensive enough in the *E*2 transition state, the electrophilic substitution (at C-β), as well as the nucleophilic substitution (at C-α) coupled with it, might involve inversion. This would produce a *syn*-planar stereospecificity."[6,8] Alternatively, ion-pairing may be important and the anion may attack close to the 'onium ion preferentially (see **3**).[4] The reason why *trans*-olefins only are formed by a *syn*-mechanism is, however, not yet clear.

The *E*2 reactions of medium- and large-ring cycloalkyl bromides have also been investigated.[9] With potassium *tert*-butoxide in *tert*-butyl alcohol the

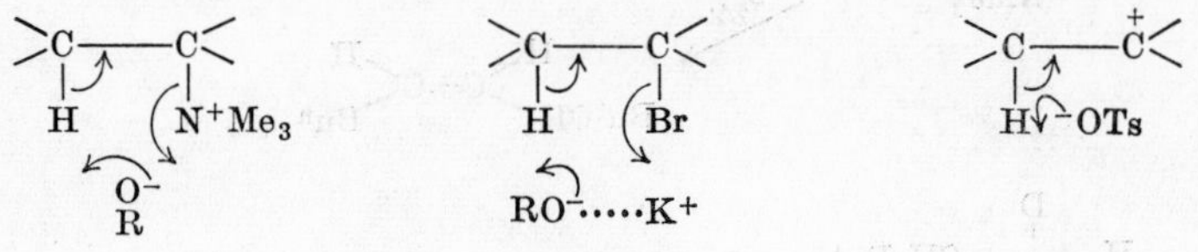

(3) **(4)** **(5)**

results were similar to those obtained with the cycloalkylammonium salts;[2,3] thus the rates of elimination to yield *cis*-olefin showed a dependence on ring size characteristic of an *anti*-elimination and those to yield *trans*-olefin showed one characteristic of a *syn*-elimination. With potassium ethoxide in ethanol, however, both reactions showed a dependence on ring size characteristic of an *anti*-mechanism. It was suggested that the *syn*-elimination involved the $RO^- \ldots K^+$ ion pair (see **4**) and that its greater importance in *tert*-butyl alcohol resulted from this solvent's being more favourable for ion-pair formation than ethanol is. If this explanation is correct the rate of the *syn*-elimination would presumably depend on the cation of the alkali-metal alkoxide used, but whether this is so has not been reported so far.

Heating the conformationally stabilized 4,4,7,7-tetramethylcyclodecyl toluene-*p*-sulphonate in acetic acid, dimethylformamide, pyridine, dimethyl sulphoxide, or collidine yields the mixture of olefins shown in reaction (3), the *trans*-olefin (**7**) always predominating (*ca.* 75%). Experiments with specifically deuterated toluene-*p*-sulphonate showed that at least 95% of

[9] J. Závada, J. Krupička, and J. Sicher, *Chem. Commun.*, **1967**, 66.

this olefin was formed by a *syn*-elimination. The *cis*- (**8**) and *trans*-olefin (**9**) were also formed predominantly by *syn*-eliminations, but too little of the *cis*-olefin (**6**) was formed to allow the steric course of its formation to be determined. Similar behaviour was observed with the isomeric toluene-*p*-sulphonate

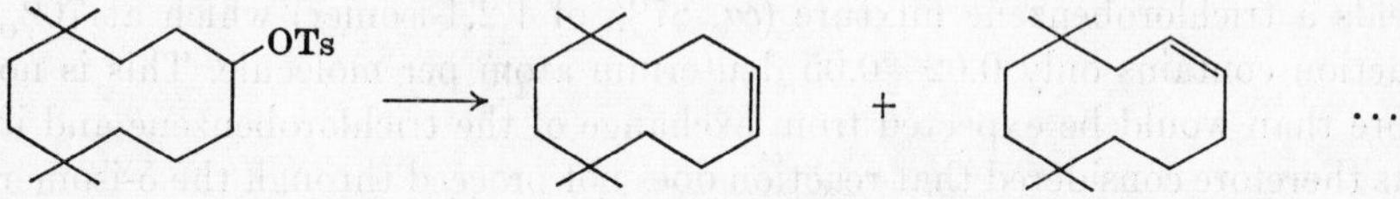

...(3)

cis (**6**) *cis* (**8**)
trans (**7**) *trans* (**9**)

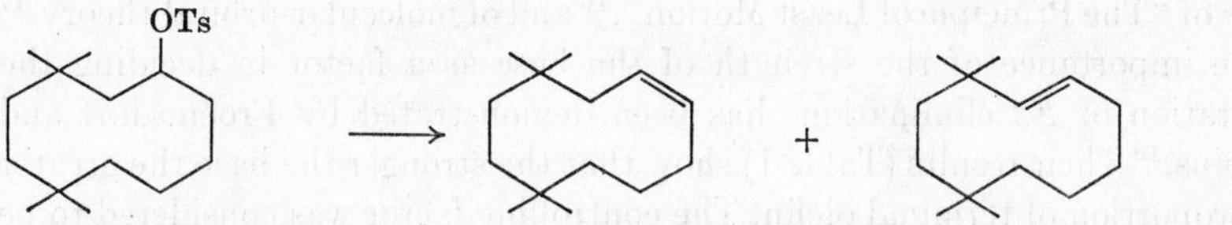

...(4)

of reaction (4), which occurred predominantly by *syn*-elimination. The formation of both the *cis*- and the *trans*-olefin by a *syn*-mechanism contrasts with the reactions just discussed. Possibly ion pairs are involved, the toluene-*p*-sulphonate counter-ion removing the β-proton (see **5**).[10]

Some *syn*-elimination occurs in the Hofmann degradation of the menthyltrimethylammonium ion (**10**) which yields olefin containing 0—27% of

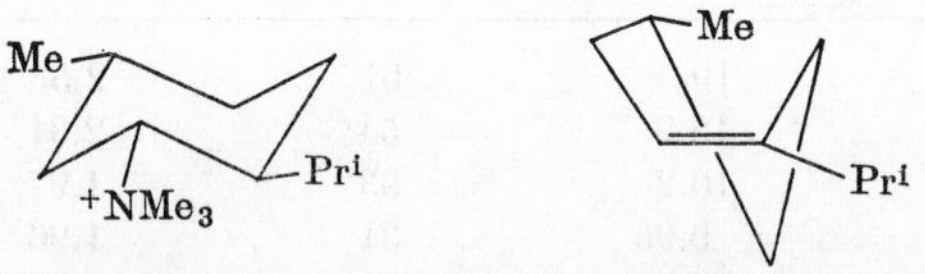

(**10**) (**11**)

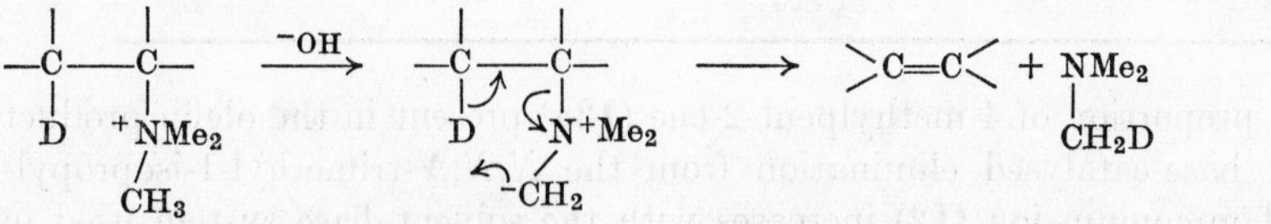

...(5)

menth-3-ene (**11**), the amount of *syn*-elimination depending on the conditions. It was shown by deuterium-labelling experiments (reactions 5) that 8—15% of the menth-3-ene results from an ylide mechanism, and the balance was thought to be formed by a concerted *E*1*cB*-like *syn*-*E*2 mechanism.[11]

[10] M. Svoboda, J. Závada, and J. Sicher, *Coll. Czech. Chem. Commun.*, **32**, 2104 (1967).
[11] M. A. Baldwin, D. V. Banthorpe, A. G. Loudon, and F. D. Waller, *J. Chem. Soc.*, *B*, **1967**, 509.

The Hofmann elimination of *N*,*N*,*N*-trimethyl-3-*exo*-deuterio-2-*exo*-norbornylammonium hydroxide also proceeds by a *syn*-mechanism, to yield deuterium-free norbornene.[12] This reaction thus resembles the *E*2 reaction of 2-*exo*-norbornyl bromide.[13, 14]

The dehydrochlorination of β-hexachlorocyclohexane by NaOMe in MeOD yields a trichlorobenzene mixture (*ca.* 87% of 1,2,4-isomer) which at 70% reaction contains only 0.02—0.05 deuterium atom per molecule. This is no more than would be expected from exchange of the trichlorobenzene and it was therefore considered that reaction does not proceed through the δ-isomer as suggested previously. Instead a *syn*-elimination, possibly proceeding via a boat transition state was suggested.[15]

The stereochemical course of elimination reactions has been discussed in terms of "The Principle of Least Motion",[16] and of molecular-orbital theory.[17]

The importance of the strength of the base as a factor in deciding the orientation of *E*2 eliminations has been demonstrated by Froemsdorf and Robbins.[18] Their results (Table 1) show that the stronger the base the greater the proportion of terminal olefin. The controlling factor was considered to be the greater C–N stretching in the reactions with the stronger bases, which makes hydrogen acidity a more important factor in determining the relative rate of elimination into each branch (see refs. 25*a* and 25*b*, however).

Table 1. Olefin distributions in the *E*2 reactions of *sec*-butyl toluene-*p*-sulphonate at 55° in dimethyl sulphoxide solution.

Base	pK_a in water	1-ene (%)	2-ene *trans*/*cis*
$KOBu^t$	19	61	2.53
KOEt	15.9	54	2.34
KOC_6H_4OMe-*p*	10.2	33	1.97
KOC_6H_5	9.95	31	1.96
$KOC_6H_4NO_2$-*p*	7.14	16	1.81
$KOC_6H_4NO_2$-*o*	7.23	16	1.85

The proportion of 4-methylpent-2-ene (**13a**) present in the olefin product of the base-catalysed elimination from the *N*,*N*,*N*-trimethyl-1-isopropylpropylammonium ion (**12**) increases with the solvent–base system used in

[12] J. L. Coke and M. P. Cooke, *J. Am. Chem. Soc.*, **89**, 2779 (1967).
[13] H. Kwart, T. Takeshita, and J. L. Nyce, *J. Am. Chem. Soc.*, **86**, 2606 (1964).
[14] See *Organic Reaction Mechanisms*, **1966**, 105.
[15] J. Hine, R. D. Weimar, P. B. Langford, and O. B. Ramsay, *J. Am. Chem. Soc.*, **88**, 5522 (1966).
[16] J. Hine, *J. Am. Chem. Soc.*, **88**, 5525 (1966).
[17] W. T. Dixon, *Chem. Ind. (London)*, **1967**, 789; see also *Tetrahedron Letters*, **1967**, 2531.
[18] D. H. Froemsdorf and D. M. Robbins, *J. Am. Chem. Soc.*, **89**, 1737 (1967).

the order ethanol–ethoxide, isopropyl alcohol–isopropoxide, *tert*-butyl alcohol–*tert*-butoxide. It was thought[19] that a steric effect, and not greater carbanionic character of the transition state, was probably responsible for the

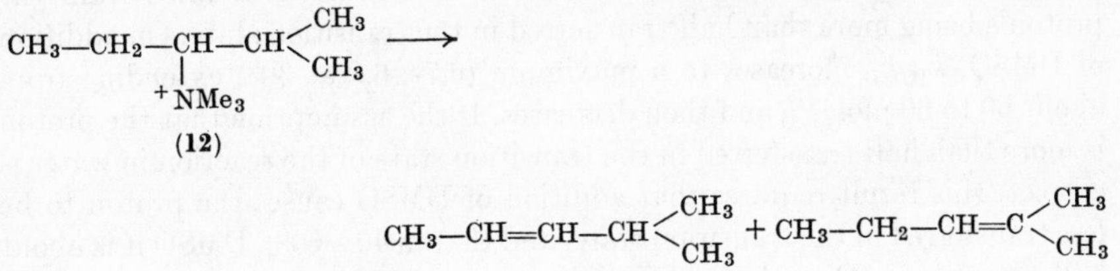

(12) (13a) (13b)

low proportion of 2-methylpent-2-ene (**13b**) formed with *tert*-butoxide, since unpublished deuterium isotope and substituent effects with other eliminations indicated less carbanionic character with *tert*-butyl alcohol–*tert*-butoxide than with ethanol–ethoxide.

Last year an elegant demonstration of the concerted nature of $E2$ reactions was reported by Willi who showed that an increase in the leaving group tendency of the arenesulphonyloxy-group of a series of 2,2-diphenylethyl arenesulphonates was accompanied by an increase in the β-deuterium isotope effect.[20] This year a complementary investigation has been reported by Fraser and Hoffmann[21] who showed that the ratio k_{OTs}/k_{Br} for the $E2$ reaction of a series of substituted phenethyl compounds, $ArCH_2CH_2X$ (X = Br or OTs) increased with the acidity of the β-hydrogen atoms. Thus when $Ar = p\text{-}MeO \cdot C_6H_4$, $k_{OTs}/k_{Br} = 0.15$; but when $Ar = p\text{-}NO_2 \cdot C_6H_4$, $k_{OTs}/k_{Br} = 1.57$. The ratio k_{OTs}/k_{Br}, which was considered to be a measure of the degree of C–X bond-breaking in the transition state, therefore increases with the degree of C_β–H bond-breaking.[21]

A detailed investigation of the effect of dimethyl sulphoxide (DMSO)[22] on the activation parameters, ρ-constant, primary deuterium isotope effect, and primary sulphur isotope effect of the elimination reaction of *S,S*-dimethyl-phenethylsulphonium bromide with hydroxide ion in aqueous DMSO has been reported.[23, 24] On changing the solvent from water to 85 moles % of DMSO the rate increases by 10^7 as a result of a decrease in $\Delta H^\ddagger$ which is only partly compensated for by a decrease in $\Delta S^\ddagger$. On going from water to 19.4

19 I. N. Feit and W. H. Saunders, *Chem. Commun.*, **1967**, 610.
20 See *Organic Reaction Mechanisms*, **1966**, 708.
21 G. M. Fraser and H. M. R. Hoffmann, *J. Chem. Soc., B*, **1967**, 265.
22 For a review of the solvent properties of dimethyl sulphoxide, see D. Martin, A. Weise, and H.-J. Niclas, *Angew. Chem. Intern. Ed. Engl.*, **6**, 318 (1967).
23 A. F. Cockerill, *J. Chem. Soc., B*, **1967**, 964.
24 A. F. Cockerill and W. H. Saunders, *J. Am. Chem. Soc.*, **89**, 4985 (1967).

mole % of DMSO the ρ-constant increases from 2.110 to 2.536, but further addition of DMSO causes only a slight additional increase up to 60.3 mole %, the highest concentration used. The value of k_H/k_D in water, 5.92 at 30°, is less than the predicted maximum and was considered to arise from the proton's being more than half transferred in the transition state. On addition of DMSO, k_H/k_D increases to a maximum (6.7—6.8 at 30°) extending from about 50 to 60 mole % and then decreases. If the assumption that the proton is more than half transferred in the transition state of the reaction in water is correct, this result requires that addition of DMSO causes the proton to be less transferred in the transition state, and that in 40—60% DMSO it is about half transferred. Dimethyl sulphoxide presumably increases the basicity of the ^-OH ion and it was thought reasonable that this should reduce the degree of proton transfer in the transition state "since the stronger base can exert from a greater distance enough pull on the proton to remove it"[25a] (see also ref. 25*b*). The sulphur isotope effect, k_{32}/k_{34}, is 1.0074 in water at 40° and decreases to 1.0011 in 7 mole % DMSO, which was interpreted as indicating a decrease in the amount of C–S bond-stretching in the transition state. The overall effect of adding DMSO is therefore that both C–H and C–S bond-stretching are reduced in the transition state, which becomes more "reactant like".

In contrast to these results there is little variation in k_H/k_D and ρ for the reaction of the phenethyl bromide with potassium *tert*-butoxide in *tert*-butyl alcohol–DMSO mixtures on increasing the DMSO concentration, despite marked enhancement in rate. It was therefore considered that the structure of the transition state remains essentially constant.[25a]

The large secondary isotope effects for the reactions of DO^- in D_2O *versus* HO^- in H_2O for the elimination reactions of $PhCH_2CH_2N^+Me_3$, $PhCH_2CH_2S^+Me_2$, $PhCH_2CH_2N^+Me_2Ph$, and p-$ClC_6H_4CH_2CH_2N^+Me_3$ ($k_{DO^-}/k_{HO^-} = 1.79$, 1.57, 1.62, and 1.73, respectively) suggest transition states in which there is extensive proton transfer to the base.[25b]

The primary β- (4.5 EtOH; 7.5 *tert*-BuOH) and the secondary α- (1.14 EtOH; 1.15 *tert*-BuOH) deuterium isotope effects for the base-catalysed elimination reaction of cyclohexyl toluene-*p*-sulphonate are normal, but the secondary β-effects are rather large (1.33 EtOH; 1.52 *tert*-BuOH). Possible contributing factors are changes in hybridization and the greater electron-repelling character of deuterium than of hydrogen.[26]

Calculations of primary hydrogen isotope effects have been reported.[27]

The relative effectiveness of thioethoxide and methoxide ions in promoting

[25a] A. F. Cockerill, S. Rottschaefer, and W. H. Saunders, *J. Am. Chem. Soc.*, **89**, 901 (1967).
[25b] L. J. Steffa and E. R. Thornton, *J. Am. Chem. Soc.*, **89**, 6149 (1967).
[26] K. T. Finley and W. H. Saunders, *J. Am. Chem. Soc.*, **89**, 898 (1967).
[27] R. A. More O'Ferrall and J. Kouba, *J. Chem. Soc., B*, **1967**, 985.

elimination from compounds $PhCH_2CMe_2X$ depends on X. When X is Cl, a good leaving group, k_{EtS^-}/k_{MeO^-} is high (6.5) but it is low when X is $MeSO_2$, a poor one, and has an intermediate value (0.8) when X is SMe_2, a moderately good one. These results are explained readily in terms of Hudson's theory of nucleophilicity. With the poor leaving group the partial bond between the base and the eliminating hydrogen in the transition state is strong, and the energy released on its formation, which will be greater with MeO^- than with EtS^-, is an important factor and hence MeO^- is more effective. When X is a good leaving group this partial bond is weaker and the energy released on its formation is much less important, and the other factors which determine nucleophilic reactivity are relatively more important. These are the energy required to desolvate the nucleophile and partially to remove an electron from an outer shell, both of which favour EtS^- over MeO^-. There appears to be no need then to invoke a merged mechanism to explain these and similar results.[28a] The effect of substituents on the rate of elimination of α,α-dimethylphenethyl chloride have been determined.[28b]

A merged mechanism has been proposed to explain the observations that the rate ratios of the *E*2 reactions, k_{OTs}/k_{Br} (*E*2), of ethyl, *n*-propyl, and isobutyl toluene-*p*-sulphonate and bromide with *tert*-butoxide ion in *tert*-butyl alcohol and with ethoxide ion in ethanol (cations unspecified) are very low (0.01—0.2), but those for the concurrent S_N2 reactions, k_{OTs}/k_{Br} (S_N2) are high, and a rise in the latter is accompanied by a fall in the former. It was suggested[29a] that the reactions proceed through a common species (**14**) which was considered to be, not an intermediate, but "one array of reactants or a plateau on the way to products"![29a]

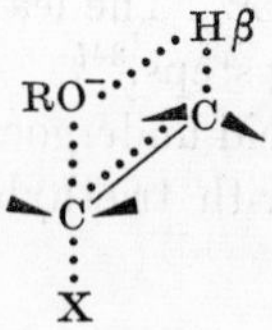

(**14**)

The rates of the piperidine-catalysed elimination from compounds (**15a**) and (**15b**) are similar, and the rate for compound (**15a**) is changed only slightly

(**15a**)

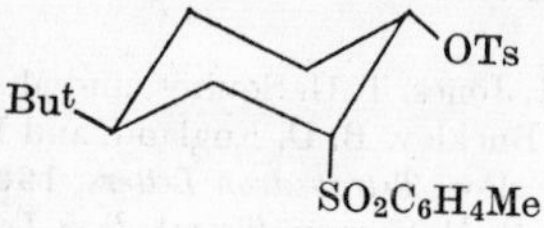

(**15b**)

[28a] J. F. Bunnett and E. Bacciochi, *J. Org. Chem.*, **32**, 11 (1967).
[28b] L. F. Blackwell, A. Fischer, and J. Vaughan, *J. Chem. Soc., B*, **1967**, 1084.
[29a] G. M. Fraser and H. M. R. Hoffmann, *J. Chem. Soc., B*, **1967**, 425.

when the toluene-*p*-sulphonyloxy leaving group is changed to methanesulphonyloxy. Heterolysis of the carbon–oxygen bond can therefore only play a minor part in the rate-determining step. However, little if any deuterium exchange occurs when the reactions are carried out in a deuterium-containing solvent, and there is no epimerization at the sulphone carbon. According to the authors,[29b] it is not possible to conclude that a carbanion is formed as a discrete intermediate.

The variation in the rate of the base-promoted elimination reactions of phenethyl bromide and ethylene bromide with solvent composition in methanol– and ethanol–water mixtures have been measured in an attempt to find if this could be used to distinguish between the *E*2 and *E*1c*B* mechanisms.[30] The validity of the conclusion that, if exchange between the β-hydrogen of a compound $R_2CH \cdot CR'_2X$ and the solvent occurs concurrently with an elimination, then the mechanism of the latter must be *E*1c*B* has been discussed and Breslow's criticism[31] of this criterion has been challenged.[32]

The iodide-promoted 1,4-elimination of 1,4-di-iodobut-2-ene is about 4 times slower than the analogous 1,2-elimination from 1,2-di-iodoethane.[33]

When aqueous solutions of their salts are heated *erythro*- and *threo*-2,3-dibromo-2,3-dimethylsuccinic acid undergo concurrent dehydrohalogenation and dehalogenative decarboxylation. The former is stereospecific and thought to proceed by an *E*2 mechanism; the latter is non-stereospecific and thought to proceed by an *E*1 mechanism.[34a]

Further examples have been reported of unimolecular reactions for which the proportion of elimination depends on the leaving group. Thus the solvolysis of α,α-dimethylbenzyl chloride in ethanol yields 11% of olefin, but the corresponding thionbenzoate yields 85%. The leaving group must take part in at least one of the product-forming steps.[34b]

2-*tert*-Butylperoxypropionic acid undergoes the elimination reaction shown in equation (6) on treatment with triethylamine in chlorobenzene at 25°.

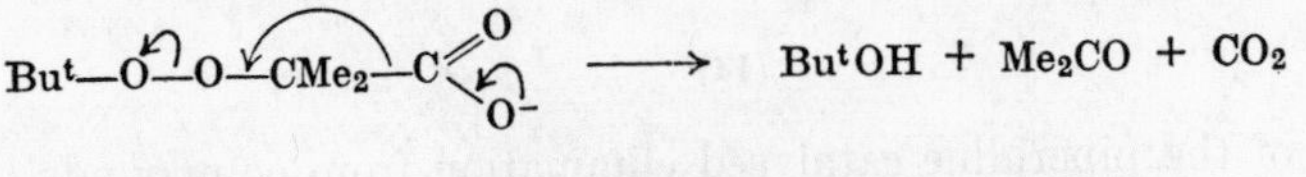

····(6)

Since the spontaneous rupture of the peroxide bond would be very slow under these conditions it was thought that the whole process was concerted.[35]

[29b] W. M. Jones, T. G. Squires, and M. Lynn, *J. Am. Chem. Soc.*, **89**, 318 (1967).
[30] P. D. Buckley, B. D. England, and D. J. McLennan, *J. Chem. Soc., B*, **1967**, 98.
[31] R. Breslow, *Tetrahedron Letters*, **1964**, 399.
[32] H. M. R. Hoffmann, *Tetrahedron Letters*, **1967**, 4393.
[33] R. F. Hudson, J. Arendt, and A. Mancuso, *J. Chem. Soc., B*, **1967**, 1069.
[34a] C. Rappe and K. Andersson, *Acta Chem. Scand.*, **21**, 1741 (1967).
[34b] R. L. Buckson and S. G. Smith, *J. Org. Chem.*, **32**, 634 (1967).
[35] W. H. Richardson and R. S. Smith, *J. Am. Chem. Soc.*, **89**, 2230 (1967).

In the presence of potassium hydroxide, 9-fluorenyl toluene-*p*-sulphonate in methanol–acetone (1:9, v/v) undergoes a bimolecular elimination, designated $E_{CO}2$, to yield 9-fluorenone and potassium toluene-*p*-sulphinate. The proportion of this reaction decreases if the proportion of methanol in the solvent is increased with concomitant increase in the S_N2 reaction.[36]

A detailed investigation of the elimination reaction across the C–N bond of one of the diastereoisomeric 2,3-bis-(*N*,*N*-difluoroamino)butanes (**16**) in aqueous diglyme has been reported. The reaction involves the elimination of

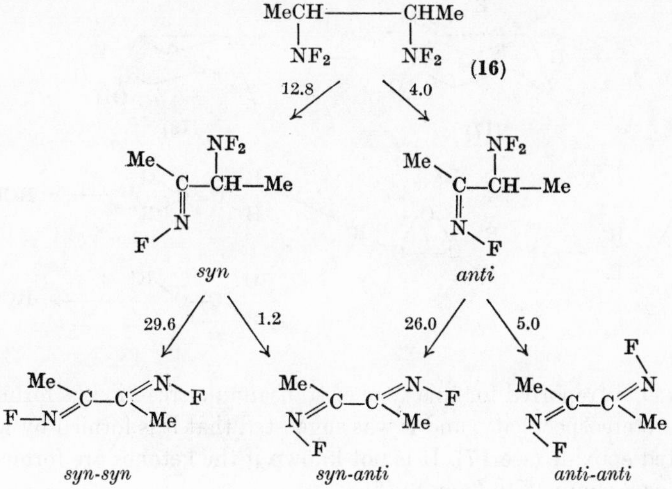

(**16**)

2 moles of HF and it was possible to isolate the separate steps, and to determine the configuration of both the mono- and the di-dehydrofluorinated products. At 50° the rate constants for the individual steps had the values shown in Scheme 1 (10^5k in sec^{-1}).[37]

The deamination of *cis*- and *trans*-2,3-dimethylaziridine by difluoroamine proceeds stereospecifically to yield *cis*- and *trans*-but-2-ene, respectively. It is claimed that orbital-symmetry theory predicts a non-stereospecific reaction and that therefore the theory as now constituted is not applicable to three-membered rings.[38, 39]

The products of the reactions of epoxides with strongly basic but poorly nucleophilic bases (e.g., $LiNEt_2$) have been rationalized as resulting from competing α- and β-eliminations. With acyclic and cyclic epoxides which can easily achieve the correct conformation for the *anti*-β-elimination the product

[36] G. W. Cowell, A. Ledwith, and D. G. Morris, *J. Chem. Soc., B,* **1967**, 697.
[37] S. K. Brauman and M. E. Hill, *J. Am. Chem. Soc.*, **89**, 2131 (1967).
[38] J. P. Freeman and W. H. Graham, *J. Am. Chem. Soc.*, **89**, 1761 (1967).
[39] See *Organic Reaction Mechanisms*, **1966**, 71.

is an allylic alcohol (e.g., eqn. 7). With medium-ring epoxides it is more difficult to achieve this conformation, the overall reaction is slower, and the products frequently include compounds formed by α-elimination. Thus cycloheptene oxide yield cycloheptadiene, cycloheptanone, cyclohept-2-enol, and *cis*-bicyclo[4.1.0]heptan-*endo*-2-ol (**18**). The last of these, like every other

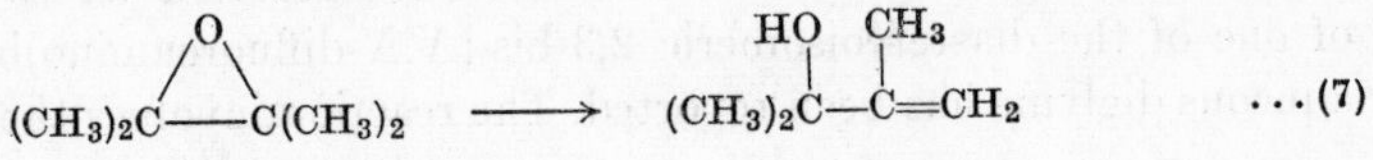

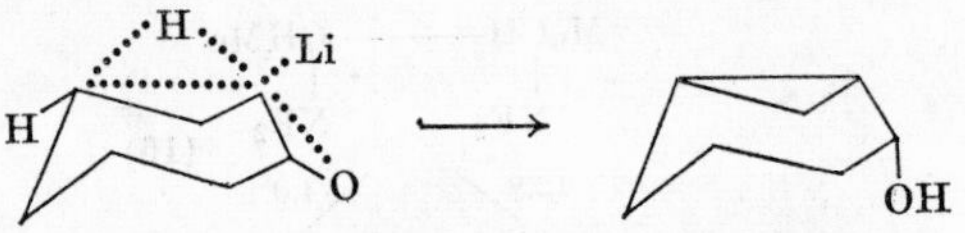

(17) (18)

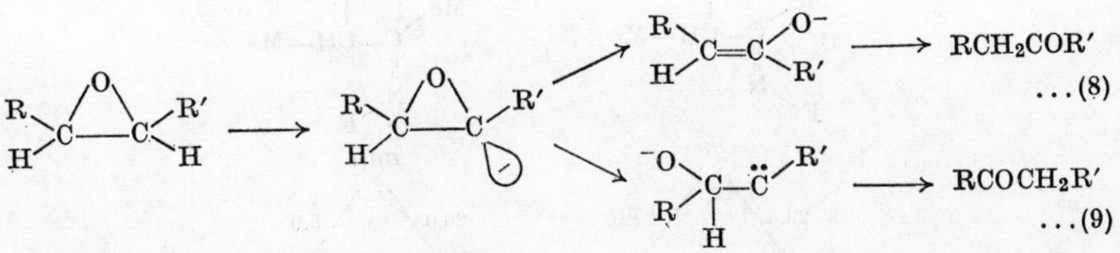

product so far reported for this type of transannular reaction, is formed with complete stereospecificity, and it was suggested that it is formed by way of a metallated epoxide (see **17**). It is not known if the ketones are formed by α- (eqn. 9) or β-elimination (eqn. 8).[40]

Epoxides of acyclic olefins which lack a β-hydrogen atom may also undergo α-elimination. Thus *trans*-di-*tert*-butylethylene oxide yields the diastereoisomeric cyclopropylmethyl alcohols (eqn. 10), the insertion this time taking place non-stereospecifically. The intermediate carbene or carbenoid species

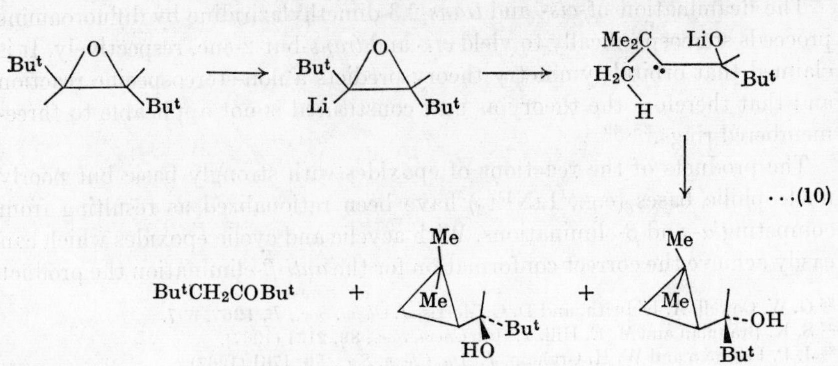

40 J. K. Crandall and L.-H. Chang, *J. Org. Chem.*, **32**, 435 (1967).

can also add intramolecularly to a double bond to yield a cyclopropane (eqn. 11). The other products of this reaction are the alcohol (**20**), formed by

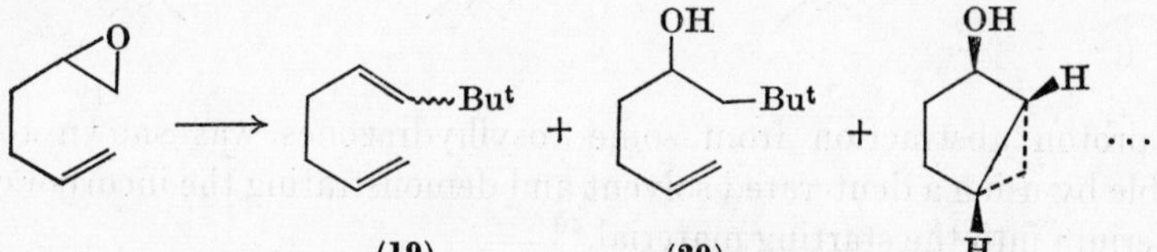

...(11)

(**19**) (**20**)

nucleophilic attack, and the diene (**19**), which is probably also formed from a carbene.[41] This is analogous to the formation of *trans*-di-*tert*-butylethylene from *tert*-butylethylene oxide and *tert*-butyl-lithium (eqn. 12). The last step

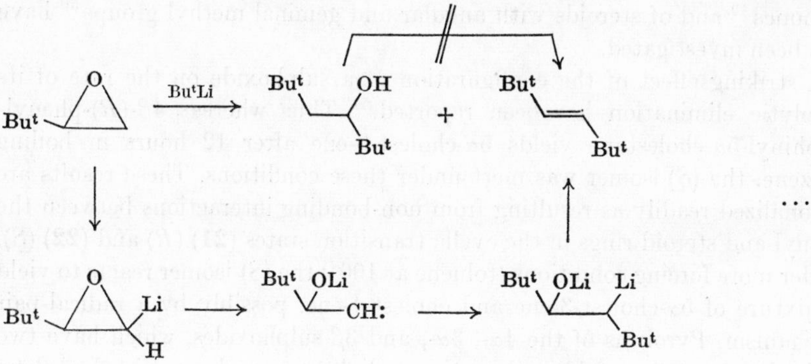

...(12)

in this reaction is the β-elimination of lithium oxide, the driving force for which was considered to be its high heat of formation (142 kcal $mole^{-1}$).[42]

α-Eliminations from organolithium compounds have been reviewed.[43]

racemic- and *meso*-2,3-Dibromobutane undergo stereospecific *trans*-elimination on treatment with phenyl-lithium in ether. This is either a concerted process involving attack of the phenyl anion on bromine or possibly a radical process in which stereochemistry is preserved by bridging by a bromine.[44]

Tosylhydrazones of aliphatic ketones containing an α-hydrogen undergo a ready elimination with alkyl-lithiums to yield olefins (eqn. 13).[45,46] The

[41] J. K. Crandall and L.-H. Chang, Lin, *J. Am. Chem. Soc.*, **89**, 4526 (1967).
[42] J. K. Crandall and L.-H. Chang, Lin, *J. Am. Chem. Soc.*, **89**, 4527 (1967).
[43] G. Köbrich, A. Akhtar, F. Ansari, W. E. Breckoff, H. Büttner, W. Drischel, R. H. Fischer, K. Flory, H. Frölich, W. Goyert, H. Heinemann, L. Hornke, H. R. Merkle, H. Trapp, and W. Zündorf, *Angew. Chem. Intern. Ed. Engl.*, **6**, 41 (1967).
[44] H. J. S. Winkler and H. Winkler, *Ann. Chem.*, **705**, 76 (1967).
[45] R. H. Shapiro and M. J. Heath, *J. Am. Chem. Soc.*, **89**, 5734 (1967).
[46] G. Kaufman, F. Cook, H. Shechter, J. Bayless, and L. Friedman, *J. Am. Chem. Soc.*, **89**, 5736 (1967).

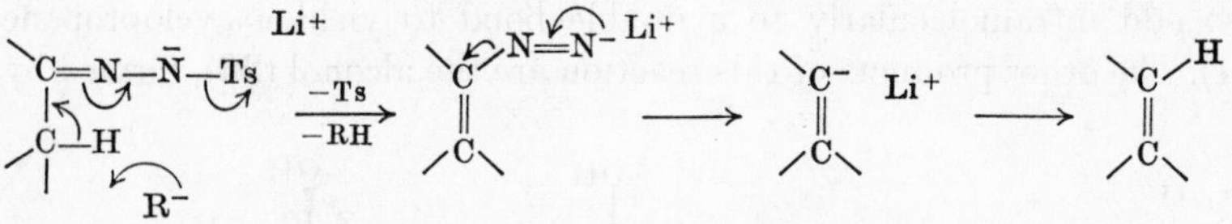

. . . (13)

initial proton abstraction from some tosylhydrazones was shown to be reversible by using a deuterated solvent and demonstrating the incorporation of deuterium into the starting material.[46]

Dehydrogenation of *cis*-1,2-dideuterioacenaphthene by 2,3-dichloro-5,6-dicyanobenzoquinone and tetrachloro-*o*-benzoquinone proceeds with 77.7% and 62.9% of *cis*-elimination, respectively. To account for these results it was proposed[47] that the previously suggested[48] mechanism should be modified to include ion-pairs as well as free ions. The dehydrogenations of steroidal quinones[49] and of steroids with angular and geminal methyl groups[50] have also been investigated.

A striking effect of the configuration of a sulphoxide on the rate of its pyrolytic elimination has been reported.[51] Thus whereas 4β-(*R*)-phenylsulphinyl-5α-cholestane yields 5α-cholest-3-ene after 12 hours in boiling benzene, the (*S*)-isomer was inert under these conditions. These results are rationalized readily as resulting from non-bonding interactions between the phenyl and steroid rings in the cyclic transition states (**21**) (*R*) and (**22**) (*S*). Under more forcing conditions (toluene at 100°) the (*S*)-isomer reacts to yield a mixture of 5α-cholest-3-ene and cholest-4-ene, possibly by a radical-pair mechanism. Pyrolysis of the 4α-, 3α-, and 3β-sulphoxides, which have two *cis*-hydrogen atoms, yields a (sometimes, slight) preponderance of that olefin expected from a consideration of non-bonding interactions between the phenyl and the steroid rings. Thus although the 4α-(*R*)-sulphoxide gave more cholest-4-ene than 5α-cholest-3-ene as expected (cf. **23** and **24**), the ratio was only 54:46. An additional factor here may be the difficulty in forming a double bond exocyclic to the six-membered B-ring. Details have been published of Goldberg and Sahli's investigation of the pyrolysis of (+)(*R*)- and (−)(*S*)-*trans*-4-methylcyclohexyl toluene-*p*-sulphoxide.[52, 53] The pyrolysis of cyclo-octyl and cyclodecyl (*S*)-hydratropates yield optically inactive *trans*-cycloalkenes.[54]

The proportions of the different alkenes formed on pyrolysis of sulphoxides

47 B. M. Trost, *J. Am. Chem. Soc.*, **89**, 1847 (1967).
48 See L. M. Jackman, *Adv. Org. Chem.*, **2**, 329 (1960).
49 A. B. Turner and H. J. Ringold, *J. Chem. Soc., C*, **1967**, 1720.
50 H. Dannenberg and H.-H. Keller, *Ann. Chem.*, **702**, 149 (1967).
51 D. N. Jones and M. J. Green, *J. Chem. Soc., C*, **1967**, 532.
52 S. I. Goldberg and M. S. Sahli, *J. Org. Chem.*, **32**, 2059 (1967).
53 See *Organic Reaction Mechanisms*, **1966**, 117.
54 O. Červinka, J. Budilová, and M. Daněček, *Coll. Czech. Chem. Commun.*, **32**, 2381 (1967).

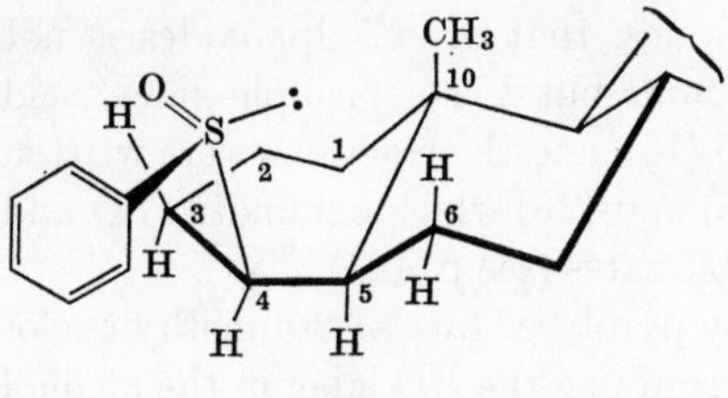

(21)

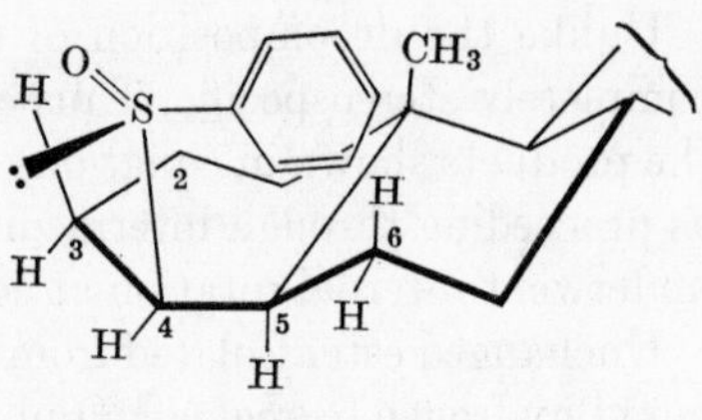

(22)

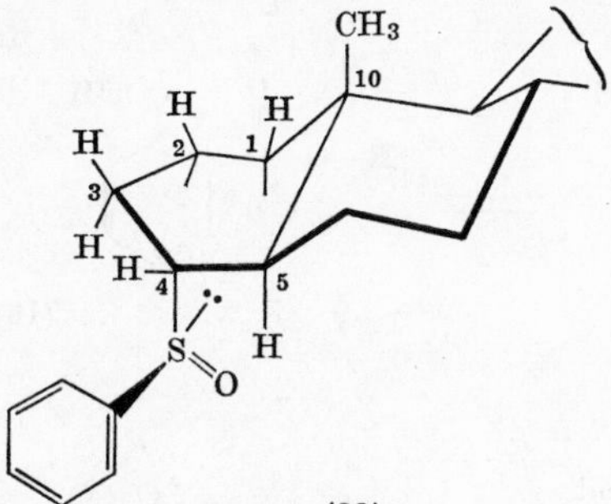

(23)

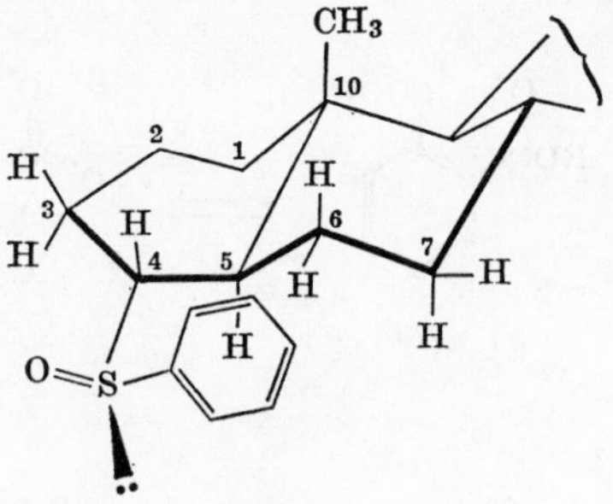

(24)

have been determined and compared with those formed on pyrolysis of analogous amine oxides, acetates, xanthates, and amides.[55]

The relative rates of pyrolysis of cyclopentyl, cyclohexyl, and cycloheptyl phenyl sulphoxides at 129.6° are 25:1:120, i.e., very similar to those of the corresponding cycloalkyldimethylamine 1-oxides.[56, 57]

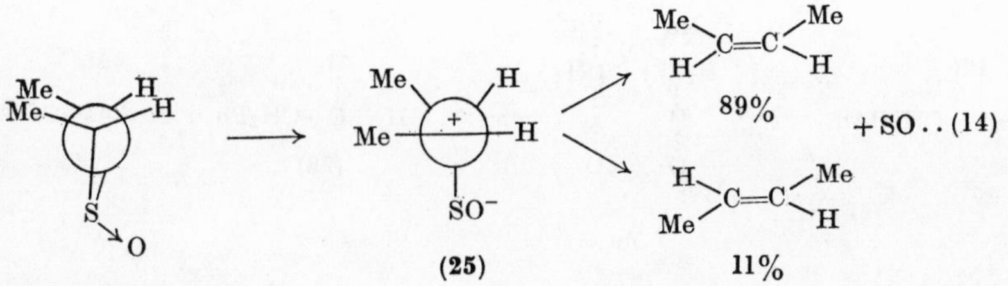

(25)

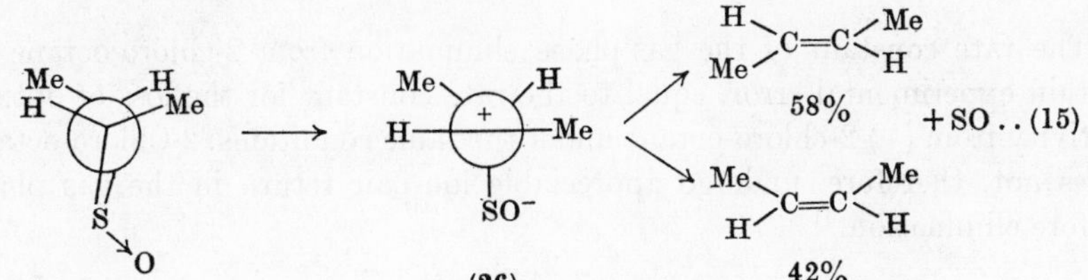

(26)

[55] D. W. Emerson, A. P. Craig, and I. W. Potts, *J. Org. Chem.*, **32**, 102 (1967).
[56] J. L. Kice and J. D. Campbell, *J. Org. Chem.*, **32**, 1631 (1967).
[57] See *Organic Reaction Mechanisms*, **1966**, 104, Fig. 1.

Unlike the decomposition of episulphones, that of episulphoxides is not completely stereospecific. Thus *cis*- and *trans*-but-2-ene episulphoxides yield the products shown in equations (14) and (15), and the reactions were written as proceeding through intermediates (**25**) and (**26**) which eliminated SO and underwent internal rotation at comparable rates (see p. 127).[58]

Unchanged ester isolated from partially pyrolysed *trans*-1,2-dimethylcyclohexyl hydrogen [*carbonyl*-^{18}O]phthalate contains the ^{18}O label in the alcohol portion, which suggests that the reaction proceeds by an ion-pair mechanism (eqn. 16).[59]

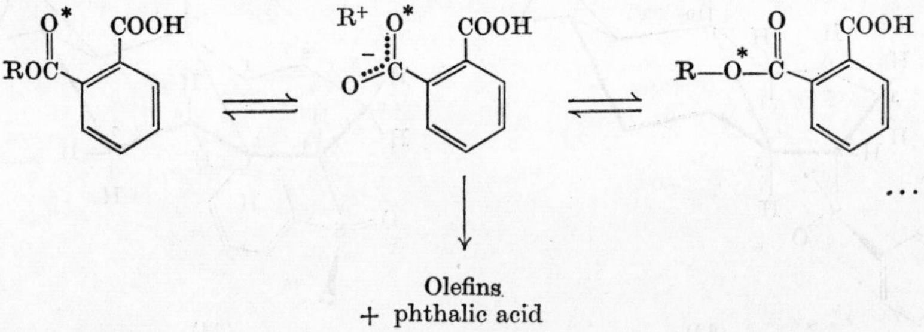

...(16)

The pyrolysis of neophyl (β,β-dimethylphenethyl) acetate proceeds with phenyl migration to form more of the less stable non-conjugated olefin (**28**) than of its isomer. A mechanism involving a seven-membered cyclic transition state (**27**) was therefore preferred to one involving ion pairs.[60]

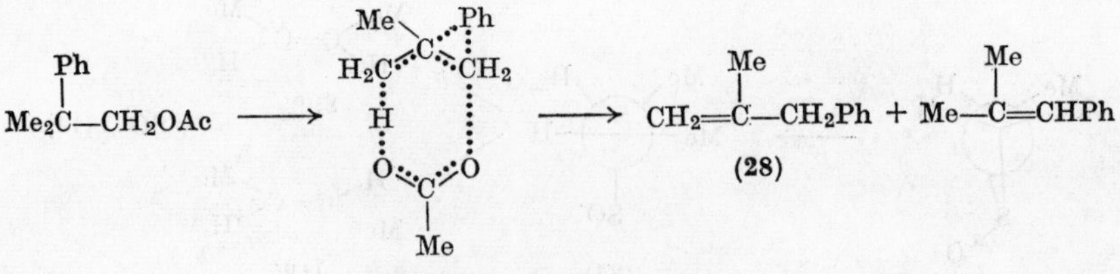

(**27**) (**28**)

The rate constant of the gas-phase elimination from 2-chloro-octane is, within experimental error, equal to the rate constant for the loss of optical activity from (+)-2-chloro-octane under the same conditions. 2-Chloro-octane does not, therefore, undergo appreciable ion-pair return in the gas phase before elimination.[61]

[58] G. Hartzell and J. N. Paige, *J. Org. Chem.*, **32**, 459 (1967).
[59] K. G. Rutherford and R. M. Ottenbrite, *Can. J. Chem.*, **45**, 679 (1967).
[60] H. Kwart and D. P. Hoster, *Chem. Commun.*, **1967**, 1155.
[61] C. J. Harding, A. Maccoll, and R. A. Ross, *Chem. Commun.*, **1967**, 289.

Six-centred transition states in the gas-phase elimination from alkyl halides and related reactions have been discussed.[62]

Gas-phase pyrolyses of the following compounds have also been investigated: *sec*-butyl bromide,[63] 3-bromopentane,[64] ethyl iodide,[65] isopropyl iodide (at very low pressures),[66] *tert*-butyl and ethyl chloride,[67,68] 1-ethoxyethyl chloride,[69] 7,7-dichlorobicyclo[4.1.0]heptane,[70] 2,2-dichloro-butane and -propane,[71] 3-chlorobut-1-ene and 3-chloro-2-methylbut-1-ene,[72] ethyl trimethylacetate,[73] *tert*-butyl alcohol,[74] *n*-butyl vinyl ether,[75] 3,3-dimethyloxetane,[76] organosilicon compounds,[77] cyclohexa-1,4-diene,[78] and 1-methylcyclohexa-1,4-diene.[79] Other gas-phase reactions that have been investigated include HBr-catalysed decomposition of trimethylacetic acid,[80] HI-catalysed decomposition of propan-2-ol,[81] and decomposition of chemically activated ethyl fluoride[82] and vibrationally excited 1,1,2-trifluoroethane.[83]

Further work has been reported on the preferential mode of electron-impact-induced eliminations.[84–86] The elimination of H_2S from pentane-1-thiol, unlike the elimination of H_2O from alcohols, proceeds with almost equal amounts of 1,3- and 1,4-elimination.[84] *n*-Butyl fluoride, in contrast to *n*-butyl chloride, reacts by both 1,3- and 1,4-elimination, and *n*-hexyl fluoride reacts predominantly (74%) in a seven-membered ring 1,5-elimination.[85]

62 C. J. Harding, A. G. Loudon, A. Maccoll, P. G. Rodgers, R. A. Ross, S. K. Wong, J. Shapiro, E. S. Swinbourne, V. R. Stimson, and P. J. Thomas, *Chem. Commun.*, **1967**, 1187.
63 M. R. Bridge and J. L. Holmes, *J. Chem. Soc., B*, **1967**, 1008.
64 N. Capon, A. Maccoll, and R. A. Ross, *Trans. Faraday Soc.*, **63**, 1152 (1967).
65 T. Yamada, H. Miyakawa, and A. Amano, *Kogyo Kagaku Zasshi*. **69**, 1268 (1966).
66 S. W. Benson and G. N. Spokes, *J. Am. Chem. Soc.*, **89**, 2525 (1967).
67 H. Heydtmann, *Chem. Phys. Lett.*, **1**, 105 (1967).
68 K. A. Holbrook and A. R. W. Marsh, *Trans. Faraday Soc.*, **63**, 643 (1967).
69 R. L. Failes and V. R. Stimson, *Australian J. Chem.*, **20**, 1553 (1967).
70 O. Nefedow, N. Nowizkaja, and A. Iwaschenko, *Ann. Chem.*, **707**, 217 (1967).
71 B. C. Young and E. S. Swinbourne, *J. Chem. Soc., B*, **1967**, 1181.
72 P. J. Thomas, *J. Chem. Soc., B*, **1967**, 1238.
73 J. T. D. Cross and V. R. Stimson, *Australian J. Chem.*, **20**, 177 (1967).
74 S. Sunner and S. Thoren, *Acta Chem. Scand.*, **21**, 565 (1967).
75 T. O. Bamkole and E. V. Emovon, *J. Chem. Soc., B*, **1967**, 523.
76 G. F. Cohoe and W. D. Walters, *J. Phys. Chem.*, **71**, 2327 (1967).
77 S. J. Band, I. M. T. Davidson, C. A. Lambert, and I. L. Stephenson, *Chem. Commun.*, **1967**, 723; I. M. T. Davidson, M. R. Jones, and C. Pett, *J. Chem. Soc., B*, **1967**, 937.
78 S. W. Benson and R. Shaw, *Trans. Faraday Soc.*, **63**, 985 (1967).
79 H. M. Frey and D. H. Lister, *J. Chem. Soc., A*, **1967**, 509.
80 J. T. D. Cross and V. R. Stimson, *J. Chem. Soc., B*, **1967**, 88.
81 R. L. Failes and V. R. Stimson, *Australian J. Chem.*, **20**, 1143 (1967).
82 J. A. Kerr, A. W. Kirk, B. V. O'Grady, and A. F. Trotman-Dickenson, *Chem. Commun.*, **1967**, 365.
83 J. T. Bryant, B. Kirtman, and G. O. Pritchard, *J. Phys. Chem.*, **71**, 1960 (1967); see also G. O. Pritchard and R. L. Thommarson, *ibid.*, p. 1674.
84 A. M. Duffield, W. Carpenter, and C. Djerassi, *Chem. Commun.*, **1967**, 109.
85 W. Carpenter, A. M. Duffield, and C. Djerassi, *Chem. Commun.*, **1967**, 1022.
86 See *Organic Reaction Mechanisms*, **1966**, 122.

The following reactions have also been investigated: stannous chloride-promoted debromination of tetrabromoethane,[87] sodium selenide-promoted debromination of *meso*- and racemic 2,3-dibromobutane,[88] bromide-promoted debromination of dimethyl *meso*-dibromosuccinate,[89] and iodide-promoted dechloromercuration of *cis*- and *trans*-2-chlorovinylmercuric chloride;[90] base-promoted conversion of 4-acetoxy-4-methylpentan-2-one into 4-methylpent-3-en-2-one,[91] and of 2-pyrazolines into cyclopropanes;[92] Hofmann eliminations;[93] dehydration of alcohols;[94] β- and γ-elimination reactions of *N,N,N*-trimethyl-(2-benzyl-3-phenylpropyl)ammonium iodide with sodamide in liquid ammonia;[95] $E2$ and S_N2 reactions of ω-chloroalkylbenzene,[96] β-substituted ethyl chlorides,[97] and 1,2-dichloroethane;[98] thermal decomposition of metal *O,O*-dialkyl phosphorodithioates,[99] imidates,[100] nitronic acids,[101] *tert*-alkyl *N*-arylcarbamates,[102] and β-cholestanyl *S*-methyl xanthate;[103] dehydrogenation and disproportionation of cyclohexene;[104, 105] and base-catalysed eliminations from steroidal toluene-*p*-sulphonates.[106]

87 W. K. Kwok and S. I. Miller, *Can. J. Chem.*, **45**, 1161 (1967).
88 M. Prince and B. W. Bremmer, *J. Org. Chem.*, **32**, 1655 (1967).
89 F. Badea, T. Constantinescu, A. Juvara, and C. Nenitzescu, *Ann. Chem.*, **706**, 20 (1967).
90 A. N. Nesmeyanov, A. E. Borisov, and I. S. Savel'eva, *Izv. Akad. Nauk SSSR, Ser. Khim.*, **1966**, 1498.
91 L. R. Fedor, *J. Am. Chem. Soc.*, **89**, 4479 (1967).
92 W. M. Jones, P. O. Sanderfer, and D. G. Baarda, *J. Org. Chem.*, **32**, 1367 (1967).
93 H. W. Bersch and D. Schon, *Arch. Pharm.*, **299**, 1039 (1966); **300**, 82 (1967); A. T. Babayan, G. T. Martirosyan, and D. V. Grigovyan, *Zh. Org. Khim.*, **2**, 1974 (1966); *Chem. Abs.*, **66**, 75436t (1967).
94 L. de Mourgues, F. Peyron, and Y. Trambouze, *Rev. Chim. (Bucharest)*, **17**, 583 (1966); *Chem. Abs.*, **66**, 104535g (1967); J. Herling, N. C. Sih, and H. Pines, *J. Org. Chem.*, **31**, 4085 (1967); J. Herling and H. Pines, *ibid.*, p. 4088; N. C. Sih and H. Pines, *ibid.*, p. 4092; H. Pines, W. F. Fry, N. C. Sih, and C. T. Goetschel, *ibid.*, p. 4094; T. Shono, T. Yoshimura, and R. Oda, *J. Org. Chem.*, **32**, 1088 (1967); M. Kraus and K. Kochloefl, *Coll. Czech. Chem. Commun.*, **32**, 2329 (1967); L. de Mourgues, F. Peyron, Y. Trambouze, and M. Prettre, *J. Catal.*, **7**, 17 (1967).
95 C. L. Bumgardner and H. Iwerks, *J. Am. Chem. Soc.*, **88**, 5518 (1966).
96 K. Okamoto, T. Kita, and H. Shingu, *Bull. Chem. Soc. Japan*, **40**, 1908 (1967).
97 K. Okamoto, T. Kita, K. Araki, and H. Shingu, *Bull. Chem. Soc. Japan*, **40**, 1913 (1967).
98 K. Okamoto, H. Matsuda, H. Kawasaki, and H. Shingu, *Bull. Chem. Soc. Japan*, **40**, 1917 (1967).
99 J. J. Dickert and C. N. Rowe, *J. Org. Chem.*, **32**, 647 (1967).
100 N. P. Marullo, C. D. Smith, and J. F. Terapane, *Tetrahedron Letters*, **1966**, 6279.
101 V. A. Tartakovskii, S. L. Ioffe, and S. S. Novikov, *Zh. Org. Khim.*, **3**, 628 (1967); *Chem. Abs.*, **67**, 43198w (1967).
102 M. P. Thorne, *Can. J. Chem.*, **45**, 2538 (1967).
103 L. Pan, T. N. Andersen, and H. Eyring, *J. Phys. Chem.*, **71**, 2258 (1967).
104 P. Tetenyi and K. Schnachter, *Acta Chim. Acad. Sci. Hung.*, **50**, 129 (1966).
105 S. Carrà and V. Ragaini, *Tetrahedron Letters*, **1967**, 1079.
106 L. Levisalles and J.-P. Pète, *Bull. Soc. Chim. France*, **1967**, 3747.

CHAPTER 5

Addition Reactions

Electrophilic Reactions

Addition of halogens. Fluorine adds to *cis*- and *trans*-propenylbenzene in non-polar solvents at very low temperatures; the addition, which is more stereospecific than that of chlorine, is predominantly *cis*. A four-centre molecular complex is proposed as intermediate; this largely collapses to the *cis*-adduct but also gives an "open" carbonium ion, rather than a fluoronium ion, which leads to the *trans*-adduct and, in methanol, the observed methoxyfluoropropylbenzenes.[1]

Poutsma[2] has reviewed the mechanism of the addition of chlorine to olefins with particular reference to competition between polar and radical mechanisms in non-polar media. In a reinvestigation of the addition of chlorine to butadiene, he has observed and separated a radical mechanism as well as a polar mechanism. The former involves addition of chlorine atoms at a rate not particularly greater than that for terminal olefins; the latter, observed in the presence of radical inhibitors such as oxygen, is slower than that of but-1-ene. Under normal dark conditions in the absence of inhibitors with neat olefin, the radical pathway predominates. Both mechanisms gave 3,4-dichlorobut-1-ene and *trans*-1,4-dichlorobut-2-ene.[3] The very rapid, stereospecific 1,2-*cis*-addition of chlorine to cyclo-octatetraene has been shown to proceed through the homoaromatic 8-chlorohomotropylium cation (Scheme 1); in both steps, attack by chlorine is *endo*, possibly because π-overlap between the orbitals at $C_{(1)}$ and $C_{(7)}$ of the ion is substantially greater on the side opposite the $C_{(8)}$ bridge.[4]

The kinetics and products of addition of chlorine to methyl *trans*-cinnamate in acetic acid at 25° in the absence and presence of lithium acetate, chloride, and perchlorate, and of perchloric acid, have been determined. The reaction, which is not acid-catalysed, involves attack by a highly polarized chlorine molecule on the double bond (**1**) to give an ion-pair which exists in two readily interconverted conformations (**2**) and (**3**). The dichlorides are formed by ion-pair collapse and, in the presence of lithium chloride, by capture of the two carbonium ions by external chloride. The chloro-acetates are formed by

[1] R. F. Merritt, *J. Am. Chem. Soc.*, **89**, 609 (1967).
[2] M. L. Poutsma, *Science*, **157**, 997 (1967).
[3] M. L. Poutsma, *J. Org. Chem.*, **31**, 4167 (1966).
[4] G. Boche, W. Hechtl, H. Huber, and R. Huisgen, *J. Am. Chem. Soc.*, **89**, 3344 (1967); R. Huisgen, G. Boche, and H. Huber, *ibid.*, p. 3345.

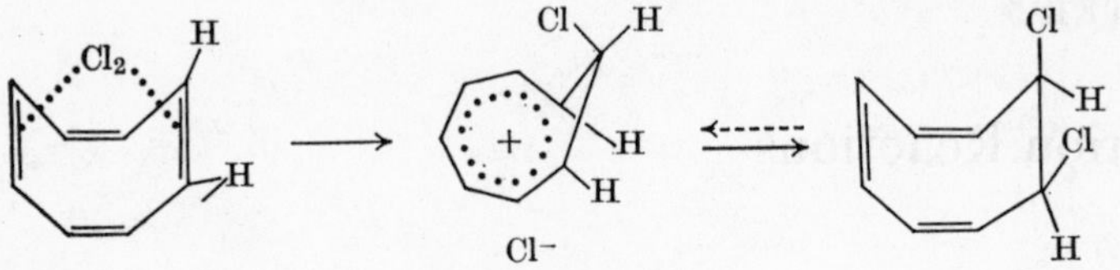

Scheme 1

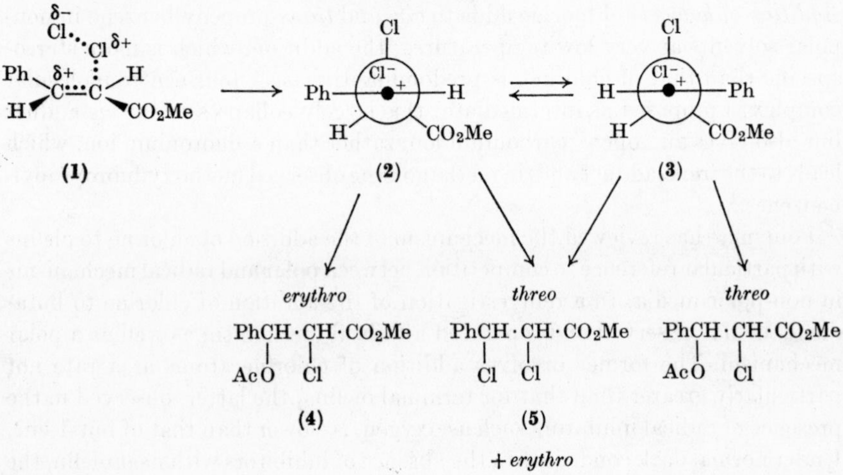

capture of (**2**) and (**3**) by solvent, a process enhanced by added electrolyte. The products of *cis*-addition of chlorine (**5**) and *trans*-addition of solvent (**4**) predominate.[5]

Naphthalene and its methyl derivatives undergo both substitution and addition when treated with sulphuryl chloride, much as with chlorine. The reaction of 1-methylnaphthalene with chlorine in acetic acid also led to substitution and addition; reaction paths were discussed.[6]

Transannular reactions result when chlorine, from lead tetrachloride, is added to *cis*-cyclo-octene giving *cis*- and *trans*-1,4- and *trans*-1,2-dichlorocyclo-octane,[7] whilst addition of chlorine itself gives the *trans*-1,2-dichloride.[8]

The dependence of the ratio of substitution to addition upon solvent in the reaction of chlorine with 1,1-dichloroethylene,[9] the reaction between chlorine

[5] M. C. Cabaleiro and M. D. Johnson, *J. Chem. Soc., B,* **1967**, 565.

[6] P. B. D. de la Mare and H. Suzuki, *J. Chem. Soc., C,* **1967**, 1586; G. Cum, P. B. de la Mare, and M. D. Johnson, *ibid.*, p. 1590.

[7] P. W. Henniger, L. J. Dukker, and E. Havinga, *Rec. Trav. Chim.*, **85**, 1177 (1966).

[8] N. L. Allinger and L. A. Tushaus, *Tetrahedron*, **23**, 2051 (1967).

[9] I. V. Bodrikov, Z. S. Smolyan, and G. A. Korchagina, *Zh. Vses. Khim. Obshchest*, **11**, 708 (1966); *Chem. Abs.*, **66**, 54910c (1967).

water and ethylene in a wetted-wall column,[10] and the addition of chlorine to 1,3-dichloropropene and to *N*,*N*,*N*-trimethylallylammonium perchlorate[11] have also been studied.

Full details[12] of the kinetics of addition of bromine to substituted styrenes which provided non-stereochemical evidence for cyclic bromonium ion intermediates[13] have appeared. The rather limited evidence for such ions, and the probability of a spectrum of intermediates ranging from α-halogeno-carbonium ions to completely symmetrical halonium ions, were stressed.[12] The addition of bromine to *cis*- and *trans*-stilbene in 1,2-dichloroethane, chlorobenzene, and carbon tetrachloride was of the first order in stilbene and second order in bromine. With tetrabutylammonium tribromide as the source of bromine the reaction was of the first order in each reactant. The *cis*-isomer reacted faster than the *trans*, and the rate increased with the polarity of the solvent, consistently with an electrophilic mechanism[14] (see also ref. 15). When the olefins $Ph_2ArC{\cdot}CR{=}CH_2$ (R = H or Me; Ar = Ph or *p*-C_6H_4OMe) were treated with bromine in various solvents migration of Ar was observed, its extent depending upon the nature of Ar and the solvent; rearrangement was often extensive, showing that the initially formed carbonium-bromonium ion is powerfully stabilized by interaction with the neighbouring aryl group.[16] Bicyclo[2.2.0]hexa-2,5-diene ("Dewar" benzene) reacted with bromine to give a mixture of dibromides and with *m*-chloroperbenzoic acid to give an epoxide without aromatization; thus the cyclobutyl-to-homoallyl cation rearrangement (eqn. 1) does not occur, presumably for orbital-symmetry reasons.[17]

The addition of "bromine fluoride", i.e., *N*-bromoacetamide and hydrogen

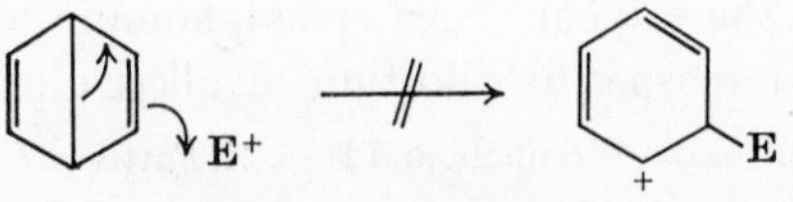

. . . . (1)

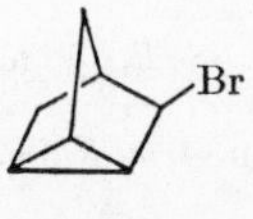

(6)

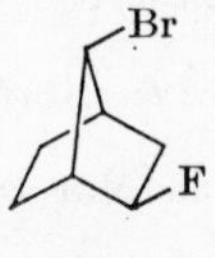

(7)

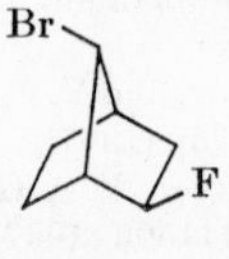

(8)

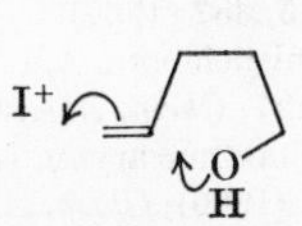

(9)

10 P. W. Dun and T. Wood, *J. Appl. Chem.* (*London*), **17**, 53 (1967).

11 Y. A. Serguchev and E. A. Shilov, *Ukr. Khim. Zh.*, **33**, 173 (1967); *Chem. Abs.*, **67**, 10913x (1967).

12 K. Yates and W. V. Wright, *Can. J. Chem.*, **45**, 167 (1967).

13 *Organic Reaction Mechanisms*, **1965**, 104.

14 R. E. Buckles, J. L. Miller, and R. J. Thurmaier, *J. Org. Chem.*, **32**, 888 (1967).

15 G. Heublein and E. Mlejnek, *Z. Chem.*, **7**, 340 (1967).

16 R. O. C. Norman and C. B. Thomas, *J. Chem. Soc., B*, **1967**, 598.

17 E. E. van Tamelen and D. Carty, *J. Am. Chem. Soc.*, **89**, 3922 (1967).

fluoride in ether, to norbornene gave the products (**6**), (**7**), and (**8**), all of which could readily result from the expected initial intermediate formed by electrophilic *exo*-attack by bromine.[18] The addition of bromine to several α,β- and β,γ-unsaturated sulphones,[19] to *cis*-3-methylcyclohex-4-ene-*cis*-1,2-dicarboxylic acid,[20] to 5,5-dichloroallylacetic acids,[21] to diaroylethylenes,[22] and to 1,1-diarylethylenes[23] has also been investigated.

Another example of participation by the hydroxyl group in the addition of iodine to an unsaturated alcohol has been reported. The relative rates of addition of iodine in aqueous potassium iodide at 0° to $CH_2{=}CH(CH_2)_nOH$ were 1:2:200:35 for $n = 1, 2, 3,$ and 4; when $n = 3$ the product was tetrahydro-2-iodomethylfuran (cf. **9**).[24] Neighbouring-group participation by sulphinyl-oxygen in the reaction of iodine with bicyclo[2.2.1]hept-5-ene derivatives has again been demonstrated.[25] The polar, dark, addition of iodine to cyclohexene is second-order reaction in a variety of solvents, but in certain (organic chloride) solvents the reaction rate decreased with increasing temperature.[26] The addition of iodine isocyanate to dienes and acetylenes,[27] and steric factors in the iodolactonization of cyclohexenedicarboxylic acids,[28] have also been investigated.

Related additions. Further evidence of kinetic complexity in the addition of 2,4-dinitrobenzenesulphenyl halides in solvents of low polarity has been reported. Reaction of the bromide with cyclohexene is first-order in each reactant in chloroform, while in the less polar benzene the order is greater than one in bromide and the reaction is accelerated by the product. Again this is explained by rate-determining formation of an episulphonium ion, with developing charge in the transition state stabilized by bromide and the product more than by the solvent.[29] An episulphonium ion intermediate is also proposed in the stereospecific addition of alkylsulphenyl chlorides to

[18] F. H. Dean, D. R. Marshall, E. W. Warnhoff, and F. L. M. Pattison, *Can. J. Chem.*, **45**, 2279 (1967).

[19] S. Shanmuganathan and V. Subramanian, *Trans. Faraday Soc.*, **63**, 911 (1967); *Indian J. Chem.*, **5**, 387 (1967).

[20] A. S. Onishchenko, A. L. Shabanov, and V. F. Kucherov, *Izv. Akad. Nauk SSSR, Ser. Khim.*, **1967**, 127; *Chem. Abs.*, **66**, 115173p (1967).

[21] G. M. Shakhnazaryan, L. A. Saakyan, V. A. Garibyan, and M. T. Dangyan, *Arm. Khim. Zh.*, **19**, 815 (1966); *Chem. Abs.*, **67**, 11150h (1967).

[22] M. J. Janssen, F. Wiegman, and H. J. Kooreman, *Tetrahedron Letters*, **1966**, 6375.

[23] J. E. Dubois and W. V. Wright, *Tetrahedron Letters*, **1967**, 3101.

[24] D. L. H. Williams, *Tetrahedron Letters*, **1967**, 2001.

[25] M. Cinquini, S. Colonna, and F. Montanari, *J. Chem. Soc., C*, **1967**, 1213; see *Organic Reaction Mechanisms*, **1966**, 127.

[26] C.-S. Chao, R.-R. Lii, C.-D. Lee, and M.-G. Kung, *Hua. Hsueh*, **1966**, 127; *Chem. Abs.*, **67**, 53512h (1967).

[27] B. E. Grimwood and D. Swern, *J. Org. Chem.*, **32**, 3665 (1967).

[28] V. I. Staninets, E. A. Shilov, and E. B. Koryak, *Dokl. Akad. Nauk SSSR*, **172**, 874 (1967); *Chem. Abs.*, **66**, 94602k (1967).

[29] D. S. Campbell and D. R. Hogg, *J. Chem. Soc., B*, **1967**, 889.

3,4-dihydro-2*H*-pyran.[30] The rate of addition of toluene-*p*-sulphenyl chloride to aliphatic acetylenes is sensitive to the acetylene structure but insensitive to (non-*ortho*) substituents in the aromatic ring, as shown last year for aromatic acetylenes. The reactions were insensitive to ultraviolet irradiation, and no evidence of a radical mechanism was found.[31] Isobutane-1-sulphenic acid has been characterized and shown to add to ethyl acrylate and methyl propiolate to give sulphoxides;[32] acetylthiosulphenyl chloride, $CH_3COSSCl$, also adds to olefins.[33a] Thiocyanogen chloride adds to olefins in a two-step, kinetically controlled reaction in which the first step is addition of ^+SCN to form a cyclic sulphonium ion.[33b]

The addition of hypochlorous acid to highly hindered olefins does not give oxygen-containing products but leads to replacement of hydrogen by chlorine, by proton loss from a chloronium ion and by a free-radical chain mechanism.[34] A striking selectivity in the conversion of polyolefins into bromohydrins by *N*-bromosuccinimide has been demonstrated; reaction at a "terminal" double bond (actually 2,3) is much faster than at an "internal" double bond, possibly because of shielding of the latter by coiling of the molecule.[35] In the stereospecific formation of bromohydrins from olefins and *N*-bromosuccinimide in DMSO containing water, attack on the bromonium ion is exclusively by DMSO, as shown by ^{18}O-labelling.[36] The addition of nitrosyl fluoride to fluoroalkenes,[37] and of nitrosyl chloride to allene to give, abnormally, 1,3-dichloroacetone,[38] has been described.

Iodine azide, prepared *in situ* in methyl cyanide, adds stereospecifically *trans* to a variety of unsaturated systems; electrophilic addition via a cyclic iodonium ion is proposed.[39] However, the addition of diboron tetrachloride to olefins and acetylenes is stereospecifically *cis*; this strongly suggests the four-centre mechanism (eqn. 2) required for maximum overlap of the p_π-orbital and the vacant *p*-orbitals on boron.[40] Tetrafluorohydrazine adds two NF_2 groups 1,4 to cyclo-octatetraene and to 6,6-diphenylfulvene.[41]

30 M. J. Baldwin and R. K. Brown, *Can. J. Chem.*, **45**, 1195 (1967).
31 A. Dondoni, G. Modena, and G. Scorrano, *Ric. Sci. Rend., Sez. A*, **6**, 665 (1964); *Chem. Abs.*, **65**, 18447 (1966).
32 J. R. Shelton and K. E. Davis, *J. Am. Chem. Soc.*, **89**, 718 (1967).
33a W. H. Mueller and P. E. Butler, *J. Org. Chem.*, **32**, 2925 (1967).
33b R. G. Guy and I. Pearson, *Chem. Ind. (London)*, **1967**, 1255.
34 S. Marmor and J. G. Maroski, *J. Org. Chem.*, **31**, 4278 (1966).
35 E. E. van Tamelen and K. B. Sharpless, *Tetrahedron Letters*, **1967**, 2655.
36 D. R. Dalton and D. G. Jones, *Tetrahedron Letters*, **1967**, 2875.
37 B. L. Dyatkin, E. P. Mochalina, R. A. Bekker, S. R. Sterlin, and I. L. Knunyants, *Tetrahedron*, **23**, 4291 (1967).
38 H. G. Peer and A. Schors, *Rec. Trav. Chim.*, **86**, 167 (1967).
39 F. W. Fowler, A. Hassner, and L. A. Levy, *J. Am. Chem. Soc.*, **89**, 2077 (1967).
40 R. W. Rudolph, *J. Am. Chem. Soc.*, **89**, 4216 (1967); M. Zeldin, A. R. Gatti, and T. Wartik, *ibid.*, p. 4217.
41 T. S. Cantrell, *J. Org. Chem.*, **32**, 911 (1967).

The addition of a variety of reagents to *cis,trans*-cyclodeca-1,5-diene (**10**) has been studied. Concerted additions, e.g., of diimide, diborane, ozone, peracids, and the Simmons–Smith reagent, show a high degree of selectivity for the *trans*-double bond, whilst reagents which add stepwise, e.g., bromine,

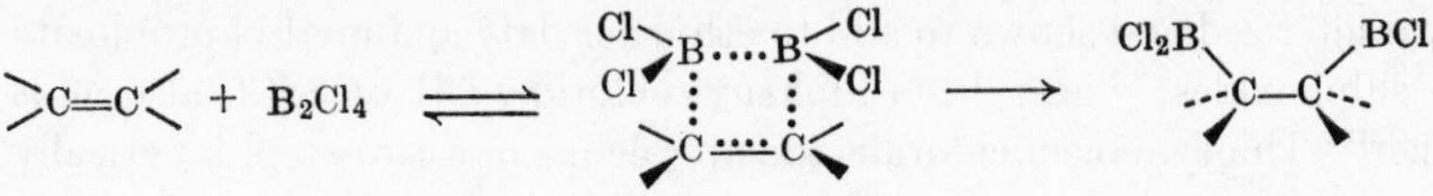

...(2)

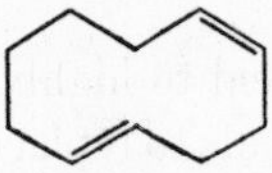

(**10**)

lead tetra-acetate, methanesulphenyl chloride, and mercury salts, give *cis*-decalins by transannular reactions. Thus (**10**) may have diagnostic value in olefin addition mechanisms.[42] Various ionic additions to the double bond of benzonorbornadiene have been investigated further, and rationalized by participation by the benzene ring.[43] A quantitative rule for polar addition to olefins has been proposed[44] and additions to fluoro-olefins have been reviewed.[45]

Addition of hydrogen halides. The generally accepted participation of vinyl carbonium ions in polar additions to acetylenes has received further support. The reaction of 1-phenylpropyne with hydrogen chloride in acetic acid is of the third order – first in the acetylene and second in hydrogen chloride. The major product (*ca.* 70%) was 1-chloro-*cis*-1-phenylpropene (**14**), but the *trans*-isomer (**16**) and the *cis*- and *trans*-acetoxy-compounds (**15**), hydrolysed to propiophenone in a secondary reaction, were also found. To account for the reaction order, the stereochemistry of the products, and particularly the variation of the products with hydrogen chloride concentration, the multistep mechanism shown was required. The rate-determining step is formation of an ion-pair between the hydrogen dichloride ion and a linear vinyl carbonium ion (where the carbonium *p*-orbital is conjugated with the phenyl ring). Intermediate between the two intimate ion pairs (**11**) and (**13**) [which collapse to (**14**) and (**16**), respectively] is a solvent-separated ion pair (**12**) with the hydrogen dichloride ion not directly associated with either face of the car-

42 J. G. Traynham, G. R. Franzen, G. A. Knesel, and D. J. Northington, *J. Org. Chem.*, **32**, 3285 (1967).

43 J. W. Wilt, G. Gutman, W. J. Ranus, and A. R. Zigman, *J. Org. Chem.*, **32**, 893 (1967).

44 M.-S. Chiang, S. Li, C.-C. Chang, and C.-H. Liu, *K'o Hsueh T'ung Pao*, **17**, 345 (1966); *Chem. Abs.*, **66**, 28168x (1967).

45 O. Paleta, *Chem. Listy*, **60**, 1363 (1966).

bonium ion; this species is the precursor of the enol acetates. The observed *cis*-stereochemistry is the same as that observed in similar reactions of olefins, suggesting that vinylic carbonium ions behave much as do their alkyl analogues.[46]

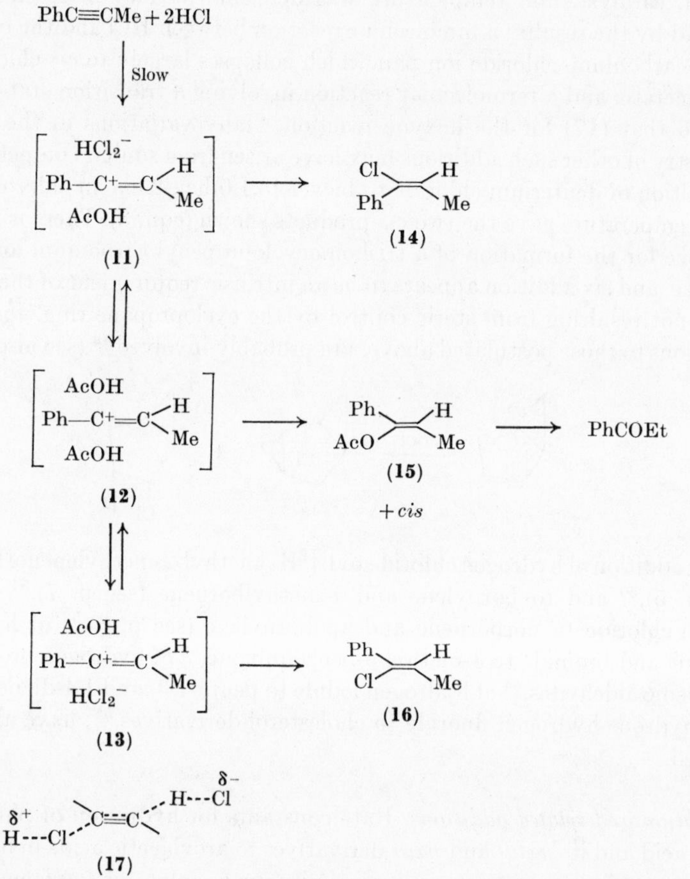

In contrast with this addition of hydrogen chloride to 1-phenylpropyne, addition to hex-3-yne was slower and gave almost exclusively the *trans*-adduct. From the high fractional order in hydrogen chloride and the effect of tetramethylammonium chloride as catalyst, a mechanism involving prior coordination of the triple bond with HCl and rate-determining attack on this by HCl, or acetic acid, in a transition state such as (**17**) was proposed.[47] A similar termolecular mechanism, in competition with a bimolecular one, has been

[46] R. C. Fahey and D. J. Lee, *J. Am. Chem. Soc.*, **88**, 5555 (1966).
[47] R. C. Fahey and D.-J. Lee, *J. Am. Chem. Soc.*, **89**, 2780 (1967).

proposed by Fahey and Monahan[48] to account for the kinetics and products of the addition of HCl to cyclohexene in acetic acid. Cyclohexyl acetate was formed by a mainly *trans*-addition, but cyclohexyl chloride was formed by both *cis*- and *trans*-additions, and the dependence of these on the concentration of HCl, catalyst, and temperature was determined. Two mechanisms are required by the results: a bimolecular reaction between HCl and the olefin to give a carbonium–chloride ion pair which collapses largely to *cis*-chloride or *trans*-acetate, and a termolecular reaction involving a transition state analogous to that (**17**) for the hexyne reaction. Many variations in the stereochemistry of other such additions may have arisen from similar competition.[48]

Addition of deuterium chloride to bicyclo[3.1.0]hex-2-ene in chloroform at room temperature gave the two *cis*-products shown (eqn. 3). There is thus no evidence for the formation of a trishomocyclopropenyl carbonium ion intermediate, and *cis*-addition appears to be an intrinsic requirement of the double bond, not resulting from steric control by the cyclopropane ring. Ion pairs, analogous to those postulated above, are probably involved[49] (see also p. 48).

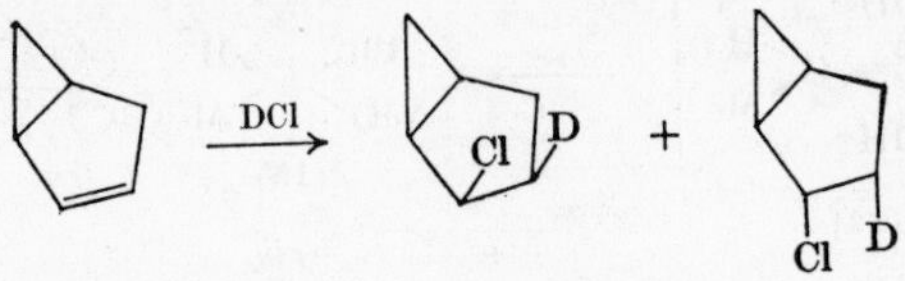

.... (3)

The addition of hydrogen chloride to 1-[$^2H_3$]methyl-2-methylenenorbornane (see p. 5),[50] and to bornylene and 1-methylbornene (see p. 7),[51] of deuterium chloride to norbornene and apobornylene (see p. 6),[52] of hydrogen chloride and bromide to 4-methyl-*o*-benzoquinone,[53] of hydrogen bromide to acetylenic aldehydes,[54] of hydrogen iodide to penta-1,3- and 1,4-diene,[55a] and of anhydrous hydrogen fluoride to cholesterol derivatives,[55] have also been studied.

Hydration and related additions. Rate constants for hydration of phenylpropiolic acid and its *meta*- and *para*-derivatives to aroylacetic acids in 62—70% sulphuric acid show strict proportionality to h_0, plots of $\log k$ against H_0 having essentially unit slope, and they are quite sensitive to electron-donating

[48] R. C. Fahey and M. W. Monahan, *Chem. Commun.*, **1967**, 936.
[49] P. K. Freeman, F. A. Raymond, and M. F. Grostic, *J. Org. Chem.*, **32**, 24 (1967).
[50] H. C. Brown and K.-T. Liu, *J. Am. Chem. Soc.*, **89**, 466 (1967).
[51] H. C. Brown and K.-T. Liu, *J. Am. Chem. Soc.*, **89**, 3898 (1967).
[52] H. C. Brown and K.-T. Liu, *J. Am. Chem. Soc.*, **89**, 3900 (1967).
[53] L. Horner and T. Burger, *Ann. Chem.*, **708**, 105 (1967).
[54] L. F. Chelpanova and G. Bondarev, *Zh. Organ. Khim.*, **2**, 1561; *Chem. Abs.*, **66**, 64790c (1967).
[55a] K. W. Egger and S. W. Benson, *J. Phys. Chem.*, **71**, 1933 (1967).
[55b] J.-C. Jacquesy, R. Jacquesy, and J. Levisalles, *Bull. Soc. Chim. France*, **1967**, 1649.

substituents in the phenyl ring. Hydration is about 4 times faster in H_2O–H_2SO_4 than in D_2O–D_2SO_4. The ρ-value was determined together with activation parameters for *p*-chlorophenylpropiolic acid. The results were most consistent with rate-determining protonation of the α-carbon to give a vinyl carbonium ion (eqn. 4).[56]

$$\text{Ar—C}\equiv\text{C—CO}_2\text{H} + \text{H}^+ \xrightarrow{\text{Slow}} \text{Ar—}\overset{+}{\text{C}}\text{=CH}\cdot\text{CO}_2\text{H} \rightarrow \rightarrow \text{ArCOCH}_2\text{CO}_2\text{H} \quad ..(4)$$

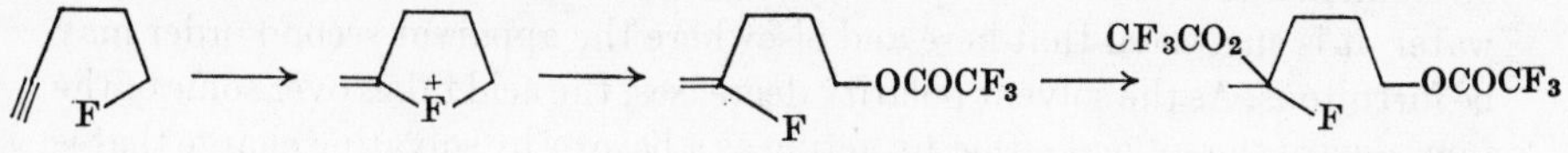

(18) (19)

The hydrolysis of cyanoketen dimethyl acetal to methyl cyanoacetate is general acid-catalysed and its rate-determining step, as in the hydration of isobutene and styrenes and the hydrolysis of vinyl ethers, is addition of a proton to the carbon–carbon double bond (see p. 323).[57] The same has been demonstrated for the hydration of α-methylstyrenes.[58]

The rates of acid-catalysed hydration of mesityl oxide have been measured at pressures up to 1360 atm. It was inferred from the volume of activation that two water molecules are involved in the transition state, i.e., that proton-transfer from a hydronium ion is concerted with formation of the new carbon–oxygen bond. The acid-catalysed addition of methanol is mechanistically similar.[59] The acid-catalysed addition of water and of methanol to ψ-ionone[60] and of water to crotonaldehyde[61a] and to methyl 2-nitrocrotonate[61b] have been investigated.

Evidence for fluorine participation in the form of a fluoronium ion intermediate or transition state has been presented for the first time. On addition of trifluoroacetic acid to 5-fluoropent-1-yne the diacetate (**19**) (15%) was formed as shown, together with the normal addition products. That the proposed initial adduct (**18**) could give (**19**), unlike the analogous chloro-compound for example, was shown to be entirely reasonable since 2-fluoropropene reacted with trifluoroacetic acid very much faster than propene or 2-chloro- or 2-bromo-propene. The last result provides a striking illustration of the effect of fluorine in stabilizing an adjacent carbonium ion by 2*p*–2*p*

[56] D. S. Noyce, M. A. Matesich, and P. E. Peterson, *J. Am. Chem. Soc.*, **89**, 6225 (1967).
[57] V. Gold and D. C. A. Waterman, *Chem. Commun.*, **1967**, 40.
[58] J.-C. Simandoux, B. Torck, M. Hellin, and F. Coussemant, *Tetrahedron Letters*, **1967**, 2971.
[59] J. J. Scott and K. R. Brower, *J. Am. Chem. Soc.*, **89**, 2682 (1967).
[60] B. C. Roest, J. U. Veenland, and T. J. de Boer, *Tetrahedron*, **23**, 3071 (1967).
[61a] V. A. Terent'ev, V. S. Markevich, and N. E. Shtivel, *Zh. Org. Khim.*, **3**, 452 (1967); *Chem. Abs.*, **67**, 2558v (1967).
[61b] V. M. Belikov and Y. N. Belokon, *Izv. Akad. Nauk SSSR, Ser. Khim.*, **1967**, 1054; *Chem. Abs.*, **67**, 72943r (1967).

overlap.[62] Isobutene reacts rapidly with trifluoroacetic acid in ethylene dichloride at ordinary temperatures to give *tert*-butyl trifluoroacetate; the reaction is of the first order in olefin and of the second order in acid. The boron trifluoride-catalysed addition of acetic acid to but-2-ene follows a similar rate law with the acid concentration replaced by that of the 1:1 acid–boron trifluoride complex. The quadratic power in acid agrees with other acid-catalysed reactions in aprotic solvents of intermediate polarity; as this order often increases as the solvent polarity decreases and as it decreases to unity in water, it is suggested that here and elsewhere the apparent second order may be fortuitous. As the solvent polarity decreases, the acid takes over some of the functions of the solvent either by acting as a base or by solvating charge that is developed in the transition state.[63] The kinetics of the vapour-phase addition of some alkanoic acids to acetylene over zinc acetate–charcoal were studied by a flow method.[64] And the mechanism of the Prins reaction has been considered further.[65]

The active hydroboration reagent prepared from α-pinene and diborane was represented as di-isopinocampheylborane.[66] It has now been suggested that the tetra- and tri-substituted dimers are the important hydroborating species for *cis*- and for *trans*-1,2-disubstituted olefins, respectively. A careful conformational analysis of the olefin–diborane interactions then predicts more successfully the chiralities of the alcohols formed.[67] Optically active [1-^{2}H]-butan-1-ol has been prepared by partial asymmetric hydroboration of *cis*-[1-^{2}H]but-1-ene and di-isopinocampheylborane. To account for the stereochemistry of the butanol, Brown's four-centre mechanism for hydroboration has been slightly modified: the olefin–borane is considered to be a triangular π-complex (**20**) with the transition state very similar to this.[68]

In the hydroboration of substituted styrenes the relative proportions of isomeric products formed by attack by boron at the α- and the β-positions are explained by stabilization of the two transition states (**21**) and (**22**).[69] The directive effect in hydroboration of a carboxylate group,[70] an allylic hydroxyl group,[71] and a cyclopropyl group,[72] and the temperature-dependence of the

62 P. E. Peterson and R. J. Bopp, *J. Am. Chem. Soc.*, **89**, 1283 (1967).
63 G. A. Latrémouille and A. M. Eastham, *Can. J. Chem.*, **45**, 11 (1967).
64 S. Do-Chen and M. Kraus, *Coll. Czech. Chem. Commun.*, **32**, 2972 (1967).
65 S. Watanabe, *Chiba Daigaku Kogakubu Kenkyu Hokoku*, **17**, 17 (1966); *Chem. Abs.*, **65**, 16848 (1966).
66 H. C. Brown, N. R. Ayyangar, and G. Zweifel, *J. Am. Chem. Soc.*, **84**, 4341 (1964).
67 D. R. Brown, S. F. A. Kettle, J. McKenna, and J. M. McKenna, *Chem. Commun.*, **1967**, 667.
68 A. Streitweiser, L. Verbit, and R. Bittman, *J. Org. Chem.*, **32**, 1530 (1967).
69 H. C. Brown and R. L. Sharp, *J. Am. Chem. Soc.*, **88**, 5851 (1966); see also J. Klein, E. Dunkelblum, and M. A. Wolff, *J. Organometal. Chem.*, **7**, 377 (1967).
70 J. Klein, E. Dunkelblum, and D. Avrahami, *J. Org. Chem.*, **32**, 935 (1967).
71 J. Klein and E. Dunkelblum, *Tetrahedron Letters*, **1966**, 6047.
72 S. Nishida, I. Moritani, K. Ito, and K. Sakai, *J. Org. Chem.*, **32**, 939 (1967).

hydroboration of (+)-car-3-ene[73] have been investigated. Irradiation of vinylboranes results in the transfer of one alkyl group from boron to the adjacent carbon, giving high yields of the substituted olefin; this was rationalized in terms of an initial, fast, base- or solvent-assisted migration of the alkyl group, followed by deboronation to the *cis*-olefin (eqn. 5).[74]

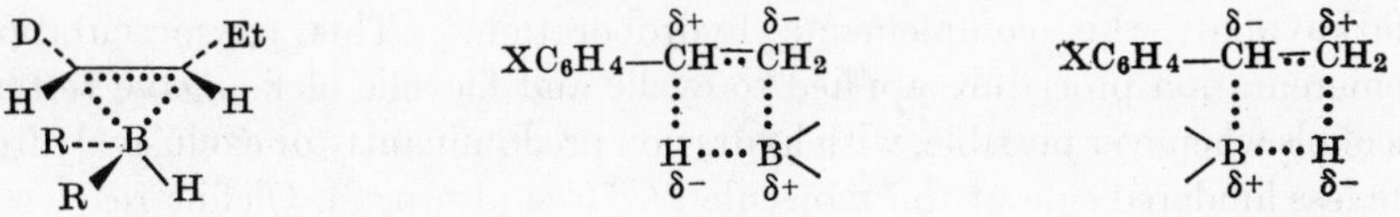

(20) **(21)** **(22)**

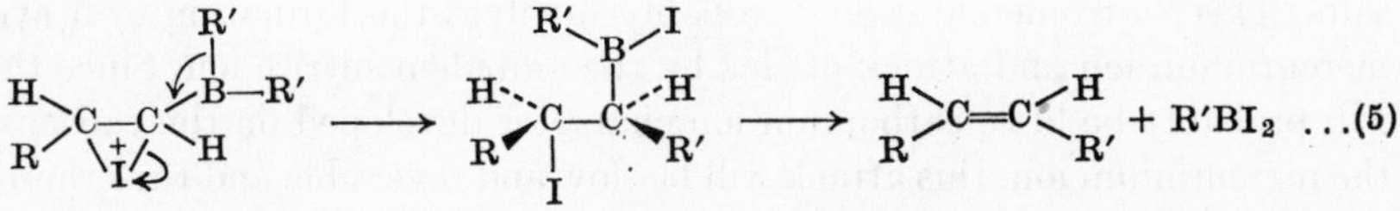

Allene and its five methyl derivatives react rapidly at 25° with mercuric acetate in methanol to form one or more adducts, which have all been characterized. Attack of methyl- and dimethyl-substituted allenic π-bonds gave exclusively monoadducts with the acetoxymercuri-group on the central carbon atom. Attack at unsubstituted π-bonds, however, occurred only at the terminal carbon to give an allylmercury compound which rapidly reacted further to give the 1,3-bis(acetoxymercuri)ketal. The order of reactivity of dimethyl, monomethyl, and unsubstituted π-bonds is 16:4:1; tetramethylallene reacts 10^4 times faster than tetramethylethylene. The results were best interpreted in terms of a stable, σ-bonded cyclic mercurinium ion (**23**) [rather than a π-complex (**24**)] with the ion formed reversibly before the slow step and not subject to steric control. Opening of the ion by methanol is both rate- and product-determining and is stereospecifically *trans*. This opening is very sensitive to the steric influence of groups on the other end of the allene, see (**23**).[75] Very similar results were found in the oxymercuration of cyclonona-1,2-diene, cyclonona-1,2,6-triene, and penta-2,3-diene with ethanolic mercuric

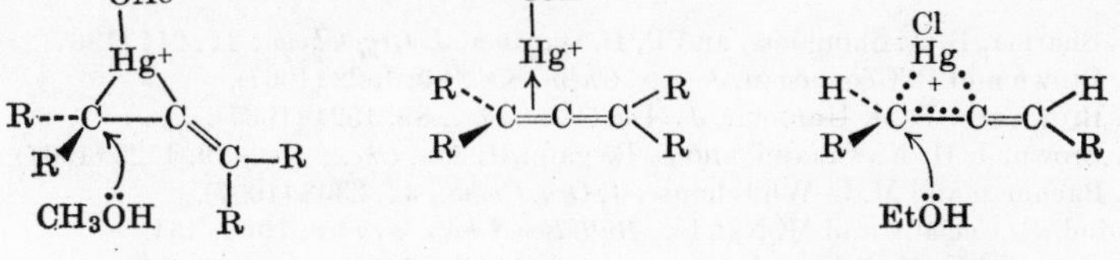

(23) **(24)** **(25)**

[73] W. Cocker, P. V. R. Shannon, and P. A. Staniland, *J. Chem. Soc., C*, **1967**, 915.
[74] G. Zweifel, H. Arzoumanian, and C. C. Whitney, *J. Am. Chem. Soc.*, **89**, 3652 (1967).
[75] W. L. Waters and E. F. Kiefer, *J. Am. Chem. Soc.*, **89**, 6261 (1967).

chloride, which proceeded by opening of the intermediate mercury complex at the outer carbon (**25**). The same orientation was observed for the addition of hydrogen chloride and of water to these compounds.[76]

A convenient procedure for the hydration of double bonds is provided by mercuration with mercuric acetate in aqueous tetrahydrofuran, followed by reduction with sodium borohydride. Since the overall hydration is Markovnikov this complements hydroboration.[77] This oxymercuration–demercuration procedure applied to cyclic and bicyclic olefins gave tertiary alcohols whenever possible, with hydration predominantly or exclusively from the less hindered side of the molecule[78, 79] (see also p. 8). Olefins react with solutions containing mercuric and nitrite ions to give 2-nitroalkylmercuric salts. This "nitromercuration" probably involves the formation of a cyclic mercurinium ion and attack of this by the ambident nitrite ion. Since there will probably be little carbonium ion character developed on the carbons of the mercurinium ion, this attack will be slow and reversible and thus thermodynamically controlled.[80] The addition of mercuric acetate to methyl oleate has also been investigated.[81] The relative proportions of the two cinnamic acids formed in the hydrocarboxylation of monosubstituted diphenylacetylenes with nickel carbonyl and aqueous acetic acid in methanol have been measured. These can be correlated with substituent σ-constants for *meta*- and *para*-derivatives; Markovnikov addition is observed. *ortho*-Substituents cause preferential formation of the α-arylcinnamic acid regardless of their electronic properties. Possible mechanisms are discussed.[82] Cobalt carbonyl-catalysed hydrocarboxylation of various dienes has been reported.[83]

Epoxidation. In spite of trenchant criticism last year[84] the suggestion of a 1,3-dipolar mechanism for peracid epoxidation of olefins has been maintained. It is claimed that 1,3-dipolar reagents without a double bond in the sextet structure, such as the proposed active reagent in epoxidation, $RC^{+}(OH)OO^{-}$, should not necessarily show large rate differences for reactions with cyclohexene and norbornene. However, the proposed 1,2-dioxolan intermediate has been shown to be much too stable to be involved in epoxidations and it is now replaced by a transition state of similar geometry.[85a] The clearest distinc-

76 R. K. Sharma, B. A. Shoulders, and P. D. Gardner, *J. Org. Chem.*, **32**, 241 (1967).
77 H. C. Brown and P. Geoghegan, *J. Am. Chem. Soc.*, **89**, 1522 (1967).
78 H. C. Brown and W. J. Hammar, *J. Am. Chem. Soc.*, **89**, 1524 (1967).
79 H. C. Brown, J. H. Kawakami, and S. Ikegami, *J. Am. Chem. Soc.*, **89**, 1525 (1967).
80 G. B. Bachman and M. L. Whitehouse, *J. Org. Chem.*, **32**, 2303 (1967).
81 A. Tubul, E. Ucciani, and M. Naudet, *Bull. Soc. Chim. France*, **1967**, 464.
82 C. W. Bird and E. M. Briggs, *J. Chem. Soc., C*, **1967**, 1265.
83 N. M. Bogoradovskaya, N. S. Imyanitov, and D. M. Rudkovskii, *Zh. Prikl. Khim.*, **39**, 2807 (1966); *Chem. Abs.*, **66**, 85213g (1967).
84 *Organic Reaction Mechanisms*, **1966**, 136.
85a H. Kwart, P. S. Starcher, and S. W. Tinsley, *Chem. Commun.*, **1967**, 335.

tion between Kwart's mechanism and that normally accepted has thus disappeared. For a further criticism of the 1,3-dipolar addition mechanism see ref. 85*b*, footnote 8.

The retarding effects of *gem*-dimethyl groups on the rate of epoxidation of cyclopentene and methylenecyclobutane, and on the direction of epoxidation of some methylenecyclohexanes, in various weakly polar solvents have been measured. The (small) changes in rate and direction with solvent are attributed partly to less favourable non-bonded interactions between the hydrocarbon substituents and the solvating solvent in the transition state, since these differences are greatest in polar halogenated solvents.[86] 3-Oxo-Δ^4-steroids without polar substituents reacted with alkaline hydrogen peroxide to give the β-epoxide, but with a polar substituent at position 17 or in the side chain appreciable amounts of the α-epoxide were formed. This was rationalized in terms of the accepted mechanism (reversible nucleophilic addition of the hydroperoxide anion to the double bond, followed by slow displacement of hydroxide ion) by considering the effect of the polar substituent on the negative charge of the enolate group. In the transition state for α-epoxide formation the electrostatic attractive force passes through the steroid molecule and is therefore maximal; for the β-isomer it passes through the higher dielectric of the solvent.[87]

Epoxidation of unhindered methylenecyclohexanes with peracids occurred predominantly by axial attack, whilst more hindered olefins underwent predominantly equatorial attack. The possibly more bulky reagent in alkaline hydrogen peroxide–benzonitrile (peroxybenzimidic acid) gave predominantly equatorial attack in all cases.[88] The rates of reaction of *cis*- and *trans*-1,2-dineopentylethylene with *m*-chloroperbenzoic acid and of related olefins with perbenzoic acid showed that the steric effect of a neopentyl group in these reactions is less than in reactions involving attack at carbonyl-carbon; models show this to be reasonable.[89] The importance of sp^2–sp^2 torsional strain as a contributor to the high reactivity of medium-ring *trans*-double bonds, as measured for example by epoxidation rates, has been discussed.[90] Other topics investigated are the reaction of sulphur atoms with olefins to give episulphides,[91] participation by the 6α-hydroxyl group in the α-epoxidation of $\Delta^{9,10}$-steroids,[92] stereochemistry of epoxidation of

85*b* B. Rickborn and J. H.-H. Chan, *J. Org. Chem.*, **32**, 3576 (1967).

86 R. C. Ewins, H. B. Henbest, and M. A. McKervey, *Chem. Commun.*, **1967**, 1085.

87 H. B. Henbest and W. R. Jackson, *J. Chem. Soc., C*, **1967**, 2459; H. B. Henbest, W. R. Jackson, and I. Malunowicz, *ibid.*, p. 2469.

88 R. G. Carlson and N. S. Behn, *J. Org. Chem.*, **32**, 1363 (1967).

89 M. S. Newman, N. Gill, and D. W. Thomson, *J. Am. Chem. Soc.*, **89**, 2059 (1967).

90 F. H. Allen, E. D. Brown, D. Rogers, and J. K. Sutherland, *Chem. Commun.*, **1967**, 1116.

91 E. M. Lown, E. L. Dedio, O. P. Strausz, and H. E. Gunning, *J. Am. Chem. Soc.*, **89**, 1056 (1967).

92 J.-C. Guilleux and M. Mousseron-Canet, *Bull. Soc. Chim. France*, **1967**, 24.

7-*tert*-butoxynorbornadiene,[93] kinetics of the reaction of *tert*-butyl hydroperoxide with isobutene and α-methylstyrene,[94] and epoxidation of alkylidenecyanoacetic esters with hydrogen peroxide catalysed by sodium tungstate or trisodium phosphate.[95]

Nucleophilic Additions

The reactivities of a wide range of amines in addition to *p*-tolyl vinyl sulphone are very sensitive to their steric requirement, much as in nucleophilic aromatic substitution, another reaction that takes place at sp^2-carbon. However, primary amines were less reactive than secondary amines with ethanol as solvent, and this was attributed to their greater degree of solvation, through hydrogen-bonding by both amino-hydrogen atoms. In benzene solution the reactions were of the first order in sulphone and also in amine in ethanol but of the second order in amine. Thus a second molecule of amine, or one of ethanol, is involved in the transition state and therefore the concerted proton transfer (**26**) is proposed.[96a] However, considerable anionic character was shown to be developed on the carbon adjacent to the sulphonyl group, by a Hammett treatment of the addition of *tert*-butylamine to nuclear-substituted aryl vinyl sulphones. Thus $C_{(\beta)}$–N bond-formation must be well advanced in the transition state, and the great sensitivity of the reaction to the bulk of both amine and olefin is understandable.[96b] The latter was shown by replacement of an α- and a β-H by methyl, which reduced the reactivity towards piperidine 6500- and 840-fold, respectively. Entropies of activation are very negative, supporting the highly ordered transition state (**26**).[96c]

In the reaction of secondary amines with methyl propiolate and acetylenedicarboxylate both *cis*- and *trans*-addition are observed;[97] both isomers arise from the same intermediate (**27**), which by intramolecular proton shift gives *cis*-addition (**28**), and by intermolecular protonation gives both adducts, but mainly the *trans* (**29**). The kinetics of these reactions were studied for dilute solutions in aprotic solvents where only the *cis*-adducts were formed. The reactions were second-order, and the rate increased with solvent polarity, as is to be expected if formation of the zwitterion (**27**) is rate-determining. Aziridine was only about one-hundredth as reactive as piperidine.[98] Measurement of addition rates for amino-acids, peptides, and proteins to αβ-unsaturated compounds (mostly acrylonitrile) has been extended to solutions in an

[93] G. W. Klumpp, A. H. Veefkind, W. L. de Graaf, and F. Bickelhaupt, *Ann. Chem.*, **706**, 47 (1967).
[94] V. L. Antonovskii and Y. D. Emelin, *Neftekhimiya*, **6**, 733 (1966); *Chem. Abs.*, **66**, 28165u (1967).
[95] M. Igarashi and H. Midorikawa, *J. Org. Chem.*, **32**, 3399 (1967).
[96] S. T. McDowell and C. J. M. Stirling, *J. Chem. Soc., B*, **1967**, (*a*) 343, (*b*) 348, (*c*) 351.
[97] R. Huisgen, B. Giese, and H. Huber, *Tetrahedron Letters*, **1967**, 1883.
[98] B. Giese and R. Huisgen, *Tetrahedron Letters*, **1967**, 1889.

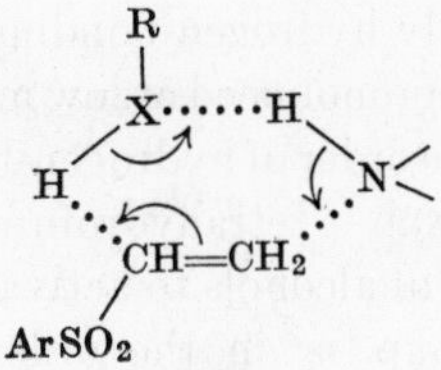

(26)

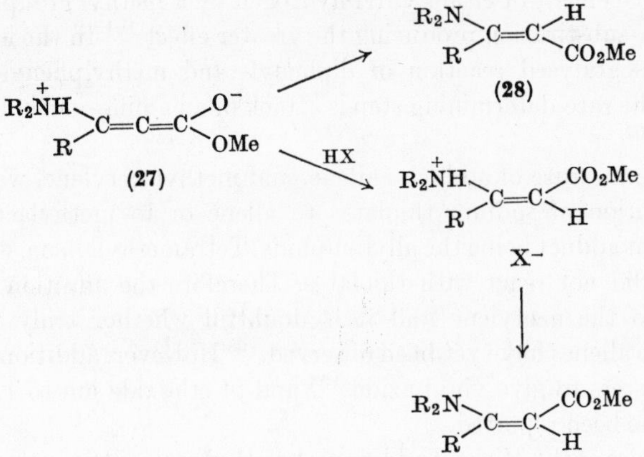

(27) **(28)** **(29)**

aqueous buffer–DMSO mixture where rates are more sensitive to pK changes than in aqueous solution. The results were rationalized in terms of participation of DMSO in acid–base equilibria, changes in ground- and transition-state stabilization and in amine hydrogen-bonding (since DMSO is a strong hydrogen-bond acceptor only).[99]

The change in volume of activation with solvent has been measured for the addition of ammonia and of thiophenoxide ion to mesityl oxide. A zwitterionic transition state is postulated for the former, and for the latter the new carbon–sulphur bond appears to be nearly complete in the transition state.[59] The kinetics of the methoxide ion-catalysed addition of methanol to acrylonitrile have been measured in mixtures of methanol with DMSO, DMF, DMF–dioxan, dioxan, tetrahydrofuran, and benzene. The rate was of the first order in acrylonitrile, of the first order in metal methoxide, and proportional to $1/[\mathrm{MeOH}]^n$ where n depends on the aprotic solvent. The rate increased with increasing concentration of the aprotic solvent since this increased the concentration of free methoxide ions by suppressing the methanol–methoxide equilibrium, and by specifically desolvating methanol–solvated methoxide

[99] M. Friedman, *J. Am. Chem. Soc.*, **89**, 4709 (1967).

ions by strongly hydrogen-bonding the methanol. This enhancement of rate was especially pronounced at low methanol concentrations. The rate constants increased in the order of hydrogen-bonding capacity of the aprotic solvent, i.e., benzene < dioxan < tetrahydrofuran < DMF < DMSO.[100] In the base-catalysed addition of alcohols to activated olefins the rate-enhancing effect of the activating group is in the order $CONHR < CONH_2 < CONR_2 < CO_2R < SO_2NR_2 < CN < SO_2R < COR < {}^+PR_3$ where R is the same lower alkyl group. Replacement of either vinyl hydrogen by a methyl group reduces the rate, with α-substitution producing the greater effect.[101] In the low-temperature amine-catalysed reaction of diphenyl- and methylphenyl-keten with methanol the rate-determining step is attack of an amine–alcohol complex on the keten.[102]

The same mixture of adducts, allene, and methylacetylene, were obtained in the addition of sodium thiolates to allene or to methylacetylene, the predominant adduct being the allyl sulphide. Tetramethylallene, which cannot isomerize, did not react with thiolates. Therefore the addition observed is probably to the acetylene and it is doubtful whether truly nucleophilic additions to allenes have yet been observed.[103] However, addition of azide ion to allenic esters to give vinyl azides[104] and of ethoxide ion to 1,1-diphenylallene[105] has been reported.

The kinetics of the Michael addition of diethyl malonate to ethyl cinnamate were consistent with the generally accepted mechanism of pre-equilibrium formation of malonate anion and slow addition of this to the double bond.[106] The kinetics of the reactions of hydroxylamine and *O*-methylhydroxylamine with cytidine suggests that the product, uridine 4-oxime, can be formed directly and not only by prior addition of the hydroxylamine to the 4,5-double bond.[107] The direct formation of deoxyuridine 4-oxime from deoxycytidine has also been demonstrated.[108] Phenyl-lithium adds to cyclopropene with greater than 99% stereospecificity, to give *cis*-2-phenylcyclopropyl-lithium.[109] The addition of non-carbon nucleophiles to cyclopropenes has been further studied.[110]

Benkeser has reviewed his work on addition reactions of silane.[111] The very

100 B.-A. Feit, J. Sinnreich, and A. Zilkha, *J. Org. Chem.*, **32**, 2570 (1967).
101 R. N. Ring, G. C. Tesoro, and D. R. Moore, *J. Org. Chem.*, **32**, 1091 (1967).
102 A. Tille and H. Pracejus, *Chem. Ber.*, **100**, 196 (1967).
103 W. H. Mueller and K. Griesbaum, *J. Org. Chem.*, **32**, 856 (1967).
104 P. Beltrame, D. Pitea, A. Marzo, and M. Simonetta, *J. Chem. Soc., B*, **1967**, 71.
105 G. R. Harvey and K. W. Ratts, *J. Org. Chem.*, **31**, 3907 (1966).
106 K. T. Finley, D. R. Call, G. W. Sovocool, and W. J. Hayles, *Can. J. Chem.*, **45**, 571 (1967).
107 N. K. Kochetkov, E. I. Budowsky, E. D. Sverdlov, R. P. Shibaeva, V. N. Shibaev, and G. S. Monastirskaya, *Tetrahedron Letters*, **1967**, 3253.
108 P. D. Lawley, *J. Mol. Biol.*, **24**, 75 (1967).
109 J. G. Welch and R. M. Magid, *J. Am. Chem. Soc.*, **89**, 5300 (1967).
110 T. C. Shields and P. D. Gardner, *J. Am. Chem. Soc.*, **89**, 5425 (1967).
111 R. A. Benkeser, *Pure Appl. Chem.*, **13**, 133 (1966).

rapid reaction of dimethylphenyl-, methyldiphenyl-, and triphenyl-silyl-lithium with 1,1-diphenylethylene is of first order in each reactant and probably involves nucleophilic addition of the monomeric silyl-lithium.[112] Stereospecific platinum-catalysed hydrosilation of oct-1-ene,[113] the addition of di-isobutylaluminium hydride to isomeric *n*-undecenes,[114] and the dicobalt octacarbonyl-catalysed addition of silicon hydrides to terminal olefins,[115] have been reported. The kinetics and stereochemistry of addition of trialkyltin hydrides (hydrostannation) to electrophilic acetylenes have been studied; substituent, solvent, and isotope effects support an ionic mechanism in which rate-determining nucleophilic transfer of the hydride hydrogen is well advanced in the transition state (**30**) and is followed by *trans*-addition of the tin cation to the vinyl carbon.[116]

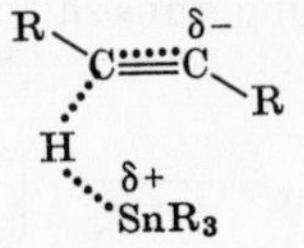

(**30**)

The following nucleophilic additions of thiols and their anions have been investigated: additions to 2-methylbut-1-en-3-yne,[117] of butanethiol to cyclopentenes,[118] of thiolate anions to vinylacetylene,[119] and of thiols to 1,5-diacetylenes.[120] Other reports include: influence of solvent and base-concentration on the site of cyanoethylation of 2-naphthol;[121] an unusual double Michael reaction where isophorone acts, in turn, as donor and acceptor;[122] stereospecific addition of lithium diphenylarsenide to alkynes;[123] "addition" of hydrogen cyanide to isobutene in the presence of a cuprous halide to give *tert*-butyl isocyanide–cuprous complexes;[124] sodium- and

112 A. G. Evans, M. L. Jones, and N. H. Rees, *J. Chem. Soc., B*, **1967**, 961.
113 L. H. Sommer, K. W. Michael, and H. Fujimoto, *J. Am. Chem. Soc.*, **89**, 1519 (1967).
114 F. Asinger, B. Fell, and F. Theissen, *Chem. Ber.*, **100**, 937 (1967).
115 A. J. Chalk and J. F. Harrod, *J. Am. Chem. Soc.*, **89**, 1640 (1967).
116 A. J. Leusink, H. A. Budding, and J. W. Marsman, *J. Organometal. Chem.*, **9**, 285 (1967); A. J. Leusink, H. A. Budding, and W. Drenth, *ibid.*, p. 295.
117 T. L. Jacobs and A. Mihailovski, *Tetrahedron Letters*, **1967**, 2607.
118 M. Procházka, L. Streinz, and V. Všetečka, *Coll. Czech. Chem. Commun.*, **32**, 3799 (1967).
119 E. N. Prilezhaeva and V. N. Petrov, *Izv. Akad. Nauk SSSR, Ser. Khim.*, **1966**, 1494; *Chem. Abs.*, **66**, 54757h (1967); V. N. Petrov, G. M. Andrianova, and E. N. Prilezhaeva, *Izv. Akad. Nauk SSSR, Ser. Khim.*, **1966**, 2180; *Chem. Abs.*, **66**, 85217m (1967).
120 A. A. Petrov and M. P. Forost, *Zh. Organ. Khim.*, **2**, 1178, 1358 (1966); *Chem. Abs.*, **66**, 37065p, 54751b (1967).
121 K. H. Takemura, K. E. Dreesen, and M. E. Petersen, *J. Org. Chem.*, **32**, 3412 (1967).
122 H. L. Brown, G. L. Buchanan, A. F. Cameron, and G. Ferguson, *Chem. Commun.*, **1967**, 399.
123 A. M. Aguiar, T. G. Archibald, and L. A. Kapicak, *Tetrahedron Letters*, **1967**, 4447.
124 S. Otsuka, K. Mori, and K. Yamagami, *J. Org. Chem.*, **31**, 4170 (1966).

potassium-catalysed addition of 4-methyl- and 4-propyl-pyridine to butadiene;[125] the relative rates of reaction of water and potassium cyanide with substituted benzylidenemalononitriles;[126a] and the addition of primary and secondary amines to activated acetylenes.[126b]

Radical Additions

Elad has reviewed photochemically initiated, homolytic addition reactions of olefins.[127]

Two new sets of results on thiol additions have been discussed in terms of partial bridging in the intermediate radicals. Co-oxidation of substituted thiophenols with indene (leading to hydroxy-sulphoxides) gives both *cis*- and *trans*-addition, but the proportion of *cis*-adduct falls when the thiophenol contains electron-releasing substituents. Such substituents (e.g., *p*-methyl) were considered to favour bridging, thus shifting the supposed equilibrium (6)

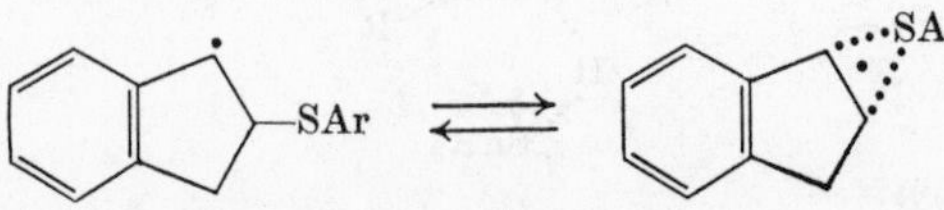

...(6)

to the right and therefore giving more of the *trans*-adduct.[128] All four stereoisomeric adducts (**32**—**35**) are obtained from the chloro-olefin (**31**) and thioacetic acid;[129] however, the product distribution varies markedly with temperature and thiol concentration. The proportion of the products (**32** and **33**) resulting from axial attack of thiyl radicals is largest from reactions conducted at low temperature and with high thiol concentration; presumably this reflects a kinetic preference for axial attack by thiyl radical, with successful interception of the intermediate adduct radical. Because thiyl addition is reversible, higher temperatures or lower thiol concentrations favour product formation from the thermodynamically preferred equatorial adduct radical [giving (**34**) and (**35**)]. More important, however, is the variation of the proportions of (**32**) and (**33**), and of (**34**) and (**35**). The first pair show a strong preference for *trans*-addition, particularly at low temperature. Whilst this clearly points to partial bridging, very similar results are obtained with a wide range of thiols, some of which might be expected to have a much greater, or much smaller, propensity for bridging. Also puzzling is the fact that the ratio of yields (**34**):(**35**) (= *ca.* 2:1) is relatively insensitive to reaction conditions and is substantially larger than that for other thiol additions to (**31**).

[125] H. Pines and J. Oszczapowicz, *J. Org. Chem.*, **32**, 3183 (1967).
[126a] R. B. Pritchard, C. E. Lough, J. B. Reesor, H. L. Holmes, and D. J. Currie, *Can. J. Chem.*, **45**, 775 (1967).
[126b] C. H. McMullen and C. J. M. Stirling, *J. Chem. Soc., B*, **1966**, 1217, 1221.
[127] D. Elad, *Fortschr. Chem. Forsch.*, **7**, 528 (1967).
[128] H. H. Szmant and J. J. Rigau, *Tetrahedron Letters*, **1967**, 3337.
[129] N. A. LeBel and A. DeBoer, *J. Am. Chem. Soc.*, **89**, 2784 (1967).

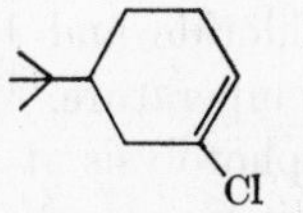

(31)

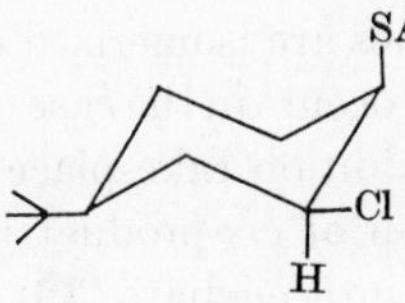

(32)

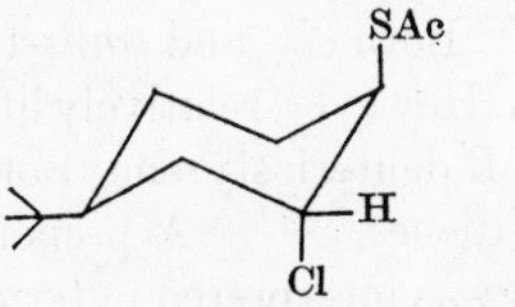

(33)

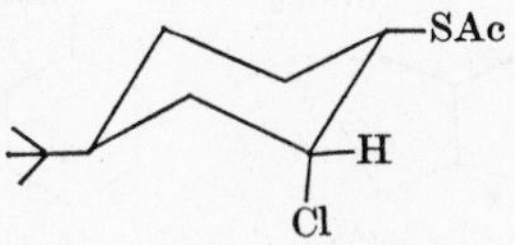

(34)

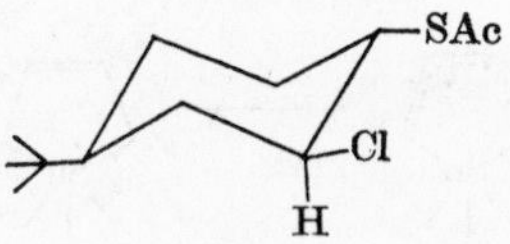

(35)

There was no evidence for reversal of thiyl radical addition to *trans*-Δ^2-octalin (**36**).[130] This is an unusually strained cyclohexene, and relief of strain in forming the adduct radical was considered responsible for the irreversible nature of this step. The preference (90%) for axial addition was attributed to

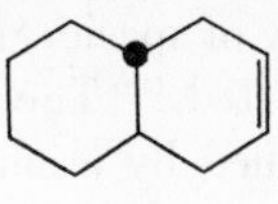

(36)

the fact that the adduct radical leading to the axial product is formed in a chair conformation, whilst its epimer is produced in a twist-boat conformation.

Addition of thiols to allylic halides can give rearranged products.[131] Thus allyl chloride with an alkanethiol gives a mixture of (**37**) and (**38**). The more

$$CH_2{:}CHCH_2Cl + RSH \rightarrow \underset{\textbf{(37)}}{RSCH_2CH_2CH_2Cl} + \underset{\textbf{(38)}}{RSCH_2CHClCH_3}$$

reactive arenethiols give only the aryl counterpart of (**37**); but, even with arenethiols, allyl bromide gives exclusively rearranged products. The rearrangement was attributed to loss of a halogen atom from the intermediate adduct radical, followed by electrophilic addition of HX to the resulting allyl sulphide derivative. This was supported by isolation of a crossed adduct of HX to a substituted allyl sulphide present in a reaction mixture.

130 E. S. Huyser, H. Benson, and H. J. Sinnige, *J. Org. Chem.*, **32**, 622 (1967).
131 D. N. Hall, *J. Org. Chem.*, **32**, 2082 (1967).

A complex between allyl chloride and HBr has been detected spectroscopically at liquid-nitrogen temperature.[132] This is partially dissociated at 100°K, at which temperature photolysis at the absorption maximum of the complex will initiate chain addition.

Two groups have reported on the reversible addition of stannyl radicals to olefins.[133] Both *cis*- and *trans*-1-deuterioalk-1-enes are isomerized during the addition. However, relatively little isomerization occurs in the case of the more reactive *β*-deuteriostyrene. Both 1,2- and 1,4-addition take place with conjugated dienes.[134, 135] A preponderant formation of *cis*-product in the 1,4-addition was interpreted in terms of the bridged intermediate (**39**). The rapid

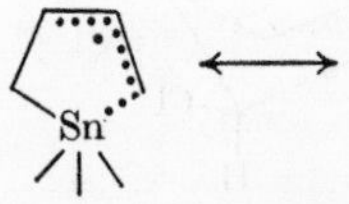

(**39**)

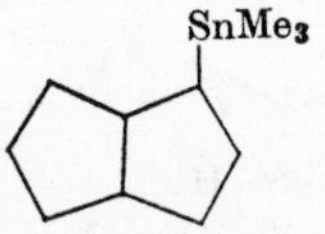

(**40**)

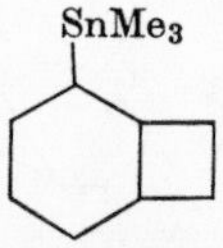

(**41**)

chain transfer by tin hydrides leads to almost exclusive 1,2-adduct formation from cyclo-octa-1,5-diene at 36°, but at higher temperatures transannular addition competes more favourably, giving significant yields of compound (**40**). At 175° a further product was detected, and this appears to be the alternative transannular adduct (**41**).[135]

Radical additions of a variety of species, RH, across the carbonyl group of perfluoro-ketones has been reported.[136] The major product of alkane addition is a tertiary alcohol, accompanied by a small amount of ether. Rapid and reversible alkyl radical addition at oxygen is believed to occur, but the greater reactivity of an alkoxyl radical in hydrogen abstraction is dominant in product formation. On the other hand, the principal product from addition of benzaldehyde is a benzoate ester.[137]

$$R^F_2CO + R\cdot \rightleftharpoons R^F_2\dot{C}{-}OR \xrightarrow{RH} R^F_2CHOR$$

$$R^F_2CRO\cdot \xrightarrow{RH} R^F_2CROH$$

An interesting addition, which is probably quite general, is the formation of nitroxides from nitrones (reaction **7**).[138]

132 E. I. Yakovenko, G. B. Sergeev, and M. M. Rakhimov, *Dokl. Akad. Nauk SSSR*, **171**, 890 (1966); *Chem. Abs.*, **66**, 94530k (1967).

133 W. P. Neumann, H. J. Albert, and W. Kaiser, *Tetrahedron Letters*, **1967**, 2041; H. G. Kuivila and R. Sommer, *J. Am. Chem. Soc.*, **89**, 5616 (1967).

134 W. P. Neumann and R. Sommer, *Ann. Chem.*, **701**, 28 (1967).

135 R. H. Fish, H. G. Kuivila, and I. J. Tyminski, *J. Am. Chem. Soc.*, **89**, 5861 (1967).

136 E. G. Howard, P. B. Sargeant, and C. G. Krespan, *J. Am. Chem. Soc.*, **89**, 1422 (1967).

137 W. H. Urry, A. Nishihara, and J. H. Y. Niu, *J. Org. Chem.*, **32**, 347 (1967).

138 M. Iwamura and N. Inamoto, *Bull. Chem. Soc. Japan*, **40**, 703 (1967).

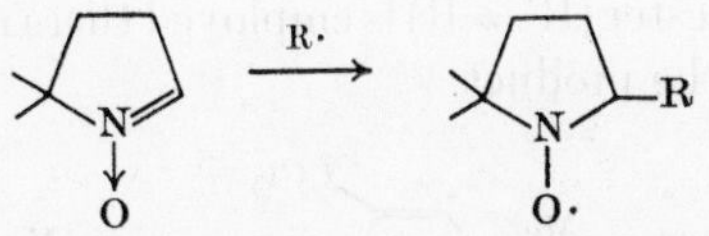

...(7)

Poutsma has commented further on the competition between radical and ionic chlorination of olefins,[139] including an extension to conjugated dienes. The rate of addition of Cl· to dienes is comparable to that of addition to mono-olefins, reflecting the low selectivity of chlorine atoms (cf. p. 131).

Two new examples of additions to 1,6-dienes have been reported,[140] in which the initial adduct radical can undergo intramolecular cycloaddition. The now familiar preference for formation of five-membered rings obtains.

Intramolecular thiol addition has been examined by Surzur *et al.*,[141] and some typical results are annexed. The proportions of products did, however, vary somewhat depending on the method of initiation.

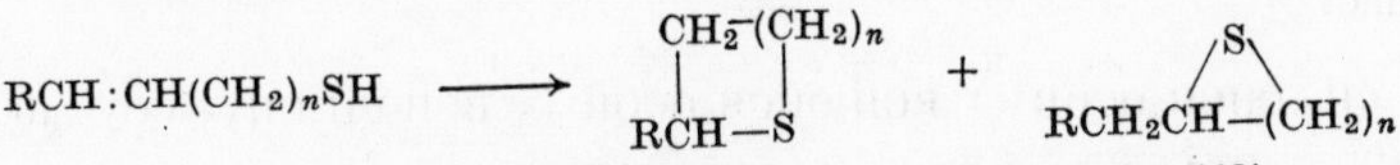

(42) **(43)**

Proportion of monocyclic products from **(42)** *and* **(43)**

R	H	H	Et	H
n	2	3	3	4
42:43	99:< 1	72:28	30:70	15:85

The cyclization of (**44**) with butyl-lithium appears to involve homolytic addition to the acetylene,[142] as a vinyl proton signal shows strongly in the

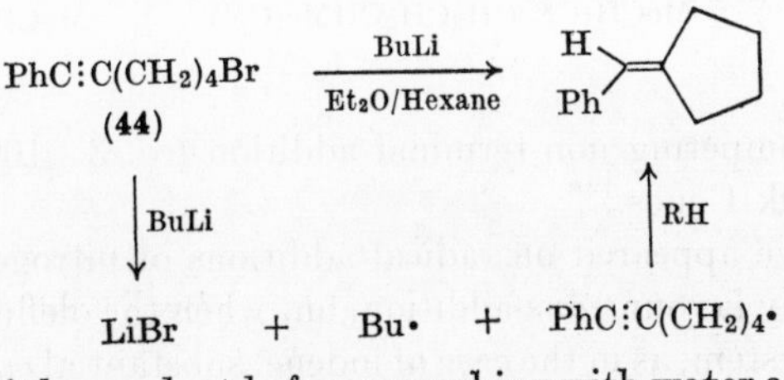

NMR spectrum of the product before quenching with water and, furthermore, only 10% of deuterium is incorporated when quenching is effected with D_2O. Products arising from butyl radicals were also detected. Cyclization of the propargylic ester (**45**) with carbon tetrachloride[143] has a parallel in the homolytic reaction of hept-1-yne with carbon tetrachloride.[144] In the former

139 M. L. Poutsma, *Science*, **157**, 997 (1967); *J. Org. Chem.*, **31**, 4167 (1966); see also *Organic Reaction Mechanisms*, **1965**, 105, 120.

140 N. O. Brace, *J. Org. Chem.*, **32**, 2711 (1967); S. F. Reed, *ibid.*, p. 3675.

141 J.-M. Surzur, M.-P. Crozet, and C. Dupuy, *Compt. Rend., Ser. C*, **264**, 610 (1967).

142 H. R. Ward, *J. Am. Chem. Soc.*, **89**, 5517 (1967).

143 E. I. Heiba and R. M. Dessau, *J. Am. Chem. Soc.*, **89**, 2238 (1967).

144 E. I. Heiba and R. M. Dessau, *J. Am. Chem. Soc.*, **89**, 3772 (1967).

case, if an asymmetric ester (R′ ≠ R) is employed there is appreciable retention of optical activity in the product.

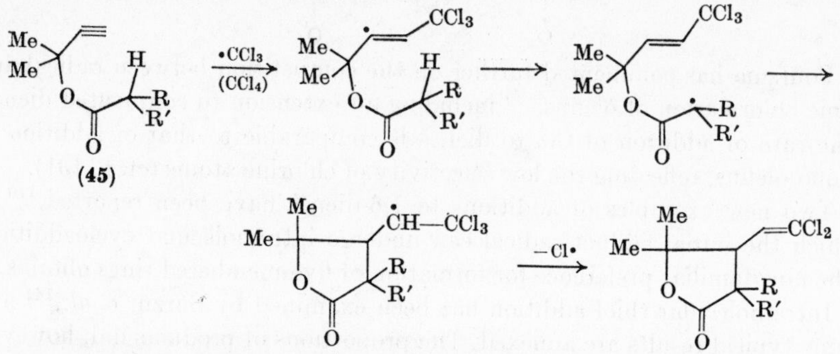

(45)

The gas-phase thermolysis of allyl esters involves the addition–elimination sequence:[145]

$$CH_2{:}CHCH_2OCOR \xrightarrow{R\cdot} RCH_2\dot{C}HCH_2OCOR \rightarrow RCH_2CH{:}CH_2 + CO_2 + R\cdot$$

In the electrolytic hydrodimerization of methacrylonitrile, some 10% of the unsymmetrical product **(46)** accompanies the symmetrical dimer.[146] If the mechanism shown is assumed, this superficially surprising result is in fact predicted by the theoretical calculations of Figeys.[147] Russian workers have

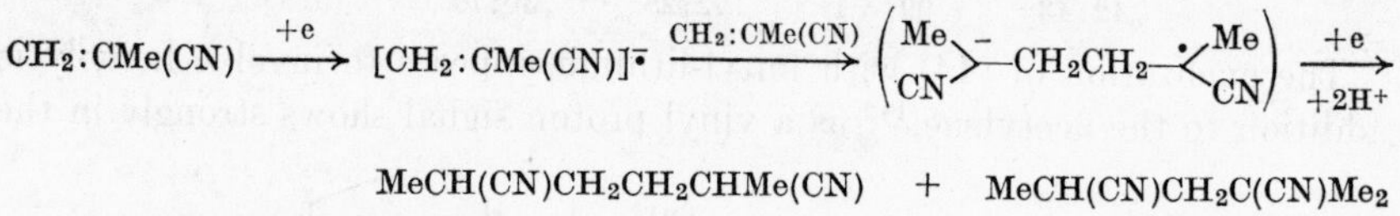

(46)

also reported competing non-terminal addition (*ca.* 3—10%) of acetone or acetonitrile to alk-1-enes.[148]

New data have appeared on radical additions of nitrogen dioxide. Steric factors commonly favour *trans*-addition, but when the olefin has an extensive flat π-electron system, as in the case of indene, substantial *cis*-addition occurs, perhaps reflecting complex formation between indene and N_2O_4.[149] The major product isolated from the reaction of N_2O_4 with α-bromostyrene in ether was ω-nitroacetophenone.[150] Related cases were also reported.

145 R. Louw and E. C. Kooyman, *Rec. Trav. Chim.*, **86**, 147 (1967).

146 G. C. Jones and T. H. Ledford, *Tetrahedron Letters*, **1967**, 615.

147 M. Figeys, *Tetrahedron Letters*, **1967**, 2237.

148 G. I. Nikishin, M. G. Vinogradov, and R. V. Kereselidze, *Zh. Org. Khim.*, **2**, 1918 (1966); *Chem. Abs.*, **66**, 75468e (1967).

149 H. Shechter, J. J. Gardikes, T. S. Cantrell, and G. V. D. Tiers, *J. Am. Chem. Soc.*, **89**, 3005 (1967).

150 S. V. Vasil'ev and O. T. Burdelev, *Zh. Org. Khim.*, **2**, 1961, 1965 (1966); *Chem. Abs.*, **66**, 94473u, 94474v (1967).

Neale and Marcus have explored the addition of *N*-chloroamines to a wide variety of unsaturated molecules in acetic–sulphuric acid mixtures.[151] Under these conditions, addition (of $R_2\overset{\cdot+}{N}H$) is strongly preferred to hydrogen abstraction although, when R is sufficiently large, intramolecular (Hofmann–Loeffler) hydrogen abstraction does compete effectively. Photolytic additions of chloramine itself[152] and of *N*-chlorourethane[153] have also been reported.

There have been further reports of radical additions to chlorinated norbornadienes,[154] including new examples of vinyl migration and interception of the cyclopropylcarbinyl intermediate. Whilst electronic and steric effects are clearly both important factors in determining the product composition, these are still not fully understood.

An extensive study of relative reactivities of a series of olefins towards addition of methyl mercaptoacetate has been reported.[155] Here again, traces of products arising from non-terminal addition to terminal olefins were detected. The results were consistent with the assignment of electrophilic character to thiyl radicals. A competitive approach to the relative reactivities of compounds RH (related to malonic ester) in their additions to oct-1-ene has also been reported.[156] Typically, ethyl cyanoacetate, the most reactive addend studied, is *ca.* 12 times more reactive than diethyl malonate, but still has only one-eighth of the reactivity of carbon tetrachloride.

New examples of homolytic transannular addition[157] and hydrogen transfer[158] have been noted in the cyclo-octene series.

Further data on the addition of $BrCH(CN)_2$ have also appeared,[159] and homolytic additions of cyclopentanone and cyclohexanone to terminal olefins have been noted.[160] Cyclo-octane adds to reactive olefins such as acrylic acid, in the presence of radical initiators,[161] and iodoacetic acid participates in a free-radical reaction with terminal olefins, to give unsaturated acids.[162] Unsubstituted carboxylic acids have been added (at their α-positions) to

151 R. S. Neale, *J. Org. Chem.*, **32**, 3263 (1967); R. S. Neale and N. L. Marcus, *ibid.*, p. 3273.

152 Y. Ogata, Y. Izawa, and H. Tomioka, *Tetrahedron*, **23**, 1509 (1967).

153 K. Schrage, *Tetrahedron*, **23**, 3033, 3039 (1967).

154 C. K. Alden and D. I. Davies, *J. Chem. Soc., C*, **1967**, 1017, 2007; D. I. Davies and P. J. Rowley, *ibid.*, pp. 2245, 2249; see also *Organic Reaction Mechanisms*, **1965**, 118, 119; **1966**, 143.

155 J. I. G. Cadogan and I. H. Sadler, *J. Chem. Soc., B*, **1966**, 1191.

156 J. I. G. Cadogan, D. H. Hey, J. T. Sharp, *J. Chem. Soc., B*, **1967**, 803; see also F. Asinger, B. Fell, and H.-H. Vogel, *Tetrahedron Letters*, **1967**, 3867.

157 A. C. Cope, M. A. McKervey, and N. M. Weinshenker, *J. Am. Chem. Soc.*, **89**, 2932 (1967).

158 J. G. Traynham and T. M. Couvillon, *J. Am. Chem. Soc.*, **89**, 3205 (1967).

159 P. Boldt, L. Schulz, and J. Etzemüller, *Chem. Ber.*, **100**, 1281 (1967); P. Boldt and L. Schulz, *Tetrahedron Letters*, **1967**, 4351; see also *Organic Reaction Mechanisms*, **1966**, 295.

160 G. I. Nikishin and G. V. Somov, *Zh. Org. Khim.*, **3**, 299 (1967); *Chem. Abs.*, **66**, 115028v (1967).

161 E. van Bruggen, H. Boelens, and F. Rijkens, *Rec. Trav. Chim.*, **86**, 958 (1967).

162 N. Kharasch, P. Lewis, and R. K. Sharma, *Chem. Commun.*, **1967**, 435.

allylcyclohexanes, to give naphthenic acids in yields of up to 30%;[163] CCl_3–SCl has been added to phenoxyethylene,[164] and *tert*-butyl peroxide initiates the addition of ethanol to β-pinene to give (**47**),[165] together with the corresponding ketone.

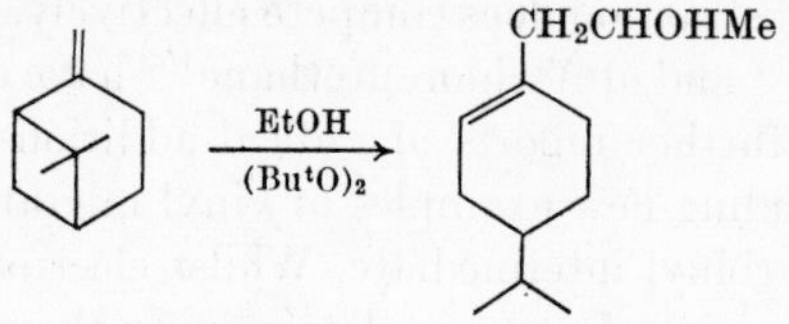

(**47**)

In a flow system, both hydroxy-radical adducts of allyl alcohol have been detected by ESR spectroscopy. At higher alcohol concentrations, 1:2 adduct radicals were observed.[166]

Methanol has been added to perfluorocyclopentene,[167] and chlorination of perfluorobut-2-ene has been studied.[168]

Tetrafluorohydrazine gives (**48**) with diphenylacetylene at 80°.[169] Because of the high reactivity of vinyl thioacetate, only a 2:1 adduct could be obtained from thioacetic acid and acetylene itself.[170]

$$F_2N—CPhF—CPh{=}NF$$

(**48**)

The gas-phase reactions of tetrafluorohydrazine with vinyl halides,[171] of dimethylamino-radicals from tetramethyltetrazene with ethylene,[172] and of bromotrichloromethane with fluorinated propenes[173] have been examined. Gas-phase addition of alkyl radicals to allene[174] and propyne[175] both occur at

163 S. I. Sadykh-Zade and R. A. Dzhalilov, *Dokl. Akad. Nauk, Azerb. SSR*, **22**, 17 (1966); *Chem. Abs.*, **66**, 94475w (1967).

164 E. F. Kolmakova, A. V. Kalabina, Y. K. Maksyutin, and L. N. Spiridonova, *Zh. Org. Khim.*, **2**, 2048 (1966); *Chem. Abs.*, **66**, 75472b (1967).

165 R. Lalande, B. Paskoff, and M. Cazaux, *Compt. Rend., Ser. C*, **264**, 1083 (1967).

166 P. Smith and P. B. Wood, *Can. J. Chem.*, **45**, 649 (1967); see also W. E. Griffiths, G. F. Longster, J. Myatt, and P. F. Todd, *J. Chem. Soc., B*, **1967**, 530.

167 R. F. Stockel and M. T. Beachem, *J. Org. Chem.*, **32**, 1658 (1967).

168 M. E. Roselli and H. J. Schumacher, *Anales Assoc. Quim. Arg.*, **53**, 23 (1965); *Chem. Abs.*, **66**, 37131g (1967).

169 R. C. Petry, C. O. Parker, F. A. Johnson, T. E. Stevens, and J. P. Freeman, *J. Org. Chem.*, **32**, 1534 (1967).

170 N. P. Petukhova and E. N. Prilezhaeva, *Zh. Org. Khim.*, **2**, 1947 (1966); *Chem. Abs.*, **66**, 75471g (1967).

171 A. J. Dijkestra, J. A. Kerr, and A. F. Trotman-Dickenson, *J. Chem. Soc., A*, **1967**, 864.

172 A. Good and J. C. J. Thynne, *J. Chem. Soc., B*, **1967**, 684.

173 J. M. Tedder and J. C. Walton, *Trans. Faraday Soc.*, **63**, 2678 (1967).

174 R. R. Getty, J. A. Kerr, and A. F. Trotman-Dickenson, *J. Chem. Soc., A*, **1967**, 979.

175 R. R. Getty, J. A. Kerr, and A. F. Trotman-Dickenson, *J. Chem. Soc., A*, **1967**, 1360.

the terminal carbon atom, and a technique has been developed for measuring the rates of alkyl-radical addition to olefins.[176]

A re-examination of the radical-chain bromination of ethylene has been reported.[177]

Diels–Alder Reactions

Valuable reviews on the mechanism of the Diels–Alder reaction[178a] and its use in heterocyclic syntheses[178b] have appeared.

In recent years the very general preference for *endo*-addition in the Diels–Alder reaction has been ascribed to secondary attractive forces operating between centres not bonded in the product, and this has been formulated in terms of orbital symmetry.[179] Herndon and Hall[180] have now suggested that these secondary relationships may be relatively unimportant and that the transition-state stabilities may be governed by a less subtle factor, namely, the geometrical overlap of the π-orbitals at the bond-forming primary centres. They calculate overlap integrals for the dimerization of cyclopentadiene based on transition-state models with the two planes parallel for the *exo*-addition and at 60° for the *endo*-addition. When all orbital interactions are considered, the *endo*-transition state is the more stable by 0.079β and the *primary* interactions account for over 90% of this difference. A reasonable value for the resonance integral ($\beta = 60$ kcal mole^{-1}) leads to an energy difference of 4.7 kcal mole^{-1}, to be compared with the experimental value[181] of 4.4 kcal mole^{-1}; a longer interaction distance would give a smaller energy difference, but it is claimed that the geometrical primary effect which favours *endo*-addition is still the most important factor, even if a highly unsymmetrical transition state is chosen. This effect also rationalizes the observed preference for *endo*-addition of mono-olefinic dienophiles such as cyclopropene and cyclopentene, not accounted for previously, and it predicts an even higher stereospecificity for the reaction of cyclopentadiene with cyclopentene than with itself; it will be very interesting to see if this is confirmed experimentally.[180] It was shown that *exo*- and *endo*-dicyclopentadiene both undergo thermal unimolecular dissociation into monomer without detectable competition from isomerization,[181] this agreeing with the dissociation–recombination mechanisms for this isomerization.[182] The LCAO MO approximation has been

[176] R. J. Cvetanović and R. S. Irwin, *J. Chem. Phys.*, **46**, 1694 (1967).
[177] T. Berces and Z. G. Szabo, *Acta Chim. Acad. Sci. Hung.*, **53**, 55 (1967).
[178a] J. Sauer, *Angew. Chem. Intern. Ed. Engl.*, **6**, 16 (1967).
[178b] J. Harner, "1,4-Cycloaddition Reactions", Academic Press, New York, 1967.
[179] R. Hoffmann and R. B. Woodward, *J. Am. Chem. Soc.*, **87**, 2046 (1965).
[180] W. C. Herndon and L. H. Hall, *Tetrahedron Letters*, **1967**, 3095.
[181] W. C. Herndon, C. R. Grayson, and J. M. Manion, *J. Org. Chem.*, **32**, 526 (1967); see also V. A. Mironov, T. M. Fadeeva, A. U. Stepanyants, and A. A. Akhrem, *Izv. Akad. Nauk SSSR, Ser. Khim.*, **1966**, 2213; *Chem. Abs.*, **66**, 85252u (1967).
[182] *Organic Reaction Mechanisms*, **1965**, 126; **1966**, 150.

applied to the Diels–Alder reaction[183] and to the electronic structure of the π-complex formed between diene and dienophile.[184] The bond order, p_{14}, between the termini of a diene system correlates well with the *para*-localization energy which is considered to be a reliable index of Diels–Alder reactivity. Hence the static index, p_{14}, which is readily calculated even for large molecules, also provides a useful guide to this reactivity.[185]

Kinetic data on the dimerization of butadiene to 4-vinylcyclohexene, the pyrolysis of cyclo-octa-1,5-diene to give butadiene and 4-vinylcyclohexene, and the reverse of these reactions are shown to be consistent with a common biradical intermediate, the octa-1,8-diyl-2,6-dienyl radical (**49**). The diradical mechanism was therefore favoured for these reactions.[186]

Diels–Alder reactions of *o*-benzoquinones can give two adducts, of type (**50**) and (**51**). The latter are the predominant products of kinetic control, but

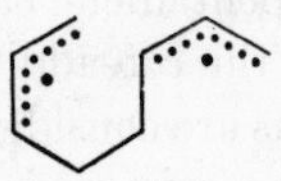
(49)

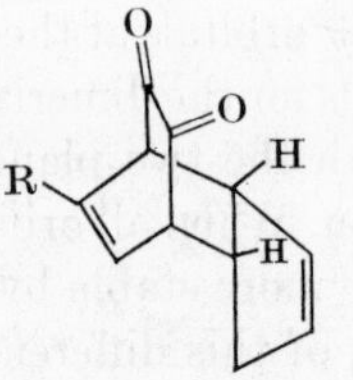

(50)

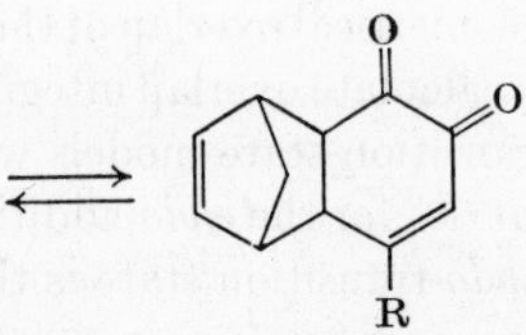

(51)

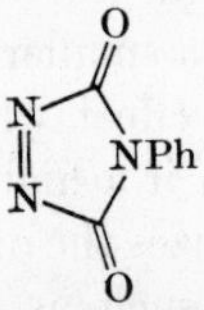

(52)

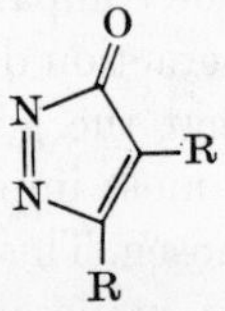

(53)

these rearrange thermally to the more stable adducts (**50**). This conversion is an intramolecular Cope rearrangement and not a retro-Diels–Alder reaction followed by recombination, as is shown by failure to trap cyclopentadiene.[187] The effect of changes in solvent, temperature, and Lewis acid-catalysis on asymmetric induction in the Diels–Alder reaction of cyclopentadiene with acrylic and fumaric esters of optically active alcohols has been investigated;[188a] aluminium chloride-catalysis of methyl acrylate–diene condensations have been studied further.[188b]

[183] W. C. Herndon and L. H. Hall, *Theor. Chim. Acta*, **7**, 4 (1967).
[184] P. Markov and N. Tyutyulkov, *Monatsh. Chem.*, **97**, 1229 (1966).
[185] I. R. Epstein, *Trans. Faraday Soc.*, **63**, 2085 (1967).
[186] S. W. Benson, *J. Chem. Phys.*, **46**, 4920 (1967).
[187] M. F. Ansell, A. F. Gosden, and V. J. Leslie, *Tetrahedron Letters*, **1967**, 4537.
[188a] J. Sauer and J. Kredel, *Tetrahedron Letters*, **1966**, 6359.
[188b] T. Inukai and T. Kojima, *J. Org. Chem.*, **32**, 869, 872 (1967).

cis-Azo-compounds have recently been shown to be much more reactive dienophiles than are their *trans*-isomers.[189] 4-Phenyl-1,2,4-triazole-3,5-dione (**52**) is particularly reactive with a wide variety of dienes, and its dienophilic reactivity is enhanced preferentially over that for additive substitution;[190] it is a more powerful dienophile than tetracyanoethylene.[191] Oxidation of suitable pyrazolinones with lead tetra-acetate in the presence of dienes gave the pyrazolones (**53**) which were trapped as their Diels–Alder adducts; adducts were obtained with dienes, such as cyclohexadiene and cycloheptatriene, that do not react with, for example, ethyl azodicarboxylate.[192]

Dicyanoacetylene adds 1,4 to benzene and its methyl derivatives at about 180°, to give bicyclo[2.2.2]octatrienes. The reaction is strongly catalysed by aluminium chloride; a 1:1 complex of the dicyanide and aluminium chloride is probably an intermediate. [2.2]Paracyclophane reacts with dicyanoacetylene to give 1:1 and 2:1 adducts (its first reported Diels–Alder adducts) at relatively low temperatures (120°), presumably because of a high ground-state energy.[193]

2,3-Homotropone added maleic anhydride and *N*-phenylmaleimide across the diene system, i.e., without involving the 2,3-cyclopropyl bond.[194] 2-(*para*-Substituted phenyl)furans underwent uncatalysed addition of tetracyanoethylene and of dimethyl acetylenedicarboxylate; the rate of the latter reaction was shown to be relatively insensitive to the substituents, in accord with a concerted cycloaddition.[195] The cyclopentadiene–maleic anhydride *endo*-adduct, stereospecifically labelled with ^{14}C in one carbonyl group, was synthesized by Roberts and his co-workers[196] to investigate the supposed "internal" mechanism for its isomerization to the *exo*-isomer, before it was found that this mechanism was not operative here. Ethylenetetracarboxylic dianhydride is formed in solution in a retro-Diels–Alder reaction of its adducts with 9,10-diethoxy- and 9,10-dimethoxy-anthracene.[197] Diels–Alder reactions of indene,[198] naphthacene, [199] and β-*trans*-ocimene,[200] and of dichloromaleic anhydride with cyclic dienes under ultraviolet

[189] See B. T. Gillis, ref. 178*b*, Chapter 6.
[190] R. C. Cookson, S. S. H. Gilani, and I. D. R. Stevens, *J. Chem. Soc., C*, **1967**, 1905.
[191] J. Sauer and B. Schröder, *Chem. Ber.*, **100**, 678 (1967).
[192] B. T. Gillis and R. Weinkam, *J. Org. Chem.*, **32**, 3321 (1967).
[193] E. Ciganek, *Tetrahedron Letters*, **1967**, 3321.
[194] L. A. Paquette and O. Cox, *Chem. Ind.* (*London*), **1967**, 1748.
[195] D. C. Ayres and J. R. Smith, *Chem. Commun.*, **1967**, 886.
[196] U. Scheidegger, J. E. Baldwin, and J. D. Roberts, *J. Am. Chem. Soc.*, **89**, 894 (1967).
[197] J. Sauer, B. Schröder, and R. Wiemer, *Chem. Ber.*, **100**, 306 (1967); J. Sauer, B. Schröder, and A. Mielert, *ibid.*, p. 315.
[198] C. F. Huebner, P. L. Strachan, E. M. Donoghue, N. Cahoon, L. Dorfman, R. Margerison, and E. Wenkert, *J. Org. Chem.*, **32**, 1126 (1967).
[199] J. S. Meek, F. M. Dewey, and M. W. Hanna, *J. Org. Chem.*, **32**, 69 (1967).
[200] K. T. Joseph and G. S. K. Rao, *Tetrahedron*, **23**, 519 (1967).

irradiation[201] have been investigated. A Diels–Alder mechanism was proposed for the formation of tetramers from liquid allene.[202]

Other Cycloaddition Reactions

1,2-*Cycloadditions*. The mechanism of cycloaddition of alkenes, such as tetracyano-, 1,1-difluoro-, and 1,1-dichlorodifluoro-ethylene, to give cyclobutanes,[203] and of the cycloaddition of unconjugated dienes,[204] has been reviewed.

The 1,2-cycloaddition of 4-phenyl-1,2,4-triazole-3,5-dione (**52**) to bicyclo-[2.1.0]pentane-5-spirocyclopropane (**54**), to give (**56**) proceeds stereospecifically with inversion at $C_{(1)}$ and $C_{(4)}$, as shown by deuterium-labelling. Therefore the diradical (**55**) is assumed to be an intermediate.[205]

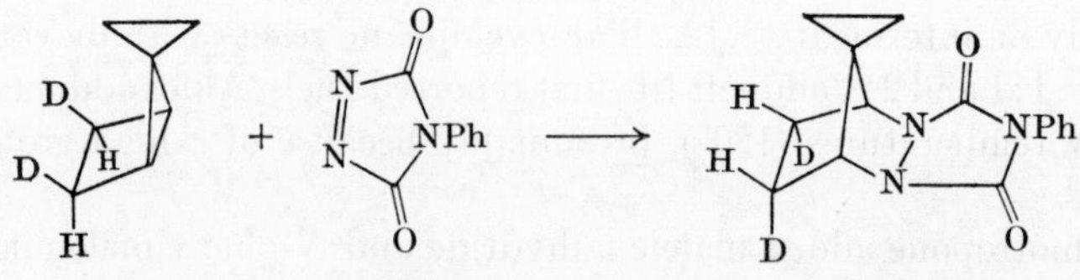

(54) **(52)** **(56)**

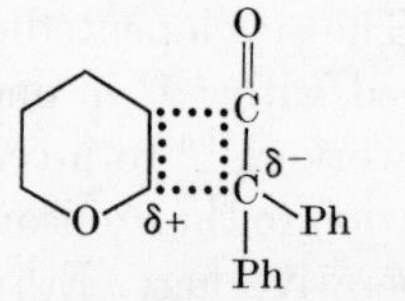

(57)

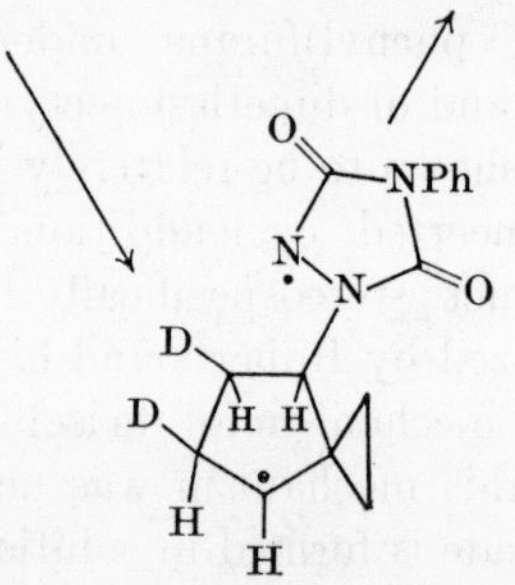

(55)

A number of 1,2-cycloadditions of the carbon–carbon double bond of ketens to nucleophilic olefins, such as vinyl ethers and enamines, have been reported recently.[206] For diphenylketen and dihydropyran, this reaction is of the second order, has a large negative entropy of activation, and is relatively insensitive to solvent, the rate increasing with decreasing solvent polarity. Thus the cycloaddition is a nearly concerted, one-step process in which

[201] H.-D. Scharf, *Tetrahedron Letters*, **1967**, 4231.
[202] B. Weinstein and A. H. Fenselau, *J. Org. Chem.*, **32**, 2278 (1967).
[203] P. D. Bartlett, *Nucleus*, **43**, 251 (1966).
[204] I. Tabushi and K. Fujita, *Yuki Gosei Kagaku Kyokai Shi.*, **25**, 10 (1967); *Chem. Abs.*, **66**, 75425p (1967).
[205] W. R. Roth and M. Martin, *Tetrahedron Letters*, **1967**, 4695.
[206] See, for example, *Organic Reaction Mechanisms*, **1966**, 155.

electron-releasing substituents in the olefin stabilize some charge separation in the transition state (**57**).[207] In agreement with this, it was found that dichloroketen is more reactive than dimethylketen in 1,2-cycloaddition to unactivated olefins, chlorine being better able to stabilize the incipient carbanion than is methyl.[208] The course of addition of the cyclopropene double bond of a triafulvene to enamines has been rationalized by invoking 1,2-cycloadducts as intermediates.[209] The cycloaddition of chlorotrifluoro- and 1,1-dichlorodifluoro-ethylene to allene,[210] and the stereochemistry of the cycloaddition of sulphonyl isocyanates to enol ethers[211a] and of the rhodium-catalysed dimerization of norbornadiene,[211b] have also been described.

1,3-*Cycloadditions*. A considerable amount of work has been devoted to 1,3-dipolar cycloadditions again this year. The rearrangement, with loss of nitrogen, which occurred when the norbornene–phenyl azide adduct was heated with phenyl isocyanate[212] has been studied further and evidence is presented for a 1:1 intermediate complex, possibly of structure (**58**). Baldwin[213] proposes a wider definition of cycloaddition to embrace reactions, such as this one, in which a small molecule is eliminated: "cycloadditions are chemical transformations giving at least one product having at least two new bonds as constituents of a new ring." Diethyl phosphorazidate also adds to norbornene to give the triazoline (**59**) in a second-order reaction. Decomposition of (**59**) proceeded in two consecutive first-order reactions via a ring-opened dipolar intermediate.[214] The addition of substituted phenyl azides to olefins and acetylenes obeyed the Hammett equation; the rate for phenyl

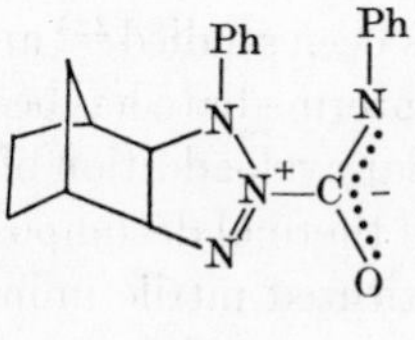

(**58**)

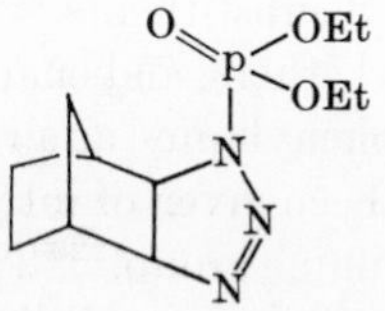

(**59**)

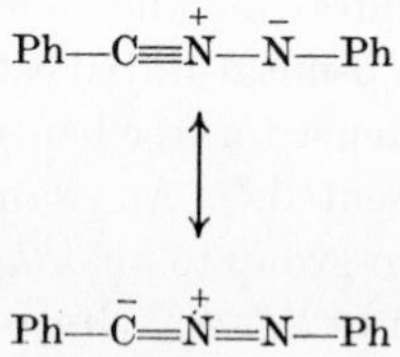

(**60**)

azide with 1-cyclopentenylpyrrolidine depended only slightly on the solvent polarity, and the reaction appears to be a concerted cycloaddition.[215]

[207] W. T. Brady and M. R. O'Neal, *J. Org. Chem.*, **32**, 612 (1967).
[208] W. T. Brady and O. H. Waters, *J. Org. Chem.*, **32**, 3703 (1967).
[209] J. Ciabattoni and E. C. Nathan, *J. Am. Chem. Soc.*, **89**, 3081 (1967).
[210] D. R. Taylor and M. R. Warburton, *Tetrahedron Letters*, **1967**, 3277.
[211a] F. Effenberger and G. Kiefer, *Angew. Chem. Intern. Ed. Engl.*, **6**, 951 (1967).
[211b] T. J. Katz and N. Acton, *Tetrahedron Letters*, **1967**, 2601.
[212] *Organic Reaction Mechanisms*, **1965**, 129.
[213] J. E. Baldwin, *J. Org. Chem.*, **32**, 2438 (1967).
[214] K. D. Berlin, L. A. Wilson, and L. M. Raff, *Tetrahedron*, **23**, 965 (1967).
[215] R. Huisgen, G. Szeimies, and L. Möbius, *Chem. Ber.*, **100**, 2494 (1967).

On the basis of kinetic measurements[216, 217] and MO calculations[216] a 1,3-dipolar cycloaddition mechanism is considered most reasonable for the reaction of substituted benzonitrile oxides with arylacetylenes, though the relatively high entropy of activation suggests a less rigid transition state than usual. Electrophilic catalysis of a 1,3-dipolar cycloaddition has been demonstrated for the first time in the reactions of benzonitrile oxide with nitriles, aldehydes, and ketones in the presence of boron trifluoride etherate.[218] Nitrile oxides also add to the N=S bond of *N*-sulphinylanilines, but the dependence of rate on substituents in the sulphinylaniline ring differs from that for the other cycloadditions and a two-step mechanism – nucleophilic attack by the nitrile oxide oxygen on sulphur followed by ring closure – was proposed.[219] Benzonitrile oxide forms a cycloadduct with methylenetriphenylphosphorane.[220]

Many further reports on nitrile imines as 1,3-dipoles have appeared. The competition of benzonitrile phenylimine (**60**) for a wide range of dipolarophiles, taken in pairs, has been measured quantitatively;[221] relative rates varied by 10^5. Substituent effects on the free enthalpy of activation are additive. The strong activation of dipolarophiles by conjugation was explained by stabilization of partial charges and increased polarizability. The carbon terminus of the nitrile imine was much more sensitive than the nitrogen to the bulk of the dipolarophile.[221] The cycloaddition of benzonitrile phenylimine to unconjugated olefins to give 1,3-diphenyl-2-pyrazolines was stereospecifically *cis*, and to norbornene was specifically *exo*.[222] With styrenes 1,3,5-triphenyl-2-pyrazolines were formed, the orientation being unchanged by electron-withdrawing and -releasing aryl substituents.[223] The orientation in additions to α,β-unsaturated esters, vinyl ethers, and enamines has been studied[224] and evidence for free benzonitrile phenylimine as a reaction intermediate has been presented.[225] An example has been given of intramolecular cycloaddition of a nitro-group to an *ortho*-nitrile imine group.[226] The rate of thermal decomposition of 2,5-diaryltetrazoles to nitrogen and diaryl-substituted nitrile imines has a Hammett $\rho = +1.16$ for substituents in the 2-aryl group and $\rho = -0.23$

216 P. Beltrame, C. Veglio, and M. Simonetta, *J. Chem. Soc., B*, **1967**, 867.
217 A. Dondoni, *Tetrahedron Letters*, **1967**, 2397.
218 S. Morrocchi, A. Ricca, and L. Velo, *Tetrahedron Letters*, **1967**, 331.
219 P. Beltrame, A. Comotti, and C. Veglio, *Chem. Commun.*, **1967**, 996.
220 R. Huisgen and J. Wulff, *Tetrahedron Letters*, **1967**, 917.
221 A. Eckell, R. Huisgen, R. Sustmann, G. Wallbillich, D. Grashey, and E. Spindler, *Chem. Ber.*, **100**, 2192 (1967).
222 R. Huisgen, H. Knupfer, R. Sustmann, G. Wallbillich, and V. Weberndörfer, *Chem. Ber.*, **100**, 1580 (1967).
223 J. S. Clovis, A. Eckell, R. Huisgen, R. Sustmann, G. Wallbillich, and V. Weberndörfer, *Chem. Ber.*, **100**, 1593 (1967).
224 R. Huisgen, R. Sustmann, and G. Wallbillich, *Chem. Ber.*, **100**, 1786 (1967).
225 J. S. Clovis, A. Eckell, R. Huisgen, and R. Sustman, *Chem. Ber.*, **100**, 60 (1967).
226 R. Huisgen and V. Weberndörfer, *Chem. Ber.*, **100**, 71 (1967).

for substituents in the 5-aryl group, suggesting an unsymmetrical transition state (**61**) for the 1,3-cycloelimination reaction.[227]

The addition of 2-diazopropane to allenic esters occurs at the more electrophilic (α,β) double bond, but the orientation was very dependent on the γ-substituents, as shown in Scheme 3. When $R^1 = R^2 =$ Me the transition state for formation of the expected adduct (**62**) is more severely crowded than that for (**63**).[228] Vinyldiazomethane undergoes cycloaddition with various

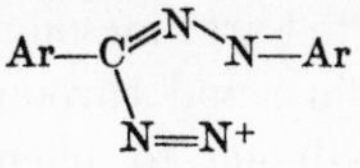

(61)

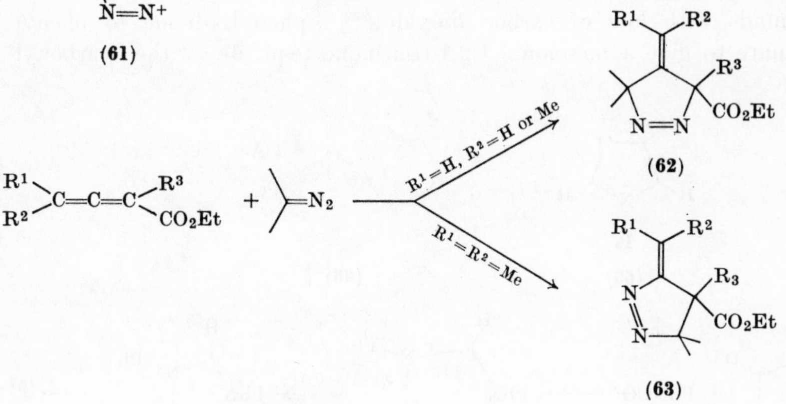

Scheme 3

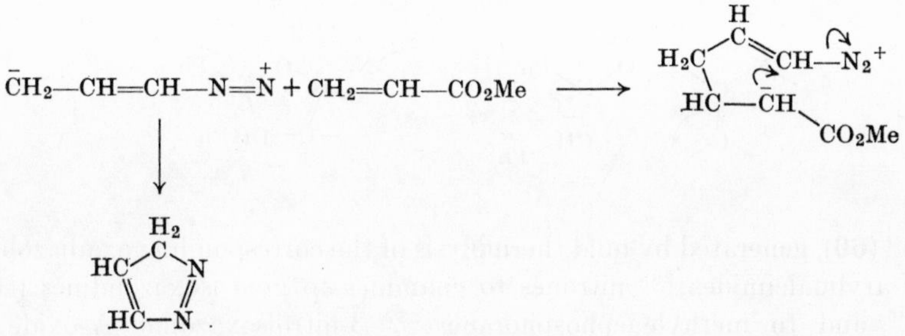

...(8)

(64)

olefins to give vinyl pyrazolines, cyclopropanes, and dienes. With methyl acrylate a small amount of methyl cyclopent-2-enecarboxylate was formed and was shown not to arise from other products. Thus the participation of a 1,5-dipolar structure is indicated, either as shown in equation (8) or by 1,5-cycloaddition followed by loss of nitrogen from the seven-membered

[227] J. E. Baldwin and S. Y. Hong, *Chem. Commun.*, **1967**, 1136.
[228] S. D. Andrews and A. C. Day, *Chem. Commun.*, **1967**, 902.

ring.[229] In the absence of olefins, vinyldiazomethane cyclizes to the 3*H*-pyrazole (**64**) which rapidly isomerizes to 1*H*-pyrazole. This reaction is light-catalysed at a wavelength to which the diazo-compound is transparent, and it was suggested that 3*H*-pyrazole is the primary absorbing species.[230] The related ring closure of a vinyl azide to a triazole has been observed for the first time.[231] The steric effects observed in the formation of pyrazolines from diazomethane and unsaturated esters accord with the concerted cycloaddition mechanism.[232]

Other 1,3-dipolar cycloadditions which have been reported are: mesoionic oxazolones (**65**), as azomethine ylides, to acetylenes, olefins, and carbonyl compounds with loss of carbon dioxide;[233] 3-phenylsydnone to phenyl isocyanate to give a mesoionic 1,2,4-triazolone (eqn. 9);[234] the "carbonyl

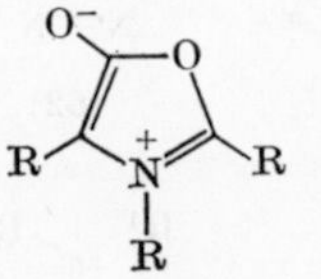

(65)

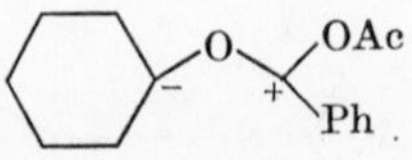

(66)

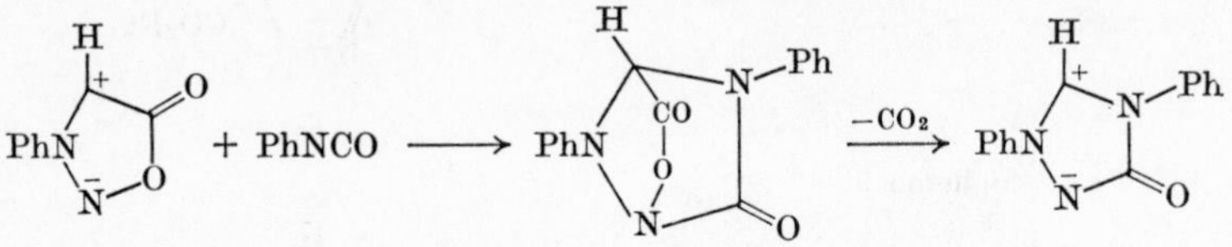

. . . (9)

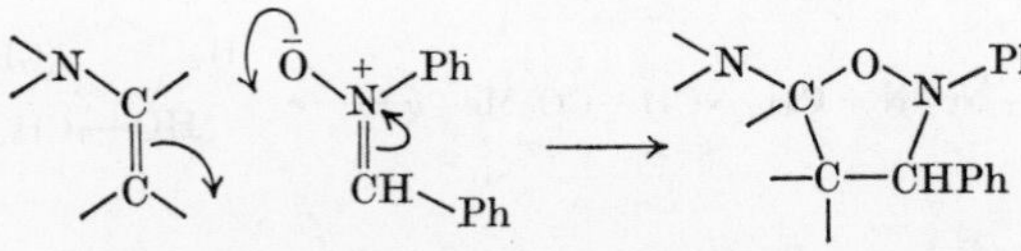

. . . (10)

ylide" (**66**), generated by mild thermolysis of the corresponding oxadiazoline, to *N*-arylmaleimides;[235] nitrones to enamines to give isoxazolidines (eqn. 10)[236] and to methylenephosphoranes;[237] 3-nitroisoxazoline *N*-oxide to

229 I. Tabushi, K. Takagi, M. Okano, and R. Oda, *Tetrahedron*, **23**, 2621 (1967).
230 A. Ledwith and D. Parry, *J. Chem. Soc., B*, **1967**, 41.
231 J. S. Meek and J. S. Fowler, *J. Am. Chem. Soc.*, **89**, 1967 (1967).
232 A. Ledwith and Y. Shih-Lin, *J. Chem. Soc., B*, **1967**, 83.
233 R. Huisgen, E. Funke, F. C. Schaefer, H. Gotthardt, and E. Brunn, *Tetrahedron Letters*, **1967**, 1809.
234 H. Kato, S. Sato, and M. Ohta, *Tetrahedron Letters*, **1967**, 4261.
235 P. Rajagopalan and B. G. Advani, *Tetrahedron Letters*, **1967**, 2689.
236 O. Tsuge, M. Tashiro, and Y. Nishihara, *Tetrahedron Letters*, **1967**, 3769; Y. Nomura, F. Furusaki, Y. Takeuchi, *Bull. Chem. Soc. Japan*, **40**, 1740 (1967).
237 J. Wulff and R. Huisgen, *Angew. Chem. Intern. Ed. Engl.*, **6**, 457 (1967).

styrene;[238] fulminic acid to carbon–carbon double bonds;[239] and formaldoxime to α,β-unsaturated cyanides and esters to give isoxazolidines.[240] A 1,3-dipolar cycloaddition has been proposed to explain the formation of 1,2,4-triazoles from benzaldehyde azine and strong base.[241]

The properties of strained double bonds[242] and of allenes[243] including their addition and cycloaddition reactions have been reviewed.

Other cycloadditions and cyclizations. Huisgen and his co-workers have also reported a number of 1,4-cycloadditions: of the 1,4-dipole formed from isoquinoline and dimethyl acetylenedicarboxylate to phenyl isocyanate, diethyl mesoxalate, and dimethyl azodicarboxylate in a two-step mechanism;[244] of two molecules of 3,4-dihydroisoquinoline (an azomethine) to electrophilic double bonds,[245] and to isothiocyanates and carbon disulphide;[246] the same addition to α,β-unsaturated ketones has also been reported.[247] Heterocyclic *N*-sulphinylamines undergo 1,4-cycloaddition to norbornene, to give derivatives of (**67**), and to ethoxyacetylene.[248] The mixed products of reaction of phenylacetylene and phenyldiimide are those expected from a radical mechanism and not simply *trans*-stilbene, the product of a concerted *cis*-1,4-cycloaddition.[249] pH is shown to be an important factor in the hydrogenation of hex-1-ene, phenylacetylene, and nitrobenzene over a platinum catalyst; the rate, mechanism, selectivity, and direction of hydrogenation are all affected.[250a] Di-(α-bromobenzyl) ketone and sodium iodide or benzyl α-chlorobenzyl ketone and alkaline alumina gave diphenylcyclopropanone, or a tautomer such as the carbonium ion–enolate dipole which added to cyclopentadiene or to furan to give the *cis*- and *trans*-adducts (eqn. 11).[250b]

Sulphur dioxide adds in a novel 1,6-cycloaddition to give 2,7-dihydrothiepin 1,1-dioxide (**68**).[251] Details have now appeared[252] of the cyclodimerization of *N*-substituted azepines; the more stable dimers are formed by

238 V. A. Tartakovskii, A. A. Onishchenko, and S. S. Novikov, *Izv. Akad. Nauk SSSR, Ser. Khim.*, **1967**, 177; *Chem. Abs.*, **66**, 104460d (1967).

239 R. Huisgen and M. Christ, *Angew. Chem. Intern. Ed. Engl.*, **6**, 456 (1967).

240 M. Ochiai, M. Obayashi, and K. Morita, *Tetrahedron*, **23**, 2641 (1967).

241 J. T. A. Boyle and M. F. Grundon, *Chem. Commun.*, **1967**, 1137.

242 N. S. Zefirov and V. I. Sokolov, *Russian Chem. Rev.*, **36**, 87 (1967).

243 D. R. Taylor, *Chem. Rev.*, **67**, 317 (1967).

244 R. Huisgen, K. Herbig, and M. Morikawa, *Chem. Ber.*, **100**, 1107 (1967); R. Huisgen, M. Morikawa, K. Herbig, and E. Brunn, *ibid.*, p. 1094.

245 M. Morikawa and R. Huisgen, *Chem. Ber.*, **100**, 1616 (1967).

246 R. Huisgen, M. Morikawa, D. S. Breslow, and R. Grashey, *Chem. Ber.*, **100**, 1602 (1967).

247 C. Szántay and L. Novák, *Chem. Ber.*, **100**, 2038 (1967).

248 H. Beecken, *Chem. Ber.*, **100**, 2159 (1967).

249 R. Fuchs, *Tetrahedron Letters*, **1967**, 2419.

250a A. M. Sokol'skaya and S. M. Reshetnikov, *Probl. Kinetiki i Kataliza Akad. Nauk SSSR.* **11**, 158 (1966); *Chem. Abs.*, **66**, 2029s (1967).

250b R. C. Cookson, M. J. Nye, and G. Subrahmanyam, *J. Chem. Soc., C*, **1967**, 473.

251 W. L. Mock, *J. Am. Chem. Soc.*, **89**, 1281 (1967).

252 A. L. Johnson and H. E. Simmons, *J. Am. Chem. Soc.*, **89**, 3191 (1967).

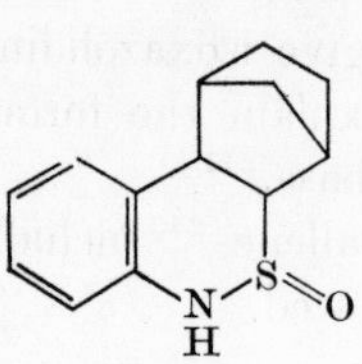

(67)

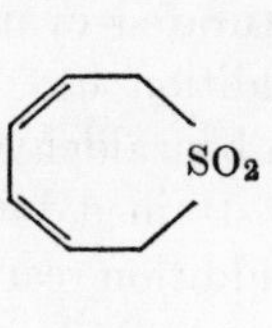

(68)

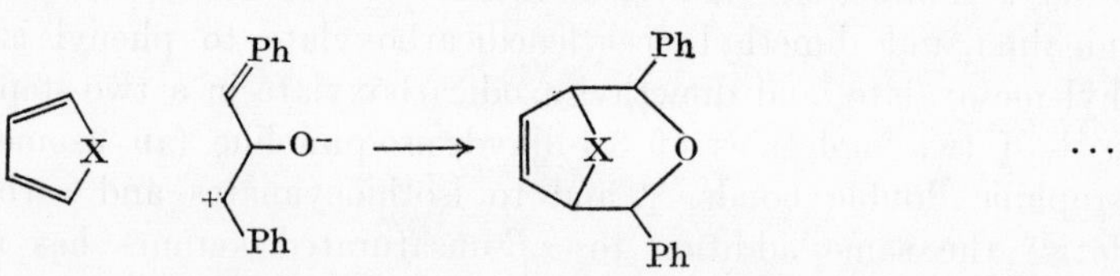

. . . (11)

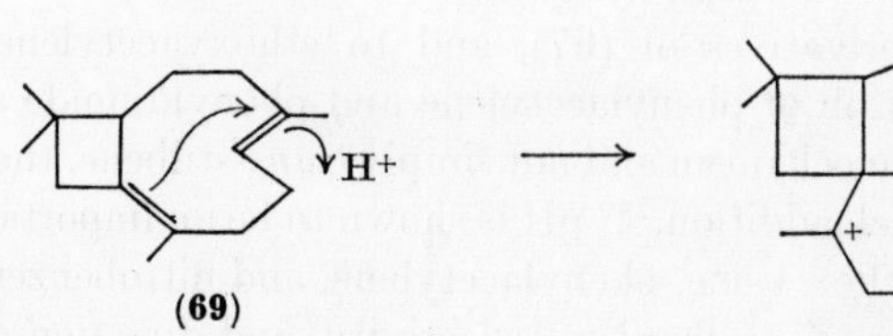

(69) **(70)**

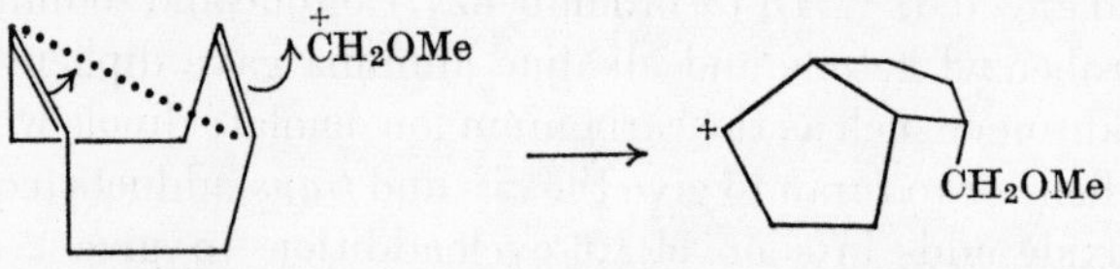

(71) **(72)**

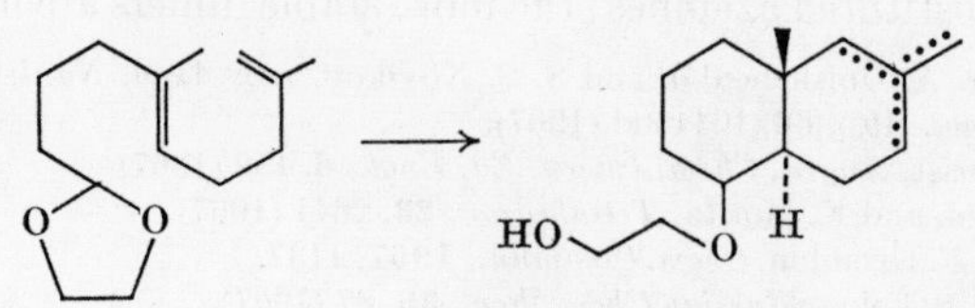

(73) **(74)**

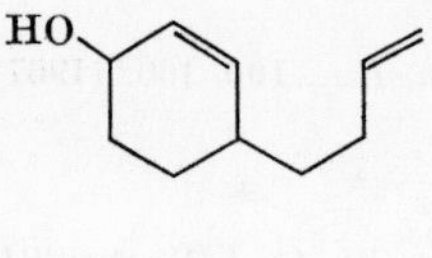

(75)

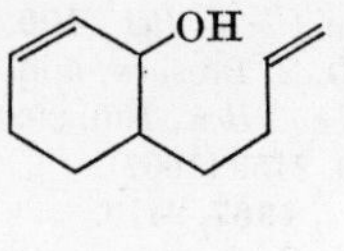

(76)

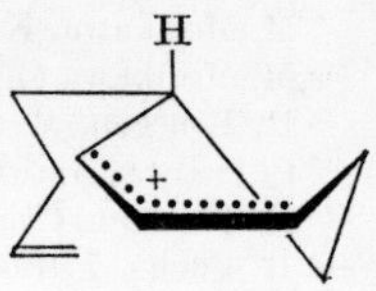

(77)

subsequent rearrangement of the kinetically controlled, symmetry-allowed dimers.[253] The cyclotrimerization of acetylenes has been studied further.[254]

A key step in the proposed mechanism for the acid-catalysed conversion of caryophyllene into neoclovene was the cyclization (**69**) → (**70**); this has now been supported by the demonstration that when the carbonium ion (**70**) was independently generated from the corresponding alcohol under the same conditions it gave neoclovene in high yield.[255] In the boron trifluoride-catalysed addition of methoxymethyl acetate to *cis,cis*-cyclo-octa-1,5-diene the stereochemistry of the complex mixture of products suggests that the electrophile approaches almost exclusively *exo* (**71**) to give the 2-*endo*-substituted *cis*-bicyclo[3.3.0]octylcarbonium ion (**72**).[256]

Johnson and van Tamelen have continued to investigate biogenetic-type cyclizations.[257] The stereospecific cyclization of an acyclic triene acetal to the tricyclic system of natural configuration by very mild treatment with stannic chloride was reported last year.[257] It has now been shown that this reaction can produce a fused system with an angular methyl group and that the stereochemical course is dictated by the configuration of the olefinic double bonds. Thus *trans*-olefin (**73**) gave the *trans*-octalins (**74**), and the corresponding *cis*-olefin gave the *cis*-octalins.[258] Cyclization of the terminal epoxide of geranylgeranyl acetate to the tricyclic terpenoid has been effected in the laboratory with the same reagent.[259] A revised structure for the major 2-octalol formed from either cyclohexenol (**75**) or (**76**) by brief treatment with formic acid is now consistent with quasi-axial attack by the butenyl side chain on the allylic cation (**77**) to give the most favourable orbital overlap in the transition state.[260]

253 *Organic Reaction Mechanisms*, **1966**, 157.
254 P. Chini, A. Santambrogio, and N. Palladino, *J. Chem. Soc., C*, **1967**, 830, 836; V. O. Reikhsfel'd, T. Y. Gindorf, and S. Vyrbanov, *Zh. Org. Chem.*, **3**, 270 (1967); *Chem. Abs.*, **66**, 115076j (1967).
255 T. F. W. McKillop, J. Martin, W. Parker, and J. S. Roberts, *Chem. Commun.*, **1967**, 162.
256 I. Tabushi, K. Fujita, and R. Oda, *Tetrahedron Letters*, **1967**, 3815.
257 Cf. *Organic Reaction Mechanisms*, **1966**, 158.
258 W. S. Johnson, A. van der Gen, and J. J. Swoboda, *J. Am. Chem. Soc.*, **89**, 170 (1967).
259 E. E. van Tamelen and R. G. Nadeau, *J. Am. Chem. Soc.*, **89**, 176 (1967).
260 W S. Johnson and K. E. Harding, *J. Org. Chem.*, **32**, 478 (1967).

CHAPTER 6

Nucleophilic Aromatic Substitution

Further evidence has been presented this year for Bunnett's two-step intermediate complex mechanism (e.g., Scheme 1) for nucleophilic aromatic substitution, particularly from the investigations of base-catalysis by Zollinger, Bernasconi, and their co-workers.

The rates of the base-catalysed reaction of 2,4-dinitrophenyl phenyl ether with benzylamine and *N*-methylbenzylamine in aqueous dioxan have been measured as a function of sodium hydroxide and of amine concentration. The

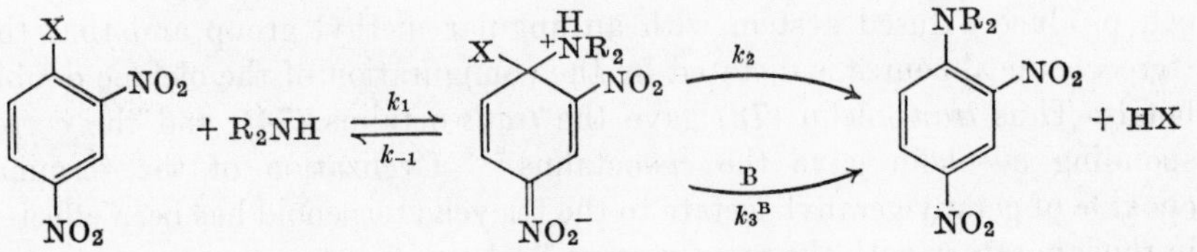

Scheme 1

secondary-amine reaction is strongly catalysed by sodium hydroxide and moderately so by itself, whilst the primary-amine reaction is independent of hydroxide concentration and the second-order rate constants decrease slightly with increasing amine concentration. In these and related reactions the relative catalytic effectiveness of OH^- and H_2O (k_3^{OH}/k_2) increases considerably with increasing basicity of the leaving group (in accord with earlier observations that reactions with good leaving groups are not base-catalysed whilst those with poor-leaving groups are), but decreases only slightly with increased basicity of the attacking amine. The results were analysed in terms of Scheme 1, though full mechanistic discussion was deferred.[1] A very similar investigation with morpholine as nucleophile, where the results are typical of those with other secondary amines, has also been reported. At high hydroxide concentration, hydrolysis of the product to 2,4-dinitrophenol had to be allowed for. Other things being equal, secondary amines seem to be more nucleophilic than primary amines towards aromatic carbon.[2] The reaction of 1-fluoro-2,4-dinitrobenzene with piperidine in benzene is catalysed by dimethyl sulphoxide (DMSO) about as strongly as by 1,4-diazabicyclo[2.2.2]octane (DABCO) and much more strongly than by the stronger base pyridine. Thus

[1] C. F. Bernasconi, *J. Org. Chem.*, **32**, 2947 (1967).
[2] C. F. Bernasconi and P. Schmid, *J. Org. Chem.*, **32**, 2953 (1967).

the DMSO is not simply acting as a base-catalyst. The dependence of rate upon piperidine concentration also changes on addition of DMSO, and the reaction of 1-chloro-2,4-dinitrobenzene, which is normally insensitive to base-catalysis (but see below), is also accelerated by DMSO. It is proposed that the dipolar S–O bond of DMSO specifically solvates the dipolar tetrahedral intermediate (with the negative charge on the *o*-nitro-group) of Scheme 1.[3] The rates of reaction of these two halides with *p*-anisidine in benzene, with and without pyridine and DABCO as catalysts, were also measured. Reaction of the fluoro-compound is catalysed strongly by pyridine and very strongly by DABCO, and is of more than the second order in *p*-anisidine as a result of base-catalysis. The same effects were observed with the chloro-compound but to a much smaller extent; thus base-catalysis in an aminolysis of 1-chloro-2,4-dinitrobenzene appears to have been unequivocally demonstrated for the first time.[4] Entirely analogous base-catalysis was demonstrated for the reactions of 1-fluoro-2,4-dinitrobenzene with benzylamine and *N*-methylbenzylamine[5] and with morpholine.[6] The last reaction is considerably more sensitive to pyridine and DABCO catalysis than that of piperidine, indicating enhanced sensitivity to base-catalysis with decreasing base strength of the nucleophile.[6]

The need for caution in generalizing about the order of mobility of the halogens in nucleophilic aromatic substitution (usually $F \gg Cl \sim Br \sim I$ for activated substrates), and for due regard to mechanism, was shown by comparing the rates of reaction of *o*- and *p*-fluoro- and -chloro-nitrobenzenes with piperidine in benzene. The *p*-fluoronitrobenzene reaction is second-order in piperidine, being catalysed by it, whilst that of the *ortho*-isomer is not. The role of piperidine in catalysing the decomposition of the tetrahedral complex in the *para* case is presumably usurped by the *o*-nitro-group and therefore, with the *o*-nitro-compounds, the ratio (k_{ArF}/k_{ArCl}) of the second-order rate constants extrapolated to zero piperidine concentration is not a measure of the relative mobility of fluorine and chlorine. What is observed here is the nitro-group-assisted elimination of the ammonium proton and the fluoride ion. When the relative mobility was properly computed, for the *para*-compounds, fluorine was found to be much less reactive than chlorine. In dipolar aprotic and protic solvents the fluoride was much more reactive than the chloride, but only the former reactions were catalysed by the solvent.[7]

In previous studies of the reaction of phenoxide ion with 1-chloro-2,4-dinitrobenzene in aqueous or alcoholic solvents, phenol has been added to suppress the equilibium conversion of phenoxide into phenol. Since this change in the medium could change the reactivity of the phenoxide ion, the

[3] C. F. Bernasconi, M. Kaufmann, and H. Zollinger, *Helv. Chem. Acta*, **49**, 2563 (1966).
[4] C. F. Bernasconi and H. Zollinger, *Helv. Chim. Acta*, **49**, 2570 (1966).
[5] C. F. Bernasconi and H. Zollinger, *Helv. Chim. Acta*, **50**, 3 (1967).
[6] G. Becker, C. F. Bernasconi, and H. Zollinger, *Helv. Chim. Acta*, **50**, 10 (1967).
[7] F. Pietra and F. Del Cima, *Tetrahedron Letters*, **1967**, 4573.

effect of added phenol on the reaction of sodium phenoxide with 1-chloro-2,4-dinitrobenzene in methanol at 35° was investigated in detail. There was a linear relation between the observed second-order rate constant and the ratio [MeOH]/[PhOH] from which the phenoxide ion rate constant can be calculated; however, this linearity did not extend to high [PhOH] where the reaction was retarded. This retardation was attributed to the formation of the biphenoxide ion (**1**) which would be less nucleophilic than phenoxide, and some support for this was provided by the absence of this rate suppression with the bulkier *o*-tolyl oxide ion.[8]

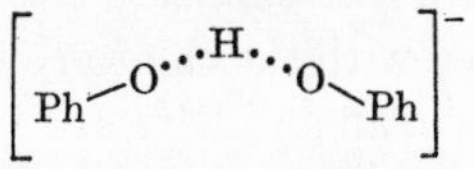

(**1**)

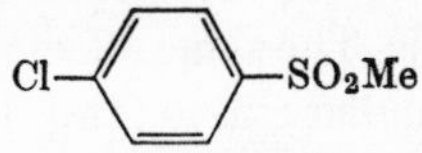

(**2**)

When equimolecular amounts of *p*-chloro(methylsulphonyl)benzene (**2**) and potassium cyanide were boiled in DMSO an equimolecular mixture of *p*-bis(methylsulphonyl)benzene and *p*-dicyanobenzene (but no 1-cyano-4-methylsulphonylbenzene) was formed in high yield; (**2**) was unchanged in boiling DMSO. Presumably cyanide displaces chlorine and the methanesulphonyl group to give the dicyanobenzene, and then the methanesulphinate anion displaces chlorine to give the disulphone.[9] *p*-Nitrosophenol is rapidly converted into its alkyl ethers in alcohols containing acid, presumably because of the powerfully activating effect of the protonated nitroso-group ($-N{=}O^{+}-H \leftrightarrow -N^{+}-OH$) in nucleophilic substitution. For the same reason the product ethers react rapidly with primary aromatic amines to give *p*-nitrosodiphenylamines.[10]

The radiochemical exchange reaction between iodide ion and 1-iodo-2,4-dinitrobenzene in methanol has been reinvestigated and shown, contrary to earlier work but in accord with expectation, to be overall of the second order, being of the first order in each reagent, and the activation energy has been redetermined.[11] The effect of the dielectric constant of the solvent on the isotopic exchange between iodide and 1-iodo-2,4-dinitro-benzene and -naphthalene,[12] and the isotopic exchange between 1-chloro-2,4,6-trinitrobenzene and $Li^{36}Cl$ in ethanol,[13] have also been studied.

The bimolecular reaction rates of 1-chloro-, 1-ethoxy-, and 1-methoxy-2,4-

[8] C. L. Liotta and R. L. Karelitz, *J. Org. Chem.*, **32**, 3090 (1967).
[9] P. ten Haken, *Tetrahedron Letters*, **1967**, 759.
[10] J. T. Hays, E. H. de Butts, and H. L. Young, *J. Org. Chem.*, **32**, 153 (1967); J. T. Hays, H. L. Young, and H. H. Espy, *ibid.*, p. 158.
[11] F. H. Kendall and J. Miller, *J. Chem. Soc., B*, **1967**, 119.
[12] C. A. Marcopoulos, *Z. Phys. Chem. (Leipzig)*, **234**, 297 (1967); see also *Radiochim. Acta*, **5**, 121 (1966).
[13] P. H. Gore, D. F. C. Morris, and T. J. Webb, *Radiochim. Acta*, **6**, 122 (1966).

dinitrobenzene with aqueous sodium hydroxide have been measured at 25° and a unique J_- acidity function has been shown to be defined by the results.[14] The equilibrium constant, $K = [OEt^-][H_2O]/[OH^-][EtOH]$, was calculated for mixtures of water and ethanol by measuring the simultaneous rates of nucleophilic displacement by water and by ethanol for certain aromatic substrates.[15] Further results comparing a polarizable nucleophile (thiomethoxide ion) with a poorly polarizable one (methoxide) have been presented.[16] Lamm has reviewed, in English, his own and related work on nucleophilic aromatic substitution.[17] Shein and his co-workers have reported more results on methylsulphonyl and trifluoromethylsulphonyl as activating groups, in the reaction of 1-chloro-2,4-bis(trifluoromethylsulphonyl)benzene and related compounds with sodium methoxide and with ammonia,[18] of 1-chloro- and 1-alkoxy-2,4-bis(trifluoromethylsulphonyl)benzenes with sodium alkoxides,[19] of 1-chloro- and 1-alkoxy-2,4-bis(methylsulphonyl)benzenes with sodium alkoxides,[20] and of *p*-chloro(trifluoromethylsulphonyl)benzene with sodium alkoxides.[21] The kinetics of the following reactions have also been investigated: 1-halogeno-2,4-dinitronaphthalenes with ammonia,[22] *p*-chloronitrobenzene with trimethylamine,[23] 1-fluoro-2,4-dinitrobenzene with aqueous sodium hydroxide,[24] and 1,2,4,5-tetrachlorobenzene with sodium hydroxide in ethylene glycol.[25]

The effect of a range of *meta*-substituents on nucleophilic substitution in 3-substituted 1-fluoro-5-nitrobenzene[26] has been investigated; also that of the following: the nature of the leaving group in nucleophilic aromatic substitution,[27] changes in concentration, polarity of solvent, and polar and steric factors on the reactions of 2,4-dinitrophenyl benzenesulphonate with

[14] C. H. Rochester, *J. Chem. Soc., B*, **1967**, 1076.

[15] J. R. Aspland, A. Johnson, and R. H. Peters, *J. Chem. Soc., B*, **1967**, 1041.

[16] L. Di Nunno and P. E. Todesco, *Tetrahedron Letters*, **1967**, 2899.

[17] B. Lamm, *Svensk Kem. Tidskr.*, **88**, 583 (1966).

[18] S. M. Shein, V. A. Ignatov, L. A. Kozorez, and L. F. Cherbatyuk, *Zh. Obshch. Khim.*, **37**, 114 (1967); *Chem. Abs.*, **67**, 21322e (1967).

[19] S. M. Shein and M. I. Krasnosel'skaya, *Zh. Obshch. Khim.*, **36**, 2148 (1966); *Chem. Abs.*, **66**, 94447p (1967).

[20] S. M. Shein and A. D. Khmelinskaya, *Izv. Sib. Otd. Akad. Nauk SSSR, Ser. Khim. Nauk*, **1966**, 89; S. M. Shein, A. D. Khmelinskaya, V. V. Litvak, and N. K. Danilova, *ibid.*, p. 116; *Chem. Abs.*, **67**, 81642u, 81643v (1967).

[21] S. M. Shein, M. I. Krasnosel'skaya, and V. N. Boiko, *Zh. Obshch. Khim.*, **36**, 2141 (1966); *Chem. Abs.*, **66**, 94446n (1967).

[22] R. C. Krauss and J. D. Reinheimer, *Can. J. Chem.*, **45**, 77 (1967).

[23] H. Suhr, *Ann. Chem.*, **701**, 101 (1967).

[24] H. D. Harlan and P. H. Franke, *Tex. J. Sci.*, **18**, 402 (1966); *Chem. Abs.*, **67**, 32088j (1967).

[25] J. Kaszubska, *Zeszyty Nauk Politech. Szczecin, Chem.*, No. 6, 33 (1965); *Chem. Abs.*, **65**, 18441 (1966).

[26] C. W. L. Bevan, J. Hirst, and S. J. Una, *Nigerian J. Sci.*, **1**, 27 (1966); *Chem. Abs.*, **66**, 28153p (1967).

[27] R. V. Vizgert and I. M. Ozdrovskaya, *Reakts. Sposobnost. Org. Soedin, Tartu. Gos. Univ.*, **3**, 35 (1966); *Chem. Abs.*, **66**, 115055b (1967).

aromatic amines;[28] the retarding effect of methyl groups on the rate of reaction of 4-chloropyridine and *p*-chloronitrobenzene with benzylamine,[29] and the copper-catalysed reaction of 8-bromo- and 8-iodo-naphthoic acid with aqueous piperidine.[30]

Volumes of activation have been calculated for the reaction of 1-chloro-2,4-dinitrobenzene with *n*- and *tert*-butylamine.[31]

Meisenheimer Complexes

PMR spectroscopy continues to be used effectively for studying the structure of these and related complexes.[32] Servis has reported and discussed in full his PMR results on Meisenheimer complexes,[33] described last year.[32] Somewhat surprisingly, triethylamine reacts with 2,4,6-trinitroanisole in DMSO to give the zwitterionic complex (**3**).[33] Further PMR studies have confirmed the structures of the σ-complexes formed between methoxide ion or acetonate ion and 1-substituted 2,4-dinitro-, 2,6-dinitro-, and 2,4,6-trinitro-benzenes, 1-substituted 2,4-dinitronaphthalenes, and 10-substituted 9-nitroanthracenes.[34] These well-established 1,1-dialkoxycyclohexadienylide structures from trinitroanisole and phenetole have now been demonstrated for the less stable dinitroanisole and phenetole complexes;[35] these symmetrical σ-complexes are probably intermediates in methoxyl exchange reactions.[32] Further evidence that $C_{(1)}$ is tetrahedral comes from a re-examination of the PMR spectrum of the spiro-anion (**4**) derived from 1-(2-hydroxyethoxy)-2,4-

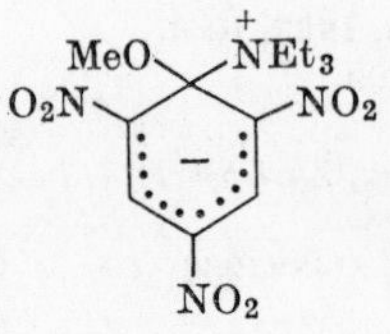

(**3**)

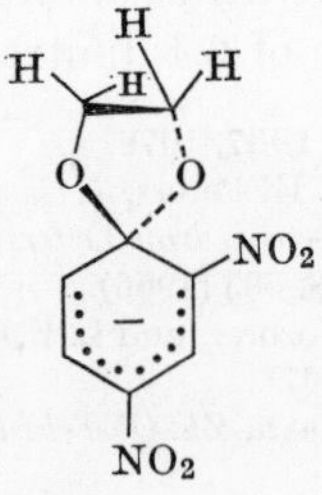

(**4**)

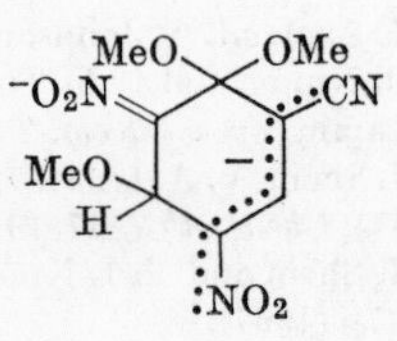

(**5**)

[28] I. M. Ozdrovskaya and R. V. Vizgert, *Reakts. Sposobnost. Org. Soedin. Tartu. Gos. Univ.*, **3**, 146 (1966); *Chem. Abs.*, **67**, 2529m (1967).

[29] N. I. Kudryashova, N. V. Khromov-Borisov, and T. D. Fedorova, *Zh. Organ. Khim.*, **2**, 1663 (1966); *Chem. Abs.*, **66**, 64781a (1967).

[30] V. N. Lisitsyn and L. A. Didenko, *Zh. Org. Chem.*, **3**, 103 (1967); *Chem. Abs.*, **66**, 94810b (1967).

[31] N. I. Prokhorova, B. S. El'yanov, and M. G. Gonikberg, *Izv. Akad. Nauk SSSR, Ser. Khim.*, **1967**, 263; *Chem. Abs.*, **67**, 32223z (1967).

[32] See *Organic Reaction Mechanisms*, **1965**, 137; **1966**, 168.

[33] K. L. Servis, *J. Am. Chem. Soc.*, **89**, 1508 (1967).

[34] R. Foster, C. A. Fyfe, P. H. Emslie, and M. I. Foreman, *Tetrahedron*, **23**, 227 (1967).

[35] W. E. Byrne, E. J. Fendler, J. H. Fendler, C. E. Griffin, *J. Org. Chem.*, **32**, 2506 (1967).

dinitrobenzene; the dioxolane-methylene protons conform to the required A_2B_2 system.[36]

Zollinger and his co-workers have presented a critical assessment of the structure of Meisenheimer complexes and of their role in nucleophilic aromatic substitution.[37–39] The PMR spectra of the stable σ-complexes of 2,4-dinitro-, 2,4,6-trinitro-, and 2-cyano-4,6-dinitro-anisole with methoxide are described; the 1:1 complex from the cyano-compound reacts further with methoxide in concentrated solution to give the 1:2 complex anion (**5**). The interpretation of the spectroscopic data, and also HMO calculations (of charge order, π-electron density, and atom–atom polarizability) concur with accumulation of the additional negative charge on the nitro-groups, the π-electron charge density in the ring actually decreasing on complex formation.[37] Similar spectroscopy shows that tetraethylammonium azide and picric acid derivatives also react to give π-complexes; Arrhenius parameters were calculated for the methyl picrate reaction.[38] With *N*-methyl-*N*,2,4,6-tetranitroaniline, 1-chloro-2,4,6-trinitrobenzene, and 1-chloro-2-cyano-4,6-dinitrobenzene and aromatic amines, however, electron donor–acceptor (π) complexes are formed.[39]

Complex formation between 1,3,5-trinitrobenzene and aliphatic amines in DMSO was examined spectroscopically and by electrical conductance measurements. Ammonia, primary, and secondary amines interact significantly whilst tertiary amines do not (cf. trinitroanisole and triethylamine, above). An equilibrium is established between one trinitrobenzene and two amine molecules:

$$\text{TNB} + 2\text{RNH}_2 \rightleftharpoons [\text{TNB}\cdot\text{NHR}]^- + \text{RNH}_3^+$$

suggesting that proton removal from the 1:1 complex ammonium ion is necessary for stabilization.[40] Amine-catalysed hydrogen exchange in 1,3,5-trinitrobenzene appears to proceed by general base-catalysed hydrogen abstraction to give the trinitrophenyl anion, followed by neutralization of this with D_2O.[41] 4-Nitrobenzyl cyanide ionizes in basic media by loss of a methylene-proton.[42]

With potassium hydroxide in methanol at room temperature, 6-substituted 2,4-dinitrophenyl and 2,4-dinitro-1-naphthyl thiocarbamate and iminothiocarbonate give highly coloured *p*-quinonoid potassium 1,1-dimethoxy-*aci*-nitro-salts.[43]

36 C. E. Griffin, E. J. Fendler, W. E. Byrne, and J. H. Fendler, *Tetrahedron Letters*, **1967**, 4473.
37 P. Caveng, P. B. Fischer, E. Heilbronner, A. L. Miller, and H. Zollinger, *Helv. Chim. Acta*, **50**, 848 (1967).
38 P. Caveng and H. Zollinger, *Helv. Chim. Acta*, **50**, 861 (1967).
39 P. Caveng and H. Zollinger, *Helv. Chim. Acta*, **50**, 866 (1967).
40 M. R. Crampton and V. Gold, *J. Chem. Soc., B*, **1967**, 23.
41 E. Buncel and E. A. Symons, *Chem. Commun.*, **1967**, 771.
42 M. R. Crampton, *J. Chem. Soc., B*, **1967**, 85.
43 M. Pianka and J. D. Edwards, *J. Chem. Soc., C*, **1967**, 2290.

Substitution in Polyfluoro-aromatic Compounds

The greater reactivity towards amines of halogen *ortho* rather than *para* to a nitro-group, and the reverse relationship with alkoxides, are commonly attributed to hydrogen-bonding in the transition state of the amine reactions. There are exceptions to this, however, and these are now being investigated with polyfluoro-aromatic substrates. The ratio of the amount of displacement by various nucleophiles of the 4- and the 6-fluorine atom of tetrafluoronitrobenzenes with a 2-substituent have been determined, and the variations have been ascribed to a steric effect.[44] Pentafluoro-(methylthio)- and -(arylthio)-benzenes reacted with ethoxide and with thiophenoxide ions in ethanol exclusively at the position *para* to sulphur; the sulphur group is activating in the order p-Cl·C_6H_4S > PhS > p-Me·C_6H_4S > MeS > H, i.e., in the order of its ability to stabilize the negative charge.[45] Similar stabilization was invoked to explain why 4,5,6,7-tetrafluorobenzo[*b*]thiophen with sodium methoxide in methanol gave mainly the trifluoro-6-methoxy derivative;[46] 1,2,3,4-tetrafluorodibenzofuran, however, underwent nucleophilic substitution of the fluorine *meta* to oxygen.[47] The ease of nucleophilic displacement of the fluorines in tetrafluoropyrimidines is, as expected from that for chloropyrimidines, 4- and 6- > 2- $\gg$ 5-.[48] The prediction that fluorine in pentachlorofluorobenzene would be more reactive than in hexafluorobenzene has now been confirmed for reaction with methoxide ion; thus pentachlorophenyl is more activating than pentafluorophenyl.[49] The reaction of N-polyfluorophenyl-substituted N-oxides with primary and secondary amines gave predominant displacement of *ortho*-fluorine; this was ascribed to hydrogen-bonding between the amine and the N-oxide group.[50]

Other reactions studied were those of hexafluorobenzene, pentafluorobenzene, and pentachlorofluorobenzene with methoxide ion,[51] of hexafluorobenzene with sodium cyanide in methanol which gave a small amount of 1,4-dicyano-2,3,5,6-tetramethoxybenzene,[52] of pentafluorobenzenes with sodium pentafluorophenoxide,[53a] and of tetrafluoroanthraquinone with ammonia and aliphatic amines.[53b]

44 J. Burdon, D. R. King, and J. C. Tatlow, *Tetrahedron*, **23**, 1347 (1967).
45 J. M. Birchall, M. Green, R. N. Haszeldine, and A. D. Pitts, *Chem. Commun.*, **1967**, 338.
46 G. M. Brooke and M. A. Quasem, *Tetrahedron Letters*, **1967**, 2507.
47 P. J. N. Brown, R. Stephens, and J. C. Tatlow, *Tetrahedron*, **23**, 4041 (1967).
48 R. E. Banks, D. S. Field, and R. N. Haszeldine, *J. Chem. Soc., C*, **1967**, 1822.
49 J. Miller and H. W. Yeung, *Australian J. Chem.*, **20**, 379 (1967).
50 M. Bellas, D. Price, and H. Suschitzky, *J. Chem. Soc., C*, **1967**, 1249.
51 V. A. Sokolenko, L. V. Orlova, and G. G. Yakobson, *Izv. Sib. Otd. Akad. Nauk SSSR, Ser. Khim. Nauk*, **1966**, 113; *Chem. Abs.*, **67**, 32106p (1967).
52 B. J. Wakefield, *J. Chem. Soc., C*, **1967**, 72.
53a R. J. DePasquale and C. Tamborski, *J. Org. Chem.*, **32**, 3163 (1967).
53b E. P. Fokin, V. A. Loskutov, and A. V. Konstantinova, *Izv. Sib. Otd. Akad. Nauk SSSR, Ser. Khim. Nauk*, **1966**, 110; *Chem. Abs.*, **67**, 63609j (1967).

Heterocyclic Systems

The rates of reaction of 2- and 4-chloroquinoline with piperidine in aprotic solvents parallel the dielectric constant of the solvent, but are enhanced in methanol, especially with the 4-chloro-compound, presumably because of hydrogen-bond stabilization of the transition state. With the 2-chloro-isomer this hydrogen-bonding is less effective because of partial internal neutralization of the developing charges.[54] This difference in hydrogen-bond stabilization is also used to rationalize the relative rates of piperidino-dechlorination of 2-chloro-4-substituted and 4-chloro-2-substituted quinolines,[55] and the much larger steric, rate-retarding, effect of an 8-*tert*-butyl substituent in methanol as solvent than in aprotic solvents.[56] The same effect has been demonstrated for the rate differences in the piperidino-dechlorination of 2- and 6-alkyl-4-chloropyrimidines in toluene and in ethanol.[57] In all these reactions the nucleophile, piperidine, is too weakly protic to interact significantly with the heteroatom; however, such interaction does occur with toluene-*p*-thiol, where nucleophilic displacement of chlorine is preceded by a rapid equilibration:

$$ArSH + QCl \rightleftharpoons ArS^- + HQ^+Cl$$

$$ArS^- + HQ^+Cl \rightarrow HQ^+SAr + Cl^-$$

This mechanism explains why the non-catalysed reaction of a chloroquinoline with the thiol in methanol is faster than the reactions involving the thiophenoxide anion or the chloro-*N*-methylquinolinium ion with the neutral form of the other reactant; and why electron-withdrawing substituents are, abnormally, deactivating.[58]

3-Methylpyridine and *o*-tolyl-lithium in ether react to give substantial amounts of 1,2,5,6-tetrahydro-3-methyl-2-(*o*-tolyl)pyridine, presumably by disproportionation of the corresponding dihydro-intermediate; thus evidence

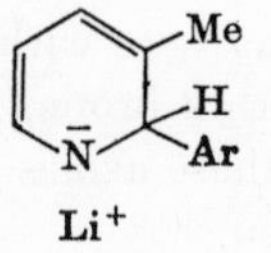

(6)

is provided for the intervention of the σ-complex (**6**).[59] The phenylation of other 3-substituted pyridines has also been reported.[60] In contrast to pentafluoropyridine, pentachloropyridine suffers displacement of both the 2- and

54 G. Illuminati, G. Marino, and G. Sleiter, *J. Am. Chem. Soc.*, **89**, 3510 (1967).
55 F. Genel, G. Illuminati, and G. Marino, *J. Am. Chem. Soc.*, **89**, 3516 (1967).
56 M. Calligaris, G. Illuminati, and G. Marino, *J. Am. Chem. Soc.*, **89**, 3518 (1967).
57 M. Calligaris, P. Linda, and G. Marino, *Tetrahedron*, **23**, 813 (1967).
58 G. Illuminati, P. Linda, and G. Marino, *J. Am. Chem. Soc.*, **89**, 3521 (1967).
59 R. A. Abramovitch and G. A. Poulton, *Chem. Commun.*, **1967**, 274, 564.
60 R. A. Abramovitch and G. A. Poulton, *J. Chem. Soc., B*, **1967**, 267.

the 4-chlorine atom by nucleophiles, larger nucleophiles favouring the more open 2-position;[61, 62] the importance of solvent has been demonstrated here also. Pentachloropyridine 1-oxide is even more susceptible to nucleophilic attack and the α-positions are now almost exclusively involved; thus secondary amines gave 2,6-disubstituted products in nearly quantitative yield at room temperature, and even under more vigorous conditions no 2,4- or 2,4,6-trisubstituted product was observed;[62] this selectivity results, we suggest, from hydrogen-bonding between the *N*-oxide group and the secondary amine.

Methyl- and phenyl-sulphinyl and -sulphonyl groups are readily displaced by nucleophiles from the 2- and the 4-position of pyrimidines; second-order rate constants for the reaction with *n*-pentylamine in DMSO show that methyl-sulphonyl and -sulphinyl and phenylsulphonyl are slightly better leaving groups than chlorine and $> 10^5$ better than the parent thioethers.[63] The methylsulphonyl group is also very readily displaced from pyridines, pyridazines, and pyrazines,[64] and their benzo-derivatives.[65] Comparative reactivities for all six monochlorodiazabenzenes with *p*-nitrophenoxide in methanol have been measured, an excess of *p*-nitrophenol being added to suppress methanolysis (though note ref. 8 of this Chapter).[66] The reactivity order for the displacement of chlorine from the chloropyridazine 1-oxides by piperidine or sodium ethoxide was $5 > 3 > 6 > 4$.[67] The activating effect of the *N*-oxide function for nucleophilic displacement of chlorine by ammonia and amines in pyrazines has been qualitatively demonstrated.[68] Reaction of 4-chloro-2-phenyl[4-^{14}C]pyrimidine with potassamide in liquid ammonia gave 4-amino-2-phenyl[4-^{14}C]pyrimidine, demonstrating a normal displacement rather than the ring-opening mechanism observed with the 5-methoxy-derivative.[69]

Kinetics of the second-order reactions of various bromo-*N*-methyltetrazoles, -triazoles, and -imidazoles with piperidine in ethanol show, as expected, that the reactivity increases with the number of "double bound" nitrogen atoms. Comparison with 2-bromopyridine shows that two or three such nitrogens are required in these azoles to overcome the electron-release from the "singly bound" nitrogen.[70]

Other reactions studied were the hydrolysis of 9-chloroacridine[71] and its

[61] W. T. Flowers, R. N. Haszeldine, and S. A. Majid, *Tetrahedron Letters*, **1967**, 2503.
[62] S. M. Roberts and H. Suschitzky, *Chem. Commun.*, **1967**, 893.
[63] D. J. Brown and P. W. Ford, *J. Chem. Soc., C*, **1967**, 568.
[64] G. B. Barlin and W. V. Brown, *J. Chem. Soc., B*, **1967**, 648.
[65] G. B. Barlin and W. V. Brown, *J. Chem. Soc., B*, **1967**, 736.
[66] T. L. Chan and J. Miller, *Australian J. Chem.*, **20**, 1595 (1967).
[67] S. Sako and T. Itai, *Chem. Pharm. Bull. (Tokyo)*, **14**, 269 (1966).
[68] B. Klein, E. O'Donnell, and J. Auerbach, *J. Org. Chem.*, **32**, 2412 (1967).
[69] H. W. van Meeteren and H. C. van der Plas, *Rec. Trav. Chim.*, **86**, 567 (1967).
[70] G. B. Barlin, *J. Chem. Soc., B*, **1967**, 641.
[71] A. Ledochowski, *Roczniki Chem.*, **41**, 717 (1967); *Chem. Abs.*, **67**, 53204j (1967).

methyl- and methoxy-derivatives,[72] the butylaminolysis of substituted 2- and 4-methoxy- and -methylthio-pyrimidines,[73] the self-quaternization of 3-chloro-6-methylpyridazine,[74] the reactions of chloropurines and their 9-methyl derivatives with sodium ethoxide and with piperidine,[75a] nucleophilic substitution on 6-chloropurine ribonucleoside,[75b] the covalent hydration of quinazolines and triazanaphthalenes,[76] and the reactions of halogenonitrothiophens with sodium thiophenoxide and with piperidine.[77]

Nucleophilic substitution in pyridines[78a] and in purines[78b] and the hydration of heteroaromatic compounds[79] have been reviewed. Some substitutions via hetarynes are described in the last section of this Chapter.

Other Reactions

Suschitzky has reviewed his use of the labilization, towards nucleophilic attack, of aromatic fluorine by the diazonium and related groups as a means of diagnosing the formation of such groups during the decomposition of various aryl radical precursors.[80] A 4-methyl group was found to decrease 4- to 6-fold the rate of decomposition of naphthalene-1-diazonium salts in aqueous solution, and this was attributed to σ–π hyperconjugation of the methyl group with the naphthalene ring.[81] The decomposition of benzenediazonium chloride is unimolecular in acidic solution, but in weakly acid or neutral solution the diazonium salt couples with the phenol produced.[82] Benzenediazonium ions react with aldehyde arylhydrazones in neutral solution at the methine carbon to give, initially, bis(arylazo)methanes, rather than at nitrogen to give tetrazenes (eqn. 1).[83]

$$\mathrm{RCH{=}N\cdot NHAr} + \mathrm{PhN_2^+} \longrightarrow \mathrm{RCH}\begin{cases}\mathrm{N{=}NAr}\\ \mathrm{N{=}NPh}\end{cases} \longrightarrow \mathrm{RC}\begin{cases}\mathrm{{=}N\cdot NHAr}\\ \mathrm{N{=}NPh}\end{cases} \qquad (1)$$

Excitation of the absorption band near 280 mμ of fluorobenzene, chlorobenzene, and anisole in the presence of good nucleophiles, such as piperidine

[72] A. Ledochowski, *Roczniki Chem.*, **40**, 2015 (1966); *Chem. Abs.*, **67**, 43066b (1967).
[73] D. J. Brown and R. V. Foster, *Australian J. Chem.*, **19**, 2321 (1966).
[74] H. Lund and S. Gruhn, *Acta Chem. Scand.*, **20**, 2637 (1966).
[75a] G. B. Barlin, *J. Chem. Soc.*, *B*, **1967**, 954.
[75b] B. T. Walsh and R. Wolfenden, *J. Am. Chem. Soc.*, **89**, 6221 (1967).
[76] J. W. Bunting and D. D. Perrin, *J. Chem. Soc.*, *B*, **1967**, 950; W. L. F. Armarego and J. I. C. Smith, *ibid.*, p. 449.
[77] D. Spinelli, C. Dell'Erba, and G. Guanti, *Ann. Chim.* (*Rome*), **55**, 1252, 1260 (1965).
[78a] R. A. Abramovitch and J. G. Saha, *Adv. Heterocyclic Chem.*, **6**, 274 (1965).
[78b] J. H. Lister, *Adv. Heterocyclic Chem.*, **6**, 11 (1966).
[79] A. Albert, *Angew. Chem. Intern. Ed. Engl.*, **6**, 919 (1967).
[80] H. Suschitzky, *Angew. Chem. Intern. Ed. Engl.*, **6**, 596 (1957).
[81] J. Reichel and B. Demian, *Rev. Roumaine Chim.*, **11**, 1241 (1966).
[82] M. Matrka, Z. Ságner, V. Chmátal, V. Štěrba, and M. Veselý, *Coll. Czech. Chem. Commun.*, **32**, 1462 (1967).
[83] A. F. Hegarty and F. L. Scott, *J. Org. Chem.*, **32**, 1957 (1967).

and alkoxide ions, resulted in displacement of the substituent by the nucleophile. Yields were usually very low and there was much polymerization. Neither radicals nor benzyne appears to be involved in the displacement reaction; an ionic mechanism was proposed in which the nucleophile attacks the electron-deficient centres in the excited state of the aromatic compound (e.g., eqn. 2). The quantum yield was very low, implying that deactivation of the excited states was much faster than reaction with nucleophile ($k_{-1} \gg k_2$). In agreement with this it was shown that the rate of formation of 1-phenylpiperidine increased with piperidine concentration.[84] Nitroanilines were formed when nitrobenzenes were irradiated in liquid ammonia in quartz vessels. Nitrobenzene gave *o*- and *p*-nitroaniline, the latter predominating; this photosubstitution was much faster than that of *p*-chloronitrobenzene.[85] Irradiation of 3-bromopyridine in aqueous sodium hydroxide gave 3-hydroxypyridine in a zero-order reaction; however, *m*-bromonitrobenzene did not react.[86] Irradiation of 4-nitroveratrole with methylamine in water gave *N*-methyl-2-methoxy-4-nitroaniline, as the major product, in a zero-order reaction.[87] Details of the light-catalysis of the reaction of 4-nitropyridine 1-oxide with piperidine in ethanol have appeared; the quantum yield varied with piperidine concentration in accord with a rate-determining displacement of the nitro-group by piperidine from the excited *N*-oxide.[88] Photolysis of

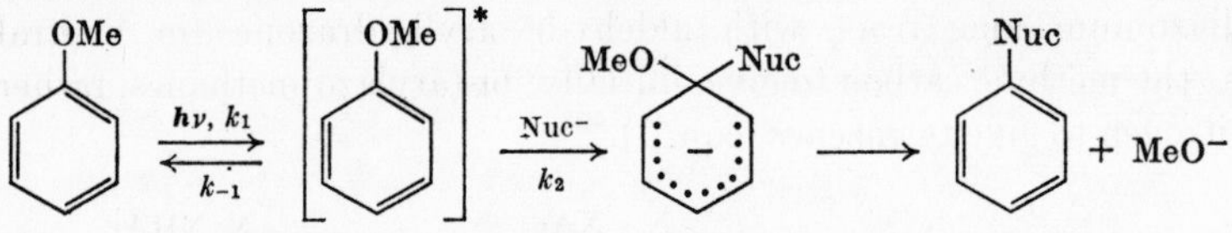

(2)

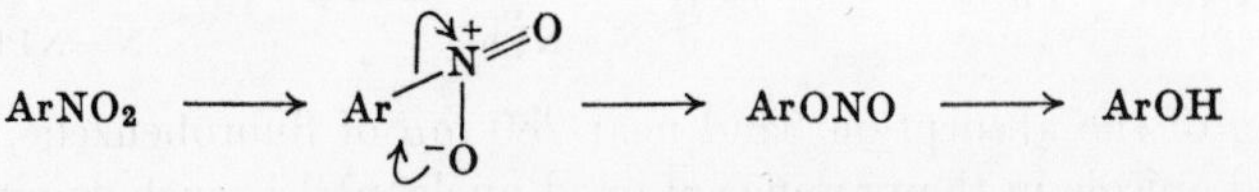

(3)

pyridine and quinoline 1-oxides with sterically hindered 4-nitro-groups, such as 3,5-dimethyl-4-nitropyridine 1-oxide, in alcohols gave the corresponding 4-hydroxy-compound; the 4-nitrite esters were proposed as intermediates (eqn. 3).[89]

The Ullmann reaction of 2,3-di-iodo-1-nitrobenzene gave a biphenylene, a biphenyl, and a tetraphenylene; the proposed copper aryl (or aryl radical) intermediate could give 3-nitrobenzyne, and hence the biphenylene by

[84] J. A. Barltrop, N. J. Bunce, and A. Thompson, *J. Chem. Soc., C*, **1967**, 1142.
[85] A. van Vliet, M. E. Kronenberg, and E. Havinga, *Tetrahedron Letters*, **1966**, 5957.
[86] G. H. D. van der Stegen, E. J. Poziomek, M. E. Kronenberg, and E. Havinga, *Tetrahedron Letters*, **1966**, 6371.
[87] M. E. Kronenberg, A. van der Heyden, and E. Havinga, *Rec. Trav. Chim.*, **86**, 254 (1967).
[88] R. M. Johnson and C. W. Rees, *J. Chem. Soc., B*, **1967**, 15.
[89] C. Kaneko, I. Yokoe, and S. Yamada, *Tetrahedron Letters*, **1967**, 775.

dimerization.[90] Catalytic effects in the coupling of aryl halides have been commented upon, and the customary view that Ullmann coupling necessarily requires cuprous oxide has been questioned.[91] The rates of the reaction of 3-chloro- and 3-bromo-pyridine and of *p*-chloro- and *p*-bromo-benzoate ions with ammonia in the presence of cuprous acetate or chloride are proportional to $[ArX][Cu^+][NH_3]^\alpha$, where α varies from 0 to 1.[92]

An extensive review on the hydrated electron,[93] e^-_{aq}, and a review on its application in analytical chemistry,[94] have appeared. The temperature-dependence of a number of reactions of e^-_{aq} with a variety of substrates has been measured; although the specific rates vary over a range of 10^5, the activation energies were constant at 3.5 ± 0.4 kcal $mole^{-1}$, the activation energy for diffusion in water. Thus reactions of e^-_{aq} require no activation energy in excess of that for diffusion, and entropy effects must be responsible for the rate differences.[95] The relative reaction rates in ice at 77°K have been shown to parallel quantitatively those in water at 300°K, and thus to support the significance of entropy effects.[96] Solvated electrons have now been generated electrolytically in a highly proton-donating solvent (ethanol–hexamethylphosphoramide) and have been used to reduce the benzene ring.[97] Other aspects of the chemistry of e^-_{aq} have also been studied.[98]

Benzyne and Related Intermediates

An excellent monograph on dehydrobenzenes and cycloalkynes has appeared.[99] The relative stability of arynes and cycloalkynes has been estimated quantitatively by dehydrochlorination of the appropriate aryl or vinyl chloride with lithium piperidide in the presence of phenyl-lithium (Scheme 2).[100] The competition between phenyl-lithium (k_1) and lithium piperidide (k_2) for the intermediate was measured, the stability of the intermediate being assumed to parallel its selectivity. The ratio, k_1/k_2, decreases,

90 K. Iqbal and R. C. Wilson, *J. Chem. Soc., C*, **1967**, 1690.
91 R. G. R. Bacon, S. G. Seeterram, and O. J. Stewart, *Tetrahedron Letters*, **1967**, 2003.
92 F. M. Vainshtein, E. I. Tomilenko, and E. A. Shilov, *Dokl. Akad. Nauk SSSR*, **170**, 1077 (1966); *Chem. Abs.*, **66**, 18400c (1967).
93 D. C. Walker, *Quart. Rev.*, **21**, 79 (1967); see also S. R. Logan, *J. Chem. Educ.*, **44**, 344 (1967).
94 E. J. Hart, *Record. Chem. Progr.*, **28**, 25 (1967).
95 M. Anbar, Z. B. Alfassi, and H. Bregman-Reisler, *J. Am. Chem. Soc.*, **89**, 1263 (1967).
96 L. Kevan, *J. Am. Chem. Soc.*, **89**, 4238 (1967).
97 H. W. Sternberg, R. E. Markby, I. Wender, and D. M. Mohilner, *J. Am. Chem. Soc.*, **89**, 186 (1967).
98 W. L. Waltz, A. W. Adamson, and P. D. Fleischauer, *J. Am. Chem. Soc.*, **89**, 3923 (1967); G. V. Buxton, F. S. Dainton, and G. Thielens, *Chem. Commun.*, **1967**, 201; R. R. Dewald and R. V. Tsina, *Chem. Commun.*, **1967**, 647; W. C. Gottschall and E. J. Hart, *J. Phys. Chem.*, **71**, 2102 (1967); B. Cercek and M. Ebert, *Trans. Faraday Soc.*, **63**, 1687 (1967); D. C. Walker, *Can. J. Chem.*, **45**, 807 (1967); G. V. Buxton, F. S. Dainton, and M. Hammerli, *Trans. Faraday Soc.*, **63**, 1191 (1967); M. Anbar and E. J. Hart, *J. Phys. Chem.*, **71**, 3700 (1967).
99 R. W. Hoffmann, "Dehydrobenzenes and Cycloalkynes", Academic Press, New York, 1967.
100 L. K. Montgomery and L. E. Applegate, *J. Am. Chem. Soc.*, **89**, 5305 (1967).

as expected, from cyclo-octyne to cyclopentyne (although only by a factor of 10); that for benzyne is between cyclopentyne and cyclohexyne.[100] The importance of steric effects in this type of competition has been stressed.[101] The greater selectivity of 9,10-phenanthryne than of benzyne or naphthalyne in competitive experiments with diethylamine and di-isopropylamine was attributed to steric hindrance in the reaction of the bulkier amine by the *peri*-hydrogen atoms (see **7**). There was effectively no difference between the

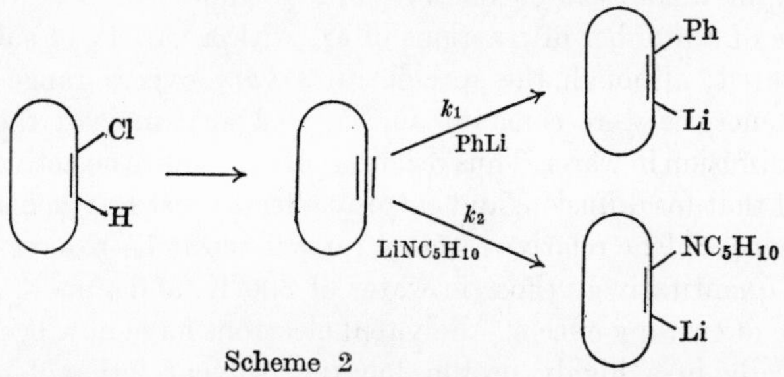

Scheme 2

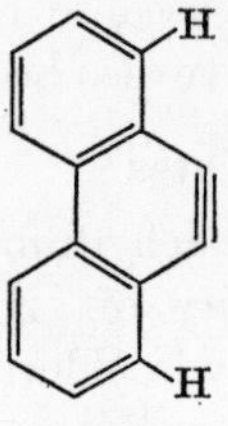

(7)

arynes in competition for diethylamine and piperidine. The 1-position of 1,2-naphthalyne was also shown to be hindered to attack by di-isopropylamine.[101] This contrasts with Huisgen's earlier explanation of selectivity of the aryne bond increasing with decreasing bond length. The proportion of elimination–addition involved in the reaction of 4-halogenopyridines with lithium amides has been determined.[102]

Good evidence for the intermediacy of benzyne in the photolysis of *o*-di-iodobenzene is provided by the formation of *o*-deuterioanisole in CH_3OD as solvent; an aryl radical would have abstracted an α-hydrogen atom.[103] The products of decomposition of benzenediazonium-2-carboxylate (**8**) in benzene have been shown to be sensitive to silver ion contamination introduced in the normal preparation of (**8**). As $[Ag^+]$ was increased, the yields of biphenylene and benzobicyclo[2.2.2]octatriene decreased and of benzocyclo-octatetraene

[101] T. Kauffmann, H. Fischer, R. Nürnberg, M. Vestweber, and R. Wirthwein, *Tetrahedron Letters*, **1967**, 2911, 2917.
[102] T. Kauffmann and R. Nürnberg, *Chem. Ber.*, **100**, 3427 (1967).
[103] N. Kharasch and R. K. Sharma, *Chem. Commun.*, **1967**, 492.

(**9**) and biphenyl increased markedly. Friedman[104] proposed the formation of a benzyne–Ag^+ complex which is more electrophilic than benzyne and attacks benzene to give (**10**) rather than the normal 1,2- and 1,4-addition product. Decomposition of the diazonium carboxylate in benzene has now been shown to give the 2:1 benzyne–benzene adduct (**11**) probably by further addition of benzyne to benzobicyclo[2.2.2]octatriene.[105] Full details have appeared of the

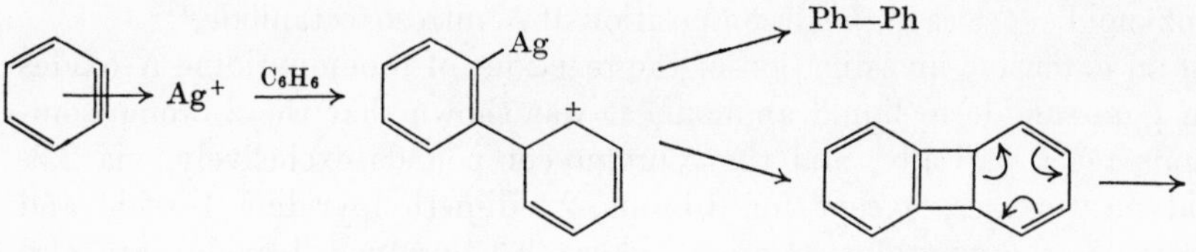

(9)

(10)

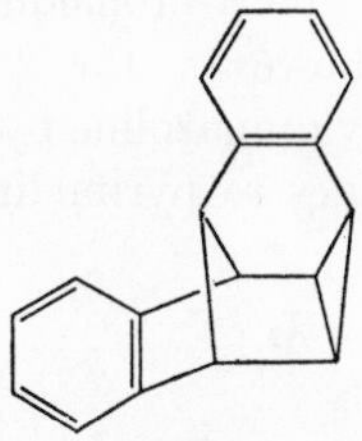

(11)

pyrolysis of phthalic anhydride and *o*-sulphobenzoic anhydride at 700—1100°/0.1—10 mm. to give biphenylene and triphenylene by way of benzyne. The yield of biphenylene from these and related compounds correlates roughly with the instability of the parent molecular ion in the mass spectrometer.[106]

There was no evidence for the formation of benzyne in the mass-spectral fragmentation or the pyrolysis (800°) of *o*-phenylene sulphite.[107] From the mass spectrometry of benzyne, generated by vacuum-pyrolysis of di-(*o*-iodophenyl)mercury, its ionization potential was found to be 0.26 eV higher than that of benzene; the enthalpy of formation of benzyne was calculated to be 118 ± 5 kcal $mole^{-1}$.[108]

From the nature of the products it was suggested that benzyne is an intermediate in the pyrolysis (690°) of acetylene; butadiyne could be formed and

104 L. Friedman, *J. Am. Chem. Soc.*, **89**, 3071 (1967).
105 M. Stiles, U. Burckhardt, and G. Freund, *J. Org. Chem.*, **32**, 3718 (1967).
106 R. F. C. Brown, D. V. Gardner, J. F. W. McOmie, and R. K. Solly, *Australian J. Chem.*, **20**, 139 (1967).
107 D. C. DeJongh, R. Y. Van Fossen, and C. F. Bourgeois, *Tetrahedron Letters*, **1967**, 271.
108 H.-F. Grützmacher and J. Lohmann, *Ann. Chem.*, **705**, 81 (1967).

then undergo cycloaddition with another molecule of acetylene.[109] At 400° biphenylene gave tetraphenylene almost quantitatively by dimerization of the 2,2′-biphenylene diradical; in the gas phase, where bimolecular reactions are minimized, this diradical fragments extensively to benzyne.[110]

Arynes are formed in good yield from aryl halides and potassium *tert*-butoxide with or without an inert solvent.[111] Further evidence has been obtained, by trapping experiments, for the participation of benzyne or "benzynoid" species in the decomposition of *N*-nitrosoacetanilide.[112]

In an extensive investigation of the reactions of bromopyridine *N*-oxides with potassamide in liquid ammonia it was shown that the 2-bromo-compounds react partially, and the 3-bromo-compounds exclusively, via 2,3-pyridyne 1-oxides, except for 3-bromo-2,5-dimethylpyridine 1-oxide and 3-bromo-5-ethoxypyridine 1-oxide where 3,4-pyridyne 1-oxides are also involved. The 4-bromo-compounds react by the normal addition–elimination mechanism.[113] In the same reaction 3- and 4-bromoquinoline gave 3- and 4-aminoquinoline via the quinolyne; however, the 2-bromo-isomer gave 2-aminoquinoline, together with 2-methylquinazoline by a rearrangement[114] (Scheme 3) analogous to that of pyridines to pyrimidines reported earlier.

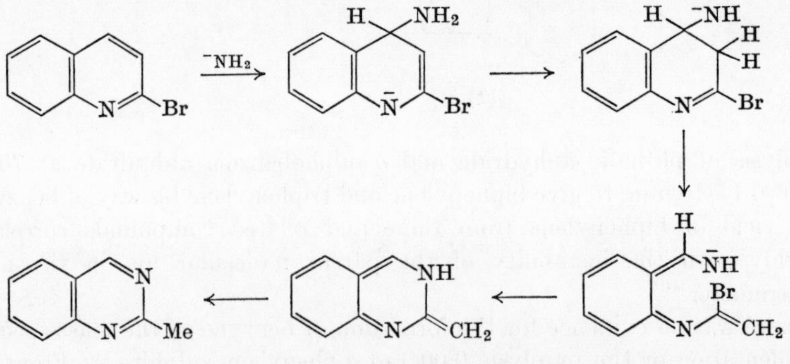

Scheme 3

Pentachloropyridine with *n*-butyl-lithium in ether gave mainly 2,3,5,6-tetrachloro-4-pyridyl-lithium which, when heated in aromatic hydrocarbons, gave the 1,4-adducts of 2,5,6-trichloro-3,4-pyridyne.[115] This pyridyne, like its

109 E. K. Fields and S. Meyerson, *Tetrahedron Letters*, **1967**, 571.
110 D. F. Lindow and L. Friedman, *J. Am. Chem. Soc.*, **89**, 1271 (1967).
111 J. I. G. Cadogan, J. K. A. Hall, and J. T. Sharp, *J. Chem. Soc., C*, **1967**, 1860.
112 D. L. Brydon, J. I. G. Cadogan, D. M. Smith, and J. B. Thomson, *Chem. Commun.*, **1967**, 727.
113 R. J. Martens and H. J. Den Hertog, *Rec. Trav. Chim.*, **86**, 655 (1967).
114 H. J. den Hertog and D. J. Buurman, *Rec. Trav. Chim.*, **86**, 187 (1967).
115 J. D. Cook and B. J. Wakefield, *Tetrahedron Letters*, **1967**, 2535.

trifluoro-analogue,[116] was not trapped by furan. 4,5,6-Trichloro-2,3-pyridyne could not be generated and trapped in the same way.[115] 4,5-Dehydro-1-methylimidazole was shown to participate in the conversion of 5-halogeno-1-methylimidazoles into the 4- and 5-amino-derivatives; with a pyrrolidine–piperidine mixture the product ratio was independent of the nature of the halogen.[117]

When acridine was treated with diazotized anthranilic acid in a large volume of methylene chloride a small amount of 9-dichloromethyl-10-phenylacridan (**12**) was formed, by incorporation of the solvent.[118] Mechanisms have been tentatively proposed for the formation of thioanisole, thiophenol, and diphenyl sulphide from benzyne and dimethyl disulphide[119] and carbon disulphide.[120]

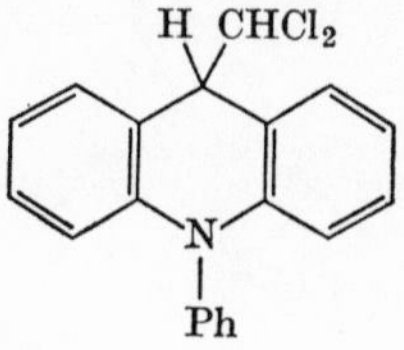

(12)

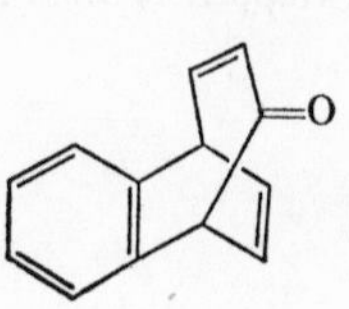

(13)

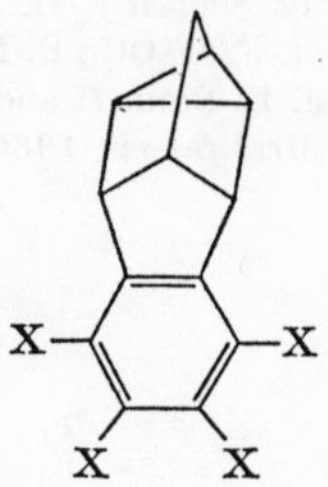

(14)

Benzyne reacts with *N*-phenylsydnone to give 2-phenylindazole,[121] with tropone by 1,4-addition to give 6,7-benzobicyclo[3.2.2]nona-3,6,8-trien-2-one (**13**),[122] with cyclo-octyne to give 1,2,3,4,5,6-hexahydrocyclo-octa[*m*]-phenanthrene,[123] with α-methylstyrene,[124] with 7-dehydrocholesteryl methyl ether and cyclohexa-1,3-diene,[125a] and with DMSO.[125b] It was proposed that tetrachloro- and tetrafluoro-benzyne added across the 2,3- and 2,6-positions (to give **14**) of bicycloheptadiene in a non-concerted and a concerted process, respectively.[126] Oxidation of 1-aminonaphthotriazoles with lead tetra-acetate gave the corresponding dibenzobiphenylenes, the reaction

[116] R. D. Chambers, F. G. Drakesmith, J. Hutchinson, and W. K. R. Musgrave, *Tetrahedron Letters*, **1967**, 1705.
[117] T. Kauffmann, R. Nurnberg, J. Schulz, and R. Stabba, *Tetrahedron Letters*, **1967**, 4273.
[118] B. H. Klanderman, *Tetrahedron Letters*, **1966**, 6141.
[119] I. Tabushi, K. Okazaki, and R. Oda, *Tetrahedron Letters*, **1967**, 3591.
[120] I. Tabushi, K. Okazaki, and R. Oda, *Tetrahedron Letters*, **1967**, 3827.
[121] A. Y. Lazaris, *Zh. Organ. Khim.*, **2**, 1322 (1966); *Chem. Abs.*, **66**, 65426a (1967).
[122] J. Ciabattoni, J. E. Crowley, and A. S. Kende, *J. Am. Chem. Soc.*, **89**, 2778 (1967).
[123] V. Franzen and H.-I. Joschek, *Ann. Chem.*, **703**, 90 (1967).
[124] E. Wohlthuis and W. Cady, *Angew. Chem. Intern. Ed. Engl.*, **6**, 555 (1967).
[125a] I. F. Eckhard, H. Heaney, and B. A. Marples, *Tetrahedron Letters*, **1967**, 4001.
[125b] M. Kise, T. Asari, N. Furukawa, and S. Oae, *Chem. Ind.* (*London*), **1967**, 276; H. H. Szmant and S. Vázquez, *ibid.*, p. 1000.
[126] H. Heaney and J. M. Jablonski, *Tetrahedron Letters*, **1967**, 2733.

involving dimerization of naphthalyne.[127] Hoffmann and Sieber have published full details of their generation of 1,8-dehydronaphthalene.[128] Bromocyclo-octatetraene and potassium *tert*-butoxide gave the 1,2-dehydroderivative, as shown by its dimerization and by several trapping experiments,[129a] and dehydrobullvalene has been generated and trapped similarly.[129b]

Details of Bunnett and Happer's generation of *o*-halogenophenyl anions have appeared.[130]

[127] J. W. Barton and S. A. Jones, *J. Chem. Soc., C*, **1967**, 1276.

[128] R. W. Hoffmann and W. Sieber, *Ann. Chem.*, **703**, 96 (1967); see *Organic Reaction Mechanisms*, **1965**, 148.

[129a] A. Krebs and D. Byrd, *Ann. Chem.*, **707**, 66 (1967).

[129b] G. Schröder, H. Röttele, R. Merényi, and J. F. M. Oth, *Chem. Ber.*, **100**, 3527 (1967); J. F. M. Oth, R. Merényi, H. Röttele, and G. Schröder, *ibid.*, p. 3538.

[130] J. F. Bunnett and D. A. R. Happer, *J. Org. Chem.*, **32**, 2701 (1967); see *Organic Reaction Mechanisms*, **1966**, 184.

CHAPTER 7

Radical and Electrophilic Aromatic Substitution

Radical Substitution

Volume II of "Advances in Free Radical Chemistry" (see Chapter 9, ref. 2) contains reviews by Hey and by Abramovitch on inter- and intra-molecular radical substitution. A review dealing with substitution—including free-radical substitution—in heterocyclic molecules has also appeared,[1] and Suschitzky has surveyed the literature on ion-pair formation in aryl radical precursors, particularly as detected by nucleophilic fluorine displacement in *p*-fluorophenylazo-derivatives,[2] e.g.:

$$FC_6H_4N{:}NOAc \rightleftarrows [FC_6H_4N_2^{+}OAc^{-}] \rightleftarrows [AcOC_6H_4N_2^{+}F^{-}] \rightleftarrows AcOC_6H_4N{:}NF \rightarrow AcOC_6H_4\cdot$$

The "nitro-group effect", whereby the yields of biphenyl and benzoic acid from the decomposition of benzoyl peroxide in benzene are greatly improved by the presence of aromatic nitro-compounds, was discussed last year[3] in terms of interception of phenylcyclohexadienyl radicals by the nitro-compound. An alternative interpretation suggests that a small quantity of nitro-compound is reduced to nitroso-compound which then scavenges a phenyl radical to form a stable aryl phenyl nitroxide radical.[4] This then maintains a relatively high stationary-state concentration and oxidizes phenylcyclohexadienyl radicals before they can dimerize or otherwise lead to complex products. The concentration of nitroxide is maintained by reoxidation of arylphenylhydroxylamine by molecular benzoyl peroxide. This molecule-induced decomposition was supported by observation of a rapid reaction between benzoyl peroxide and diphenylhydroxylamine in benzene at relatively low temperatures. Whilst the evidence for the new mechanism (Scheme 1) (page 184) seems strong for reactions when traces of nitrosobenzene or diphenylhydroxylamine are added in place of the nitro-compound, the stable radical observed (ESR) when nitro-compounds are employed has not yet been identified with certainty.

The dominant role of stable radical intermediates has been discussed before

[1] R. A. Abramovitch and J. G. Saha, *Adv. Heterocyclic Chem.*, **6**, 229 (1966).
[2] H. Suschitzky, *Angew. Chem. Intern. Ed. Engl.*, **6**, 596 (1967).
[3] See *Organic Reaction Mechanisms*, **1966**, 188.
[4] G. R. Chalfont, D. H. Hey, K. S. Y. Liang, and M. J. Perkins, *Chem. Commun.*, **1967**, 367; see also D. H. Hey, K. S. Y. Liang, M. J. Perkins, and G. H. Williams, *J. Chem. Soc., C*, **1967**, 1153.

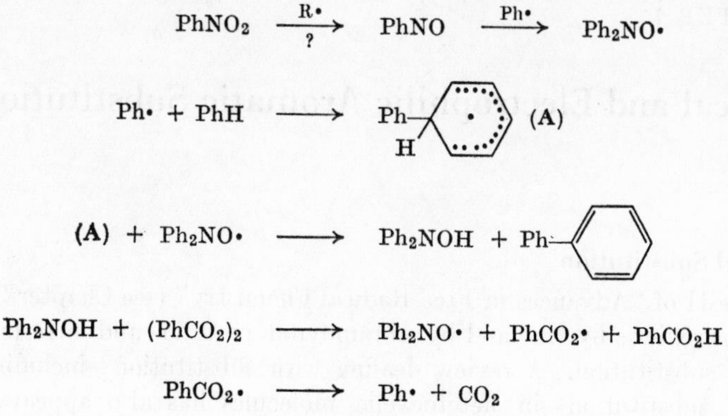

$$PhNO_2 \xrightarrow[?]{R\cdot} PhNO \xrightarrow{Ph\cdot} Ph_2NO\cdot$$

$$Ph\cdot + PhH \longrightarrow Ph\underset{H}{\diagup}\!\!\langle\cdot\rangle \quad (A)$$

$$(A) + Ph_2NO\cdot \longrightarrow Ph_2NOH + Ph\!-\!\langle\rangle$$

$$Ph_2NOH + (PhCO_2)_2 \longrightarrow Ph_2NO\cdot + PhCO_2\cdot + PhCO_2H$$

$$PhCO_2\cdot \longrightarrow Ph\cdot + CO_2$$

Scheme 1

in these pages,[5] particularly in the context of phenylation with nitroso-acetanilide (the Hey reaction). The previously discussed mechanism for this reaction involved the phenyldiazotate (PhN:NO·) radical in this crucial role.[5] New ESR data on the stable radical in this reaction involve ^{15}N labelling and have led to a revised assignment of the nitrogen coupling constants; that next to oxygen is *ca*. 12 Gauss.[6] An alternative interpretation of these spectra

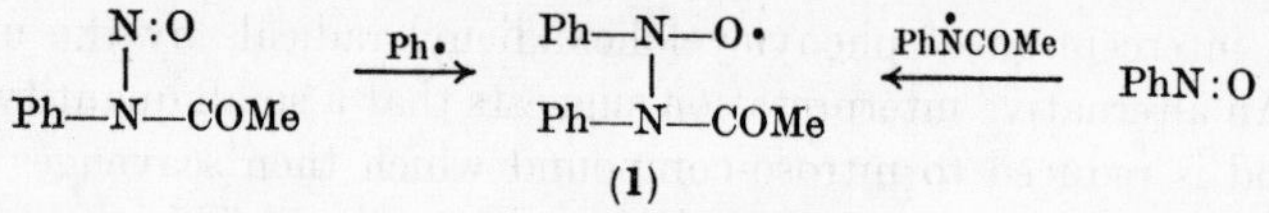

$$\underset{Ph-N-COMe}{\overset{N:O}{|}} \xrightarrow{Ph\cdot} \underset{Ph-N-COMe}{\overset{Ph-N-O\cdot}{|}} \xleftarrow{Ph\overset{\cdot}{N}COMe} PhN:O$$
$$(1)$$

is in terms of structure (1).[4,7] This is considered to arise by scavenging of phenyl radicals by the nitroso-group of the nitroso-amide, and support for the new structure was obtained when the same radical was generated by irradiation of a benzene solution of *N*-bromoacetanilide in the presence of nitroso-benzene.[7] An alternative mechanistic scheme[5] for the Hey reaction was therefore suggested[7] with (1) instead of PhN:NO· acting as the stable radical.

It appears that nitrosoacetanilide may also behave as a source of benzyne, for in the presence of a good benzyne trap, such as tetracyclone, characteristic benzyne adducts are observed.[8] The scavenging of benzyne in this system was markedly dependent on the scavenger, and it was postulated that free benzyne may not be involved, but that the elements of benzyne might be transferred to the tetracyclone in a concerted fashion. The identity of the "benzynoid" (cf. "carbenoid") was not determined, but it was pointed out that neither of

[5] See *Organic Reaction Mechanisms*, **1965**, 154; **1966**, 188.
[6] G. Binsch, E. Merz, and C. Rüchhardt, *Chem. Ber.*, **100**, 247 (1967).
[7] G. R. Chalfont and M. J. Perkins, *J. Am. Chem. Soc.*, **89**, 3054 (1967).
[8] D. L. Brydon, J. I. G. Cadogan, D. M. Smith, and J. B. Thomson, *Chem. Commun.*, **1967**, 727.

the free-radical schemes discussed for homolytic phenylation[5, 7] could at the same time accommodate benzyne production. This is not too surprising since radical reactions are notoriously susceptible to modification or catalysis by reactive impurities (as, for example, in the nitro-group effect discussed above) and, in any case, if a benzynoid species is involved, formation of this would presumably precede the genesis of free radicals. By way of a final comment we suggest consideration of species (**2**) as a possible benzynoid. This is usually regarded as the transition state of the initial rearrangement of the nitroso-amide, but such [1.3]sigmatropic rearrangements are normally symmetry-forbidden and it seems possible that structure (**2**) might represent an energy minimum resulting from intramolecular nucleophilic attack by nitroso-oxygen on what is an unusually electrophilic amide carbon. Certainly, (**2**) has all the structural requirements of a benzyne precursor (effective leaving group; strong base adjacent to *ortho*-proton).

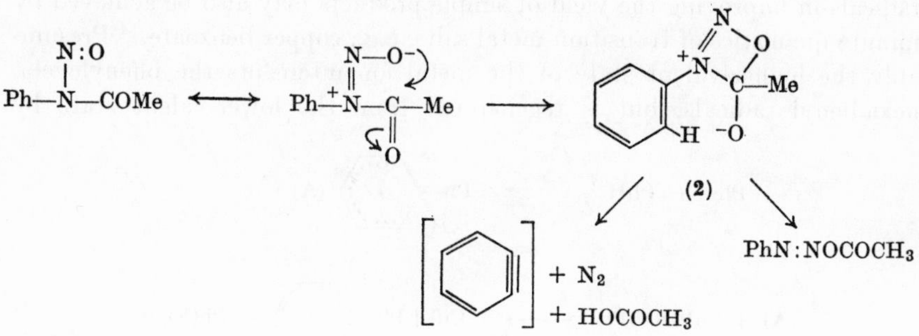

The most detailed analysis to date of the uncatalysed thermal reaction of benzoyl peroxide with benzene has appeared, close scrutiny being given to secondary reaction products.[9] The results were combined with existing product and kinetic data from the literature and subjected to computer analysis.[10] Order-of-magnitude rate constants for some 100 elementary steps adequately accommodated the empirical results. Application of these rate constants to the decomposition of phenylazotriphenylmethane in benzene quantitatively confirmed the dominant role ascribed[11] to stable triphenylmethyl radicals in this reaction. ESR experiments have also demonstrated that a relatively high concentration of triphenylmethyl radicals prevails during the reaction.[12]

It has been found that *N*-phenyl-*N'*-tosyloxydiimide *N*-oxide gives biaryls

[9] D. F. DeTar, R. A. J. Long, J. Rendleman, J. Bradley, and P. Duncan, *J. Am. Chem. Soc.*, **89**, 4051 (1967).
[10] D. F. DeTar, *J. Am. Chem. Soc.*, **89**, 4058 (1967).
[11] See *Organic Reaction Mechanisms*, **1965**, 155.
[12] I. P. Gragerov and A. F. Levit, *Zh. Org. Khim.*, **3**, 550 (1967); *Chem. Abs.*, **67**, 21303z (1967).

on pyrolysis (> 80°) in aromatic solvents.[13] A mechanism was suggested involving the phenyldiazotate radical:

$$PhN{=}NOTs \longrightarrow \cdot OTs + Ph{-}N{\overset{O}{=}}N \longrightarrow PhN{=}NO\cdot$$

$$\text{then,} \quad PhN{=}NO\cdot + RH \longrightarrow PhN{=}NOH \longrightarrow Ph\cdot$$

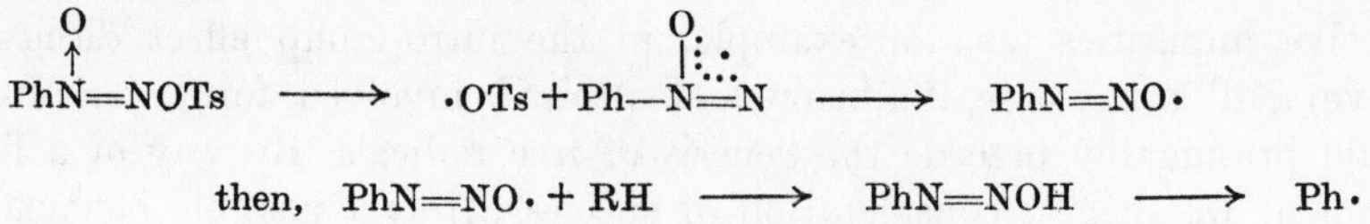

However, in view of the above discussion, other possibilities may have to be considered. A stable radical (if one is formed in this system) could be produced by radical addition at N':

$$O{\leftarrow}NPh{:}NOTs \overset{R\cdot}{\rightarrow} \cdot O{-}NPh{-}NR{-}OTs$$

In the phenylation of benzene with benzoyl peroxide, the effect of nitroxide radicals in improving the yield of simple products may also be achieved by minute quantities of transition-metal salts, e.g., copper benzoate.[14] Presumably the higher-valent state of the metal ion intercepts the phenylcyclohexadienyl radicals, but is regenerated from the lower-valent state by

$$Ph\cdot + PhH \longrightarrow Ph{-}\!\!\!\overset{\bullet}{\underset{H}{\bigcirc}}\!\!\! \quad (A)$$

$$(A) + Cu^{II}OCOPh \longrightarrow Cu^{I} + Ph{-}\!\!\!\bigcirc\!\!\! + PhCO_2H$$

$$Cu^{I} + (PhCO_2)_2 \longrightarrow Cu^{II}OCOPh + PhCO_2\cdot$$

$$PhCO_2\cdot \longrightarrow Ph\cdot + CO_2$$

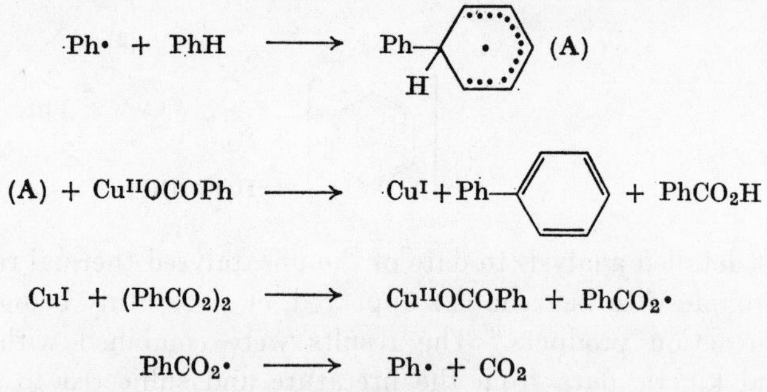

Scheme 2

$$Ph\cdot$$

$$\overset{\bullet}{\bigcirc}{-}X$$

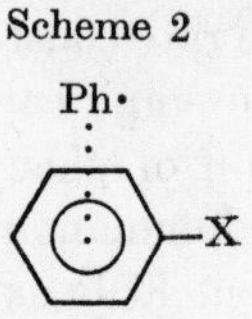

(3)

molecular peroxide (Scheme 2). With substituted benzenes, high yields of biaryls were again obtained and no significant change in isomer composition was observed. It was pointed out that this argued against a relatively stable π-complex (3), because oxidation of (3) to a cationic π-complex by electron

[13] E. A. Dorko and T. E. Stevens, *Chem. Commun.*, **1966**, 871, 925.
[14] D. H. Hey, K. S. Y. Liang, and M. J. Perkins, *Tetrahedron Letters*, **1967**, 1477.

transfer to a metal ion should give quite different isomer distribution typical of electrophilic attack.

Phenyl benzoate is a by-product of the benzoyl peroxide–benzene reaction, produced in part by addition of benzoyloxy-radicals to benzene. Several facts suggest that this addition is reversible and new results from Simamura and his colleagues provide strong support for this view.[15] These workers found that photolysis of benzoyl peroxide in benzene at room temperature provides some 10% of phenyl benzoate. Scavenging experiments with iodine showed this to be a cage product. However, in the presence of oxygen the yield of ester rose to 50%, consistently with the mechanism of Scheme 3. At the temperature of the experiment, decarboxylation of benzoyloxy-radicals would be slower, and complex-formation with benzene more pronounced, than in the thermal reaction at 80°. Benzoyloxylation of anthracene by

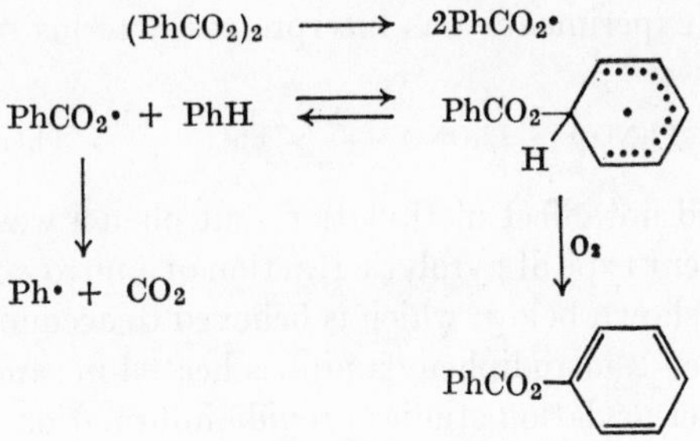

Scheme 3

benzoyl peroxide has also been re-examined, as well as the effect of sulphur dioxide on this reaction.[16a]

Lead tetrabenzoate gives products very similar to those from benzoyl peroxide when its decomposition in boiling benzene is promoted by a radical initiator.[16b] Perhaps decomposition is induced by phenylcyclohexadienyl radicals, for no comparable decomposition can be promoted in carbon tetrachloride at 80°.

Decomposition of acyl-1,3-diaryltriazenes in nitrobenzene at 95° leads to homolytic arylation of the solvent in good yield, but 1,3-acyl migration competes with decomposition of an unsymmetrical diaryltriazene.[17] Photolysis of benzenediazonium fluoroborate in a rigid matrix generates phenyl radicals, which have been detected by ESR spectroscopy.[18]

[15] T. Nakata, K. Tokumaru, and O. Simamura, *Tetrahedron Letters*, **1967**, 3303.
[16a] H. Takeuchi, T. Nagai, and N. Tokura, *Bull. Chem. Soc. Japan*, **40**, 2375 (1967).
[16b] N. A. Maier and Y. A. Ol'dekop, *Dokl. Akad. Nauk SSSR*, **172**, 349 (1967); *Chem. Abs.*, **66**, 94513g (1967).
[17] D. Y. Curtin and J. D. Druliner, *J. Org. Chem.*, **32**, 1552 (1967).
[18] M. Sukigara and S. Kikuchi, *Bull. Chem. Soc. Japan*, **40**, 461 (1967).

High-temperature (400—500°) arylation of benzene and chlorobenzene has been studied in the gas-phase, with azobenzene as a radical source.[19] The isomer distribution of chlorobiphenyls was *ortho* 22%, *meta* 52%, and *para* 26%, which is substantially different from the values from solution chemistry at lower temperatures. The figures were considered to reflect the relative thermodynamic stabilities of the isomeric cyclohexadienyl radicals under the reaction conditions. An alternative source of phenyl radicals was also found in the reaction between carbon tetrachloride and benzene. At the high temperatures employed, it was considered that a chlorine atom abstracted hydrogen from benzene.

Also in the gas-phase at high temperature, it has been noted that nitrobenzene constitutes a source of phenyl radicals, though in these experiments the surprising result was that substitution patterns closely paralleled those obtained in solution chemistry.[20] The formation of substantial amounts of phenol in these experiments was interpreted in terms of a nitro-inversion process:

$$PhNO_2 \rightarrow PhONO \xrightarrow{-NO} PhO\cdot \xrightarrow{RH} PhOH$$

Nitromethane did not effect methylation, but phenol was again produced.[21]

A rather different type of pyrolytic reaction of a nitro-compound is typified by the sequence shown below, which is believed to account for the formation of phenazine when 2-nitrodiphenylamine is heated in sand.[22]

Numerous other arylation studies provide information on the phenylation of protonated heterocycles,[23] substituted thiazoles[24] and pyridines,[25] and

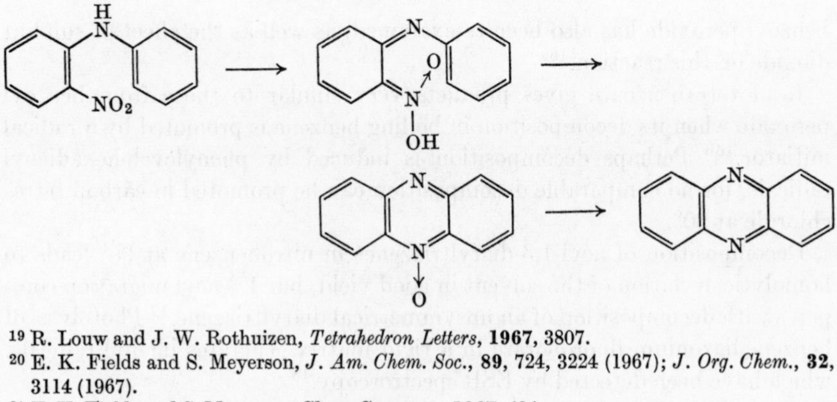

19 R. Louw and J. W. Rothuizen, *Tetrahedron Letters*, **1967**, 3807.

20 E. K. Fields and S. Meyerson, *J. Am. Chem. Soc.*, **89**, 724, 3224 (1967); *J. Org. Chem.*, **32**, 3114 (1967).

21 E. K. Fields and S. Meyerson, *Chem. Commun.*, **1967**, 494.

22 H. Suschitzky and M. E. Sutton, *Tetrahedron Letters*, **1967**, 3933.

23 H. J. M. Dou and B. M. Lynch, *Bull. Soc. Chim. France*, **1966**, 3815, 3820.

24 H. J. M. Dou, G. Vernin, and J. Metzger, *Compt. Rend., C*, **263**, 1243, 1310 (1966); **264**, 336 (1967); *Tetrahedron Letters*, **1967**, 2223.

25 J.-M. Bonnier, J. Court, and M. Gelus, *Compt. Rend., Ser. C*, **263**, 262 (1966); **264**, 1023 (1967); J.-M. Bonnier, J. Court, and T. Fay, *Bull. Soc. Chim. France*, **1967**, 1204.

methylnaphthalenes.[26] The high reactivity of [2.2]paracyclophane towards phenylation (33 relative to benzene) reflects the strain in that molecule.[27] Partial rate factors have also been obtained for substitution of toluene by perfluorophenyl radicals obtained by photolysis of pentafluoroiodobenzene,[28] and new data have also been provided for thienylation by 2- and 3-thienyl radicals obtained by photolysis of 2-iodothiophen[29] and thermolysis of 3-thenoyl peroxide,[30] respectively. Decomposition of the thenoyl peroxide in benzene gave a small quantity of biphenyl which, it was suggested, may have arisen as shown (where Ar = C_4H_3S—).

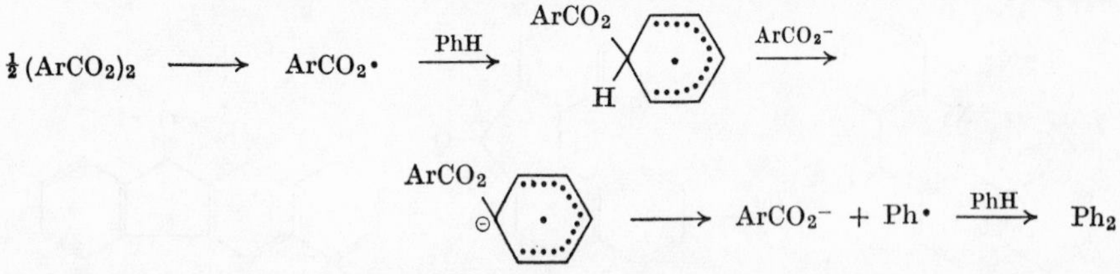

Two new reports of production of aryl radicals by photolysis of aryl chlorides or bromides have appeared,[31] and photolysis of the bromopyridinium salt (**4**) results in internuclear cyclization.[32] Internuclear cyclization by the Grignard method[33] has been employed to synthesize phenanthrenes[34] and

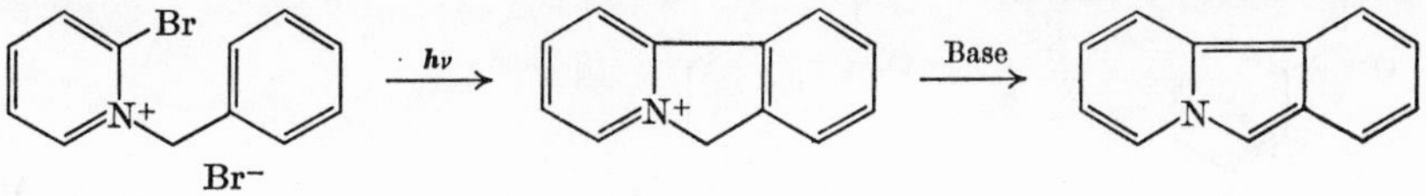

(**4**)

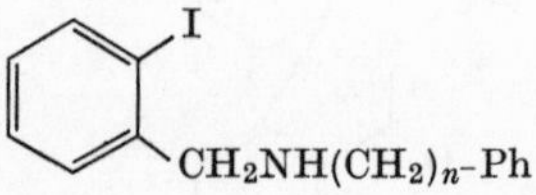

$n = 1$ or 2

(**5**)

[26] J.-M. Bonnier, M. Gelus, and J. Rinaudo, *Compt. Rend., Ser. C*, **264**, 541 (1967); *Tetrahedron Letters*, **1967**, 627; J.-M. Bonnier, J. Rinaudo, and C. Bouvier, *Bull. Soc. Chim. France*, **1967**, 4067.

[27] S. C. Dickerman and N. Milstein, *J. Org. Chem.*, **32**, 852 (1967).

[28] J. M. Birchall, R. Hazard, R. N. Haszeldine, and W. W. Wakalski, *J. Chem. Soc., C*, **1967**, 47.

[29] L. Benati and M. Tiecco, *Boll. Sci. Fac. Chim. Ind. Bologna, Suppl.*, **24**, 225 (1966); *Chem. Abs.*, **67**, 90193a (1967).

[30] D. Mackay, *Can. J. Chem.*, **44**, 2881 (1966).

[31] E. Latouska and T. Latowski, *Roczniki Chem.*, **40**, 1977 (1966); M. A. Chel'tsova and G. I. Nikishin, *Izv. Akad. Nauk SSSR, Ser. Khim.*, **1967**, 456; *Chem. Abs.*, **67**, 21296z (1967).

[32] A. Fozard and C. K. Bradsher, *J. Org. Chem.*, **32**, 2966 (1967).

[33] See *Organic Reaction Mechanisms*, **1966**, 190.

[34] B. F. Bonini and M. Tiecco, *Gazz. Chim. Ital.*, **96**, 1792 (1966).

dibenzothiophen,[35] and in aqueous acid solvents seven- and eight-membered ring heterocycles have been obtained in quite good yield by photolysis of iodoamines (**5**).[36]

New data on intramolecular arylation by copper-catalysed decomposition of appropriately substituted diazonium fluoroborates in acetone point unambiguously to a radical rather than a cationic mechanism. Thus cyclization of (**6**) gives (**7**) and (**8**) in proportions essentially identical with those obtained in other radical cyclizations,[37a] and (**9**) gives (**10**) as a major by-product which it is difficult to envisage as anything other than a radical coupling product.[37b]

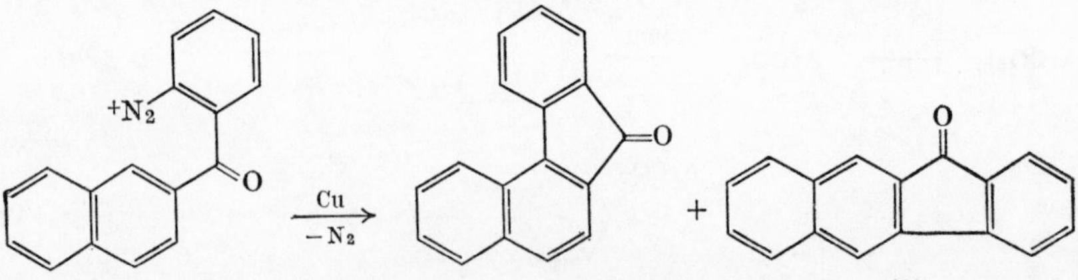

(**6**) (**7**) (**8**)

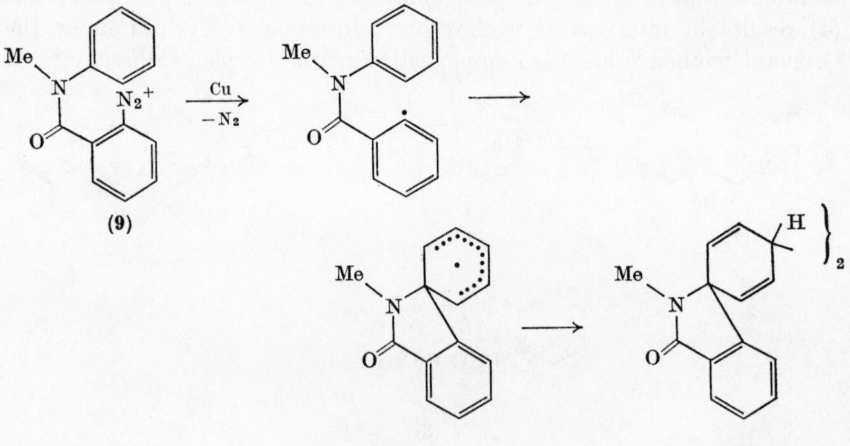

(**9**)

(**10**)

The oxidation of 2′-substituted biphenyl-2-carboxylic acids by $Pb(OAC)_4$ to yield benzocoumarins appears to proceed by a homolytic mechanism.[37c]

There have also been reports of radical alkylation of 4-cyanopyridine,[38]

[35] G. Nespoli and M. Tiecco, *Bol. Sci. Fac. Chim. Ind. Bologna, Suppl.*, **24**, 239 (1967); *Chem. Abs.*, **67**, 73467a (1967).

[36] P. W. Jeffs and J. F. Hansen, *J. Am. Chem. Soc.*, **89**, 2798 (1967).

[37a] R. A. Abramovitch and A. Robson, *J. Chem. Soc., C*, **1967**, 1101.

[37b] D. H. Hey, C. W. Rees, and A. R. Todd, *J. Chem. Soc., C*, **1967**, 1518.

[37c] D. I. Davies and C. Waring, *J. Chem. Soc., C*, **1967**, 1639.

[38] H. D. Eilhauer and G. Reckling, *Arch. Pharm.*, **299**, 891 (1966).

methylation of pyridine and picolines,[39] and gas-phase methylation of benzene (150°).[40] An anodic cyanation of naphthalene has been noted, though the mechanism could not be defined,[41] and radical reactions of the triazine (**11**) have been examined.[42] Benzoyloxylation of (**11**) occurs at the positions marked by arrows, but oxidation by diphenylpicrylhydrazyl gives a dimer, apparently by abstraction of aromatic hydrogen.

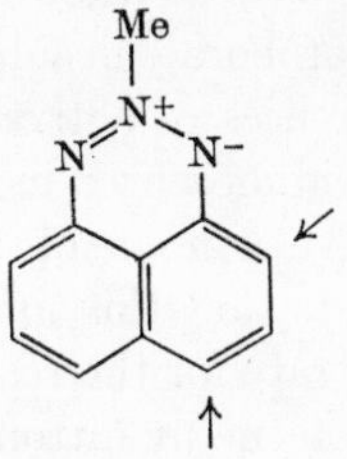

(**11**)

Hydroxylation of anisole employing ^{18}O-labelled hydrogen peroxide as source of HO• provided no evidence for displacement of the methoxyl group.[43] The hydroxylation which occurs when an oxygen-free aqueous solution of sodium anthraquinone-2-sulphonate is irradiated has been attributed to electron-transfer from photoexcited anthraquinone to a ground-state molecule, and oxidation of OH^- to HO• by the resulting radical cation.[44] Trimethylsilylation of anisole is brought about by photolysis of bis(trimethylsilyl)-mercury,[45] and dimethylamination of anisole and other reactive aromatic compounds by *N*-chlorodimethylamine in acid has been elaborated.[46] These reactions involve the radical cation $Me_2HN\cdot^+$; it has been confirmed that the neutral radical, $Me_2N\cdot$, does not participate in aromatic substitution.[47]

Substitution into thiophen by thiyl radicals occurs at the 2-position.[48]

New molecular-orbital calculations have been carried out on the relative ease of attack of hydrogen atoms on substituted benzenes.[49]

39 R. A. Abramovitch and K. Kenaschuk, *Can. J. Chem.*, **45**, 509 (1967).
40 A. D. Malievskii and N. M. Emanuel, *Dokl. Akad. Nauk SSSR*, **169**, 1342 (1966); *Chem. Abs.*, **66**, 65163n (1967).
41 K. Koyama, T. Susuki, and S. Tsutsumi, *Tetrahedron*, **20**, 2665, 2675 (1967).
42 H. Beecken and P. Tavs, *Ann. Chem.*, **704**, 172 (1967).
43 L. G. Shevchuk and N. A. Vysotskaya, *Zh. Organ. Khim.*, **2**, 1229 (1966); *Chem. Abs.*, **66**. 64741n (1967).
44 A. D. Broadbent, *Chem. Commun.*, **1967**, 382; G. O. Phillips, N. W. Worthington, J. F. McKellar, and R. R. Sharpe, *ibid.*, p. 835.
45 C. Eaborn, R. A. Jackson, and R. Pearce, *Chem. Commun.*, **1967**, 920.
46 F. Minisci, R. Galli, M. Cecene, and V. Trabucchi, *Chim. Ind.* (*Milan*), **48**, 1147 (1966); F. Minisci, R. Galli, and M. Cecere, *ibid.*, p. 1324.
47 R. E. Jacobson, K. M. Johnston, and G. H. Williams, *Chem. Ind.* (*London*), **1967**, 157.
48 Y. A. Gol'dfarb, G. P. Pokhil, and L. I. Belen'kii, *Dokl. Akad. Nauk SSSR*, **167**, 823 (1966).
49 P. V. Schastnev and G. M. Zhidomirov, *Kinetika y Katalyz*, **8**, 203 (1967); *Chem. Abs.*, **67**, 43108s (1967).

Electrophilic Substitution

Reviews discuss substitution in purines,[50a] and in heterocyclic systems generally,[1, 50b] also in polyalkylbenzenes,[51] and in strongly deactivated systems.[52] The directing effects of aryloxy- and alkoxy-substituents have been surveyed,[53] as have general substituent effects in electrophilic attack on substituted pyridines.[54]

There do not appear to have been any new contributions to the mechanism of nitration with nitronium fluoroborate in sulpholane.[55] However, it has been noted that the relative reactivities of activated substrates to nitration by nitric acid in 68% sulphuric acid measured in competition with benzene have a limiting value of about 40.[56] An estimate of the concentration of NO_2^+ under the reaction conditions suggests that this limiting value corresponds to the diffusion-controlled encounter rate for the reactive substrate and NO_2^+.

Nitration of 9-deuterioanthracene by nitronium fluoroborate in sulpholane shows a pronounced isotope effect ($k_H/k_D = 2.6$), which is consistent with rate-determining destruction of an initially formed adduct (**12**).[57] The adduct was observed spectroscopically and is relatively stable. As the isotope effect

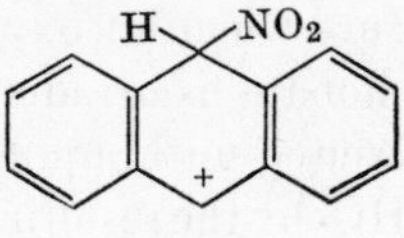

(12)

was measured by competition between attack at C-9 and C-10, rather than by kinetic measurement, it would appear, rather surprisingly, that addition of NO_2^+ to anthracene must be reversible. In naphthalene, however, deprotonation is much faster than addition of NO_2^+, for the α-nitronaphthalene formed from 1,4-dideuterionaphthalene under similar reaction conditions was substituted equally in the deuterated and the non-deuterated ring.

Bonner and Hancock[58] have continued their work on nitrations in carbon tetrachloride. They find that nitration of activated species by N_2O_4 involves oxidation of the initial nitroso-derivative, not by nitric acid formed during the reaction as older studies would imply, but by unreacted N_2O_4. Indeed

50a J. H. Lister, *Adv. Heterocyclic Chem.*, **6**, 1 (1966).
50b A. R. Katritzky and C. D. Johnson, *Angew. Chem. Intern. Ed. Engl.*, **6**, 608 (1967).
51 E. Baciocchi and G. Illuminati, *Progr. Phys. Org. Chem.*, **5**, 1 (1967).
52 J. H. Ridd, "Aromaticity", The Chemical Society, London, 1967.
53 G. Kohnstam and D. L. H. Williams in "The Chemistry of the Ether Linkage", S. Patai, ed., Interscience, London, 1967, p. 132.
54 R. A. Abramovitch and J. G. Saha, *Adv. Heterocyclic Chem.*, **6**, 229 (1966).
55 See *Organic Reaction Mechanisms*, **1965**, 160; **1966**, 193.
56 R. G. Coombes, R. B. Moodie, and K. Schofield, *Chem. Commun.*, **1967**, 352.
57 H. Cerfontain and A. Telder, *Rec. Trav. Chim.*, **86**, 371 (1967).
58 T. G. Bonner and R. A. Hancock, *Chem. Commun.*, **1967**, 780.

nitric acid was an inefficient oxidizing agent for the nitroso-compounds in question.

Spectra of intermediates observed during the reaction of nitrosylsulphuric acid with aromatic compounds have been attributed to π-complexes of the form $ArH \cdot NO^+$ and $(ArH)_2NO^+$.[59]

The effect of increased pressure on the nitration of *tert*-butylbenzene is to reduce the partial rate factor for *para*-substitution.[60] The additivity principle does not adequately predict the isomer distributions in nitrations of polyalkylated benzenes[61] and, alone amongst transition metals investigated, mercury ions were found to affect the isomer distribution in nitrations of toluene.[62] Other workers have examined (or re-examined) orientational effects in the nitrations of 2.3-dimethoxynaphthalene,[63] β-trifluoromethylstyrene,[64] and 1,4-dialkyl-2-nitrobenzenes.[65] In the last of these, substitution is predominantly at the 3- and the 6-position when alkyl is methyl or ethyl, but when alkyl is *tert*-butyl steric effects, reinforced by buttressing, direct attack principally to the 5-position.

Recent results on the nitration of the dimethylphenylsulphonium ion have been elaborated, together with those for $[PhSeMe_2]^+$.[66] Hartshorn and Ridd have determined the secondary kinetic isotope effects for *meta*- and *para*-nitration of $PhND_3^+$.[67] The values of k_H/k_D were 1.08 and 1.18 per hydrogen atom, respectively. These figures were apparently greater than observed for $PhNMe_2D^+$, and it was concluded that there is no conformational preference for optimum isotope effect. In our opinion the accuracy of the determinations did not fully justify this conclusion; it seems quite possible that the preferred conformation of the dimethylanilinium ion is such as to minimize the secondary isotope effect if this is of a hyperconjugative nature, and experiments with molecules incorporating the positive nitrogen in a rigid ring system might be particularly instructive. The isotope effects were distinguished from solvent isotope effects by the discovery that hydrogen exchange at nitrogen in the concentrated acid is slower than nitration. Thus it was possible to nitrate $PhNH_3^+$ in D_2SO_4. This result clearly establishes that it is not an equilibrium proportion of free base which is nitrated.

59 Z. J. Allan, J. Podstata, D. Šnobl, and J. Jarkovský, *Coll. Czech. Chem. Commun.*, **32**, 1449 (1967).

60 T. Asano, R. Goto, and A. Sera, *Bull. Chem. Soc. Japan*, **40**, 2208 (1967).

61 A. Fischer, J. Vaughan, and G. J. Wright, *J. Chem. Soc., B*, **1967**, 368.

62 T. Osawa, T. Yoshida, and K. Namba, *Kogyo Kayaku Kyokaishi*, **27**, 162 (1966); *Chem. Abs.* **67**, 21326j (1967).

63 C. W. J. Chang, R. E. Moore, and P. J. Scheuer, *J. Chem. Soc., C*, **1967**, 840.

64 L. M. Yagupol'skii, Y. A. Fialkov, and A. G. Panteleimonov, *Zh. Obshch. Khim.*, **36**, 2127 (1966); *Chem. Abs.*, **66**, 75773a (1967).

65 C. D. Johnson and M. J. Northcott, *J. Org. Chem.*, **32**, 2029 (1967).

66 H. M. Gilow and G. L. Walker, *J. Org. Chem.*, **32**, 2580 (1967); see *Organic Reaction Mechanisms*, **1965**, 165.

67 S. R. Hartshorn and J. H. Ridd, *Chem. Commun.*, **1967**, 133.

Nitration and acetylation of [18]annulene and a tridehydro[18]annulene have been noted,[68] as has nitration of *trans*-15,16-dihydro-15,16-dimethylpyrene (**13**).[69] Electrophilic substitution of the *meta*-cyclophane (**14**) gives

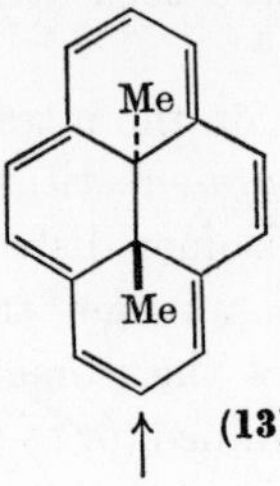

(**13**)

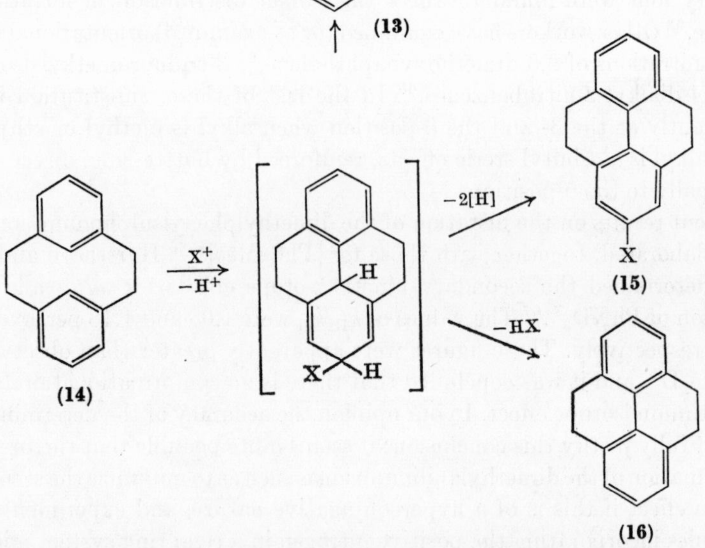

(**14**) (**15**) (**16**)

(**15**) or (**16**) depending on the oxidizing properties of the reaction medium.[70]

A melt of mixed inorganic nitrates to which a persulphate has been added acts as an effective nitrating medium for benzene or other volatile aromatic compounds which may be introduced in the vapour phase:[71]

$$NO_3^- + S_2O_7^{2-} \rightleftarrows 2SO_4^{2-} + NO_2^+$$

A report is available of ^{18}O tracer studies of simultaneous hydroxylation and nitration of aromatic compounds with pernitrous acid ("HNO_4").[72]

68 I. C. Calder, P. J. Garratt, H. C. Longuet-Higgins, F. Sondheimer, and R. Wolovsky, *J. Chem. Soc., C*, **1967**, 1041.

69 J. B. Phillips, R. J. Molyneux, E. Sturm, and V. Boekelheide, *J. Am. Chem. Soc.*, **89**, 1704 (1967); see also V. Boekelheide and T. Miyasaka, *ibid.*, p. 1709.

70 M. Fujimoto, T. Sato, and K. Hata, *Bull. Chem. Soc. Japan*, **40**, 600 (1967); T. Sato, M Wakabayashi, Y. Okamura, T. Amada, and K. Hata, *ibid.*, p. 2363.

71 R. B. Temple, C. Fay, and J. Williamson, *Chem. Commun.*, **1967**, 966.

72 N. A. Vysotskaya and A. E. Brodsky, *Abh. Deut. Akad. Wiss. Berlin, Kl. Chem., Geol., Biol.*, **1964**, 653; *Chem. Abs.*, **67**, 21232a (1967).

A new example of a substantial primary isotope effect ($k_H/k_D = 3.6$) has been observed in bromination of 1,3,5-tri-*tert*-butylbenzene.[73] The hydrogen abstraction step is presumably largely rate-determining. The same research group has examined the effect of varying substituents at the point of attack.[74] Thus very small steric effects, but substantial polar ones, are evident in the rates shown for dienone formation from (**17**).

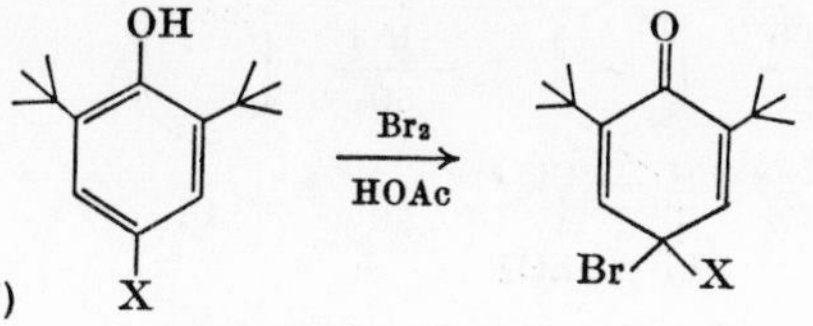

(**17**)

X	H	Me	Bu^t	Br
k (l. mole^{-1} sec^{-1}, 25°)	4.79	1.28	1.30	$< 10^{-4}$

de la Mare and El Dousouqui have confirmed the existence of an isotope effect in the bromination of PhOD in acetic [^{2}H]acid,[75] and have reaffirmed their contention that the most plausible explanation for this ($k_H/k_D = ca.\ 1.8$) is the weakening of the O–D bond in the transition state.

Conditions for selective *ortho*-bromination of phenols at low temperatures have been found, and it was tentatively suggested that aryl hypobromites may be key intermediates.[76]

The equilibrium

$$Hg(OAc)_2 + I_2 \rightleftarrows AcOI + IHgOAc$$

provides a source of the electrophilic iodinating agent acetyl hypoiodite.[77] The kinetics of iodination of pentamethylbenzene with this system indicate that the hypoiodite is much less electrophilic than acetyl hypobromite prepared analogously.[78]

New results on chlorination by chloramine-T lead to the conclusion that the reactive chlorinating agent here is not hypochlorous acid but dichloramine-T.[79]

de la Mare's group have obtained new information on competing chlorine substitution in, and addition to, the naphthalene ring.[80] The disappearance of naphthalene shows an inverse isotope effect ($k_H/k_D = 0.85$), possibly associated with rehybridization in the initial addition to form the σ-complex.[81] Chlorine

73 E. Baciocchi, G. Illuminati, G. Sleiter, and F. Stegel, *J. Am. Chem. Soc.*, **89**, 125 (1967).
74 E. Baciocchi and G. Illuminati, *J. Am. Chem. Soc.*, **89**, 4017 (1967).
75 P. B. D. de la Mare and O. M. H. El Dusouqui, *J. Chem. Soc., B*, **1967**, 251.
76 D. E. Pearson, R. D. Wysong, and C. V. Breder, *J. Org. Chem.*, **32**, 2358 (1967).
77 E. M. Chen, R. M. Keefer, and L. J. Andrews, *J. Am. Chem. Soc.*, **89**, 428 (1967).
78 See *Organic Reaction Mechanisms*, **1965**, 167.
79 T. Higuchi, K. Ikeda, and A. Hussain, *J. Chem. Soc., B*, **1967**, 546; T. Higuchi and A. Hussain, *ibid.*, p. 549.
80 G. Cum, P. B. D. de la Mare, J. S. Lomas, and M. D. Johnson, *J. Chem. Soc., B*, **1967**, 244.
81 P. B. D. de la Mare and J. S. Lomas, *Rec. Trav. Chim.*, **86**, 1082 (1967).

substitution and addition were again observed when employing sulphuryl chloride as chlorinating agent,[82] and chlorination of anisole with this reagent was also studied.[83] The reaction is of the first order in both species, and the possibility of attack by molecular sulphuryl chloride was discussed.

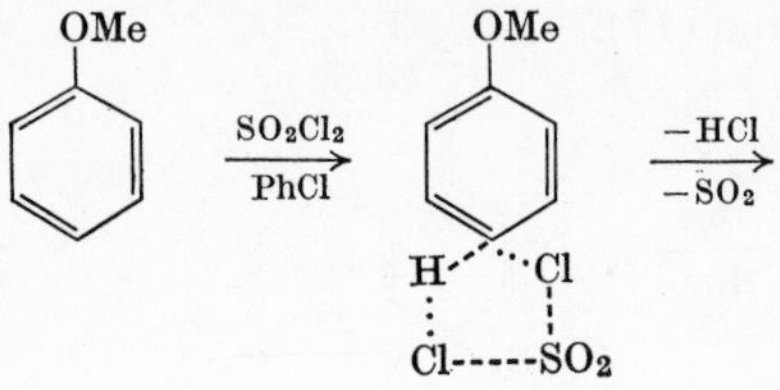

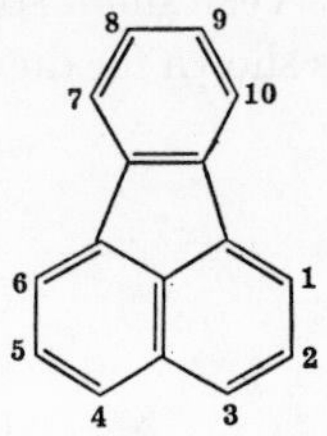

(18)

The rate of bromination of aniline co-ordinated to cobalt is roughly midway between those of the free base and its conjugate acid.[84] The rates of iodination of metal complexes of 8-hydroxyquinoline-5-sulphonic acid show a very large spread (*ca.* 10^5) which can almost entirely be accommodated in variations in the pre-exponential factor.[85]

A nitro-group in position 2 of fluoranthene (**18**) directs bromine substitution to position 9.[86] The rates of bromination of polymethylbenzenes in trifluoroacetic acid show a reaction constant ρ of −16.7, indicating an extreme sensitivity to charge distribution in the transition state.[87] The unique properties of this solvent were further reflected by comparison of the rate of bromination of mesitylene with that in acetic acid. In the former solvent the rate is lower by a factor of 10^6, yet the dielectric constants of the two acids are very similar.

Iodination of a number of methylphenols has been studied as models for the reaction of tyrosine,[88] and the kinetics of bromination of *p*-dimethoxybenzene in nitrobenzene show the reaction to be of the third order.[89]

New molecular-orbital calculations pertain to the bromination of polynuclear aromatic compounds,[90] and alkyl-, alkoxy-, and amino-benzenes.[91]

New experimental results are available on sulphonation of naphthalene,[92]

[82] R. Bolton, P. B. D. de la Mare, and H. Suzuki, *Rec. Trav. Chim.*, **85**, 1206 (1966); P. B. D. de la Mare and H. Suzuki, *J. Chem. Soc., C*, **1967**, 1586; G. Cum, P. B. D. de la Mare, and M. D. Johnson, *ibid.*, p. 1590.

[83] R. Bolton and P. B. D. de la Mare, *J. Chem. Soc., B*, **1967**, 1044.

[84] N. K. Chawla, D. G. Lambert, and M. M. Jones, *J. Am. Chem. Soc.*, **89**, 557 (1967).

[85] R. C. McNutt and M. M. Jones, *J. Inorg. Nucl. Chem.*, **29**, 1415 (1967).

[86] E. H. Charlesworth and A. J. Dolenko, *Can. J. Chem.*, **45**, 96 (1967).

[87] P. Alcais, F. Rothenberg, and J.-É. Dubois, *J. Chim. Phys.*, **63**, 1443 (1966).

[88] W. E. Mayberry, *Biochemistry*, **6**, 1320 (1967).

[89] K. V. Seshadri and R. Ganesan, *Current Sci.* (*India*), **34**, 429 (1965); *Chem. Abs.*, **66**, 10312f (1967).

[90] L. Altschuler and E. Berliner, *J. Am. Chem. Soc.*, **88**, 5837 (1966).

[91] J.-E. Dubois and J.-P. Doucet, *Tetrahedron Letters*, **1967**, 3413.

[92] H. Cerfontain and A. Telder, *Rec. Trav. Chim.*, **86**, 527 (1967).

alkylbenzenes,[93] and chlorobenzene[94] and on the exchange of sulphonate groups between aromatic sulphonic acids and sulphuric acid.[95]

An interesting example of stereoselectivity was observed in the Friedel–Crafts alkylation of benzene by optically active 4-valerolactone, catalysed by aluminium chloride.[96] After allowance for the slow racemization of the valerolactone under the reaction conditions, the actual substitution step was found to proceed with nearly 50% net inversion of configuration. The mechanism of this process was envisaged as shown. The internal ion-pair complex of lactone with aluminium chloride can racemize by bond rotation and subsequently revert to starting material (to give racemized lactone), or it can attack a benzene molecule. Consistent with this type of mechanism was the observation that dilution with carbon disulphide led to increased racemization.

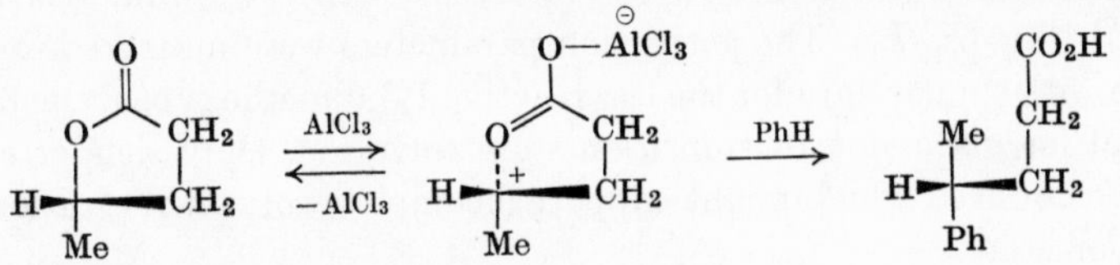

An *X*-ray study of the benzoyl chloride–antimony pentachloride complex shows that co-ordination is through oxygen.[97]

Other reports of alkylation or acylation describe alkylation of fluorene,[98] chloromethylation of methylbenzenes,[99] and stannic chloride-catalysed polycondensation of benzyl chloride.[100] There is no acetyl exchange between acetophenone and acetyl chloride under Friedel–Crafts conditions.[101] Acid-catalysed alkylation of phenols by olefins gives a high *ortho*:*para* ratio;[102] heterogeneous catalysis of alkylation of benzene–toluene mixtures by alkyl bromides over zeolites gives relative reactivities consistent with Brown's selectivity relationship, which is normally applied to homogeneous reactions.[103]

93 H. de Vries and H. Cerfontain, *Rec. Trav. Chim.*, **86**, 873 (1967).

94 C. W. F. Kort and H. Cerfontain, *Rec. Trav. Chim.*, **86**, 865 (1967).

95 V. F. Grechanovskii and N. T. Maleeva, *Zh. Obshch. Khim.*, **36**, 1189 (1966); *Chem. Abs.*, **65**, 16805 (1966).

96 J. I. Brauman and A. J. Pandell, *J. Am. Chem. Soc.*, **89**, 5421 (1967).

97 R. Weiss and B. Chevrier, *Chem. Commun.*, **1967**, 145.

98 O. G. Akperov, S. T. Akhmedov, and M. A. Salimov, *Uch. Zap. Azerb. Gos. Univ., Ser. Khim. Nauk*, **1965**, 43.

99 G. S. Mironov, M. I. Farberov, V. D. Shein, and I. I. Bespalova, *Zh. Organ. Khim.*, **2**, 1639 (1966); *Chem. Abs.*, **66**, 115198a (1967).

100 D. B. V. Parker, W. G. Davies, and K. D. South, *J. Chem. Soc., B*, **1967**, 471.

101 M. Frangopol, A. Genunche, N. Negoiță, P. T. Frangopol, and A. T. Balaban, *Tetrahedron*, **23**, 841 (1967).

102 N. M. Karavaev, S. A. Dmitmev, K. I. Zimina, E. I. Kazakov, K. D. Korenev, G. G. Kotova, and O. N. Tsvetkov, *Dokl. Akad. Nauk SSSR*, **173**, 832.

103 P. B. Venuto, *J. Org. Chem.*, **32**, 1272 (1967).

Intermolecular transfer of isopropyl groups between phenols, catalysed by aluminium chloride, has been noted.[104]

Several interesting papers deal with hydrogen exchange, and direct physical measurements have been made on the proton adducts of fluorobenzene[105] and anthracene.[106] In the latter case, a ^{13}C–H spin-coupling constant of 127.5 cps confirms the presence of tetrahedral hybridization at C-9.

Hydrogen exchange in polymethylbenzenes has been used to study the acidity behaviour of Lewis acid/acetic acid mixtures.[107]

New work on the exchange reactions of 1,3,5-trimethoxybenzene employs the three hydrogen isotopes.[108] Primary and secondary isotope effects were measured for hydrogen at the nuclear site undergoing exchange. There was no measurable effect resulting from isotopic substitution at positions not undergoing exchange. The results were consistent with the Swain relationship: $(k_H/k_D)^{1.442} = (k_H/k_T)$. The activation parameters were measured, and comparison with similar data for the less reactive 1,3-dimethoxybenzene provided material for discussion of transition-state solvation. Only slender evidence could be adduced which might support incomplete solvent relaxation at the transition state.

The reactions of azulene constitute a fruitful area for investigation. Isotope effects for the individual steps (protonation and deprotonation) at position 1 were measured and found to reach a maximum ($k_H/k_D = 9.5$) with H_3O^+ as the proton source.[109] The azulenium ion and H_3O^+ have similar pK_a's and this result agrees, then, with the prediction of a maximum isotope effect for proton transfer between bases of the same strength. Rather different behaviour has, however, been found for the iodination of azulene.[110] Removal of the proton is rate-determining, but the magnitude of the isotope effect depends on steric effects in the base, as well as its strength. Thus from relative base strengths, deprotonation of the iodine–azulene σ-complex should show a smaller isotope effect with 2,4,6-trimethylpyridine than with pyridine itself: in practice, the values for k_H/k_D are 6.5 and 2.0, respectively. The large value obtained with the hindered base was associated with proton tunnelling.

The first excited singlet state of azulene has been shown to be a weaker base than is azulene itself,[111] and it has been pointed out that azulenes constitute

104 R. Lamartine and R. Perrin, *Compt. Rend.*, **264**, 1337 (1967).
105 G. A. Olah and T. E. Kiovsky, *J. Am. Chem. Soc.*, **89**, 5692 (1967).
106 V. A. Koptyug, I. S. Isaev, and A. I. Rezvukhin, *Tetrahedron Letters*, **1967**, 823; see also V. A. Koptyug, V. A. Bushmelev, and T. N. Gerasimova, *Zh. Organ. Khim.*, **3**, 140 (1967).
107 A. P. Sannikov, E. Z. Vtyanskaya, P. P. Alikhanov, and A. P. Shatenshtein, *Zh. Obshch. Khim.*, **36**, 2036 (1966).
108 A. J. Kresge and Y. Chiang, *J. Am. Chem. Soc.*, **89**, 4411 (1967); A. J. Kresge, Y. Chiang, and Y. Sato, *ibid.*, p. 4418.
109 L. C. Gruen and F. A. Long, *J. Am. Chem. Soc.*, **89**, 1287 (1967); J. L. Longridge and F. A. Long, *ibid.*, p. 1292.
110 E. Grovenstein and F. C. Schmalstieg, *J. Am. Chem. Soc.*, **89**, 5084 (1967).
111 R. Hagen, E. Heilbronner, W. Meier, and P. Seiler, *Helv. Chim. Acta*, **50**, 1523 (1967).

convenient models with which to examine new electrophilic reactions, because of the ease with which the transformation may be followed spectroscopically.[112]

Kendall and his colleagues have extended their examination of hydrogen-transfer in substituted dimethylanilines.[113] In the naphthalene series a revised treatment[114] of transmission of substituent effects gives an excellent correlation of the results previously obtained[115] for detritiation of substituted 1-tritionaphthalenes. Other publications deal with hydrogen exchange in aryl derivatives of trivalent phosphorus,[116] *meso*-exchange in porphyrins, [117] and acid-catalysed exchange at boron in carboranes (with base, exchange is at carbon).[118] A new correlation of exchange in aromatic hydrocarbons with calculated atom localization energies has also appeared.[119]

A very rapid hydrogen exchange in the *ortho*- and *para*-positions of dibenzylmercury (which gives σ^+ for p-$CH_2HgCH_2Ph = -1.14$) has been attributed to pronounced carbon–metal hyperconjugation.[120]

The rate of detritiation of 1-tritiotriptycene (see formula for numbering) has been compared with those of 9,10-dihydro-1-tritioanthracene and 3-tritio-*o*-xylene.[121] Extrapolation from the last two rates to a predicted rate for triptycene gives a value considerably too large. On the other hand, the

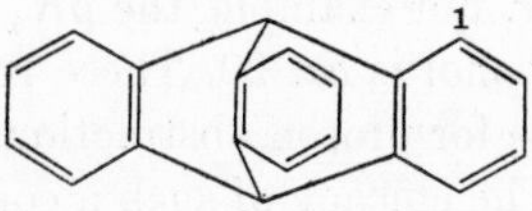

predicted value for 2-tritiotriptycene is lower than the experimental figure. These results were discussed in terms of ring strain in triptycene and of the Mills–Nixon effect.[122]

It has been mentioned above that, in concentrated sulphuric acid, nitration of anilinium ions is faster than hydrogen exchange at nitrogen. The ring-hydrogen atoms also exchange more rapidly than NH-hydrogen atoms when an activating group (e.g., methoxyl) is also present.[123] Results are not yet

[112] W. Treibs, *Chem. Ztg. Chem. App.*, **90**, 691 (1966).
[113] A. C. Ling and F. H. Kendall, *J. Chem. Soc., B*, **1967**, 440, 445; see *Organic Reaction Mechanisms*, **1966**, 201.
[114] K. C. C. Bancroft and G. R. Howe, *Tetrahedron Letters*, **1967**, 4207.
[115] See *Organic Reaction Mechanisms*, **1965**, 169.
[116] E. A. Yakovleva, E. N. Tsvetkov, D. I. Lobanov, M. I. Kabachnik, and A. I. Shatenshtein, *Dokl. Akad. Nauk SSSR*, **170**, 1103 (1966); *Chem. Abs.*, **66**, 37038g (1967).
[117] R. Bonnett, I. A. D. Gale, and G. F. Stephenson, *J. Chem. Soc., C*, **1967**, 1168.
[118] V. N. Setkina, I. G. Malakhova, V. I. Stanko, A. I. Klimova, and L. I. Zakharkin, *Izv. Akad. Nauk SSSR, Ser. Khim.*, **1966**, 1678; *Chem. Abs.*, **66**, 64757x (1967).
[119] C. Parkanyi, K. Bocek, Z. Dolejsek, and R. Zahradnik, *Abh. Deut. Akad. Wiss. Berlin, Kl. Chem., Geol., Biol.*, **1964**, 657; *Chem. Abs.*, **66**, 37051f (1967).
[120] W. Hanstein and T. G. Traylor, *Tetrahedron Letters*, **1967**, 4451.
[121] R. Taylor, G. J. Wright, and A. J. Homes, *J. Chem. Soc., B*, **1967**, 780.
[122] See *Organic Reaction Mechanisms*, **1965**, 162.
[123] J. R. Blackborow and J. H. Ridd, *Chem. Commun.*, **1967**, 132.

available for the much slower exchange on the unsubstituted anilinium ions. Acid-catalysed exchange was relatively rapid at nitrogen when the activating group was situated *meta* to the positive nitrogen and this may probably be rationalized by the sequence shown; it was much faster than that for the *N*,*N*-dimethylanilinium ion under otherwise identical conditions.

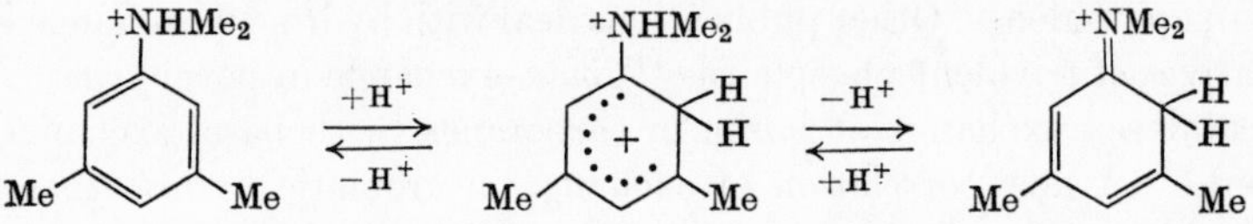

Protonated *p*-methoxybenzophenones appear to undergo ring-hydrogen exchange in strong acid exclusively via the deprotonated species, unlike the diarylethylenes discussed last year.[124]

Aspects of base-catalysed hydrogen exchange in aromatic systems have been discussed;[125, 126] in particular it has been pointed out that low k_H/k_D ratios for exchange of protons in aromatic hydrocarbons, catalysed by *tert*-butoxide in dimethyl sulphoxide, are not necessarily inconsistent with rate-determining proton-transfer. For example, the pK_a of toluene (CH_3) is *ca.* 40 whilst that of *tert*-butyl alcohol is *ca.* 20. These results imply a highly unsymmetrical transition state for proton abstraction, for which a k_H/k_D value of *ca.* 2 can be estimated. The efficacy of such a comparatively weak base in this exchange reaction is attributable to the ease with which butoxide ions are desolvated in the dimethyl sulphoxide medium, giving a high base strength as measured by H_-. Bases that are stronger (as measured by pK_a of the conjugate acid), e.g., lithium cyclohexylamide, give much larger k_H/k_D ratios for the same exchange reaction (see also p. 107).

Exchange of aromatic hydrogen by heterogeneous catalysis is well known.[127] Examples of homogeneous catalysis by noble-metal salts have also been documented recently.[128]

Whilst carbonium ions appear to be generated by electrolysis of acetonitrile solutions containing an alkyl iodide and lithium perchlorate as supporting electrolyte, similar electrolysis of aryl iodides leads to diaryliodonium salts.[129]

[124] G. L. Eian and C. A. Kingsbury, *J. Org. Chem.*, **32**, 1864 (1967); see *Organic Reaction Mechanisms*, **1966**, 200.

[125] J. Massicot and F. Zonszajn, *Bull. Soc. Chim. France*, **1967**, 2206; J. Massicot, *ibid.*, p. 2204; E. Buncel and A. W. Zabel, *J. Am. Chem. Soc.*, **89**, 3082 (1967); E. Buncel and E. A. Symons, *Chem. Commun.*, **1967**, 771.

[126] J. R. Jones, *Chem. Commun.*, **1967**, 710.

[127] For recent examples see: B. D. Fisher and J. L. Garnett, *Australian J. Chem.*, **19**, 2299 (1966); C. G. MacDonald and J. S. Shannon, *ibid.*, **20**, 297 (1967).

[128] J. L. Garnett and R. J. Hodges, *Chem. Commun.*, **1967**, 1001; *J. Am. Chem. Soc.*, **89**, 4546 (1967).

[129] L. L. Miller and A. K. Hoffmann, *J. Am. Chem. Soc.*, **89**, 593 (1967).

$$RI \xrightarrow{-e} RI\cdot^{+} \rightarrow R^{+} + I\cdot$$

$$R^{+} + MeCN \rightarrow RN{=}\overset{+}{C}Me \xrightarrow{H_2O} RNHCOMe$$

The reactive intermediate was considered to be $ArI\cdot^{+}$, and it was argued that this, rather than $Ar\overset{+}{I}OH$, may also be the reactive species when diaryliodonium salts are generated in a more conventional manner from ArI:O, Ar′H, and sulphuric acid. New work on the latter reaction contradicts an earlier report that reaction with Ar′H (= PhMe or PhCl) gives exclusively *para*-substitution.[130]

An extensive study of aromatic mercuration has been reported.[131] Mercuration by mercuric acetate, catalysed by perchloric acid, can involve $Hg(OAc)_2$, $[HgOAc^{+}\ ClO_4^{-}]$ ion pair, and $[Hg^{2+}2ClO_4^{-}]$ as mercurating agents. The reaction of $[HgOAc^{+}\ ClO_4^{-}]$ with ArH in acetic acid leads to $ArHgClO_4$, because there is no autocatalysis by $HClO_4$ which would be liberated if the alternative product ArHgOAc were formed. An isotope effect of $k_H/k_D = 6$ showed that proton loss is rate-determining; possibly the proton is lost to acetate in a concerted elimination of acetic acid as illustrated.

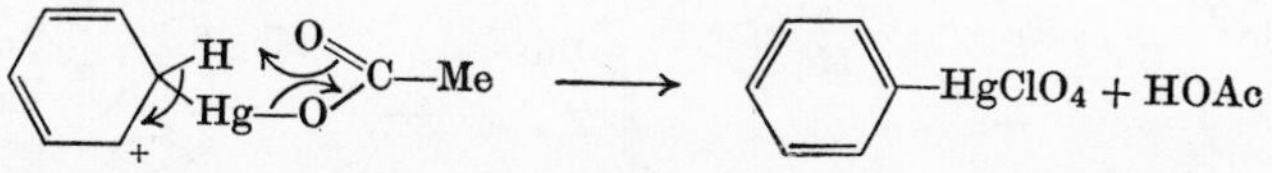

The exchange of mercury between arylmercury compounds in benzene solution and metallic mercury may involve initial electron transfer to the bulk metal, for encounter of the organic species with the metal surface has been shown to be much more effective in the exchange process than is encounter with dissolved mercury atoms.[132]

Isotopic mercury exchange between phenylmercuric chloride and mercuric chloride is of the first order in both reagents;[133] it was discussed in terms of an intermediate:

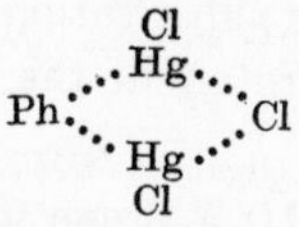

Perhaps some similar representation may be appropriate to mercuridestannylation:

$$XC_6H_4SnEt_3 + Hg(OAc)_2 \rightarrow XC_6H_4HgOAc + AcOSnEt_3$$

130 D. J. Le Count and J. A. W. Reid, *J. Chem. Soc., C*, **1967**, 1298.
131 A. J. Kresge, M. Dubeck, and H. C. Brown, *J. Org. Chem.*, **32**, 745 (1967); A. J. Kresge and J. F. Brennan, *ibid.*, p. 752; A. J. Kresge and H. C. Brown, *ibid.*, p. 756.
132 M. M. Kreevoy and E. A. Walters, *J. Am. Chem. Soc.*, **89**, 2986 (1967).
133 T. A. Smolina, C. Mieh-ch'u, and O. A. Reutov, *Izv. Akad. Nauk SSSR, Ser. Khim.*, **1966**, 413.

in view of the good rate correlation with σ rather than σ^+ ($\rho = -3.5$).[134] Protodemercuration[135] and iododemercuration[136] have also received attention.

Protodestannylation has been examined under conditions of base catalysis. The rate-determining step was considered to be either carbanion formation from a pentaco-ordinate tin complex or direct nucleophilic displacement:[137]

$$XC_6H_4SnMe_3 + OH^- \xrightarrow{\text{Aq.MeOH}} XC_6H_5 + HOSnMe_3$$

Kovacic's group have extended their investigations of amination by "σ-substitution",[138] and of oxygenation by peroxides.[139]

Amination of 2,6-disubstituted phenoxide ions with chloramine leads to azepinones, apparently by bimolecular nucleophilic displacement of chlorine from nitrogen. To probe this type of displacement, substituent effects in the phenoxide ion were examined, and the preferential formation of (**20**) from (**19**) suggests that normal steric factors come into play.[140]

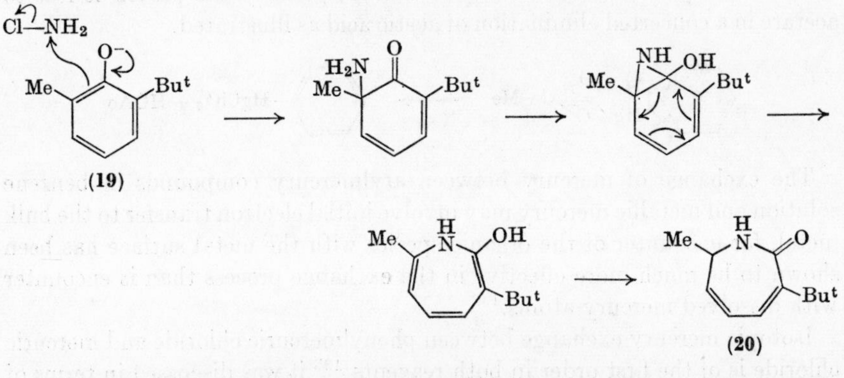

(**19**)

(**20**)

Enzymic hydroxylations are frequently accompanied by a shift of the proton from the site undergoing substitution to the position *ortho* to it, and consequent retention of this proton in the molecule. In a model system,

134 H. Hashimoto and Y. Morimoto, *J. Organometal. Chem.*, **8**, 271 (1967).

135 I. P. Beletskaya, A. E. Myshkin, and O. A. Reutov, *Zh. Organ. Khim.*, **2**, 2086 (1966); *Chem. Abs.*, **66**, 94448q (1967); I. P. Beletskaya, A. E. Myshkin, and O. A. Reutov, *Izv. Akad. Nauk SSSR, Ser. Khim.*, **1967**, 238, 245; *Chem. Abs.*, **67**, 43086h, 43087j (1967).

136 O. Itoh, H. Taniguchi, A. Kawabe, and K. Ichikawa, *Kogyo Kagaku Zasshi*, **69**, 913 (1966); *Chem. Abs.*, **65**, 19951 (1966).

137 C. Eaborn, H. L. Hornfeld, and D. R. M. Walton, *J. Chem. Soc., B*, **1967**, 1036.

138 P. Kovacic and A. K. Harrison, *J. Org. Chem.*, **32**, 207 (1967); P. Kovacic, K. W. Field, P. D. Roskos, and F. V. Scalzi, *ibid.*, p. 585; P. Kovacic and R. J. Hopper, *Tetrahedron*, **23**, 3965, 3977 (1967); see *Organic Reaction Mechanisms*, **1966**, 199.

139 M. E. Kurz and P. Kovacic, *J. Am. Chem. Soc.*, **89**, 4960 (1967); see *Organic Reaction Mechanisms*, **1966**, 199.

140 L. A. Paquette and W. C. Farley, *J. Am. Chem. Soc.*, **89**, 3595 (1967).

p-deuterioacetanilide reacts with trifluoroperacetic acid (a source of "OH^+") to give *p*-hydroxyacetanilide retaining some of the deuterium label.[141] This suggests that cationoid intermediates are involved in both this and the enzymic proton migrations. Such a mechanism finds support in the retention of *ca.* 20% of the deuterium in *p*-chlorophenol obtained by acid-catalysed dehydration of (**21**).[142] The mechanism may involve a pinacolic deuteride

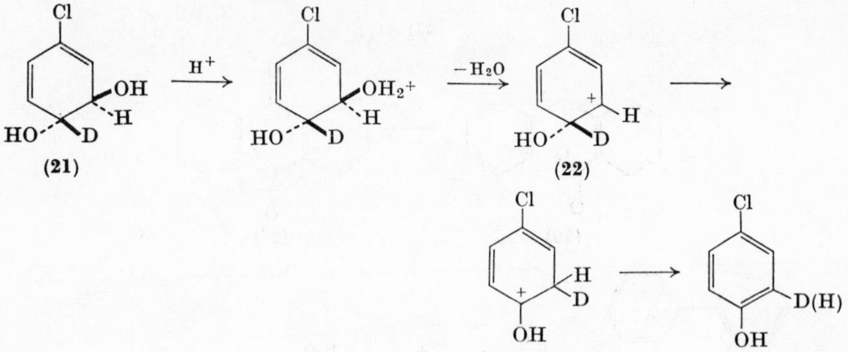

(**21**) (**22**)

shift, though it seems surprising that migration rather than deprotonation should occur if (**22**) is an intermediate, and the stereochemistry is wrong for a deuteride shift concerted with the elimination of water.

An intermediate from the reaction between dimethylaniline and tetracyanoethylene is regarded as the zwitterion (**23**),[143] though Hall and his

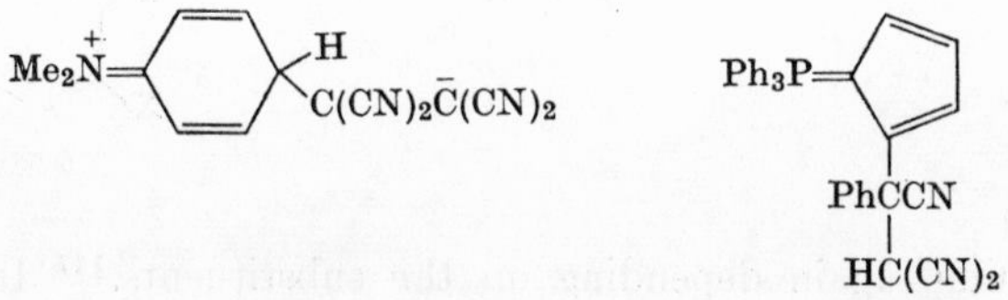

(**23**) (**24**)

colleagues present good evidence that a comparable product from cyclopentadienylidenetriphenylphosphorane and tricyanovinylbenzene has structure (**24**).[144]

Cyclization of anils (**25**) of *β*-keto-acid amides in sulphuric acid proceeds exclusively by attack by C-1 on ring A (the Conrad–Limpach quinoline

[141] D. Jerina, J. Daly, W. Landis, B. Witkop, and S. Udenfriend, *J. Am. Chem. Soc.*, **89**, 3347 (1967).
[142] D. M. Jerina, J. W. Daly, and B. Witkop, *J. Am. Chem. Soc.*, **89**, 5488 (1967).
[143] P. G. Farrell, J. Newton, and R. F. M. White, *J. Chem. Soc., B*, **1967**, 637.
[144] C. W. Rigby, E. Lord, and C. D. Hall, *Chem. Commun.*, **1967**, 714.

synthesis), and not C-3 on ring B (the Knorr reaction).[145] In a study of cyclization of substituted *o*-benzoylbenzoic acids, a dominating feature was found to be the effect of substituents on the position of the equilibrium (**26**) ⇄ (**27**).[146] In cyclizations (**28**) → (**29**) either step (b) or step (c) may be

(**25**)

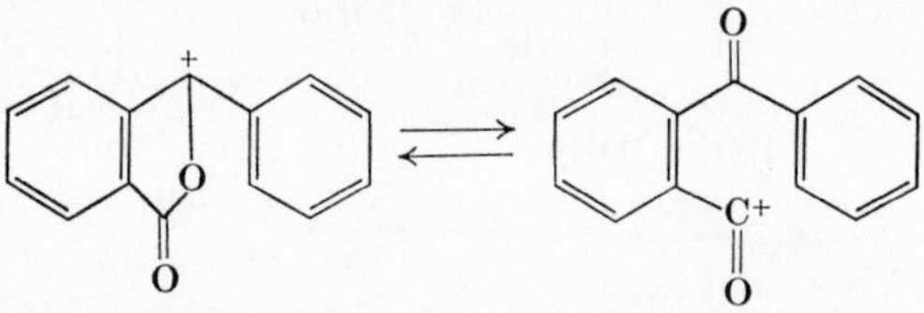

(**26**) (**27**)

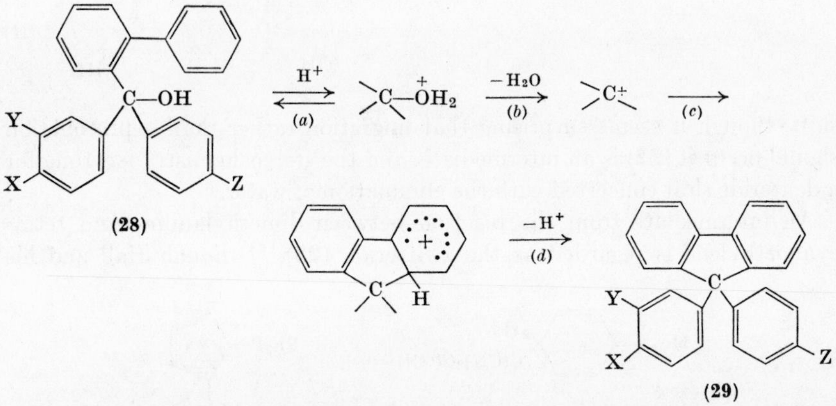

(**28**) (**29**)

rate-determining, again depending on the substituents.[147] In the former circumstance, $\rho = -2.5$ (correlation with σ), in the latter $\rho = +2.8$ (correlation with σ^+).

A new Friedel–Crafts reaction resulting in ring-closure to indanones employs carbon monoxide and an appropriate polyfunctional halide and is exemplified by the formation of (**30**) (p. 205).[148]

Copper-catalysed decomposition of (**9**) was discussed in terms of a radical mechanism (p. 190). Decomposition in dilute sulphuric acid at 70° is believed

145 J. Moszew, *Zesz. Nauk Uniw. Jagiellon., Pr. Chem.*, No. 10, 7 (1965); *Chem. Abs.*, **66**, 28128j (1967).
146 D. S. Noyce and P. A. Kittle, *J. Org. Chem.*, **32**, 2459 (1967).
147 H. Hart and E. A. Sedor, *J. Am. Chem. Soc.*, **89**, 2342 (1967).
148 H. A. Bruson and H. L. Plant, *J. Org. Chem.*, **32**, 3356 (1967).

to be ionic, and new products from alkoxyl derivatives can be rationalized only by a cationic mechanism, e.g., (**31**) → (**32**) + (**33**).[149]

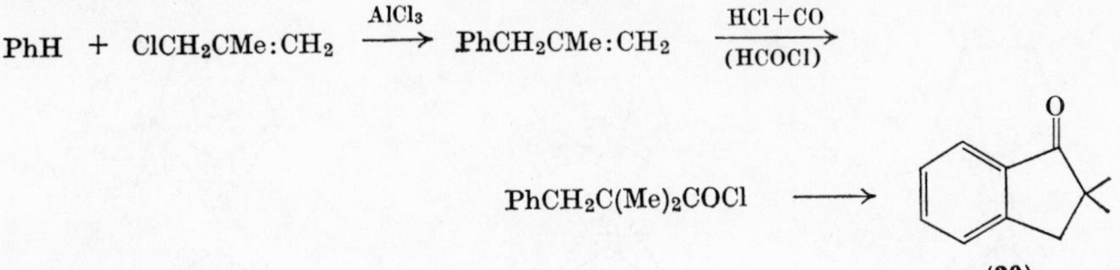

(30)

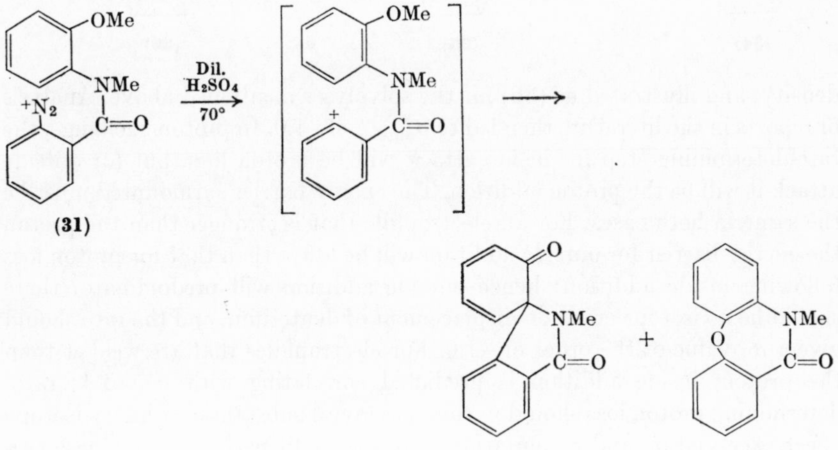

(31) **(32)** **(33)**

Several new reports on electrophilic reactions of metallocenes and related compounds[150] are dominated by Mangravite and Traylor's most interesting dissection[151] of the reactions of ferrocene into those involving "inside" or "outside" attack. In an extension of the results on C–M hyperconjugation, discussed above for the case M = Hg,[120] it was found that σ^+ for a *para*-ferrocenyl substituent = –0.72. Exchange on the ferrocene nucleus was also measured, and gave $\sigma^+_{\alpha\text{-Fer}} = -1.1$. This contrasts with a $\sigma^+_{\alpha\text{-Fer}}$ of –1.4 obtained from solvolysis of ferrocenylmethyl halides. The divergent values led to the

[149] D. H. Hey, J. A. Leonard, C. W. Rees, and A. R. Todd, *J. Chem. Soc., C*, **1967**, 1513.

[150] E. O. Fischer, M. von Foerster, C. G. Kreiter, and K. E. Schwarzhans, *J. Organometal. Chem.*, **7**, 113 (1967); A. N. Nesmeyanov, D. N. Kursanov, V. N. Setkina, N. V. Kislyakov, N. E. Kolobova, and K. N. Anisimov, *Izv. Akad. Nauk SSSR, Ser. Khim.*, **1966**, 944; F. S. Yakushin, V. N. Setkina, E. A. Yakovleva, A. I. Shatenshtein, and D. N. Kursanov, *ibid.*, **1967**, 206; *Chem. Abs.*, **66**, 94426f (1967); G. P. Sollott, and W. R. Peterson, *J. Am. Chem. Soc.*, **89**, 5054 (1967); M. D. Rausch and R. A. Genetti, *ibid.*, p. 5502; G. R. Knox, I. G. Morrison, P. L. Pauson, M. A. Sandhu, and W. E. Watts, *J. Chem. Soc., C*, **1967**, 1853.

[151] J. A. Mangravite and T. G. Traylor, *Tetrahedron Letters*, **1967**, 4457, 4461.

concept of electron interaction from different sides of a cyclopentadiene ring; "inside" (**34**) and "outside" (**35**), illustrated for electrophilic substitution. The σ^+ of -1.4 corresponds to σ^+_{inside} attack at the site of highest electron

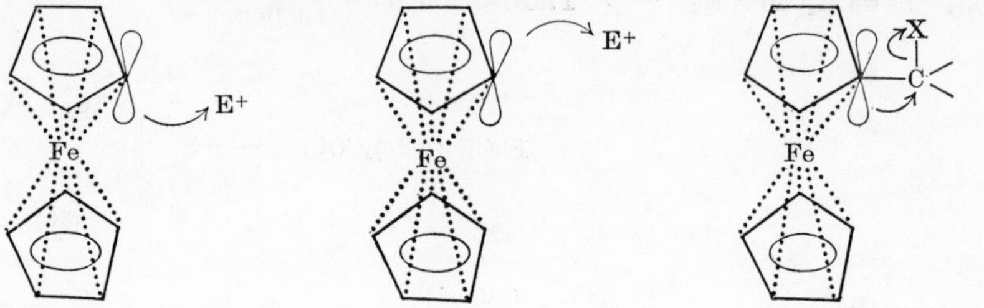

(**34**) (**35**) (**36**)

density, and illustrated as (**36**) for the solvolyses mentioned above. Analysis of reports in the literature then led to $\sigma^+_{\text{outside}}$ of -1.0. In proton exchange, the rate-determining step for inside attack will be proton loss, but for outside attack it will be the proton addition. The energy barrier surmounted must be the same in both cases. For an electrophile that is stronger than the proton the energy barrier for outside addition will be lower than that for proton loss following inside addition; hence outside addition will predominate; there should be no isotope effect on displacement of deuterium, and the rate should give a σ^+ value of the order of -1.0. For electrophiles that are weaker than the proton, inside addition is predicted correlating with $\sigma^+ = 1.4$; rate-determining proton loss should in this case give a substantial primary isotope effect. Acetylation and mercuration of ferrocene fit perfectly into these two reaction categories.

Extensive new work deals with the exchange behaviour in strong acids of pyridines (as free base or conjugate acid),[152] of their *N*-oxides,[153] and of pyrimidones,[154] pyridones, pyrones, and thiapyrones.[155] The relative reactivities of the last three molecules (as free bases) compare with those of pyrrole, furan, and thiophen. No common H_+ function correlated the second protonation of dibasic aminopyridines, phenylhydrazines, and diaza-heterocyclic compounds in strong acid.[156]

[152] J. A. Zoltewicz and C. L. Smith, *J. Am. Chem. Soc.*, **89**, 3358 (1967); C. D. Johnson, A. R. Katritzky, B. J. Ridgewell, and M. Viney, *J. Chem. Soc., B*, **1967**, 1204; C. D. Johnson, A. R. Katritzky, and M. Viney, *ibid.*, 1211; G. P. Bean, C. D. Johnson, A. R. Katritzky, B. J. Ridgewell, and A. M. White, *ibid.*, p. 1219.

[153] G. P. Bean, P. J. Brignell, C. D. Johnson, A. R. Katritzky, B. J. Ridgewell, H. O. Tarhan, and A. M. White, *J. Chem. Soc., B*, **1967**, 1222.

[154] G. E. Wright, L. Bauer, and C. L. Bell, *J. Heterocyclic Chem.*, **3**, 440 (1966).

[155] P. Bellingham, C. D. Johnson, and A. R. Katritzky, *J. Chem. Soc., B*, **1967**, 1226; *Chem. Commun.*, **1967**, 1047.

[156] P. J. Brignell, C. D. Johnson, A. R. Katritzky, N. Shakir, H. O. Tarhan, and G. Walker, *J. Chem. Soc., B*, **1967**, 1233.

Nitration of pyridine *N*-oxide gives the 4-nitro-derivative, but in strong acid pyridine *N*-oxides with electron-donating substituents in the α-position are nitrated, as their conjugate acids, in a β-position.[157] Nitrations of bicyclic *N*-oxides have also been examined.[158]

Anomalous behaviour is found in the nitration of 4-phenylpyrimidine.[159] In mixed nitric and sulphuric acids, nitration is in the *ortho*- and *meta*-positions of the phenyl ring; possibly the high *ortho* attack is associated with intramolecular transfer of NO_2^+ following initial co-ordination with nitrogen at position 3. Substitution occurs additionally in the *para*-position when the solvent is trifluoroacetic anhydride; in this solvent some of the nitration may involve an initial adduct (**37**), and related adducts were in fact isolated when acetic anhydride was used.

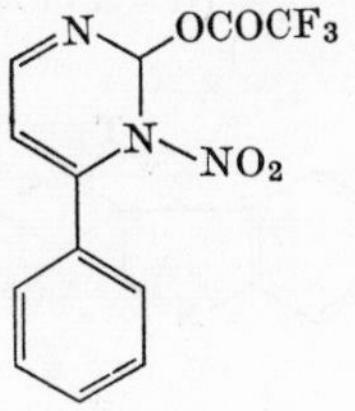

(37)

Base-catalysed hydrogen exchange in pyridine *N*-oxides[160] involves carbanion formation, as established by the trapping of the carbanion by carbonyl compounds.[161]

Several papers compare the reactivities of pyrroles, furans, and thiophens towards electrophilic attack.[162] These fall in the general order pyrrole $\gg$ thiophen $\geqslant$ furan $\gg$ benzene. Selenophen lies close to thiophen and furan in its reactivity as regards mercuration.[163]

Also reported are base-catalysed hydrogen exchange in furan ($\alpha \gg \beta$),[164]

157 C. D. Johnson, A. R. Katritzky, N. Shakir, and M. Viney, *J. Chem. Soc., B*, **1967**, 1213.
158 M. Hamana and T. Nagayoshi, *Chem. Pharm. Bull.* (*Tokyo*), **14**, 319 (1966); I. Suzuki, T. Nakashima, and N. Nagasawa, *ibid.*, p. 816.
159 B. M. Lynch and L. Poon, *Can. J. Chem.*, **45**, 1431 (1967).
160 R. A. Abramovitch, G. M. Singer, and A. R. Vinutha, *Chem. Commun.*, **1967**, 55; J. A. Zoltewicz and G. M. Kauffman, *Tetrahedron Letters*, **1967**, 337.
161 R. A. Abramovitch, M. Saha, E. M. Smith, and R. T. Coutts, *J. Am. Chem. Soc.*, **89**, 1537 (1967).
162 K. Schwetlick, K. Unverferth, and R. Mayer, *Z. Chem.*, **7**, 58 (1967); S. Clementi, F. Genel, and G. Marino, *Chem. Commun.*, **1967**, 498; P. Linda and G. Marino, *ibid.*, p. 499; *Tetrahedron*, **23**, 1739 (1967).
163 Y. K. Yur'ev, M. A. Gal'bershtam, I. I. Kandror, *Khim. Geterotsikl. Soedin.*, **1966**, 897; *Chem. Abs.*, **66**, 94453n (1967).
164 A. I. Shatenshtein, A. G. Kamrad, I. O. Shapiro, Y. I. Ranneva, and S. Hillers, *Khim. Geterotsikl. Soedin.*, **1966**, 643; *Chem. Abs.*, **66**, 85184y (1967).

acetylation of deactivated pyrroles,[165] nitration (in the 4-position) of 5-amino-pyrazole,[166] iodination of imidazole co-ordinated to nickel,[167] degradation of the imidazole ring of histidine by iodine,[168] iodination of aminochromes and adrenochrome,[169] hydrogen exchange in polyazaindenes,[170] and nitration of the dihydrodiazepinium perchlorates (**38**).[171]

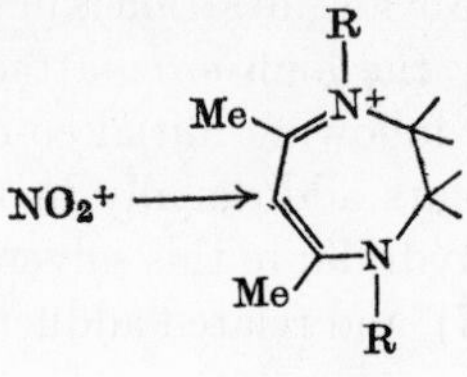

(**38**)
(R = H or Me)

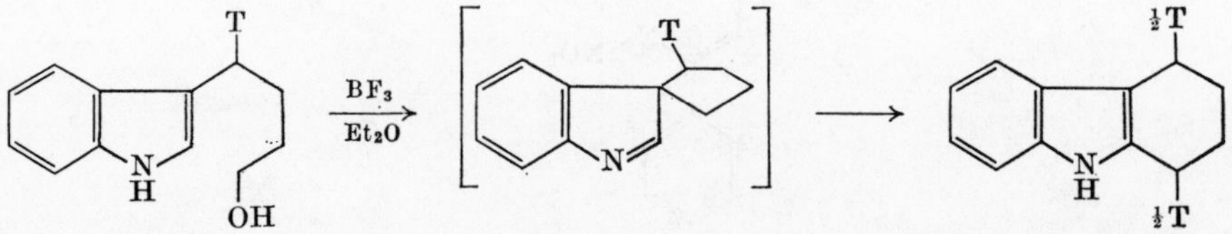

Finally, mention must be made of the elegant study by Jackson and Smith,[172] who show by the tritium-labelling experiment formulated that electrophilic substitution in 3-substituted indoles may (and probably normally does) involve initial addition at the 3-position followed by rearrangement.

[165] H. J. Anderson and C. W. Huang, *Can. J. Chem.*, **45**, 897 (1967).
[166] H. Dorn and H. Dilcher, *Ann. Chem.*, **707**, 141 (1967).
[167] D. G. Lambert and M. M. Jones, *J. Am. Chem. Soc.*, **88**, 5537 (1966).
[168] L. Schutte and E. Havinga, *Rec. Trav. Chim.*, **86**, 385 (1967).
[169] G. L. Mattock and D. L. Wilson, *Can. J. Chem.*, **45**, 1721, 2473 (1967).
[170] W. W. Paudler and L. S. Helmick, *Chem. Commun.*, **1967**, 377.
[171] C. Barnett, *Chem. Commun.*, **1967**, 637.
[172] A. H. Jackson and P. Smith, *Chem. Commun.*, **1967**, 264; see also M. Ahmed and B. Robinson, *J. Chem. Soc., B*, **1967**, 411.

CHAPTER 8

Molecular Rearrangements

Aromatic Rearrangements

Balaban and Fărcaşiu have made the important observation that the known rearrangement of polycyclic aromatic hydrocarbons by aluminium chloride occurs also with naphthalene. When [1-^{14}C]naphthalene was heated in benzene

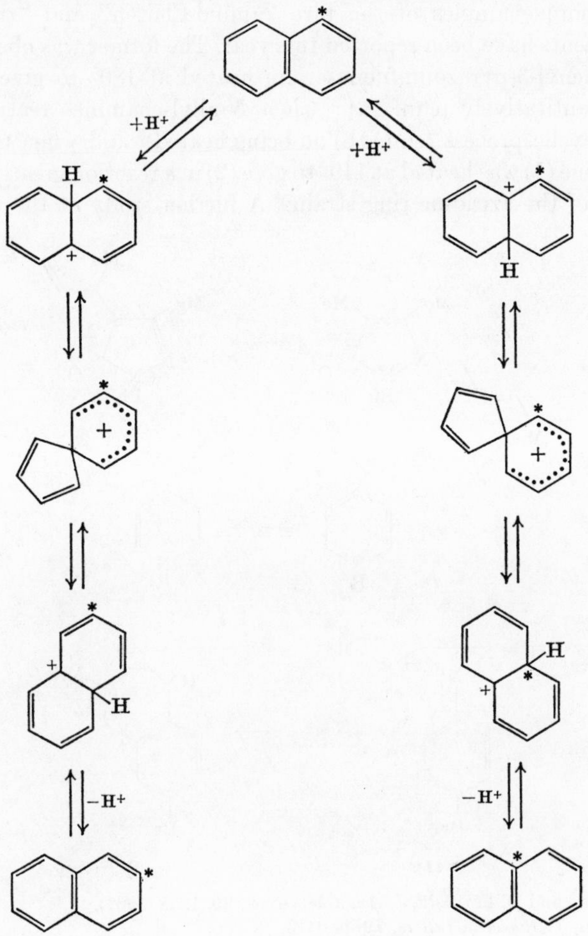

Scheme 1

for 2 hours at 60° with anhydrous aluminium chloride and then recovered, the total activity was conserved but approximately statistical distribution of the label between all the carbon atoms resulted. The proposed mechanism (Scheme 1) involves protons formed from aluminium chloride and atmospheric moisture. Protonation of the reactive α-positions does not lead to rearrangement, but protonation of $C_{(9)}$ or $C_{(10)}$ followed by 1,2-shifts gives a symmetrical cation in which a reversed 1,2-shift causes the observed scrambling. Obviously this isomerization under such mild conditions is potentially significant for electrophilic substitution of naphthalenes generally. The term automerization was proposed for isomerizations such as this where the structure is overall unchanged.[1]

A few more examples of the rare "amino-Claisen" and "thio-Claisen" rearrangements have been reported this year. The former was observed when 2-allyl-1-phenyl-3-pyrazolin-5-ones were heated at 180° to give the 4-allyl isomers quantitatively (eqn. 1a),[2] when *N*-allyl-enamines rearranged in a concerted cyclic process (eqn. 1b) on being heated,[3] and when the 1-aryl-2-vinylaziridine (**1**) was heated at 140° to give (**2**) in a reaction greatly facilitated by release of the aziridine ring strain.[4] A further study of the thio-Claisen

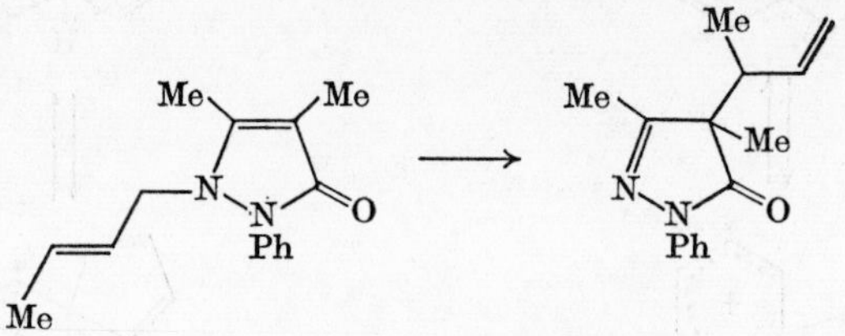

...(1a)

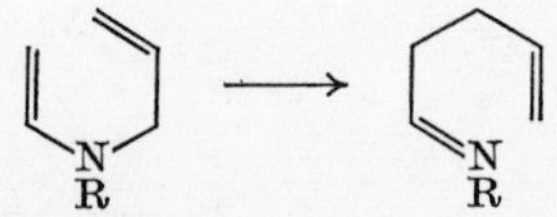

...(1b)

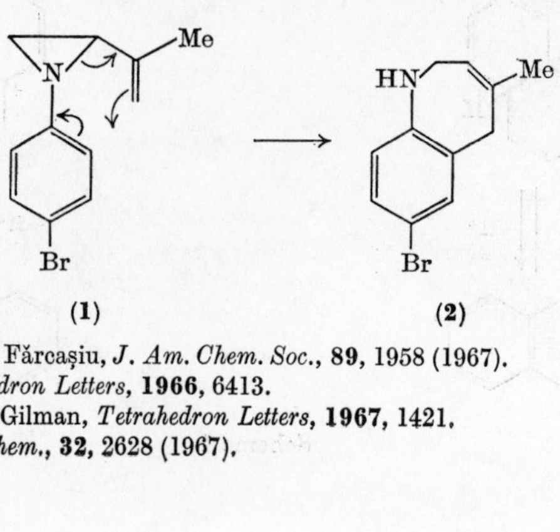

(1) (2)

[1] A. T. Balaban and D. Fărcaşiu, *J. Am. Chem. Soc.*, **89**, 1958 (1967).
[2] Y. Makisumi, *Tetrahedron Letters*, **1966**, 6413.
[3] R. K. Hill and N. W. Gilman, *Tetrahedron Letters*, **1967**, 1421.
[4] P. Scheiner, *J. Org. Chem.*, **32**, 2628 (1967).

rearrangement[5] now suggests that the two cyclic products (the thiacoumaran and the thiachroman) arise from different intermediates, a thiiran and an *o*-allylthiophenol. Possible roles for the solvent, quinoline, necessary for Claisen rearrangement were considered.[6] However, this rearrangement has been shown to proceed without complication, even in the absence of basic solvents, with allyl 4-quinolyl sulphides (**3**) to give single products in high yield; these products (**4**) were formed by subsequent cyclization of the 3-allyl-4(1*H*)-quinolinethiones.[7]

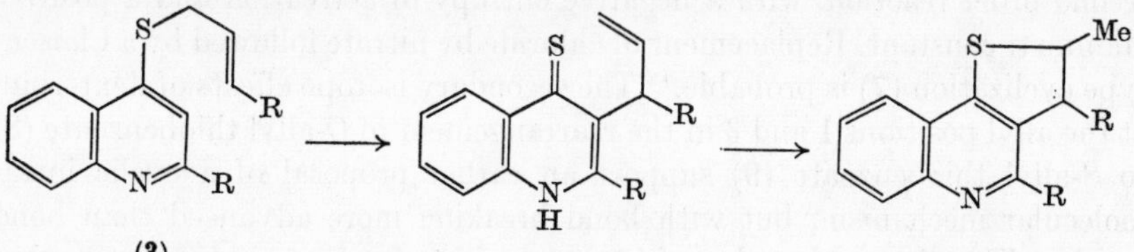

(**3**) (**4**)

Further evidence for the mechanism of the "abnormal" Claisen rearrangement reported earlier,[8] involving a homodienyl 1,5-hydrogen shift in a spirodienone intermediate, has been produced.[9] The rates of the ready interconversions of *cis*- and *trans*-*o*-(1,3-dimethylallyl)phenol (**5**) and of deuterium incorporation showed that every *cis*-**5** molecule converted into *trans*-**5** incorporates one deuterium atom into the allyl side chain at $C_{(3)}$. Thus the proton transfer is stereospecific, as required by the 1,5-homodienyl shift.[9]

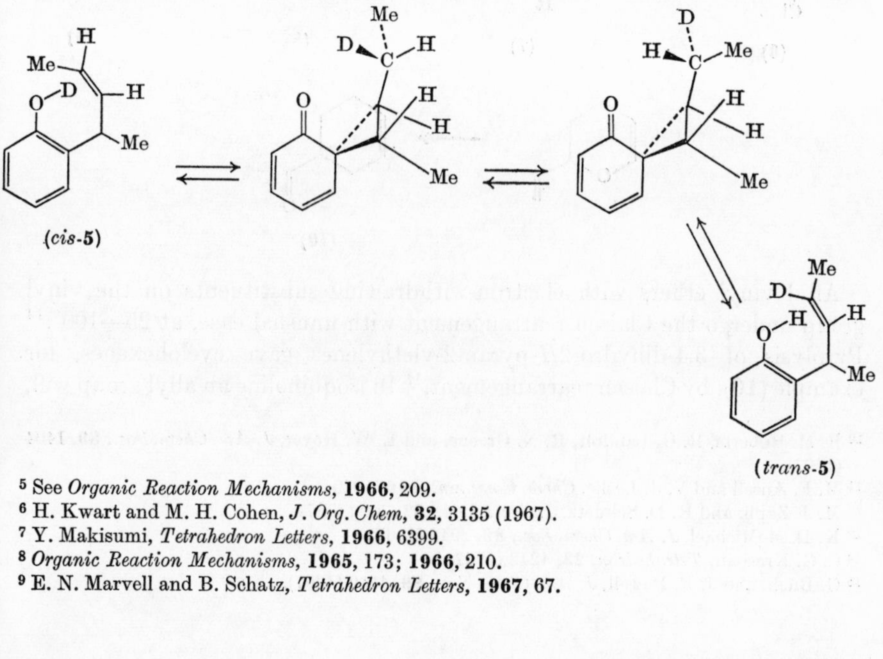

(*cis*-**5**) (*trans*-**5**)

[5] See *Organic Reaction Mechanisms*, **1966**, 209.
[6] H. Kwart and M. H. Cohen, *J. Org. Chem*, **32**, 3135 (1967).
[7] Y. Makisumi, *Tetrahedron Letters*, **1966**, 6399.
[8] *Organic Reaction Mechanisms*, **1965**, 173; **1966**, 210.
[9] E. N. Marvell and B. Schatz, *Tetrahedron Letters*, **1967**, 67.

Details have appeared[10] of Roberts and Landolt's work, reported earlier,[8] on the rearrangement of acylalkylcyclopropanes; the relations of this to the abnormal Claisen rearrangement and to other 1,5-hydrogen shifts were discussed.

The 1:1 adduct of *o*-chloranil and dimethylbutadiene is an allylic *o*-dienone and as such undergoes the retro-Claisen rearrangement shown (**6**) on very mild heating.[11] *para*-Substituted phenyl chloroformates reacted with silver nitrate in methyl cyanide, to give 4-substituted 2-nitrophenols in high yield, in a second-order reaction, with a negative entropy of activation and a positive Hammett constant. Replacement of chloride by nitrate followed by a Claisen-type cyclization (**7**) is probable.[12] The secondary isotope effects of deuterium at the allyl positions 1 and 3 in the rearrangement of *O*-allyl thiobenzoate (**8**) to *S*-allyl thiobenzoate (**9**) support an earlier proposal of a cyclic intramolecular mechanism, but with bond breaking more advanced than bond making. The diagnostic value of secondary isotope effects in locating such a transition state along the reactant–product coordinate is assessed.[13]

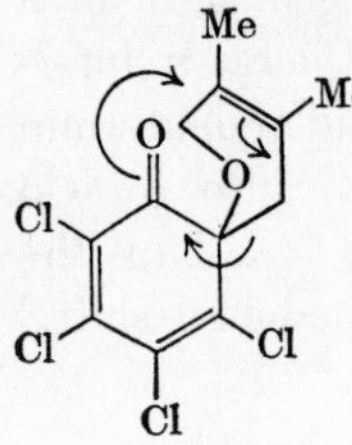

(**6**)

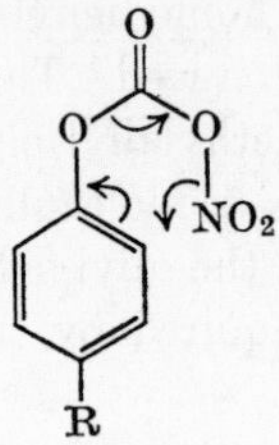

(**7**)

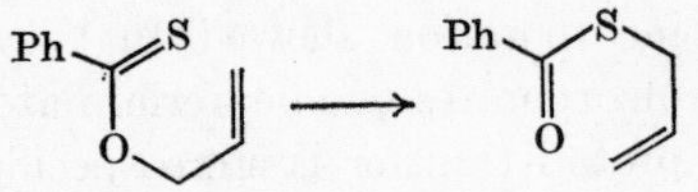

(**8**) (**9**)

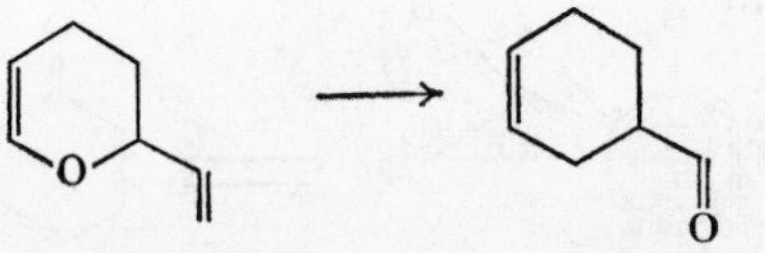

(**10**)

Allyl vinyl ethers with electron-withdrawing substituents on the vinyl group undergo the Claisen rearrangement with unusual ease, at 25—100°.[14] Pyrolysis of 3,4-dihydro-2*H*-pyran-2-ylethylenes gave cyclohexenes, for example (**10**), by Claisen rearrangement.[15] In isoquinoline an allyl group will,

10 R. M. Roberts, R. G. Landolt, R. N. Greene, and E. W. Heyer, *J. Am. Chem. Soc.*, **89**, 1404 (1967).
11 M. F. Ansell and V. J. Leslie, *Chem. Commun.*, **1967**, 949.
12 M. J. Zabik and R. D. Schuetz, *J. Org. Chem.*, **32**, 300 (1967).
13 K. D. McMichael, *J. Am. Chem. Soc.*, **89**, 2943 (1967).
14 C. G. Krespan, *Tetrahedron*, **23**, 4243 (1967).
15 G. Büchi and J. E. Powell, *J. Am. Chem. Soc.*, **89**, 4559 (1967).

as expected, migrate from $C_{(1)}$ to N and from $C_{(3)}$ to $C_{(4)}$, but not from $C_{(3)}$ to N.[16] Other Claisen rearrangements that have been reported include the *ortho*-rearrangement of symmetrical[17] and unsymmetrical[18] 1,4-diaryloxy-*trans*-but-2-enes, of 4-(phenoxymethyl)-3-chromene,[19] of allyl *p*-halogenophenyl ethers,[20] and of various heterocyclic systems;[21] also, stereoselectivity in the *para*-Claisen rearrangement has been described.[22]

Full details have appeared[23] of Miller's work on the interconversion of the remarkably stable di-*tert*-butylcyclohexadienones which have the structure of Claisen rearrangement intermediates.[5] A further example of the great sensitivity of the course of the dienone–phenol rearrangement to small structural changes in the dienone,[24] and other dienone–phenol rearrangements,[25] have been reported.

In a study of the acid-catalysed reaction, i.e., benzidine rearrangement and disproportionation, of hydrazobenzenes with bulky *para*-substituents, Shine and his co-workers have questioned the π-complex and the polar transition-state mechanisms for the rearrangement. The former mechanism claims that steric hindrance by *para*-substituents is a factor in determining the order in acid, and this is shown not to be so. The latter mechanism does not accommodate disproportionation, except as a separate concurrent reaction. In the present reaction the (large) relative contribution of disproportionation was not altered significantly by changing the initial hydrazobenzene concentration, showing that the two reactions follow the same kinetic law. This is most simply explained by invoking a common intermediate and it was suggested that the intermediateless, polar transition-state mechanism should be modified to the extent of incorporating an intermediate which can rearrange or react with another molecule of hydrazobenzene.[26] These results have been confirmed and extended by Ingold and his co-workers, who have also observed one-proton disproportionation for the first time, showing that the kinetic dichotomy of rearrangement extends to disproportionation. It is proposed

[16] H. Win and H. Tieckelmann, *J. Org. Chem.*, **32**, 59 (1967).

[17] B. S. Thyagarajan, K. K. Balasubramanian, and R. B. Rao, *Tetrahedron.* **23**, 3205 (1967).

[18] B. S. Thyagarajan, K. K. Balasubramanian, and R. B. Rao, *Tetrahedron*, **23**, 3533 (1967).

[19] B. S. Thyagarajan, K. K. Balasubramanian, and R. B. Rao, *Tetrahedron*, **23**, 1893 (1967).

[20] J. Mirek, *Zesz, Nauk Uniw. Jagiellon Pr. Chem.*, No. 9, 77 (1964); *Chem. Abs.*, **66**, 37116f (1967).

[21] J. K. Elwood and J. W. Gates, *J. Org. Chem.*, **32**, 2956 (1967); for a review, see *Adv. Heterocyclic Chem.*, **8**, 143 (1967).

[22] B. S. Thyagarajan, K. K. Balasubramanian, and R. B. Rao, *Chem. Ind.* (*London*), **1967**, 401.

[23] B. Miller, *J. Am. Chem. Soc*, **89**, 1685 (1967).

[24] K. H. Bell, *Tetrahedron Letters*, **1967**, 397.

[25] M. Heller, R. H. Lenhard, and S. Bernstein, *J. Am. Chem. Soc.*, **89**, 1911 (1967); A. I. Brodskii, V. D. Pokhodenko, N. N. Kalibabchuk, and V. S. Kuts, *Dokl. Akad. Nauk SSSR*, **172**, 122 (1967); *Chem. Abs.*, **66**, 85255x (1967).

[26] H. J. Shine and J. T. Chamness, *J. Org. Chem.*, **32**, 901 (1967); H. J. Shine and J. P. Stanley, *ibid.*, p. 905.

that the required intermediate, which must be formed irreversibly after the rate-determining step, could have the quinonoid ring-linked, but still protonated, structure (e.g., **11**). This could then take part in the necessary redox

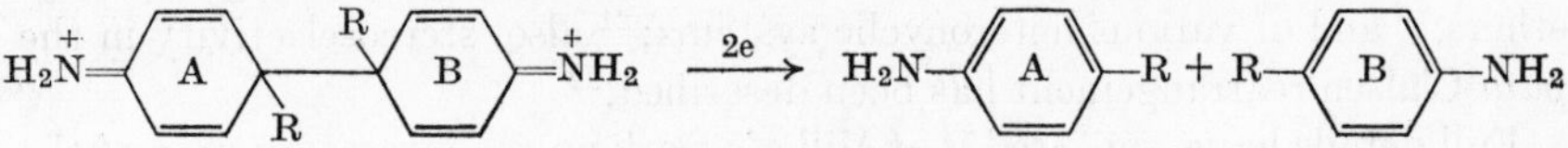

(11)

process in a rapid step.[27] The rearrangement of hydrazobenzene-2,2′-dicarboxylic acid, which gave only the benzidine,[28] the effect of pressure on the rate of rearrangement of 2,2′-hydrazotoluene,[29] and the benzidine rearrangement generally[30] have been described.

It is now clear that all the steps in the rearrangement of phenylsulphamic acid to sulphanilic acid are intermolecular, since that of orthanilic to sulphanilic acid was shown to be intermolecular by using [^{35}S]sulphuric acid.[31] The diazoaminobenzene rearrangement has also been shown to be intermolecular by ^{15}N-labelling.[32] Other aromatic rearrangements studied are the Smiles,[33] Orton,[34] and Wallach[35] rearrangements and the aluminium chloride-catalysed isomerization of methyldiphenylmethanes[36] and of halogen-substituted carboxylic and sulphonic acids,[37] and the sulphuric acid-catalysed isomerization of toluenesulphonic acids[38] and of naphthalenesulphonic acids.[39]

A monograph on aromatic rearrangements has appeared.[40]

Cope and Related Rearrangements: Valence-bond Isomerization

Geometrical specificity in the Cope rearrangement of 1,5-dienes was elegantly demonstrated, and used to prove the chair-like conformation of the transition

[27] D. V. Banthorpe, A. Cooper, and C. K. Ingold, *Nature*, **216**, 232 (1967).
[28] M. Kurihara and N. Yoda, *Bull. Chem. Soc. Japan*, **40**, 2429 (1967).
[29] J. Osugi, M. Sasaki, and I. Onishi, *Rev. Phys. Chem. Japan*, **36**, 100 (1966); *Chem. Abs.*, **67**, 43163f, 90204e (1967).
[30] V. O. Lukashevich, *Tetrahedron*, **23**, 1317 (1967).
[31] F. L. Scott and W. J. Spillane, *Tetrahedron Letters*, **1967**, 2707.
[32] S. Weckherlin and W. Lüttke, *Ann. Chem.*, **700**, 59 (1966).
[33] V. N. Drozd and T. Y. Frid, *Zh. Org. Khim.*, **3**, 373 (1967); *Chem. Abs.*, **67**, 2586c (1967); V. N. Drozd and V. I. Sheichenko, *Zh. Org. Khim.*, **3**, 554 (1967); *Chem. Abs.*, **67**, 10965r (1967).
[34] J. M. W. Scott and J. G. Martin, *Can. J. Chem.*, **44**, 2901 (1966).
[35] C. S. Hahn, K. W. Lee, and H. H. Jaffé, *J. Am. Chem. Soc.*, **89**, 4975 (1967).
[36] G. A. Olah and J. A. Olah, *J. Org. Chem.*, **32**, 1612 (1967).
[37] A. M. Komagorov, V. P. Chzhu, and V. A. Koptyug, *Izv. Sib. Otd. Akad. Nauk SSSR, Ser. Khim. Nauk*, **1966**, 93; *Chem. Abs.*, **66**, 37108e (1967).
[38] A. C. M. Wanders, H. Cerfontain, and C. W. F. Kort, *Rec. Trav. Chim.*, **86**, 301 (1967).
[39] V. A. Koptyug and S. A. Shkol'nik, *Zh. Org. Khim.*, **2**, 1870 (1966); *Chem. Abs.*, **66**, 54796v (1967).
[40] H. J. Shine, "Aromatic Rearrangements", Elsevier, Amsterdam, 1967.

state, by Doering and Roth six years ago.[41] Hill and Gilman[42] have now demonstrated optical stereospecificity in this rearrangement in complete accord with the earlier results. *trans*-3-Methyl-3-phenylhepta-1,5-diene (**12**) rearranged quantitatively at 250° to the mixture of *cis*- and *trans*-heptadienes shown, which were not interconverted at this temperature. Rearrangement of (dextrorotatory) (*R*)-(**12**) gave optically active products (optical yield 94—96%), with opposite configuration at the newly induced asymmetric centre. This nearly quantitative transfer of asymmetry provides strong additional evidence for a concerted cyclic rearrangement, and the 87:13 preference for the conformation with the phenyl group equatorial supports predictions made on the basis of cyclohexane-type conformational analysis of the transition state.[42]

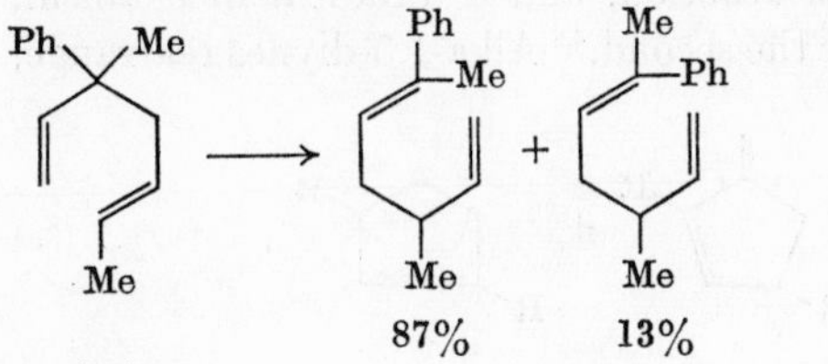

(**12**)

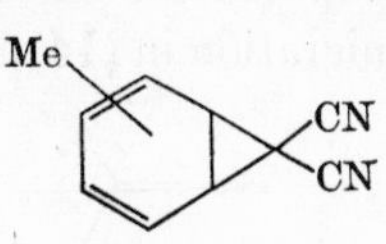

(**13**)

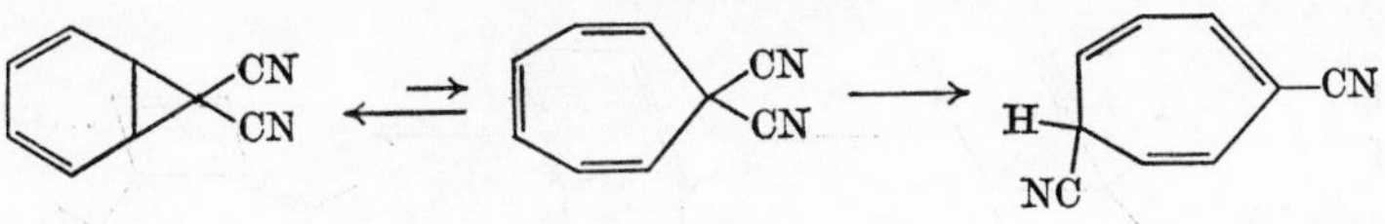

....(2)

Following the demonstration[43] that *x*,7,7-trialkylcycloheptatrienes are interconverted by skeletal rearrangements involving valence isomerization to norcaradienes at about 300°, Berson and his co-workers[44] have shown that methyl-substituted 7,7-dicyanonorcaradienes (**13**) undergo the same re-arrangement, but under very much milder conditions (above 55°). The drastic reduction in activation energy (*ca.* 18 kcal mole^{-1}) is attributed to the absence of the preliminary (endothermic) Cope rearrangement to form the norcara-diene, and to the weakening of the cyclopropane bonds by the cyano-groups. Isotopic labelling experiments were again consistent with a circumambulatory mechanism.[44] Details of Ciganek's work on the addition of dicyanocarbene to aromatic compounds have appeared. The greater stability of the norcardiene over the cycloheptatriene structure with these 7,7-dicyano-compounds is possibly explained by the greater NC—C—CN bond angle and hence smaller

41 W. von E. Doering and W. R. Roth, *Tetrahedron*, **18**, 67 (1962).
42 R. K. Hill and N. W. Gilman, *Chem. Commun.*, **1967**, 619.
43 *Organic Reaction Mechanisms*, **1965**, 179; **1966**, 217.
44 J. A. Berson, F. W. Grubb, R. A. Clark, D. R. Hartter, and M. R. Willcott, *J. Am. Chem. Soc.*, **89**, 4076 (1967).

dipole–dipole repulsion between the cyano-groups, or by the formation of an internal charge-transfer complex between one cyano-group and the diene system.[45] However, introduction of three phenyl groups also stabilizes the norcaradiene structure; the valence-bond tautomerization of 2,5,7-triphenylnorcaradiene, the first simple example with a hydrogen on $C_{(7)}$, has been investigated.[46] Thermolysis of 7,7-dicyanonorcaradiene gave phenylmalononitrile and 3,7-dicyanocycloheptatriene. Kinetic and other evidence suggests that the latter was formed by an intramolecular 1,5-cyano-shift in the tautomeric cycloheptatriene (reactions 2).[47]

Alk-1-en-5-ynes undergo a reversible Cope rearrangement at 340° to give 1,2,5-alkatrienes which, in turn, cyclize to 3- and 4-methylenecyclopentene (reaction 3). From the effect of methyl groups on the rate a concerted mechanism is proposed for the first reaction, and a diradical mechanism, with hydrogen migration in (**14**), for the second.[48] Alka-1,5-diynes rearrange,

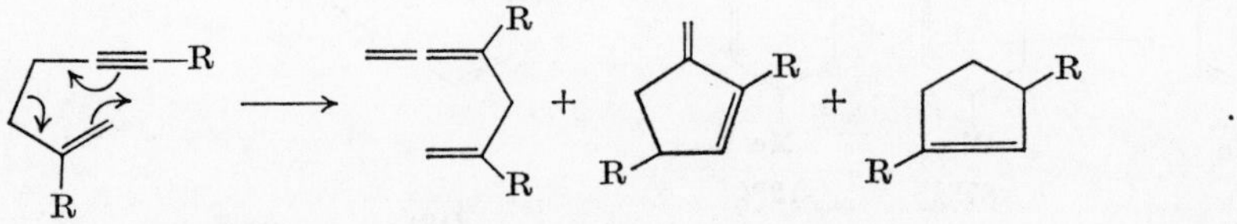

...(3a)

R=H or Me

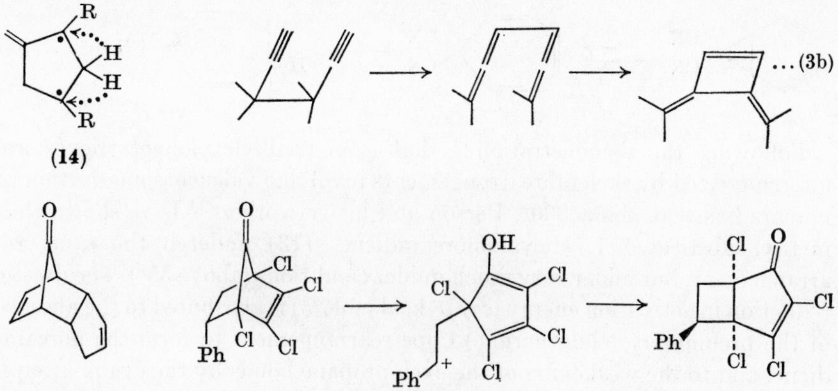

...(3b)

(**14**) (**15**) (**16**) (**17**) (**18**)

on being heated, to dimethylenecyclobutenes by a conrotatory process; a likely mechanism is Cope rearrangement to the bisallene followed by intramolecular allene dimerization (reactions 3b).[49]

45 E. Ciganek, *J. Am. Chem. Soc.*, **89**, 1454 (1967).
46 T. Mukai, H. Kubota, and T. Toda, *Tetrahedron Letters*, **1967**, 3581.
47 E. Ciganek, *J. Am. Chem. Soc.*, **89**, 1458 (1967).
48 W. D. Huntsman, J. A. De Boer, and M. H. Woosley, *J. Am. Chem. Soc.*, **88**, 5846 (1966).
49 W. D. Huntsman and H. J. Wristers, *J. Am. Chem. Soc.*, **89**, 342 (1967).

Dicyclopentadien-8-one (**15**) has been isolated and shown to rearrange at its melting point and in polar solvents to the 1-oxo-isomer[50] (cf. ref. 51); the rearrangement is strongly catalysed by Lewis acids. On treatment with sulphuric acid the *exo*- and *endo*-bicycloheptene (**16**) gave the bicycloheptenone (**18**) probably via the carbonium ion (**17**).[52] Similar isomerization of the corresponding pentaphenyl compounds (Ph for Cl in **16**) gave the corresponding product. Since both *exo*- and *endo*-isomers gave the same product, a concerted 1,3-allylic shift of the benzylic carbon cannot be occurring and the open carbonium ion is probably involved. An alternative product of cyclization

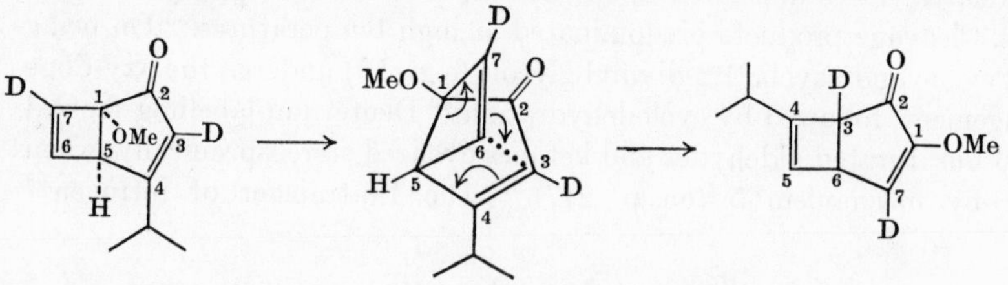

. . . (4)

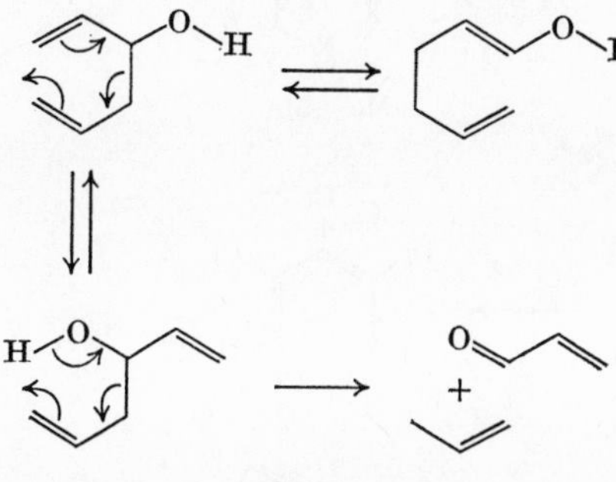

Scheme 2

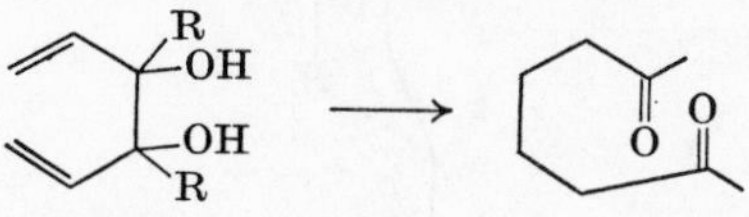

(**19**)

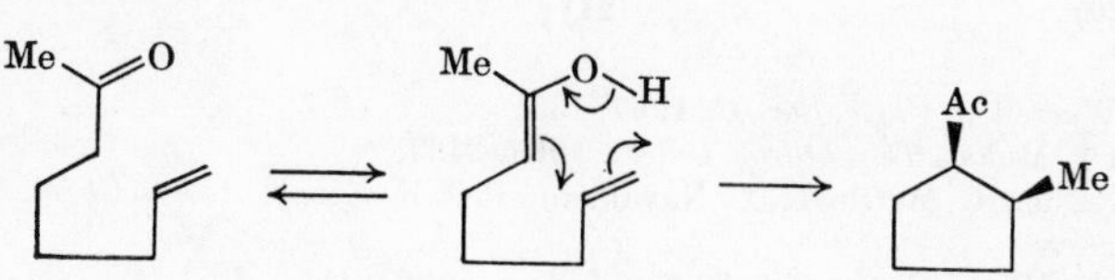

. . . (5)

[50] R. C. Cookson, J. Hudec, and R. O. Williams, *J. Chem. Soc., C*, **1967**, 1382.
[51] R. B. Woodward and T. J. Katz, *Tetrahedron*, **5**, 70 (1959).
[52] L. S. Besford, R. C. Cookson, and J. Cooper, *J. Chem. Soc., C*, **1967**, 1385.

of this carbonium ion was isolated when the bridgehead phenyl groups were replaced by methyl.[53]

Thermal rearrangement of 1,4- into 3,7-disubstituted bicyclo[3.2.0.]hepta-dienones has been shown by deuterium labelling to proceed by a Cope mechanism (reaction 4, p. 217).[54]

In an extensive review of the "oxy-Cope" rearrangement the products of vapour-phase thermolysis of ten methyl-substituted hexa-1,5-dien-3-ols were all accounted for by two competing concerted reactions with six-atom cyclic transition states, a Cope rearrangement and a 1,5-hydrogen shift (Scheme 2), whose relative rates depended upon the stabilities of the appropriate con-formers. Cleavage products predominated at high temperatures.[55] On being heated, cyclic and acyclic 1,2-divinyl glycols (e.g., **19**) undergo the oxy-Cope rearrangement, followed by cyclodehydration.[56] Deuterium-labelling showed that 5,6-unsaturated aldehydes and ketones cyclized stereospecifically, when heated, by mechanism 5 (on p. 217).[57] The 1,5-transfer of nitrogen[58]

. . . (6)

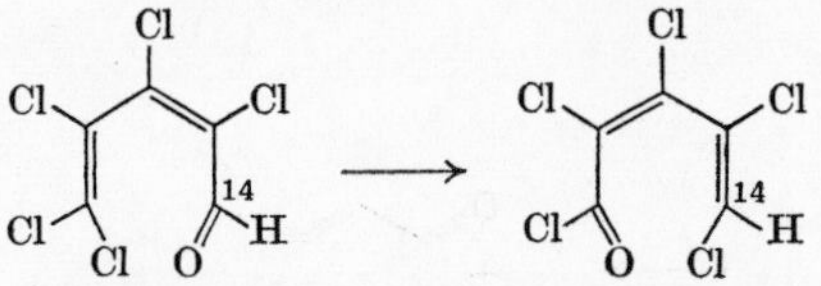

. . . (7)

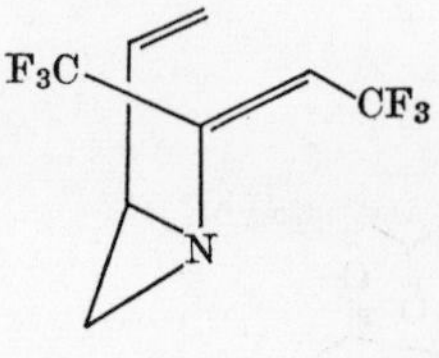

(20)

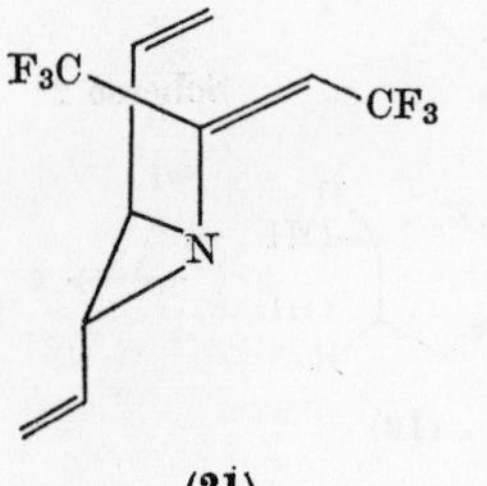

(21)

53 R. C. Cookson and D. C. Warrell, *J. Chem. Soc., C*, **1967**, 1391.
54 T. Miyashi, M. Nitta, and T. Mukai, *Tetrahedron Letters*, **1967**, 3433.
55 A. Viola, E. J. Iorio, K. K. Chen, G. M. Glover, U. Nayak, and P. J. Kocienski, *J. Am. Chem. Soc.*, **89**, 3462 (1967).
56 E. Brown, P. Leriverend, and J.-M. Conia, *Tetrahedron Letters*, **1966**, 6115.
57 F. Rouessac, P. Le Perchec, and J.-M. Conia, *Bull. Soc. Chim. France*, **1967**, 818; J.-M. Conia, *Ind. Chim. Belge*, **32**, 413 (1967); R. Bloch and J.-M. Conia, *Tetrahedron Letters*, **1967**, 3409.
58 H. W. Bersch and D. Schon, *Arch. Pharm.*, **300**, 82 (1967).

(eqn. 6) and of oxygen[59] (eqn. 7) via six-atom cyclic transition states has also been demonstrated.

The divinylaziridines (**20**) and (**21**) isomerize, by the Cope mechanism, at the same rate. In (**21**) the inverting *N*-vinyl group is always *cis* to another vinyl group, as required for the Cope rearrangement. Since (**20**) isomerizes as fast as (**21**), inversion about the ring nitrogen must be much faster than valence tautomerization.[60] Other examples of valence tautomerization by Cope rearrangement have been reported with diazepines,[61] triazepines,[62] bicyclo-[6.1.0]nona-2,4,6-triene,[63] hepta-1,2,6-triene which gave 3-methylenehexa-1,5-diene,[64] vinylbicyclo[3.2.1]oct-2-enes,[65] cyclo-octa-1,3,5-trienes,[66] and caryophyllenes.[67]

Subsequent work by Newman and Lala has shown that the thermolyses reported last year,[68] which were considered to involve bicyclo-[3.2.1] and -[3.3.1] transition states are, in fact, more complex than this.[69]

The concept of molecular-orbital symmetry conservation has been extended to certain transition metal-catalysed reactions and the very intriguing suggestion is made[70] that metals containing suitable orbitals are "capable of rendering otherwise forbidden cycloaddition reactions allowed by providing a template of atomic orbitals" through which the appropriate electron pairs may flow. Striking examples of this phenomenon could be the enormous increase in rate of the thermally forbidden isomerization of quadricyclane to norbornadiene in the presence of various rhodium, palladium, and platinum complexes,[71] and the dramatic catalysis by metal ions of fused-ring cyclobutene-to-butadiene isomerizations.[72] For example, dibenzotricyclo-octadiene (**22**) isomerizes via (**23**) slowly at 180°, but in the presence of molar quantities of silver fluoroborate the isomerism is complete in 10 seconds at room temperature. It was proposed that the metal ion (Ag^{+},Cu^{+}) forms a π-complex in which

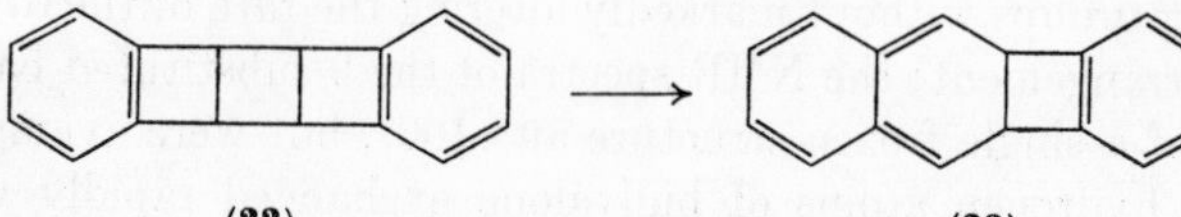

(**22**) (**23**)

[59] A. Roedig, G. Märkl, F. Frank, R. Kohlhaupt, and M. Schlosser, *Chem. Ber.*, **100**, 2730 (1967).
[60] E. L. Stogryn and S. J. Brois, *J. Am. Chem. Soc.*, **89**, 605 (1967).
[61] M. A. Battiste and T. J. Barton, *Tetrahedron Letters*, **1967**. 1227.
[62] H. W. Heine and J. Irving, *Tetrahedron Letters*, **1967**, 4767.
[63] T. L. Burkoth, *J. Org. Chem.*, **31**, 4259 (1966); W. Grimme, *Chem. Ber.*, **100**, 113 (1967).
[64] H. M. Frey and D. H. Lister, *J. Chem. Soc., A*, **1967**, 26.
[65] J. M. Brown, *Chem. Commun.*, **1967**, 638.
[66] M. Kröner, *Chem. Ber.*, **100**, 3163, 3172 (1967).
[67] G. Ohloff, G. Uhde, and K. H., Schulte-Elte, *Helv. Chim. Acta*, **50**, 561 (1967).
[68] *Organic Reaction Mechanisms*, **1966**, 224—225.
[69] M. S. Newman and L. M. Lala, *J. Org. Chem.*, **32**, 3225 (1967).
[70] F. D. Mango and J. H. Schachtschneider, *J. Am. Chem. Soc.*, **89**, 2484 (1967).
[71] H. Hogeveen and H. C. Volger, *J. Am. Chem. Soc.*, **89**, 2486 (1967).
[72] W. Merk and R. Pettit, *J. Am. Chem. Soc.*, **89**, 4788 (1967).

the sterically preferred, but normally forbidden, disrotatory process is now allowed, and the internal strain of the cyclobutenes can then be rapidly released.[72] Another example is provided by the valence isomerization of hexamethylprismane to hexamethyl-Dewar-benzene and to hexamethylbenzene; the latter predominates (30:1) in the thermal isomerization whilst the former, a "forbidden" product, predominates (20:1) under transition-metal catalysis.[73]

A novel type of valence tautomerization is demonstrated by the temperature-dependence of the NMR spectrum of tetracarbonyl(tetramethylallene)-iron; the $Fe(CO)_4$ unit is co-ordinated to one double bond but is moving from one π-M.O. to the other, orthogonal, one.[74] Valence tautomerism has also been investigated in tricarbonylcyclo-octatetraeneruthenium,[75] tricarbonyl-(methylcyclo-octatetraene)iron,[76] and tricarbonyl-*N*-(ethoxycarbonyl)azepineiron.[77]

The cyclization of a tetraene to a cyclo-octatriene is a conrotatory process in accord with orbital-symmetry predictions.[78] A wide range of tropones were decarbonylated to the corresponding benzenes by pyrolysis (500—900°), the most likely path being by valence tautomerism to the bicyclo[4.1.0]-system.[79] Bicyclo[6.2.0]deca-2,4,6,9-tetraene readily and quantitatively isomerized to *trans*-4a,8a-dihydronaphthalene.[80]

Valence tautomerization,[81] electrocyclic reactions,[82] and the stereochemistry of the transition state of various multicentre reactions[83] have been reviewed.

Rational syntheses of bullvalene, bullvalone (**24**), barbaralane (**26**), and barbaralone, which afford independent synthetic evidence of their structure, and a study of their "fluxional" character have been described by Doering and his co-workers.[84] The carbonyl group of barbaralone was replaced by various other groups without markedly altering the rate of the divinylcyclopropane rearrangement; the NMR spectra of the 9-substituted barbaralanes were those of a single frozen structure at –100°, but were averaged at 25°. Two of the hydrogen atoms of bullvalone exchanged rapidly with dilute

73 H. Hogeveen and H. C. Volger, *Chem. Commun.*, **1967**, 1133; H. C. Volger and H. Hogeveen, *Rec. Trav. Chim.*, **86**, 830 (1967).
74 R. Ben-Shoshan and R. Pettit, *J. Am. Chem. Soc.*, **89**, 2231 (1967).
75 M. I. Bruce, M. Cooke, M. Green, and F. G. A. Stone, *Chem. Commun.*, **1967**, 523.
76 F. A. L. Anet, *J. Am. Chem. Soc.*, **89**, 2491 (1967).
77 H. Günther and R. Wenzl, *Tetrahedron Letters*, **1967**, 4155.
78 E. N. Marvell and J. Seubert, *J. Am. Chem. Soc.*, **89**, 3377 (1967).
79 T. Mukai, T. Nakazawa, and T. Shishido, *Tetrahedron Letters*, **1967**, 2465.
80 S. Masamune, C. G. Chin, K. Hojo, and R. T. Seidner, *J. Am. Chem. Soc.*, **89**, 4804 (1967).
81 J. Wolters, *Chem. Weekblad*, **62**, 588 (1966).
82 C. Horig, *Z. Chem.*, **7**, 298 (1967); O. Červinka and O. Kříž, *Chem. Listy*, **61**, 1036 (1967).
83 W. R. Roth, *Chem. Weekblad*, **63**, 9 (1966).
84 W. von E. Doering, B. M. Ferrier, E. T. Fossel, J. H. Hartenstein, M. Jones, G. Klumpp, R. M. Rubin, and M. Saunders, *Tetrahedron*, **23**, 3943 (1967).

alkaline deuterium oxide, as expected; there then follows further exchange of three, four, and finally all ten hydrogen atoms, presumably by successive divinylcyclopropane rearrangements of the enolic tautomer (**25**). This complete exchange is strong structural evidence for the bullvalene rearrangement, leading to random distribution of all the carbon atoms.[84] To the three interconvertible $C_{10}H_{10}$ hydrocarbons, including bullvalene, mentioned last

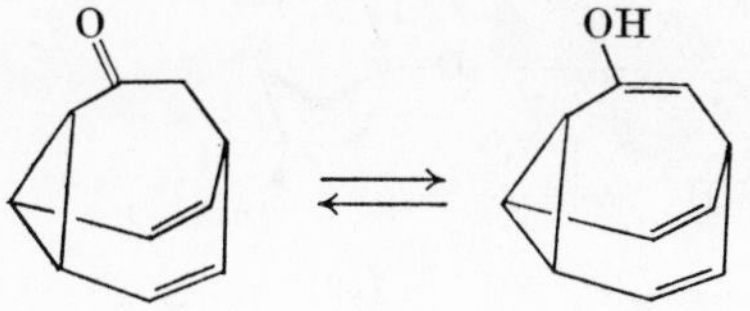

(**24**) (**25**)

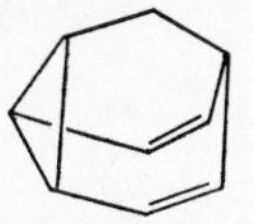

(**26**)

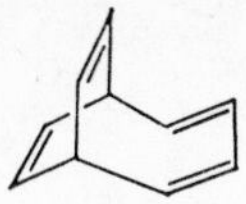

(**27**)

year,[85] a fourth has been added: bicyclo[4.2.2]deca-2,4,7,9-tetraene (**27**) was formed by irradiation of 4a,8a-dihydronaphthalene and, in turn, was photoisomerized to bullvalene.[86]

A methoxybullvalene has been synthesized and its fluxional nature shown by its temperature-dependent NMR spectrum.[87] Possible mechanisms have been considered for thermal isomerizations of a bridged homotropylidene system.[88] Details of the crystal and molecular structure of a silver–bullvalene complex have appeared.[89]

A comparison of the rates of homolysis of the 1,7- and 1,5-bonds in bicyclo[3.2.0]hept-2-en-*endo*-6-yl acetate (**28**) would provide an intramolecular measure of the inhibition of normal allylic stabilization in the transition state for breaking a bond lying in the π-nodal plane. This has now been done with the 3-deuterio-derivative by showing that it rearranges, at 298° in decalin, to (**29**), which requires 1,7-bond fission, at least 30 times faster than it rearranges by the alternative 1,5-bond fission. The total energy-deficit associated with breaking the bond in the π-nodal plane is thus estimated to be at least 6 kcal mole^{-1}, i.e., nearly half of the normal allylic stabilization.[90] This is in very good agreement with the value estimated by Willcott and Goerland[91] by comparing the activation energy for conversion of bicyclo[3.2.0]heptadiene (**30**) into cycloheptatriene and of bicyclo[3.2.0]hept-2-ene (**31**) into cycloheptadiene. The difference in activation energy, $45.5 - 39.5 = 6$ kcal mole^{-1}, is the energy gain in forming an allylic radical with minimum overlap. The

[85] *Organic Reaction Mechanisms*, **1966**, 218.
[86] M. Jones and L. T. Scott, *J. Am. Chem. Soc.*, **89**, 150 (1967); W. von E. Doering and J. W. Rosenthal, *Tetrahedron Letters*, **1967**, 349.
[87] L. A. Paquette, T. J. Barton, and E. B. Whipple, *J. Am. Chem. Soc.*, **89**, 5481 (1967).
[88] J. N. Labows, J. Meinwald, H. Röttele, and G. Schröder, *J. Am. Chem. Soc.*, **89**, 612 (1967).
[89] J. S. McKechnie, M. G. Newton, and I. C. Paul, *J. Am. Chem. Soc.*, **89**, 4819 (1967).
[90] J. A. Berson and R. S. Wood, *J. Am. Chem. Soc.*, **89**, 1043 (1967).
[91] M. R. Willcott and E. Goerland, *Tetrahedron Letters*, **1966**, 6341.

difference in activation energy for the thermal ring opening of (**31**) and for cyclobutene itself is 13 kcal mole^{-1}. Now the monocyclic ring will open by conrotation, as required by orbital symmetry, but the bicyclic olefin must open by disrotation (to give the *cis,cis*-cyclic 1,3-diene) and thus the energy

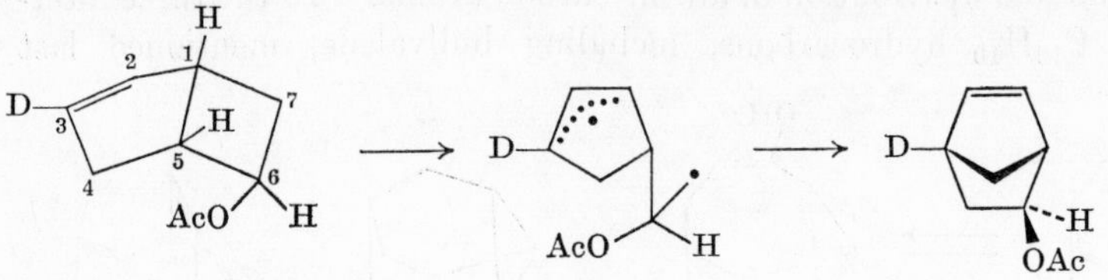

(**28**) (**29**)

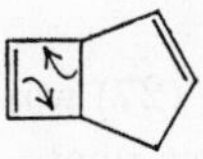

(**30**)

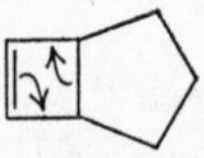

(**31**)

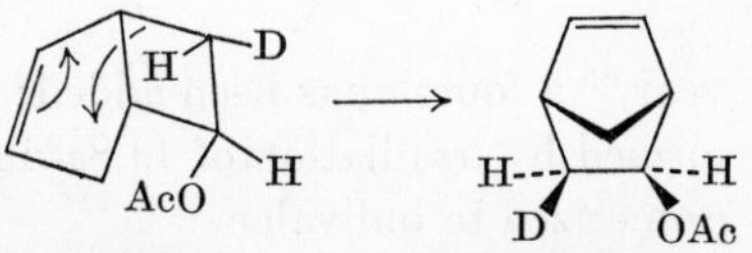

(**32**) (**33**)

difference between the allowed and the forbidden process is 13 kcal mole^{-1}.[91]

In a concerted thermal 1,3-sigmatropic rearrangement a migrating atom, such as the deuterium-bearing carbon in (**32**), might use both lobes of an antisymmetric orbital and thereby achieve a suprafacial, rather than the direct antarafacial process. This would necessarily be accompanied by inversion of configuration of the migrating group, and such an inversion has now been demonstrated in a system where the rearrangement is forced to be suprafacial. Thus in the rearrangement, at 307° in decalin, of (**32**) to (**33**) the deuterium originally *trans* to the acetate becomes *cis* to it.[92]

When 2-deuteriovinylcyclopropane (**34**) was heated at 360° the stereospecificity at the deuterium-labelled site was lost at least 5 times faster than the compound was converted into cyclopentene. Since the former process involves opening of the cyclopropane ring to the trimethylene radical it is likely, though not proved, that the vinylcyclopropane-to-cyclopentene rearrangement also proceeds through this diradical intermediate.[93] The major products of the pyrolysis of (**35**) arise from homolysis of the peripheral rather than the radial bond to give the diradical (**36**); the activation energy is about 8.5 kcal mole^{-1} lower than for vinylcyclopropane itself, presumably because of the extra strain in the spiropentane system.[94] The kinetics and secondary deuterium isotope effects in the pyrolysis of 3-vinyl-1-pyrazoline (**37**) to vinylcyclopropane and cyclopentene suggest the participation of a similar

[92] J. A. Berson and G. L. Nelson, *J. Am. Chem. Soc.*, **89**, 5503 (1967).
[93] M. R. Willcott and V. H. Cargle, *J. Am. Chem. Soc.*, **89**, 723 (1967).
[94] J. J. Gajewski, *Chem. Commun.*, **1967**, 920.

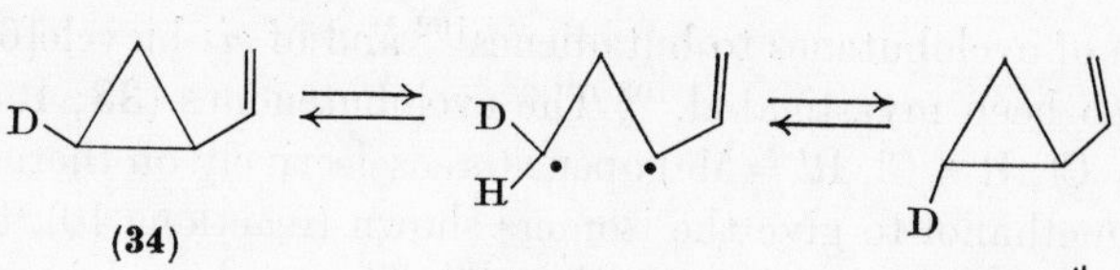

(34)

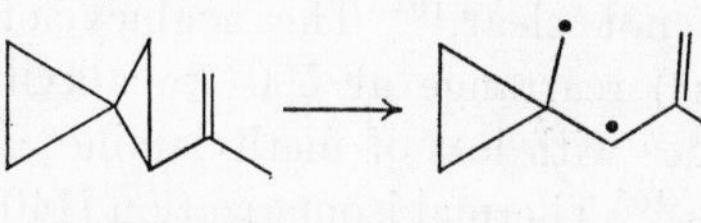

(35) (36)

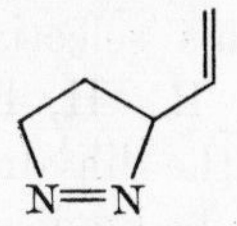

(37)

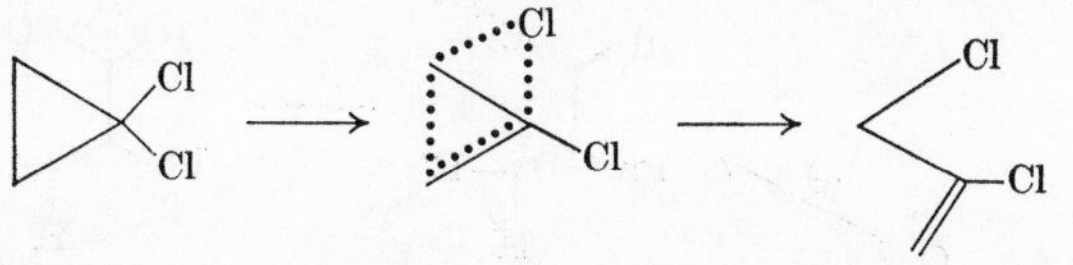

...(8)

nitrogen-free diradical.[95] A vinylcyclopropane-to-cyclopentene rearrangement was proposed as a likely mechanism for the formation of cyclopentadienes from 1,1-dibromo-2-vinylcyclopropanes and methyl-lithium.[96]

Product[97] and kinetic[98] studies of the thermal isomerization of 1,1-dichlorocyclopropanes show the reaction to involve intramolecular migration of a chlorine atom through a unimolecular transition state (reaction 8). Ring-opening to give the more stable trimethylene biradical is precluded since this would not give the observed products. The thermal rearrangement of tetra-arylcyclopropenes to triarylindenes and the non-rearrangement of an arylcyclopropane to an indane have been demonstrated.[99]

The degenerate thermal rearrangement of 1,2-dimethylenecyclobutane at 250—300° was demonstrated by the labelling shown for reaction (9); allene

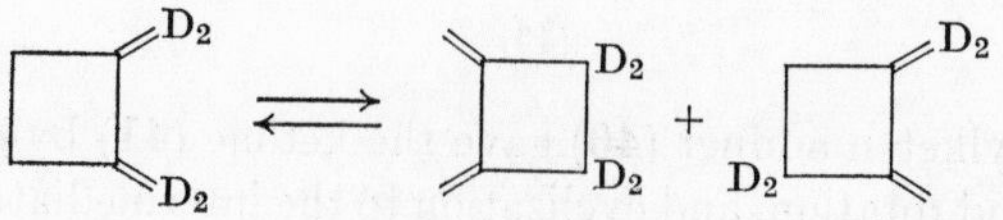

...(9)

was not formed and does not dimerize at these temperatures; the tetramethylene diradical is a possible intermediate.[100] Cyclization of a vinylallene to a cyclobutene was demonstrated by the thermal equilibration of 1-methyl-3-methylenecyclobutene and 2-methylpenta-1,3,4-triene.[101] The thermal

[95] R. J. Crawford and D. M. Cameron, *Can. J. Chem.*, **45**, 691 (1967).
[96] L. Skattebøl, *Tetrahedron*, **23**, 1107 (1967).
[97] R. Fields, R. N. Haszeldine, and D. Peter, *Chem. Commun.*, **1967**, 1081.
[98] K. A. W. Parry and P. J. Robinson, *Chem. Commun.*, **1967**, 1083.
[99] M. A. Battiste, B. Halton, and R. H. Grubbs, *Chem. Commun.*, **1967**, 907.
[100] J. J. Gajewski and C. N. Shih, *J. Am. Chem. Soc.*, **89**, 4532 (1967); W. von E. Doering and W. R. Dolbier, *ibid.*, p. 4534.
[101] E. Gil-Av and J. Herling, *Tetrahedron Letters*, **1967**, 1.

isomerization of cyclobutenes to butadienes[102] and of *cis*-bicyclo[6.2.0]dec-9-enes have also been investigated.[103] The cyclobutenones (**38**; R = R′ = Cl; R = Me, R′ = Cl; R = Cl, R′ = Me) open stereoselectively on thermolysis and pyrolysis in methanol to give the isomers shown (reactions 10), though the reason for this selectivity is not clear.[104] The azabicycloheptenes (**39**; R = R′ = Et; R = H, R′ = Bu^n) rearrange at 500° to alkylbenzenes, presumably via the dihydroazepines with loss of methylamine; again the ring opening must be by disrotation.[105] Thermal isomerization (140—190°) of the

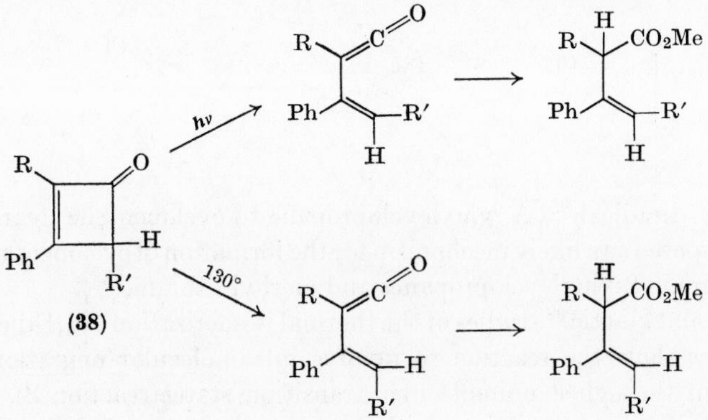

(**38**)

. . . (10)

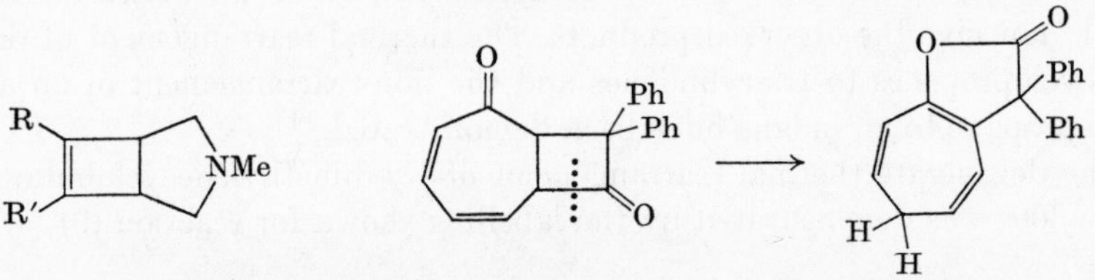

(**39**) (**40**) (**41**)

tropone–diphenylketen adduct (**40**) gave the ketone (**41**) by cleavage of the bond shown, bond rotation, and cyclization to the intermediate lactone which suffers a 1,5-hydrogen shift; the first step is probably rate-determining.[106]

The thermal isomerization of 2,3-dicyanoquadricyclone[107a] and of methylenequadricyclanes,[107b] and the catalysed synthesis[108a] and Cope rearrangement[108b] of *cis*-1,2-divinylcyclobutane, have also been studied.

102 H. M. Frey, D. C. Montague, and I. D. R. Stevens, *Trans. Faraday Soc.*, **63**, 372 (1967); H. M. Frey, B. M. Pope, and R. F. Skinner, *ibid.*, p. 1166.
103 P. Radlick and W. Fenical, *Tetrahedron Letters*, **1967**, 4901.
104 J. E. Baldwin and M. C. McDaniel, *J. Am. Chem. Soc.*, **89**, 1537 (1967).
105 R. F. Childs and A. W. Johnson, *J. Chem. Soc., C*, **1967**, 874.
106 A. S. Kende, *Tetrahedron Letters*, **1967**, 2661.
107a J. R. Edman, *J. Org. Chem.*, **32**, 2920 (1967).
107b H. Prinzbach and J. Rivier, *Tetrahedron Letters*, **1967**, 3713.
108 H. Heimbach and W. Brenner, *Angew. Chem. Intern. Ed. Engl.*, **6**, (*a*) 800, (*b*) 800 (1967).

Intramolecular Hydrogen Migrations

The hydroxypleiadenone (**42**) rearranged with alkali-metal *tert*-butoxides in DMSO to the isomer (**43**). The primary deuterium-isotope effect, metal-cation effects, the Arrhenius parameters, and the retention of deuterium when the compound deuterated on the hydroxyl-bearing carbon rearranges, are consistent with rate-determining transannular hydride transfer, via transition state (**44**).[109] The driving force for this rearrangement appears to be that the chlorination prevents maximum conjugation of the carbonyl group with the naphthalene ring, for the methyl analogue rearranges similarly.[109]

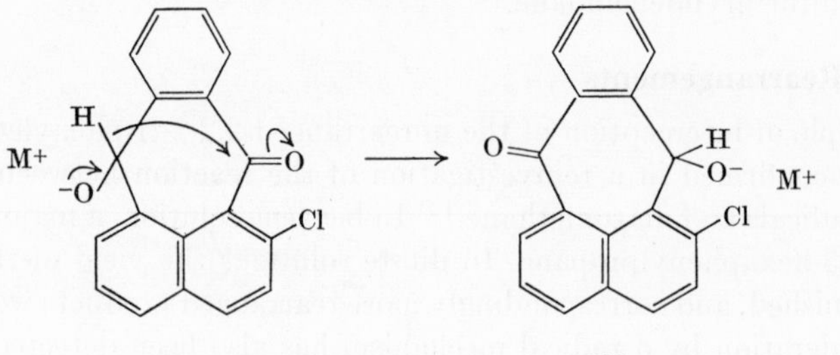

(**42**) (**43**)

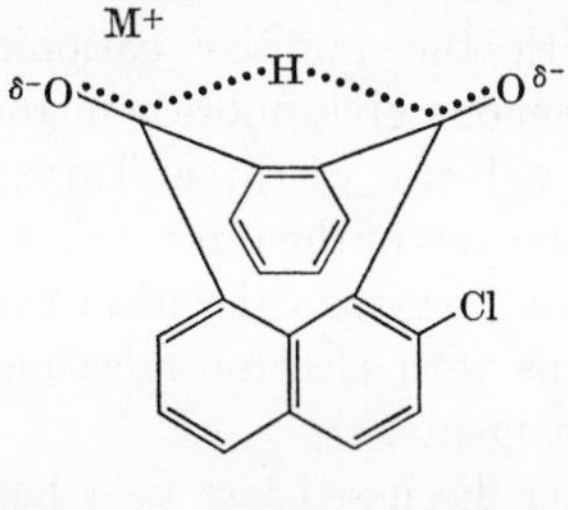

(**44**)

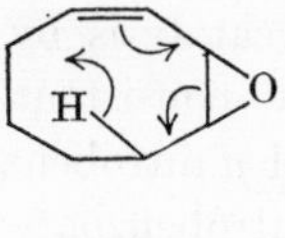

(**45**)

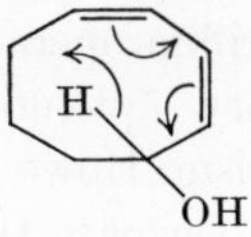

(**46**)

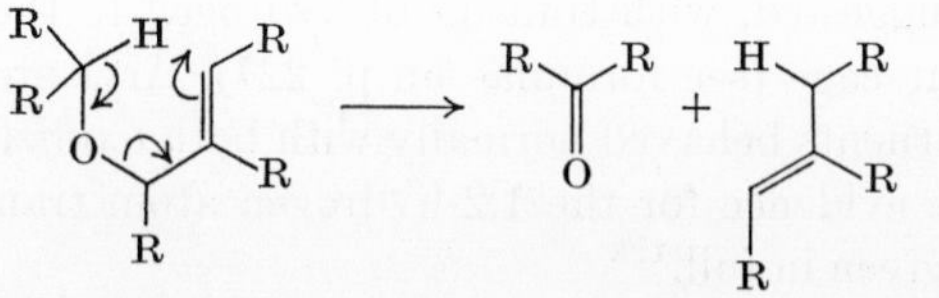

. . . (11)

A 1,5-homodienyl shift has been demonstrated to occur in epoxycycloalkenes, where the customary cyclopropane ring is replaced by an epoxide ring, again stressing the relative unimportance of the nuclei concerned; thus, 3,4-epoxycyclo-octene (**45**) gave, *inter alia*, *cis,cis*-3-oxa-1,4-cyclononadiene.[110] When heated, cyclo-octa-2,4- and -3,5-dienol are interconverted by 1,5-hydrogen shifts (e.g., **46**) and are irreversibly converted into cyclo-oct-3-enone through its enol.[111] Valence-bond tautomerization and 1,5-hydrogen shifts have also been investigated with cyclo-octatrienes,[112] methylcyclo-

[109] P. T. Lansbury and F. D. Saeva, *J. Am. Chem. Soc.*, **89**, 1890 (1967).
[110] J. K. Crandall and R. J. Watkins, *Tetrahedron Letters*, **1967**, 1717.
[111] J. K. Crandall and L.-H. Chang, *J. Org. Chem.*, **32**, 532 (1967).
[112] D. A. Bak and K. Conrow, *J. Org. Chem.*, **31**, 3958 (1966).

heptatrienes,[113] triphenylbenzocycloheptatrienes,[114] and *p*-(dimethylaminophenyl)cycloheptatrienes.[115] The thermal isomerizations of *cis*-hexa-1,3-diene[116] and *cis*-[1,1-$^2H_2$]penta-1,3-diene[117] are both 1,5-hydrogen shifts involving cyclic six-atom hydrogen-bridged transition states.

On pyrolysis at about 500°, allyl ethers decompose through a similar transition state to give a carbonyl compound and a propene, with allylic inversion (eqn. 11). In accord with this, the reaction rate is insensitive to structural variation.[118] Thermal sigmatropic rearrangement of hydrogen, methyl, and ethyl have been observed in macrocyclic pyrrolic nickel complexes.[119] Base-catalysed rearrangement of 3-furfuryl-1-furfurylideneindene gave 1,3-difurfurylideneindane.[120]

Radical Rearrangements

An example of interception of the unrearranged 2,2,2-triphenylethyl radical has been confirmed in a reinvestigation of the reaction between triphenylmethyl radicals and diazomethane.[121] In benzene solution, a major product is 1,1,1,3,3,3-hexaphenylpropane. In dilute solutions the yield of this product was diminished, and correspondingly more rearranged products were isolated.

Aryl migration by a radical mechanism has also been detected in certain Baeyer–Villiger oxidations of unsymmetrically substituted benzophenones.[122] With trifluoroperacetic acid and catalysis by trifluoroacetic acid, relative migratory aptitudes are entirely consistent with the familiar cationic mechanism. However, oxidation of *p*-nitrobenzophenone gives more *p*-nitrophenyl benzoate than phenyl *p*-nitrobenzoate by a factor of three. This is clearly inconsistent with a cationic mechanism; homolysis of the intermediate was suggested, with transfer of hydrogen to the acetoxy-radical within the solvent cage (see formulae on p. 227). Aryl groups with electron-releasing substituents behaved normally with both oxidizing agents.

The evidence for the 1,2-hydrogen atom transfer discussed last year has been given in full.[123]

Several new results relate to the chemistry of cyclopropylcarbinyl radicals. Hydrogen-abstraction from dicyclopropylcarbinol gives radical (**47**)[124] which

[113] K. W. Egger, *J. Am. Chem. Soc.*, **89**, 3688 (1967).
[114] W. Tochtermann, G. Schnabel, and A. Mannschreck, *Z. Naturforsch.*, **21**b, 897 (1966).
[115] A. P. ter Borg, H. Kloosterziel, and Y. L. Westphal, *Rec. Trav. Chim.*, **86**, 474 (1967).
[116] H. M. Frey and B. M. Pope, *J. Chem. Soc.*, *A*, **1966**, 1701.
[117] W. R. Roth and J. König, *Ann. Chem.*, **699**, 24 (1966).
[118] R. C. Cookson and S. R. Wallis, *J. Chem. Soc.*, *B*, **1966**, 1245.
[119] R. Grigg, A. W. Johnson, K. Richardson, and K. W. Shelton, *Chem. Commun.*, **1967**, 1192.
[120] R. Ahlberg and G. Bergson, *Arkiv Kemi*, **27**, 59 (1967).
[121] D. B. Denney and N. F. Newman, *J. Am. Chem. Soc.*, **89**. 4692 (1967).
[122] J. C. Robertson and A. Swelim, *Tetrahedron Letters*, **1967**, 2871.
[123] T. Nagai, K. Nishitomi, and N. Tokura, *Bull. Chem. Soc. Japan*, **40**, 1183 (1967); see *Organic Reaction Mechanisms*, **1966**, 232.
[124] D. C. Neckers and A. P. Schaap, *J. Org. Chem.*, **32**, 22 (1967).

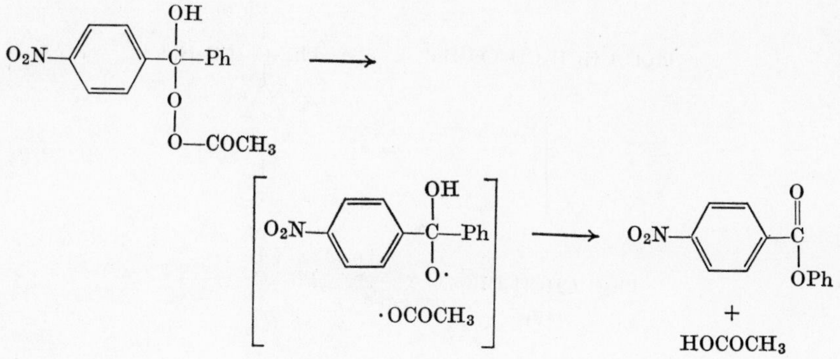

undergoes similar rearrangements to those of the monocyclopropyl analogues reported previously.[125] Thiol addition to vinylcyclopropane gives a cyclopropylcarbinyl radical (**48**) in which rearrangement competes with chain transfer.[126]

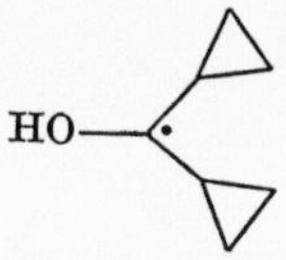

(**47**)

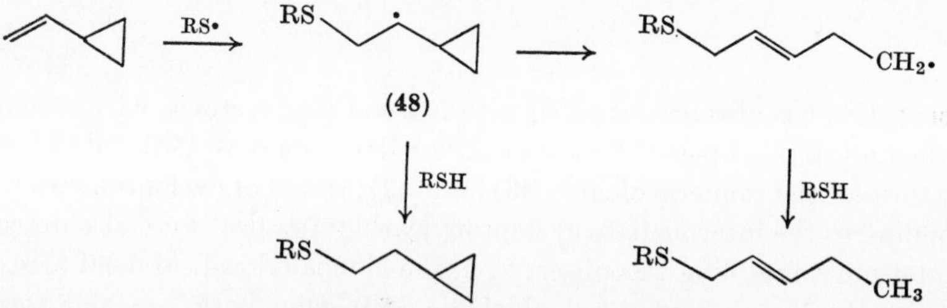

Substituted *tert*-butyl peracetates have been employed as sources of the radicals (**49**) and (**50**) (see p. 228).[127] Both give rearranged and unrearranged products (**51** and **52**) on decomposition in a variety of hydrogen-donor solvents. The precursor of (**49**) gives the same proportions of (**51**) and (**52**) with triethyltin hydride irrespective of the hydride concentration. This points to a rapid equilibration between (**49**) and (**50**); this point is substantiated by the observation that the deuterated perester (**53**) gives (**51**) which is equally deuterated at positions 3 and 4. The vinyl migration implied in this last

[125] See *Organic Reaction Mechanisms*, **1965**, 187; **1966**, 229.

[126] I. S. Lishanskii, A. M. Guliev, A. G. Zak, O. S. Fomina, and A. S. Khachaturov, *Dokl. Akad. Nauk SSSR*, **170**, 1084 (1966); *Chem. Abs.*, **66**, 28131e (1967).

[127] T. A. Halgren, M. E. H. Howden, M. E. Medof, and J. D. Roberts, *J. Am. Chem. Soc.*, **89**, 3051 (1967).

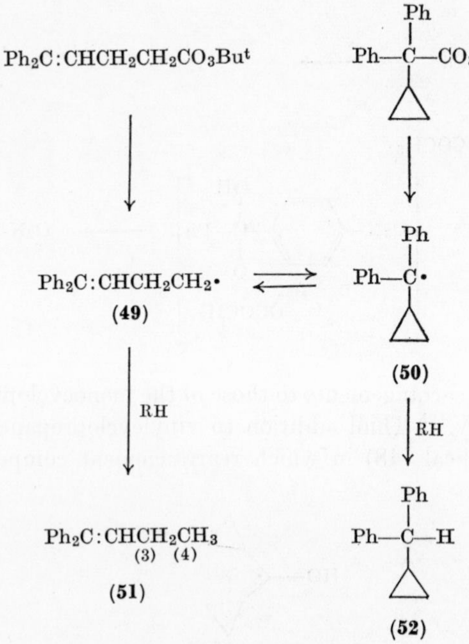

$Ph_2C{:}CHCH_2CD_2CO_3Bu^t$

(53)

observation has also been studied with less complex systems, with aldehydes as the radical precursers.[128] Thus decarbonylation of both (**54**) and (**55**) gives mixtures of the isomeric olefins (**56**) and (**57**); traces of cyclopropanes corresponding to the intermediate cyclopropylcarbinyl radical were also detected. Deuterium scrambling was observed in the homoallyl radical itself (**58**), and the specifically *cis*-deuterated aldehyde (**59**) gave both *cis*- and *trans*-1-deuteriobut-1-ene. The latter observation was interpreted to mean that the cyclopropylcarbinyl radical (**60**) is sufficiently long-lived for rotation about the carbon–carbon bond to occur. Furthermore, at high aldehyde concentrations, there is a slight increase in the proportion of *cis*-deuterated product, suggesting interception of the homoallyl radical before complete equilibration can occur.

New kinetic data obtained for the decomposition of a variety of precursors of cyclopropyl-substituted carbinyl radicals that undergo only partial (or even negligible) opening of the three-membered ring show rate enhancements that strongly implicate a resonance-stabilization of the radical by the cyclo-

[128] L. K. Montgomery, J. W. Matt, and J. R. Webster, *J. Am. Chem. Soc.*, **89**, 923 (1967); L. K. Montgomery and J. W. Matt, *ibid.*, pp. 934, 3050.

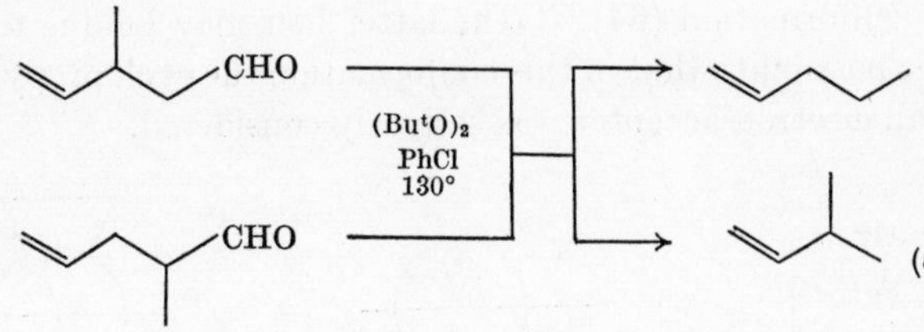

(54) (55) (56) (57)

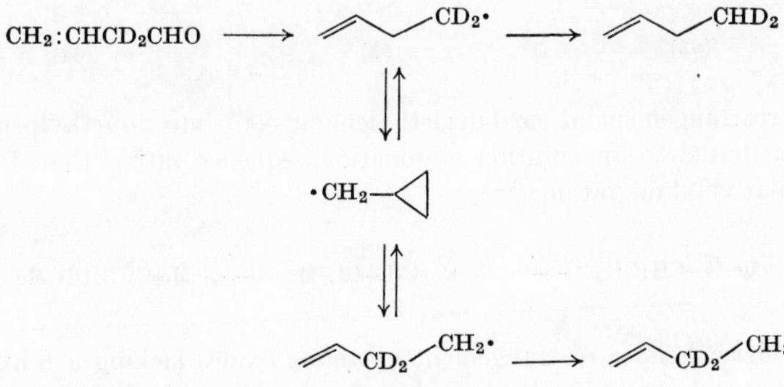

(58)

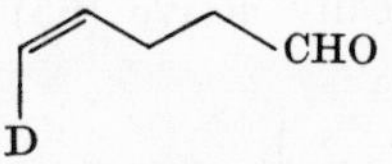

(59)

(60)

propyl substituent.[129] Thus the azo-compound (**61**) undergoes homolysis some 240 times more rapidly than does azobisisobutyronitrile under comparable conditions, yet in this case almost 80% of the radical source can be

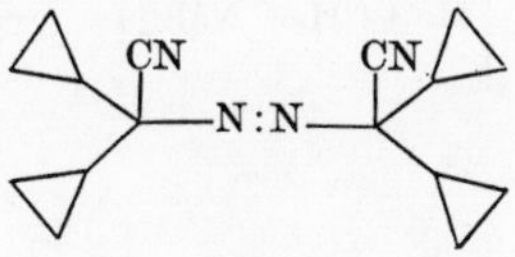

(**61**)

accounted for in the form of products which retain the cyclopropyl substituents intact.

Some interesting ESR data have appeared regarding the preferred conformations of cyclopropyl-substituted radical anions. Semidione (**62**)[130] and the radical cation of 9-cyclopropylanthracene[131] prefer the bisected conformation (**63**). However, the corresponding cyclopropylanthracene radical

[129] J. C. Martin, J. E. Schultz, and J. W. Timberlake, *Tetrahedron Letters*, **1967**, 4629.
[130] G. A. Russell and H. Malkus, *J. Am. Chem. Soc.*, **89**, 160 (1967).
[131] N. L. Bauld, R. Gordon, and J. Zoeller, *J. Am. Chem. Soc.*, **89**, 3948 (1967).

anion prefers conformation (**64**).[131] The latter fact may be due to torsional effects, but the possibility that in this conformation the cyclopropyl unit may conjugate as an electron-acceptor was seriously considered.

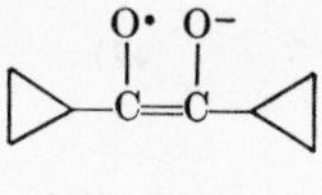

(62)

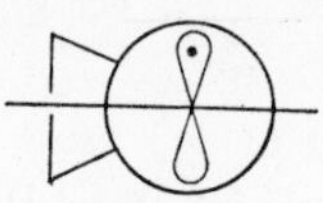

(63)

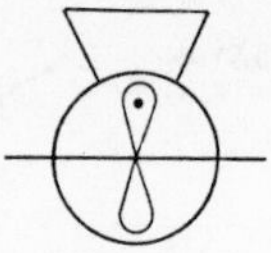

(64)

The rearrangement of *tert*-butylethylene at 500° into 2-methylpent-2-ene was attributed to an addition–elimination sequence rather than to intramolecular vinyl migration:[132]

$$\mathrm{Me_3C{-}CH{:}CH_2 \xrightarrow{Me\cdot} Me_3C{-}\dot{C}H{-}CH_2Me \xrightarrow{-Me\cdot} Me_2C{:}CHCH_2Me}$$

The Meisenheimer rearrangement of amine oxides lacking a β-hydrogen atom is considered to involve homolysis to give nitroxide radicals, and these relatively stable species have now been detected by ESR spectroscopy in the decomposition of (**65**).[133] When optically active (**65**) was employed, no

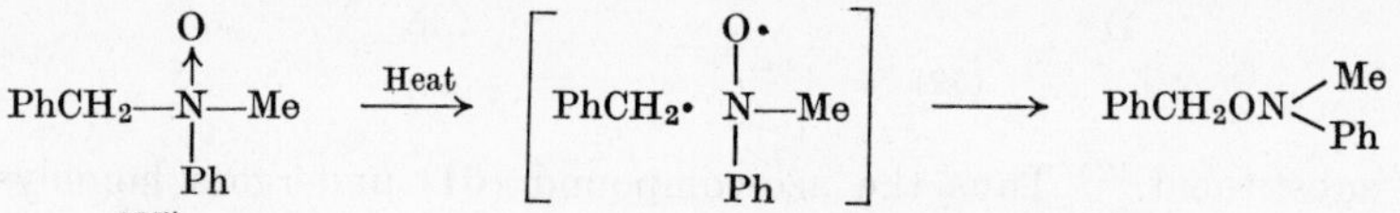

(65)

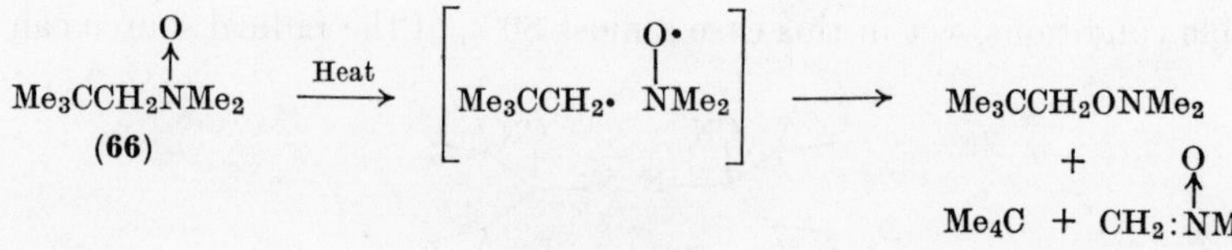

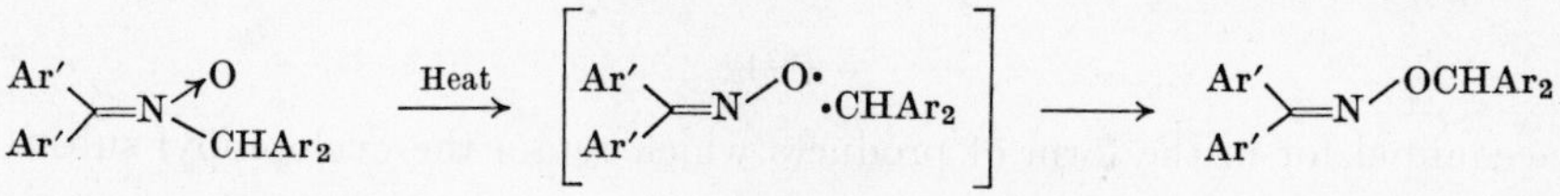

(67) **(68)**

racemization of recovered starting material was detected, indicating that the initial step is irreversible. A similar rearrangement of the non-benzylic *N*-oxide (**66**) occurs also under relatively mild conditions (E_a *ca.* 32 kcal

[132] H. Ladenheim and W. Bartok, *J. Am. Chem. Soc.*, **89**, 1786 (1967).
[133] U. Schöllkopf, U. Ludwig, M. Patsch, and W. Franken, *Ann. Chem.*, **703**, 77 (1967).

mole^{-1}), confirming the dominant role of nitroxide stability in the initial decomposition.[134]

A similar radical mechanism has been proposed for the rearrangement of nitrone (**67**). This accords with the nature of by-products and the direct observation (ESR) of radical intermediates.[135] The radicals were not identified, but strong candidates would seem to be nitroxides formed by radical addition to (**67**) (see p. 150), rather than the much less stable iminoxyl intermediate (**68**).

The rearrangement of *N*-chloroacetanilide under homolytic conditions has been further examined. The homolytic component of the reaction proved difficult to isolate, as hydrogen chloride which is liberated in side reactions catalysed the ionic rearrangement. Furthermore, peroxide catalysts were converted into hypochlorites which also cause ionic chlorination.[136]

New examples of 1,2-halogen migrations have been noted,[137] and an old controversy on the interpretation of the stereospecificity of iodine exchange in *trans*-di-iodoethylene has received a critical re-examination;[138] conclusions had been drawn regarding the geometry of the intermediate radical, but these seem to be of little generality because of the effects of α-substitution and of bridging by iodine.

There was no evidence for bromine participation in the rate, or products, of decomposition of *tert*-butyl ω-bromoperoxyhexanoate,[139] though the possibility of iodine bridging cannot be ruled out in the interesting peroxide-induced eliminations of iodine from 1,3-di-iodopropane and 1,5-di-iodopentane to give cyclopropane and cyclopentane, respectively.[140]

Other instances of bridging in radical intermediates, and examples of radical rearrangement, will be found in the section on radical additions (p. 148).

Heterocyclic Rearrangements

The Woodward–Hoffmann predictions that the thermal isomerization of cyclopropyl anion to allyl anion is conrotatory and that the photochemical isomerization is disrotatory has been elegantly confirmed by Huisgen and his co-workers who proved the stereospecificity of the isomerization of aziridine to azomethine ylide (Scheme 3). Thermolysis (conrotatory) of the *cis*-ester and photolysis (disrotatory) of the *trans*-ester gave the *trans*-1,3-dipole, trapped

134 J. I. Brauman and W. A. Sanderson, *Tetrahedron*, **23**, 37 (1967).
135 E. J. Grubbs, J. A. Villarreal, J. D. McCullough, and J. S. Vincent, *J. Am. Chem. Soc.*, **89**, 2234 (1967).
136 J. Coulson, G. H. Williams, and K. M. Johnston, *J. Chem. Soc., B*, **1967**, 174.
137 P. S. Juneja and E. M. Hodnett, *J. Am. Chem. Soc.*, **89**, 5685 (1967); B. G. Yasnitskii, *Zh. Org. Khim.*, **3**, 789, 800 (1967); *Chem. Abs.*, **67**, 43141x, 43142y (1967).
138 R. M. Noyes, D. E. Applequist, S. W. Benson, D. M. Golden, and P. S. Skell, *J. Chem. Phys.*, **146**, 1221 (1967).
139 W. S. Trahanovsky and M. P. Doyle, *J. Org. Chem.*, **32**, 146 (1967).
140 L. Kaplan, *J. Am. Chem. Soc.*, **89**, 1753, 4566 (1967).

with dimethyl acetylenedicarboxylate, and vice versa.[141] Deamination of *cis*- and *trans*-2,3-dimethylaziridine with difluoramine was stereospecific, giving *cis*- and *trans*-but-2-ene, respectively; however, the significance of the

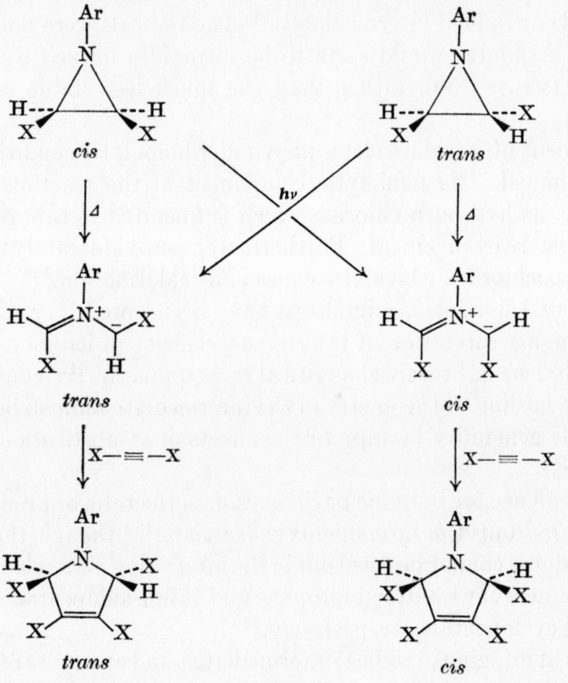

[X = CO_2Me]

Scheme 3

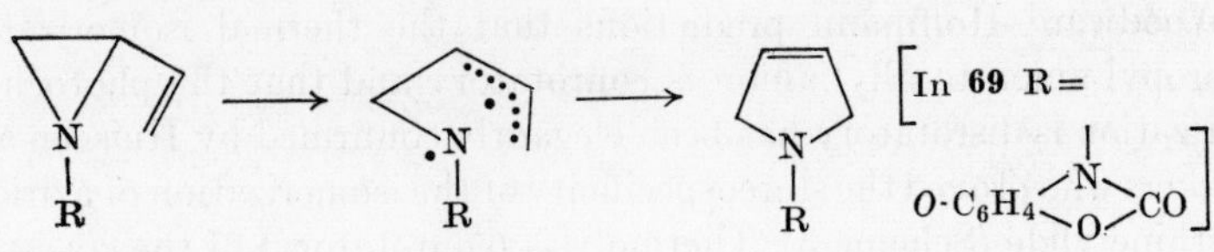

(69)

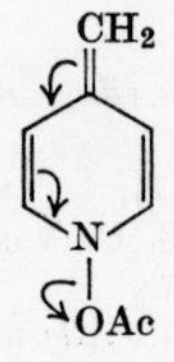

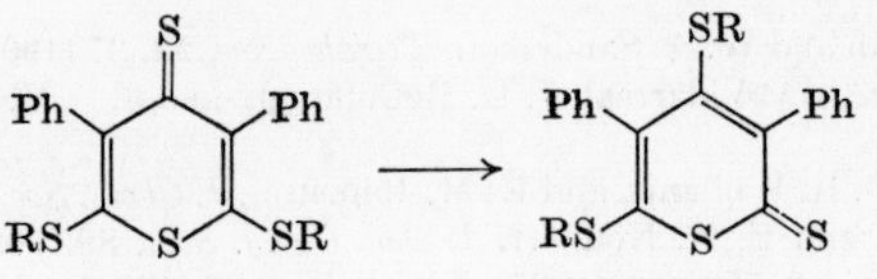

(70) (71) (72)

[141] R. Huisgen, W. Scheer, and H. Huber, *J. Am. Chem. Soc.*, **89**, 1753 (1967).

Woodward–Hoffmann rules for this is not yet clear (see also p. 123).[142] A thermal rearrangement of vinylaziridine to 3-pyrroline analogous to that of vinylcyclopropane to cyclopentene has been reported; substantially lower temperatures were required for the new reaction, probably because of the lower activation energy for formation of the allylic hydrazino-diradical (**69**).[143] The isomerization of 1-aroylaziridines into oxazolines has been reported.[144]

In the reaction of 4-picoline *N*-oxide with acetic anhydride the generation of picolyl cations rather than radicals has been clearly demonstrated by using as solvents anisole, benzonitrile, and a mixture of the two. In anisole a mixture of the three picolylanisoles consistent with electrophilic substitution was formed, and in benzonitrile attack on nitrogen and not in the ring (known to occur readily with radicals) was observed. In a mixture, substitution occurred in the anisole ring only, though benzonitrile is more susceptible to radical attack. Although the pyridyl cation could be formed in a side reaction it seems likely to be an intermediate in the rearrangement of *N*-oxide derived from (**70**).[145] The reactions of 4-substituted pyridine *N*-oxides with *p*-nitrobenzene-sulphenyl and -sulphinyl chloride[146] and of 2-picoline *N*-oxide with acetic anhydride have[147] also been investigated.

Thermal rearrangement of the 4-thiothiapyrones (**71**; R = various alkyl) to the 2-thio-isomers (**72**) was shown to be intermolecular, contrary to previous report, and probably to involve nucleophilic dealkylation by the 4-thio-sulphur.[148] Treatment of the dihydropyridine (**73**) with potassium hydrogen sulphide gave a compound (**74**) by ring expansion of the conjugate base to an azepine, attack of this by the hydrogen sulphide ion, followed by an internal Michael addition.[149] Some reactions of 3,5-diacetyl-4-chloromethyl-1,4-dihydro-2,6-lutidine (**73**; Ac for CO_2Me) have been described.[150] When heated in benzene the azepine (**75**) undergoes the interesting and extensive rearrangement to the fulvene (**76**) almost quantitatively; the mechanism shown was proposed.[151]

Several rearrangements of oxygen and nitrogen heterocycles of the type (**77**) → (**78**) have been collated.[152] Pyrolysis of 3,5-diaryl-1,2,4-oxadiazoles gave a cyanide and an isocyanate by the proposed scheme (reactions 12); thus

[142] J. P. Freeman and W. H. Graham, *J. Am. Chem. Soc.*, **89**, 1761 (1967).
[143] R. S. Atkinson and C. W. Rees, *Chem. Commun.*, **1967**, 1232.
[144] H. W. Heine and M. S. Kaplan, *J. Org. Chem.*, **32**, 3069 (1967).
[145] T. Cohen and G. L. Deets, *J. Am. Chem. Soc.*, **89**, 3939 (1967).
[146] S. Oae and K. Ikura, *Bull. Chem. Soc. Japan*, **40**, 1420 (1967).
[147] B. K. Varma and A. B. Lal, *J. Indian Chem. Soc.*, **43**, 613 (1966).
[148] H. J. Teague and W. P. Tucker, *J. Org. Chem.*, **32**, 3140 (1967).
[149] J. Ashby and U. Eisner, *J. Chem. Soc., C*, **1967**, 1706.
[150] R. C. Allgrove and U. Eisner, *Tetrahedron Letters*, **1967**, 499.
[151] R. F. Childs, R. Grigg, and A. W. Johnson, *J. Chem. Soc., C*, **1967**, 201.
[152] A. J. Boulton, A. R. Katritzky, and A. M. Hamid, *J. Chem. Soc., C*, **1967**, 2005.

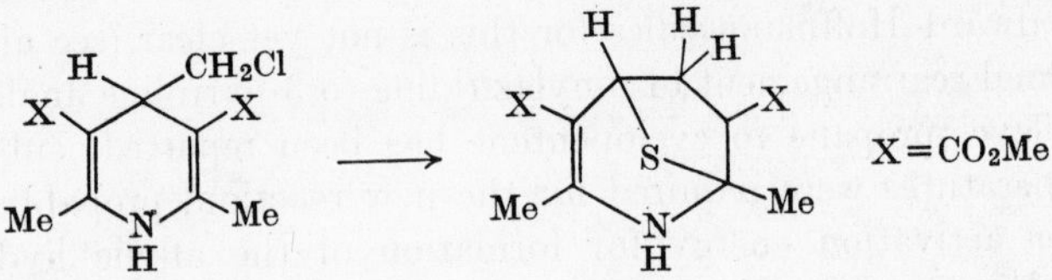

(73) (74)

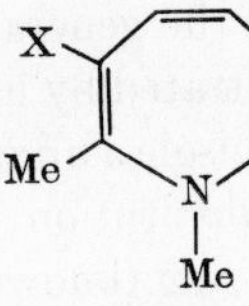

(75)

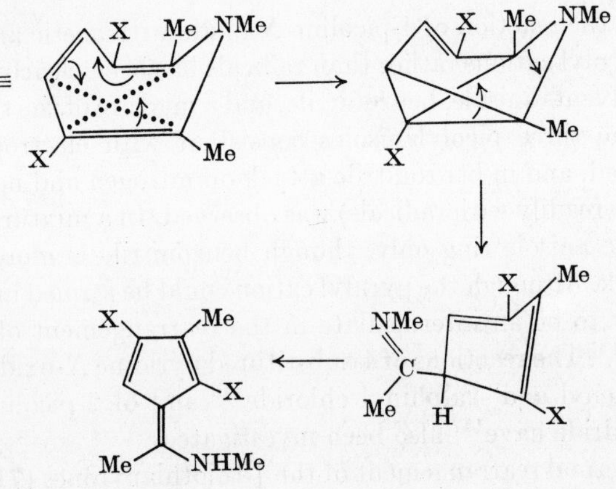

(76)

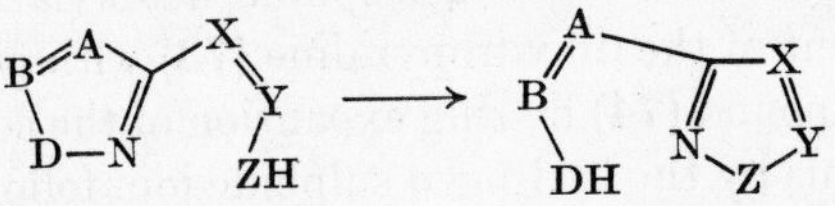

(77) (78)

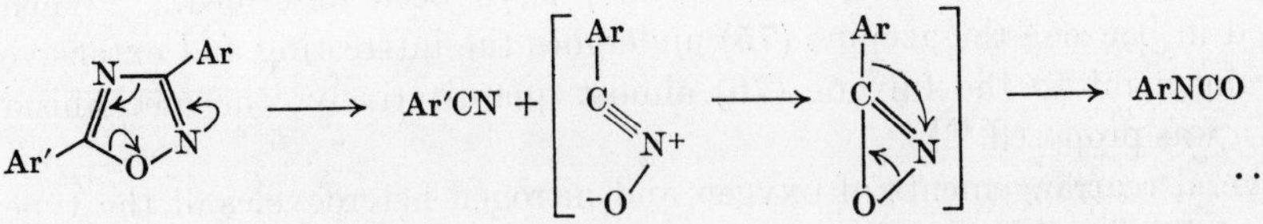

. . . (12)

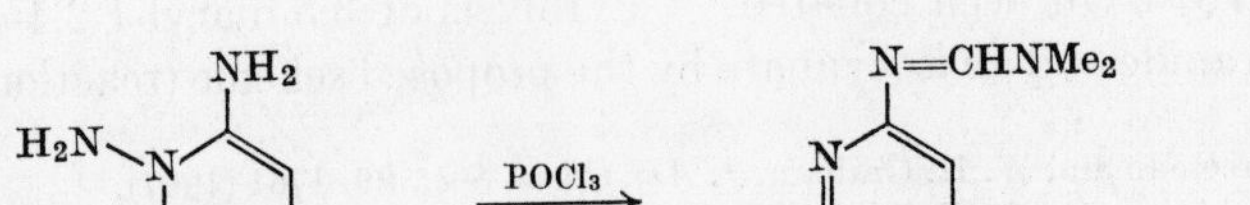

(79) (80)

the initial fragmentation is not to the aroylnitrene which would also have given the isocyanate.[153]

Further studies of the Dimroth and related isomerizations have been

[153] C. Ainsworth, *J. Heterocyclic Chem.*, **3**, 470 (1966).

reported.[154] The novel rearrangement of (**79**) to (**80**) by phosphorus oxychloride and dimethylformamide appears to involve the participation of the doubly bonded sulphur as a neighbouring group.[155] The very rapid rearrangement of the dihydrodiazepinone (**81**) to the aminopyridine (**82**) by sodium hydroxide was of the first order in hydroxide and was not subject to general base-catalysis. The favoured mechanism was nucleophilic attack at $C_{(7)}$ to give the carbinolamine in the rate-determining step, followed by fragmentation of the $C_{(3)}$ carbanion to the acyclic intermediate which rapidly cyclizes.[156]

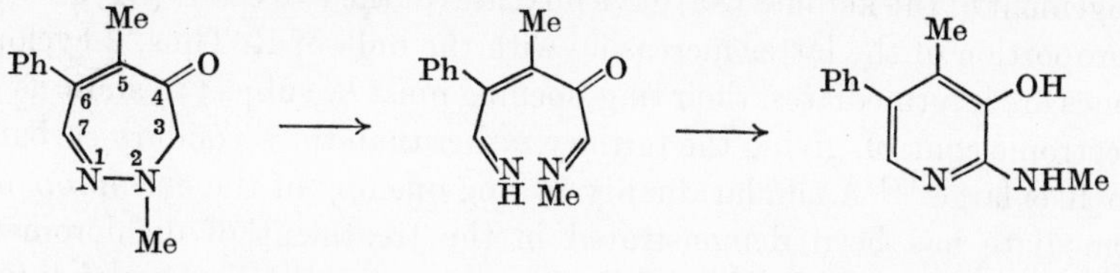

(**81**) (**82**)

Many other interesting rearrangements related to this have been described and rationalized.[157] The suggestion that the acid-catalysed rearrangement of *O*-aryloximes to benzofurans occurs by a Fischer indole-type mechanism has been supported by the isolation of *o*-hydroxyaryl ketimine salts which were proposed intermediates.[158] Details have been published of Gassman and Fox's work on alkyl migration to electron-deficient nitrogen.[159]

Other heterocyclic rearrangements which have been reported include: the tautomeric rearrangement of 4-alkylthiol-4-en-2-ones,[160] conversion of 4-formyl-2,3-dimethyl-1-phenyl-5-pyrazolone into 4-acetyl-2-methyl-1-phenyl-5-pyrazolone,[161] the base-catalysed exchange of benzoyl and acetyl groups in 2-acetylimino-3-phenacylthiazolidine,[162] further tautomerism of substituted benzofuroxans,[163] the Fischer indole synthesis,[164] the formation of *N*-phenylpyridinium chloride from 1,7-diphenyl-1,7-diazahepta-1,3,5-triene,[165] general Lewis-acid catalysis of 5-nitro-2-pyridone $O \rightarrow N$-glycosyl

[154] D. J. Brown and M. N. Paddon-Row, *J. Chem. Soc., C*, **1967**, 903, 1856, 1928; D. J. Brown and B. T. England, *ibid.*, p. 1922; J. C. Parham, J. Fissekis, and G. B. Brown, *J. Org. Chem.*, **32**, 1151 (1967).

[155] E. C. Taylor and R. W. Morrison, *J. Org. Chem.*, **32**, 2379 (1967).

[156] J. A. Moore, H. Kwart, G. Wheeler, and H. Bruner, *J. Org. Chem.*, **32**, 1342 (1967).

[157] J. A. Moore, R. L. Wineholt, F. J. Marascia, R. W. Medeiros, and F. J. Creegan, *J. Org. Chem.*, **32**, 1353 (1967); J. M. Eby and J. A. Moore, *ibid.*, p. 1346.

[158] A. Mooradian and P. E. Dupont, *Tetrahedron Letters*, **1967**, 2867.

[159] P. G. Gassman and B. L. Fox, *J. Am. Chem. Soc.*, **89**, 338 (1967).

[160] A.-B. Hörnfeldt, *Acta Chem. Scand.*, **21**, 673 (1967).

[161] M.-P. Sinh and N. P. Buu-Hoï, *Bull. Soc. Chim. France*, **1967**, 802.

[162] G. R. Bedford, P. Doyle, M. C. Southern, and R. W. Turner, *Chem. Commun.*, **1967**, 155.

[163] A. J. Boulton, A. R. Katritzky, M. J. Sewell, and B. Wallis, *J. Chem. Soc., B*, **1967**, 914.

[164] R. J. Owellen, J. A. Fitzgerald, B. M. Fitzgerald, D. A. Welsh, D. M. Walker, and P. L. Southwick, *Tetrahedron Letters*, **1967**, 1741.

[165] E. N. Marvell, G. Caple, and I. Shahidi, *Tetrahedron Letters*, **1967**, 277.

rearrangement,[166] conversion of 3-amino-2-bromoquinoline by potassamide into 3-cyanoindole and 2-isocyanobenzyl cyanide,[167] of 4,6-disubstituted pyrimidines into 3,5-disubstituted isoxazoles,[168] ring contractions of pyrimidines with hydrazine,[169] and a new Cope-type rearrangement with tetrahydroharmans.[170]

Other Rearrangements

Several papers have appeared on the Favorskii and related reactions. Rearrangement of the ketones (**83**) gave mixtures of the two esters (**84**) and (**85**), the proportion of the latter increasing with the bulk of R. Thus, if cyclopropanones are intermediates, their ring-opening must be subject to steric as well as electronic control, giving the tertiary rather than the secondary carbanion when R is large.[171] A similar duality in ring opening of the cyclopropanone intermediate has been demonstrated in the treatment of dichloromethyl ketones by sodium methoxide.[172] Rearrangement of 2-bromocyclobutanone to cyclopropanecarboxylic acid has been studied further; the semibenzilic acid mechanism for the sodium carbonate-catalysed reaction was supported, and the occurrence of silver nitrate-catalysis was explained by a "push-pull"

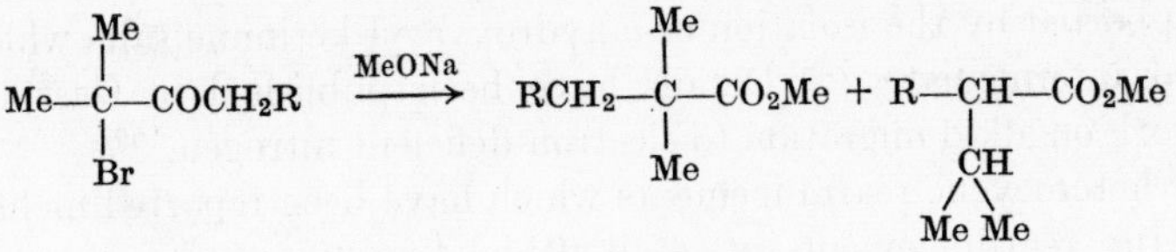

(**83**) (**84**) (**85**)

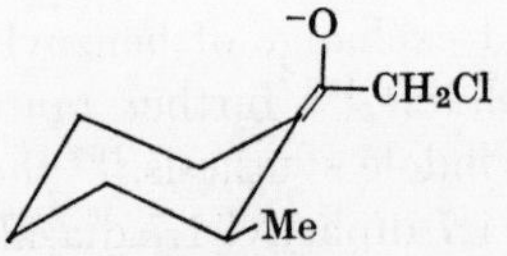

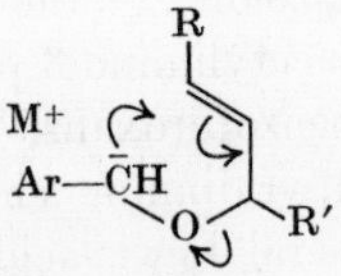

(**86**) (**87**)

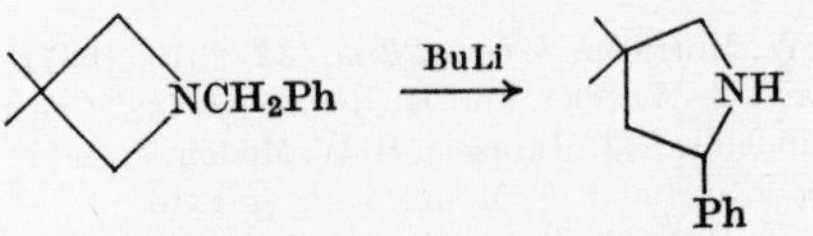

(**88**) (**89**)

166 D. Thacker and T. L. V. Ulbricht, *Chem. Commun.*, **1967**, 122.
167 H. J. den Hertog and D. J. Buurman, *Tetrahedron Letters*, **1967**, 3657.
168 T. Kato, H. Yamanaka, and N. Yasuda, *J. Org. Chem.*, **32**, 3593 (1967).
169 H. C. van der Plas and H. Jongejan, *Tetrahedron Letters*, **1967**, 4385.
170 E. Winterfeldt and W. Franzischka, *Chem. Ber.*, **100**, 3801 (1967).
171 C. Rappe and L. Knutsson, *Acta Chem. Scand.*, **21**, 2205 (1967).
172 N. Schamp and W. Coppens, *Tetrahedron Letters*, **1967**, 2697.

mechanism.[173] In the Favorskii rearrangement of 2-chloro-, 2-bromo-, and 2-iodo-cholestan-3-one with sodium ethoxide the order of reactivity was I > Br > Cl, a primary deuterium isotope effect was observed, and interruption experiments on the deuterated halides showed that the hydrogen was rapidly lost from $C_{(2)}$ before the rearrangement.[174] The *cis-* and *trans*-isomers of chloromethyl 2-methylcyclohexyl ketone gave the same mixture of ester products with sodium methoxide; a planar intermediate, presumably the enolate anion (**86**), must therefore precede the rearrangement step.[175] The reactions of pulegone epoxides with sodium ethoxide have, however, been rationalized in terms of stereospecific Favorskii rearrangements,[176] and the rearrangement of methyl 6,6-dichloro-3,3-dimethyl-5-oxohexanoate has been described.[177] Rearrangements of α,α-dichlorobenzyl sulphones and dichloromethyl sulphones by means of base proceed by initial formation of the chloro-episulphone, and then further elimination of hydrogen chloride, to give thiiren dioxides which decompose to acetylenes, vinyl chlorides, and α,β-unsaturated sulphonic acids.[178] The Ramberg–Bäcklund rearrangement has been reviewed.[179]

In agreement with a report last year[180] the Wittig rearrangement of 9-allyloxyfluorenes proceeds with complete inversion of the allyl group; the same applies to allyl quinaldinyl ethers, with either sodamide in liquid ammonia or butyl-lithium in tetrahydrofuran, thus supporting the S_N2' mechanism (**87**).[181] Treatment of 1-benzyl-3,3-dimethylazetidine (**88**) with ethereal butyl-lithium gave 4,4-dimethyl-2-phenylpyrrolidine (**89**) and an acyclic product, thus demonstrating a Wittig rearrangement of an amine for the first time. A possible mechanism is ring opening of the benzyl carbanion followed by cyclization on to carbon.[182]

Sommelet–Hauser-type rearrangements have been studied further this year. Rearrangement of 2,2,3-triphenylpropyl-lithium to 1,1,3-triphenylpropyl-lithium in the presence of [α-^{14}C]benzyl-lithium results in incorporation of ^{14}C into the product, suggesting an elimination–addition mechanism (eqn. 13). The relative reactivities of organolithium reagents towards 1,1-diphenylethylene are PhLi < EtLi, *n*-BuLi ⋘ iso-PrLi < $PhCH_2Li$ and,

173 C. Rappe and L. Knutsson, *Acta Chem. Scand.*, **21**, 163 (1967).
174 H. R. Nace and B. A. Olsen, *J. Org. Chem.*, **32**, 3438 (1967).
175 H. O. House and F. A. Richey, *J. Org. Chem.*, **32**, 2151 (1967).
176 G. W. K. Cavill and C. D. Hall, *Tetrahedron*, **23**, 1119 (1967); W. Reusch and P. Mattison, *ibid.*, p. 1953.
177 N. Schamp, *Bull. Soc. Chim. Belges*, **76**, 400 (1967).
178 L. A. Paquette and L. S. Wittenbrook, *J. Am. Chem. Soc.*, **89**, 4483 (1967); L. A. Paquette, L. S. Wittenbrook, and V. V. Kane, *ibid.*, p. 4487.
179 F. G. Bordwell, in "Organosulfur Chemistry", ed. M. J. Janssen, Interscience, 1967 Chapter 16.
180 *Organic Reaction Mechanisms*, **1966**, 241.
181 Y. Makisumi and S. Notzumoto, *Tetrahedron Letters*, **1966**, 6393.
182 A. G. Anderson and M. T. Wills, *J. Org. Chem.*, **32**, 3241 (1967).

consistently with this sequence, isopropyl-lithium was the only organolithium besides benzyl-lithium to trap the intermediate of reaction (13).[183] In contrast, the rearrangement of 2,2,2-triphenylethyl-lithium to 1,1,2-triphenylethyl-lithium appears to be intramolecular since in the presence of [^{14}C]-phenyl-lithium or -benzyl-lithium there is no incorporation of radioactive phenyl or benzyl into the products.[183] A mechanism involving phenyl-lithium and 1,1-diphenylethylene in a cage seems unlikely since phenyl-lithium adds only slowly to 1,1-diphenylethylene under the reaction conditions. A mechanism involving a transition state (**90**) was preferred to one in which the diphenylethylene was bound as a ligand to the lithium in a lithium cation–phenyl anion ion-pair, since it was found that the migratory aptitude of the 4-biphenylyl group was 24.5 times that of the 3-biphenylyl group. It was thought that whereas a *p*-phenyl group would not stabilize a phenyl anion it would stabilize an intermediate or transition state of type (**90**).[184]

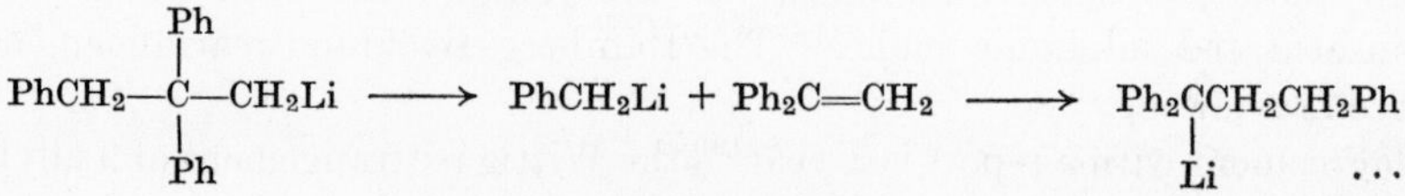

...(13)

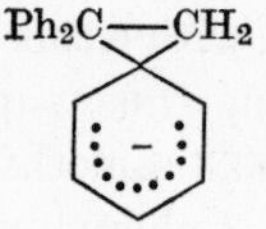

(90)

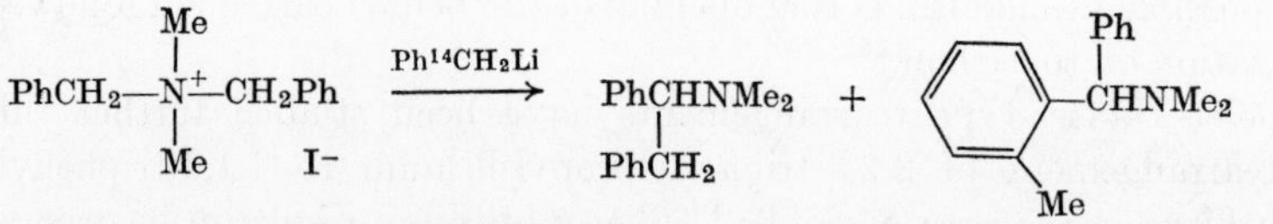

...(14)

The Stevens and Sommelet rearrangements of dibenzyldimethylammonium halide in the presence of [α-^{14}C]benzyl-lithium also yield products into which there has been no incorporation of ^{14}C and hence proceed by intramolecular mechanisms (reaction 14).[183]

When *N*,*N*,*N*-trimethyl-α-phenylneopentylammonium chloride (**91**) was treated with various bases under different conditions the *para*-product (**92**) was formed in low yield (0—9%); (**92**) is not formed from the *ortho*-rearrangement or other products. This first example of a *para*-Sommelet–Hauser rearrangement provides support for an ion-pair, rather than an S_N2' mechanism, although the ion-pair could not be trapped. The absence of *para*-products has previously been used as an argument against the ion-pair

[183] E. Grovenstein and G. Wentworth, *J. Am. Chem. Soc.*, **89**, 1852 (1967).
[184] E. Grovenstein and G. Wentworth, *J. Am. Chem. Soc.*, **89**, 2348 (1967).

mechanism.[185] However, the cyclic mechanism (**93**) was favoured over an elimination–addition process for the rearrangement of the dianions of *ortho*-substituted benzyltrimethylammonium salts with potassamide.[186] The

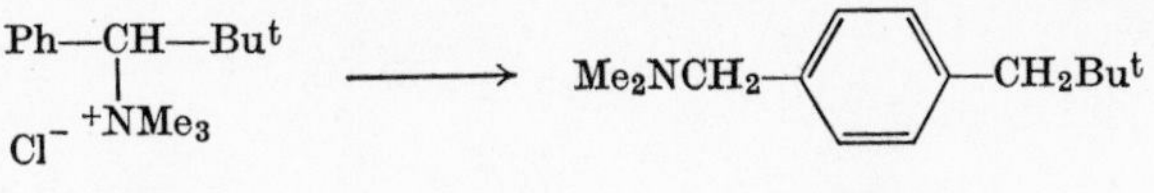

(**91**) (**92**)

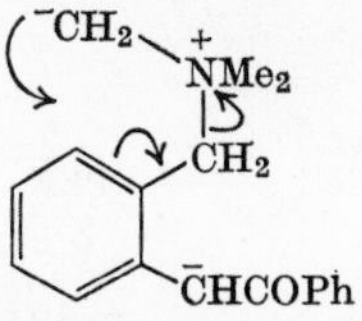

(**93**)

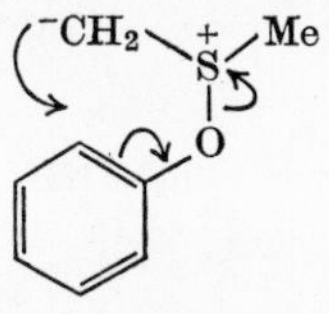

(**94**)

principal products of the acid-catalysed reaction of phenols with dicyclohexylcarbodi-imide in DMSO are the *o*-[(methylthio)methyl]phenols probably formed by the analogous mechanism (**94**).[187] Rearrangements of other benzylammonium salts[188] and the conversion of *N*-benzyl-*N*-methylaniline oxide into *O*-benzylhydroxylamine[189] have been investigated.

The sodium salt of perfluoroacetanilide and diethyl acetylenedicarboxylate gave the normal adduct together with the rearranged product (**95**); the intramolecular mechanism on p. 240 was proposed.[190] Following up their demonstration of a 1,2-anionic rearrangement of organosilylhydrazines in which trimethylsilyl groups migrate from one nitrogen to the other, West and his co-workers[191] have now shown that the dianions (but *not* the monoanions) of 1,1-diphenyl- and 1-methyl-1-phenyl-hydrazine undergo a similar though very much slower rearrangement (reactions 15). The tendency towards the rearrangement shown in equation (16) increases with increasing electron-release from *para*-substituents in Ar; ^{14}C-labelling showed that it is the aryl group *trans* to chlorine which migrates.[192]

A benzilic acid rearrangement promoted by metal chelation under extremely mild conditions represents the first *in vitro* example of a molecular rearrangement involving group migration effected by a metal template. 2,2′-Pyridil (**96**)

[185] S. H. Pine, *Tetrahedron Letters*, **1967**, 3393.
[186] K. P. Klein, R. L. Vaulx, and C. R. Hauser, *J. Am. Chem. Soc.*, **88**, 5802 (1966).
[187] M. G. Burdon and J. G. Moffatt, *J. Am. Chem. Soc.*, **88**, 5855 (1966).
[188] A. R. Lepley and A. G. Guimanini, *J. Org. Chem.*, **32**, 1706 (1967); A. R. Lepley and T. A. Brodof, *ibid.*, p. 3234; K. P. Klein and C. R. Hauser, *ibid.*, **31**, 4276 (1966).
[189] U. Schöllkopf, U. Ludwig, M. Patsch, and W. Franken, *Ann. Chem.*, **703**, 77 (1967).
[190] G. M. Brooke and R. J. D. Rutherford, *Chem. Commun.*, **1967**, 318.
[191] R. West, H. F. Stewart, and G. R. Husk, *J. Am. Chem. Soc.*, **89**, 5050 (1967).
[192] G. Köbrich, H. Trapp, and I. Hornke, *Chem. Ber.*, **100**, 961 (1967).

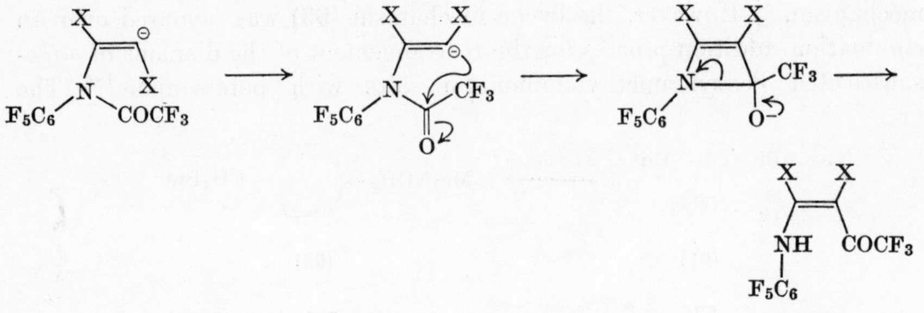

(95)

$$Ph_2N\cdot NH_2 \xrightarrow{MeLi} \underset{Li^+}{Ph_2N\cdot \bar{N}H} \xrightarrow{MeLi} \underset{2Li^+}{Ph_2N\cdot N^{=}} \longrightarrow \underset{2Li^+}{Ph\bar{N}\cdot \bar{N}Ph} \xrightarrow{H_2O} PhNH\cdot NHPh \quad \ldots(15)$$

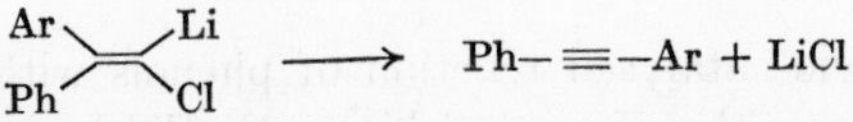

...(16)

with hydrated nickel or cobalt acetate in methanol gave the 2,2′-pyridilic acid complexes (**97**) in almost quantitative yield. Chelation is a prerequisite for reaction under these conditions since benzil did not react, and hydrated sodium acetate is not a catalyst.[193] The optically active α-hydroxy-imine (**98**) rearranged to amino-ketones with methyl and with phenyl migration; the latter proceeds with almost total retention of optical activity in agreement with an intramolecular cyclic mechanism.[194] Migration of a benzyl group has been observed for the first time in a tertiary ketol rearrangement (eqn. 17).[195] Base-induced rearrangements of α-epoxy-ketones[196] and a benzilic acid-type rearrangement of a nitro-olefin have been reported.[197]

In the cyclization of *o*-phenylbenzophenone oxime in polyphosphoric acid to give 6-phenylphenanthridine the possibility of formation of the iminium ion (**99**) and direct ring closure of this, without a Beckmann rearrangement, was considered but disproved by labelling of the *o*-phenyl with methyl. Thus the accepted rearrangement of the oxime polyphosphate with concerted aryl migration is supported.[198] Beckmann rearrangement of 2-methoxycyclohexanone oxime with thionyl chloride gave 1-chloro-5-cyanopentyl methyl ether, as shown by its interception with various nucleophiles.[199] The rearrange-

193 D. St. C. Black, *Chem. Commun.*, **1967**, 311.
194 S. Yamada, H. Mizuno, and S. Terashima, *Chem. Commun.*, **1967**, 1058.
195 D. Y. Curtin and A. C. Henry, *J. Org. Chem.*, **32**, 847 (1967).
196 G. R. Treves, H. Stange, and R. A. Olofson, *J. Am. Chem. Soc.*, **89**, 6257 (1967).
197 T. J. De Boer and J. C. Van Velzen, *Rec. Trav. Chim.*, **86**, 107 (1967).
198 P. T. Lansbury and R. P. Spitz, *J. Org. Chem.*, **32**, 2623 (1967).
199 M. Ohno and I. Terasawa, *J. Am. Chem. Soc.*, **88**, 5683 (1966).

ments of arylglyoxylonitrile oxime tosylate,[200] and of organic fluoroamines[201] and fluoroimines,[202] have been reported and the sulphuric acid-catalysed Beckmann rearrangements of alicyclic ketoximes have been reviewed.[203]

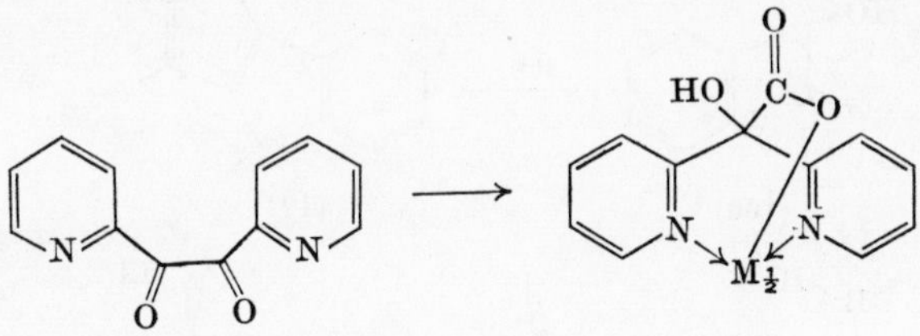

(96) **(97)**

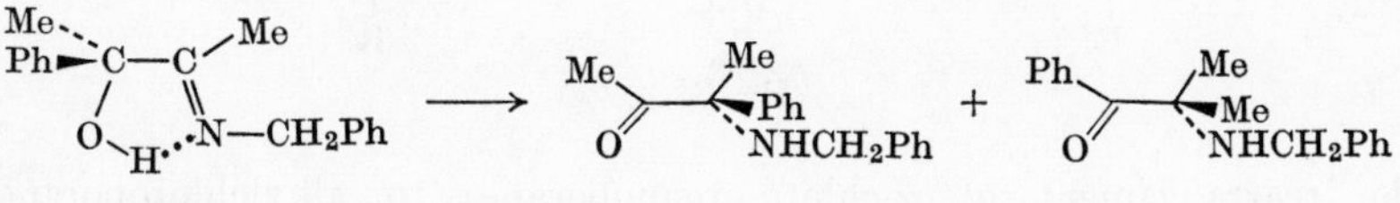

(98)

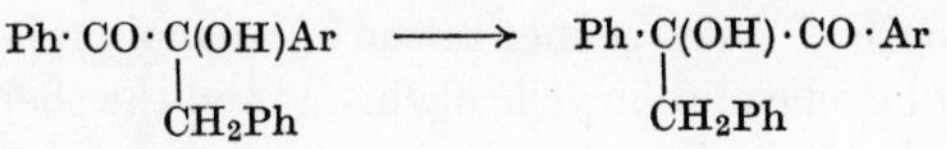

. . . (17)

Ar = p-ClC_6H_4

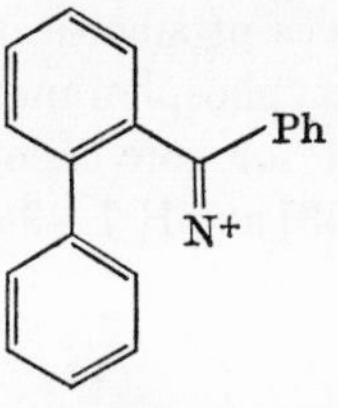

(99)

The interesting acid-catalysed rearrangement of the benzobicyclo[2.2.2]-octadienol (**100**) to (**101**) has been reported and rationalized.[204] 2,2,4,4-Tetramethylpentan-3-one isomerized in sulphuric acid to 2,2,3,3-tetramethylpentan-4-one; a pinacolone-type rearrangement involving the migration of a methyl and then a *tert*-butyl group was proposed.[205] The rearrangement of pentan-3-one to pentan-2-one by phosphoric acid,[206] and of tetrahydro-2,2-dimethyldicyclopentadiene to 1,3-dimethyladamantane by aluminium

200 T. E. Stevens, *J. Org. Chem.*, **32**, 670 (1967).
201 T. E. Stevens and W. H. Graham, *J. Am. Chem. Soc.*, **89**, 182 (1967).
202 T. E. Stevens, *Tetrahedron Letters*, **1967**, 3017.
203 M. I. Vinnik and N. G. Zarakhani, *Russian Chem. Rev.*, **36**, 51 (1967).
204 A. C. G. Gray, T. Kakihana, P. M. Collins, and H. Hart, *J. Am. Chem. Soc.*, **89**, 4556 (1967).
205 J.-É. Dubois and P. Bauer, *Bull. Soc. Chim. France*, **1967**, 1156.
206 A. Fry and W. H. Corkern, *J. Am. Chem. Soc.*, **89**, 5894 (1967).

chloride,[207] and migratory aptitudes in the pinacol rearrangement[208] have been studied.

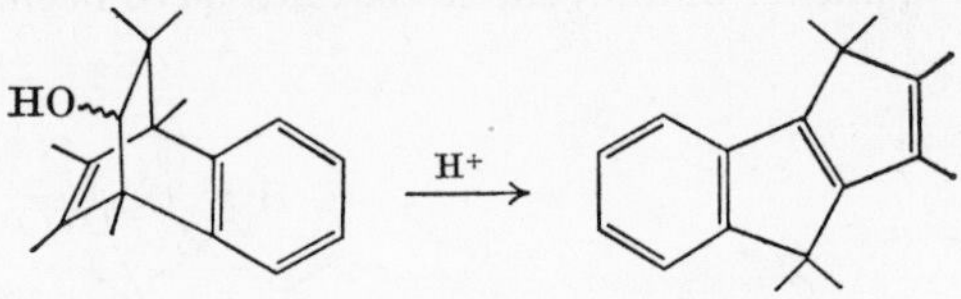

(100) **(101)**

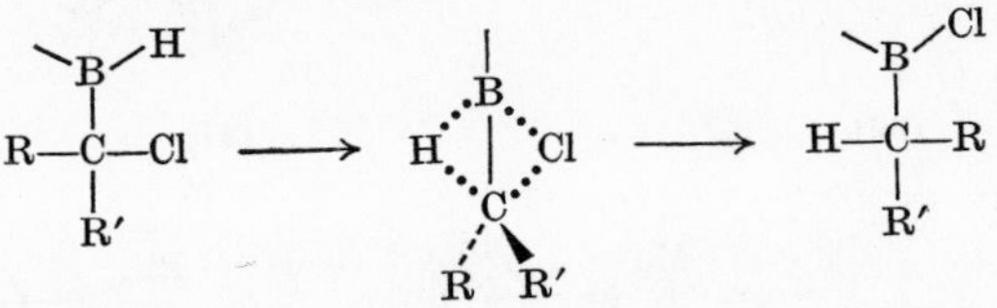

...(18)

The rearrangement of α-chloro-organoboranes to alkylchloroboranes proceeds with complete inversion at carbon; the intramolecular mechanism shown (18) was favoured.[209] The isomerization of organoboranes derived from the hydroboration of cyclic and bicyclic olefins[210] and the thermal isomerization of α-branched trialkylboranes[211] were also studied. Phosphines, $Ph_2P—CR_2CH{=}CH_2$, undergo allylic rearrangement at 210° in the presence, but not in the absence, of traces of silanes or silanols;[212] rearrangements during the solvolyses of pentaoxyphosphoranes have been reported.[213]

The first-order rate constant for conversion of 3-ethoxycyclohex-2-enol (**102**) into cyclohex-2-enone (**103**) at pH 7—9 at 25° was proportional to the

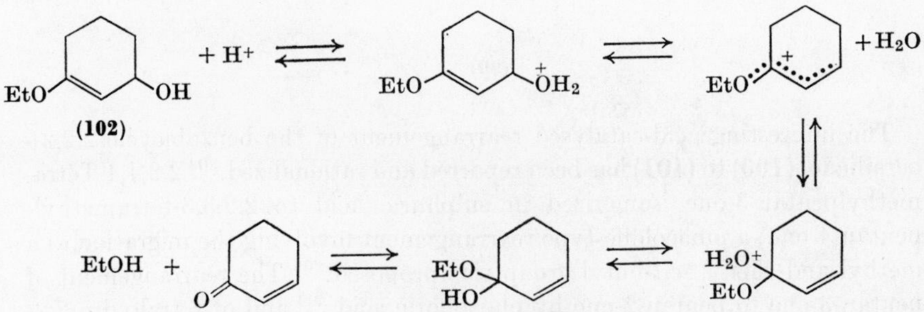

(102) **(103)**

207 D. J. Cash and P. Wilder, *Tetrahedron Letters*, **1966**, 6445.
208 K. Matsumoto, R. Goto, A. Sera, and T. Asano, *Nippon Kagaku Zasshi*, **87**, 1076 (1966); *Chem. Abs.*, **66**, 64844y (1967).
209 D. J. Pasto and J. Hickman, *J. Am. Chem. Soc.*, **89**, 5608 (1967).
210 H. C. Brown and G. Zweifel, *J. Am. Chem. Soc.*, **89**, 561 (1967).
211 F. M. Rossi, P. A. McCusker, and G. F. Hennion, *J. Org. Chem.*, **32**, 450 (1967).
212 M. P. Savage and S. Trippet, *J. Chem. Soc., C*, **1967**, 1998.
213 F. Ramirez, S. B. Bhatia, A. V. Patwardhan, and C. P. Smith, *J. Org. Chem.*, **32**, 3547 (1967).

hydrogen-ion concentration; in ^{18}O-enriched water the ketone produced was enriched in ^{18}O, and the ethanol was not. The allylic rearrangement mechanism shown was therefore proposed.[214] In a further study of the Westphalen rearrangement[215] it has been shown that the products depend markedly on the nature of the 3β-substituent, but the rates depend much more on that of the 6β-substituent.[216]

Other rearrangements which have been studied include: the Amadori rearrangement;[217] the Wolff rearrangement of α-acetoxy-diazoketones;[218] an anomalous Neber rearrangement;[219] the Schmidt reaction;[220] the Hofmann bromoamide rearrangement;[221] pyrolysis of phenylhydrazones;[222] interconversion of *meso*- and racemic 2,3-dimethyl-2,3-diphenylsuccinonitrile;[223] rearrangement of cumyl perbenzoate;[224] isomerization of menaquinones[225] and of perfluoro-α,ω-bisazomethines;[226] and ring expansions of the *cis*-perhydro-8-methylindene system.[227]

Reinvestigation of the rates of thermal *cis*-to-*trans* isomerization of a wide range of substituted azobenzenes, starting with pure *cis*-isomers has led to rejection of earlier results and of their interpretation. The rates are rather insensitive to substituents and even less so to solvent; all *para*-substituents, regardless of their nature, increase the rate. A linear transition state was proposed in which one or both of the nitrogen atoms rehydridize to the *sp*-state, with the π-bond intact and the non-bonded electron pair in a *p*-orbital.[228]

A low energy barrier for rotation about a carbon–carbon double bond flanked by electron-releasing and -withdrawing groups has been demonstrated by NMR spectroscopy.[229] Investigation of geometrical and positional isomerization of olefins included: cyclohexa-1,4- to -1,3-diene,[230] iodine-catalysed isomerization of hept-1-, -2-, and -3-ene,[231] bromine-catalysed isomerization

214 M. Stiles and A. L. Longroy, *J. Org. Chem.*, **32**, 1095 (1967).
215 *Organic Reaction Mechanisms*, **1965**, 189.
216 A. Fischer, M. J. Hardman, M. P. Hartshorn, D. N. Kirk, and A. R. Thawley, *Tetrahedron*, **23**, 159 (1967).
217 F. Micheel, S. Degener, and I. Dijong, *Ann. Chem.*, **701**, 233 (1967).
218 F. W. Bachelor and G. A. Miana, *Tetrahedron Letters*, **1967**, 4733.
219 K. R. Henery-Logan and T. L. Fridinger, *J. Am. Chem. Soc.*, **89**, 5724 (1967).
220 H. W. Moore and H. R. Shelden, *J. Org. Chem.*, **32**, 3603 (1967); J. Mirek, *Zesz. Nauk Uniw. Jagiellon.*, *Pr. Chem.*, No. 10, 61 (1965); *Chem. Abs.*, **66**, 37125h (1967).
221 K. M. Joshi and K. K. Shah, *J. Indian Chem. Soc.*, **43**, 481 (1966).
222 W. D. Crow and R. K. Solly, *Australian J. Chem.*, **19**, 2119 (1966).
223 L. I. Peterson, *J. Am. Chem. Soc.*, **89**, 2677 (1967).
224 N. V. Yablokova, V. A. Yablokov, and O. F. Rachkova, *Reacts. Sposobnostorg. Soedin., Tartu. Gos. Univ.*, **3**, 223 (1966); *Chem. Abs.*, **67**, 2670a (1967).
225 P. Cohen and P. Mamont, *Bull. Soc. Chim. France*, **1967**, 1164.
226 P. H. Odgen and R. A. Mitsch, *J. Am. Chem. Soc.*, **89**, 5007 (1967).
227 G. Di Maio, *Tetrahedron*, **23**, 2291 (1967).
228 E. R. Talaty and J. C. Fargo, *Chem. Commun.*, **1967**, 65.
229 Y. Shvo, E. C. Taylor, and J. Bartulin, *Tetrahedron Letters*, **1967**, 3259.
230 T. Yamaguchi, T. Ono, K. Nagai, C. C. Sin, and T. Shirai, *Chem. Ind. (London)*, **1967**, 759.
231 K. W. Egger, *J. Am. Chem. Soc.*, **89**, 504 (1967).

of terminal olefins,[232] the friedelene–oleanene rearrangement,[233] nucleophile-catalysed isomerizations,[234] isomerization of butene over alumina and silica–alumina catalysts,[235] ethoxide ion-catalysed isomerization of 1,4- to 1,3-diyne,[236] and transition-metal-catalysed isomerizations.[237]

232 H. G. Meunier and P. I. Abell, *Tetrahedron Letters*, **1967**, 3633.
233 R. M. Coates, *Tetrahedron Letters*, **1967**, 4143.
234 Z. Grünbaum, S. Patai, and Z. Rappoport, *J. Chem. Soc., B*, **1966**, 1133.
235 J. W. Hightower and W. K. Hall, *J. Am. Chem. Soc.*, **89**, 778 (1967).
236 I. M. Mathai, H. Taniguchi, and S. I. Miller, *J. Am. Chem. Soc.*, **89**, 115 (1967).
237 B. Fell, P. Krings, and F. Asinger, *Chem. Ber.*, **99**, 3688 (1966); H. A. Tayim and J. C. Bailar, *J. Am. Chem. Soc.*, **89**, 3420 (1967); R. G. Miller, *ibid.*, p. 2785; I. I. Moiseev, S. V. Pestrikov, and L. M. Sverzh, *Izv. Akad. Nauk SSSR, Ser. Khim.*, **1966**, 1866; *Chem. Abs.*, **66**, 64859g (1967); I. I. Moiseev and S. V. Pestrikov, *Dokl. Akad. Nauk SSSR*, **171**, 151 (1966); *Chem. Abs.*, **66**, 37112b (1967).

CHAPTER 9

Radical Reactions

The Gomberg Centenary Symposium Lectures have now been published,[1] and further valuable reviews are collected in Volume II of "Advances in Free Radical Chemistry"[2] and Volume IV of "Progress in Reaction Kinetics".[3] The last of these contains an interesting section by Vassil'ev on the use of chemiluminescence in kinetic studies: references to research papers on chemiluminescence are collected in this volume in Chapter 13 (p. 419). Other reviews deal with hydrogen abstraction,[4a] stable phenoxy radicals,[4b] photo-oximation,[5] reactions under high pressure,[6] metal-ion redox catalysis of radical reactions,[7] the importance of π-complexing and hydrogen bonding in radical reactions,[8] the study of conformation and structure by ESR spectroscopy,[9] the reactions of halogenated methyl radicals,[10a] the decomposition of organo-mercury compounds in solution,[10b] and the kinetics of homogeneous gas-phase reactions.[11] A useful general review of recent developments in free-radical chemistry,[12] and a discussion (in Russian) of the role of amines as inhibitors,[13] have also appeared. Also of general interest are two, rather different, attempts to separate polar and radical-stabilizing effects in modified Hammett treatments of homolytic reactions,[14, 15] and a comparison of liquid-phase and vapour-phase reactions of free radicals.[16a] The last of these emphasizes that even the least polar solvents can play a major role; and therefore, to make

1 *Pure Appl. Chem.*, **15**, Part 1 (1967).
2 G. H. Williams, ed., Logos Press and Academic Press, London and New York, 1967.
3 G. Porter, ed., Pergamon, London and New York, 1967.
4a R. S. Davidson, *Quart. Rev.*, **21**, 249 (1967).
4b E. R. Altwicker, *Chem. Rev.*, **67**, 475 (1967).
5 M. Pape, *Fortschr. Chem. Forsch.*, **7**, 559 (1967).
6 W. J. le Noble, *Progr. Phys. Org. Chem.*, **5**, 207 (1967).
7 J. K. Kochi, *Record Chem. Progr.*, **27**, 207 (1966); *Science*, **155**, 415 (1967).
8 A. L. Buchachenko and O. P. Sukhanova, *Russian Chem. Rev.*, **36**, 192 (1967).
9 D. H. Geske, *Progr. Phys. Org. Chem.*, **4**, 125 (1967).
10a J. M. Tedder and J. C. Walton. *Progr. Reaction Kinetics*, **4**, 37 (1967).
10b K. C. Bass, *Organometallic Rev.*, **1**, 391 (1966).
11 W. Schirmer, *Z. Chem.*, **7**, 249 (1967).
12 W. G. Bentrude, *Ann. Rev. Phys. Chem.*, **18**, 283 (1967).
13 E. V. Vasil'ev and G. F. Bebkikh, *Vestn. Mosk. Univ.*, Ser. 11, **21**, 50 (1966); *Chem. Abs.*, **66**, 85182w (1967).
14 T. Yamamoto and T. Otsu, *Chem. Ind. (London)*, **1967**, 787; T. Yamamoto, *Bull. Chem. Soc. Japan*, **40**, 642 (1967).
15 A. D. Jenkins in ref. 2, p. 139.
16a F. R. Mayo, *J. Am. Chem. Soc.*, **89**, 2654 (1967).

any quantitative assessment of solvent effect, comparison should be with the vapour-phase reaction. Another text dealing with the applications of ESR in chemistry has appeared,[16b] and the contributions to a symposium on ESR spectroscopy have been published,[16c] as well as a useful review.[16d]

Although more than six hundred papers and abstracts have been scanned for this Chapter and the free-radical sections of Chapters 5, 7, and 8, it seems fair to record that during 1967 there have been very few contributions that add significantly to our understanding of fundamental free-radical behaviour.

Production of Radicals

One area of significant interest concerns the geometry of free radicals, and several new results relate to the formation and stereochemical integrity of vinyl radicals.[17] Two groups have found new evidence for partial retention of configuration in reactive substrates. Sargent and Browne[18] generated vinyl radicals from 3-chloro-*cis*(or *trans*)-hex-3-ene by electron transfer from sodium naphthalenide. In ethereal solvents the major component of the resulting hex-3-ene mixture is always the *trans*-isomer. However, the proportion of *cis*-product is significantly greater from the *cis*-starting material. Similar stereoselectivity has been found by Singer and Kong[19] who employed peresters as their radical precursors: in cumene at 110° the proportions of *cis*- and *trans*-β-chlorostyrene clearly reflect the identity of the starting material. The stereoselectivity diminishes as the reaction mixture is diluted with benzene. However, in the limit the two precursors do not give the same mixture of chlorostyrenes. This was attributed to the formation from (**2**) of a carboxylate radical–cumene complex that can undergo direct decarboxylation to the *cis*-olefin. Complex-formation may also occur with the carboxylate radical from (**1**), but the limiting composition of styrenes from this perester seems unexceptional.

The ratio of disproportionation to recombination of unsubstituted vinyl radicals formed in the photolysis of vinyl iodide has been determined as 0.34,[20] and new ESR results for vinyl radicals in a neon matrix at 4°K fail to distinguish between a linear structure and rapidly equilibrating bent radicals such as are implied in Scheme 1.[21]

The formation of cyclohexyl acetate on decomposition of acetyl peroxide

16b P. B. Ayscough, "Electron Spin Resonance in Chemistry", Methuen, London, 1967.
16c *J. Phys. Chem.*, **71**, 1—214 (1967).
16d R. O. C. Norman and B. C. Gilbert, *Adv. Phys. Org. Chem.*, **5**, 53 (1967).
17 See *Organic Reaction Mechanisms*, **1966**, 249.
18 G. D. Sargent and M. W. Browne, *J. Am. Chem. Soc.*, **89**, 2788 (1967).
19 L. A. Singer and N. P. Kong, *J. Am. Chem. Soc.*, **89**, 5251 (1967); see also L. A. Singer and N. P. Kong, *Tetrahedron Letters*, **1967**, 643.
20 P. C. Roberge and J. A. Herman, *Can. J. Chem.*, **45**, 1361 (1967).
21 P. H. Kasai and E. B. Whipple, *J. Am. Chem. Soc.*, **89**, 1033 (1967).

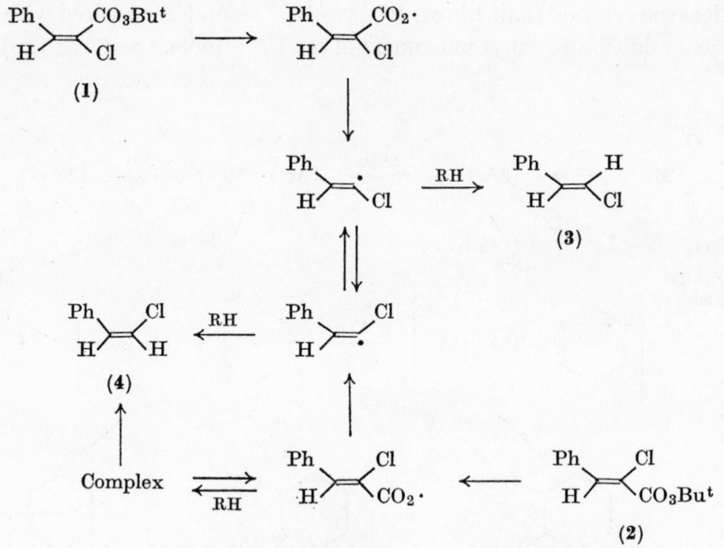

Scheme 1

		Composition of styrene fraction	
		(**3**) (%)	(**4**) (%)
In cumene	from (**1**)	76	24
	from (**2**)	13	87
Dil. soln. of cumene in benzene	from (**1**)	66	34
	from (**2**)	37	63

in cyclohexene has been re-examined.[22] Carbonyl-^{18}O-labelled peroxide was used and, contrary to an earlier report, the label was completely randomized in the cyclohexyl acetate produced. This could not be interpreted in terms of induced decomposition, and radical-scavenging experiments showed that the ester was not a cage product. An acyloxyl radical–solvent complex was again proposed (see above), and indeed this was considered to be the only form in which the acetoxyl radical escaped the solvent cage. Two new cage products (yield unaffected by scavengers) were also detected (**5** and **6**). The overall interpretation of product formation is shown in Scheme 2. Formulation of the acetoxyl radical complex as π rather than σ is consistent with the results of competitive acetate formation from several olefins as correlated with their π-basicity. Formation of a π-complex could be sufficiently rapid to compete

[22] J. C. Martin, J. W. Taylor, and E. H. Drew, *J. Am. Chem. Soc.*, **89**, 129 (1967).

with decarboxylation (half-life *ca.* 10^{-10} to 10^{-9} seconds, compared with rates of radical addition to vinyl monomers of *ca.* 10^5 l. mole^{-1} sec^{-1} or less).

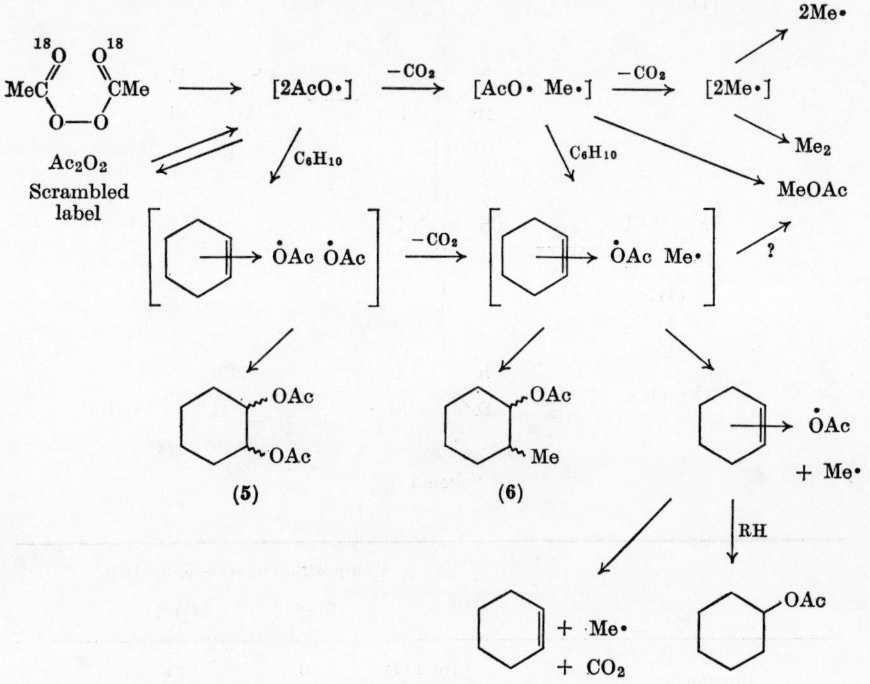

Scheme 2

Cage formation of ethane in Scheme 2 finds an interesting analogy in the formation of methyldiphenylcyclopropane (*ca.* 2%) from the peroxide (**7**) when the latter decomposes in boiling carbon tetrachloride.[23] The non-

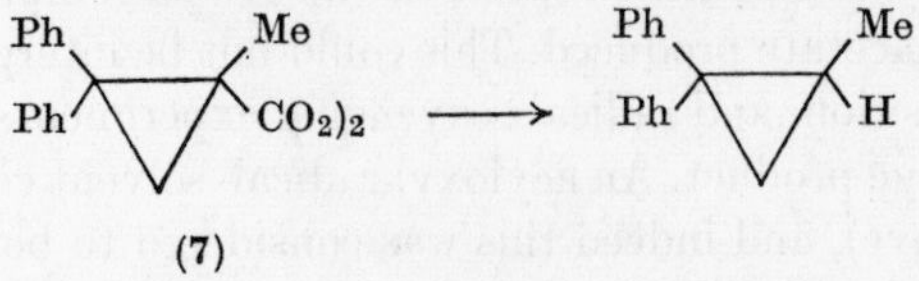

halogenated cyclopropane was considered to be formed by cage disproportionation. It could not be eliminated by radical scavengers, and when optically active peroxide was employed, the cyclopropane retained some 40% of its optical purity. In hydrogen-donor solvents the same product, obtained in much higher yield, was essentially racemic.

[23] H. M. Walborsky and C.-J. Chen, *J. Am. Chem. Soc.*, **89**, 5499 (1967).

The rate of decomposition of *tert*-butyl perdeuteriopivalate (**8**) gives a k_H/k_D value of 1.02 per deuterium atom, agreeing well with calculations based

$$(CD_3)_3CCO \cdot O \cdot O \cdot C(CH_3)_3$$

(**8**)

on a hyperconjugative model.[24] An incidental implication of these calculations is that a trigonal carbon radical suffers less destabilization on distortion from coplanarity than does a corresponding carbonium ion.[25]

Walling and Waits have re-examined the effect of pressure on the decomposition of di-*tert*-butyl peroxide in the light of work on cage recombination of oxygen radicals.[26] The rate falls with increasing pressure, and also in solvents of high viscosity, both of which conditions favour cage recombination.

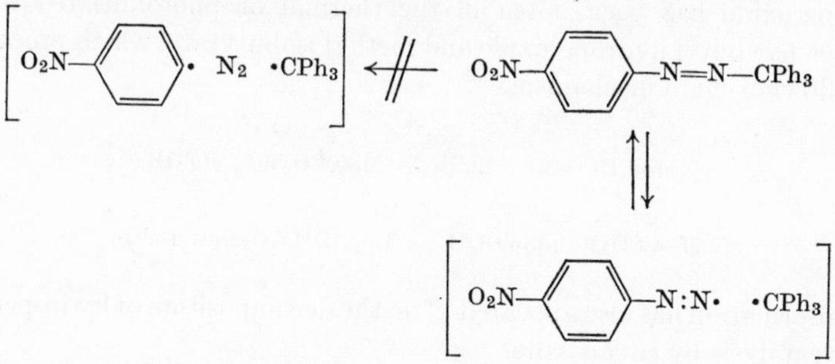

An analysis of viscosity effects on cage reactions has been given by Pryor and Smith[27] who studied the effect of this variable on the rate of decomposition of (*p*-nitrophenylazo)triphenylmethane. A regular rate decrease with increasing solvent viscosity was interpreted in terms of cleavage of single bonds, as it was considered unlikely that cage recombination of two radicals and molecular nitrogen would occur. A related analysis of cage recombination was employed by Neuman and Behar[28] in selecting *tert*-butyl phenylperacetate with which to probe the effect of pressure on the rate of peroxide decomposition. This compound is known to undergo concerted two-bond homolysis, and it was therefore assumed that cage recombination could be neglected: a small retardation was observed at high pressures, giving a volume of activation of *ca.* +12 ml mole^{-1} in both cumene and chlorobenzene.

The decompositions of a series of ω-iodoacyl peroxides, $(I \cdot [CH_2]_n \cdot CO_2)_2$, in various solvents have been examined.[29] A δ-lactone was obtained with $n = 4$,

[24] T. Koenig and R. Wolf, *J. Am. Chem. Soc.*, **89**, 2948 (1967).
[25] See *Organic Reaction Mechanisms*, **1965**, 211, 212.
[26] C. Walling and H. P. Waits, *J. Phys. Chem.*, **71**, 2361 (1967).
[27] W. A. Pryor and K. Smith, *J. Am. Chem. Soc.*, **89**, 1741 (1967).
[28] R. C. Neuman and J. V. Behar, *J. Am. Chem. Soc.*, **89**, 4549 (1967).
[29] R. G. Woolford and R. N. Gedye, *Can. J. Chem.*, **45**, 291 (1967).

whose formation was attributed to radical-induced decomposition of the peroxide by abstraction of iodine followed by intramolecular displacement on oxygen.

In a review of perester decomposition,[30] Rüchardt has summarized neighbouring-group effects in the homolytic decomposition of these compounds. Also of general interest are the Hoffmann molecular-orbital calculations pertinent to peroxide decompositions, which have been reported by a Japanese group.[31]

Numerous other investigations of peroxide,[32] hydroperoxide,[33] and peracid[34] decompositions have been reported, including a number on peroxide homolysis induced by nitrogenous bases.[35a] Nucleophilic attack on peroxide oxygen has also been reviewed.[35b] A brief report of the decomposition of hydroperoxides by nitric oxide has appeared,[36a] and free-radical displacement on a hydroperoxide has been noted in the thermal or photoinitiated reaction between *tert*-butyl hydroperoxide and methyl isobutyrate, which proceeds by the following chain mechanism:[36b]

$$Me_2CHCO_2Me + Bu^tO\cdot \rightarrow Me_2\dot{C}CO_2Me + Bu^tOH$$

$$Me_2\dot{C}CO_2Me + Bu^tOOH \rightarrow Me_2C(OH)CO_2Me + Bu^tO\cdot$$

New information has also appeared[36c] on the decomposition of hydroperoxides under catalysis by cobalt salts.

30 C. Rüchardt, *Fortschr. Chem. Forsch.*, **6**, 251 (1966).
31 T. Yonezawa, H. Kato, and O. Yamamoto, *Bull. Chem. Soc., Japan* **40**, 307 (1967).
32 C. Rüchardt and H. Böck, *Chem. Ber.*, **100**, 654 (1967); G. Greig and J. C. J. Thynne, *Trans. Faraday Soc.*, **63**, 1369, 2196 (1967); C. Leggett and J. C. J. Thynne, *ibid.*, p. 2504; R. Rado, *Chem. Listy*, **61**, 785 (1967); T. Suehiro, H. Tsuruta, and S. Hibino, *Bull. Chem. Soc. Japan*, **40**, 674 (1967).
33 L. Dulog and A. Sanner, *Tetrahedron Letters*, **1966**, 6353; W. J. Maguire and R. C. Pink, *Trans. Faraday Soc.*, **63**, 1097 (1967); S. D. Razumovskii and Y. N. Yur'ev, *Neftekhimiya*, **6**, 737 (1966); *Chem. Abs.*, **66**, 28232 (1967).
34 J. Drimus and C. Matasa, *Bull. Soc. Chim. France*, **1966**, 3994; A. V. Andrianova, E. A. Kuz'mina, V. A. Shushunov, and M. K. Shchennikova, *Tr. Khim. Tekhnol.*, **1965**, 85; *Chem. Abs.*, **66**, 54856q (1967).
35a H.-T. Feng and K.-Y. Ch'iu, *K'o Hsueh Tung Pao*, **17**, 166 (1966); *Chem. Abs.*, **66**, 37161s (1967); N. M. Beileryan, F. O. Karapetyan, and O. A. Chatykyan, *Dokl. Akad. Nauk Arm. SSR*, **43**, 108 (1966); *Chem. Abs.*, **66**, 37179d (1967); S. Kashino, Y. Mugino, and S. Hasegawa, *Bull. Chem. Soc. Japan*, **40**, 2004 (1967); B. M. Sogomonyan, N. M. Beileryan, and O. A. Chaltykyan, *Armyansk. Khim. Zh.*, **19**, 391 (1966); *Chem. Abs.*, **66**, 2017q (1967); K. Maruyama, T. Otsuki, and T. Iwao, *J. Org. Chem.*, **32**, 82 (1967); G. Sosnovsky and E. H. Zaret, *Chem. Ind. (London)*, **1967**, 1297.
35b E. J. Behrman and J. O. Edwards, *Progr. Phys. Org. Chem.*, **4**, 93 (1967).
36a J. R. Shelton and R. F. Kopczewski, *J. Org. Chem.*, **32**, 2908 (1967).
36b D. J. Trecker and R. S. Foote, *Chem. Commun.*, **1967**, 841.
36c M. K. Shchennikova and E. A. Kuz'mina, *Tr. Khim. Khim. Tekhnol.*, **1966**, 29; *Chem. Abs.*, **67**, 72953u (1967); E. A. Kuz'mina, V. A. Shushunov, and M. K. Shchennikova, *Zh. Obshch. Khim.*, **37**, 81 (1967); *Chem. Abs.*, **67**, 11005q (1967).

The Wieland reaction of triphenylmethyl radicals with symmetrically substituted diaroyl peroxides shows a pronounced polar character in the rate-determining step.[37] Hammett σ correlation gives $\rho = 1.45$, which is consistent with charge transfer to the peroxide. The decomposition of diacyl peroxides by organotin hydrides has also been studied;[38] with a trialkyltin hydride, benzoyl peroxide gives the corresponding trialkyltin benzoate. Experiments with carbonyl-^{18}O-labelled peroxide gave carbonyl-labelled product, indicating that induced decomposition by stannyl radicals occurs by attack at peroxide oxygen.

New facts about ester pyrolysis[39] and the reactions of multivalent iodine compounds[40] have been reported.

The pyrolysis of benzophenone azine does not proceed with loss of nitrogen, but by N–N bond fission, and subsequent reactions of $Ph_2C{:}N\cdot$ as indicated.[41] Oxidation of phenylhydrazones with manganese dioxide gives a spectrum of products interpretable in terms of the intermediate (**9**),[42] and ESR spectra

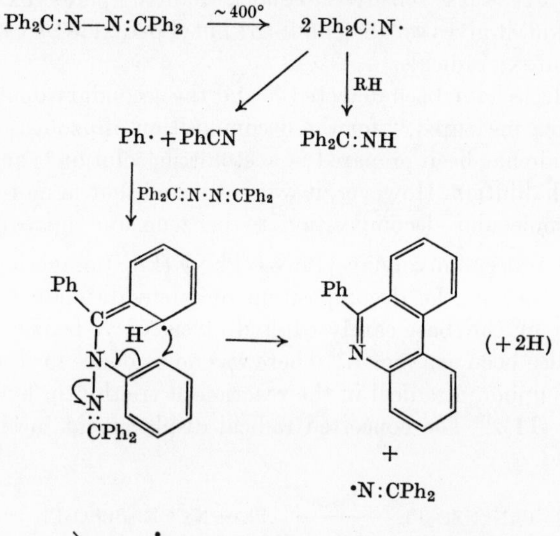

(**9**)

[37] T. Suehiro, A. Kanoya, H. Hara, T. Nakahama, M. Omori, and T. Komori, *Bull. Chem. Soc. Japan*, **40**, 668 (1967).

[38] W. P. Neumann, K. Rübsamen, and R. Sommer, *Chem. Ber.*, **100**, 1063 (1967); W. P. Neuman and K. Rübsamen, *ibid.*, p. 1621.

[39] R. Louw and E. C. Kooyman, *Rec. Trav. Chim.*, **86**, 1041 (1967); see also Chapter 5, ref. 145.

[40] O. A. Ptitsyna, S. I. Orlov, and O. A. Reutov, *Izv. Akad. Nauk SSSR, Ser. Khim.*, **1966**, 1947; *Chem. Abs.*, **66**, 75433q (1967); J. E. Leffler and J. L. Story, *J. Am. Chem. Soc.*, **89**, 2333 (1967).

[41] S. S. Hirsch, *J. Org. Chem.*, **32**, 2433 (1967).

[42] I. Bhatnagar and M. V. George, *J. Org. Chem.*, **32**, 2252 (1967).

observed when hydrazines react with anhydrides have been attributed to radical cations of the type represented by **(10a)** and **(10b)**.[43]

$$R-\overset{O}{\overset{\|}{C}}-O-\overset{O}{\overset{\|}{C}}-R \;+\; R'_2N-N< \;\rightleftharpoons\; >N-\overset{+}{N}R'_2-\overset{O^-}{\overset{|}{\underset{|}{C}}}-R$$
$$\phantom{R-C-O-C-R \;+\; R'_2N-N<}OCOR$$

$$RCO\cdot \;+\; >N-\overset{+\cdot}{N}R'_2$$
(10a)

$$>N-\overset{+}{N}R'_2COR \;+\; RCO_2^-$$

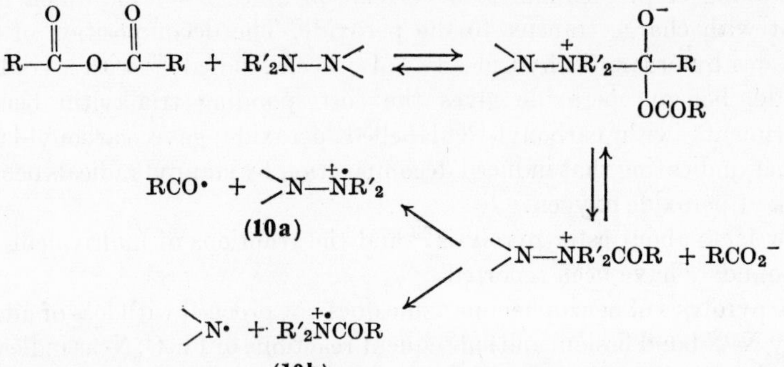

$$>N\cdot \;+\; R'_2\overset{+\cdot}{N}COR$$
(10b)

A convenient low-temperature source of *tert*-butoxy radicals is di-*tert*-butyl hyponitrite.[44] It is less sensitive to shock than the peroxyoxalate but, like that compound, it gives *ca.* 10% of di-*tert*-butyl peroxide by cage recombination of the butoxy radicals.

Solvent effects have been detected,[45] and the secondary deuterium isotope effect has been measured,[46] for the decomposition of azobisisobutyronitrile. Phenyldi-imide has been prepared in acetonitrile solution[47] and is relatively stable at high dilution. However, it was suggested that in more concentrated solutions bimolecular decomposition to benzene and nitrogen involved a radical cage process involving [PhN₂· PhṄNH₂]. Competing radical and anionic processes in the decomposition of phenyldi-imide obtained as an intermediate in the base-catalysed hydrolysis of N-benzoyl-N'-phenyldi-imide have also been examined.[48] There was no evidence for the formation of the benzenesulphonyl radical in the reaction of triethyltin hydride with the azosulphone **(11)**.[49] The concerted radical displacement mechanism shown was suggested.

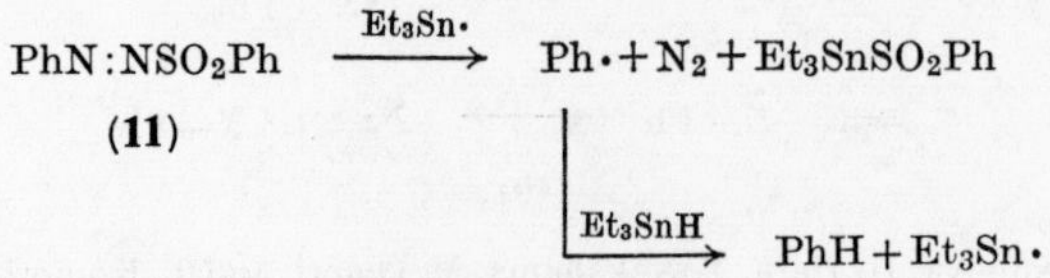

$$PhN:NSO_2Ph \;\xrightarrow{Et_3Sn\cdot}\; Ph\cdot + N_2 + Et_3SnSO_2Ph$$
(11)
$$\xrightarrow{Et_3SnH}\; PhH + Et_3Sn\cdot$$

[43] S. F. Nelsen, *J. Am. Chem. Soc.*, **88**, 5666 (1966).

[44] H. Kiefer and T. G. Traylor, *Tetrahedron Letters*, **1966**, 6163.

[45] L. M. Andronov and G. E. Zaikov, *Kinet. Katal.*, **8**, 270 (1967); *Chem. Abs.*, **67**, 81662a (1967).

[46] S. Rummel, H. Hubner, and P. Krumbiegel, *Z. Chem.*, **7**, 351 (1967).

[47] P. C. Huang and E. M. Kosower, *J. Am. Chem. Soc.*, **89**, 3910, 3911 (1967).

[48] R. W. Hoffmann and G. Guhn, *Chem. Ber.*, **100**, 1474 (1967); see also J. F. Bunnett and D. A. R. Happer, *J. Org. Chem.*, **32**, 2701 (1967).

[49] W. P. Neumann and H. Lind, *Angew. Chem. Intern. Ed. Engl.*, **6**, 76 (1967).

Several interesting results have been obtained with biradicals, not the least intriguing being the theoretical prediction that incorporation of two allyl radicals into a spiroheptane structure (as **12**) should produce a species with a singlet ground state.[50]

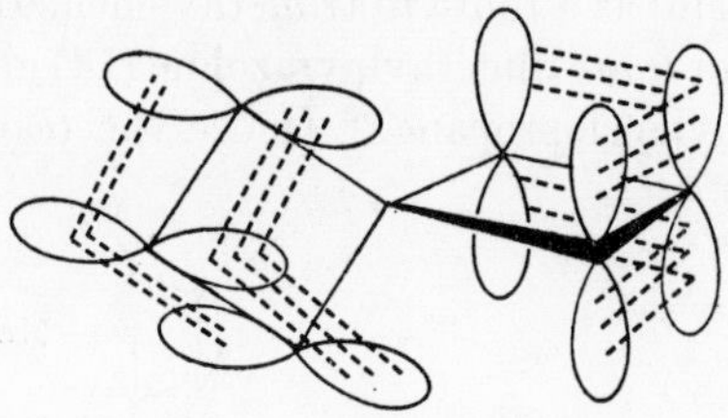

(12)

Several new papers deal with trimethylenemethane. The ESR spectrum of this triplet species has now been obtained in experiments involving photolysis of 3-methylenecyclobutanone,[51] and Skell and Doerr have generated it by the reaction of 3-iodo-2-(iodomethyl)propene with potassium vapour.[52] The products of this reaction include 1,4-dimethylenecyclohexane and *p*-xylene, both presumably formed by dimerization. A similar reaction with 1,3-dichloroacetone produced no dimeric products, but gave predominantly carbon monoxide and ethylene by way of a cyclopropanone.[53] This is consistent with

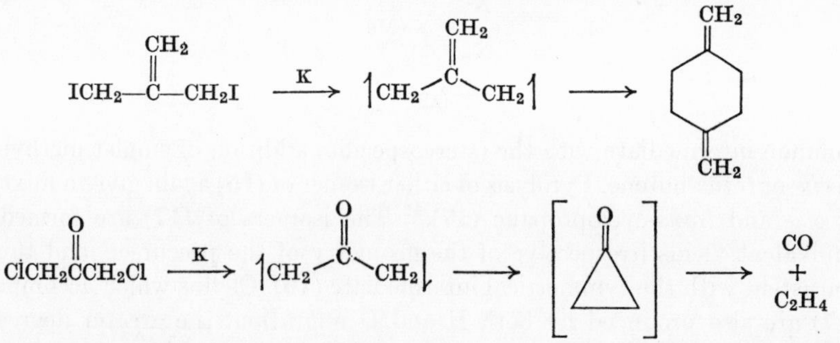

the prediction that the intermediate species should in the latter case have a singlet ground state. The experimental arrangement for this work has been described separately.[54]

New molecular-orbital calculations on trimethylenemethane[55] suggest that

[50] R. Hoffmann, A. Imamura, and G. D. Zeiss, *J. Am. Chem. Soc.*, **89**, 5215 (1967); see also H. E. Simmons and T. Fukunaga, *ibid.*, p. 5208.

[51] P. Dowd and K. Sachdev, *J. Am. Chem. Soc.*, **89**, 715 (1967).

[52] P. S. Skell and R. G. Doerr, *J. Am. Chem. Soc.*, **89**, 3062, 4688 (1967).

[53] R. G. Doerr and P. S. Skell, *J. Am. Chem. Soc.*, **89**, 4684 (1967).

[54] P. S. Skell, E. J. Goldstein, R. J. Petersen, and G. L. Tingey, *Chem. Ber.*, **100**, 1442 (1967).

[55] W. T. Borden, *Tetrahedron Letters*, **1967**, 259.

thermal and photochemical decompositions of (**13**) should produce the biradical in different electronic states. The calculations also predict that as triplet methylenecyclopropane correlates with a high-energy state of trimethylenemethane, photosensitized reaction of the cyclopropane would therefore be unsuccessful as a route to trimethylenemethane.

Heating either *cis*- or *trans*-dimethylpyrazoline (**14**) gives a mixture of both *cis*- and *trans*-dimethylcyclopropane.[56] Hence the reaction cannot share a

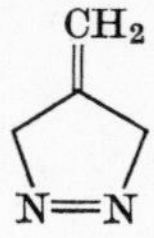

(**13**)

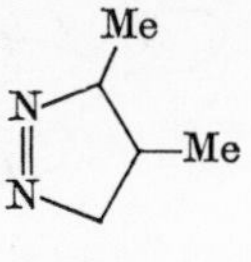

(**14**)

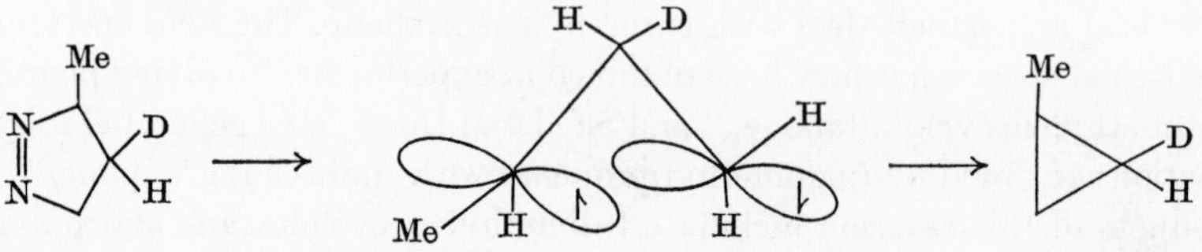

(**15**) (**16**) (**17**)

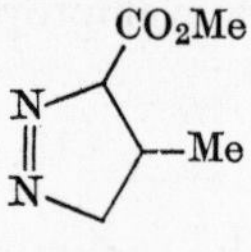

(**18**)

common intermediate with the (stereospecific) addition of singlet methylene to *cis*- or *trans*-butene. Pyrolysis of either isomer of (**15**) again gives a mixture of *cis*- and *trans*-cyclopropane (**17**).[57] The isomers of (**17**) are formed in equivalent yields irrespective of the geometry of the precursor, and this is consistent with the symmetrical intermediate (**16**). Olefins which accompany (**17**) are also produced by both H and D migration, the greater degree of protium migration being due to a kinetic isotope effect. This contrasts with the results obtained by using the methyl 4-methylpyrazoline-3-carboxylate (**18**),[58] for which there is some evidence that the migration of hydrogen is concerted with loss of nitrogen.

Pyrolysis of (**19**) or (**20**) gives mixtures of the four dienes (**21**)—(**24**), consistently with conformational isomerization in the diradical intermediates,[59] but pyrolysis of (**25**) gives more of the inverted product (**26**) than of

[56] R. J. Crawford and L. H. Ali, *J. Am. Chem. Soc.*, **89**, 3908 (1967).
[57] R. J. Crawford and G. L. Erickson, *J. Am. Chem. Soc.*, **89**, 3907 (1967).
[58] D. E. McGreer and W.-S. Wu, *Can. J. Chem.*, **45**, 461 (1967).
[59] W. R. Roth and M. Martin, *Tetrahedron Letters*, **1967**, 3865.

(**27**) by a factor of three.[60] Photolysis is less stereoselective, and, indeed, at –70° this gives (**27**) as the major product. The results with compound (**25**) are therefore inconsistent with a 1,3-diradical of sufficiently long life to equilibrate, and also inconsistent with one having the planar geometry represented by (**16**).

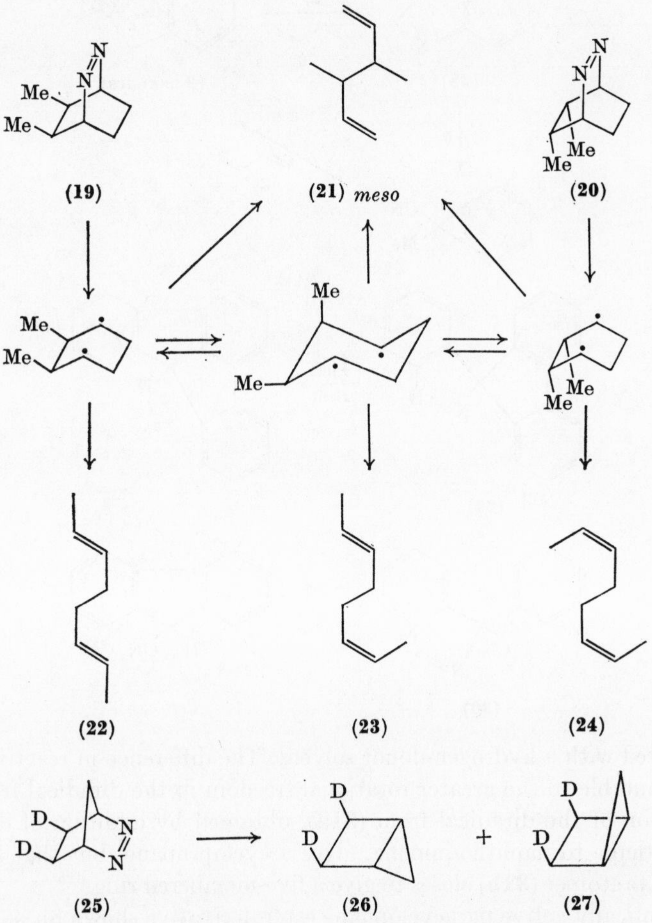

A triplet state ESR spectrum has been obtained from an intermediate formed on photolysis of (**28**) in a rigid matrix at 77°K.[61] This is presumably the spectrum of the precursor of the benzocyclopropene which has previously been obtained by photolysis of (**28**). The intermediate may be intercepted by

[60] W. R. Roth and M. Martin, *Ann. Chem.*, **702**, 1 (1967).
[61] G. L. Closs, L. R. Kaplan, and V. I. Bendall, *J. Am. Chem. Soc.*, **89**, 3376 (1967).

a reactive diene, both on photolysis of (**28**) and on warming the benzocyclopropene. Intramolecular recyclization occurs on thermal decomposition of the indazoles (**29**) and (**30**).[62a] Only in the former case can an intermediate be

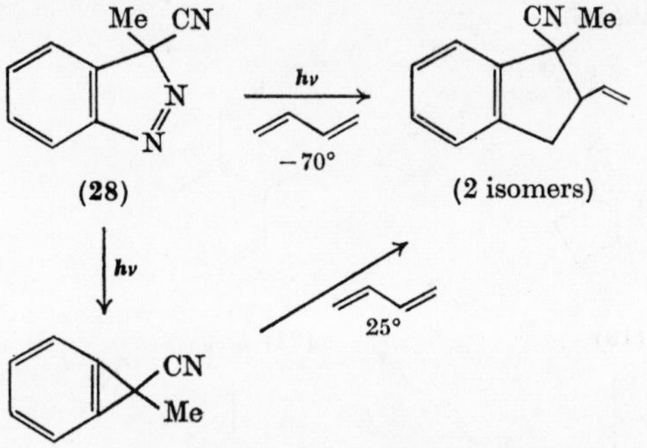

(**28**) (2 isomers)

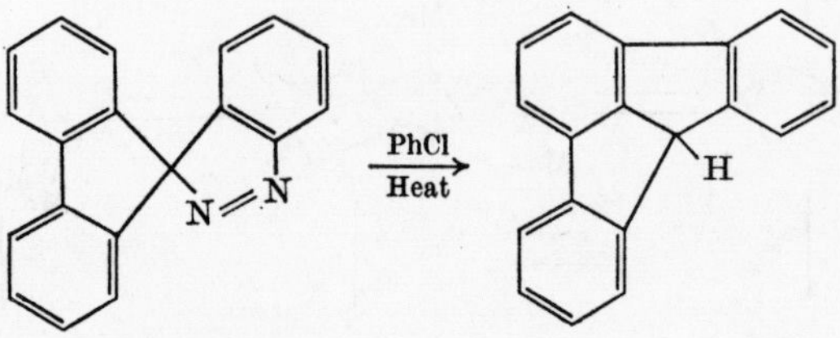

(**29**)

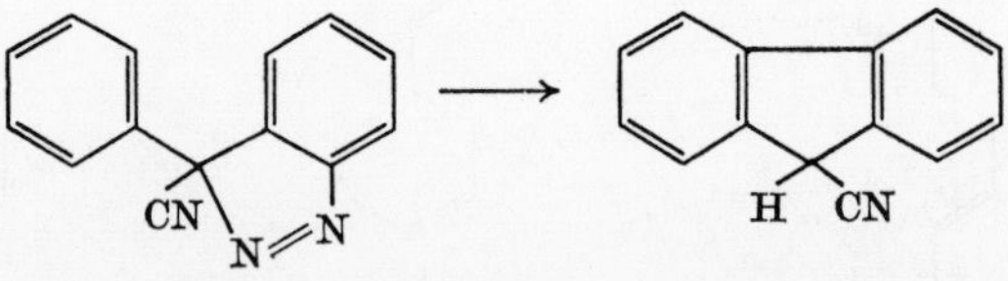

(**30**)

intercepted with a hydrogen-donor solvent. The difference in reactivity may be attributable to the greater rotational freedom in the diradical from (**30**). Cyclization of the diradical from (**31a**), obtained by addition of diphenyldiazomethane to naphthoquinone, gives a cyclopropane, but that from the phenolic tautomer (**31b**) closes to give a five-membered ring.[62b]

The optically active paracyclophane ester derivative shown on p. 257 may be racemized by heat. The mechanism of racemization involves a diradical which may be intercepted by, for example, a hydrogen-donor solvent to give a 4,4′-dimethylbibenzyl.[63]

The photolysis of azobisisobutyronitrile in a rigid glass at 77°K has been

[62a] G. Baum, R. Bernard, and H. Shechter, *J. Am. Chem. Soc.*, **89**, 5307 (1967).
[62b] P. G. Jones, *Chem. Commun.*, **1966**, 894.
[63] H. J. Reich and D. J. Cram, *J. Am. Chem. Soc.*, **89**, 3078 (1967).

found to produce radical pairs in which there is considerable spin interaction, and a typical ESR triplet spectrum is obtained.[64] The radicals are about 6 Å apart.

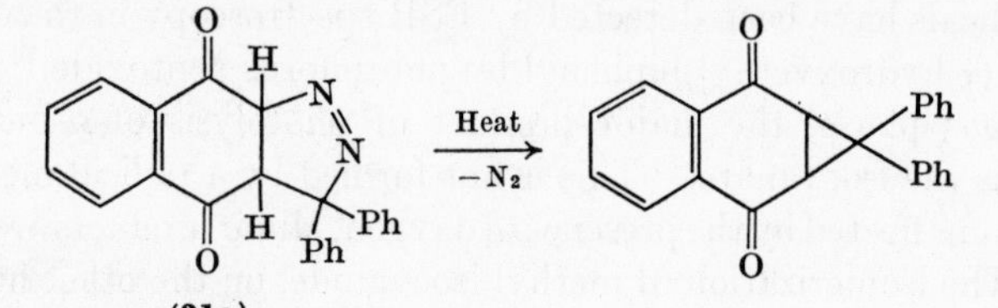

(31a)

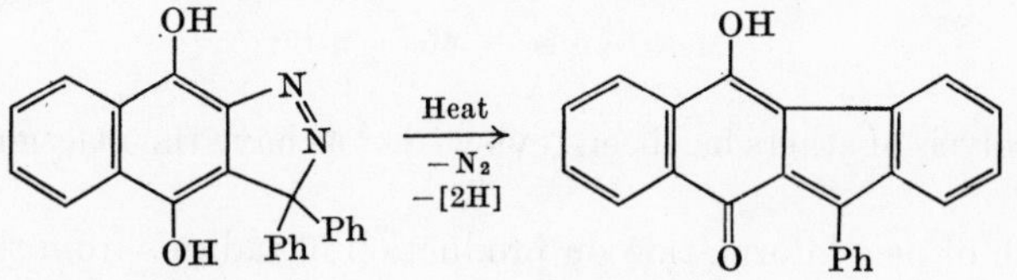

(31b)

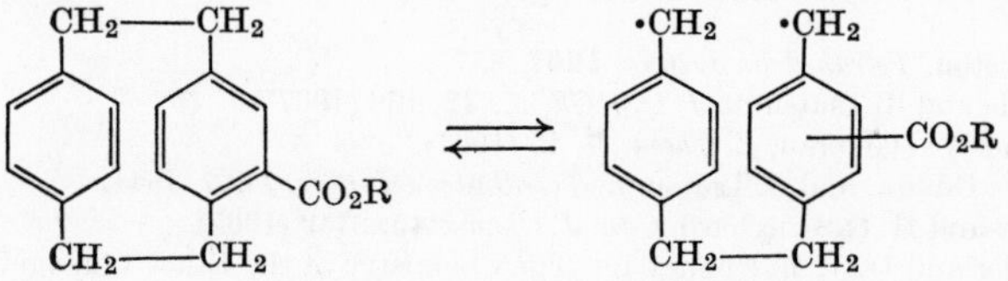

The kinetics of processes at electrodes have been surveyed,[65] and an electrochemical method for detecting rapid reactions of electrolytically produced radical anions has been described.[66] No Kolbe dimers are formed in the electrolysis of α-bromo-acids.[67] Reaction products in methanol include substantial quantities of methyl ester; one possible explanation advanced to account for this involved anodic oxidation of un-ionized acid:

$$RCO_2H \xrightarrow{-2e} RCO^+ + {}^+OH$$
$$RCO^+ \xrightarrow{MeOH} RCO_2Me$$

Electrolysis of $ArCH_3$ in acetic acid containing tetramethylammonium nitrate gave products attributable to benzyl radical formation.[68] Hydrogen abstraction was suggested to be by the $NO_3\cdot$ radical formed by preferential

[64] M. C. R. Symons, *Nature (London)*, **213**, 1226 (1967).
[65] B. E. Conway, *Progr. Reaction Kinetics*, **4**, 399 (1967).
[66] J. A. Harrison, *Tetrahedron Letters*, **1966**, 6457.
[67] R. G. Woolford, J. Soong, and W. S. Lin, *Can. J. Chem.*, **45**, 1837 (1967).
[68] S. D. Ross, M. Finkelstein, and R. C. Petersen, *J. Am. Chem. Soc.*, **89**, 4088 (1967).

discharge of nitrate ions. Oxidative phenol coupling has been reported in electrolysis of hydroxyacetophenones,[69] and radical reactions have been described in the electrolysis of selected organomercury compounds.[70]

Free radicals have been detected by ESR spectroscopy in the dehydration of 4,4′-bis-(1-hydroxyethyl)biphenyl by phosphorus pentoxide.[71]

In the gas phase, the major product of photolysis of *n*-butyl cyanate (BuOCN) is the isocyanate.[72] This is not formed by a radical mechanism, as the yield is unaffected by the presence of oxygen. Minor products are scavenged, however. The isomerization of methyl isocyanide, on the other hand, to give acetonitrile, does involve methyl radicals:[73]

$$\text{MeNC} + \text{Me}\cdot \rightarrow \text{Me}\cdot + \text{MeCN}$$

The pyrolysis of ethers has been reviewed,[74] as have the reactions of alkoxy radicals.[75]

A wealth of new information on production of radicals from the pyrolysis or photolysis of small molecules in the gas-phase has been published, discussion of which is beyond the scope of the present volume.[76]

69 K. M. Johnston, *Tetrahedron Letters*, **1967**, 837.
70 K. Yosahida and S. Tsutsumi, *J. Org. Chem.*, **32**, 468 (1967).
71 K. Lohs and W. Damerau, *Z. Chem.*, **7**, 15 (1967).
72 M. Hara, Y. Odaira, and S. Tsutsumi, *Tetrahedron Letters*, **1967**, 1641.
73 D. H. Shaw and H. O. Pritchard, *Can. J. Chem.*, **45**, 2749 (1967).
74 K. J. Laidler and D. J. McKenney in "The Chemistry of the Ether Linkage", ed. S. Patai, Interscience, London, 1967, p. 167.
75 P. Gray, R. Shaw, and J. C. J. Thynne, *Progr. Reaction Kinetics*, **4**, 63 (1967).
76 See, for example: J. C. Amphlett and E. Whittle, *Trans. Faraday Soc.*, **63**, 80 (1967); R. J. Akers and J. J. Throssell, *ibid.*, p. 124; J. M. Brown and B. A. Thrush, *ibid.*, p. 630; R. D. Giles, L. M. Quick, and E. Whittle, *ibid.*, p. 662; M. I. Christie and M. A. Voisey, *ibid.*, p. 2459; J. M. Tedder and J. C. Walton, *ibid.*, p. 2464; E. R. Morris and J. C. J. Thynne, *ibid.*, p. 2470; J. A. Kerr and A. C. Lloyd, *ibid.*, p. 2480; J. C. Amphlett and E. Whittle, *ibid.*, p. 2695; M. I. Christie and M. A. Voisey, *ibid.*, p. 2702; A. Good and J. C. J. Thynne, *ibid.*, p. 2708, 2720; J. O. Terry and J. H. Futrell, *Can. J. Chem.*, **45**, 2327 (1967); M. Krech and S. J. W. Price, *ibid.*, p. 157; P. Gray and A. Jones, *ibid.*, p. 333; R. J. Kominar, M. G. Jacko, and S. J. Price, *ibid.*, p. 575; M. C. Lin and K. J. Laidler, *ibid.*, p. 1315; A. Kato and R. J. Cvetanović, *ibid.*, p. 1845; L. F. Loucks and K. J. Laidler, *ibid.*, p. 2763, 2767, 2785, 2795; L. F. Loucks, *ibid.*, p. 2775; M. C. Lin and K. J. Laidler, *ibid.*, **44**, 2927 (1966); L. Batt and F. R. Cruickshank, *J. Phys. Chem.*, **71**, 1836 (1967); C. E. Waring and R. Pellin, *ibid.*, p. 2044; T. J. Houser and B. M. H. Lee, *ibid.*, p. 3422; J. A. Kerr and A. C. Lloyd, *Chem. Commun.*, **1967**, 164; J. M. Edwards and M. I. Christie, *ibid.*, p. 789; D. G. Dalgleish and J. H. Knox, *ibid.*, **1966**, 917; K. J. Laidler and M. T. H. Liu, *Proc. Roy. Soc., A*, **297**, 635 (1967); A. S. Kallend, J. H. Purnell, and B. C. Shurlock, *ibid., A*, **300**, 120 (1967); G. A. Hughes and L. Phillips, *J. Chem. Soc., A*, **1967**, 894; S. E. Braslavsky, J. Grotewold, and E. A. Lissi, *ibid., B*, **1967**, 414; J. M. Hay and D. Lyon, *ibid., B*, **1967**, 970; W. E. Morganroth and J. G. Calvert, *J. Am. Chem. Soc.*, **88**, 5387 (1966); G. E. Zaikov, A. A. Vichutinskii, and Z. K. Maizus, *Kinet. Katal.*, **8**, 675 (1967); *Chem. Abs.*, **67**, 90321 (1967); F. Maria, G. Acs, and Z. Szabo, *Acta Chem. Acad. Sci. Hung.*, **50**, 263 (1966); P. Molyneux, *Tetrahedron*, **22**, 2929 (1966); G. M. Come, M. Dzierzynski, R. Martin, and M. Niclause, *Compt. Rend., Ser. C*, **264**, 548 (1967); K. Scherzer, W. Petzerling, H. Tilgner, and G. Geiseler, *Z. Physik. Chem. (Frankfurt)*, **51**, 155 (1966); F. Marta and L. Seres, *Ber. Bunsenges. Phys. Chem.*, **70**, 921

Photolysis of $CF_2{=}N{-}N{=}CF_2$ generates $CF_2{=}N\cdot$ radicals which have been trapped by combination with $CF_3\cdot$ or by addition to tetrafluoroethylene.[77] Vacuum-ultraviolet photolysis of simple aliphatic amines at 77°K causes loss of a hydrogen atom from nitrogen, and ESR spectra of several simple amino-radicals were recorded.[78]

Pyrolysis studies of cyclohexa-1,3-diene show that concerted loss of hydrogen occurs much less readily than from the 1,4-isomer ($\Delta E_a > 9$ kcal mole^{-1}) as predicted by orbital-symmetry considerations. Indeed, benzene is formed from the 1,3-isomer by a radical-chain process.[79]

A frequent justification for recent photochemical and pyrolysis studies has been the comparison of the behaviour of the energy-rich molecules formed, with the behaviour of those obtained by electron-bombardment in a mass spectrometer. To this end, Hedaya *et al.* have developed the technique of flash vacuum pyrolysis;[80] in their apparatus a sample of volatilized organic material is passed through a pyrolysis cell at *ca.* 900° (the "radical gun") and the pyrolysate is directed immediately onto the walls of a Dewar vessel at liquid-nitrogen temperature. Thus benzyl bromide gives benzyl radicals which were condensed in a rare-gas matrix and examined spectroscopically. Also, pyrolysis of allyl phenyl ether gives products which contain a cyclopentadiene ring, having been formed by decarbonylation of phenoxyl radicals, a process known to occur in the mass spectrometer.

Reactions of Radicals

Radical abstraction and displacement processes. The results of Walling[81a] and K. U. Ingold[81b] and their co-workers on the radical-chain chlorination of toluene by *tert*-butyl hypochlorite have been set out in detail; butoxyl radicals are responsible for hydrogen abstraction, and Walling and Mintz[82] have summarized substituent effects on the ease of hydrogen abstraction by this species. Some typical results are given in the following Table; these were obtained at 0° except where noted; the rate of abstraction of a methyl hydrogen in "Alkyl CH_3" is taken as unity.

(1966); L. M. Andronov and Z. K. Maizus, *Izv. Akad. Nauk SSSR, Ser. Khim.*, **1967, 519;** *Chem. Abs.*, **67**, 53341b (1967); G. E. Zaikov, Z. K. Maizus, and N. M. Emanuel, *Dokl. Akad. Nauk SSSR*, **173**, 859 (1967); *Chem. Abs.*, **67**, 53335c (1967); D. E. Hoare and D. A. Whytock, *Can. J. Chem.*, **45**, 2741, 2841 (1967); see also Chapter 13.

77 R. A. Mitsch and P. H. Ogden, *Chem. Commun.*, **1967**, 59; P. H. Ogden and R. A. Mitsch, *J. Am. Chem. Soc.*, **89**, 3868 (1967).

78 S. G. Hadley and D. H. Volman, *J. Am. Chem. Soc.*, **89**, 1053 (1967).

79 S. W. Benson and R. Shaw, *J. Am. Chem. Soc.*, **89**, 5351 (1967).

80 E. Hedaya and D. McNeil, *J. Am. Chem. Soc.*, **89**, 4213 (1967); C. L. Angell, E. Hedaya, and D. McLeod, *ibid.*, p. 4214.

81 (*a*) C. Walling and V. P. Kurkov, *J. Am. Chem. Soc.*, **89**, 4895 (1967); (*b*) D. J. Carlsson and K. U. Ingold, *J. Am. Chem. Soc.*, **89**, 4885, 4891 (1967); see also *Organic Reaction Mechanisms*, **1966**, 258.

82 C. Walling and M. J. Mintz, *J. Am. Chem. Soc.*, **89**, 1515 (1967).

Relative rates of hydrogen abstraction of a methylene hydrogen in RCH_2Me by $Bu^tO\cdot$ at 0°C.

R =	Alkyl	Phenyl	Vinyl	Cl	OAlkyl	COAlkyl	{CN, CO_2Alkyl, CO_2H}
	13	45	61	5*	78	3.8	0.67*

* From competition experiments at 40°.

Contrary to earlier reports it has been found that protonated dialkylamino-radicals will participate in intermolecular hydrogen abstraction, and that they do so with high selectivity.[83] Thus in acetic/sulphuric acid mixtures, chloramines may be employed as highly selective chlorinating agents. Proportions of monochloro-derivatives from methyl hexanoate and *N*-chlorodimethylamine in this solvent system are as follows:

MeOCO—	CH_2—	CH_2—	CH_2—	CH_2—	CH_3
	0	4.7	13.5	78	3.9 %

However, to attribute the difference between this result and that for abstraction by chlorine atoms obtained by Tedder[84] as entirely due to the selectivity of $Me_2HN\cdot^+$ seems unsatisfactory, as in the solvent system used in the present study the ester probably exists almost entirely in a protonated form. This would presumably have a marked polar influence on the selectivity, and indeed just such a polar effect has been observed in hydrogen abstraction by atomic chlorine from carboxylic acids dissolved in liquid HF.[85]

Radical-chain chlorination by chlorine dioxide involves abstraction by both Cl· and ClO·.[86] Numerous other studies of chlorination include that of isovaleronitrile with various reagents (Cl_3CSO_2Cl, Cl_3CSCl, SCl_2, S_2Cl_2),[87] of alkylchlorosilanes,[88] of carboranes (by displacement of H from boron),[89] of *n*-propylbenzene,[90] of chloropropanes,[91] of chloroethanes,[92] and of ethane

[83] F. Minisci, R. Galli, A. Galli, and R. Bernardi, *Tetrahedron Letters*, **1967**, 2207. Similar results have been obtained for bromination by Me_2NBr: F. Minisci, R. Galli, and R. Bernardi, *Chem. Commun.*, **1967**, 903.

[84] See *Organic Reaction Mechanisms*, **1965**, 205.

[85] J. Kollonitsch, G. A. Doldouras, and V. F. Verdi, *J. Chem. Soc., B*, **1967**, 1093.

[86] D. D. Tanner and N. Nychka, *J. Am. Chem. Soc.*, **89**, 121 (1967).

[87] J. Rouchaud and A. Bruylants, *Bull. Soc. Chim. Belges*, **76**, 50 (1967).

[88] Y. Nagai, N. Machida, H. Kono, and T. Migita, *J. Org. Chem.*, **32**, 1194 (1967).

[89] L. I. Zakharkin, V. I. Stanko, and A. I. Klimova, *Izv. Akad. Nauk SSSR, Ser. Khim.*, **1966**, 1946; *Chem. Abs.*, **66**, 76054d (1967).

[90] I. Azad and A. Guillemonat, *Compt. Rend., Ser. C*, **264**, 720 (1967).

[91] Y. I. Rotshtein, B. E. Krasotkina, and N. G. Sokolovskaya, *Zh. Organ. Khim.*, **2**, 1539 (1966); *Chem. Abs.*, **66**, 64778e (1967).

[92] T. Migita, M. Kosugi, and Y. Nagai, *Bull. Chem. Soc. Japan*, **40**, 920 (1967); C. Cillien, P. Goldfinger, G. Huybrechts, and G. Martens, *Trans. Faraday Soc.*, **63**, 1631 (1967).

itself.[93] Photochlorination of bicyclo[2.2.0]hexane gives a mixture of rearranged and unrearranged products including the unrearranged bridgehead derivative.[94] Benzonorbornene gives almost entirely the *exo*-chloro-derivative (**32**),[95] which can perhaps be rationalized best in terms of the torsional arguments on p. 1.

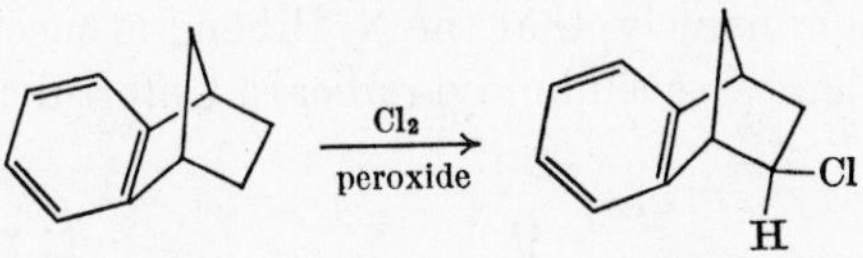

(**32**)

Chlorination of propyne with *tert*-butyl hypochlorite gives 3-chloropropyne but no allenic chloride.[96] Allene behaves in the same way and, surprisingly, competition experiments with toluene show that α-hydrogen abstraction from propyne, and vinyl-hydrogen abstraction from allene, occur with equal facility. A new approach has also been made to determine the relative stabilizing effects of alkenyl and alkynyl substituents on a carbon radical.[97] It was argued that previous comparisons were based on exothermic hydrogen-abstraction reactions and therefore reflected ground-state properties rather than stability of the radicals formed. The new work examines the decomposition of a series of appropriately substituted peresters which undergo two-bond fission. Although analysis of the kinetic data suggested that the allylic radical from (**33**) was more stable than its propargylic counterpart from (**34**) by some

RCO_3Bu^t	R = MeCH:CHCH$_2$—	(**33**)
	MeC⫶CCH$_2$—	(**34**)
	PhCH:CHCH$_2$—	(**35**)
	PhC⫶CCH$_2$—	(**36**)

4 kcal mole^{-1}, there is a rate difference of only 8. This is because of the unfavourable entropy factor for the allylic radical, which has to become coplanar for maximum orbital overlap. Extension of the work to peresters (**35**) and (**36**) revealed that in the olefin, but not the acetylene, a further gain in resonance stabilization was achieved by introducing the phenyl substituent.

It is now generally considered that the bromine atom, and not the succinimidyl radical, is responsible for hydrogen abstraction in allylic bromination by *N*-bromosuccinimide. Hedaya and his co-workers, in seeking an authentic source of succinimidyl radical, have examined *N*,*N′*-bisuccinimide and the

93 T. Migita, M. Kosugi, and Y. Nagai, *Yuki Gosei Kagaku Kyokai Shi*, **24**, 1237 (1966); *Chem. Abs.*, **66**, 54740x (1967).
94 R. Srinivasan and F. I. Sonntag, *Tetrahedron Letters*, **1967**, 603.
95 J. W. Wilt, G. Gutman, W. J. Ranus, and A. R. Zigman, *J. Org. Chem.*, **32**, 893 (1967).
96 M. C. Caserio and R. E. Pratt, *Tetrahedron Letters*, **1967**, 91.
97 M. M. Martin and E. B. Sanders, *J. Am. Chem. Soc.*, **89**, 3777 (1967).

perester (**37**).[98] Bisuccinimide proved to be extremely stable, and from the data obtained it was concluded that the bond strength of N–H in succinimide may be as high as 100 kcal mole^{-1}. This is much larger than previous estimates, and we would point out that it renders suspect one of the few non-circumstantial arguments against a succinimidyl radical chain for brominations by *N*-bromosuccinimide: namely, that the N–H bond in succinimide is too weak for allylic abstraction by succinimidyl radicals. Unfortunately, even (**37**) was

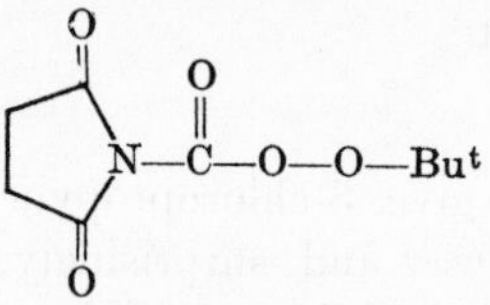

(37)

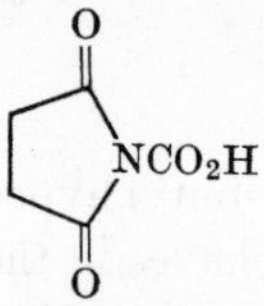

(38)

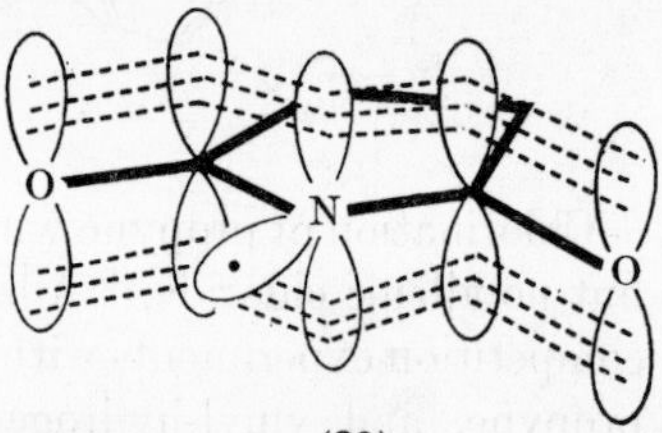

(39)

not a satisfactory source of succinimidyl. Although in hydrogen-donor solvents succinimide was formed, this may largely have resulted from abstraction by carboxyl radical to give (**38**), with subsequent decarboxylation. The apparent lack of stabilization in the succinimidyl radical was discussed in terms of MO calculations favouring the σ-structure (**39**). The reaction of nitric oxide with *N*-bromophthalimide to give the *N*-nitroso-compound[36a] would seem to be rationalized best in terms of a phthalimido radical intermediate.

Evidence favours a bromine atom chain in radical bromination by (**40**),[99a]

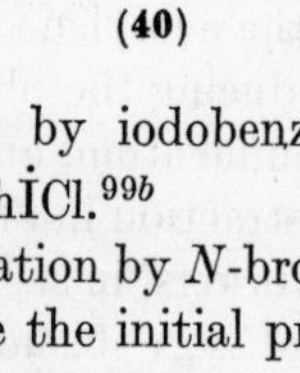

(40)

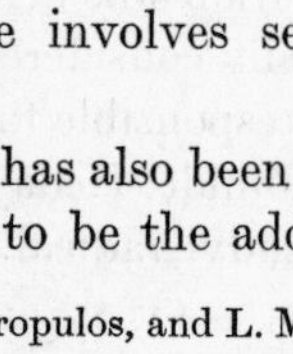

(41)

but chlorination by iodobenzene dichloride involves selective hydrogen abstraction by PhİCl.[99b]

Allylic bromination by *N*-bromoacetamide has also been re-examined, and with cyclohexene the initial product proves to be the adduct (**41**).[100] This

[98] E. Hedaya, R. L. Hinman, V. Schomaker, S. Theodoropulos, and L. M. Kyle, *J. Am. Chem. Soc.*, **89**, 4875 (1967).

[99a] B. R. Kennedy and K. U. Ingold, *Can. J. Chem.*, **45**, 2632 (1967); see also V. D. Pokhodenko and N. N. Kalibabchuk, *Zh. Organ. Khim.*, **2**, 1397 (1966); *Chem. Abs.*, **66**, 54804w (1967).

[99b] D. D. Tanner and P. B. Van Bostelen, *J. Org. Chem.*, **32**, 1517 (1967).

[100] S. Wolfe and D. V. C. Awang, *J. Am. Chem. Soc.*, **89**, 5287 (1967).

seems to be formed by initial radical disproportionation to give *N*,*N*-dibromoacetamide which then adds to the olefin (by an undefined mechanism).

Radical bromination of stearic acid carrying a bromine substituent between C-5 and C-17 gives a vicinal dibromide as the major product.[101]

Examination of the gas-phase reaction between iodobenzene and HI has led to a new estimate of the dissociation energy of the Ph–H bond of 112 kcal $mole^{-1}$.[102a] This is higher than former values, but the earlier data on abstraction can be reconciled in terms of addition-abstraction mechanisms involving cyclohexadienyl radicals. The gas-phase iodination of alkanes has also been examined.[102b]

Photolysis of cyclohexane solutions of tetracovalent phosphorus compounds having a P–Cl bond promotes a radical-chain substitution to give *P*-cyclohexyl derivatives,[103] and chloroformylation of [2.2]paracyclophane in a bridge position proceeds without ring-opening.[104]

Photo-oximation of bicyclo[2.2.2]octane deuterated at one bridgehead position proceeded in high yield (> 70%) without loss of deuterium.[105] As chlorination of this compound gives substantial bridgehead attack (*ca.* 50%),

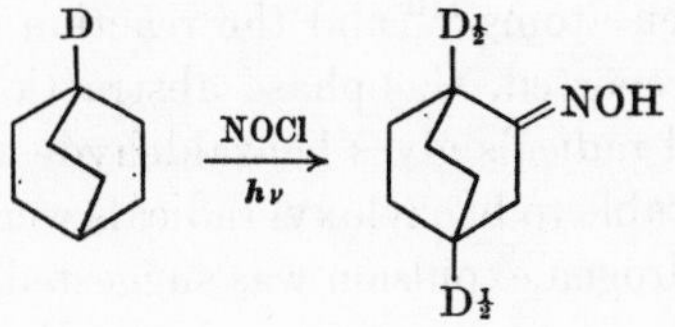

photo-oximation apparently cannot involve hydrogen abstraction by chlorine atoms. A possible alternative involves direct attack by photoexcited NOCl.

New results confirm the correlation of rate of hydrogen abstraction by *tert*-butoxyl radicals from substituted toluenes with σ^+ and not σ.[106] The radical source was *tert*-butyl peroxyoxalate in dilute solution in a Freon, in chlorobenzene, or in acetonitrile. The ρ-values showed a small but unmistakable solvent-dependence. A solvent-complexing effect has also been reported in the competition between addition and allylic abstraction in the reaction between trichloromethanesulphonyl chloride and cyclohexene.[107] Comparison with bromotrichloromethane reactions indicated relatively little hydrogen abstraction, and the results were interpreted in terms of the equilibrium

101 E. Ucciani, F. Morot-Sir, and M. Naudet, *Bull. Soc. Chim. France*, **1967**, 1913.
102a A. S. Rodgers, D. M. Golden, and S. W. Benson, *J. Am. Chem. Soc.*, **89**, 4578 (1967).
102b J. H. Knox and R. G. Musgrave, *Trans. Faraday Soc.*, **63**, 2201 (1967).
103 E. Müller and H. G. Padeken, *Chem. Ber.*, **100**, 521 (1967).
104 E. Hedaya and L. M. Kyle, *J. Org. Chem.*, **32**, 197 (1967).
105 E. Müller and U. Trense, *Tetrahedron Letters*, **1967**, 2045.
106 H. Sakurai and A. Hosomi, *J. Am. Chem. Soc.*, **89**, 458 (1967); see also *Organic Reaction Mechanisms*, **1966**, 257.
107 E. S. Huyser and L. Kim, *J. Org. Chem.*, **32**, 618 (1967).

shown. Allylic abstraction was considered to be due to the free sulphonyl radical, and a rather selective addition of trichloromethyl radicals due to bimolecular reaction between sulphonyl-radical complex and a second molecule of olefin.

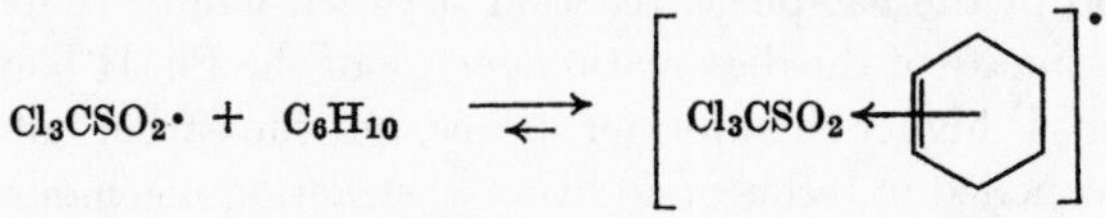

Cheng and Szwarc[108] have sought solvent effects on the reactions of methyl and trifluoromethyl radicals by probing the effect on reactivity of dilution with an inert fluorocarbon solvent. However, no significant change in product ratios was detected.

Hay[109a] has attempted to correlate A-factors for hydrogen abstraction by Me·, and also for unimolecular homolyses, with Hückel calculations for radicals. Abstraction of hydrogen from ethylenediamine and *N*-deuterated ethylenediamine has been examined,[109b] and the abstraction of hydroxylic hydrogen by methyl radicals has received further attention.[109c] Fluorine abstraction by hydrogen atoms,[110] and the reaction of CF_3· with trichlorosilane,[111a] have been reported. Gas-phase abstraction of methyl hydrogen from anisole by methyl radicals gives benzaldehyde as a major product.[111b] As no products attributable to benzyloxyl radicals were detected, a concerted rearrangement and hydrogen expulsion was suggested:

$$PhOCH_2\cdot \rightarrow PhCHO + H\cdot$$

Several new ESR investigations of reactions of hydroxyl radicals with various substrates in a flow system have yielded information on radicals formed from acetamide and formamide[112a] (giving carbon and nitrogen radicals respectively), esters[112b] (α-abstraction from the alcohol group), amino-acids,[112c] and sugars.[112d] The effect of pH on carbohydrate degradation by Fenton's reagent has also been examined.[113] Hydroxy-radical oxidation of

[108] W. J. Cheng and M. Szwarc, *J. Phys. Chem.*, **71**, 2726 (1967).
[109a] J. M. Hay, *J. Chem. Soc., B*, **1967**, 1175.
[109b] P. Gray and A. A. Herod, *Trans. Faraday Soc.*, **63**, 2489 (1967).
[109c] V. Keler, N. V. Kazanskaya, and I. V. Berezin, *Vestn. Mosk. Univ. Ser. II. Khim.*, **21**, 29 (1966); *Chem. Abs.*, **66**, 10383n (1967).
[110] P. M. Scott and K. R. Jennings, *Chem. Commun.*, **1967**, 700.
[111a] T. N. Bell and B. B. Johnson, *Australian J. Chem.*, **20**, 1545 (1967).
[111b] M. F. R. Mulcahy, B. G. Tucker, D. J. Williams, and J. R. Wilmshurst, *Australian J. Chem.*, **20**, 1155 (1967).
[112] (*a*) P. Smith and P. B. Wood, *Can. J. Chem.*, **44**, 3085 (1966); (*b*) A. R. Metcalfe and W. A. Waters, *J. Chem. Soc., B*, **1967**, 340; (*c*) W. A. Armstrong and W. G. Humphreys, *Can. J. Chem.*, **45**, 2589 (1967); (*d*) P. J. Baugh, O. Hinojosa, and J. C. Arthur, *J. Phys. Chem.*, **71**, 1135 (1967).
[113] B. Larsen and O. Smidsrød, *Acta Chem. Scand.*, **21**, 552 (1967).

phenylacetic acid in a flow system gives a spectrum of the benzyl radical, probably by the mechanism shown here.[114a] In acidic solution, oxidation of oximes to nitroxides also appears to involve initial addition as indicated,[114b] and new information on the radicals formed by oxidation of hydroxamic acids has been published.[114c]

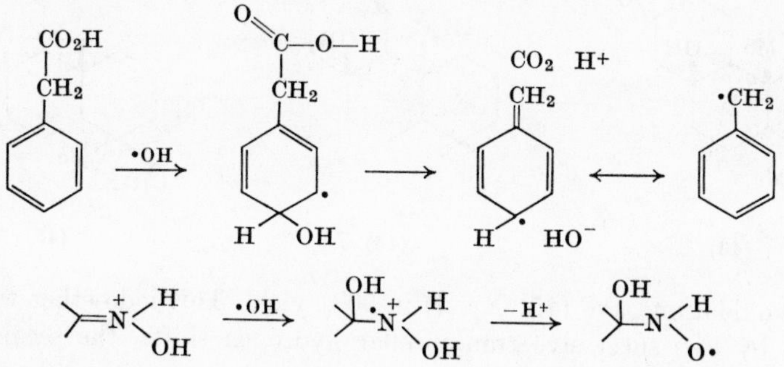

Oxidation of anthracene with periodic acid in aqueous acetic acid gives anthraquinone, but from pyrene 1,1′-bipyrenyl was obtained. The latter product was thought to be formed from the 1-pyrenyl radical for which an ESR spectrum was recorded.[114d]

Intramolecular radical abstraction can be of considerable preparative value, as for example in nitrite photolyses. It has now been found that the discrepancy between the behaviour of nitrites on photolysis and on solution-phase pyrolysis is attributable to a ready acid-catalysed reaction in the pyrolysis.[115a] In the thermal gas-phase reaction, free-radical behaviour was found that was closely akin to the photochemical results. The nitrite photolysis has been extended to the alkaloid field,[115b] and it has been reported that photolysis of a steroidal 6β-nitrate can functionalize the 19-methyl (though in very poor yield).[115c]

Photolytic or thermal rearrangement of *N*-chlorosulphonamides (**42**) results in γ- and some δ-chlorination, the latter process signifying an unusually favourable seven-membered ring transition state in the hydrogen transfer.[116]

$$R[CH_2]_4SO_2NR'Cl \rightarrow RCH_2CHCl[CH_2]_2SO_2NHR' + RCHCl[CH_2]_3SO_2NHR'$$

(**42**)

[114a] R. O. C. Norman and R. J. Pritchett, *J. Chem. Soc., B*, **1967**, 926; cf. *Organic Reaction Mechanisms*, **1966**, 260.
[114b] J. Q. Adams, *J. Am. Chem. Soc.*, **89**, 6022 (1967).
[114c] D. F. Minor, W. A. Waters, and J. V. Ramsbottom, *J. Chem. Soc., B*, **1967**, 180.
[114d] A. J. Fatiadi, *Chem. Commun.*, **1967**, 1087.
[115a] D. H. R. Barton, G. C. Ramsay, and D. Wege, *J. Chem. Soc., C*, **1967**, 1915.
[115b] H. Suginome, N. Sato, and T. Masamune, *Tetrahedron Letters*, **1967**, 1557.
[115c] B. W. Finucane, J. B. Thomson, and J. S. Mills, *Chem. Ind.* (*London*), **1967**, 1747.
[116] M. Okahara, T. Ohashi, and S. Komori, *Tetrahedron Letters*, **1967**, 1629.

Finally, an interesting case of intramolecular abstraction has been reported in the lead tetra-acetate–iodine oxidation of (**43**).[117] In addition to the expected ether (**45**; X = H), a second product was found which was identified as (**45**; X = I). It proved possible to isolate the intermediate (**44**) which on

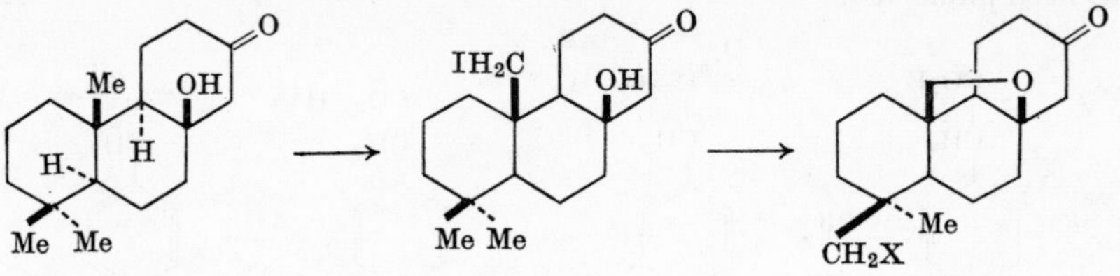

(**43**) (**44**) (**45**)

further oxidation gave (**45**; X = I) in 50% yield. This iodo-ether was thus formed by two successive transannular hydrogen shifts, the second from carbon to carbon.

Oxygen and nitrogen radicals. There have been numerous publications dealing with radical oxidations with molecular oxygen. These include oxidation of chloroform,[118] of chloroacetyl chloride,[119] and chloroacetaldehyde,[120] and co-oxidation of benzaldehyde with cyclohexene[121] and with α-methylstyrene.[122] Also reported are new data on the oxidation of alkenes[123] and cycloalkenes[124] and of benzene.[125] Howard and K. U. Ingold have examined the autoxidation of 1,4-dihydrobenzene and 1,4-dihydronaphthalene in chlorobenzene and have found the large rate constant of 10^9 mole^{-1} sec^{-1} for the termination reaction:[126]

$$2HOO\cdot \rightarrow O_2 + H_2O_2$$

There are also reports of inhibition of the oxidation of ethylbenzene by ferrocene,[127] of new results for oxidation of cumene,[128] and of ESR studies of

[117] E. Wenkert and B. L. Mylari, *J. Am. Chem. Soc.*, **89**, 174 (1967).
[118] S. Kawai, *Yakugaku Zasshi*, **86**, 1125 (1966); *Chem. Abs.*, **66**, 104551j (1967).
[119] B. G. Yasnits'kii, O. B. Dol'berg, and G. I. Kovalenko, *Dopov. Akad. Nauk Ukr. SSR*, **1967**, 67; *Chem. Abs.*, **67**, 21311a (1967).
[120] B. G. Yasnitskii and A. P. Zaitsev, *Zh. Organ. Khim.*, **2**, 1022 (1966); *Chem. Abs.*, **65**, 18450 (1966).
[121] T. Ikawa, H. Tomizawa, and T. Yanagihara, *Can. J. Chem.*, **45**, 1900 (1967).
[122] E. Niki and Y. Kamiya, *Bull. Chem. Soc. Japan*, **40**, 583 (1967).
[123] D. E. Van Sickle, F. R. Mayo, E. S. Gould, and R. M. Arluck, *J. Am. Chem. Soc.*, **89**, 977 (1967).
[124] D. E. Van Sickle, F. R. Mayo, R. M. Arluck, and M. G. Syz, *J. Am. Chem. Soc.*, **89**, 967 (1967).
[125] H. Hotta, *Bull. Chem. Soc. Japan*, **40**, 687 (1967).
[126] J. A. Howard and K. U. Ingold, *Can. J. Chem.*, **45**, 785, 793 (1967).
[127] L. M. Postnikov, E. M. Tochina, and V. Y. Shlyapintokh, *Dokl. Akad. Nauk SSSR*, **172**, 651 (1967).
[128] J. R. Thomas, *J. Am. Chem. Soc.*, **89**, 4872 (1967).

peroxy radicals formed from polymers,[129] and the interaction of oxygen and triphenylmethyl radicals.[130]

Bartlett and Guaraldi[131] have now found that, at temperatures lower than −90° in methylene chloride solution, di-*tert*-butyl tetroxide exists in stable equilibrium with two butylperoxy radicals.[131] At higher temperatures oxygen is lost, and combination of butoxy and butylperoxy radicals gives di-*tert*-butyl trioxide, stable up to −30°. New information has also appeared on di-, tri-, and tetra-sulphides.[132]

There have been several studies of phenoxy radicals, both in the context of inhibition of oxidation[133] and of oxidative coupling of phenol.[134] It has been found that in oxygen-free conditions and in the presence of a catalytic quantity of tri-*tert*-butylphenoxy radical the phenoxyphenol (**46**) undergoes "redistribution" to give monomeric and trimeric, etc., species (cf. p. 268).[135]

An interesting NMR study has separated the rates of the identity reaction A of (**47**) and the dissociation process, B.[136] Diamagnetic line-broadening of the aromatic proton/quinonoid proton signal gave information about process A (presumably a radical cage reaction), and paramagnetic broadening of the acetyl signal provided information on process B.

Among new data on lead tetra-acetate oxidation of alcohols[137, 138] is that of Starnes[137] who finds that oxidation of aryldiphenylcarbinols gives products of preferred aryl migration when aryl is *p*-nitrophenyl, consistently with a radical but not a cationic intermediate. When aryl is *p*-methoxyphenyl, relative extents of migration depend on solvent; for example, in acetonitrile, migration of the methoxyphenyl group was strongly favoured, consistently with an incipient oxygen cation as the migration terminus.

129 J. C. W. Chien and C. R. Boss, *J. Am. Chem. Soc.*, **89**, 571 (1967).
130 E. G. Janzen, F. J. Johnston, and C. L. Ayers, *J. Am. Chem. Soc.*, **89**, 1176 (1967).
131 P. D. Bartlett and G. Guaraldi, *J. Am. Chem. Soc.*, **89**, 4799 (1967); see also *Organic Reaction Mechanisms*, **1966**, 263.
132 P. M. Rao, J. A. Copeck, and A. R. Knight, *Can. J. Chem.*, **45**, 1369 (1967); T. L. Pickering, K. J. Saunders, and A. V. Tobolsky, *J. Am. Chem. Soc.*, **89**, 2364 (1967); S. Chubachi, P. K. Chatterjee, and A. V. Tobolsky, *J. Org. Chem.*, **32**, 1511 (1967).
133 M. A. DaRooge and L. R. Mahoney, *J. Org. Chem.*, **32**, 1 (1967); L. R. Mahoney, *J. Am. Chem. Soc.*, **89**, 1895 (1967); L. R. Mahoney and M. A. DaRooge, *ibid.*, p. 5619; W. H. Starnes and N. P. Neureiter, *J. Org. Chem.*, **32**, 333 (1967).
134 W. G. B. Huysmans and W. A. Waters, *J. Chem. Soc.*, *B*, **1967**, 1163; A. C. Waiss, J. A. Kuhnle, J. J. Windle, and A. K. Wiersema, *Tetrahedron Letters*, **1966**, 6251; J. Petránek, J. Pilař, and D. Doskočilova, *ibid.*, **1967**, 1979; J. D. Fitzpatrick, C. Steelink, and R. E. Hansen, *J. Org. Chem.*, **32**, 625 (1967); W. J. Mijs, O. E. van Lohuizen, J. Bussink, and L. Vollbracht, *Tetrahedron*, **23**, 2253 (1967); C.-H. Brieskorn and K. Ullmann, *Chem. Ber.*, **100**, 618 (1967); A. E. Brodskii, V. D. Pokhodenko, and L. N. Ganyuk, *Abhandl. Deut. Akad. Wiss. Berlin, Kl. Chem., Geol. Biol.*, **1964**, 635 (1963); *Chem. Abs.*, **66**, 28195d (1967).
135 D. A. Bolon, *J. Org. Chem.*, **32**, 1584 (1967).
136 D. J. Williams and R. Kreilick, *J. Am. Chem. Soc.*, **89**, 3408 (1967).
137 W. H. Starnes, *J. Am. Chem. Soc.*, **89**, 3368 (1967).
138 M. L. Mihailović, L. Živković, Z. Maksimović, D. Jeremić, Ž. Čeković, and R. Matić, *Tetrahedron*, **23**, 3095 (1967).

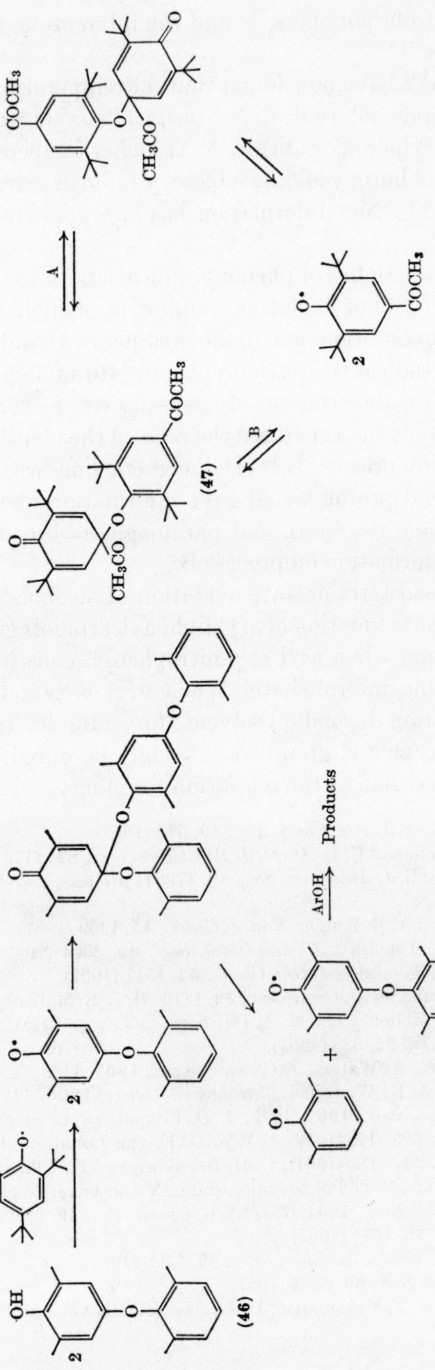

(46) (47)

Oxidation of the readily enolizable acetylacetone by ceric ions in a flow system gives the radical MeCOĊHCOMe.[139] From the proton coupling constant it seems that this is indeed best represented as a carbon radical with little delocalization onto oxygen, though arylindanediones (**48a**) are used as antioxidants. Hydrogen-abstraction from these compounds by diphenylpicrylhydrazyl has been examined.[140a]

Hydrogen-transfer from substituted anilines to indophenoxyl (**48b**) shows a marked polar effect revealed by the Hammett correlation of ease of oxidation with σ.[140b]

Thermochemical data pertinent to the stability of nitroxides give the unexpectedly high O–H bond dissociation energy of *ca.* 70 kcal mole^{-1} for two *N*-hydroxypiperidines (**49**).[141] *tert*-Butyl phenyl nitroxide disproportionates

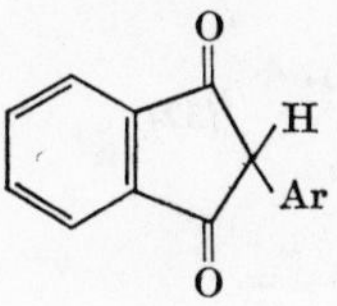

(**48a**)

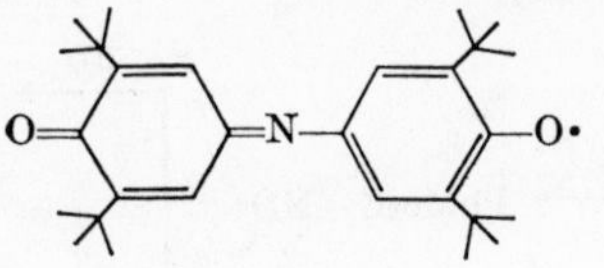

(**48b**)

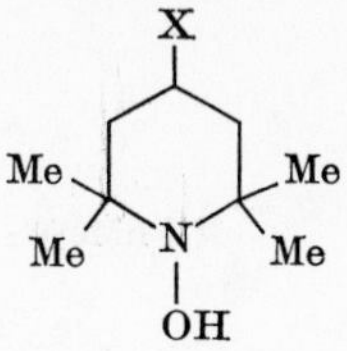

(**49**)

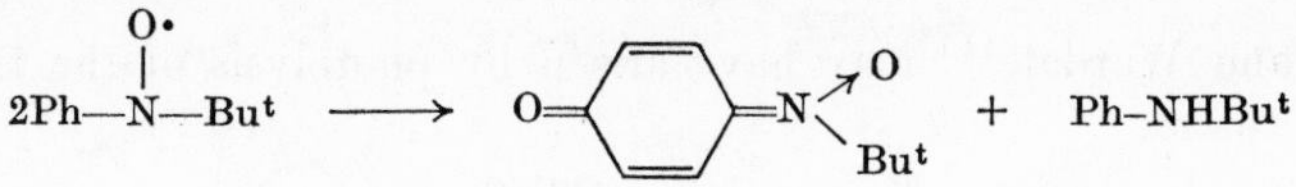

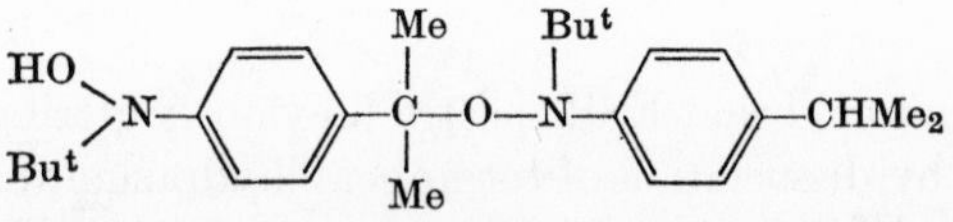

(**50**)

as shown; however, appropriate blocking groups (*p*-phenyl, *p*-*tert*-butyl, 3,5-dimethyl) have been found to inhibit this reaction to an extent such that the monomeric radical may be isolated.[142] With a *p*-isopropyl group the radicals are destroyed by hydrogen abstraction leading to a dimeric species (**50**).

[139] G. A. Russell and J. Lokensgard, *J. Am. Chem. Soc.*, **89**, 5059 (1967).

[140a] V. V. Moiseer and L. P. Zalukaev, *Zh. Organ. Chem.*, **3**, 731 (1967); *Chem. Abs.*, **67**, 53329d (1967).

[140b] V. D. Pokhodenko and V. A. Bidzilya, *Teor. Eksp. Khim.*, **2**, 691 (1966); *Chem. Abs.*, **66**, 94614r (1967).

[141] Y. A. Lebedev, E. G. Rozantsev, M. B. Neiman, and A. Y. Apin, *Zh. Fiz. Khim.*, **40**, 2340 (1966); *Chem. Abs.*, **66**, 37297r (1967).

[142] A. Calder and A. R. Forrester, *Chem. Commun.*, **1967**, 682.

The photochromic behaviour of *o*-nitrocumene is accompanied by production of an ESR signal attributed to (**51**),[143] and photolysis of organic nitrites

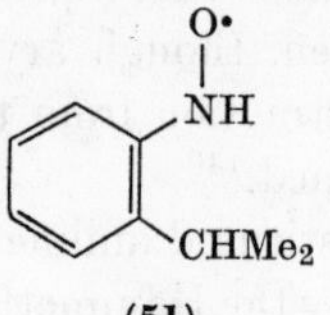

(**51**)

in cumene has been found to generate two relatively stable radicals (**52**) and (**53**).[144] It seems possible that the bistrichloromethylnitroxide reported by

$$\mathrm{RONO} \xrightarrow{h\nu} \mathrm{NO} + \mathrm{RO}\cdot \xrightarrow{\mathrm{PhCHMe_2}} \mathrm{ROH} + \mathrm{PhMe_2C}\cdot$$

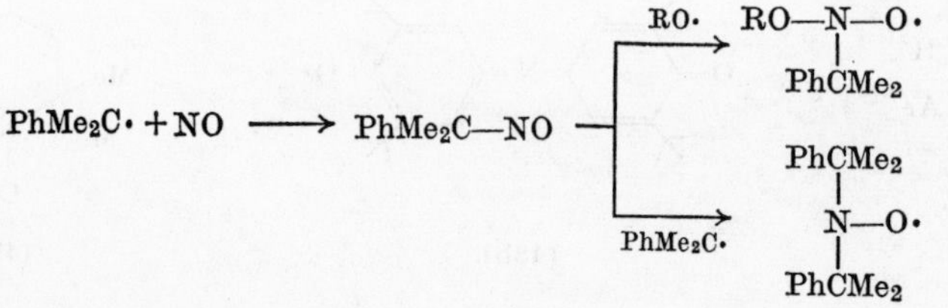

(**52**)

(**53**)

Sutcliffe and Wardale[145] may have arisen by photolysis of the trichloronitrosomethane:

$$\mathrm{Cl_3CNO} \xrightarrow{h\nu} \mathrm{NO} + \mathrm{Cl_3C}\cdot \xrightarrow{\mathrm{Cl_3CNO}} (\mathrm{Cl_3C})_2\mathrm{NO}\cdot$$

It has also been found that ESR spectra previously attributed to diphenylnitrogen formed by dissociation of tetraphenylhydrazine were in fact due to diphenylnitroxide,[146] and reversible dissociation of (**54**) in the range 90—150° gives a sulphur analogue of a nitroxide.[147]

Some interesting bisverdazyls have been prepared. For example, the radical character of (**55**) is markedly reduced at low temperature by the increased proportion of the quinonoid diamagnetic structure.[148] The corresponding 1,5-naphthalene derivative exists entirely as the diradical even at low tem-

143 E. T. Strom and J. Weinstein, *J. Org. Chem.*, **32**, 3705 (1967).
144 A. Mackor, T. A. J. W. Wajer, T. J. de Boer, and J. O. W. van Voorst, *Tetrahedron Letters*, **1967**, 385.
145 H. Sutcliffe and H. W. Wardale, *J. Am. Chem. Soc.*, **89**, 5487 (1967); cf. *Organic Reaction Mechanisms*, **1966**, 266.
146 C. Jackson and N. K. D. Patel, *Tetrahedron Letters*, **1967**, 2255.
147 J. E. Bennett, H. Sieper, and P. Tavs, *Tetrahedron*, **23**, 1697 (1967).
148 F. A. Neugebauer, H. Trischmann, and M. Jenne, *Angew. Chem. Intern. Ed. Engl.*, **6**, 362 (1967).

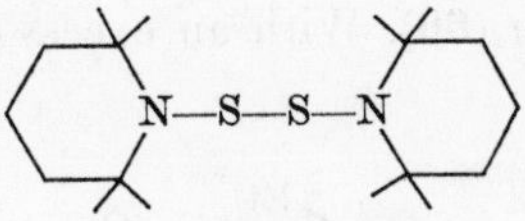

(54)

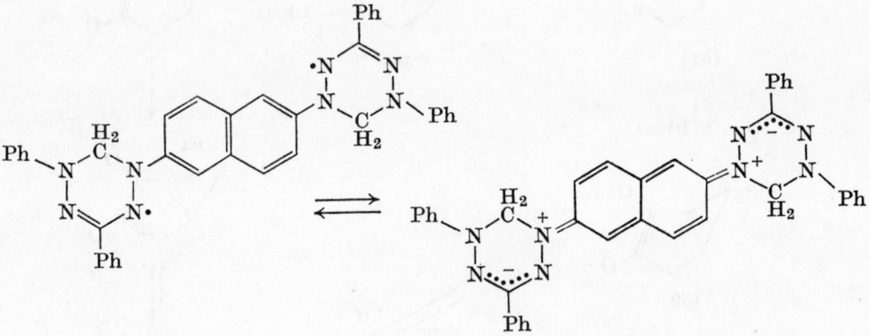

(55)

perature. Interesting results on inter- and intra-molecular association in a series of bispyridinyl biradicals have also been described.[149]

Radical ions. The anion of dibenzoylmethane reacts with sodium in tetrahydrofuran to give an extremely stable dianion radical (**56**).[150]

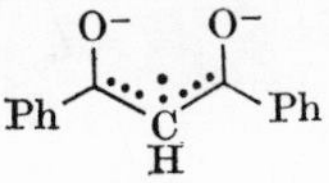

(56)

Several groups have commented on the ESR spectra of biacetyl radical anion and on the protonation of this species,[151] and an ESR technique has been employed to study proton-transfer equilibria between semiquinone radical anion and its conjugate acid.[152] Russell's group has continued its study of semidione anion radicals. This work includes the observation that oxidation of (**57**) or (**58**) with oxygen in dimethyl sulphoxide and base eventually gives the same radical species (**59**).[153] The rate of ring-inversion depends on base-strength, and may therefore involve fission of one side of the three-membered

149 E. M. Kosower and Y. Ikegami, *J. Am. Chem. Soc.*, **89**, 461 (1967); M. Itoh and E. M. Kosower, *ibid.*, p. 3655; M. Itoh and S. Nagakura, *ibid.*, p. 3959.
150 N. L. Bauld and M. S. Brown, *J. Am. Chem. Soc.*, **89**, 5413 (1967).
151 R. J. Pritchett, *Mol. Phys.*, **12**, 481 (1967); R. O. C. Norman and R. J. Prichett, *J. Chem. Soc., B*, **1967**, 378; J. R. Steven and J. C. Ward, *J. Phys. Chem.*, **71**, 2367 (1967); *Australian J. Chem.*, **20**, 2005 (1967); H. Zeldes and R. Livingston, *J. Chem. Phys.*, **47**, 1465 (1967).
152 I. C. P. Smith and A. Carrington, *Mol. Phys.*, **12**, 439 (1967).
153 G. A. Russell, P. R. Whittle, and J. McDonnell, *J. Am. Chem. Soc.*, **89**, 5515, 5516 (1967).

ring in the radical dianion (**60**). With an excess of oxygen, the *ortho*-semiquinone (**61**) is obtained.

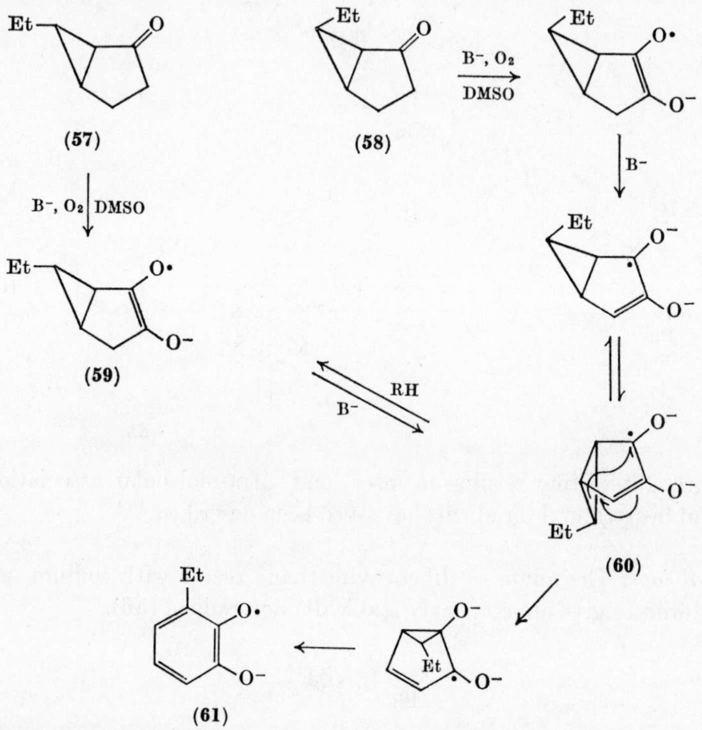

Benzophenone ketyl is methylated to give *o*- and *p*-methylbenzophenone in addition to α-methylbenzhydrol reported in the literature,[154] and oxidation of lithium 9-fluorenoxide with oxygen, yielding fluorenone ketyl, has been followed by ESR spectroscopy.[155]

Irradiation of solutions of phenyl-lithium and an easily reduced aromatic hydrocarbon, ArH, gives $ArH\cdot^-$ and phenyl radicals.[156] This constitutes a simple procedure for examining ion-pairing in $ArH\cdot^-Li^+$ as a function of solvent. Electron-transfer is probably to photoexcited ArH*. With lithium alkyls (not methyl) and anthracene, irradiation gives the anion of 9-alkylanthracene.[157a]

Lithium naphthalenide with α-amino-nitriles gives products resulting from

[154] G. O. Schenck and G. Matthias, *Tetrahedron Letters*, **1967**, 699.

[155] N. A. Sokolov, V. V. Pereshin, and G. A. Abakumov, *Zh. Obshch. Khim.*, **37**, 386 (1967); *Chem. Abs.*, **67**, 53195g (1967).

[156] H. J. S. Winkler and H. Winkler, *J. Org. Chem.*, **32**, 1695 (1967).

[157a] H. J. S. Winkler, R. Bollinger, and H. Winkler, *J. Org. Chem.*, **32**, 1700 (1967).

reactions of intermediate α-aminoalkyl radicals with the naphthalenide radical anion.[157b]

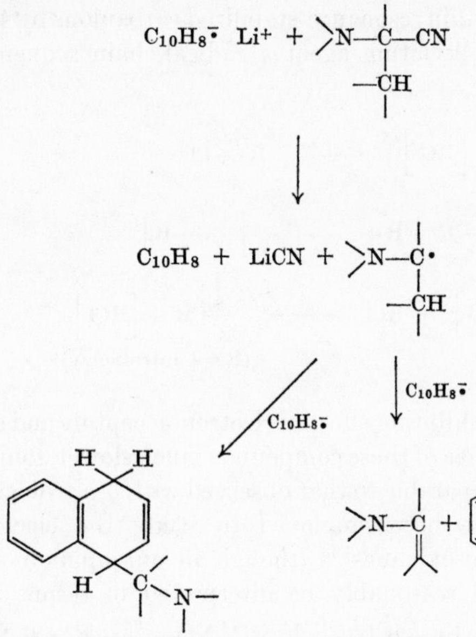

Russell *et al.* have summarized some of their work on nitrobenzene and its reduction products and have attempted to delineate conditions for observing ESR spectra of the various paramagnetic species which can be observed.[158] One ubiquitous radical previously identified as $Ph\dot{N}OH$ is now believed to have the structure PhN(O·)ONHPh. Further details on the formation of radical anions from *o*- and *p*-nitrotoluenes have also appeared.[159] In Bu^tO^-–Bu^tOH the rate-determining step is proton-abstraction. A probable source of dinitrobibenzyls in these reactions is the coupling of nitrobenzyl radical and nitrobenzyl anion in a very fast reaction. The reaction between a radical and an anion is also a key step in the mechanism proposed for peroxide-catalysed oxidation of propan-2-ol by nitrous oxide:[160]

$$N_2O^- + Me_2CHOH \rightarrow N_2 + OH^- + Me_2\dot{C}OH$$

$$Me_2CHOH + OH^- \rightleftarrows Me_2CHO^- + HOH$$

$$Me_2\dot{C}OH + Me_2CHO^- \rightarrow Me_2\dot{C}O^- + Me_2CHOH$$

$$Me_2\dot{C}O^- + N_2O \rightarrow Me_2CO + N_2O^-$$

157b C. Fabre and Z. Welvart, *Tetrahedron Letters*, **1967**, 3801.
158 G. A. Russell, E. J. Geels, F. J. Smentowski, K.-Y. Chang, J. Reynolds, and G. Kaupp, *J. Am. Chem. Soc.*, **89**, 3821 (1967).
159 G. A. Russell and E. G. Janzen, *J. Am. Chem. Soc.*, **89**, 300 (1967).
160 W. V. Sherman, *J. Am. Chem. Soc.*, **89**, 1302 (1967).

Coupling between radical and anion may be the key to biphenyl formation in the photolysis of phenyl-lithium,[161] and it has been advanced as a key step in the *C*-alkylation of certain resonance-stabilized carbanions by 4-nitrobenzyl chloride.[162] With this alkylating agent a radical-chain sequence has been proposed:

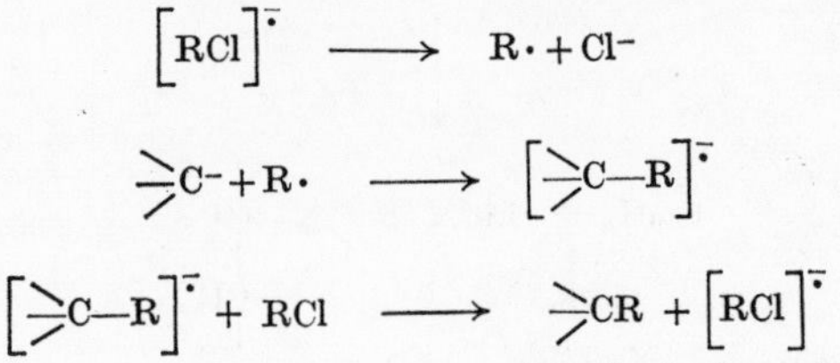

(R = 4-nitrobenzyl)

This would explain the inhibiting effect of electron-acceptors and of transition-metal salts. In the presence of these compounds much slower, ionic, alkylation takes place at a rate comparable to that observed with 3-nitrobenzyl chloride.

ESR techniques have been employed to study the electron-acceptor properties of fluoronitrosoalkanes[163] (though in our opinion many of the radicals observed could reasonably be interpreted in terms of nitroxides formed after photolysis by visible light: $R^FNO \xrightarrow{h\nu} R^F\cdot + NO$) electron-transfer between α-hydroxyalkyl radicals and nitro-compounds,[164] and the electron-transfer equilibria:[165]

$$2(PhN{:}NPh)\cdot^- \rightleftarrows (PhN{:}NPh)^{2-} + PhN{:}NPh$$

$$2(PhCOCOPh)\cdot^- \rightleftarrows (PhCOCOPh)^{2-} + PhCOCOPh$$

and also to investigate electron-exchange between benzene and its radical anion.[166] Significant differences were found between the rates of electron-transfer from (+)-hexahelicene radical anion to the two enantiomeric hexahelicenes.[167] Also reported are the dimerization of diphenylacetylene radical

161 See *Organic Reaction Mechanisms*, **1965**, 300.

162 N. Kornblum, R. E. Michel, and R. C. Kerber, *J. Am. Chem. Soc.*, **88**, 5660, 5662 (1966); G. A. Russell and W. C. Danen, *ibid.*, p. 5663.

163 V. A. Ginsburg, A. N. Medvedev, S. S. Dubov, and M. F. Lebedeva, *Zh. Obshch. Khim.*, **37**, 601 (1967); *Chem. Abs.*, **67**, 43233d (1967); V. A. Ginsburg, A. N. Medvedev, M. F. Lebedeva, M. N. Vasil'eva, and L. L. Martynova, *Zh. Obshch. Khim.*, **37**, 611 (1967); *Chem. Abs.*, **67**, 43234e (1967); V. A. Ginsburg, A. N. Medvedev, N. S. Mirzabekova, and M. F. Lebedeva, *Zh. Obshch. Khim.*, **37**, 620 (1967); *Chem. Abs.*, **67**, 43235f (1967).

164 W. E. Griffiths, G. F. Longster, J. Myatt, and P. F. Todd, *J. Chem. Soc., B*, **1967**, 533.

165 A. G. Evans, J. C. Evans, and E. H. Godden, *Trans. Faraday Soc.*, **63**, 136 (1967); A. G. Evans, J. C. Evans, and C. L. James, *J. Chem. Soc., B*, **1967**, 652.

166 G. L. Malinoski and W. H. Bruning, *J. Am. Chem. Soc.*, **89**, 5063 (1967).

167 R. Chang and S. I. Weissman, *J. Am. Chem. Soc.*, **89**, 5968 (1967).

anion,[168] proton transfer to butadiene radical anion in a solid alcohol matrix,[169] and a pulse-radiolysis study of protonation of radical anions.[170]

Miscellaneous data on free radicals. The stereochemistry of 9-decalyl free radicals has been discussed previously.[171] Greene and Lowry have examined the same problem using the chain reactions indicated to generate their

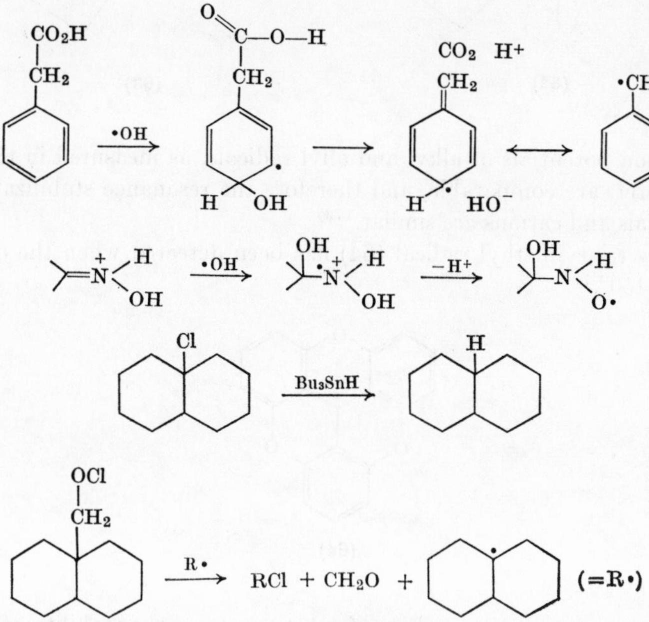

radicals.[172] The conclusions were, in general, similar. With the decalyl chloride–tributylstannane reaction, both *cis*- and *trans*-9-chlorodecalin gave the same mixture of decalins, indicating that the most stable radical had been formed before hydrogen-abstraction from the stannane. The hypochlorite reaction appears to involve much faster chain transfer, as at high concentrations (in $CFCl_3$) the product compositions are quite different. From the *trans*-hypochlorite the product ratio *trans*/*cis* for the 9-chlorodecalins is 30:1 irrespective of concentration, and this is considered to reflect the stereochemical preference for chlorine abstraction by planar *trans*-decalyl radical (**62**). The only significant point at which the interpretation diverges from that

[168] D. Dadley and A. G. Evans, *J. Chem. Soc., B,* **1967**, 418.
[169] T. Shida and W. H. Hamill, *J. Am. Chem. Soc.*, **88**, 5371 (1966).
[170] S. Arai, E. L. Tremba, J. R. Brandon, and L. M. Dorfman, *Can. J. Chem.*, **45**, 1119 (1967).
[171] See *Organic Reaction Mechanisms*, **1965**, 211.
[172] F. D. Greene and N. N. Lowry, *J. Org. Chem.*, **32**, 875, 882 (1967).

of Bartlett *et al.*[171] is in the suggestion that the radical from the *cis*-precursor may also prefer a planar conformation (**63**), but may produce more *cis*-chlorodecalin because of the different conformation of the remainder of the radical.

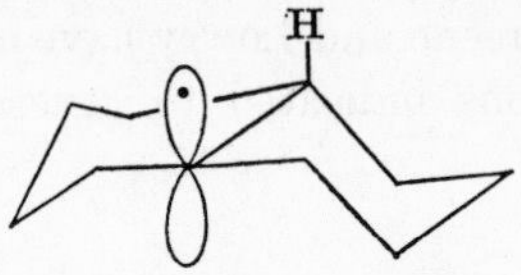

(62)

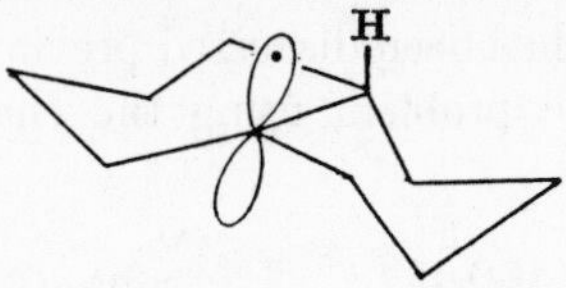

(63)

Ionization potentials of alkyl and allyl radicals, as measured in the mass spectrometer, are comparable, and therefore the resonance stabilizations of allyl radicals and cations are similar.[173]

The new triarylmethyl radical (**64**) has been detected when the dimer is heated to 150°.[174]

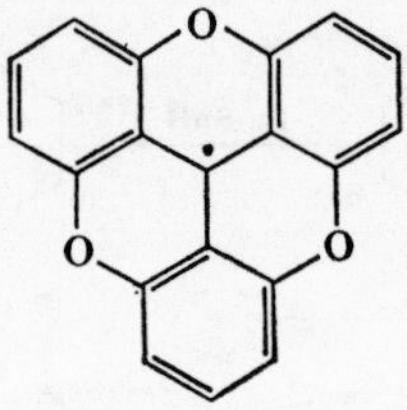

(64)

New data concerning the effects of substituents on the stability of triarylamine radical cations have been presented.[175]

Dewar's group has reported investigations of oxidation of aralkanes by single-electron transfer to trivalent manganese (see p. 425).[176] The reaction of benzoyl peroxide with cupric salts in acetic acid gives products of ligand-transfer to the phenyl radical:[177] with cupric chloride or propionate, chlorobenzene or phenyl propionate is obtained. It was argued that oxidation by electron-transfer would give phenyl acetate by interaction of phenyl cation with solvent.

N,*N*-Dichlorourethane reacts with ethers in benzene under the influence of

173 S. Pignataro, A. Cassuto, and F. P. Lossing, *J. Am. Chem. Soc.*, **89**, 3693 (1967).
174 E. Müller, A. Moosmayer, A. Rieker, and K. Scheffler, *Tetrahedron Letters*, **1967**, 3877; M. J. Sabacky, C. S. Johnson, R. G. Smith, H. S. Gutowsky, and J. C. Martin, *J. Am. Chem. Soc.*, **89**, 2054 (1967).
175 L. Hagopian, G. Kohler, and R. I. Walter, *J. Phys. Chem.*, **71**, 2290 (1967).
176 P. J. Andrulis, M. J. S. Dewar, R. Dietz, and R. L. Hunt, *J. Am. Chem. Soc.*, **88**, 5473 (1966); T. Aratani and M. J. S. Dewar, *ibid.*, p. 5479; P. J. Andrulis and M. J. S. Dewar, *ibid.*, p. 5483.
177 K. Wada, J. Yamashita, H. Hashimoto, *Bull. Chem. Soc. Japan*, **40**, 2410 (1967).

ultraviolet light to give bisurethanes (**65**), possibly by a route involving the steps indicated.[178]

$$Et_2O + R\cdot \rightarrow RH + CH_3CHO + Et\cdot$$

$$CH_3CHO + Cl_2 \rightarrow ClCH_2CHO + HCl$$

$$ClCH_2CHO + EtOCONH_2 \rightarrow \underset{\textbf{(65)}}{ClCH_2CH(NHCO_2Et)_2} + H_2O$$

The Grignard reaction of 1,4-dialkoxybut-2-ynes to give allenes appears to involve free radicals:[179]

$$ROCH_2C{\vdots}CCH_2OR \xrightarrow{R'MgBr} ROCH_2C{\vdots}CCH_2\cdot + R'\cdot + BrMgOR$$

$$R'\cdot + ROCH_2C{\vdots}CCH_2\cdot \rightarrow ROCH_2CR'{:}C{:}CH_2$$

The Grignard reaction of *N*-aryl-*N'*-tosyloxydi-imide *N*-oxides also appears to involve radicals, as a significant by-product in tetrahydrofuran incorporates a solvent molecule:[180]

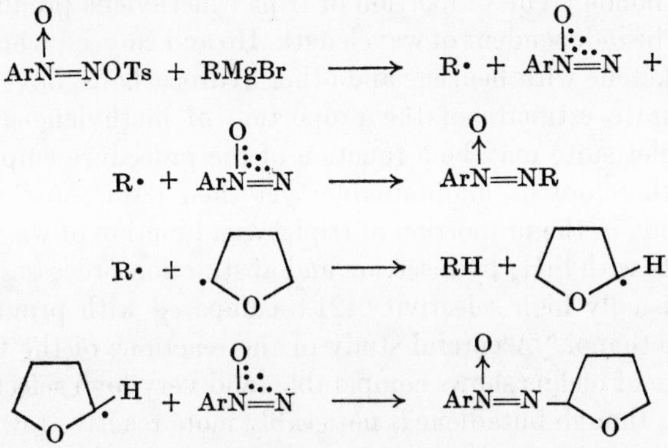

[178] T. A. Foglia and D. Swern, *Tetrahedron Letters*, **1967**, 3963.
[179] G. M. Mkryan, S. M. Gasparyan, E. A. Avetisyan, and S. L. Mndzhoyan, *Zh. Organ. Khim.*, **3**, 808 (1967); *Chem. Abs.*, **67**, 43326m (1967).
[180] T. E. Stevens, *J. Org. Chem.*, **32**, 1641 (1967).

CHAPTER 10

Carbenes and Nitrenes

A survey of the work of Köbrich's group on carbenoids of the α-halogenoalkyl-(and α-halogenoalkenyl)-lithium type has appeared,[1] as well as a review of the chemistry of valence-deficient carbene analogues such as atomic oxygen and sulphur, and univalent boron, aluminium, and phosphorus, etc.,[2a] and one on nitrenes.[2b]

Reports continue to appear concerning the gas-phase reactions of methylene generated from ketene or diazomethane, but fundamental features of these reactions are still far from clear. For example, ethyl chloride is considered to react with both singlet and triplet methylene by chlorine or hydrogen abstraction, not by insertion. The singlet shows a pronounced preference for attack at chlorine.[3] Triplet insertion reported last year[4] could not be substantiated,[5] nor could any effect of wavelength used to photolyse diazomethane be detected in the proportions of products of insertion of the resulting singlet methylene into different CH bonds.[5] The proportion of triplet methylene produced was also considered to be independent of wavelength. Ho and Noyes,[6] who studied the photolysis of ketene with benzene and other hydrocarbons, have pointed out that an accurate estimate of the proportion of methylene originally formed in the triplet state may be a function of the procedure employed to measure it, and therefore be unobtainable; yet their data show clearly a substantial variation in the proportion of triplet as a function of wavelength.

Methylene reacts with SiH_4 by insertion and abstraction processes,[7] and is inserted with unusually high selectivity (21.8 compared with primary CH) into the OH of methanol.[8] A careful study of the reactions of the two spin states with a range of olefins shows comparable (and very low) selectivity in addition reactions, though butadiene is noticeably more reactive towards the triplet species (6 times as reactive as isobutene).[9] An experimental estimate

[1] G. Köbrich, A. Akhtar, F. Ansari, W. E. Breckoff, H. Büttner, W. Drischel, R. H. Fischer, K. Flory, H. Frölich, W. Goyert, H. Heinemann, I. Hornke, H. R. Merkle, H. Trapp, and W. Zündorf, *Angew. Chem. Intern. Ed. Engl.*, **6**, 41 (1967).

[2a] O. M. Nefedov and M. N. Manakov, *Angew. Chem. Intern. Ed. Engl.*, **5**, 1021 (1966).

[2b] W. Lwowski, *Angew. Chem. Intern. Ed. Engl.*, **6**, 897 (1967).

[3] C. H. Bamford, J. E. Casson, and A. N. Hughes, *Chem. Commun.*, **1967**, 1096.

[4] See *Organic Reaction Mechanisms*, **1966**, 279.

[5] B. M. Herzog and R. W. Carr, *J. Phys. Chem.*, **71**, 2688 (1967).

[6] S.-Y. Ho and W. A. Noyes, *J. Am. Chem. Soc.*, **89**, 5091 (1967).

[7] J. W. Simons and C. J. Mazac, *Can. J. Chem.*, **45**, 1717 (1967).

[8] J. A. Kerr, B. V. O'Grady, and A. F. Trotman-Dickenson, *J. Chem. Soc., A*, **1967**, 897.

[9] S. Krzyzanowski and R. J. Cvetanović, *Can. J. Chem.*, **45**, 665 (1967).

of the energy separation between the two spin states of methylene puts the triplet *ca.* 2.5 kcal $mole^{-1}$ below the singlet.[10a] Photolysis of diazomethane in propylene oxide gives methylene insertion products together with acetone; the latter product apparently comes from rearrangement of the epoxide following energy transfer from excited diazomethane.[10b]

Skell and Engel have elaborated some earlier results on the reactions of atomic carbon[11] and, with Plonka,[12] have also looked at its behaviour with a number of oxygen compounds. Oxygen abstraction occurs, e.g., reactions (1)

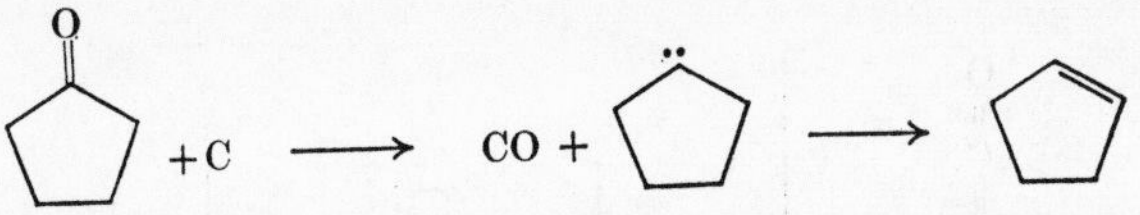

. . . (1)

$$Et_2O + C \longrightarrow CO + 2Et\cdot \longrightarrow C_2H_4 + C_2H_6 \quad \ldots(2)$$

and (2). The reaction of ${}^1C^1$ with benzene produces polymeric products, presumably because of the difficulty with which a C_7H_6 unit can proceed to simple products of low molecular weight.[13]

The collapse of cyclopropylidenes to allenes is the subject of an interesting communication by Borden[14] who, by thermodynamic and symmetry arguments, predicts that triplet cyclopropylidene cannot readily react in this fashion. Evidence from Skell's work with atomic carbon[11] supports this; the triplet adds instead to a second mole of olefin to give a spiropentane as in reaction (3).

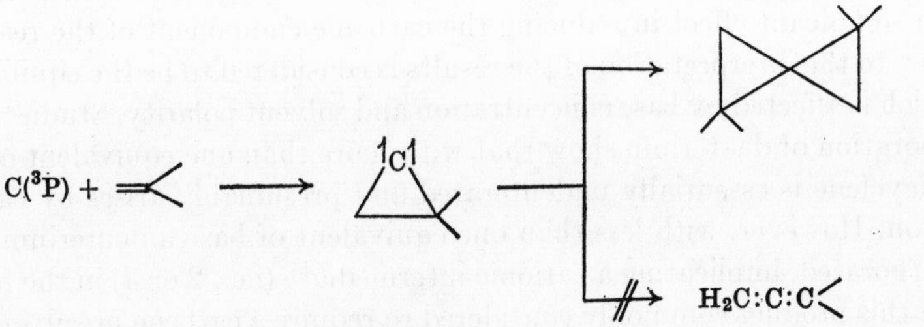

. . . (3)

A novel route to allenes, which is successful, involves the formation and collapse of diazo-heterocycles (**1**).[15a] Attempts to trap the possible carbene

[10a] M. L. Halberstadt and J. R. McNesby, *J. Am. Chem. Soc.*, **89**, 3417 (1967).
[10b] J. N. Bradley and A. Ledwith, *J. Chem. Soc., B*, **1967**, 96.
[11] P. S. Skell and R. R. Engel, *J. Am. Chem. Soc.*, **89**, 2912 (1967); see *Organic Reaction Mechanisms*, **1965**, 224.
[12] P. S. Skell, J. H. Plonka, and R. R. Engel, *J. Am. Chem. Soc.*, **89**, 1748 (1967).
[13] T. Rose, C. MacKay, and R. Wolfgang, *J. Am. Chem. Soc.*, **89**, 1529 (1967).
[14] W. T. Borden, *Tetrahedron Letters*, **1967**, 447.
[15a] R. Kalish and W. H. Pirkle, *J. Am. Chem. Soc.*, **89**, 2781 (1967).

intermediate failed. Photolysis of carbon suboxide in 1,2-dimethylcyclopropene gives a vinylacetylene, possibly as shown. There was no evidence for intramolecular insertion to form dimethyltetrahedrane.[15b].

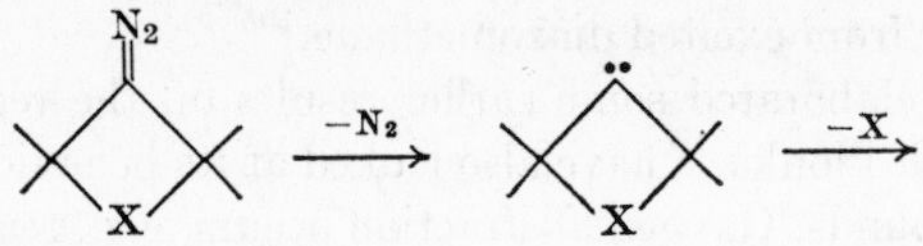

(1) $X=N_2$ or SO_2

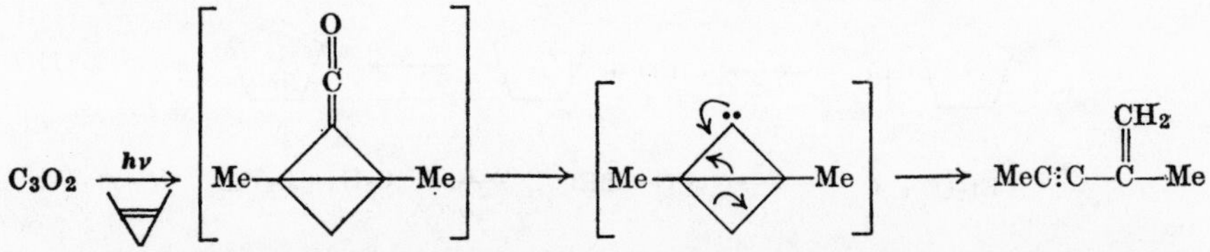

Scrutiny of the Bamford–Stevens reaction continues. Aprotic base-promoted decomposition of camphor tosylhydrazone gives camphene and tricyclene as major products, and the ratio of these (camphene : tricyclene) has been studied as a function of base and solvent polarity.[16] This ratio tends to zero as the concentration of base (sodium methoxide) is increased. A smaller trend in the same direction is observed as the solvent polarity is decreased from that of DMF to that of decalin. However, with an excess of base, addition of as much as 8% by volume of water to a reaction in diglyme (diethylene glycol dimethyl ether) still gives essentially only tricyclene; the protic solvent has no significant effect in reducing the carbenic component of the reaction. The key to the interpretation of the results is considered to be the equilibrium (4) which is affected by base concentration and solvent polarity. Studies of the incorporation of deuterium show that with more than one equivalent of base the tricyclene is essentially undeuterated and presumably arises by carbene insertion. However, with less than one equivalent of base a deuterium atom is incorporated, implicating a cationic intermediate (i.e., **2** or **3**) in the formation of this product commonly considered to require a carbene precursor. The camphene appears to arise through one of the cationic intermediates. The effect of non-sodium bases in promoting camphene formation was discussed last year[17] for aluminium and is now extended to lithium; a cationic complex (**4**) is considered to be formed.

In the proposed scheme, a detail that is not fully discussed, is the fact that sodium hydride gives results comparable to those for sodium methoxide.

[15b] H. W. Chang, A. Lautzenheiser, and A. P. Wolf, *Tetrahedron Letters*, **1966**, 6295.

[16] R. H. Shapiro, J. H. Duncan, and J. C. Clopton, *J. Am. Chem. Soc.*, **89**, 471, 1442 (1967); see also H. Babad, W. Flemon, and J. B. Wood, *J. Org. Chem.*, **32**, 2871 (1967).

[17] See *Organic Reaction Mechanisms*, **1966**, 284.

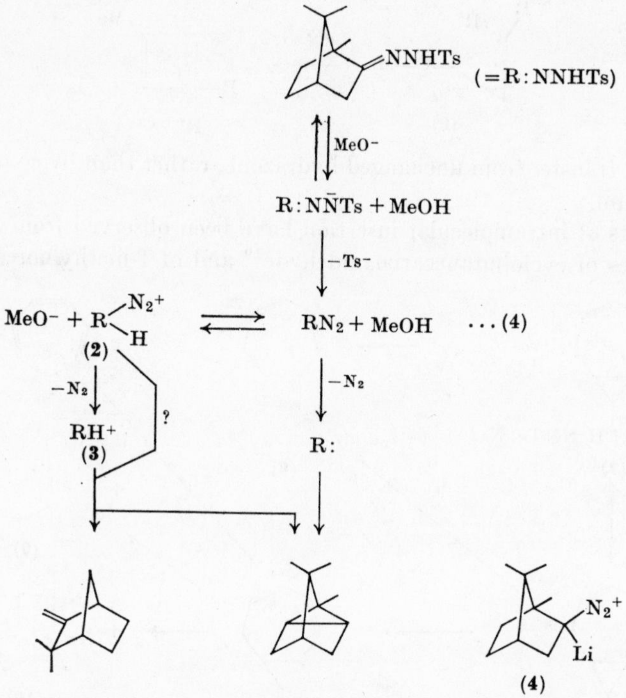

Here the proton source, which with less than a mole of base affords (**2**), must be unchanged tosylhydrazone. This has also been considered to be a reasonable explanation of the formation of cation-derived products by Biellmann and Pète[18] using an excess of lithium hydride as base in the reaction of (**5**). The French authors suggest that reaction of the hydrazone with the lithium base is a slow process, and the diazo-compound is converted into a cationic species

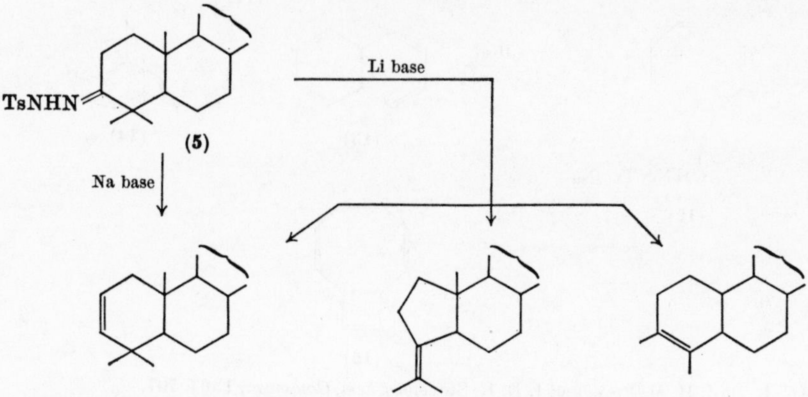

[18] J.-F. Biellmann and J.-P. Pète, *Bull. Soc. Chim. France*, **1967**, 675.

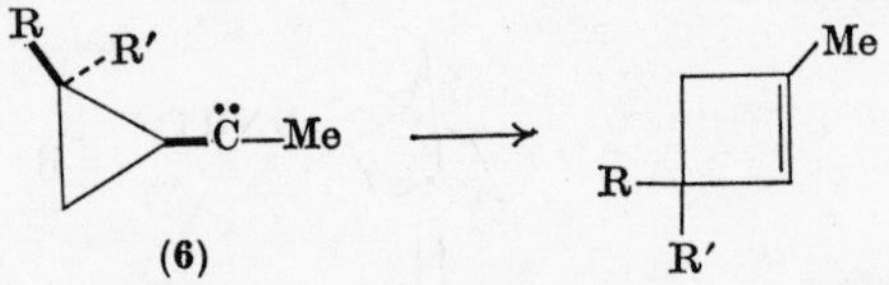

(6)

by proton transfer from unchanged hydrazone, rather than by co-ordination with lithium.

Products of intramolecular insertion have been observed from the tosylhydrazones of cyclobutanecarboxaldehyde[19] and of 1-methylnorcamphor.[20]

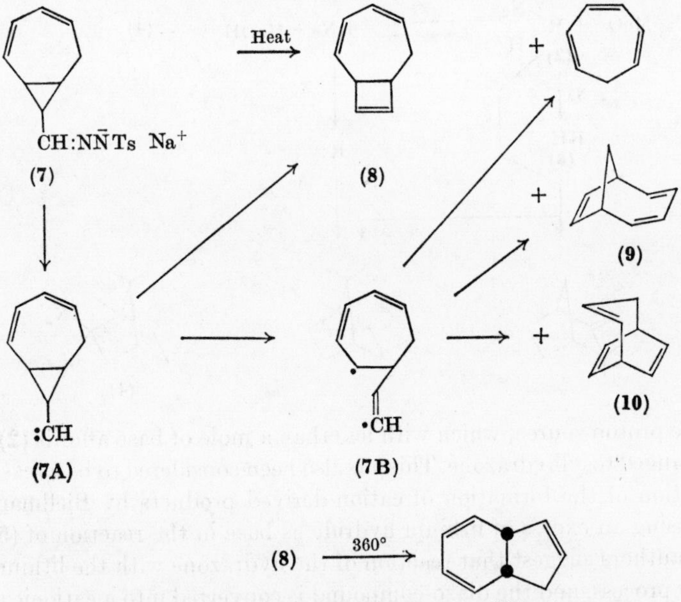

(7) (8) (9) (7A) (7B) (10) (11)

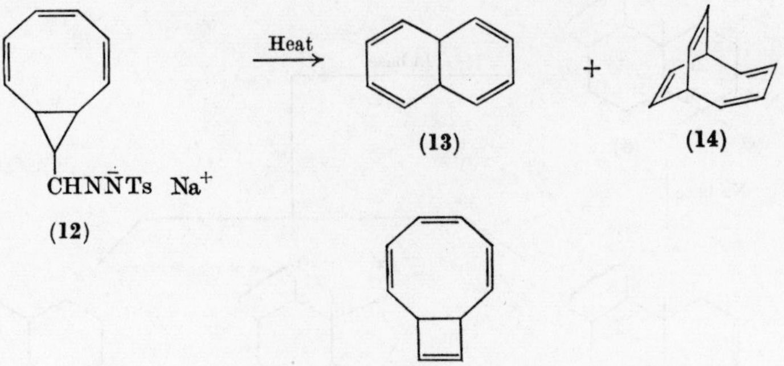

(12) (13) (14) (15)

[19] C. L. Bird, H. M. Frey, and I. D. R. Stevens, *Chem. Commun.*, **1967**, 707.
[20] D. H. Paskovich and P. W. N. Kwok, *Tetrahedron Letters*, **1967**, 2227.

The rearrangement of cyclopropylcarbenes (formed under Bamford–Stevens conditions) to cyclobutenes has been examined to assess the factors that determine which cyclopropane bond breaks. In a series of carbenes (**6**) (R and R′ are methyl or hydrogen), it is always the least substituted carbon that migrates preferentially. The effect may be of steric origin.[21]

Other interesting cyclopropylcarbene precursors have been investigated by M. Jones and his collaborators.[22] Heating (**7**) gives a mixture of hydrocarbons including the cyclobutene (**8**). Pyrolysis of (**8**) gives (**11**) and not (**9**) or (**10**). Reasonable intermediates are considered to be (**7A**) and (**7B**). Products from (**12**) include (**13**) (*cis* and *trans*) and (**14**). By analogy with the reactions of (**7**), (**13**) but not (**14**) may be derived from the expected cyclobutene (**15**) *in situ*.

Tosylhydrazone (**16**) gives (**17**) and (**18**), the latter possibly as indicated.[23a]

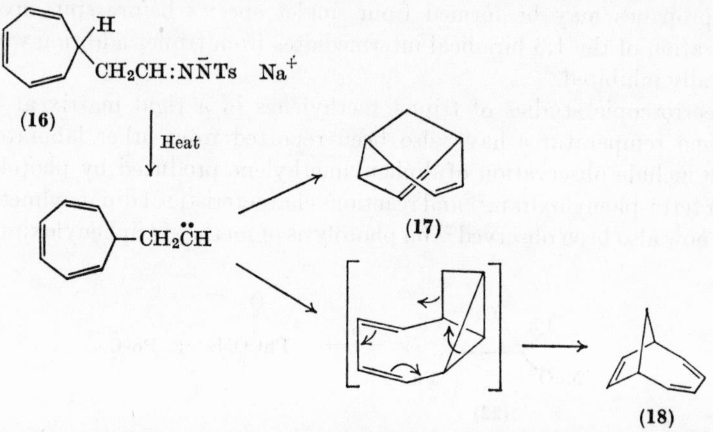

Landgrebe and Kirk[23b] have examined the competition between aryl migration and CH insertion in the carbenes formed on pyrolysis of hydrazone salts (**19**). Electron-releasing substituents in Ar favour the migration. There was no evidence for methyl migration.

$$\mathrm{ArMe_2C{-}CPh{=}N{-}\bar{N}Ts\quad Na^+}$$

(19)

Photolysis of the hydrazone sodium salt (**20**) gives products consistent with formation of the nucleophilic carbene shown.[24] Dimerization to heptafulvalene, and addition to the relatively electrophilic olefin, dimethyl fumarate,

[21] H. Krieger, S. E. Masar, and H. Ruotsalainen, *Suomen Kemistilehti, B*, **39**, 237 (1966).
[22] M. Jones and L. T. Scott, *J. Am. Chem. Soc.*, **89**, 150 (1967); M. Jones and S. D. Reich, *ibid.*, p. 3935.
[23a] H. Tsuruta, K. Kurabayashi, and T. Mukai, *Tetrahedron Letters*, **1967**, 3775.
[23b] J. A. Landgrebe and A. G. Kirk, *J. Org. Chem.*, **32**, 3499 (1967).
[24] W. M. Jones and C. L. Ennis, *J. Am. Chem. Soc.*, **89**, 3069 (1967).

were observed, although no cyclopropane derivative was formed with cyclohexene. The related dibenzo- (**21**) and tribenzo-derivatives have been generated by photolysis of the corresponding diazo-compounds; these species are, however, more closely related to diphenylmethylene, ESR spectra

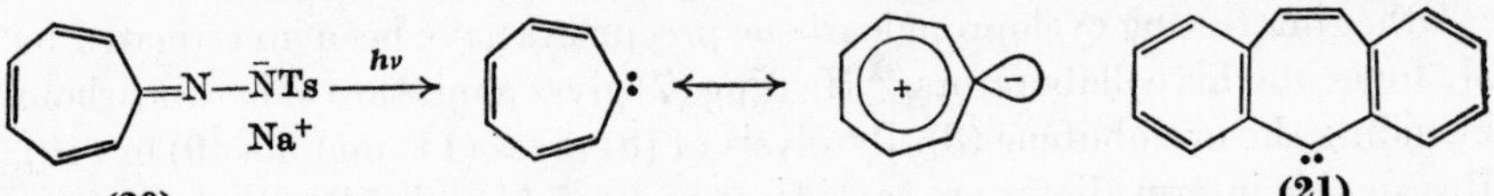

(20) (21)

showing that they have triplet ground states.[25] In view of this, their stereospecific addition to isomeric butenes[26] may seem surprising, but cyclopropane yields are low and are accompanied by typically free-radical products. The cyclopropanes may be formed from singlet species before spin inversion. Cyclization of the 1,3-biradical intermediates from triplet addition would be sterically inhibited.

Spectroscopic studies of triplet methylenes in a rigid matrix at liquid-nitrogen temperatures have also been reported from other laboratories.[27] These include observation of diphenylmethylene produced by photolysis of tri- or tetra-phenyloxiran,[28] and reactions characteristic of diphenylmethylene have now also been observed[29] on photolysis of methoxytriphenyloxiran (**22**).

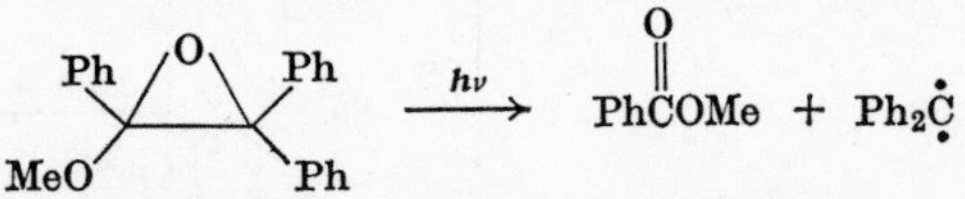

(22)

This reaction has been extended to provide sources of cyanophenylcarbene and methoxycarbonylphenylcarbene.[30] The mode of photofragmentation of the glycidic ester (**23**) led to the hypothesis that initial ring cleavage may generate the most favourable zwitterion (e.g., **24**) which then decomposes to carbonyl compound and carbene.[30]

A carbenoid reaction of ethylene oxides has been disclosed by the work of

[25] I. Moritani, S.-I. Murahashi, M. Nishino, Y. Yamamoto, K. Itoh, and N. Mataga, *J. Am. Chem. Soc.*, **89**, 1259 (1967).
[26] S.-I. Murahashi, I. Moritani, and M. Nishino, *J. Am. Chem. Soc.*, **89**, 1257 (1967).
[27] A. M. Trozzolo and W. A. Gibbons, *J. Am. Chem. Soc.*, **89**, 239 (1967); R. E. Moser, J. M. Fritsch, and C. N. Matthews, *Chem. Commun.*, **1967**, 770.
[28] A. M. Trozzolo, W. A. Yager, G. W. Griffin, H. Kristinnsson, and I. Sarkar, *J. Am. Chem. Soc.*, **89**, 3357 (1967).
[29] T. I. Temnikova and I. P. Stepanov, *Zh. Organ. Khim.*, **2**, 1525 (1966).
[30] P. C. Petrellis and G. W. Griffin, *Chem. Commun.*, **1967**, 691; P. C. Petrellis, H. Dietrich, E. Meyer, and G. W. Griffin, *J. Am. Chem. Soc.*, **89**, 1967 (1967).

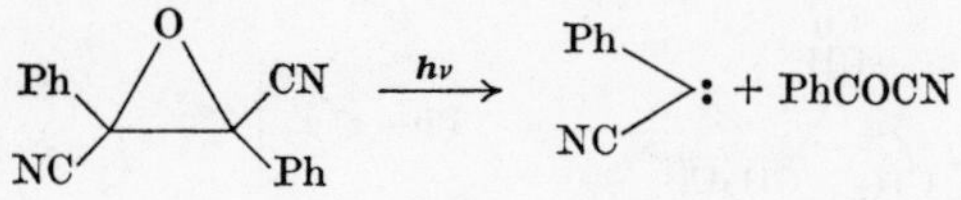

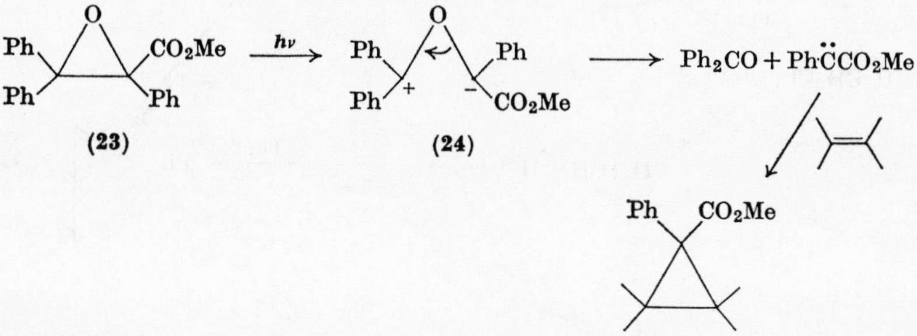

(23) (24)

Crandall *et al.* For example, α-elimination competes with β-elimination when cycloheptene oxide is treated with base, insertion products being formed.[31] In the case of epoxide (**25**), a minor product of α-elimination is the intramolecular adduct (**27**) probably formed directly from the organolithium carbenoid (**26**).[32] These reactions are discussed in greater detail in Chapter 4 (p. 125).

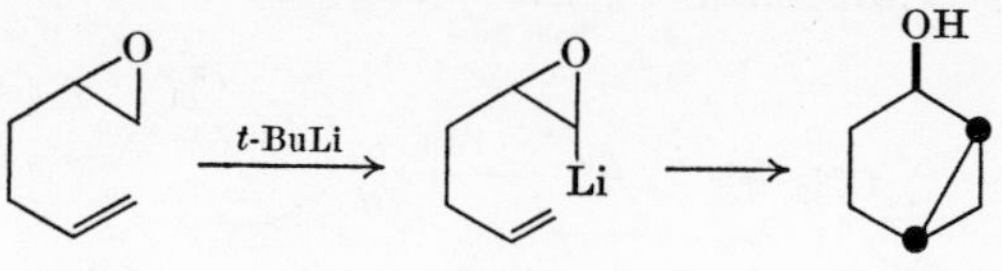

(25) (26) (27)

A carbene (or carbenoid) mechanism has also been found for the formation of phenylcyclopropane which accompanies allylbenzene formed in the reaction between allyl chloride and phenyl-lithium.[33] Mechanisms represented by (5) and (6) were distinguished by employing [1,1-$^2H_2$]allyl chloride. The resulting mixture of monodeuterated cyclopropanes (**28**) and (**29**) established the α-elimination route. A marked kinetic isotope effect was also noted, and it was confirmed that phenyl-lithium adds to cyclopropene.

The details of a kinetic study of the reaction of $PhHgCCl_2Br$ with a variety of olefins in benzene have at last appeared; they are in accord with a free carbene mechanism.[34] The rate of cyclopropanation is almost independent of olfin concentration, but it does show a slight dependence on olefin reactivity when olefins encompassing a wide range of reactivities towards dichloro-

[31] J. K. Crandall and L.-H. Chang, *J. Org. Chem.*, **32**, 435, 532 (1967).
[32] J. K. Crandall and L.-H. C. Lin, *J. Am. Chem. Soc.*, **89**, 4526 (1967).
[33] R. M. Magid and J. G. Welch, *J. Am. Chem. Soc.*, **88**, 5681 (1966).
[34] D. Seyferth, J. Y.-P. Mui, and J. M. Burlitch, *J. Am. Chem. Soc.*, **89**, 4953 (1967).

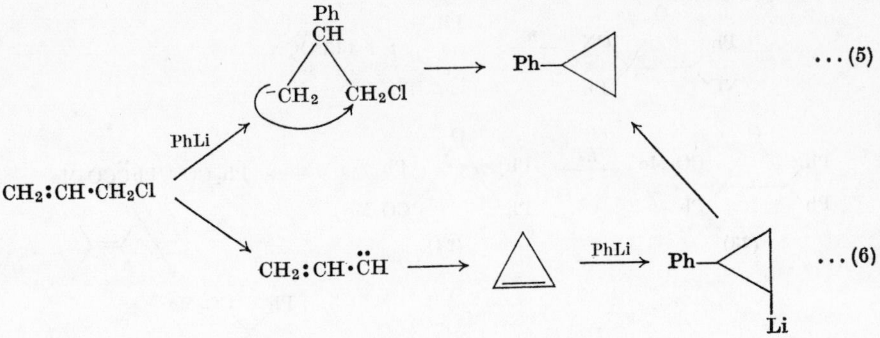

(28) (29)

carbene addition are compared. The reaction is retarded by added PhHgBr, and the results are consistent with the following scheme:

$$PhHgCCl_2Br \underset{Fast}{\overset{Slow}{\rightleftharpoons}} PhHgBr + :CCl_2$$

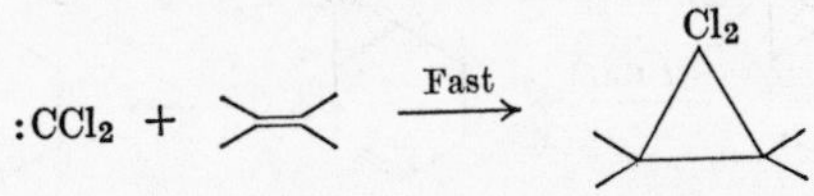

The organomercurial route has been used for a miscellany of novel halogenocarbene reactions: addition to alkenylcarboranes,[35] insertion into the strained C–Si bond of 1,1-dimethylsilacyclobutane,[36] into an Sn–Sn bond,[37] and into Si–H and Ge–H bonds.[38, 39] With the optically active silane (**30**) asymmetry was retained in the product, but insertion of diphenylmethylene and phenylnitrene gave racemic products, presumably by free-radical pathways.

The very interesting, and unexplained, observation has been made that insertion of $:CCl_2$ into C–H β to a mercury atom occurs with inversion of configuration,[40] in contrast to the retention which would have been predicted according to the discussion in last year's report.[41]

35 D. Seyferth and B. Prokai, *J. Organometal. Chem.*, 8, 366 (1967).
36 D. Seyferth, R. Damrauer, and S. S. Washburne, *J. Am. Chem. Soc.*, **89**, 1538 (1967).
37 D. Seyferth and F. M. Armbrecht, *J. Am. Chem. Soc.*, **89**, 2790 (1967).
38 D. Seyferth, J. M. Burlitch, H. Dertouzos, and H. D. Simmons, *J. Organometal. Chem.*, **7**, 405 (1967).
39 A. Ritter and L. H. Sommer, *Intern. Symp. Organosilicon Chem. Sci. Commun., Prague*, 1965, p. 279; *Chem. Abs.*, **66**, 10331 (1967).
40 J. A. Landgrebe and D. E. Thurman, *J. Am. Chem. Soc.*, **89**, 4542 (1967).
41 *Organic Reaction Mechanisms*, **1966**, 298.

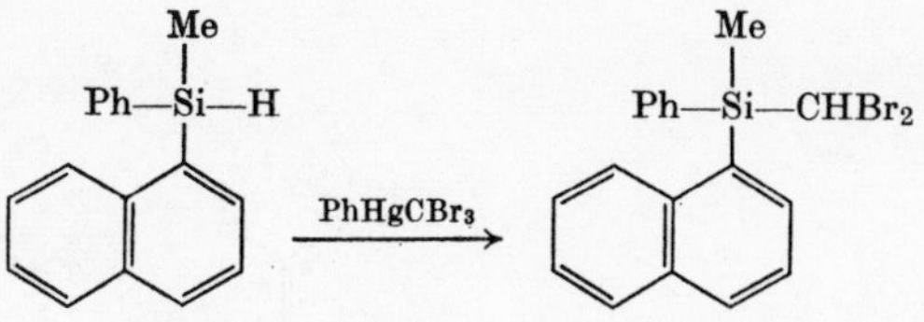

(30)

The formation of haloforms from HCl and $PhHgCX_2Br$ appears to proceed by parallel unimolecular and bimolecular paths, the latter predominating at lower temperatures. The unimolecular component is discussed in terms of protonation of the dihalogenocarbene and ion-pair collapse:[42]

$$:CX_2 + HCl \rightarrow [H\overset{+}{C}X_2 \quad Cl^-] \rightarrow HCX_2Cl$$

Phenyl(trichloromethyl)mercury reacts relatively rapidly with triethylamine, but products arising from dichlorocarbene are isolated (in poor yield). It was suggested that the high reaction rate could be explained if the carbene precursor were (**31**). The major carbene product (**32**) in fact contains two carbene fragments.[43] Results on iodide-promoted halogenocarbene formation

$PhHgN^+Et_3 \quad ^-CCl_3$ **(31)** $Et_2NCCl{:}CCl_2$ **(32)**

from (trihalogenoalkyl)metal compounds have been elaborated by Seyferth's group,[44] and new work on the Simmons–Smith reaction suggests that the carbenoid might be (**33**), present in equilibrium with (iodomethyl)zinc iodide.[45]

$$2ICH_2ZnI \rightleftharpoons (IZn)_2CH_2 + I_2CH_2$$

(33)

Methylene chloride reacts with evaporated films of magnesium or zinc to give the related $CH_2(MCl)_2$, which species transfer methylene to olefins to form cyclopropanes.

The work on α-chloroalkyl(and alkenyl)-lithium derivatives as carbenoids has been reviewed,[1] and there have been numerous new research publications

[42] D. Seyferth, J. Y.-P. Mui, L. J. Todd ,and K. V. Darragh, *J. Organometal. Chem.*, 8, 29 (1967).
[43] D. Seyferth, M. E. Gordon, and R. Damrauer, *J. Org. Chem.*, **32**, 469 (1967).
[44] D. Seyferth, M. E. Gordon, J. Y.-P. Mui, and J. M. Burlitch, *J. Am. Chem. Soc.*, **89**, 959 (1967); D. Seyferth, H. Dertouzos, R. Suzuki, and J. Y.-P. Mui, *J. Org. Chem.*, **32**, 2980 (1967); see *Organic Reaction Mechanisms*, **1965**, 227.
[45] C. Fauveau, Y. Gault, and F. G. Gault, *Tetrahedron Letters*, **1967**, 3149.

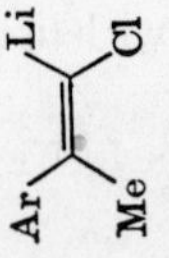

(34)

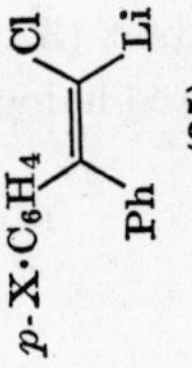

(35)

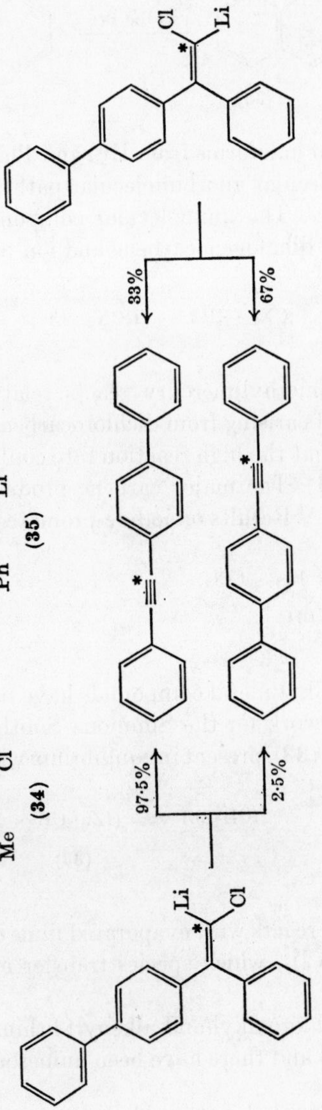

from Köbrich's group in this area.[46–49]. These include a study of geometrical isomerization and halogen exchange in (**34**),[47] and the factors affecting elimination and rearrangement in geometrical isomers of (**35**).[48] There is preference for migration of the group *trans* to the chlorine, but also for migration of the group with electron-releasing substituents (thus p-MeO > p-Me > p-Ph > H > p-Cl). Results are presented here for the case of X = p-Ph.

An interesting "inverse Stevens rearrangement" occurs when halogenoalkyl-lithium carbenoids react with triphenylboron.[49] The reaction is stereospecific, leading to the suggestion that rearrangement occurs in concert with chloride elimination.

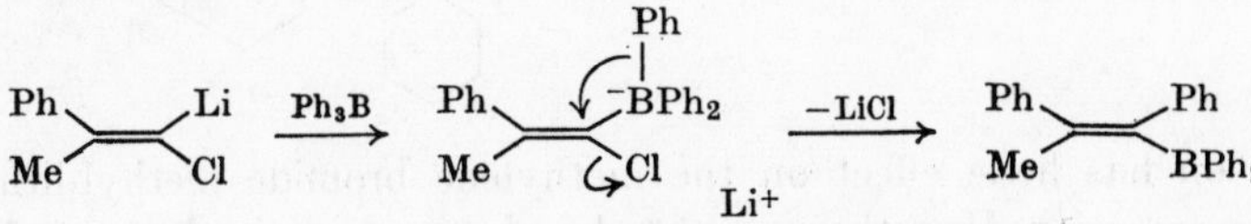

Good evidence for the equilibrium carbenoid (**36**) ⇆ carbene (**37**) has been presented by Seebach,[50] who finds, for example, that the rate of formation of (**38**) is reduced by addition of PhSLi, but increased by addition of electrophiles

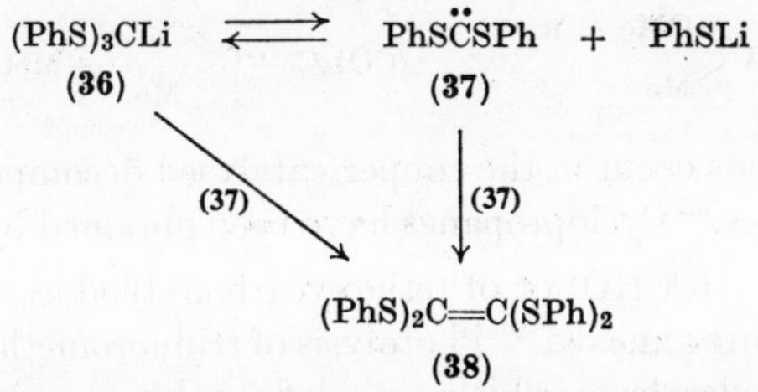

capable of scavenging PhSLi. If MeC_6H_4SLi is added, the tolyl group is incorporated into product molecules.

It is known that cyclopropanes formed by reaction of olefins with halogenomethyl-lithium carbenoids obtained by the use of methyl-lithium often include products with a methyl substituent in place of an expected halogen. The effect becomes more pronounced when the methyl-lithium has been prepared from methyl iodide, so that iodide ions are present. This salt effect has been the subject of several papers this year. Dilling and Edamura have

46 G. Köbrich, H. Heinemann, and W. Zündorf, *Tetrahedron*, **23**, 565 (1967).
47 G. Köbrich and F. Ansari, *Chem. Ber.*, **100**, 2011 (1967).
48 G. Köbrich, H. Trapp, and I. Hornke, *Chem. Ber.*, **100**, 961 (1967).
49 G. Köbrich and H. R. Merkle, *Angew. Chem. Intern. Ed. Engl.*, **6**, 74 (1967); *Chem. Ber.*, **100**, 3371 (1967).
50 D. Seebach, *Angew. Chem. Intern. Ed. Engl.*, **6**, 442 (1967).

established the importance of iodide ions in the methylene chloride–methyl-lithium–cyclohexene system,[51] and suggest that the key to methyl incorporation is sequence (7). Magid and Welch,[52] however, find that additional products are the stereoisomeric 7-iodonorcaranes, and that these can participate in a Wurtz-like reaction with methyl-lithium to form 7-methylnorcaranes. Thus alkylation of the carbenoid species may not occur.

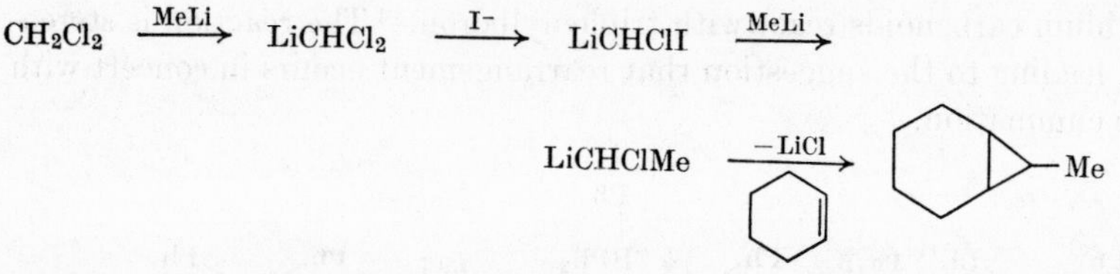

...(7)

Iodide ion has little effect on the methylene bromide–methyl-lithium–cyclohexene reaction. Here the major product is norcarane itself, formed from $LiCH_2Br$ which is a product of halogen–metal interchange.[53]

Fischer's group have examined the chemistry of transition metal–carbene complexes,[54] and have observed some interesting substituent exchange reactions (e.g., 8).

$$(CO)_5Cr\cdots C\langle^{\cdots OMe}_{Me} \xrightarrow{R_2NH} (CO)_5Cr\cdots C\langle^{\cdots NR_2}_{Me} + MeOH \qquad \ldots(8)$$

Carbenoid reactions occur in the copper-catalysed decomposition of carbohydrate diazoketones.[55] Cyclopropanes have been obtained by olefin addition of formylcarbene[56] ($H\ddot{C}CHO$) or of (ethoxycarbonyl)iodocarbene which was generated by the route annexed.[57] Photolysis of trifluoromethyldiazomethane gives trifluoromethylcarbene which, in olefin solution, gives products of

$$EtO_2CC(:N_2)\text{—}Hg\text{—}C(:N_2)CO_2Et \xrightarrow[I_2]{} I\text{—}C(:N_2)CO_2Et \rightarrow I\text{—}\ddot{C}\text{—}CO_2Et$$

stereospecific addition. In the gas phase, addition is non-stereospecific, and it was suggested that unimolecular crossing to the triplet state occurs, excess energy being accommodated by the vibrational modes of the CF_3 group.[58]

[51] W. L. Dilling and F. Y. Edamura, *Chem. Commun.*, **1967**, 183; *J. Org. Chem.*, **32**, 3492 (1967).
[52] R. M. Magid and J. G. Welch, *Tetrahedron Letters*, **1967**, 2619; *Chem. Commun.*, **1967**, 518.
[53] W. L. Dilling and F. Y. Edamura, *Tetrahedron Letters*, **1967**, 587; *J. Org. Chem.*, **32**, 3492 (1967).
[54] E. O. Fischer and R. Aumann, *Angew. Chem. Intern. Ed. Engl.*, **6**, 181 (1967); E. O. Fischer and A. Maasböl, *Chem. Ber.*, **100**, 2445 (1967); J. A. Connor and E. O. Fischer, *Chem. Commun.*, **1967**, 1024.
[55] Y. A. Zhdanov, V. I. Komilov, and G. V. Boydanova, *Carbohydrate Res.*, **3**, 139 (1966).
[56] Z. Arnold, *Chem. Commun.*, **1967**, 299.
[57] F. Gerhart, U. Schöllkopf, and H. Schumacher, *Angew. Chem. Intern. Ed. Engl.*, **6**, 74 (1967).
[58] J. H. Atherton and R. Fields, *J. Chem. Soc., C*, **1967**, 1450.

Free-radical behaviour has been observed for anthronylidene (**39**).[59] No products of radical coupling at oxygen were noted. Further work on the triplet behaviour of fluorenylidene has appeared,[60] and carbene reactions in

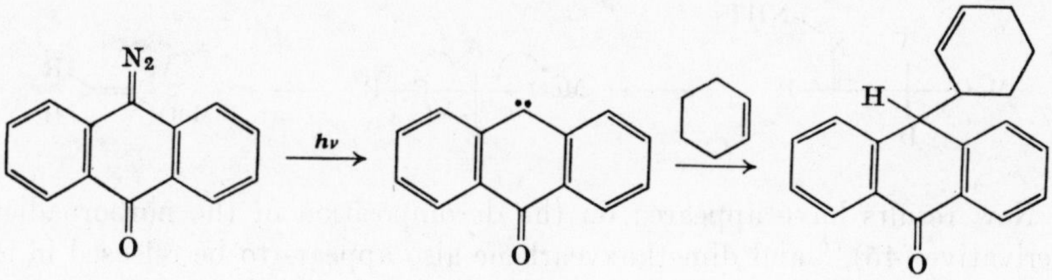

(**39**)

the photolyses of (**40**)[61] and (**41**)[62a] have been reported. Heating (**40**) with triphenylarsine gives an arsoniumcyclopentadienylide.[62b]. Heating or photo-

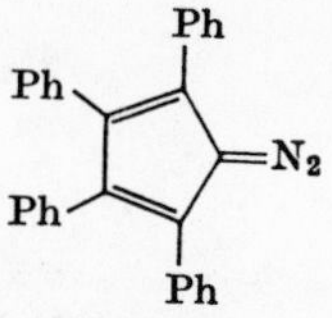

(**40**)

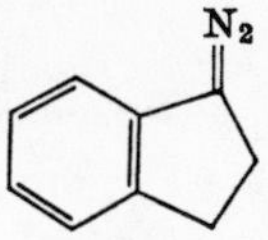

(**41**)

lysis of (**42**) gave the dimer (**44**) of aminocyanocarbene (**43**), but the intermediate could not be trapped, nor could an ESR spectrum be detected on

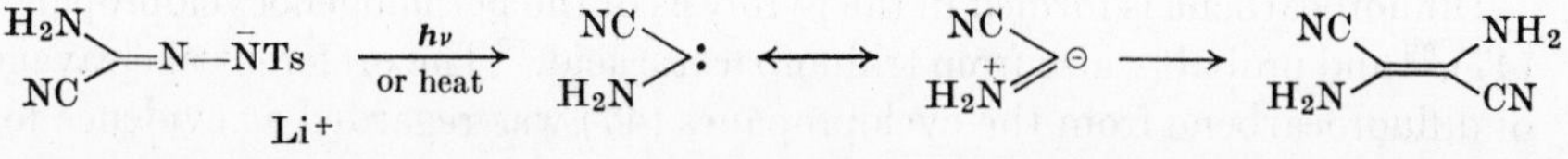

(**42**) (**43**) (**44**)

low-temperature photolysis.[63] There is probably considerable stabilization as indicated [compare (**20**) and (**81**)]. This would be accentuated by the cyano-group.

Stereospecific addition has been reported for the arylsulphonylcarbene p-MeOC$_6$H$_4$SO$_2\ddot{C}$H.[64]

[59] G. Cauquis and G. Reverdy, *Tetrahedron Letters*, **1967**, 1493.
[60] S. Murahashi, I. Moritani, and T. Nagai, *Bull. Chem. Soc. Japan*, **40**, 1655 (1967).
[61] H. Dürr and G. Scheppers, *Chem. Ber.*, **100**, 3236 (1967).
[62a] R. A. Moss and J. D. Funk, *J. Chem. Soc., C*, **1967**, 2026.
[62b] D. Lloyd and M. I. C. Singer, *Chem. Ind.* (*London*), **1967**, 510.
[63] R. E. Moser, J. M. Fritsch, T. L. Westman, R. M. Kliss, and C. N. Matthews, *J. Am. Chem. Soc.*, **89**, 5673 (1967).
[64] A. M. Van Leusen, R. J. Mulder, and J. Strating, *Rec. Trav. Chim.*, **86**, 226 (1967).

Base-promoted decompositions of the tosylhydrazones of α-alkoxy-ketones give olefins arising from preferential migration of other substituent groups (H, R, OR) on the oxygenated α-carbon, as indicated.[65]

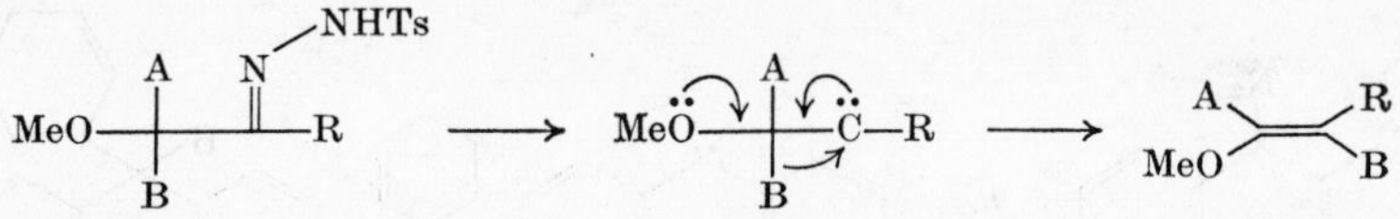

New results have appeared on the decomposition of the norbornadiene derivative (**45**),[66] and dimethoxycarbene also appears to be released in the related pyrolysis of (**46**).[67] Decomposition of certain chlorocarbons by high-frequency discharge, followed by collection of the pyrolysate at 77°K, gave a transient species whose spectrum was ascribed to $:CCl_2$.[68]

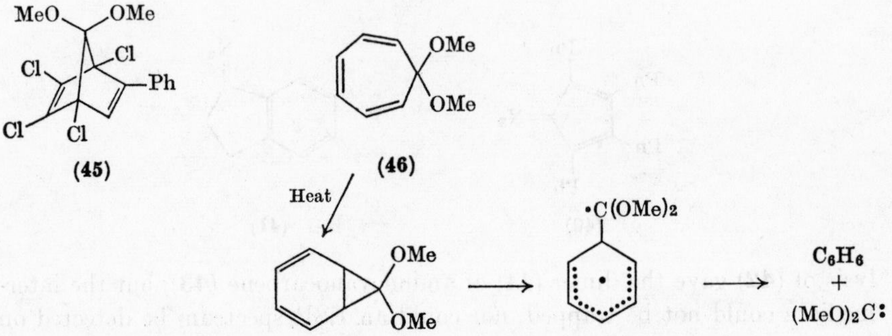

Difluorocarbene is formed in the pyrolysis of the perhalogenocyclopropanes (**47**),[69] and probably also from trifluoroacetic acid.[70] The preferential cleavage of difluorocarbene from the cyclopropanes (**47**) was regarded as evidence for

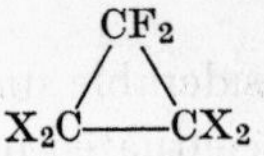

X = Cl or F $CHF_2CF_2SiF_3$

(**47**) (**48**)

the relatively great stabilization in this species. Difluorocarbene is also a probable intermediate in the easy aqueous-alkaline hydrolysis of $CHClF_2$ to

65 W. Kirmse and M. Buschhoff, *Chem. Ber.*, **100**, 1491 (1967).
66 R. W. Hoffmann and C. Wünsche, *Chem. Ber.*, **100**, 943 (1967); see *Organic Reaction Mechanisms*, **1966**, 293.
67 R. W. Hoffmann and J. Schneider, *Tetrahedron Letters*, **1967**, 4347.
68 R. Steudel, *Tetrahedron Letters*, **1967**, 1845.
69 J. M. Birchall, R. N. Haszeldine, and D. W. Roberts, *Chem. Commun.*, **1967**, 287.
70 P. G. Blake and H. Pritchard, *J. Chem. Soc., B*, **1967**, 282.

give formate,[71a] and has been observed in flash photolyses of trifluoromethyl phosphorus dihalides.[71b] Its reactions with oxygen have been examined.[71c]

Pyrolysis (150°) of the silane (**48**)[72] gives trifluoroethylene and, with alkenes or alkanes in the gas phase, products of addition or selective (*tert* ≫ *sec* ≫ *prim*) CH insertion by the carbene CHF_2CF**:**.

Chlorofluorocarbene generated by reaction (9) appears to be more selective in its behaviour towards olefins than does dichlorocarbene, again reflecting the stabilizing influence of fluorine.[73] This effect is less pronounced with the

$$(FCCl_2)_2CO \xrightarrow{BuO^-} :CClF \qquad \ldots(9)$$

less reactive olefins, possibly because the tighter transition states for reactions of these substrates experience more pronounced steric hindrance in the case of the more bulky dichlorocarbene. On the other hand, the major isomer from addition to *cis*-butene is (**49**), suggesting that the polarizability of the chlorine substituent here outweighs the effect of its bulk.[74]

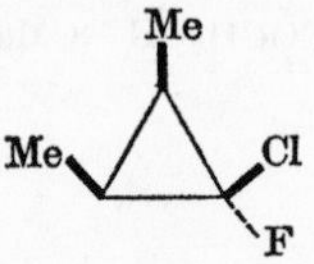

(**49**)

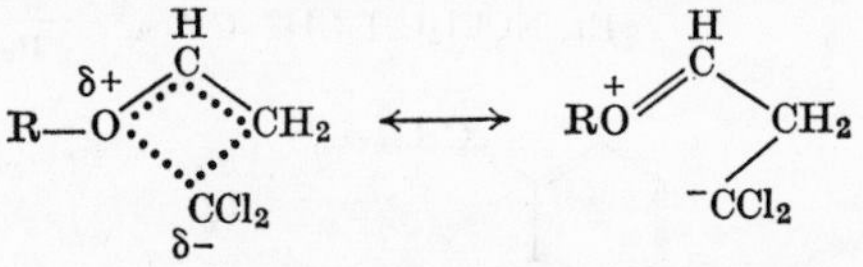

(**50**)

Comparison of the rates of dichlorocarbene additions to a series of enol ethers with rates of acidic hydrolyses suggests direct participation of the ether oxygen (e.g., as **50**).[75] Additions to allenes[76a] (with subsequent α-elimination to give cumulenes[76b]) have been reported, and addition to the indene derivatives (**51**) provides a new route to metacyclophanes (**52**).[77] Addition to allylamines gives cyclopropanes, in contrast to allyl sulphides which react by initial ylide formation.[78] Although the nitrogen ylides are probably less stable, nitrogen involvement may be a competing process (reaction 10).

[71a] T. Hayashi, *Kogyo Kagaku Zasshi*, **68**, 2002 (1965); *Chem. Abs.*, **67**, 21222x (1967).
[71b] R. G. Cavell, R. C. Dobbie, and W. J. R. Tyerman, *Can. J. Chem.*, **45**, 2849 (1967).
[71c] T. Johnston and J. Heicklen, *J. Chem. Phys.*, **47**, 475 (1967).
[72] R. N. Haszeldine and J. G. Speight, *Chem. Commun.*, **1967**, 995.
[73] R. A. Moss and R. Gerstl, *J. Org. Chem.*, **32**, 2268 (1967); *Tetrahedron*, **23**, 2549 (1967).
[74] See *Organic Reaction Mechanisms*, **1966**, 290.
[75] A. Ledwith and H. J. Woods, *J. Chem. Soc., B*, **1967**, 973; other additions of enol ethers are discussed by: W. E. Parham and R. J. Sperley, *J. Org. Chem.*, **32**, 926 (1967), and R. C. De Selms and T.-W. Lin, *Tetrahedron*, **23**, 1479 (1967).
[76a] H. G. Peer and A. Schors, *Rec. Trav. Chim.*, **86**, 161 (1967).
[76b] W. J. Ball, S. R. Landor, and N. Punja, *J. Chem. Soc., C*, **1967**, 194.
[77] W. E. Parham and J. K. Rinehart, *J. Am. Chem. Soc.*, **89**, 5668 (1967); see also W. E. Parham and J. F. Dooley, *ibid.*, p. 985.
[78] W. E. Parham and J. R. Potoski, *J. Org. Chem.*, **32**, 275, 278 (1967).

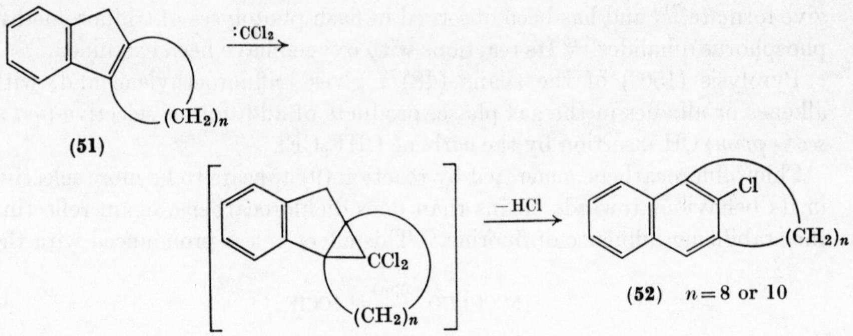

(51)

(52) $n = 8$ or 10

(isolated when $n = 4$)

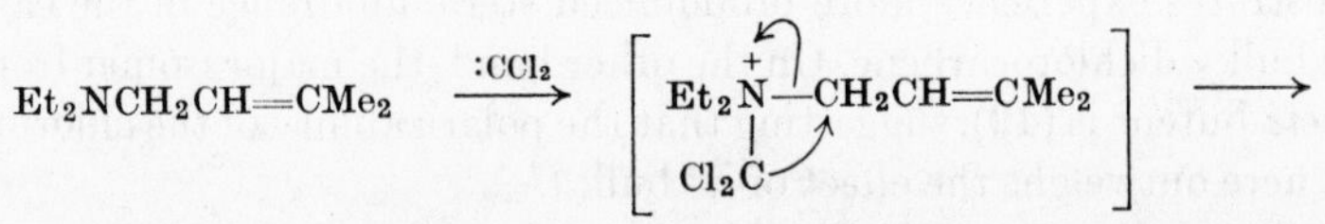

$$Et_2NCCl_2CH_2CH{=}CMe_2 \xrightarrow{H_2O} Et_2NCOCH_2CH{=}CMe_2 \quad \ldots(10)$$

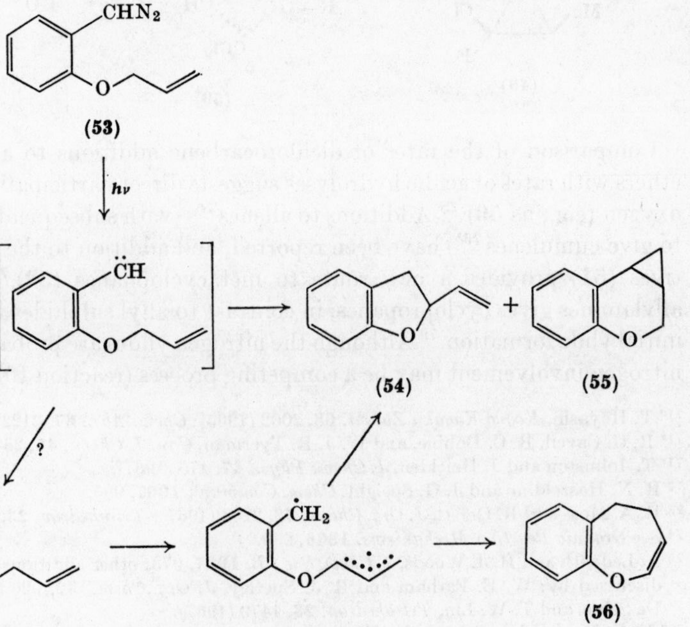

(53) (54) (55) (56)

Four isomeric products of addition of chlorocarbenoid to norbornene were reported last year.[79] Two of these were considered to derive from *endo*-attack,

[79] *Organic Reaction Mechanisms*,, 221 **1966**.

but it has now been found that they are not isomers, but are formed by *exo*-attack of bromocarbenoid formed *in situ*. The bromine arose from methyl bromide employed to generate methyl-lithium.[80]

Photolysis of the allyl ether (**53**) gives the cyclized products (**54**) and (**55**).[81] Evidence that (**56**) is also formed suggests at least some triplet component to the reaction.

The reaction (**57**) → (**58**) appears to constitute a rare example of 1,4-carbenoid addition.[82] Cuprous-catalysed reaction of diazoacetic ester with

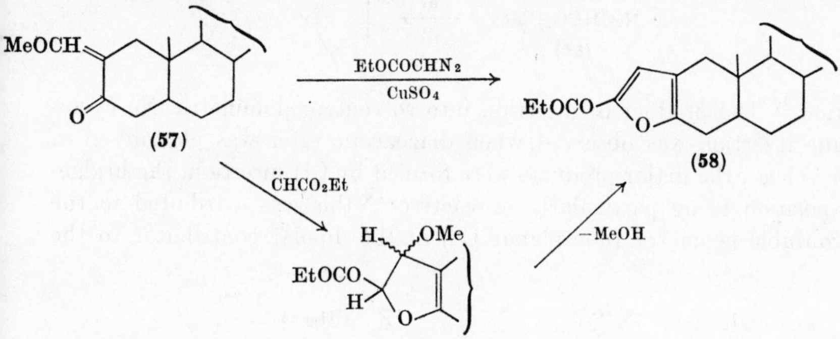

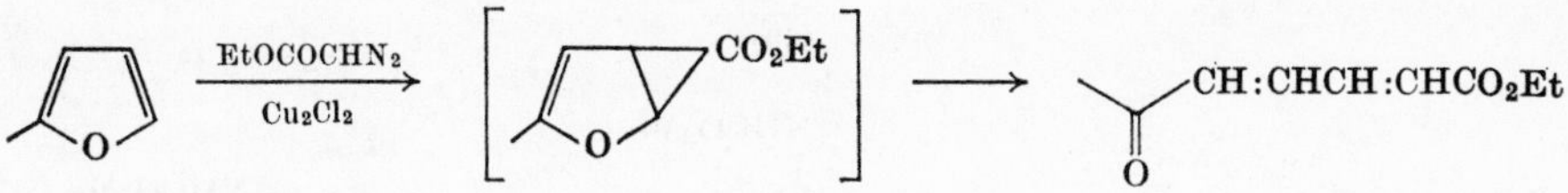

2-methylfuran may involve a steric preference for attack at the unsubstituted double bond.[83a] Photolysis of the same ester with acetylenes gives cyclopropenes,[83b] though whether this was a carbene reaction or involved photodecomposition of intermediate pyrazolines was not established.

The photolysis of dimethyl diazomalonate in the presence of olefins gives cyclopropanes. The *cis*-stereoselectivity falls on dilution with hexafluorobenzene and is absent in the photosensitized reactions. Thus bis(methoxycarbonyl)carbene appears to be capable of forming cyclopropanes from both its singlet and its triplet configuration.[84]

Kirmse's group has compared the products of metal-catalysed, and

[80] C. W. Jefford, E. H. Yen, and R. Medary, *Tetrahedron Letters*, **1966**, 6317; see also the discussion above (pp. 289—290).

[81] W. Kirmse and H. Dietrich, *Chem. Ber.*, **100**, 2710 (1967).

[82] D. L. Storm and T. A. Spencer, *Tetrahedron Letters*, **1967**, 1865.

[83a] Y. Noichi, I. Moritani, N. Obata, H. Fujita, and I. Kawanishi, *Kogyo Kagaku Zasshi*, **69**, 1491 (1966); *Chem. Abs.*, **66**, 94585g (1967).

[83b] H. Lind and A. J. Deutschman, *J. Org. Chem.*, **32**, 326 (1967).

[84] M. Jones, A. Kulczycki, and K. F. Hummel, *Tetrahedron Letters*, **1967**, 183; M. Jones, W. Ando, and A. Kulczycki, *ibid.*, p. 1391.

photolytic decomposition of diazoalkanes.[85] In the catalysed reactions olefins formed by alkyl or hydrogen migration are the major products; cyclopropanes are major products of the photolyses, being formed by intramolecular insertion into β-CH bonds. The products from ω-diazo-esters [$N_2CH(CH_2)_nCO_2Me$] are also markedly dependent on the decomposition conditions.[86] The diazo-ester (**59**) gives a small yield of lactone by δ-CH

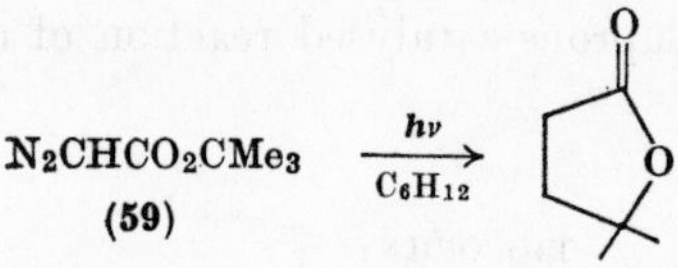

(**59**)

insertion.[87] The product of insertion into solvent predominates. No cyclopropane insertion was observed when diazoacetic ester was photolysed in nortricyclene; the major products were formed by CH insertion, the bridgehead position being particularly unreactive;[88] this was attributed to the unfavourable geometry (non-planar C^+) of the dipolar contributor to the

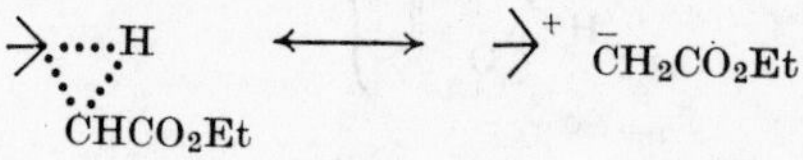

(**60**)

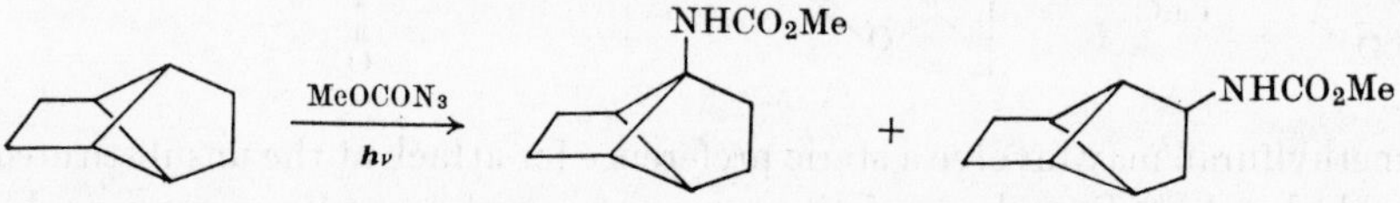

(**61**)

transition state (**60**). On the other hand, nitrene insertion has been used to obtain access to the bridgehead position in (**61**).[89] Tricyclic compounds are among the products of intramolecular insertion from (**62**)[90] and (**63**).[91] In the latter case the yields are unusually high (> 50%). No allene was found.

Carbenes generated in dimethyl sulphoxide solution are oxidized to carbonyl compounds. Thus, in DMSO, treatment of ethyl trichloroacetate with a base produces phosgene, and that of benzaldehyde tosylhydrazone gives benzaldehyde.[92]

85 W. Kirmse and K. Horn, *Chem. Ber.*, **100**, 2698 (1967).
86 S. Hauptmann and K. Hirschberg, *J. Prakt. Chem.*, **34**, 55 (1966).
87 W. Kirmse, H. Dietrich, and H. W. Bücking, *Tetrahedron Letters*, **1967**, 1833.
88 R. R. Sauers and R. J. Kiesel, *J. Am. Chem. Soc.*, **89**, 4695 (1967).
89 J. Meinwald and D. H. Aue, *Tetrahedron Letters*, **1967**, 2317.
90 W. Kirmse and K. Pöhlmann, *Chem. Ber.*, **100**, 3564 (1967).
91 C. G. Cardenas, B. A. Shoulders, and P. D. Gardner, *J. Org. Chem.*, **32**, 1220 (1967).
92 R. Oda, M. Mieno, and Y. Hayashi, *Tetrahedron Letters*, **1967**, 2363.

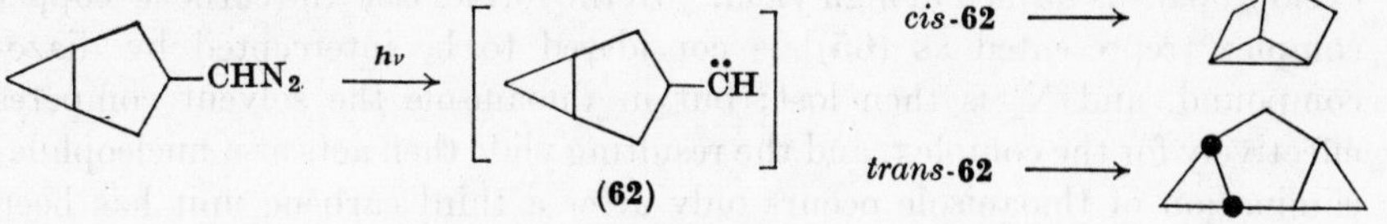

(62)

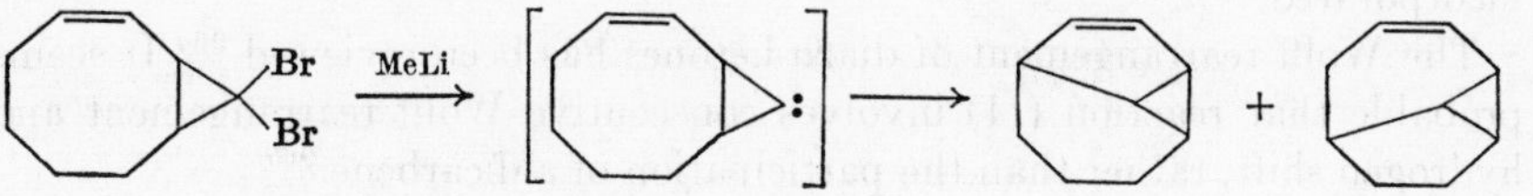

(63)

It has been found that olefins react with methylene iodide in the presence of peroxides to give cyclopropanes.[93] The probable intervention of a free-radical mechanism involving addition of $\cdot CH_2I$ to the olefin clearly bears on the mechanism of photochemical cyclopropanation by methylene iodide.[94a] However, the products of gas-phase photolysis of *gem*-di-iodoalkanes ($\lambda < 240$ mμ) include cyclopropanes which could have arisen by intramolecular carbene insertion.[94b]

The principal product of copper-catalysed decomposition of the diazoketone (**64**) in inert solvents is dipivaloylethylene (2,2,7,7-tetramethyloct-4-

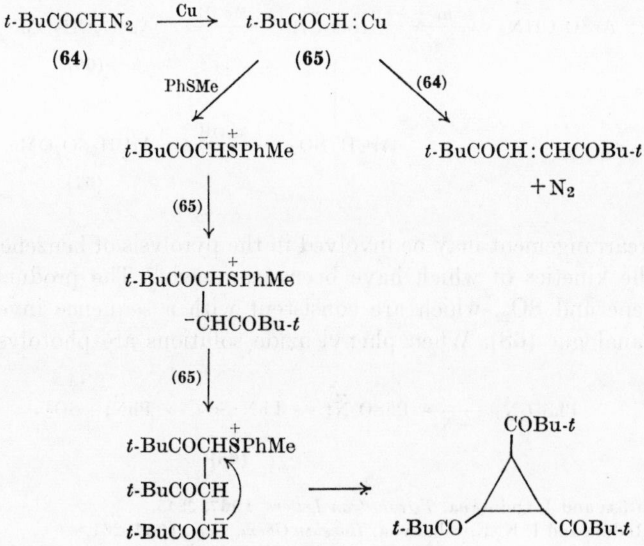

[93] L. Kaplan, *J. Am. Chem. Soc.*, **89**, 4566 (1967).
[94a] See *Organic Reaction Mechanisms*, **1965**, 228.
[94b] R. C. Neuman and R. G. Wolcott, *Tetrahedron Letters*, **1966**, 6267.

ene-3,6-dione). However, in the nucleophilic solvent thioanisole, tripivaloylcyclopropane is formed in high yield.[95] In the former case the carbene–copper complex [represented as (**65**)] is considered to be intercepted by diazo-compound, and N_2 is then lost; but in thioanisole the solvent competes effectively for the complex, and the resulting ylide then acts as a nucleophile; elimination of thioanisole occurs only after a third carbene unit has been incorporated.

The Wolff rearrangement of diazo-ketones has been reviewed.[96a] It seems probable that reaction (11) involves consecutive Wolff rearrangement and hydrogen shift, rather than the participation of a dicarbene.[96b]

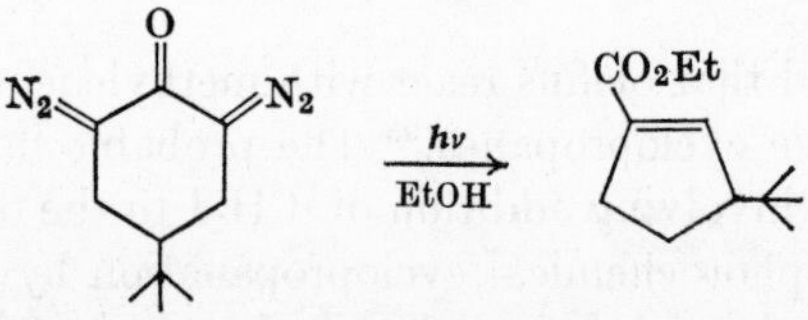

. . . (11)

Related to the Wolff rearrangement is the rearrangement of arylsulphonylcarbenes to sulfene intermediates. Electron-releasing substituents in the aryl group facilitate this, as shown by the proportions of (**66**) and (**67**).[97] A

$$ArSO_2CHN_2 \xrightarrow[-N_2]{h\nu} ArSO_2\ddot{C}H \xrightarrow{MeOH} ArSO_2CH_2OMe \quad (\mathbf{66})$$

$$ArSO_2\ddot{C}H \longrightarrow ArCH{:}SO_2 \xrightarrow{MeOH} ArCH_2SO_2OMe \quad (\mathbf{67})$$

similar rearrangement may be involved in the pyrolysis of benzenesulphonyl azide, the kinetics of which have been examined.[98] The products include azobenzene and SO_2, which are consistent with a sequence involving the sulfene analogue (**68**). When phenyl azide solutions are photolysed in the

$$PhSO_2N_3 \xrightarrow[-N_2]{} PhSO_2\ddot{N}{:} \rightarrow PhN{:}SO_2 \rightarrow Ph\ddot{N}{:} + SO_2$$

(**68**)

[95] F. Serratosa and J. Quintana, *Tetrahedron Letters,* **1967**, 2245.
[96a] L. L. Rodina and I. K. Korobitsyna, *Russian Chem. Rev.,* **1967**, 261.
[96b] R. Tasovac, M. Stefanović, and A. Stojiljković, *Tetrahedron Letters,* **1967**, 2729.
[97] R. J. Mulder, A. M. van Leusen, and J. Strating, *Tetrahedron Letters,* **1967**, 3057.
[98] G. P. Balabanov, Y. I. Dergunov, and V. A. Gal'perin, *Zh. Org. Khim.,* **2**, 1828 (1966); *Chem. Abs.,* **66**, 54848p (1967).

presence of SO_2, a sulphamic acid is formed. At first sight this suggests that the azasulfene may be reconstituted from phenyl nitrene and sulphur dioxide. However, closer inspection points to a radical sequence:[99]

$$PhN\colon \xrightarrow{RH} Ph\dot{N}H \xrightarrow{SO_2} PhNHSO_2 \xrightarrow[RH]{O_2} PhNHSO_3H$$

Copper-catalysed decomposition of benzenesulphonyl azide gives typically free-radical products in a variety of solvents.[100] The complexing of copper in these reactions was represented as in (**69**). Reaction is accelerated by dimethyl sulphoxide which gives (**70**), possibly by the 1,3-dipolar addition indicated.

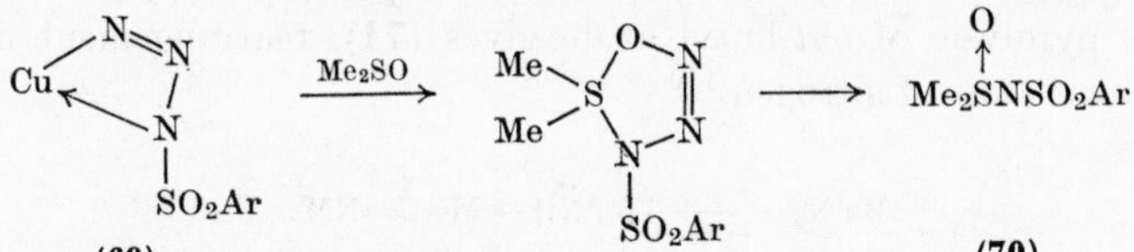

(**69**) (**70**)

Full reports of some of Lwowski's studies of (ethoxycarbonyl)nitrene have appeared,[101] and the use of an inert solvent (e.g., CH_2Cl_2) to permit singlet to triplet crossing has found new applications. For example, it has been shown that addition to benzene to give ethyl azepine-1-carboxylate occurs only from the singlet,[102] and, with a dilute solution of anthracene in chlorobenzene, 9- and 1-anthrylurethane are formed in the ratio 9:1, reflecting free-radical behaviour. At higher anthracene concentrations, increasing singlet reaction (which occurs at the most localized double bond, i.e., between $C_{(1)}$ and $C_{(2)}$) this ratio is reduced.[103] With enol acetates, reactive aziridines are formed,[104] and with cyclic ethers and acetals insertion α to oxygen predominates.[105] With aliphatic hydrocarbons, insertion into primary, secondary, and tertiary CH bonds occurs with the approximate rate ratio 1:10:32;[106] a surprising result is that the yield of urethane formed by singlet insertion into cyclohexane is actually increased by *m*-dinitrobenzene and other radical scavengers.[106, 107]

[99] T. Nagai, K. Yamamoto, and N. Tokura, *Bull. Chem. Soc. Japan*, **40**, 408 (1967).
[100] H. Kwart and A. A. Kahn, *J. Am. Chem. Soc.*, **89**, 1950, 1951 (1967).
[101] J. S. McConaghy and W. Lwowski, *J. Am. Chem. Soc.*, **89**, 2357, 4450 (1967).
[102] W. Lwowski and R. L. Johnson, *Tetrahedron Letters*, **1967**, 891.
[103] A. J. L. Beckwith and J. W. Redmond, *Chem. Commun.*, **1967**, 165; see also *Organic Reaction Mechanisms*, **1966**, 303.
[104] J. F. W. Keana, S. B. Keana, and D. Beetham, *J. Org. Chem.*, **32**, 3057 (1967).
[105] H. Nozaki, S. Fujita, H. Takaya, and R. Noyori, *Tetrahedron*, **23**, 45 (1967).
[106] D. S. Breslow, T. J. Prosser, A. F. Marcantonio, and C. A. Genge, *J. Am. Chem. Soc.*, **89**, 2384 (1967).
[107] D. S. Breslow and E. I. Edwards, *Tetrahedron Letters*, **1967**, 2123.

One possible explanation for this is that radical (or triplet) intermediates in the uninhibited reaction catalyse spin inversion of the nitrene to its triplet ground state.

Rates of azepine formation from substituted benzenes show a Hammett correlation with σ_p is -1.32. A corresponding correlation[108] for norcaradiene formation by (ethoxycarbonyl)carbene gives a ρ of -0.38. Both species therefore behave as electrophiles, but the carbene is much less discriminating. The electronic state of the carbene reaction is not certain.

Photolysis of pivaloyl azide[109a] gives products resulting from the reaction of pivaloylnitrene with solvent, as well as *tert*-butyl isocyanate which is not formed via the nitrene.

A study of the photolysis of alkyl azides and hydrazoic acid in organic solvents did not yield any evidence for nitrene participation,[109b] and whilst gas-phase pyrolysis of *tert*-butyl azide gives (**71**), rearrangement may be concerted with loss of nitrogen.[110]

$$\mathrm{Bu^tN_3} \xrightarrow[-\mathrm{N_2}]{} [\mathrm{Bu^t\ddot{N}:}] \rightarrow \mathrm{Me_2C{=}NMe}$$

(**71**)

Photosensitized decomposition of triarylmethyl azides give anils.[111] Migratory aptitudes of groups ranging from p-$NO_2C_6H_4$ to p-$CH_3C_6H_4$ are within a few percent of unity. The same is true for the unsensitized photolysis, and it was concluded that both rearrangements involve the triplet nitrene,

$$\mathrm{Ar_3CN_3} \xrightarrow[\text{Sensitizer}]{h\nu} \mathrm{Ar_2C{=}NAr + N_2}$$

and that the energy barrier for rearrangement of this species is extremely small.

Further spectroscopic studies of arylnitrenes produced by azide photolyses in a frozen matrix at 77°K have been reported,[112–114] and include observation of the quintet state of the dinitrene (**72**).[113] The corresponding dicarbene was also observed. Ultraviolet spectra of species (**73**) have been recorded,[114] and similar species were also observed from *para*-diazides with extended conjuga-

108 J. E. Baldwin and R. A. Smith, *J. Am. Chem. Soc.*, **89**, 1886 (1967).
109a G. T. Tisue, S. Linke, and W. Lwowski, *J. Am. Chem. Soc.*, **89**, 6303, 6308 (1967).
109b E. Koch, *Tetrahedron*, **23**, 1747 (1967).
110 W. Pritzkow and D. Timm, *J. Prakt. Chem.*, **32**, 178 (1966).
111 F. D. Lewis and W. H. Saunders, *J. Am. Chem. Soc.*, **89**, 645 (1967).
112 A. Reiser, G. Bowes, and R. J. Horne, *Trans. Faraday Soc.*, **62**, 3162 (1966).
113 E. Wasserman, R. W. Murray, W. A. Yager, A. M. Trozzolo, and G. Smolinsky, *J. Am. Chem. Soc.*, **89**, 5076 (1967).
114 A. Reiser, H. M. Wagner, R. Marley, and G. Bowes, *Trans. Faraday Soc.*, **63**, 2403 (1967).

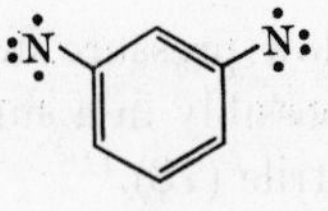

(72)

tion (e.g., **74**). Pyrolysis of *p*-diazidobenzene in decalin does not yield maleonitrile and acetylene (reaction 12) in the same way that *o*-diazides give *cis,cis*-mucononitrile.[115]

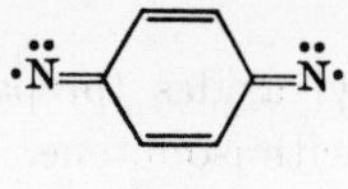

(73)

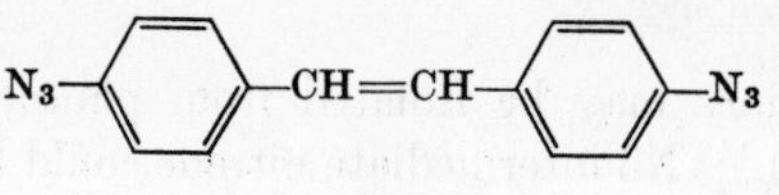

(74)

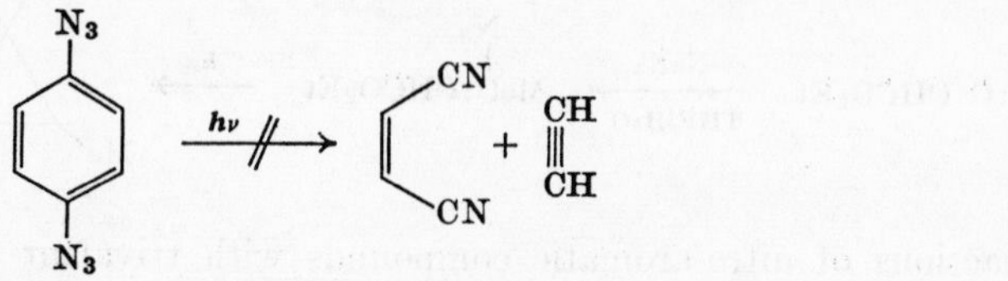

...(12)

Benzophenone-photosensitized decomposition of the geminal diazide (**75**) at 77°K gives first an azidonitrene, and then diphenylmethylene, as indicated by ESR spectrometry.[116] Photolysis of (**75**) in benzene gives the products shown. It was suggested that, as further photolysis of the tetrazole (**76**) gives only the imidazole (**77**), two stereoisomeric nitrenes may be involved.[117]

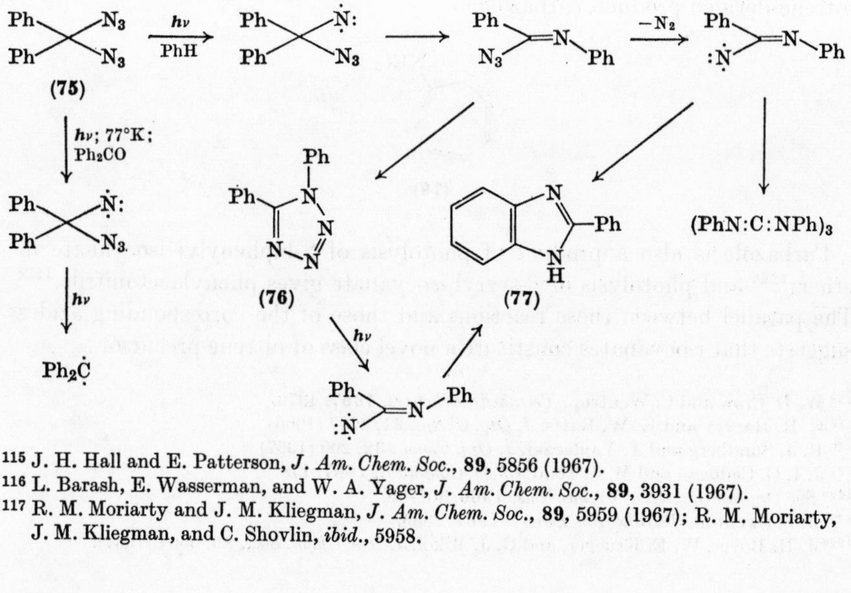

[115] J. H. Hall and E. Patterson, *J. Am. Chem. Soc.*, **89**, 5856 (1967).
[116] L. Barash, E. Wasserman, and W. A. Yager, *J. Am. Chem. Soc.*, **89**, 3931 (1967).
[117] R. M. Moriarty and J. M. Kliegman, *J. Am. Chem. Soc.*, **89**, 5959 (1967); R. M. Moriarty, J. M. Kliegman, and C. Shovlin, *ibid.*, 5958.

Pyrolysis of phenyl azide at low pressure in the vapour phase gives azobenzene, but at high pressure (possibly in a surface reaction) rearrangement gives cyclopentadiene-1-carbonitrile (**78**).[118]

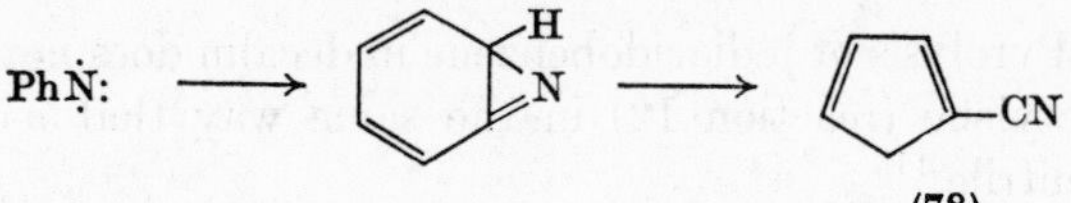

(78)

Azirines may be isolated from photolysis of vinyl azides (prepared as shown).[119] No intermediate nitrene could be trapped with isobutene.

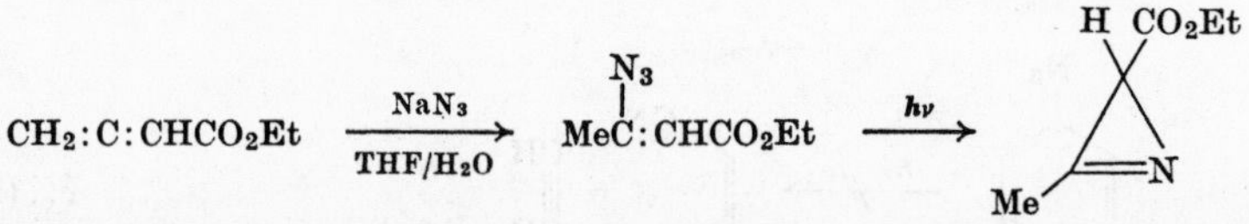

New reactions of nitro-aromatic compounds with trivalent phosphorus compounds have been reported,[120] and Cadogan and Todd[121] have presented compelling evidence that these reactions and aryl azide decompositions involve common intermediates (presumably nitrenes). For example, nitrobenzene is deoxygenated by the highly reactive diethyl methylphosphonite in diethylamine to give the azepine (**79**).[122a] A puzzling feature is that 2-nitrobiphenyl gives an azepine under these conditions at the expense of the normal nitrene-derived product, carbazole.

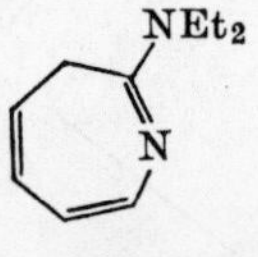

(79)

Carbazole is also a product of photolysis of 2-biphenylyl isocyanate in ether;[122b] and photolysis of β-styryl isocyanate gives phenylacetonitrile.[122c] The parallel between these reactions and those of the corresponding azides suggests that isocyanates constitute a novel class of nitrene precursor.

[118] W. D. Crow and C. Wentrup, *Tetrahedron Letters*, **1967**, 4379.
[119] G. R. Harvey and K. W. Ratts, *J. Org. Chem.*, **31**, 3907 (1966).
[120] R. J. Sundberg and T. Yamazaki, *J. Org. Chem.*, **32**, 290 (1967).
[121] J. I. G. Cadogan and M. J. Todd, *Chem. Commun.*, **1967**, 178.
[122a] See *Organic Reaction Mechanisms*, **1966**, 301—3.
[122b] J. S. Swenton, *Tetrahedron Letters*, **1967**, 2855.
[122c] J. H. Boyer, W. E. Krueger, and G. J. Mikol, *J. Am. Chem. Soc.*, **89**, 5504 (1967).

Cyanonitrene chemistry has been discussed in detail.[123] This nitrene undergoes stereospecific insertion into CH bonds, with high selectivity. In the heavy-atom solvent methylene bromide, insertion is by a non-stereospecific free-radical mechanism characteristic of the triplet ground state. Less crossing to the triplet occurs in methylene chloride solutions.

Intramolecular insertion competes with reaction with solvent when hexanoyl azide is photolysed in cyclohexane. The singlet nitrene gives both a γ- and a δ-lactam, the latter predominating.[124] From this and related work in the literature it seems that the preferred transition state for intramolecular insertion is that leading to the δ-lactam. Photosensitized decomposition of the same azide gives no lactam, but products attributable to a triplet nitrene intermediate. Photosensitized decomposition gives the triplet nitrene which does not undergo insertion reactions.

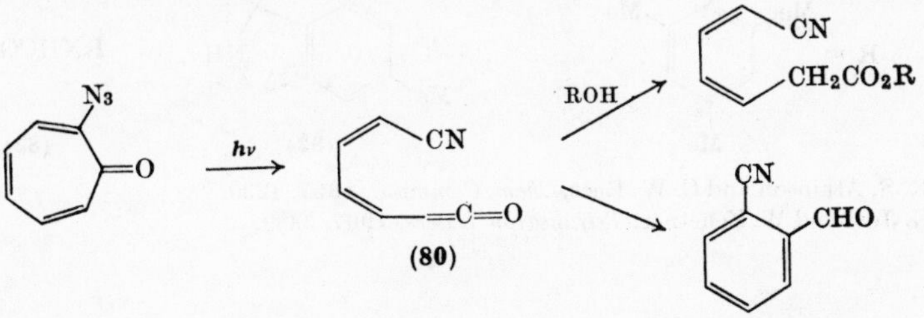

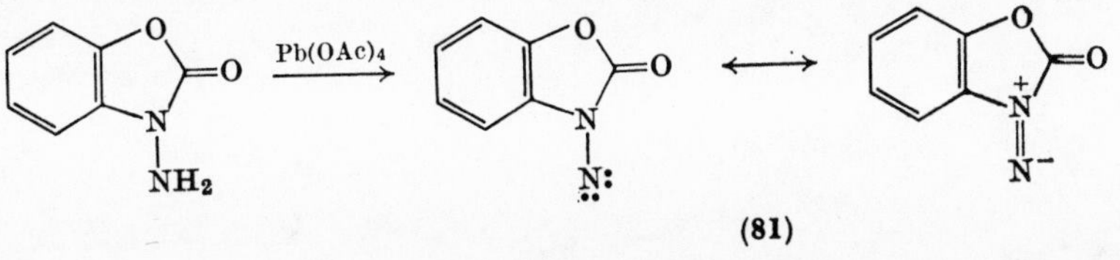

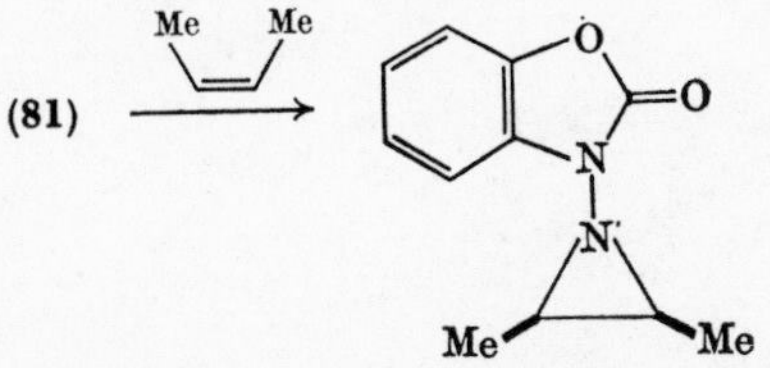

2-Azidotropone undergoes thermolysis under relatively mild conditions, suggesting that a nitrene may not precede the ketene intermediate (**80**).[125] The stabilized aminonitrene (**81**) does not fragment to benzyne, but may be

[123] A. G. Anastassiou and H. E. Simmons, *J. Am. Chem. Soc.*, **89**, 3177 (1967); A. G. Anastassiou, *ibid.*, p. 3184.

[124] I. Brown and O. E. Edwards, *Can. J. Chem.*, **45**, 2599 (1967).

[125] J. D. Hobson and J. R. Malpass, *J. Chem. Soc.*, *C*, **1967**, 1645.

trapped by stereospecific addition to olefins.[126] Related oxidation of 1-amino-oxindole gave the ring-expanded cinnolin-3-ol; no intermediate could be trapped.

The photolysis of mesitonitrile oxide in methanol gives compounds (**82**) and (**83**), possibly through an acylnitrene intermediate.[127]

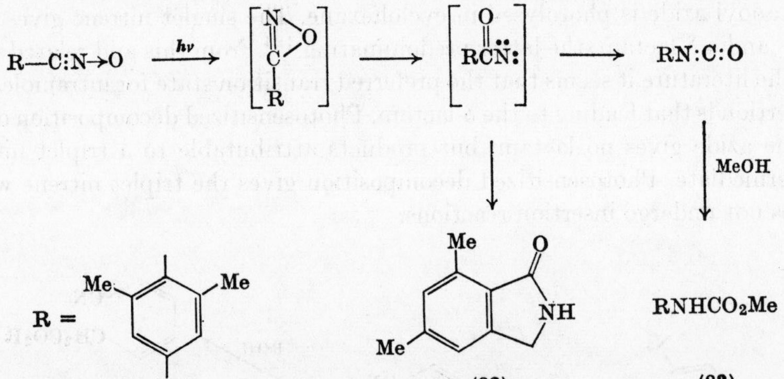

(**82**) (**83**)

[126] R. S. Atkinson and C. W. Rees, *Chem. Commun.*, **1967**, 1230.
[127] G. Just and W. Zehetner, *Tetrahedron Letters*, **1967**, 3389.

CHAPTER 11

Reactions of Aldehydes and Ketones and their Derivatives[1]

Formation and Reactions of Acetals and Ketals[2, 3]

The acid-catalysed hydrolyses of 2-aryl-4,4,5,5-tetramethyl-1,3-dioxolans (e.g., **1**) occur at about one-thousandth of the rate of those of the corresponding aryl-1,3-dioxolans (**2**) and it has been suggested that they proceed by an *A*2 mechanism.[4] This conclusion is based partly on a comparison of the kinetic parameters for the tetramethyldioxolan (determined in aqueous solution) with those for the dioxolans (as **2**) and benzaldehyde diethyl acetals (determined in 50% aqueous dioxan). Thus the entropy of activation for the

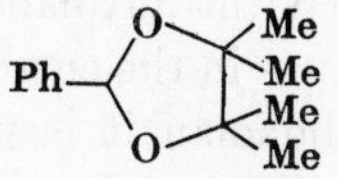

(**1**)

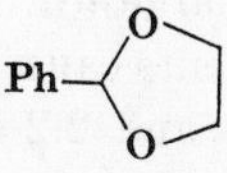

(**2**)

hydrolysis of 2-phenyl-4,4,5,5-tetramethyl-1,3-dioxolan is –14.2 e.u. and for the hydrolyses of 2-phenyl-1,3-dioxolan and benzaldehyde diethyl acetal –8.9 and +1.0 e.u., respectively. The tetramethyldioxolan was studied in dilute aqueous hydrochloric acid and the entropy of activation was calculated from the second-order rate constant determined by dividing the first-order rate constant by the acid concentration (0.1M).[5] The dioxolan and diethyl acetal were, however, studied in 50% aqueous dioxan and the entropies of activation calculated from second-order rate constants based on the pH, as determined by the glass electrode. It was considered that the change in entropies of activation calculated in these ways on going from the dioxolans to the tetramethyldioxolans indicated a change from an *A*1 to an *A*2

[1] See C. D. Gutsche, "The Chemistry of Carbonyl Compounds", Prentice Hall, Englewood Cliffs, N.J., U.S.A., 1967.

[2] E. H. Cordes, *Progr. Phys. Org. Chem.*, **4**, 1 (1967).

[3] E. Schmitz and I. Eichborn in "The Chemistry of the Ether Linkage", S. Patai, ed., Interscience, London, 1967, p. 167.

[4] T. H. Fife, *J. Am. Chem. Soc.*, **89**, 3228 (1967).

[5] This second-order rate constant probably has no significance since the rate of hydrolysis of 4,4,5,5-tetramethyl-2-(*p*-nitrophenyl)dioxolan, and hence probably that of the analogous phenyl compound, is not proportional to the acid concentration.

mechanism. Further evidence is provided by the ρ-value for the tetramethyldioxolans (–2.0) which is less negative than that for the diethyl acetals (–3.35), suggesting less carbonium ion character in the transition state; yet more evidence is the observation of buffer catalysis in the hydrolysis of 4,4,5,5-tetramethyl-2-*p*-methoxyphenyldioxolan—the first observation of buffer catalysis in the hydrolysis of an acetal but its nature was not determined. The plot of log*k* for the hydrolysis of the tetramethyl-*p*-nitrophenyl-1,3-dioxolan against H_0 is a curve,[6] and that against $\log C_{H^+}$ is a straight line of slope 1.9. The Hammett–Zucker hypothesis thus cannot be used, but the Bunnett plot yields a *w*-value of +1.9, suggesting that water acts as a nucleophile. So, although "each piece of evidence ... contains some ambiguity", the overall picture is "of a reaction in which water is participating as a nucleophile but in which either the bond being formed in the transition state is not well developed or in which an *A*1 mechanism is still making some contribution to the observed rate".[4]

In contrast to these results, the introduction of four methyl substituents into positions 4 and 5 of dioxolan and 2-methyldioxolan causes only small (5—7-fold) decreases in the rates of hydrolysis, but with 2,2-dimethyldioxolan the effect is again large (*ca.* 800-fold). With dioxolans the introduction of the methyl substituents causes little change in the entropy of activation, but with the 2-methyl- and 2,2-dimethyl-dioxolan it is made 7—9 e.u. more negative (see Table 1).[7] In this investigation Kankaanpera studied the

Table 1. Kinetic parameters for the hydrolysis of 1,3-dioxolan and its methyl derivatives.[7]

Substituents in 1,3-dioxolan	$\Delta S^{\ddagger}$	$\Delta H^{\ddagger}$ (kcal mole^{-1})	10^4k (l. mole^{-1} sec^{-1})
—	+1.4	25.6	0.0227
4,4,5,5-Tetramethyl	+2.1	26.7	0.00457
2-Methyl	+5.6	21.7	136
2,4,4,5,5-Pentamethyl	–3.8	20.0	18.9
2,2-Dimethyl	+7.9	21.0	1440
2,2,4,4,5,5-Hexamethyl	+0.5	22.7	1.77

hydrolysis of all the methyl-substituted dioxolans and a large number of dioxolans with other substituents. The hydrolysis of a large number of 1,3-dioxans has also been investigated.[8]

[6] For a review of acidity functions see M. I. Vinnik, *Russian Chem. Rev.*, **35**, 802 (1967).
[7] A. Kankaanpera, *Ann. Univ. Turku, Ser. A I*, No. 95 (1966); *Chem. Abs*,. **67**, 53495e (1967).
[8] K. Pihlaja, *Ann. Univ. Turku, Ser. A I*, No. 114 (1967); see also F. G. Riddell and M. J. T. Robinson, *Tetrahedron*, **23**, 3417 (1967).

Bruice and Piszkiewicz[9] have attempted unsuccessfully to observe intramolecular general-acid catalysis in the hydrolysis of ketals (**3**) to (**5**). This result is not surprising, however, since with all these compounds an unstable conformation would be necessary for proton-transfer to occur between the carboxyl group and the ketal oxygen. With compounds (**5**) the distance between the carboxyl and the ketal oxygen would increase on going to the transition state, a circumstance hardly likely to favour proton-transfer between them.

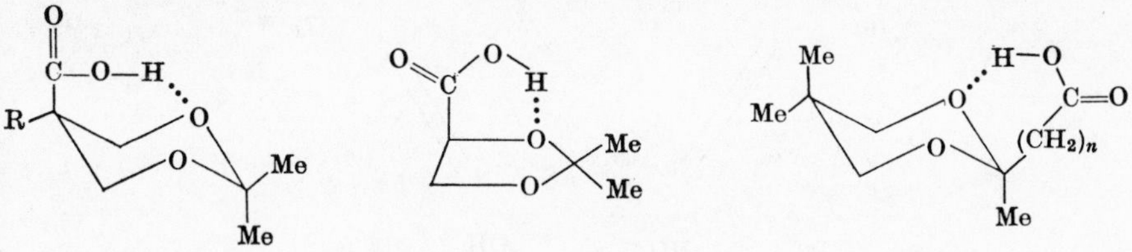

(**3**) R = Me, CH_2OH (**4**) (**5**) n = 1 or 2

The pK_a's of the conjugate acids of certain acetals and ketals have been estimated from changes in the O–D vibrational frequency when the compounds are dissolved in deuteriomethanol. The following values[10] were reported: 2,2-dimethoxypropane, –5.2; methyl orthoacetate, –6.5; methyl orthocarbonate, –8.4; *para*-substituted benzaldehyde diethyl acetals, –5.7.

The condensation of glycerol with carbonyl compounds has been investigated.[11, 12]

The entropies of activation for the hydrolysis of a series of methyl aldofuranosides are all negative. Mechanisms involving a bimolecular displacement of methanol and ring opening were considered.[13]

Replacement of the hydroxyl group at position 2 of methyl β-D-glucopyranoside by the more strongly electron-withdrawing chloro results in a rate decrease of about 30-fold and a change in the entropy of activation from +16.5 to +7.6 e.u. The latter result was interpreted as indicating an increase in the *A*2-character of the transition state, as might reasonably be expected. However, the Bunnett *w*-value, for the chloro-glycoside, –0.2, is characteristic of an *A*1 mechanism.[14]

The hydrolysis of methyl 1-thio-α-D-ribopyranoside is accompanied by anomerization and ring contraction. A reaction solution in 0.4M-hydrochloric

[9] T. C. Bruice and D. Piszkiewicz, *J. Am. Chem. Soc.*, **89**, 3568 (1967).
[10] T. Pletcher and E. H. Cordes, *J. Org. Chem.*, **32**, 2294 (1967).
[11] G. Stefanović and D. Petrović, *Tetrahedron Letters*, **1967**, 3153.
[12] See *Organic Reaction Mechanisms*, **1965**, 240; **1966**, 308.
[13] B. Capon and D. Thacker, *J. Chem. Soc., B*, **1967**, 185.
[14] E. Buncel and P. R. Bradley, *Can. J. Chem.*, **45**, 515 (1967).

acid, after refluxing for six minutes, contained, in addition to the α-pyranoside (**6**) (18.5%), β-pyranoside (**7**) (7.4%), α-furanoside (**8**) (5.8%), and β-furanoside (**9**) (15.4%). Clearly the anomerization and ring contraction must involve ring opening, but whether hydrolysis does also is not known.[15]

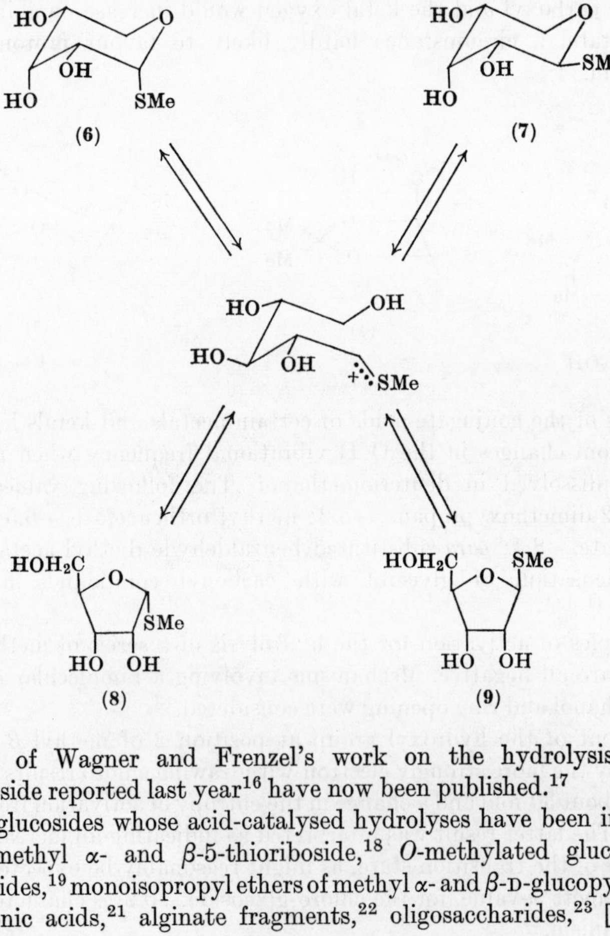

Details of Wagner and Frenzel's work on the hydrolysis 2-pyridyl β-D-glucoside reported last year[16] have now been published.[17]

Other glucosides whose acid-catalysed hydrolyses have been investigated include methyl α- and β-5-thioriboside,[18] *O*-methylated glucosides and disaccharides,[19] monoisopropyl ethers of methyl α- and β-D-glucopyranoside,[20] aldobiuronic acids,[21] alginate fragments,[22] oligosaccharides,[23] [1-^{14}C]cello-

15 C. J. Clayton, N. A. Hughes, and S. A. Saeed, *J. Chem. Soc., C*, **1967**, 644.
16 See *Organic Reaction Mechanisms*, **1966**, 311.
17 G. Wagner and H. Frenzel, *Arch. Pharm.*, **300**, 591 (1967).
18 C. J. Clayton and N. A. Hughes, *Carbohydrate Res.*, **4**, 32 (1967).
19 K. K. De and T. E. Timell, *Carbohydrate Res.*, **4**, 72 (1967).
20 J. E. Höök and B. Lindberg, *Acta Chem. Scand.*, **20**, 2363 (1966).
21 K. K. De and T. E. Timell, *Carbohydrate Res.*, **4**, 177 (1967).
22 O. Smidsrød, B. Larsen, and A. Haug, *Carbohydrate Res.*, **5**, 371 (1967).
23 A. Meller, *J. Polymer Sci.*, Part A-1, **5**, 1443 (1967).

triose,[24] and the phenyl glycoside of *N*-acetyl-D-neuraminic acid.[25] The acid-catalysed methanolysis of tri-*O*-methylamylose and tri-*O*-methylcellulose has also been investigated.[26]

Details of the determination, by *X*-ray diffraction, of the conformation of lysozyme and of a complex between lysozyme and tri-*N*-acetylchitotriose reported last year[27] have been published.[28] The mechanism that was then proposed[27] for the mode of action of lysozyme has been discussed in more detail.[29] Lowe and his co-workers[30] have shown that the hydrolyses of aryl di-*N*-acetyl-β-chitobiosides catalysed by lysozyme, although slow, follow Michaelis–Menten kinetics. K_m is almost independent of the substituent in the aglycone but k_{cat} shows a marked dependence thereon with a ρ-value of 1.2. A possible alternative mechanism was proposed (Scheme 1) involving neighbouring-group participation by the amide group of the substrate. It was thought that this would require the pyranose ring of the substrate to take up a boat conformation as shown in Scheme 1.[30,31]

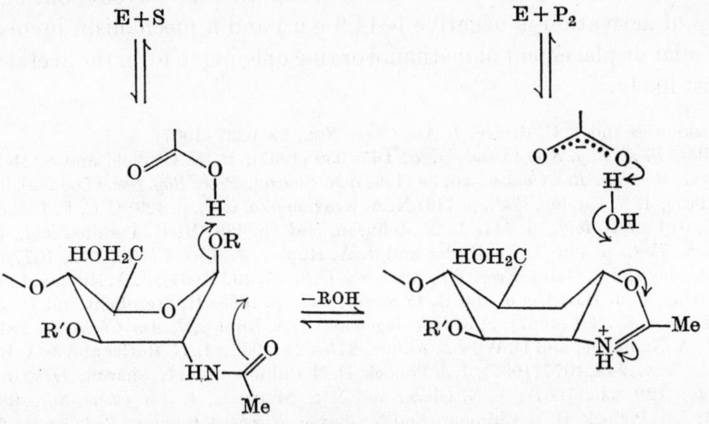

Scheme 1

Neighbouring-group participation by the 2-acetamido-group of *o*- and *p*-nitrophenyl 2-acetamido-2-deoxyglucosides has been demonstrated to occur in their netural hydrolyses, i.e., with ArO^- as leaving group, but it is

[24] M. S. Feather and J. F. Harris, *J. Am. Chem. Soc.*, **89**, 5661 (1967).
[25] P. Meindl and H. Tuppy, *Monatsh. Chem.*, **98**, 53 (1967).
[26] J. N. BeMiller and E. E. Allen, *J. Polymer Sci.*, Part A-1, **5**, 2133 (1967).
[27] See *Organic Reaction Mechanisms*, **1966**, 312—313.
[28] C. C. F. Blake, G. A. Mair, A. C. T. North, D. C. Phillips, and V. R. Sarma, *Proc. Roy. Soc.* (*London*), *B*, **167**, 365 (1967); C. C. F. Blake, L. N. Johnson, G. A. Mair, A. C. T. North, D. C. Phillips, and V. R. Sarma, *ibid.*, p. 378; D. C. Phillips, *Proc. Nat. Acad. Sci. U.S.*, **57**, 484 (1967).
[29] C. A. Vernon, *Proc. Roy. Soc.* (*London*), *B*, **167**, 389 (1967).
[30] G. Lowe, G. Sheppard, M. L. Sinnott, and A. Williams, *Biochem. J.*, **104**, 893 (1967).
[31] G. Lowe, *Proc. Roy. Soc.* (*London*), *B*, **167**, 431 (1967).

not operative in their acid-catalysed hydrolyses, i.e., with ArOH as leaving group.[32]

There have been many other investigations of lysozyme, particularly of the binding of chitin oligosaccharides and related compounds.[33]

The mechanisms of action of other glycosidases have been discussed in references 34—36.

The acid-catalysed anomerization of the methyl glucopyranosides in CD_3OD proceeds with complete exchange of the methoxyl group with the solvent.[37–39] This is consistent with the intervention of a cyclic carbonium ion (reaction 1) or a stereospecifically formed and decomposed acyclic acetal (reaction 3), but not with that of an acyclic carbonium ion (reaction 2). The mechanism (3) was excluded by the observation that the acetal (**10**) formed furanosides not pyranosides under the reaction conditions. Mechanism (1) is therefore the most likely. It is also consistent with the entropies of activation, +5 to +8 e.u. The anomerization of the methyl glucofuranosides similarly proceeds with exchange of the methoxyl group with the solvent; but here the entropy of activation is negative (–14.9 e.u.) and a mechanism involving a bimolecular displacement of methanol or ring opening to form the acetal seems the most likely.[38]

[32] D. Piszkiewicz and T. C. Bruice, *J. Am. Chem. Soc.*, **89**, 6237 (1967).

[33] P. Jollès, *Proc. Roy. Soc.* (*London*), *B*, **167**, 350 (1967); R. E. Canfield and S. McMurry, *Biochem. Biophys. Res. Comm.*, **26**, 38 (1967); N. Sharon, *Proc. Roy. Soc.* (*London*), *B*, **167**, 402 (1967); J. A. Rupley, *ibid.*, p. 416; N. A. Kravchenko, *ibid.*, p. 429; C. C. F. Blake, *ibid.*, p. 435; J. Collins, *ibid.*, p. 441; L. N. Johnson, *ibid.*, p. 439; H. R. Perkins, *ibid.*, p. 443; E. Work, *ibid.*, p. 446; L. G. Butler and J. A. Rupley, *J. Biol. Chem.*, **242**, 1077 (1967); J. A. Rupley and V. Gates, *Proc. Nat. Acad. Sci. U.S.*, **57**, 496 (1967); J. A. Rupley, L. Butler, M. Gerring, F. J. Hartdegen, and R. Pecoraro, *ibid.*, p. 1088; H. Sternlicht and D. Wilson, *Biochemistry*, **6**, 2881 (1967); F. J. Hartdegen and J. A. Rupley, *J. Am. Chem. Soc.*, **89**, 1743 (1967); A. Neuberger and D. Wilson, *Nature*, **215**, 524 (1967); L. G. Butler and J. A. Rupley, *J. Biol. Chem.*, **242**, 1077 (1967); J. J. Pollock, D. M. Chipman, and N. Sharon, *Arch. Biochem. Biophys.*, **120**, 235 (1967); A. N. Glazer and N. S. Simmons, *J. Am. Chem. Soc.*, **88**, 2335 (1966); J. J. Pollock, D. M. Chipman, and N. Sharon, *Biochem. Biophys. Res. Comm.*, **28**, 779 (1967); D. M. Chipman, V. Grisaro, and N. Sharon, *J. Biol. Chem.*, **242**, 4388 (1967); T. M. Spotwood, J. M. Evans, and J. M. Richards, *J. Am. Chem. Soc.*, **89**, 5052 (1967); F. W. Dahlquist and M. Raftery, *Nature*, **213**, 625 (1967); I. A. Cherasov and N. A. Kravchenko, *Mol. Biol.*, **1**, 381 (1967); *Chem. Abs.*, **67**, 70735a (1967); N. A. Kravchenko and Y. D. Kuznetsov, *Mol. Biol.*, **1**, 498 (1967); *Chem. Abs.*, **67**, 105495b (1967); T. Tojo, K. Hamaguchi, M. Imanishi, and T. Amano, *J. Biochem.* (*Tokyo*), **60**, 538 (1966); K. Ogasahara and K. Hamaguchi, *ibid.*, **61**, 199; S. Hara and Y. Matsushima, *ibid.*, **62**, 118; K. Ikeda, K. Hamaguchi, M. Imanishi, and T. Amano, *ibid.*, p. 315.

[34] J. E. G. Barnett, W. T. S. Jarvis, and K. A. Munday, *Biochem. J.*, **103**, 699 (1967).

[35] J. Conchie, A. J. Hay, I. Strachan, and G. A. Levvy, *Biochem. J.*, **102**, 929 (1967); J. Conchie, A. L. Gelman, and G. A. Levvy, *ibid.*, **103**, 609 (1967).

[36] A. Dahlquist, and N. G. Asp, *Biochem. J.*, **103**, 86 (1967); M. R. J. Morgan, *ibid.*, **102**, 44P (1967); O. P. Malhotra and P. M. Dey, *ibid.*, **103**, 508, 739 (1967).

[37] B. Capon, *Chem. Commun.*, **1967**, 21.

[38] B. Capon and D. Thacker, *J. Chem. Soc.*, *B*, **1967**, 1010.

[39] See also J. Świderski and A. Temeriusz, *Carbohydrate Res.*, **3**, 225 (1966); G. Wagner and H. Frenzel, *Pharmazie*, **8**, 415 (1967).

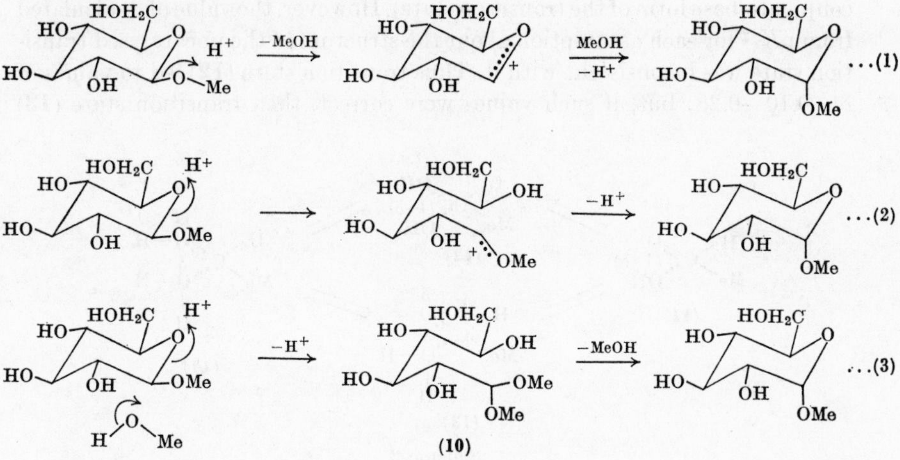

Other reactions of carbohydrates that have been investigated include the base-catalysed fission of glycosides with heterocyclic aglycones,[39] hydrolysis of the 4,6-benzylidene derivatives of glycosides of amino-sugars,[40] Fe(II)-catalysed decomposition of sinigrin,[41] and the acetolysis of acetylated aldose acetals and thioacetals.[42]

Hydration of Aldehydes and Ketones and Related Reactions[43]

The transition states for the hydration of acetaldehyde have been discussed in terms of p$K^{\ddagger}$'s, calculated from the rate constants for the hydronium-, water-, and hydroxyl-catalysed reactions by the equations:

$$pK_1{}^{\ddagger} = \log(k_{\rm H}/k_0)$$

$$pK_2{}^{\ddagger} = \log(k_0/k_{\rm OH}) + pK_{\rm w}$$

The concerted mechanism was assumed, and the transition states shown in Scheme 2 were considered. (**11**) and (**14**) were shown to be the most stable transition states for the H_3O^+- and ^-OH-catalysed reactions, but it was not possible to decide which transition state (**12** or **13**) for the water-catalysed reaction would be the more stable. Attempts were made to calculate values for δ, the difference in formal charge between transition state and reactants, from the p$K^{\ddagger}$'s, and Δm (where Δm is the difference between the number of water molecules strongly hydrogen-bonded to the acid form and to the

[40] J. Kovář and J. Jarý, *Coll. Czech. Chem. Commun.*, **32**, 854 (1967).
[41] C. G. Youngs and A. S. Perlin, *Can. J. Chem.*, **45**, 1801 (1967).
[42] N. H. Kurihara and E. P. Painter, *Can. J. Chem.*, **45**, 1467, 1475 (1967).
[43] R. P. Bell, *Advan. Phys. Org. Chem.*, **4**, 1 (1966); *Dan. Kemi*, **47**, 138 (1967).

conjugate-base form of the transition state). However, the value of δ calculated from $pK_1^‡$ for each assumption about the structure of the uncatalysed transition state was inconsistent with it. Thus transition state (**12**) led to values of $\delta = 0.10$—0.28; but, if such values were correct, then transition state (**13**)

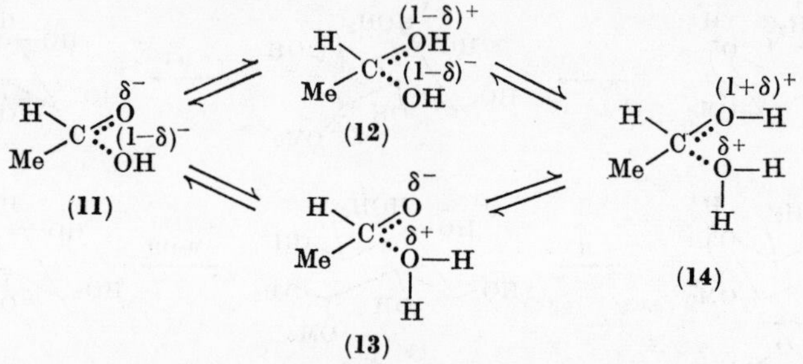

Scheme 2

would be the more stable. Similarly, transition state (**13**) led to values of $\delta = 0.78$—0.92; but these values imply that (**12**) is the more stable. It was, therefore, concluded that the original assumption of the concerted mechanism's being followed was incorrect and that the acid-catalysed reaction proceeded by a rate-determining proton-transfer. The values of δ were, however, considered to be consistent with a concerted mechanism for the water and hydroxyl ion-catalysed reactions. The catalytic constant for the former is considerably larger than the value predicted from the Brønsted equation and the value of $\Delta S^‡$ is highly negative (–38 e.u.). This was taken to indicate that a special type of concerted mechanism was being followed, in which a large number of water molecules are arranged into a more ordered structure in the transition state.[44]

The equilibrium constants for the hydration of aldehydes and ketones have been calculated from ultraviolet spectrophotometric results,[45,46] and from oxygen-17[47] and proton magnetic resonance[47–50] data. Agreement was not always obtained. For instance, one set of NMR measurements led to the conclusion that pivalaldehyde was 100% unreacted,[48] whereas another[47] led to a value of K_h of 0.20 in fairly good agreement with the spectrophotometrically determined value. The kinetics of the acid-catalysed hydration of acetaldehyde have been measured by the broadening of the ^{17}O-resonance

[44] J. L. Kurz and J. I. Coburn, *J. Am. Chem. Soc.*, **89**, 3528 (1967).
[45] J. L. Kurz, *J. Am. Chem. Soc.*, **89**, 3524 (1967).
[46] P. Greenzaid, Z. Rappoport, and D. Samuel, *Trans. Faraday Soc.*, **63**, 2131 (1967).
[47] P. Greenzaid, Z. Luz, and D. Samuel, *J. Am. Chem. Soc.*, **89**, 749 (1967).
[48] D. L. Hooper, *J. Chem. Soc., B*, **1967**, 169.
[49] V. S. Griffiths and G. Socrates, *Trans. Faraday Soc.*, **63**, 673 (1967).
[50] G. Öjelund and I. Wadsö, *Acta Chem. Scand.*, **21**, 1408 (1967).

lines of the carbonyl- and *gem*-diol oxygen atoms.[51] The hydration of pyridinecarbaldehydes catalysed by bovine erythrocyte carbonic anhydrase and by metal ions has been investigated.[52, 53]

An attempt to observe bifunctional catalysis of the hydration of acetaldehyde, the mutarotation of glucose, and iodination of acetone by the mono-anions of dicarboxylic acids was unsuccessful. The catalytic constants for the last reaction were only slightly greater than values calculated from the Brønsted relationship for the sums of the catalytic constants of the carboxyl and carboxylate groups taken individually.[54a] Dehydration of the hydrate of glyoxalatopentamminecobalt(III) has been investigated.[54b]

There have been several investigations of the mutarotation of sugars.[55–59]

The hydroxide ion-catalysed decomposition of three hemithioacetals, MeCH(OH)SR, where R = Ph, $p\text{-}NO_2C_6H_4$, or CH_3CO, have rate constants approximately 10^{10} l. mole^{-1} sec^{-1}, i.e., near the diffusion-controlled limit, and enthalpies of activation *ca.* 2 kcal mole^{-1}. The catalytic constants for catalysis by weaker bases define Brønsted lines of slope 0.80, but the points for catalysis by $^{-}$OH fall below these as required by Eigen's theory of proton-transfer reactions.[60]

The reaction of carbonyl compounds with hydrogen sulphide has been investigated.[61]

Reactions with Nitrogen Bases

The energies and entropies of activation for the hydrolysis of substituted 2-benzylidene-1,1-dimethylethylamines have been determined under a variety of conditions.[62] The complete reaction scheme is given in equations (5—8) and under steady-state conditions this yields the following expression for k_{obs}:

$$k_{obs} = \frac{k_1 k_3[H^+] + K_w k_2 k_3}{([H^+] + K_{SH})(k_{-1}[H^-] + k_{-2} + k_3)} . \qquad \ldots(4)$$

[51] P. Greenzaid, Z. Luz, and D. Samuel, *J. Am. Chem. Soc.*, **89**, 756 (1967).
[52] Y. Pocker and J. E. Meany, *Biochemistry*, **6**, 239 (1967).
[53] Y. Pocker and J. E. Meany, *J. Am. Chem. Soc.*, **89**, 631 (1967); *J. Phys. Chem.*, **71**, 3113 (1967).
[54a] G. E. Lienhard and F. H. Anderson, *J. Org. Chem.*, **32**, 2229 (1967).
[54b] H. J. Price and H. Taube, *J. Am. Chem. Soc.*, **89**, 269 (1967).
[55] H. Schmid and G. Bauer, *Z. Naturforsch.*, **6**, 21, 1009 (1966); H. Schmid and G. Bauer, *Monatsh. Chem.*, **97**, 866 (1966); H. Schmid, G. Bauer, and G. Prähauser, *ibid.*, **98**, 165 (1967); H. Schmid, *ibid.*, **98**, 2097 (1967).
[56] H. S. Isbell and C. W. R. Wade, *J. Res. Nat. Bur. Stand.*, **71A**, 137 (1967).
[57] R. E. Pincock and T. E. Kiovsky, *Chem. Commun.*, **1966**, 864.
[58] F. Grønlund and B. Andersen, *Acta Chem. Scand.*, **20**, 2663 (1966).
[59] D. Horton, J. S. Jewell, and K. D. Philips, *J. Org. Chem.*, **31**, 4022 (1966); see also D. Horton, J. B. Hughes, J. S. Jewell, K. D. Philips, and W. N. Turner, *ibid.*, **32**, 1073 (1967).
[60] R. Barnett and W. P. Jencks, *J. Am. Chem. Soc.*, **89**, 5963 (1967).
[61] S. Bleisch and R. Mayer, *Chem. Ber.*, **100**, 93 (1967).
[62] R. K. Chaturvedi and E. H. Cordes, *J. Am. Chem. Soc.*, **89**, 1230 (1967).

In dilute alkali this reduces to:

$$k_{\mathrm{obs}} = \frac{k_2 K_{\mathrm{w}}}{K_{\mathrm{SH}}} \cdot \frac{k_3}{(k_{-2} + k_3)} = \frac{k_2 K_{\mathrm{w}}}{K_{\mathrm{SH}}} \; ;$$

and hence under these conditions the measured values of the energy and entropy of activation are composite; e.g., $E_{\mathrm{obs}} = E_2 + \Delta H^{\circ}{}_{\mathrm{w}} - \Delta H^{\circ}_{\mathrm{SH}}$. The standard enthalpy and entropy changes for the dissociation of the conjugate acid of 1,1-dimethyl-2-*p*-methylbenzylideneethylamines were determined, enabling $\Delta H^{\ddagger}$ and $\Delta S^{\ddagger}$ for step 2 (eqn. 7) to be calculated. The value of $\Delta S^{\ddagger}$, −3.4 e.u., is consistent with a bimolecular reaction between the protonated

$$>\!\!=\!\!N\!-\!R \; + \; H^+ \quad \overset{K'_{\mathrm{SH}}}{\rightleftharpoons} \quad >\!\!=\!\!\overset{H}{\underset{+}{N}}\!-\!R \qquad \qquad \ldots 5$$

$$>\!\!=\!\!\overset{H}{\underset{+}{N}}\!-\!R \; + \; H_2O \quad \underset{k_{-1}}{\overset{k_1}{\rightleftharpoons}} \quad +\!\!\underset{\mathrm{OH}}{\overset{H}{N}}R \; + \; H^+ \qquad \ldots 6$$

$$>\!\!=\!\!\overset{H}{\underset{+}{N}}\!-\!R \; + \; {}^-OH \quad \underset{k_{-2}}{\overset{k_2}{\rightleftharpoons}} \quad +\!\!\underset{\mathrm{OH}}{\overset{H}{N}}R \qquad \qquad \ldots 7$$

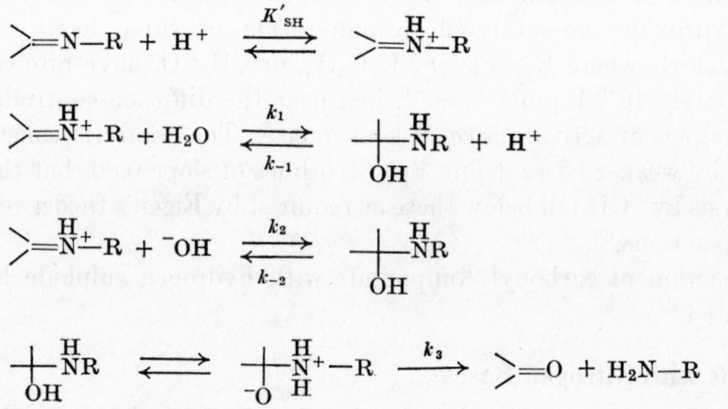

$$+\!\!\underset{\mathrm{OH}}{\overset{H}{N}}R \quad \rightleftharpoons \quad +\!\!\underset{{}^-O}{\overset{H}{N}}{}^+\!\!-\!R \quad \overset{k_3}{\longrightarrow} \quad >\!\!=\!\!O \; + \; H_2N\!-\!R \qquad \ldots 8$$

Schiff base and ${}^-OH$, and the value of $\Delta H^{\ddagger}$, 0.4 kcal mole^{-1} which is very small, suggests that there is a favourable energy term arising from electrostatic interactions between the oppositely charged species.

In slightly acidic solutions, when the Schiff bases are converted into their conjugate acids, $[H^+] \gg K_{\mathrm{SH}}$, $k_1 k_3 \gg K_{\mathrm{w}} k_2 k_3$, and $(k_{-2} + k_3) > k_{-1}[H^+]$, so that equation (9) reduces to $k_{\mathrm{obs}} = k_1$, where k_1 is the first-order rate constant for the attack of water on the protonated substrates. This yields a second-order constant on division by the concentration of water (55 mole l.$^{-1}$), which is 10^9—10^{10} less than the second-order constant for the reaction of the protonated Schiff base with ${}^-OH$. The values of $\Delta H^{\ddagger}$ varied from 12.9 to 16.6 kcal mole^{-1} and of $\Delta S^{\ddagger}$ from −31.2 to −25.8 depending on the substituent in the aryl ring.

In more strongly acid solutions, decomposition of the carbinolamine becomes rate-determining and $k_{\mathrm{obs}} = k_1 k_3 / k_{-1}[H^+]$. Step 1 (eqn. 6) is rapid and k_1/k_{-1} can be written as an equilibrium constant K_{hyd}, so that $k_{\mathrm{obs}}[H^+] = K_{\mathrm{hyd}} k_3$. The value of $\Delta S^{\ddagger}_{\mathrm{obs}}$ obtained under these conditions is thus the sum of the standard entropy change for hydration of the protonated Schiff base and the entropy of activation for decomposition of the carbinolamine, $\Delta S^{\ddagger}_{\mathrm{obs}} = \Delta S^{\circ}_{\mathrm{hyd}} + \Delta S^{\ddagger}_3$. The value of $\Delta S^{\circ}_{\mathrm{hyd}}$ was estimated from the values of

$\Delta S°$ for the hydration of acetaldehyde and for the ionization of secondary ammonium ions: $\Delta S^{\ddagger}_{0} \approx \Delta S^{\circ}_{\text{ion}} + \Delta S^{\circ}_{\text{aldehyde}}$; this yields a value near zero for $\Delta S^{\ddagger}_{3}$, which is reasonable.[62]

Activation parameters have also been measured for semicarbazone formation. The values of $\Delta H^{\ddagger}$ for attack of semicarbazide on *p*-hydroxybenzaldehyde catalysed by the hydrated proton, formic acid, and water are all near 9 kcal mole^{-1}, but the values of $\Delta S^{\ddagger}$ become progressively more negative (–16.5, –25.1, and –43.4 e.u.) as the acidity of the catalyst is decreased. The entropy of activation for the acid-catalysed dehydration of the carbinolamine from *p*-chlorobenzaldehyde was calculated from $\Delta S^{\ddagger}_{\text{obs}}$, determined under conditions where this step was slow, and the standard entropy change for the initial addition of semicarbazide. The value obtained, –16.6 e.u., is unusual for an acid-catalysed unimolecular decomposition.[63]

The kinetics of imine formation from 3-hydroxypyridine-4-carbaldehyde and alanine have been investigated by Auld and Bruice.[64] The rate constant for the dehydration of the carbinolamine intermediate is 60 times greater than with pyridine-4-carbaldehyde. It was suggested that, as in the previously reported reaction of glycine with 3-hydroxypyridine-4-carbaldehyde, this is the result of intramolecular catalysis by the phenolic hydroxyl group. The criticism of this conclusion reported last year[65] was, however, not discussed, nor were the kinetics of the reaction of 3-methoxypyridine-4-carbaldehyde investigated.

The transamination of 3-hydroxypyridine-4-carbaldehyde by alanine was also investigated.[66] The initial step is the rapid and reversible formation of the imine which then undergoes a general base-catalysed abstraction of the α-proton. Since pyridine-4-carbaldehyde does not undergo transamination, a special role was assigned to the 3-hydroxyl group. This could be because the alternative keto-enamine[67] is the reactive species, or because the intramolecularly hydrogen-bonded structure (**15**) has a partial positive charge on the azomethine nitrogen atom.

Pyridine-2-carbaldehyde, in contrast to pyridine-4-carbaldehyde, undergoes a transamination reaction with several amino-acids.[68]

The transamination of the optically active (laevorotatory) Schiff base (*S*)-(**16**) by potassium *tert*-butoxide in *tert*-butyl alcohol yields (dextrorotatory) (*R*)-(**17**) stereospecifically. Experiments with deuterated solvent

63 R. K. Chaturvedi and E. H. Cordes, *J. Am. Chem. Soc.*, **89**, 4631 (1967).
64 D. S. Auld and T. C. Bruice, *J. Am. Chem. Soc.*, **89**, 2083 (1967).
65 *Organic Reaction Mechanisms*, **1966**, 319.
66 D. S. Auld and T. C. Bruice, *J. Am. Chem. Soc.*, **89**, 2090, 2098, 4250, 4251 (1967).
67 For investigations of the tautomerism of Schiff bases see G. Dudek and E. P. Dudek, *Tetrahedron*, **23**, 3245 (1967); G. Dudek, *J. Org. Chem.*, **32**, 2016 (1967); J. W. Ledbetter, *J. Phys. Chem.*, **71**, 2351 (1967).
68 K. Asano, M. Furukawa, and N. Asano, *Chem. Pharm. Bull.* (*Tokyo*), **14**, 309 (1966).

showed that (**16**) undergoes proton exchange more rapidly than it is isomerized, and that the major pathway for isomerization is an intermolecular one. It was suggested that the reaction involves one face of an aza-allylic anion (**18**).[69]

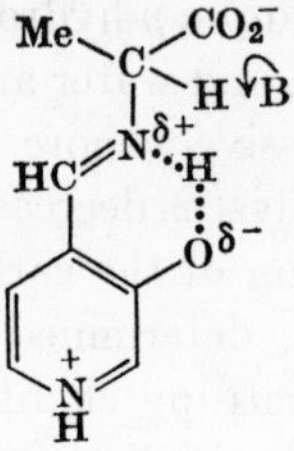

(**15**)

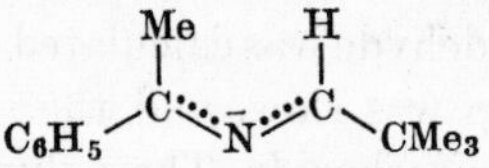

(**18**)

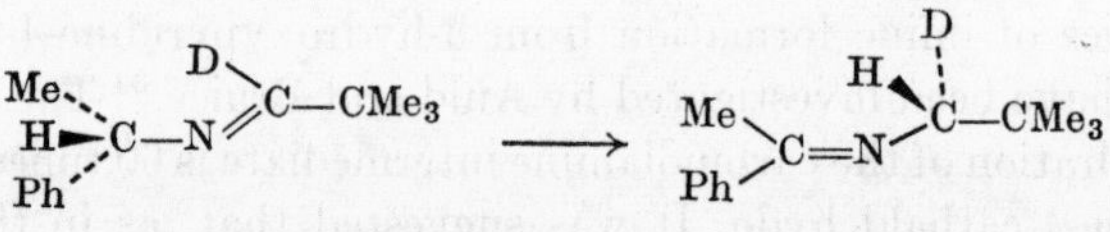

(**16**) (**17**)

Other investigations of transamination are reported in references 70 and 71.

The aldehydo-hydrogen atom of the phenylhydrazones of aldehydes without α-hydrogen is exchanged rapidly on treatment with ethanolic potassium hydroxide. Electron-withdrawing substituents in the aldehyde portion of benzaldehyde phenylhydrazone enhance the rate, but in the phenylhydrazine portion they decrease it. Benzaldehyde 1-methyl-1-phenylhydrazone undergoes no exchange under these conditions. The mechanism shown in equation (9) was proposed[72] (see also ref. 73).

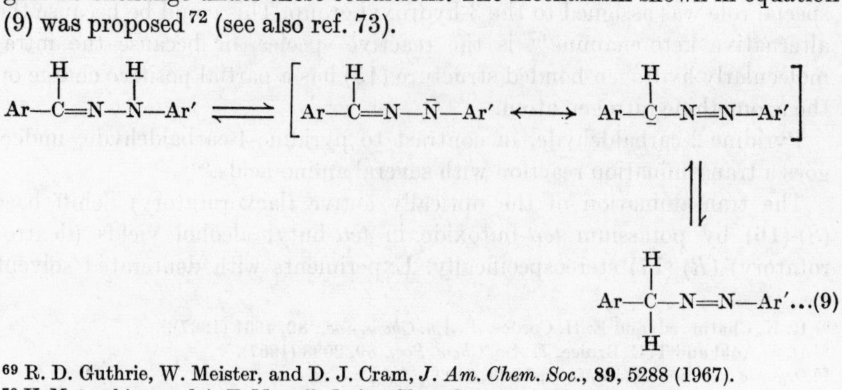

[69] R. D. Guthrie, W. Meister, and D. J. Cram, *J. Am. Chem. Soc.*, **89**, 5288 (1967).
[70] Y. Matsushima and A. E. Martell, *J. Am. Chem. Soc.*, **89**, 1322, 1331 (1967).
[71] K. Nagano and D. E. Metzler, *J. Am. Chem. Soc.*, **89**, 2891 (1967).
[72] H. Simon and W. Moldenhauer, *Chem. Ber.*, **100**, 1949 (1967).
[73] A. V. Chernova, R. R. Shagidullin, and Y. P. Kitaev, *Zh. Org. Khim.*, **3**, 916 (1967); *Chem. Abs.*, **67**, 53349k (1967).

The following reactions have also been investigated: formation of biacetyl mono- and di-oxime,[74] 2-naphthaldehyde semicarbazone,[75] aldose hydrazones,[76] osazones,[77] and glycosylamines;[78a] reaction of formaldehyde with tetrahydrofolic acid;[78b] formation and hydrolysis of some Δ^4-steroid 3-thiosemicarbazones;[79] hydrolysis of Girard hydrazones,[80] co-ordinated Schiff bases of ethylenediamine,[81] and nucleotides;[82] mutarotation of glycosylamines;[83] epimerization of lysergic acid and dihydrolysergic acid *N*-(1-hydroxyethyl)amides;[84] the Mannich reaction,[85] and reactions of the 3-ethylbenzisoxazolium ion with nucleophiles,[86] and of cyclopropanone with amines.[87]

The equilibrium constants for the formation of imines from isobutylaldehyde and aliphatic amines[88] and from salicylaldehyde and ethylamine[89] have been measured. *syn–anti*-Isomerism of imines,[90] hydrazones,[91] and *p*-benzoquinone mono-oxime[92] have been investigated.

Enolization and Related Reactions

Details have been published of Rappe's investigation of the base-catalysed halogenation of butan-2-one.[93–95] The results were considered to indicate that two different mechanisms are followed in the pH regions 5—7 and 12—14 since the ratios of 3-halogenation: 1-halogenation (7—7.5 and 0, respectively)

74 P. R. Ellefsen and L. Gordon, *Talanta*, **14**, 443 (1967).
75 H. P. Figeys and J. Nasielski, *Bull. Soc. Chem. Belges*, **75**, 601 (1966).
76 H. H. Stroh and P. Golüke, *Z. Chem.*, **7**, 60 (1967).
77 H. Simon, G. Heubach, and H. Wacker, *Chem. Ber.*, **100**, 3106 (1967); H. Simon and W. Moldenhauer, *ibid.*, p. 3121; A. Hassner, and P. Catsoulacos *Chem. Commun.*, **1967**, 121; A. Hassner and P. Catsoulacos, *Tetrahedron Letters*, **1967**, 489; L. Mester, E. Moczar, G. Vass, and A. Schimpl, *ibid.*, p. 2943.
78a K. Heyns, G. Müller, and H. Paulsen, *Ann. Chem.*, **703**, 202 (1967).
78b R. G. Kallen and W. P. Jencks, *J. Biol. Chem.*, **241**, 5845, 5851, 5864 (1966).
79 J. C. Orr, P. Carter, and L. L. Engel, *Biochemistry*, **6**, 1065 (1967).
80 M. Masui and H. Ohmori, *J. Chem. Soc., B*, **1967**, 762.
81 E. Hoyer and J. Anton, *Z. Chem.*, **7**, 197 (1967).
82 H. Venner, *Z. Physiol. Chem.*, **344**, 189 (1966).
83 T. Jasiński and K. Smiataczowa, *Z. Physik. Chem.* (*Leipzig*), **235**, 49 (1967).
84 F. Arcamone, B. Camerino, E. B. Chain, A. Ferretti, and S. Redaelli, *Tetrahedron*, **23**, 11 (1967).
85 R. Andrisano, A. S. Angeloni, P. DeMaria, and M. Tramontini, *J. Chem. Soc., C*, **1967**, 2307.
86 D. S. Kemp, *Tetrahedron*, **23**, 2001 (1967).
87 N. J. Turro and W. B. Hammond, *Tetrahedron Letters*, **1967**, 3085.
88 J. Hine and C. Y. Yeh, *J. Am. Chem. Soc.*, **89**, 2669 (1967); J. Hine and J. Mulders, *J. Org. Chem.*, **32**, 2200 (1967).
89 R. W. Green and R. J. Sleet, *Australian J. Chem.*, **19**, 2101 (1966).
90 G. Wettermark, *Arkiv Kemi*, **27**, 159 (1967); G. Wettermark, *Svensk Kem. Tidskr.*, **79**, 249 (1967).
91 A. F. Hegarty and F. L. Scott, *Chem. Commun.*, **1967**, 521.
92 R. K. Norris and S. Sternhell, *Tetrahedron Letters*, **1967**, 97.
93 C. Rappe, *Acta Chem. Scand.*, **20**, 2305 (1966).
94 C. Rappe, *Acta Chem. Scand.*, **21**, 857 (1967).
95 See *Organic Reaction Mechanisms*, **1966**, 321.

are widely different under the two sets of conditions. The precise formulation of these mechanisms is awaited with interest.

Further measurements, by means of NMR spectroscopy, of the rates of hydrogen exchange in each branch of unsymmetrical ketones have been reported.[96,97] Substituents in one methyl group of acetone affect the rate in both branches and sometimes the effect is greater in the unsubstituted branch (see Table 1).[96] It was reported in an extensive investigation of the base-catalysed hydrogen exchange of methoxyacetone that all the catalysts used, except ^-OH, catalysed exchange at the methyl position more rapidly than at the methoxymethylene position.[97]

Table 1. Rate constants for hydrogen exchange of acetone and methoxyacetone in deuterium oxide at 50° (l. mole^{-1} sec^{-1}).[96]

	Base catalysis (NaOAc)		Acid catalysis ($NaHSO_4$)	
	$10^5k_{R_1}$	$10^5k_{R_2}$	$10^5k_{R_1}$	$10^5k_{R_2}$
CH_3COCH_3[a]	0.15	0.15	10	10
$CH_3OCH_2COCH_3$[b]	0.57	1.2	0.72	1.2

[a] $R_1 = R_2 = CH_3$. [b] $R_1 = CH_3OCH_2$, $R_2 = CH_3$.

Details of the work of Warkentin and Tee on the exchange of the methyl and methylene protons of butan-2-one reported[97] last year have been published.[98]

Kinetic isotope effects for the base-catalysed ionization of *meta*- and *para*-substituted acetophenones have been obtained from their rates of bromination and the rates of detritiation of the corresponding [α-^{3}H]acetophenones. The ratio k_H/k_T increases steadily from the fastest reacting compound, *p*-nitroacetophenone ($k_H/k_T = 18.2$).[99] *ortho*-Substituted acetophenones react faster than their *para*-isomers and show lower isotope effects except *o*-methylacetophenone, for which $k_H/k_T = 19.8$ (at 25°). The Arrhenius plot for the deprotonation of this compound is curved at the lowest temperature studied and it was suggested that proton-tunnelling is important.[100] The effect of dimethyl sulphoxide on the rate of detritiation of [α^3-H]acetophenone in $Me_4N^+OH^-$–H_2O, NaOMe–MeOH, $KOBu^t$–Bu^tOH, and NaOEt–EtOH was also determined.[101]

96 A. A. Bothner-By and C. Sun, *J. Org. Chem.*, **32**, 492 (1967).
97 J. Hine, K. G. Hampton, and B. C. Menon, *J. Am. Chem. Soc.*, **89**, 2664 (1967); see *Organic Reaction Mechanisms*, **1966**, 321—322.
98 J. Warkentin and O. S. Tee, *J. Am. Chem. Soc.*, **88**, 5540 (1966); see also C. Rappe, *Acta Chem. Scand.*, **20**, 2305 (1966).
99 J. R. Jones, R. E. Marks, and S. C. Subba Rao, *Trans. Faraday Soc.*, **63**, 111 (1967).
100 J. R. Jones, R. E. Marks, and S. C. Subba Rao, *Trans. Faraday Soc.*, **63**, 993 (1967).
101 J. R. Jones and R. Stewart, *J. Chem. Soc., B*, **1967**, 1173.

The relative rates of the base-catalysed exchange of the methyl protons of acetylferrocene, (acetylcyclopentadienyl)tricarbonylmanganese, and acetophenone are 0.24:14.1:10.0 and for the acid-catalysed reaction are 4.2:2.9:4.0, respectively.[102]

The rate of iodination of ethyl pyruvate in the absence of catalysts other than water is approximately the same as that of pyruvic acid, suggesting that the latter does not react with intramolecular catalysis[103] as previously proposed.[104]

Further work on deuterium-exchange by 2-deuterioisobutyraldehyde in the presence of methylamine has been reported.[105–107] This reaction shows a term in the rate law proportional to the concentration of $MeNH_3^+$, which was explained as resulting from a base-catalysed exchange of the iminium ion, $Me_2CDCH{=}NHMe^+$.[107] A plot of the logarithms of the catalytic constants for a series of bases against the logarithms of the catalytic constants for the base-catalysed exchange of [β-^{2}H]isobutyraldehyde in the absence of methylamine was a straight line, suggesting that the base acts similarly in the two reactions.[106] 2,6-Lutidine was a particularly poor catalyst,[107] presumably as a result of steric hindrance.[108] No term in the rate law proportional to the concentration of $R_2NH_2^+$ was observed in the presence of several secondary amines;[109] this arises probably because the equilibrium constant for the formation of iminium ions from isobutyraldehyde and secondary amines is much less favourable than with methylamine owing to unfavourable steric interactions (compare **19** with **20**). Trimethylamine is a better catalyst for

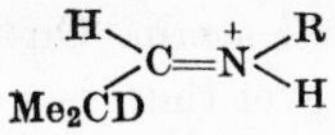

(19)

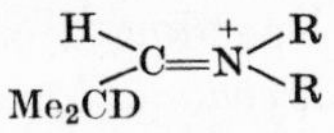

(20)

the base-catalysed exchange than dimethylamine, but *N*-methylpiperidine is poorer than piperidine, and *N*-methylmorpholine poorer than morpholine. It was suggested, surprisingly,[110] that the reactive conformations of the *N*-methyl compounds were those with the *N*-methyl groups axial.[109]

[102] N. V. Kislyakova, V. N. Setkina, and D. N. Kursanov, *Izv. Akad. Nauk SSSR, Ser. Khim.*, **1967**, 34; *Chem. Abs.*, **67**, 63458j (1967).
[103] R. P. Bell and H. F. F. Ridgewell, *Proc. Roy. Soc. (London), A*, **298**, 178 (1967).
[104] See *Organic Reaction Mechanisms*, **1965**, 248.
[105] J. Hine, F. C. Kokesh, K. G. Hampton, and J. Mulders, *J. Am. Chem. Soc.*, **89**, 1205 (1967).
[106] J. Hine, J. Mulders, J. G. Houston, and J. P. Idoux, *J. Org. Chem.*, **32**, 2205 (1967).
[107] See *Organic Reaction Mechanisms*, **1966**, 323.
[108] See *Organic Reaction Mechanisms*, **1965**, 246; cf. J. Hine, J. G. Houston, J. H. Jensen, and J. Mulders, *J. Am. Chem. Soc.*, **89**, 3085 (1967).
[109] J. Hine and J. Mulders, *J. Org. Chem.*, **32**, 2200 (1967).
[110] Cf. D. R. Brown, R. Lygo, J. McKenna, J. M. McKenna, and B. G. Hutley, *J. Chem. Soc., B*, **1967**, 1184.

Thioethoxide is about 15 times less effective a catalyst than methoxide ion for the isomerization of (−)-menthone to (+)-menthone (eqn. 10) in methanol, and about 5 times less effective for the isomerization of cholest-5-en-3-one to cholest-4-en-3-one. The thioethoxide-catalysed reactions could involve direct

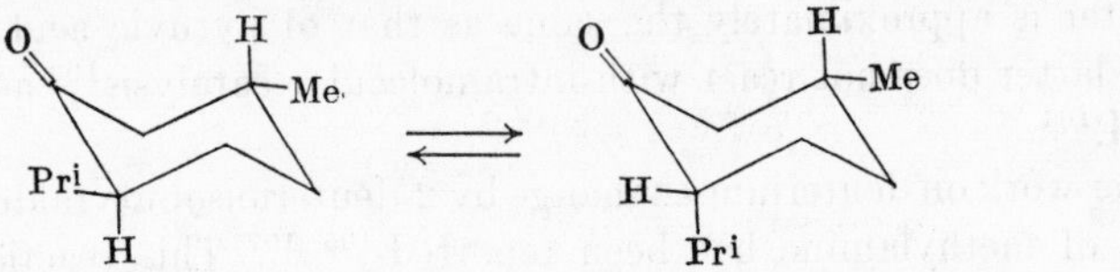

...(10)

abstraction of a proton by the thioethoxide ion to form an enolate ion, or the kinetically equivalent abstraction of a proton by methoxide with ethanethiol acting as a general acid-catalyst to yield an enol. It was tentatively suggested that since the pK_a of ethanethiol is less than that of cyclohexanone the menthone isomerization follows the latter mechanism, but that the isomerization of the (probably) more acidic cholesta-5-en-3-one may follow the former.[111]

Details have now been published[112] of the experiments which lead to the conclusion that some methyl-substituted 2-norbornanones undergo preferential exchange of the *exo*-3-hydrogen.[113] The following explanation has been offered[114] based on torsional strain. In the enolate ion from norbornanone (**22**) the dihedral angle between the $C_{(3)}$–H and $C_{(4)}$–H bonds is approximately 20°. This is increased in the transition state for protonation from the *exo*-direction (**23**) with a corresponding decrease in torsional strain. On passing to the transition state for *endo*-protonation, (**21**), however there is a decrease in the

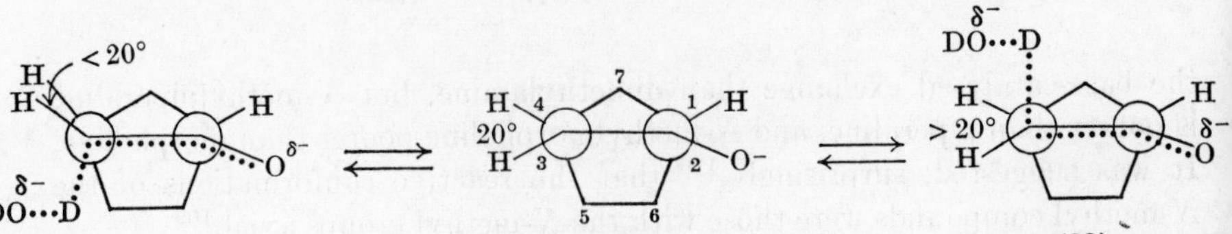

(**21**) (**22**) (**23**)

dihedral angle and hence an increase in torsional strain. "Torsional effects thus favour *exo* over *endo* attack, and, by microscopic reversibility, *exo* over *endo* departure." It should be noted, however, that α,α-dideuteriocamphor yields 21% of dideuterio-, 64% of monodeuterio-, and 15% of undeuterated com-

[111] J. F. Bunnett and L. A. Retallick, *J. Am. Chem. Soc.*, **89**, 423 (1967).
[112] A. F. Thomas, R. A. Schneider, and J. Meinwald, *J. Am. Chem. Soc.*, **89**, 68 (1967).
[113] See *Organic Reaction Mechanisms*, **1965**, 247; **1966**, 324.
[114] P. von R. Schleyer, *J. Am. Chem. Soc.*, **89**, 701 (1967).

pound on treatment with sodium hydroxide and, as pointed out,[115] this is not compatible with a more rapid exchange of the *exo*-hydrogen (see also p. 1).

The conversion of the unstable *trans*-enol* of alkyl 2-formyl-2-phenylacetates into the stable *cis*-enol has been shown by following the reaction by NMR spectroscopy to proceed through the intermediate aldehydo form (cf. 11). The reaction is acid-catalysed, with the *trans*-enol possibly catalysing its own isomerization.[116]

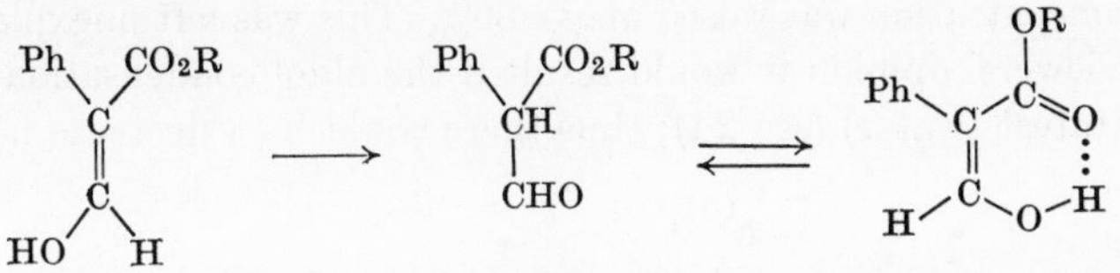

...(11)

The following reactions have also been investigated: base-catalysed bromination of acetone in the presence of high concentrations of electrolytes;[117] bromination of diethyl ketone, di-isopropyl ketone, 1-bromopropan-2-one, 2-bromopentan-3-one, 2-chloropentan-2-one, and 2-oxopropane-1,3-disulphonate,[118a] of methyl cyclohexyl, methyl cyclopentyl, and isopropyl methyl ketone,[118b] and of steroidal ketones;[118c] acid-catalysed bromination of acetaldehyde, acetone, and butan-2-one;[119] base-catalysed ionization of ethyl 2-oxocyclopentanecarboxylate;[120] and proton-transfer and tautomerization of heptane-2,4,6-trione.[121]

Equilibrium constants for the enolization of many di- and tri-carbonyl compounds have been measured.[122]

An extensive investigation of the base-catalysed aldol condensation of

* OH/CO_2R *trans*.

115 J. A. Berson, J. H. Hammons, A. W. McRowe, R. G. Bergman, A. Remanick, and D. Houston, *J. Am. Chem. Soc.*, **89**, 2597 (1967).

116 S. T. Yoffe, P. V. Petrovskii, E. I. Fedin, K. V. Vatsuro, P. S. Burenko, and M. I. Kabachnik, *Tetrahedron Letters*, **1967**, 4525.

117 J. R. Jones and S. C. Subba Rao, *Trans. Faraday Soc.*, **63**, 120 (1967).

118a R. P. Bell, G. R. Hillier, J. W. Mansfield, and D. G. Street, *J. Chem. Soc., B*, **1967**, 827.

118b M. Gaudry and A. Marquet, *Bull. Soc. Chim. France*, p. 1849.

118c A. Demarche, C. Gast, J.-C. Jacquesy, J. Levisalles, and L. Schaeffer, *Bull. Soc. Chim. France*, **1967**, 1636; R. Jacquesy and J. Levisalles, *ibid.*, **1967**, 1642.

119 P. T. McTigue and J. M. Sime, *Australian J. Chem.*, **20**, 905 (1967); see also C. F. Wells, *Trans. Faraday Soc.*, **63**, 147 (1967).

120 M. Mühlstaat, C. Weiss, O. Brede, and H. Nass, *Z. Chem.*, **6**, 467 (1966).

121 J. Stuehr, *J. Am. Chem. Soc.*, **89**, 2826 (1967).

122 N. L. Allinger, L. W. Chow, and R. A. Ford, *J. Org. Chem.*, **32**, 1994 (1967); A. Yogev and Y. Mazur, *ibid.*, p. 2162; E. W. Garbisch and J. G. Russell, *Tetrahedron Letters*, **1967**, 29; J. Osugi, T. Mizukami, and T. Tachibana, *Rev. Phys. Chem., Japan* **36**, 8 (1966); *Chem. Abs.*, **66**, 37113c (1967); M. Gorodetsky, Z. Luz, and Y. Mazur, *J. Am. Chem. Soc.*, **89**, 1183 (1967); T. B. H. McMurry and R. C. Mollan, *J. Chem. Soc., C*, **1967**, 1813; D. C. Nonhebel, *ibid.*, p. 1716; J. G. Dawber and M. M. Crane, *J. Chem. Educ.*, **44**, 150 (1967); S. Forsén, F. Merényi, and M. Nilsson, *Acta Chem. Scand.*, **21**, 620 (1967); W. Hänsel, R. Haller, and K. W. Merz, *Naturwissenschaften*, **54**, 44 (1967).

D-glyceraldehyde and of dihydroxyacetone has been reported.[123] The products are mixtures of fructose, sorbose, and dendro-ketose and so the active anion is always that derived from dihydroxyacetone. Pyridine bases are particularly effective catalysts, but those with a 2-methyl substituent, and especially 2,6-dimethyl substituents, show steric hindrance very similar to that observed in the enolization of other aldehydes and ketones.[124] The yield of hexoses formed from glyceraldehyde in aqueous pyridine fell markedly when the pyridine concentration was raised above 60%. This was left unexplained but in the Reviewers' opinion it would result if the aldol condensation required an acid catalyst (water) (see **24**); thus there could be a decrease in the rate

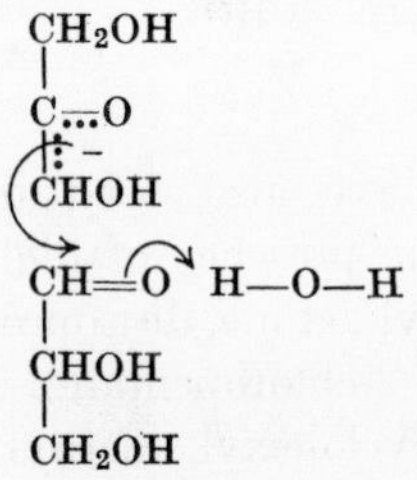

(24)

arising from the decrease in water concentration which accompanies the increase in pyridine concentration. There has also been an investigation of the interconversion of pentoses in pyridine solution.[125a]

The plot of the logarithm of the relative rate of the aldol condensation at $C_{(2)}$ of steroidal 3-ketones against the chemical shift of the 19-methyl group (relative to that of the analogous hydrocarbon) is approximately a straight line.[125b] The aldol condensation of formaldehyde and acetaldehyde,[126] and that of cyclopentanone and acetaldehyde,[127] have also been studied.

There have been several investigations relevant to the mechanism of action of aldolase.[128]

Additional evidence that the hydrolysis of enol ethers involves a rate-determining proton-transfer has been reported. This includes the observation of general acid-catalysis in the hydrolysis of ethyl vinyl ether,[129–131] and the

123 C. D. Gutsche, D. Redmore, R. S. Buriks, K. Nowotny, H. Grassner, and C. W. Armbruster, *J. Am. Chem. Soc.*, **89**, 1235 (1967).
124 See *Organic Reaction Mechanisms*, **1965**, 246.
125a M. Fedoroňko and K. Linek, *Coll. Czech. Chem. Commun.*, **32**, 2177 (1967).
125b R. Baker and J. Hudec, *Chem. Commun.*, **1967**, 891.
126 Y. Ogata, A. Kawasaki, and K. Yokoi, *J. Chem. Soc., B*, **1967**, 1013.
127 J.-E. Dubois and M. Dubois, *Tetrahedron Letters*, **1967**, 4215.
128 H. E. Sine and L. F. Hass, *J. Am. Chem. Soc.*, **89**, 1749 (1967); R. D. Kobes and E. E. Dekker, *Biochem. Biophys. Res. Commun.*, **27**, 607 (1967).
129 P. Salomaa, A. Kankaanperä, and M. Lajunen, *Acta Chem. Scand.*, **20**, 1790 (1966).
130 A. J. Kresge and Y. Chiang, *J. Chem. Soc., B*, **1967**, 53, 58.
131 See *Organic Reaction Mechanisms*, **1966**, 330.

finding that only one deuterium per mole is incorporated into the methyl group of the product, acetaldehyde, when the solvent is deuterium oxide.[130] The latter observation, and the observation that *cis*- and *trans*-alkenyl ethers are not isomerized under the conditions of hydrolysis,[132] indicate that the proton-transfer is irreversible and truly rate-determining. The isotope effects for catalysis by hydronium ion and formic acid are $(k_H/k_D)_{H^+} = 2.95$ and $(k_H/k_D)_{HCO_2H} = 6.8$.[130] The Brønsted α-value derived directly (0.51)[129] and from the variation of rate constant of the hydronium ion-catalysed reaction with D_2O concentration in D_2O–H_2O mixtures (0.65) suggests that the proton-transfer step is 0.5—0.6 complete by the time the transition state is reached.[130] This should mean that the primary isotope effect is close to the maximum which was estimated to be $(k_H/k_D)_{H^+} = 7.8$. The measured value is, however, only slightly greater than half this $(2.95/0.65 = 4.5)$, and it was therefore concluded that "the two different parts of the reaction co-ordinate which govern primary and secondary isotope effects are not closely coupled."[130]

Tracer experiments with $H_2{}^{18}O$ have demonstrated that a wide range of vinyl ethers and esters are hydrolysed in the presence of acids with vinyl–oxygen fission.[133]

The acid-catalysed hydrolysis of furan occurs about 10^4 times more slowly than that of divinyl ether and 10^6—10^7 times more slowly than that of dihydrofuran.

The rates of hydrolysis of a large number of alkyl-substituted vinyl ethers have been measured.[132–135]

The hydrolyses of vinyl ethers and acetals have been compared.[136]

A reaction, closely related to the hydrolysis of enol ethers, the hydrolysis of cyanoketen dimethyl acetal (reaction 12), has also been investigated.[137]

$$NC\cdot CH{:}C(OMe)_2 + H_2O \longrightarrow NC\cdot CH_2\cdot CO_2Me + MeOH \qquad \ldots(12)$$

This is general-acid catalysed with isotope effects $(k_H/k_D)_{H^+} = 3.0$ and $(k_H/k_D)_{AcOH} = 5.4$. The catalytic constants for catalysis by proton carboxylic acids obey the Brønsted equation with $\alpha = 0.63$, almost identical with the value obtained for catalysis by the corresponding deuteron acids $(\alpha = 0.64)$. It was suggested that a comparison of the values of α for catalysis by proton and deuteron acids could be used as a criterion for the detection of proton

132 T. Okuyama, T. Fueno, H. Nakatsuji, and J. Furukawa, *J. Am. Chem. Soc.*, **89**, 5826 (1967).
133 A. F. Rekasheva, L. A. Kiprianova, and I. P. Samchenko, *Abhandl. Deut. Akad. Wiss. Berlin, Kl. Chem. Geol. Biol.*, **1964**, 641; *Chem. Abs.*, **66**, 64759z (1967).
134 A. Kankaanperä and P. Salomaa, *Acta Chem. Scand.*, **21**, 575 (1967).
135 P. Salomaa and P. Nissi, *Acta Chem. Scand.*, **21**, 1386 (1967).
136 P. Salomaa and A. Kankaanperä, *Acta Chem. Scand.*, **20**, 1802 (1966).
137 V. Gold and D. C. A. Waterman, *Chem. Commun.*, **1967**, 40.

tunnelling, and their being equal here indicated its absence in this reaction (see p. 139).

Vinyl-interchange between benzoic acid and vinyl acetate[138] and the synthesis of steroidal enol acetates[139] have been investigated.

A *cis*-dienol ether (**25**) in aqueous acetic acid undergoes a photochemically induced hydrolysis whose rate is independent of acetic acid concentration.[140] Protonation occurs > 98% at the α-position compared with 45% in the dark reaction.[141] Similar photochemically induced hydrolysis of the *trans*-dienol ether (**26**) occurred with 30% of α-protonation while the dark reaction led to none of this product. It was considered that these results were in accord with

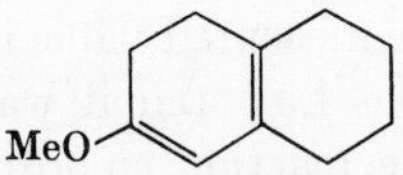

(**25**)

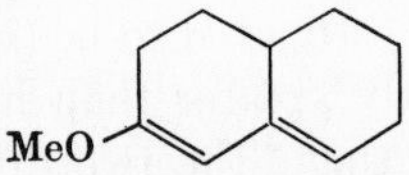

(**26**)

the charge-density changes on excitation, calculated by using the HMO approximation.[140]

The dienol (**28**), formed by the dehydration of enediol (**27**), undergoes predominantly α-protonation, to yield ketone (**29**) rather than γ-protonation to yield ketone (**30**).[142]

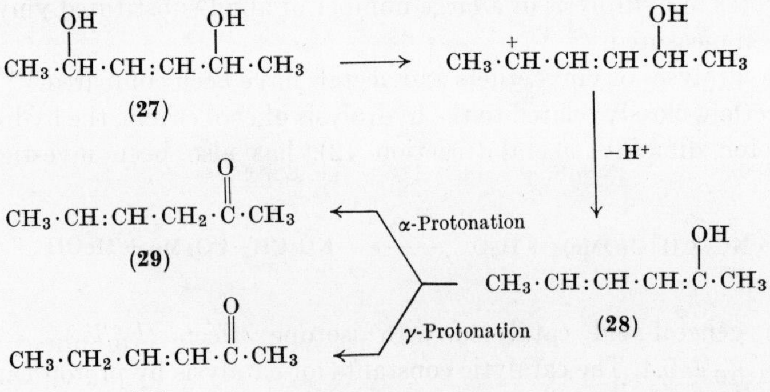

The investigation of the hydrolysis of enamines reported in 1965[143] has now been extended to more strongly acidic solutions. The rates did not show the expected independence of acid concentration, but instead decreased with

138 G. Slinckx and G. Smets, *Tetrahedron*, **22**, 3163 (1966).

139 O. R. Rodig and G. Zanati, *J. Org. Chem.*, **32**, 1423 (1967); A. J. Liston and M. Howarth, *ibid.*, p. 1034.

140 T. N. Huckerby, N. A. J. Rogers, and A. Sattar, *Tetrahedron Letters*, **1967**, 1113.

141 See *Organic Reaction Mechanisms*, **1965**, 252.

142 H. Morrison and S. R. Kurowsky, *Chem. Commun.*, **1967**, 1098.

143 See *Organic Reaction Mechanisms*, **1965**, 253.

increasing acid concentration. This was ascribed to a change in the rate-determining step from protonation of the carbon–carbon double bond (step 1) to hydration of the immonium ion (step 2) in weakly acidic solutions, and decomposition of the carbinolamine (step 3) in moderately strong acids (see Scheme 3).[144]

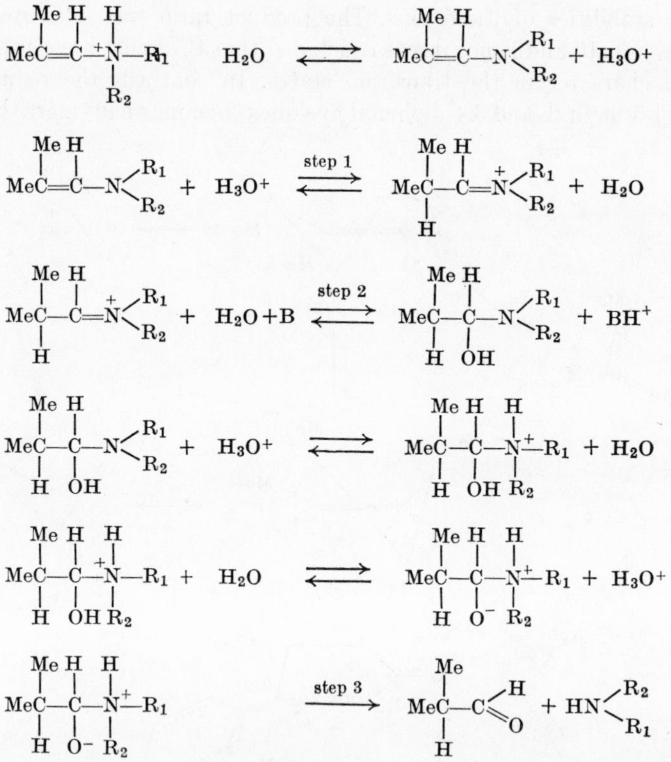

R_1, R_2 = morpholino, piperidino or pyrrolidino

Scheme 3

The following reactions of enamines have also been investigated: cyclohexanone enamines with β-nitrostyrenes[145] and phenyl vinyl sulphone;[146] equilibration of the enamines of 3-methylcyclohexanone and *trans*-2-decalone;[147] enonamine–enolimine tautomerism;[148] *cis–trans*-isomerism of

144 W. Maas, M. J. Janssen, E. J. Stamhuis, and H. Wynberg, *J. Org. Chem.*, **32**, 1111 (1967).
145 A. Risaliti, M. Forchiassin, and E. Valentin, *Tetrahedron Letters*, **1966**, 6331.
146 A. Risaliti, S. Fatutta, and M. Forchiassin, *Tetrahedron*, **23**, 1451 (1967).
147 S. K. Malhotra, D. F. Moakley, and F. Johnson, *Chem. Commun.*, **1967**, 448.
148 C. A. Grob and H. J. Wilkens, *Helv. Chim. Acta*, **50**, 725 (1967).

enamino-ketones;[149] and internal rotation about the =C–N bond of enamines.[150]

The reduction of 3,4-dimethylcyclohex-2-enone with lithium in liquid ammonia in the presence of ethanol yields a 16:84 mixture of *cis*- and *trans*-3,4-dimethylcyclohexanone. If these products were formed from a pair of anions (**31** and **32**) with $C_{(3)}$ tetrahedral, then their ratio should reflect the relative stabilities of these ions. The product ratio was estimated to be *cis*:*trans* = 1:15 and hence it was concluded that $C_{(3)}$ must have appreciable trigonal character in the transition states. In contrast, the reduction of 3-phenyl-5-methyl- and 3,4-diphenyl-cyclohex-3-enone yields more than 90%

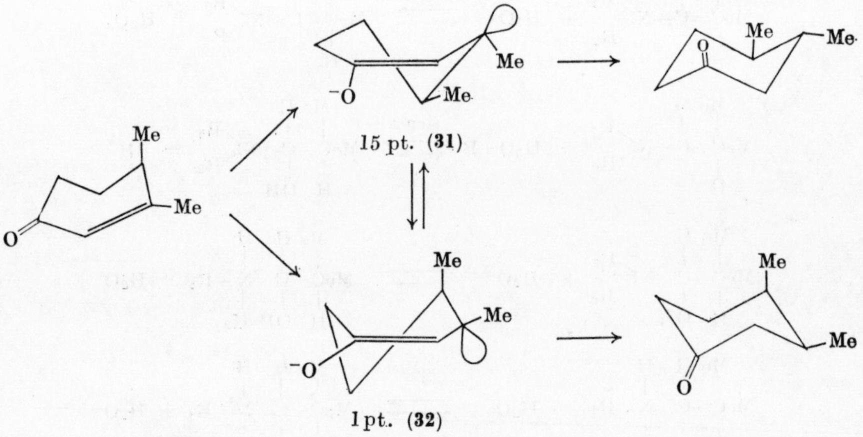

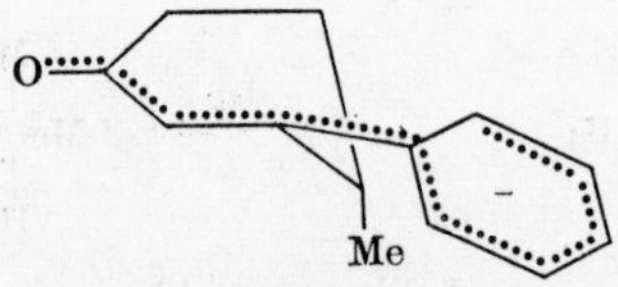

(**33**)

of the *cis*-isomers. It was suggested that the intermediate anions here are highly delocalized, have conformation (**33**), and are protonated from the axial direction.[151]

The major product of *C*-alkylation of enolate ions of type (**34**) by benzyloxymethyl halides is (**35**), formed by α-attack, when R is Me and (**36**), formed by β-attack, when R is CO_2Et. Possibly the CO_2Et group is involved directly in

[149] J. Danbrowski and J. Terpionki, *Roczniki Chem.*, **41**, 697 (1967).
[150] A. Mannschreck and V. Koelle, *Tetrahedron Letters*, **1967**, 863.
[151] S. K. Malhotra, D. F. Moakley, and F. Johnson, *Tetrahedron Letters*, **1967**, 1089.

the formation of an intermediate complex between the metal enolate and alkyl halide.[152]

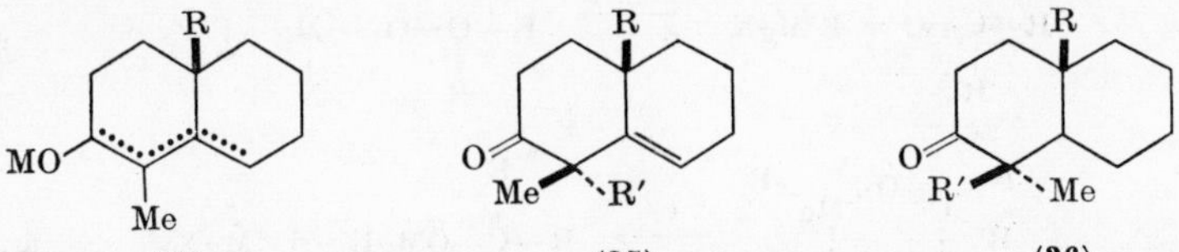

(34) **(35)** **(36)**

Predominant *O*-alkylation of the enamino-ketone (**37**) by methyl iodide occurred in hydroxylic solvents, and predominant *C*-alkylation in aprotic solvents. The *O*-alkylation was reversible and the product underwent further decomposition through nucleophilic attack by the solvent.[153]

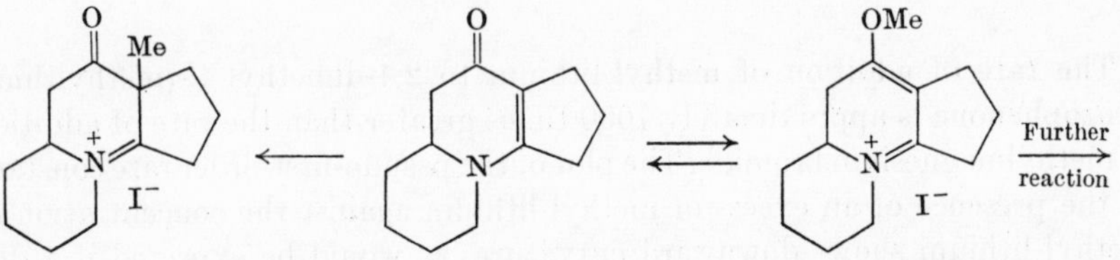

(37)

Alkylation of the metal enolates of 2,6- and 2,2-dimethylcyclohexanone has been studied.[154]

Details of the reactions of the enol–metal chelates reported last year[155] have been published.[156]

Other Reactions

The kinetics of the reaction of an excess of methylmagnesium bromide, G, with benzophenone fit a rate expression of the form:

$$k_{\text{obs}} = \frac{k_2 K[G]^2{}_0}{1 + K[G]_0},$$

where k_{obs} is the pseudo-first-order rate constant. The mechanism shown in equations (13) and (14) was proposed as consistent with this expression. The final step (eqn. 15) was postulated to explain the falling off of the pseudo-first-

152 F. H. Bottom and F. J. McQuillin, *Tetrahedron Letters*, **1967**, 1975.
153 A. I. Meyers, A. H. Reine, and R. Gault, *Tetrahedron Letters*, **1967**, 4049.
154 D. Caine and B. J. L. Huff, *Tetrahedron Letters*, **1967**, 3399.
155 See *Organic Reaction Mechanisms*, **1966**, 332—334.
156 L. Crombie, D. E. Games, and M. H. Knight, *J. Chem. Soc., C*, **1967**, 757, 763, 773, 777.

order constants at low Grignard concentrations after 50% reaction[157] (see also p. 112).

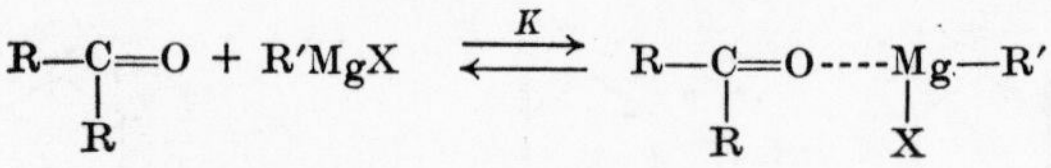

...(13)

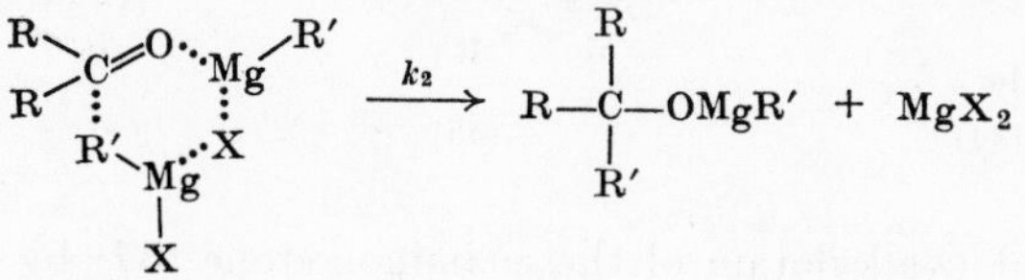

...(14)

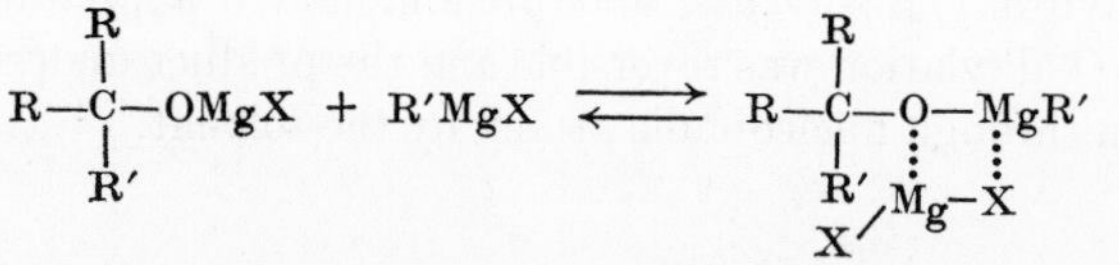

...(15)

The rate of addition of methyl-lithium to 2,4-dimethyl-4′-(methylthio)-benzophenone is approximately 7000 times greater than the rate of addition of methylmagnesium bromide. The plot of the pseudo-first-order rate constant in the presence of an excess of methyl-lithium against the concentration of methyl-lithium shows downward curvature, as would be expected if a dissociated reactive species were in equilibrium with associated forms.[158]

The stereochemistry of the additions to carbonyl groups directly bonded to asymmetric carbon atoms has been discussed.[159]

The following additions to carbonyl compounds have been investigated: nitriles and sulphones to benzophenone;[160] disodiophenylacetamide to benzaldehyde;[161] and Grignard reagents to ketones;[162] also the alkynylation of 17-keto-steroids.[163]

The reduction of ketones by aluminium hydride,[164] of cyclohexanones by lithium aluminium hydride[165] and diborane,[166] and of adamantanone by alcohol dehydrogenase,[167] and the hydrogenation of 2-methylcyclopentanone have been studied.[168]

[157] E. C. Ashby, R. B. Duke, and H. M. Neumann, *J. Am. Chem. Soc.*, **89**, 1964 (1967).
[158] S. G. Smith, *Tetrahedron Letters*, **1966**, 6075.
[159] G. J. Karabatsos, *J. Am. Chem. Soc.*, **89**, 1367 (1967).
[160] E. M. Kaiser and C. R. Hauser, *J. Am. Chem. Soc.*, **89**, 4566 (1967).
[161] D. M. von Schriltz, E. M. Kaiser, and C. R. Hauser, *J. Org. Chem.*, **32**, 2610 (1967).
[162] Y. Yasuda, N. Kawabata, and T. Tsuruta, *Kogyo Kagaku Zasshi*, **69**, 936 (1966).
[163] T. C. Miller and R. G. Christiansen, *J. Org. Chem.*, **32**, 2781 (1967).
[164] D. C. Ayres and R. Sawdaye, *J. Chem. Soc., B*, **1967**, 581.
[165] S. R. Landor and J. P. Regan, *J. Chem. Soc., C*, **1967**, 1159.
[166] J. Klein and E. Dunkelblum, *Tetrahedron*, **23**, 205 (1967).
[167] H. J. Ringold, T. Bellas, and A. Clark, *Biochem. Biophys. Res. Commun.*, **27**, 361 (1967).
[168] D. Cornet and F. G. Gault, *J. Catalysis*, **7**, 140 (1967).

There have been extensive investigations of the kinetics[169] and steric course[170–172] of the Wittig reaction and discussions of its mechanism.[169–172] A stable four-membered cyclic compound with a structure analogous to that of the possible covalent intermediate of the Wittig reaction has been reported.[173] The reaction of dimethylsulphonium benzylide with formaldehyde to form styrene oxide has been investigated.[174]

The following reactions have also been investigated: pyridoin condensation,[175] Cannizzaro reaction,[176] tautomerism of mesityl oxide,[177] addition of HCN to 3-oxo-Δ^4-steroids,[178] Tishchenko reaction of acetaldehyde with aluminium isopropoxide,[179] acid-catalysed condensation of formaldehyde with aromatic amines and arenethiols, selenols, or sulphinic acids,[180] an intramolecular diazoalkane–carbonyl reaction,[181] and thermal cyclization of unsaturated ketones.[182]

The NMR spectra of a large number of protonated carbonyl compounds[183] and protonated imines[184] have been reported.

169 C. Rüchardt, P. Panse, and S. Eichler, *Chem. Ber.*, **100**, 1144 (1967).
170 L. D. Bergelson, L. I. Barsukov, and M. M. Shemyakin, *Tetrahedron*, **23**, 2709 (1967).
171 L. V. Shubina, L. Y. Malkes, V. N. Dmitreva, and V. D. Bezuglyi, *Zh. Obshch. Khim.*, **37**, 437 (1967).
172 M. Schlosser and K. F. Christmann, *Ann. Chem.*, **708**, 1 (1967).
173 G. H. Birum and C. N. Matthews, *Chem. Commun.*, **1967**, 137.
174 M. Yoshimine and M. J. Hatch, *J. Am. Chem. Soc.*, **89**, 5831 (1967).
175 J. P. Schaefer and J. L. Bertram, *J. Am. Chem. Soc.*, **89**, 4121 (1967).
176 R. A. Khaibullina, L. G. Fatkulina, and K. R. Rustamov, *Uzbeksk. Khim. Zh.*, **10**, 54 (1966); *Chem. Abs.*, **66**, 10399q (1967).
177 G. Hesse, R. Hatz, and U. Dutt, *Chem. Ber.*, **100**, 923 (1967).
178 H. B. Henbest and W. R. Jackson, *J. Chem. Soc., C*, **1967**, 2465.
179 Y. Ogata, A. Kawasaki, and I. Kishi, *Tetrahedron*, **23**, 825 (1967).
180 I. E. Pollak and G. F. Grillot, *J. Org. Chem.*, **32**, 3101 (1967).
181 C. D. Gutsche and J. E. Bowers, *J. Org. Chem.*, **32**, 1203 (1967).
182 F. Rouessac, P. Le Perchec, and J.-M. Conia, *Bull. Soc. Chim. France*, **1967**, 818, 822, 826; J.-M., Conia and F. Leyendecker, *ibid.*, p. 830.
183 M. Brookhart, G. C. Levy, and S. Winstein, *J. Am. Chem. Soc.*, **89**, 1735 (1967); G. A. Olah, M. Calin, and D. H. O'Brien, *ibid.*, p. 3586; G. A. Olah and M. Calin, *ibid.*, p. 4736; G. A. Olah, D. H. O'Brien, and M. Calin, *ibid.*, p. 3582; D. M. Brouwer, *Chem. Commun.*, **1967**, 515; T. D. J. D'Silva and H. J. Ringold, *Tetrahedron Letters*, **1967**, 1505.
184 G. A. Olah and P. Kreienbühl, *J. Am. Chem. Soc.*, **89**, 4756 (1967).

CHAPTER 12

Reactions of Acids and their Derivatives

Carboxylic Acids[1]

Tetrahedral intermediates. Fedor and Bruice's demonstration, by kinetic means, of the intervention of an intermediate in the hydrolysis of *S*-ethyl trifluorothioacetate[2] has been complemented by an investigation of oxygen-exchange by this ester.[3] Both methods allow the partitioning of the intermediate to products and reactants to be calculated and, within experimental error, identical values were obtained. This confirms the view that exchange and hydrolysis involve the same intermediate which must be a tetrahedral addition compound.

Although the hydrolysis of *N*-acetyl-*O*-[*carbonyl*-^{18}O]cinnamoylserine amide is accompanied by oxygen-exchange, this exchange does not occur with [*carbonyl*-^{18}O]cinnamoyl-α-chymotrypsin and *p*-nitro[*carbonyl*-^{18}O]-benzoyl-α-chymotrypsin. Lack of exchange is probably because the two oxygen atoms of the tetrahedral intermediate are not equivalent in the asymmetric environment of the enzyme.[3]

Kinetic evidence has been obtained for the intervention of a tetrahedral intermediate in the hydrolysis of *S*-2,2,2-trifluoroethyl thioacetate catalysed by piperidine. The plot of k_{obs} against $[\text{Piperidine}]^2_{total}$ is a straight line with intercept, k_I, which varies with hydroxyl ion concentration according to the equation $k_I = 1.02 \times 10^4[^-OH]$. The second-order rate constant for the hydroxyl ion-catalysed reaction determined directly, in the absence of piperidine, is 64.5 l. mole^{-1} min^{-1}, i.e., about 160 times less than the apparent constant (1.02×10^4) determined in the presence of piperidine. This suggests that there is a piperidine-catalysed hydrolysis whose rate is independent of the concentration of piperidine under these conditions. The most reasonable explanation is a piperidine-catalysed attack of water on the ester, to yield a tetrahedral intermediate (T), which partitions preferentially to re-form reactants:

$$H_2O + \text{Ester} \underset{\text{pipH}^+ \cdot k_2}{\overset{\text{pip} \cdot k_1}{\rightleftharpoons}} T \xrightarrow{k_3} \text{Products.}$$

1 (*a*) General base and nucleophilic catalysis of ester hydrolysis and related reactions have been reviewed by S. L. Johnson, *Adv. Phys. Org. Chem.*, **5**, 237 (1967). (*b*) The mechanisms of hydrolysis of orthoesters has been reviewed by E. M. Cordes, *Progr. Phys. Org. Chem.*, **4**, 1 (1967).

2 See *Organic Reaction Mechanisms*, **1965**, 260.

3 M. L. Bender and H. d'A. Heck, *J. Am. Chem. Soc.*, **89**, 1211 (1967).

Steady-state analysis of this equation yields:

$$-\frac{d[\text{Ester}]}{dt} = \frac{k_1 k_3[\text{Ester}][\text{pip}][H_2O]}{k_2[\text{pipH}^+] + k_3};$$

and, when $k_2[\text{pipH}^+] \gg k_3$,

$$-\frac{d[\text{Ester}]}{dt} = \frac{k_1 k_3[\text{Ester}][\text{pip}][H_2O]}{k_2[\text{pipH}^+]} = \frac{K_a k_1 k_3[\text{Ester}][^-OH][H_2O]}{k_2 K_w}.$$

Thus $K_a k_1 k_2[H_2O]/K_w k_2 = 1.02 \times 10^4$ l. mole^{-1} min^{-1} and $k_1 k_3/k_2 = 193.6$ l. mole^{-1} min^{-1}.[4,*]

The reactions of *S*-2,2,2-trifluoroethyl thioacetate with a large number of other nucleophiles were also studied. Most of these follow a rate law of the form: Rate $= k_n$[Ester][Nuc], but that with tris[(hydroxymethyl)amino]-methane follows a law, Rate $= k_n$[Ester][Nuc] $+ k_{BH^+}$[Ester][NucH$^+$]. The second term on the right-hand side of this equation was thought to arise from a general acid-catalysed hydrolysis of the ester. Although the Brønsted plot of $\log k_n$ against pK_a of the base showed considerable scatter, there was a good correlation between $\log k_n$ and $\log k$ for the reactions of the same nucleophiles with *p*-nitrophenyl acetate, suggesting that the rate-determining step for both sets of reaction is nucleophilic attack.[4]

Hydrolysis of trifluoroacetanilide and trichloroacetanilide is specifically catalysed by hydrogen carbonate and dihydrogen phosphate ions,[5] possibly as a result of catalysis of the breakdown of the tetrahedral intermediate (cf. ref. 36, p. 340).

Details of Hibbert and Satchell's investigation of the hydrolysis of benzoyl cyanide have been published.[6] It is interesting that the reaction has now been shown to exhibit general acid-inhibition, as would be expected from the proposed mechanism and the observed general base-catalysis. Attempts to observe similar general acid-inhibition in the closely related hydrolysis of *S*-ethyl trifluorothioacetate were inconclusive.[7] The hydrolysis of acetyl and propionyl cyanide was also investigated.[8]

The hydrolysis of the 2-acetyl-3,4-dimethylthiazolium ion occurs in two easily separable steps (eqn. 1). The first, the reversible hydration to 2-(1,1-dihydroxyethyl)-3,4-dimethylthiazolium ion (*ca.* 50% complete at equilibrium) is general base-catalysed and probably proceeds as symbolized by (**1**).

* This work has now been shown to be wrong and has been retracted: N. J. Kundu and T. C. Bruice, *J. Org. Chem.*, **33**, 422 (1968).

[4] M. J. Gregory and T. C. Bruice, *J. Am. Chem. Soc.*, **89**, 2121 (1967).

[5] S. O. Eriksson and C. Holst, *Acta Chem. Scand.*, **20**, 1892 (1966).

[6] F. Hibbert and D. P. N. Satchell, *J. Chem. Soc., B*, **1967**, 653; see *Organic Reaction Mechanisms*, **1966**, 260.

[7] See *Organic Reaction Mechanisms*, **1965**, 260.

[8] F. Hibbert and D. P. N. Satchell, *J. Chem. Soc., B*, **1967**, 755.

The second, cleavage to acetic acid and dimethylthiazolium ion, is specific $^{-}$OH-catalysed, and probably proceeds as shown in equation (2).[9]

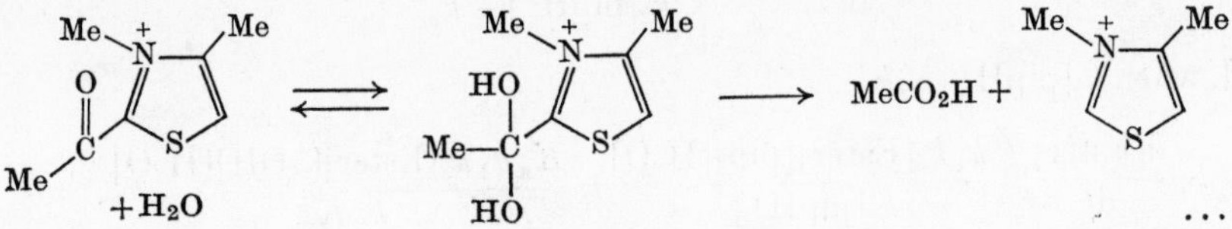

...(1)

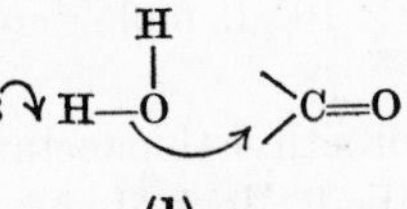

(1)

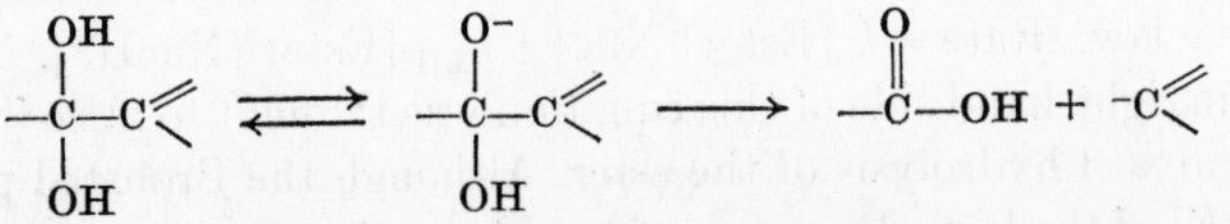

...(2)

The base-catalysed cleavage of 2-formylpropiophenone

$C_6H_5COCH(CHO)CH_3$[10]

and acetylacetone[11] have been investigated.

Intermolecular catalysis. Further extensive studies of the aminolysis of aryl esters in aqueous solution have been reported.[12, 13] The general kinetic equation is:

$$k_{obs} = k_n[N_f] + k_{gb}[N_f]^2 + k_{ga}[N_f][NH^+] + k_{OH}[N_f][OH^-]$$

where $[N_f]$ and $[NH^+]$ are the concentration of amine and its conjugate acid, and k_n, k_{gb}, k_{ga}, and k_{OH} are the rate constants for unassisted or water-assisted nucleophilic attack, apparent self-assisted general base-catalysed nucleophilic attack, nucleophilic attack assisted by the conjugate acid, and hydroxide-ion-catalysed nucleophilic attack by amine on the ester. The terms in k_{ga} and k_{gb} are more important for esters with poor leaving groups, and the relative importance of all the terms depends on the amine, the temperature, and the salt employed to maintain constant ionic strength.[12, 13] The transition states shown in (**2**)—(**5**) were proposed for the different pathways.

9 G. E. Lienhard, *J. Am. Chem. Soc.*, **88**, 5642 (1966).
10 N. Boulay, *Compt. Rend., Ser. C*, **264**, 602 (1967).
11 J.-P. Calmon and P. Maroni, *Compt. Rend., Ser. C*, **263**, 319 (1966).
12 T. C. Bruice, A. Donzel, R. W. Huffman, and A. R. Butler, *J. Am. Chem. Soc.*, **89**, 2106 (1967).
13 L. do Amaral, K. Koehler, D. Bartenbach, T. Pletcher, and E. H. Cordes, *J. Am. Chem. Soc.*, **89**, 3537 (1967).

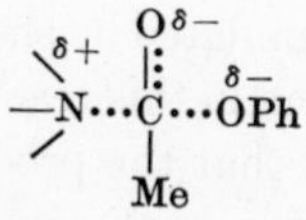

k_n (2)

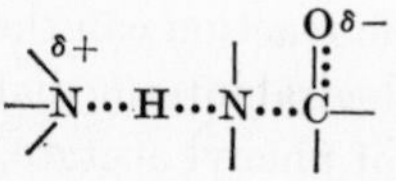

k_{gb} (3)

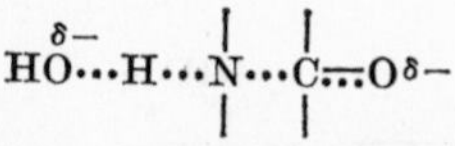

k_{OH} (5)

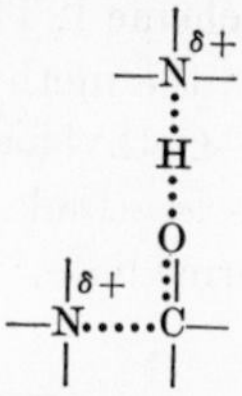

k_{ga} (4)

The catalytic constants for general acid-catalysis by carboxylic acids of the methoxyaminolysis of *p*-nitrophenyl acetate are almost independent of the pK_a of the acid, but for catalysis by ammonium ions are highly sensitive to the pK_a. Possibly the carboxylic acids act as bifunctional catalysts as shown in (**6**).[13]

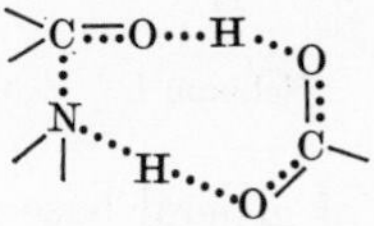

(**6**)

There is no intramolecular general base-catalysis by one amino-group of lysine or cyclohexane-*trans*-1,2-diamine in the aminolysis of *p*-nitrophenyl acetate by the others.[14]

The aminolysis of phenyl esters in dioxan and cyclohexane solution has also been investigated.[15] The reactions of phenyl dichloroacetate and difluoroacetate with primary amines in dioxan show mixed second- and third-order kinetics: $k_{obs} = k_2[\text{Amine}] + k_3[\text{Amine}]^2$, but with secondary amines, second-order kinetics are followed. The reaction of phenyl dichloroacetate with *n*-butylamine in dioxan is catalysed by triethylamine. The reactions in cyclohexane solution show third-order kinetics. In order to minimize charge separation in the transition states mechanisms were proposed in which proton transfers were concerted with nucleophilic attack.[15]

The rate of hydrolysis of phenyl *N*-methylacetimidate (**8**) from pH 9 to pH 14 is constant and the reaction is probably one of the protonated substrate with HO^-. From pH 9 to pH 2 the pH–rate profile is sigmoid with the rate proportional to the concentration of protonated substrate (**7**) (pK_a = 6.2) and the reaction is probably one of water with (**7**). Below pH 2 the rate decreases and there is probably a change in rate-determining step to breakdown of the

[14] R. W. Huffman, A. Donzel, and T. C. Bruice, *J. Org. Chem.*, **32**, 1973 (1967).
[15] A. S. A. S. Shawali, and S. S. Biechler, *J. Am. Chem. Soc.*, **89**, 3020 (1967).

tetrahedral intermediate. The reaction can therefore be formulated as shown in Scheme 1. The same tetrahedral intermediates are involved in this reaction as in the methylaminolysis of phenyl acetate, and the fact that the product is *N*-methylacetamide and phenol indicates that the slow step in the aminolysis is attack of the amine on the ester, not breakdown of the tetrahedral intermediate.[16]

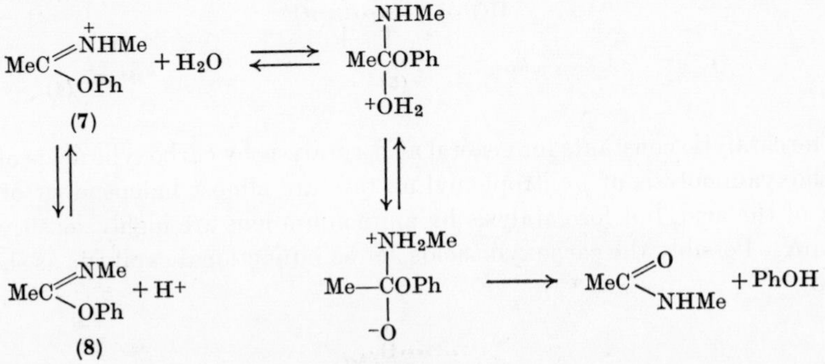

Scheme 1

An extensive investigation of general base-catalysis of the hydrolysis of alkylaminoethyl esters of carboxylic acids has been reported.[17a]

The methanolysis of *p*-nitrophenyl acetate catalysed by 2- and 2,6-disubstituted pyridines is slightly slower (< 5-fold) than predicted from their basic strengths and the Brønsted plot for pyridines without 2- and 6-substituents.[17b] It was concluded that this is consistent with the pyridine acting as a nucleophilic catalyst, but in our opinion it does not exclude the possiblity that it acts as a general base-catalyst since steric effects larger than this are known in several authentic general base-catalysed reactions (see p. 319 and ref. 18). Further, the steric effects of 2- and 6-methyl substituents on the pyridine-catalysed hydrolysis of acetic anhydride, an authentic nucleophilically catalysed reaction with similar steric requirements, are much larger than found here.[19]

The hydrolysis of ethyl oxalate in aqueous hydrochloric acid is not acid-catalysed.[20] The rate decreases about 3-fold on going from 1M- to 8M-acid and a similar decrease is brought about by the addition of lithium chloride at constant acid concentration. Similar behaviour is exhibited by ethyl *o*-nitro-

16 M. Kandel and E. H. Cordes, *J. Org. Chem.*, **32**, 3061 (1967).
17a T. Suzuki and Y. Tanimura, *Chem. Pharm. Bull. (Tokyo)*, **15**, 674 (1967).
17b A. Kirkien-Konasiewicz, A. M. Gude, M. McGraw, M. Knight, and A. Maccoll, *Chem. Ind. (London)*, **1967**, 1527.
18 *Organic Reaction Mechanisms*, **1965**, 246.
19 See V. Gold, *Progr. Stereochem.*, **3**, 194 (1962).
20 T. C. Bruice and B. Holmquist, *J. Am. Chem. Soc.*, **89**, 4028 (1967).

phenyl oxalate. Bunnett plots yield *w* values of +7.4 and +9.7, respectively, characteristic of reactions involving water acting as a proton-transfer agent, and transition state (**9**) was proposed. A plot of the catalytic constants for the

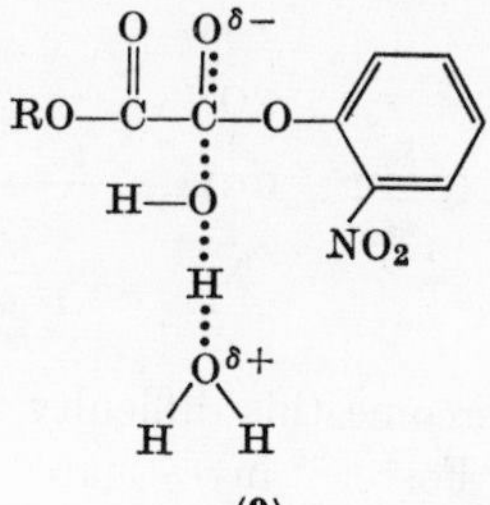

(**9**)

reaction of nucleophiles with the *o*-nitrophenyl oxalate anion against those for reaction with ethyl *o*-nitrophenyl oxalate is a straight line and the points for positively charged nucleophiles, e.g., 2-aminopyridinium ion, show none of the deviations to be expected if their reaction with the oxalate anion were enhanced by electrostatic catalysis as proposed previously.[21] The hydrolysis of *o*-nitrophenyl oxalate is catalysed by ribonuclease-A.[22a]

Nucleophiles possessing an unshared pair of electrons on an atom adjacent to the nucleophilic atom (e.g., NH_2OH, NH_2NH_2, ^-OOH) are frequently much more reactive towards carbonyl-carbon than predicted by their pK_a's from the Brønsted plot for other nucleophiles. This is termed the α-effect and it has now been shown not to operate in a general base-catalysed reaction, the ionization of nitroethane, and a nucleophilic substitution on sp^3-hybridized carbon, that of methyl iodide, for which hydrazine, *N*-methylhydrazine, hydroxylamine, and methoxylamine show no positive deviations from the Brønsted plots for other primary and secondary amines[22b] (see also p. 107).

An interesting rate decrease, possibly the result of intramolecular hydrogen bonding in the initial state, occurs in the reaction of *o*-phenylenediamine with benzoyl chloride in benzene, which is much slower than that of *p*-phenylenediamine.[23]

Imidazole-catalysis of the hydrolysis of vinyl acetate,[24] and the hydrolysis of *p*-nitrophenyl acetate catalysed by histidine derivatives[25a] and cysteine-containing peptides,[25b] have been investigated.

[21] M. L. Bender and Y.-L. Chow, *J. Am. Chem. Soc.*, **81**, 3929 (1959).
[22a] T. C. Bruice, B. Holmquist, and T. P. Stein, *J. Am. Chem. Soc.*, **89**, 4221 (1967).
[22b] M. J. Gregory and T. C. Bruice, *J. Am. Chem. Soc.*, **89**, 2327, 4400 (1967); see also ref. 11.
[23] H. S. Venkataraman, J. M. Rao, and L. Iyengar, *Indian J. Chem.*, **5**, 255 (1967).
[24] G. S. Reddy and D. G. Gehring, *J. Org. Chem.*, **32**, 2291 (1967).
[25a] I. Schmidt, W. U. Ahmad, and F. Turba, *Z. Physiol. Chem.*, **345**, 91 (1966).
[25b] I. Photaki and V. Bardakos, *Chem. Commun.*, **1967**, 275.

Determination of nucleophilicities towards carbonyl-carbon is complicated because the overall rate of reaction depends on the rate constant for nucleophilic attack, k_1, and on the partitioning of the tetrahedral intermediate, $k_3/(k_2+k_3)$ (see eqn. 3). A slow reaction may therefore result if either of these

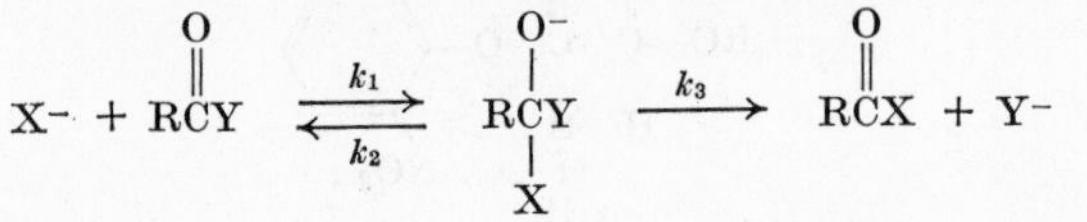

...(3)

quantities is small. To overcome this difficulty in determining the nucleophilicity of Cl^-, Jones and Foster[26] investigated chloride-exchange of ethyl chloroformate in water–acetonitrile mixtures. Here $k_2 = k_3$ and the rate constant for exchange, $k_e = k_1/2$, is 18—45 times greater than the rate constant for hydrolysis, depending on the water concentration (40—85%), and so, with this compound under these conditions, chloride ion has considerable nucleophilicity towards carbonyl-carbon.[26]

Bromide-exchange between acetyl bromide and tetra-alkylammonium bromides has been investigated.[27]

A definitive study of ester hydrolysis in aqueous sulphuric acid has been reported.[28] Three types of rate-dependence on acidity were found. Methyl, ethyl, and *n*-propyl acetate showed a rate maximum at *ca.* 50% H_2SO_4 followed by a slow increase at high sulphuric acid concentration. Isopropyl, *sec*-butyl, and benzyl acetate showed a maximum at *ca.* 50% H_2SO_4 followed by a sharp increase in rate at higher acidities. Aryl acetates showed a continuous increase with acidity over the whole range of sulphuric acid concentration (10—90%) studied. These variations of rate acidity were correlated by a modified Bunnett equation:

$$\log k_1 + H_s = r \log a_{H_2O} + \text{Constant}$$

where H_s is the acidity function appropriate to the substrate, in this case an ester. The problem that values of H_s for esters are not known was overcome by determining the ratio $[SH^+]/[S]$ for several esters which are not hydrolysed too rapidly. This is related to H_s by the equation

$$\log \frac{[SH^+]}{[S]} = -H_s + pK_{SH^+}$$

[26] L. B. Jones and J. P. Foster, *J. Org. Chem.*, **32**, 2900 (1967).

[27] K. K. Desai and B. C. Haldar, *Proc. Nucl. Radiat. Chem., Symp. Waltair, India*, **1966**, 255; *Chem. Abs.*, **67**, 21216y (1967).

[28] K. Yates and R. A. McClelland, *J. Am. Chem. Soc.*, **89**, 2686 (1967).

and since the plot of log[SH$^+$]/[S] against H_0 is a straight line

$$\log \frac{[\mathrm{SH^+}]}{[\mathrm{S}]} = -mH_0 + \text{Constant}$$

$$H_s = mH_0 + \text{Constant}$$

and

$$\log k_1 + mH_0 = r \log a_{\mathrm{H_2O}} + \text{Constant}.$$

Plots of $\log k_1 + mH_0$ against $\log a_{\mathrm{H_2O}}$ for all the esters studied showed breaks. Those for the primary alkyl and aryl esters at low acidities gave slopes ($r = ca. +2$) characteristic of an $A_{\mathrm{Ac}}2$ mechanism which changed at higher acidities to ones with slopes ($r = ca. -0.2$) characteristic of an $A_{\mathrm{Ac}}1$ mechanism. The value of r for secondary alkyl and benzyl acetates at low acidities is characteristic of an $A_{\mathrm{Ac}}2$ mechanism changing to one (*ca.* -0.5) which probably indicates an $A_{\mathrm{Alk}}1$ mechanism. The value of r at low acidities indicates an $A_{\mathrm{Ac}}2$ mechanism for all the esters involving a transition state in which there are two water molecules as (**10a**) or (**10b**).

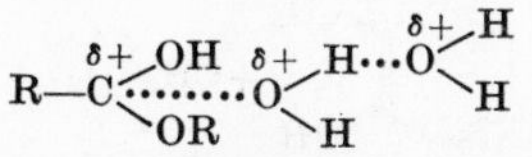

(**10a**)

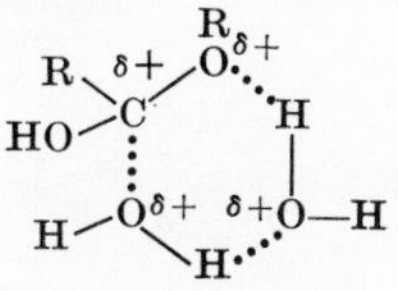

(**10b**)

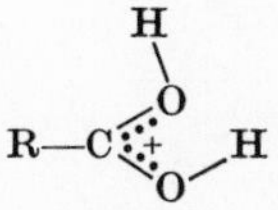

(**11**)

The acid-catalysed hydrolysis of ethyl acetate probably proceeds by attack of water on its conjugate acid protonated on the carbonyl-oxygen with rate constant *ca.* $10^{2.5}$ sec^{-1} at 25°.[29]

Esters dissolved in HSO_3F–SbF_5–SO_2 are protonated on carbonyl-oxygen since under conditions where exchange is slow there is no coupling between the α-proton of the alkyl group and the oxygen-bound proton.[30] Carboxylic acids are also protonated on carbonyl-oxygen and the NMR spectrum in HF–BF_3 at –80° shows that the two oxygen-bound protons are not equivalent but that the structure is probably as shown in (**11**).[31]

Intramolecular catalysis and neighbouring-group participation. An important investigation of the hydrolysis of aspirin and substituted aspirins has been reported by Fersht and Kirby who conclude that the enhanced rate of hydrolysis results, not from intramolecular nucleophilic catalysis, as previously

[29] R. B. Martin, *J. Am. Chem. Soc.*, **89**, 2501 (1967).
[30] G. A. Olah, D. H. O'Brien, and A. M. White, *J. Am. Chem. Soc.*, **89**, 5694 (1967).
[31] H. Hogeveen, *Rec. Trav. Chim.*, **86**, 289 (1967); G. A. Olah and A. M. White, *J. Am. Chem. Soc.*, **89**, 3591 (1967); M. Brookhart, G. C. Levy, and S. Winstein, *ibid.*, p. 1735.

supposed, but from intramolecular general base-catalysis.[32] The rates of hydrolysis of 4- and 5-substituted aspirins (see **12**) were correlated by the Jaffe four-parameter equation:

$$\log k/k_0 = \sigma_1 \rho_{acid} + \sigma_2 \rho_{phenol}$$

where σ_1 and σ_2 are the substituent constants (σ_m and σ_p) of the group in the 4- or 5-position, relative to the reacting groups in the 1- and 2-positions, and ρ_{acid} and ρ_{phenol} are the reaction constants for the effect of substituents on the acidic and phenolic groups, respectively. The values obtained were

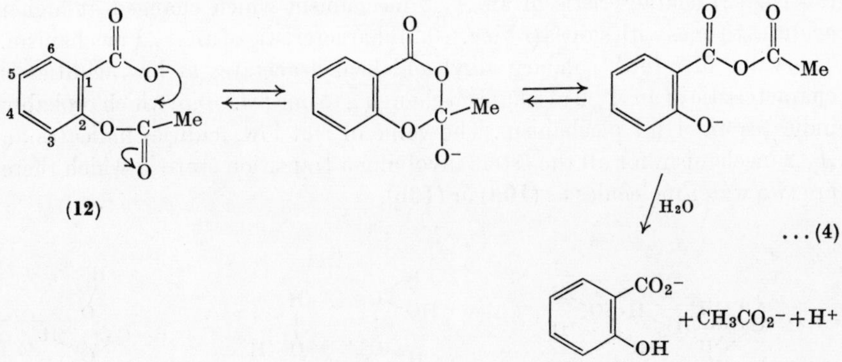

(**12**) . . . (4)

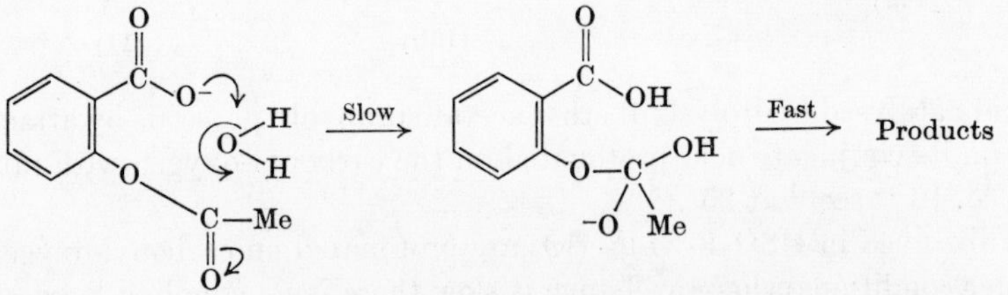

. . . (5)

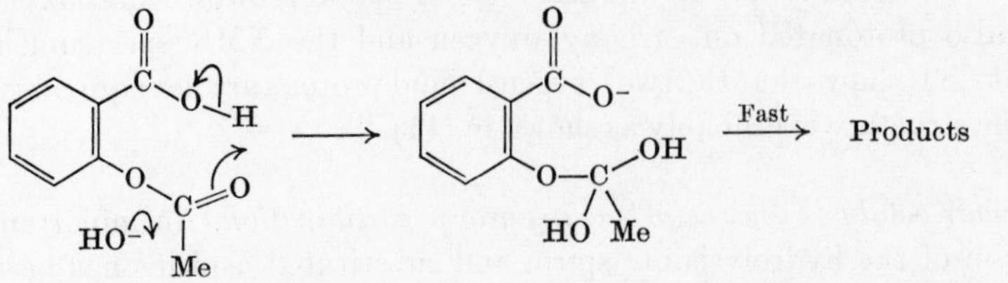

. . . (6)

$\rho_{phenol} = 0.96 \pm 0.04$ and $\rho_{acid} = -0.52 \pm 0.03$. That for ρ_{phenol} clearly excludes a mechanism involving intramolecular nucleophilic catalysis (eqn. 4) in which breakdown of the tetrahedral intermediate is rate-limiting. Several reactions

[32] A. R. Fersht and A. J. Kirby, *J. Am. Chem. Soc.*, **89**, 4853, 4857 (1967).

proceeding by this type of mechanism are known (e.g., hydrolysis of aryl hydrogen succinates and glutarates) and they all yield ρ-values > 2. A mechanism involving intramolecular nucleophilic catalysis with rate-limiting formation of the tetrahedral intermediate also seems unlikely from the value of ρ_{acid} (−0.52), as this is much less than the value of ρ (1.6—1.7) that corresponds to the β-coefficient of 0.8 found for the intermolecular reaction of nucleophiles with *p*-nitrophenyl acetate. Further, the main evidence for intramolecular nucleophilic catalysis, incorporation of ^{18}O into salicylic acid obtained from the hydrolysis of aspirin in $H_2{}^{18}O$, was shown to be incorrect and no significant incorporation could be detected. This mechanism was therefore rejected.

On the other hand, the ρ-values are consistent with intramolecular general base-catalysis (eqn. 5); ρ_{phenol} (0.96) is similar to that for the base-catalysed hydrolysis of *p*-nitrophenyl acetate (1.1 ± 0.2) and ρ_{acid} (−0.52) corresponds closely to the β-values for the general base-catalysed hydrolysis of ethyl and phenyl dichloroacetate. Further support for a bimolecular mechanism comes from the negative entropy of activation, −22.5 e.u., and the fact that the reaction is faster in aqueous alcohols than in water, since alcohols and alkoxide ions are stronger nucleophiles than water and hydroxide. This mechanism could, however, involve intramolecular acid-catalysis of attack by hydroxide ion (eqn. 6); but, since its enthalpy of activation would have to be very small (160 cal mole^{-1}) and similar catalysis for attack by other nucleophiles would be expected, but was not found, it also was rejected.

Intermolecular general base-catalysis of the hydrolysis of aspirin occurs, but this is not associated with intramolecular general acid-catalysis by the carboxyl group. The solvent isotope effect for catalysis by acetate, $k_H/k_D = 2.2$, is the same as that for the intramolecularly catalysed hydrolysis, and the entropy of activation, −30.7 e.u., is 8.2 e.u. more negative, which is approximately the entropy loss expected when two molecules come together to form a complex. The ρ-value for the phosphate-catalysed (probably general base-catalysed) hydrolysis of aspirin is 0.96, the same as ρ_{phenol} for the intramolecularly catalysed hydrolysis. This close similarity of the inter- and the intra-molecularly catalysed reactions suggests that the mechanisms are similar and therefore supports the view that intramolecular general base-catalysis is involved as shown in equation (5).

As reported last year,[33] the relative rates of the general base- and nucleophilically catalysed hydrolyses of aryl esters depends on the pK_a of the corresponding phenol. It should be possible then to observe a change in mechanism from intramolecular general base to intramolecular nucleophilic catalysis by studying an aspirin with strongly electron-withdrawing substituents. This possibility was realised by Fersht and Kirby who showed that

[33] *Organic Reaction Mechanisms*, **1966**, 338.

the hydrolysis of 3,5-dinitroaspirin in $H_2{}^{18}O$ at pH > 3 yielded 3,5-dinitrosalicylic acid with over 40% incorporation of label and that in 50% methanol–water it yielded 60% of methyl 3,4-dinitrosalicylic acid. Acetyl-3,5-dinitrosalicyloyl anhydride was proposed as an intermediate and possibly this is hydrolysed with intramolecular general base-catalysis by the ionized phenolic hydroxyl group (see **13**).[34]

The pH–rate profile for the hydrolysis of 3,5-dinitroaspirin is sigmoid at pH's below 3 suggesting that the undissociated form is being hydrolysed at an enhanced rate under these conditions, possibly as a result of intramolecular general acid catalysis.[35]

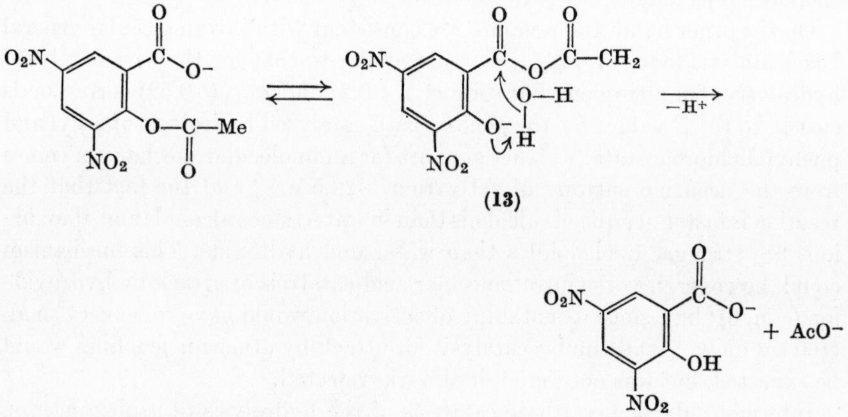

(13)

The hydrolysis of 4-hydroxybutyranilide at pH 4.90 is catalysed specifically by phosphate and hydrogen carbonate. The reaction involves neighbouring-group participation by the hydroxyl group but, in the absence of hydrogen carbonate and phosphate, breakdown of the tetrahedral intermediate (**14**) rather than the cyclization step is rate-limiting. Phosphate and hydrogen carbonate probably catalyse breakdown of (**14**) via cyclic transition states (**15**) and (**16**); consistently with this, at high concentrations of buffer the rate reaches a steady value, attributable to a change to cyclization as the rate-limiting step.[36] Intermediate (**14**) is the same as that postulated as intervening in the hydrolysis of an *N*-phenylimino-lactone discussed last year.[37]

The pH-rate profile for the hydrolysis of *p*-nitrophenyl *o*-methanesulphonamidobenzoate has a sigmoid portion in which

$$\text{rate} \propto [C_6H_4(CO_2Ar)\bar{N}SO_2 \cdot Me].$$

[34] A. R. Fersht and A. J. Kirby, *J. Am. Chem. Soc.*, **89**, 5960 (1967).
[35] A. R. Fersht and A. J. Kirby, *J. Am. Chem. Soc.*, **89**, 5961 (1967).
[36] B. A. Cunningham and G. L. Schmir, *J. Am. Chem. Soc.*, **89**, 917 (1967); B. J. Haske, M. E. Matthews, J. A. Conkling, and H. P. Perzanowski, *J. Org. Chem.*, **32**, 1579 (1967).
[37] *Organic Reaction Mechanisms*, **1966**, 354—355.

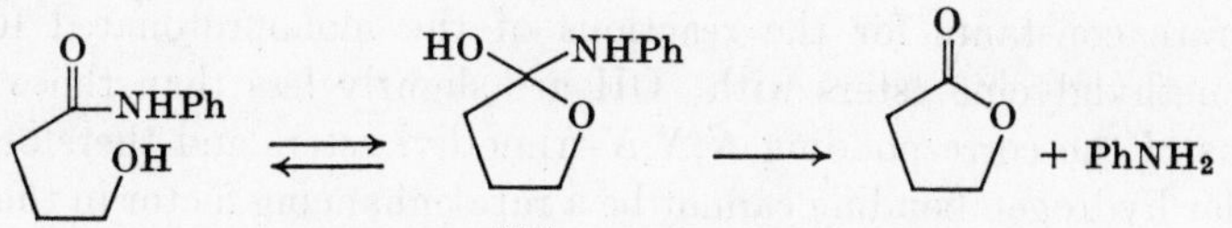

(14)

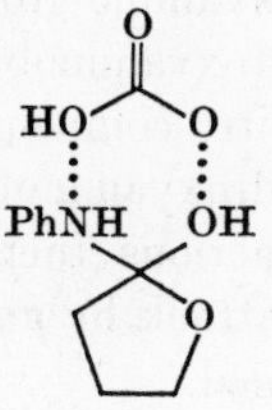

(15)

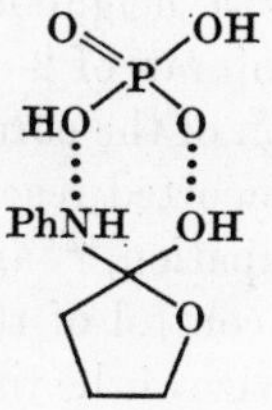

(16)

At pH 8.65 the rate is 40 times greater than that for the hydrolysis of the corresponding *para*-ester. A mechanism involving intramolecular nucleophilic catalysis was excluded by the solvent isotope effect, $k_{H_2O}/k_{D_2O} = 2.02$ and one involving general base-catalysis (eqn. 7) was preferred to one involving intramolecular general acid-catalysis because the reaction with another nucleophile (azide ion) was not accelerated as would be expected if the latter occurred.[38]

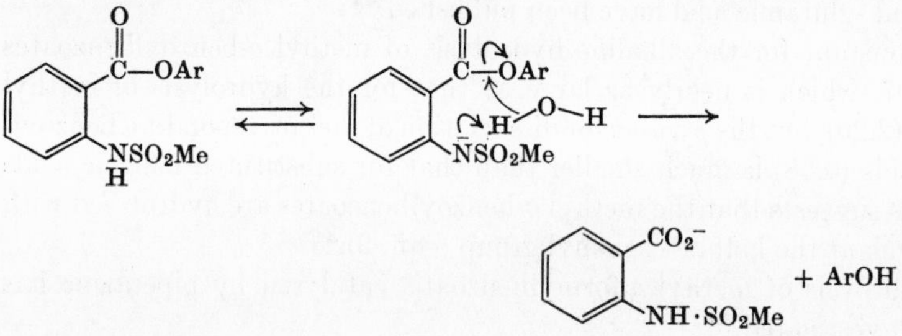

. . . (7)

The alkaline hydrolysis of methyl ester (**17**) is about four times faster than that of its *para*-isomer and after correction for steric effects the rate enhancement was calculated to be 14-fold and was ascribed to a field effect.[39]

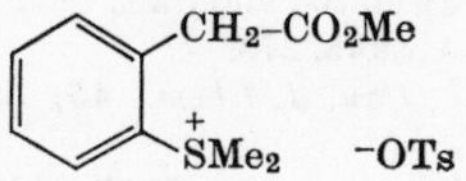

(17)

[38] F. M. Menger and C. L. Johnson, *Tetrahedron*, **23**, 19 (1967).
[39] J. Casanova, N. D. Werner, and H. R. Kiefer, *J. Am. Chem. Soc.*, **89**, 2411 (1967).

The rate constants for the reactions of the monoprotonated forms of *N*,*N*-dimethylglycine esters with ^{-}OH are slightly less than those for the reactions of the corresponding *N*,*N*,*N*-trimethyl esters and therefore intramolecular hydrogen bonding cannot be a rate-enhancing factor in the former reactions.[40]

N-Participation by a neighbouring hydroxyamino-group occurs in the reaction in tetrahydrofuran of 2-cyano-2′-hydroxyaminobiphenyl (eqn. 8)[41] generated by reduction of the corresponding nitro-compound. This contrasts with the previously reported reaction of *o*-hydroxyaminobenzonitrile which reacts with *O*-participation,[42] and these reactions therefore provide an excellent example of control of the mode of attack by an ambident neighbouring group by the size of the ring being formed.

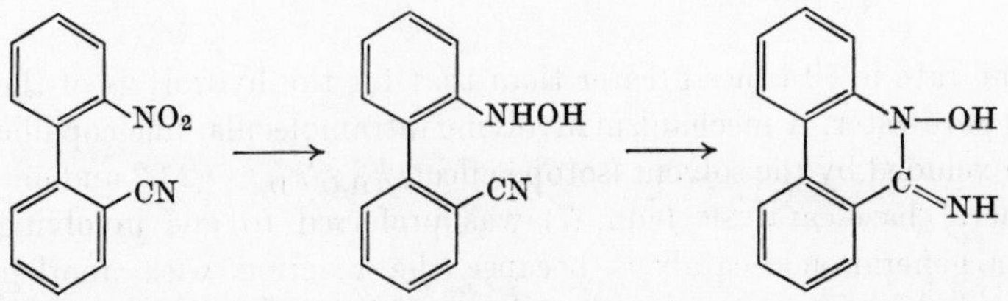

. . . (8)

Details of Capindale and Fan's investigation of the hydrolysis of *N*-benzoyl-aspartic and -glutamic acid have been published.[43]

The ρ-constant for the alkaline hydrolysis of methyl *o*-benzoylbenzoates (**18**) is 2.07, which is nearly as large as that for the hydrolysis of methyl benzoates (2.20), but the ρ-value for dissociation of the corresponding benzoyl-benzoic acids (0.58) is much smaller than that for substituted benzoic acids (1.68). This suggests that the methyl *o*-benzoylbenzoates are hydrolysed with initial attack at the ketone-carbonyl group (eqn. 9).[44]

The hydrolysis of methyl *o*-formylbenzoate, catalysed by piperazine has also been investigated.[45]

The base-catalysed racemization of the *p*-nitrophenyl esters of acyl-peptides in dichloromethane containing triethylamine proceed through an intermediate oxazolone.[46] Oxazolone formation from *O*-[(benzyloxycarbonyl-

[40] G. Aksnes and P. Frøyen, *Acta Chem. Scand.*, **21**, 1507 (1967).
[41] C. W. Muth, J. R. Elkins, M. L. DeMatte, and S. E. Chiang, *J. Org. Chem.*, **32**, 1106 (1967).
[42] See *Organic Reaction Mechanisms*, **1965**, 270.
[43] J. B. Capindale and H. S. Fan, *Can. J. Chem.*, **45**, 1921 (1967); see *Organic Reaction Mechanisms*, **1966**, 343.
[44] K. Bowden and G. R. Taylor, *Chem. Commun.*, **1967**, 1112; M. S. Newman and L. K. Lala, *Tetrahedron Letters*, **1967**, 3267.
[45] G. Dahlgren and D. M. Schell, *J. Org. Chem.*, **32**, 3200 (1967).
[46] I. Antonovics and G. T. Young, *J. Chem. Soc., C*, **1967**, 595; M. Bodanszky and A. Bodanszky, *Chem. Commun.*, **1967**, 591.

glycyl)-L-phenylalanyl]-*N*-ethylsalicylamide (**19**), as measured by the rate of racemization in DMF containing a triethylamine buffer, is specific base-catalysed.[47] Basic hydrolysis of methyl hippurate, unlike that of *p*-nitrophenyl hippurate, does not proceed with neighbouring-group participation by the amide group.[48]

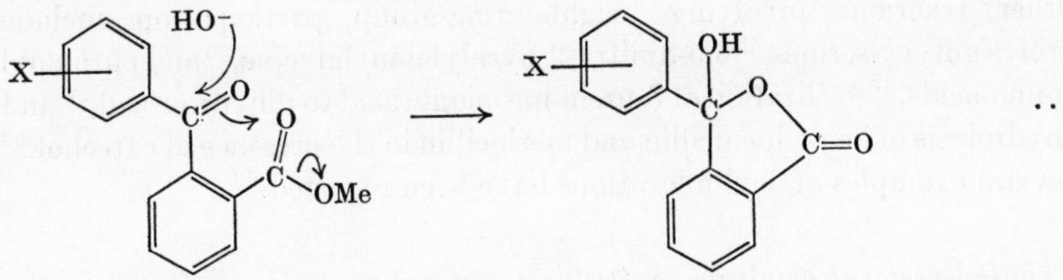

...(9)

(**18**)

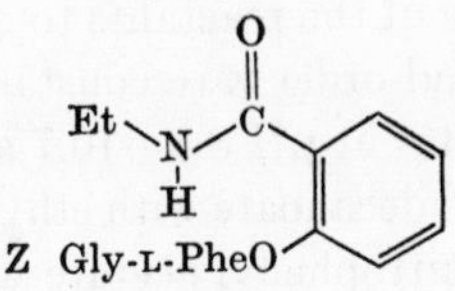

(**19**)

Competition between five- and six-membered ring formation in the Dieckmann condensation has been investigated.[49]

Kinetics of thiazole ring-opening of 3-methyl- and 2,3-dimethyl-benzo-thiazolium salts and recyclization of the resulting *o*-(*N*-acetyl-*N*-methyl-amino)thiophenols have been measured.[50]

The interactions of carboxylic acid functional groups attached to polymers have been reviewed.[51]

Acetylation of the tertiary 6a-alcohol group of (**20**) and hydrolysis of the tertiary acetate group of (**21**) proceed at enhanced rates. Mechanisms in which

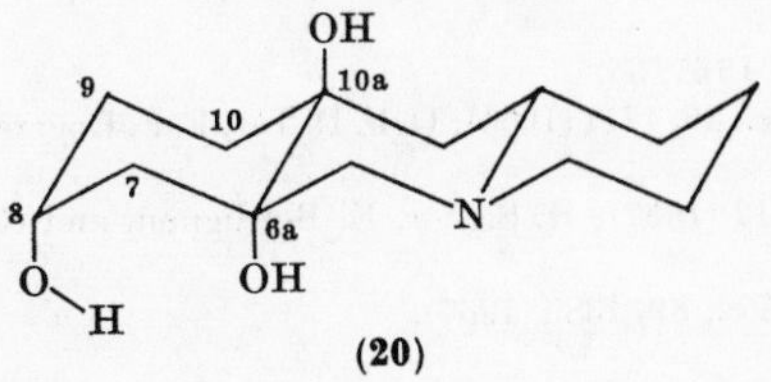

(**20**)

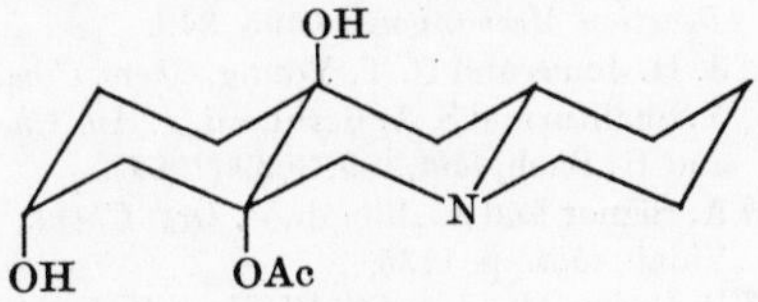

(**21**)

47 D. S. Kemp and S. W. Chien, *J. Am. Chem. Soc.*, **89**, 2745 (1967).
48 R. W. Hay and P. J. Morris, *Chem. Commun.*, **1967**, 663.
49 R. L. Augustine, Z. S. Zelawski, and D. H. Malarek, *J. Org. Chem.*, **32**, 2257 (1967).
50 H. Vorsanger, *Bull. Soc. Chim. France*, **1967**, 551, 556.
51 H. Morawetz, *Svensk Kem. Tidskr.*, **79**, 309 (1967).

the hydroxyl group acts as a general acid-catalyst and the nitrogen as a nucleophilic or general base-catalyst were considered but it is not yet possible to decide which is correct.[52]

Further work on the aminolysis of 8-quinolyl esters which probably involves intramolecular general base-catalysis has been reported.[53] Aminolysis of catechol esters probably also involves intramolecular general base-catalysis.[54]

Other reactions involving neighbouring-group participation include hydrolysis of acylserines,[55] 3,5-dinitro-2-pyridylalanylglycine,[56] and phthaloyl ω-amino-acids,[57] cyclization of 3-ureidopropionic acid to dihydrouracil,[58] and the hydrolysis of benzylpenicillin and methicillin in the presence of catechol.[59]

Several examples of acyl migrations have been reported.[60]

Association-prefaced catalysis. A striking rate enhancement is found in the reaction of *p*-nitrophenyl decanoate (4.53×10^{-6}M) with *n*-decylamine (4.08×10^{-5}M) as a result of binding of the reactants to one another through hydrophobic interactions. The second-order rate constant for the release of *p*-nitrophenol in 0.99% acetone–water at pH 8.8—10.7 and 35° is 700 times that for the reaction of *p*-nitrophenyl decanoate with ethylamine, whereas the rate constant for the reaction of *p*-nitrophenyl acetate with *n*-decylamine is only 6.8 times greater than that for the reaction with ethylamine. Product analysis shows that half the rate enhancement arises from aminolysis and half from hydrolysis. It cannot be due to incorporation of the reactants into micelles as the concentrations used were well below (10^3—10^4-fold) the critical micelle concentration, and most probably it is the result of formation of small (1:1, 1:2, 2:1, etc.) complexes between the long-chain amine and the ester.[61]

N-Stearoylhistidine (**22**) is 240 times more effective than *N*-acetylhistidine as a catalyst for the hydrolysis of the ester (**23**). Presumably reaction is facilitated by a pre-association of catalyst and substrate through hydrophobic

[52] S. M. Kupchan, J. H. Block, and A. C. Isenberg, *J. Am. Chem. Soc.*, **89**, 1189 (1967).

[53] H.-D. Jakubke, A. Voigt, and S. Burkhardt, *Chem. Ber.*, **100**, 2367 (1967); cf. *Organic Reaction Mechanisms*, **1966**, 349.

[54] J. H. Jones and G. T. Young, *Chem. Commun.*, **1967**, 35.

[55] Y. Shalitin and S. A. Bernhard, *J. Am. Chem. Soc.*, **88**, 4711 (1966); D. F. DeTar, F. F. Rogers, and H. Bach, *ibid.*, **89**, 3039 (1967).

[56] A. Signor and L. Biondi, *J. Org. Chem.*, **32**, 812 (1967); H. Signor, E. Bordignon, and G. Vidali, *ibid.*, p. 1135.

[57] P. D. Hoagland and S. W. Fox, *J. Am. Chem. Soc.*, **89**, 1389 (1967).

[58] I. G. Pojarlieff, *Tetrahedron*, **23**, 4307 (1967).

[59] M. A. Schwartz and G. R. Pflug, *J. Pharm. Sci.*, **56**, 1459 (1967).

[60] A. J. Gordon, *Tetrahedron*, **23**, 863 (1967); I. P. Freeman and I. D. Morton, *J. Chem. Soc., C*, **1966**, 1710; R. M. Vitali and R. Gardi, *Steroids*, **8**, 537 (1966); L. H. Welsh, *J. Org. Chem.*, **32**, 119 (1967); D. Y. Curtin and L. L. Miller, *J. Am. Chem. Soc.*, **89**, 637 (1967); D. Buchnea, *Chem. Phys. Lipids*, **1**, 113, 128 (1967); *Chem. Abs.*, **66**, 115265v, 115266w (1967).

[61] J. R. Knowles and C. A. Parsons, *Chem. Commun.*, **1967**, 755.

interactions. Interestingly, the catalytic efficiency of (**22**) is decreased by sodium chloride and urea.[62]

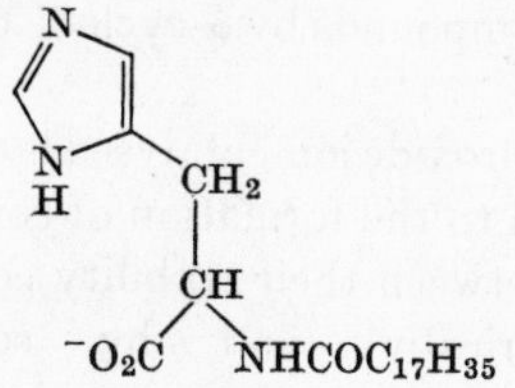

(22)

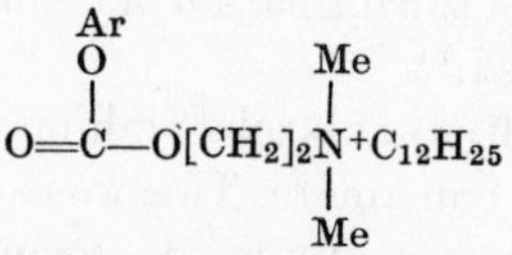

(23)

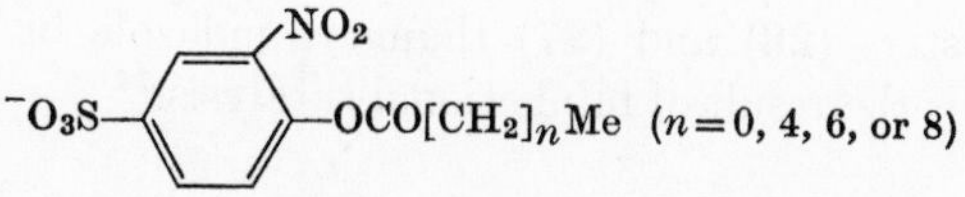

(24)

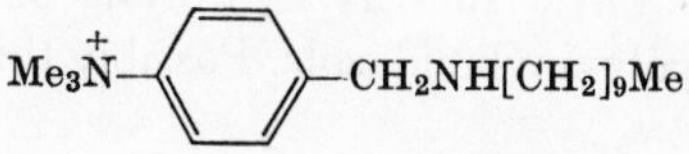

(25)

The rate of liberation of the phenol from long-chain esters (**24**) (5×10^{-5}M) is increased by the presence of long-chain amine (**25**) near its critical micelle concentration (7×10^{-3}M) according to the equation

$$k_{obs} = k_{OH}[^{-}OH] + \frac{V_m[\text{Amine}]^2}{K + [\text{Amine}]^2}.$$

The product is the acylated amine and presumably the ester is incorporated into the micelles of the amine.[63]

Laurate anion causes a decrease in the rate of the alkaline hydrolysis of *p*-nitrophenyl acetate, mono-*p*-nitrophenyl dodecanedioate, *p*-nitrophenyl octanoate, and benzoylcholine, probably as a result of incorporation of the ester functions into the interior of the micelles. On the other hand, *n*-dodecyltrimethylammonium ion causes rate increases which, for the long chain esters, start to occur below the critical micelle concentration. Possibly the esters and *n*-dodecyltrimethylammonium ions form positively charged complexes which react rapidly with hydroxide ions.[64]

The hydrolysis of other esters[65] and of aspirin[66] in the presence of micelles has also been investigated.

Details of Bender and his co-workers' investigation of catalysis of the

[62] T. E. Wagner, C. Hsu, and C. S. Pratt, *J. Am. Chem. Soc.*, **89**, 6366 (1967).
[63] T. C. Bruice, J. Katzhendler, and L. R. Fedor, *J. Phys. Chem.*, **71**, 1961 (1967).
[64] F. M. Menger and C. E. Portnoy, *J. Am. Chem. Soc.*, **89**, 4698 (1967).
[65] P. B. Sheth and E. L. Parrott, *J. Pharm. Sci.*, **56**, 983 (1967).
[66] A. G. Mitchell and J. F. Broadhedd, *J. Pharm. Sci.*, **56**, 1261 (1967).

hydrolysis of esters catalysed by cyclodextrins have been published.[67–70] Experiments with partially mesylated and methylated cyclodextrins suggest that the catalytically active hydroxyl group is that at position 2 or 3.

The kinetics of the formation of inclusion compounds by α-cyclodextrin has been investigated.[71]

Purines as well as imidazole[72] inhibit the hydroxide ion-catalysed hydrolysis of methyl *trans*-cinnamate. This was ascribed to the formation of complexes and there was reasonably good agreement between their stability constants determined kinetically, spectrophotometrically, and by solubility measurements.[73]

A copolymer of 4-vinylimidazole and *p*-vinylphenol is a more effective catalyst for the hydrolysis of esters (**26**) and (**27**) than is imidazole or polyvinylimidazole. Possibly this is the result of bifunctional catalysis.[74]

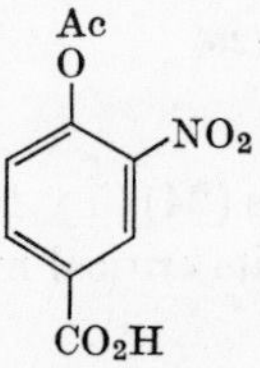

(**26**)

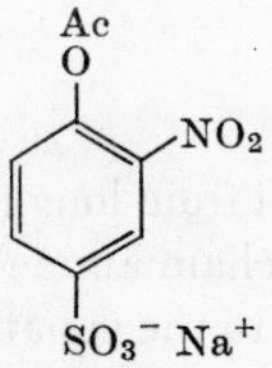

(**27**)

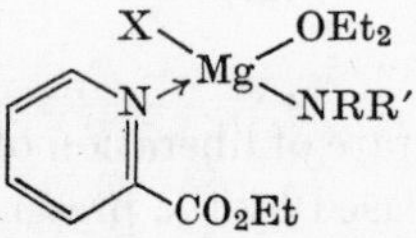

(**28**)

The aminolysis of ethyl picolinate by magnesium derivatives of amines occurs readily, and possibly reaction proceeds through a complex in which the amino-group is held close to the ester group (see **28**).[75]

Metal-ion catalysis. Nickel ions at a concentration sufficient fully to complex 2-cyano-1,10-phenanthroline increase the second-order rate constant for the reaction with hydroxide ions (which yields the corresponding amide) by a factor of 10^6. Mechanisms in which the nickel acts as an electrophilic catalyst

67 R. L. VanEtten, J. F. Sebastian, G. A. Clowes, and M. L. Bender, *J. Am. Chem. Soc.*, **89**, 3242 (1967).
68 R. L. VanEtten, G. A. Clowes, J. F. Sebastian, and M. L. Bender, *J. Am. Chem. Soc.*, **89**, 3253 (1967).
69 M. L. Bender, *Trans. N.Y. Acad. Sci.*, **29**, 301 (1967); see also M. Sherwood, *New Scientist*, **35**, 333 (1967).
70 See *Organic Reaction Mechanisms*, **1966**, 350.
71 F. Cramer, W. Saenger, and H.-Ch. Spatz, *J. Am. Chem. Soc.*, **89**, 14 (1967).
72 See *Organic Reaction Mechanisms*, **1965**, 262.
73 J. A. Mollica and K. A. Connors, *J. Am. Chem. Soc.*, **89**, 308 (1967).
74 C. G. Overberger, J. C. Salamone, and S. Yaroslavsky, *J. Am. Chem. Soc.*, **89**, 6231 (1967); H. Morawetz, C. G. Overberger, J. C. Salamone, and S. Yaroslavsky, *J. Polymer Sci.*, *B*, **4**, 609 (1966); H. Morawetz and W. R. Song, *J. Am. Chem. Soc.*, **88**, 5714 (1966).
75 R. P. Houghton and C. S. Williams, *Tetrahedron Letters*, **1967**, 3929.

(eqn. 10) and in which a ligand acts as a nucleophilic catalyst (eqn. 11) were considered and the former was preferred.[76]

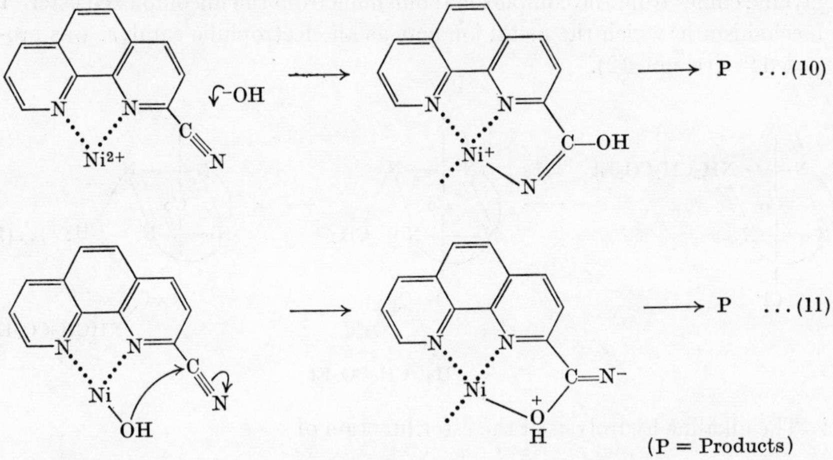

Hydrolyses of cyanopyridines are catalysed by nickel and copper oxide. The 3- and 4-nitriles yield amides, but the 2-nitrile yields a complex of the acid and metal which is itself a catalyst for the hydrolysis of pyridine-2-carboxamide. The complexes formed by nickel and copper with the 2-carboxylic acid and amide also catalyse the hydrolysis of the 2-nitrile. Presumably reaction proceeds via complex (**29**) in which the triple bond is activated to nucleophilic attack.[77]

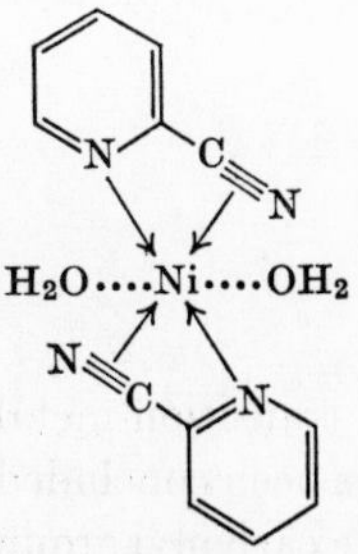

(**29**)

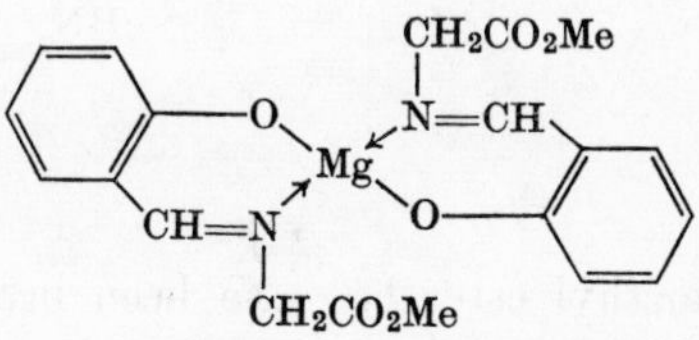

(**30**)

The ester groups of the chelate (**30**) are very reactive towards nucleophilic substitution.[78a] Hydrolysis of the ester group of ethyl glycinate-*N*,*N*-diacetic acid in the presence of Cu^{2+} and Sm^{3+} ions has been investigated.[78b]

[76] R. Breslow, R. Fairweather, and J. Keana, *J. Am. Chem. Soc.*, **89**, 2135 (1967).
[77] K. Sakai, T. Ito, and K. Watanabe, *Bull. Chem. Soc., Japan*, **40**, 1660 (1967).
[78a] R. P. Houghton, *J. Chem. Soc., C*, **1967**, 2030.
[78b] R. J. Angelici and B. E. Leach, *J. Am. Chem. Soc.*, **89**, 4605 (1967).

Ready formation of a peptide bond occurs when [Co(trien)ClglyOEt] $(ClO_4)_2$ is treated with glycine ethyl ester. Tracer studies indicate that one mole of the glycine comes from the complex and one mole from the uncomplexed ester. A mechanism in which the metal ion acts as an electrophilic catalyst was proposed[79] (see eqn. 12).

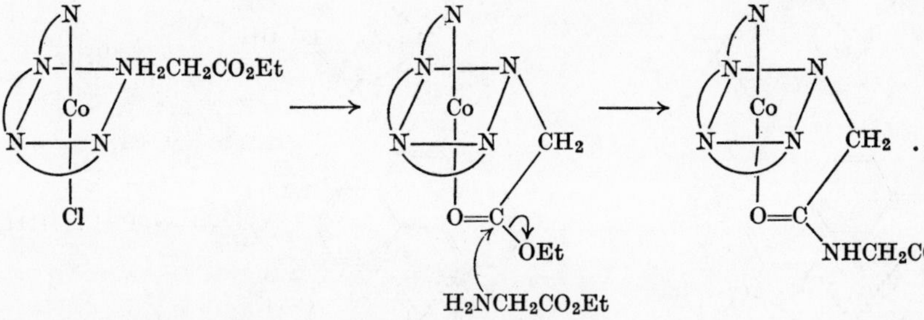

...(12)

The alkaline hydrolysis of the ester function of

cis-[Co(en)$_2$($NH_2CH_2CO_2R$)Cl]Cl_2

is about 100 times faster than that of glycine ethyl ester.[80]

The Cu(II)-catalysed alkaline hydrolysis of histidine methyl ester is thought to proceed through a complex (**31**) of the metal with the histidine that does not involve the ester group. The monoprotonated form of the ester is hydrolysed 100 times faster.[81] The metal-ion-catalysed hydrolysis of cysteine

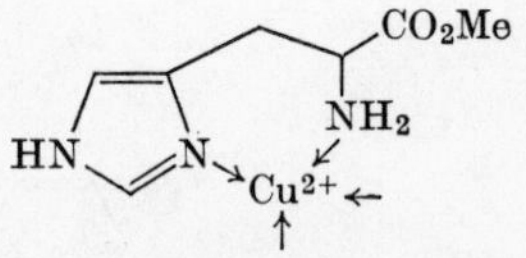

(**31**)

methyl ester has also been investigated.[82] A number of transition-metal complexes of α-amino-esters have been synthesized and it has been concluded from their IR spectra that there is no interaction between the carbonyl group and the metal.[83a]

Acetic acid and hydroxylamine undergo a nickel-ion-catalysed condensation

79 D. A. Buckingham, L. G. Marzilli, and A. M. Sargeson, *J. Am. Chem. Soc.*, **89**, 2772 (1967); D. A. Buckingham, J. P. Collman, D. A. R. Happer, and L. G. Marzilli, *ibid.*, p. 1082; D. E. Allen and R. D. Gillard, *Chem. Commun.*, **1967**, 1091.

80 R. W. Hay, M. L. Jansen, and P. L. Cropp, *Chem. Commun.*, **1967**, 621.

81 R. W. Hay and P. J. Morris, *Chem. Commun.*, **1967**, 23.

82 R. W. Hay and L. J. Porter, *Chem. Commun.*, **1967**, 653.

83a R. W. Hay and L. J. Porter, *Australian J. Chem.*, **20**, 675 (1967).

in aqueous solution which probably proceeds via an acetic acid–hydroxylamine complex.[83b]

Enzymic catalysis.[84, 85] The three-dimensional structure of toluene-*p*-sulphonyl α-chymotrypsin has been determined from an electron-density map at 2-Å resolution.[86] There is no pronounced cleft, but near the active serine there are open regions in the surface of the molecule where the structure is not closely packed. "The tosyl group bound to Ser-195 extends across one of these open regions. His-57 extends from one side pointing directly towards the sulphur atom of the tosyl group, and approaching within 5 Å of it. The position of Ser-195 and His-57 supports the accepted view that they exert a concerted action in the catalytic mechanism." His-40 is at least 13 Å away from His-57 and it is unlikely that they act concertedly, as sometimes proposed.[87]

Electron-density maps at 5-Å resolution have also been reported for chymotrypsinogen and phenylmethanesulphonyl-δ-chymotrypsin (PMS-δ-Cht) and difference maps for PMS-π-Cht *minus* PMS-δ-Cht, and PMS-δ-Cht *minus* PMS-γ-Cht. It was concluded that the gross conformations of chymotrypsinogen and of π-, δ-, and γ-chymotrypsin are very similar.[88]

Further details of Šorm and his co-workers' determination of the sequence of chymotrypsin have been published.[89]

The NMR spectrum of chymotrypsin has been discussed[90] and the reactions of chymotrypsin with *N*-acetyl-D-*p*- and -*m*-fluorophenylalanine and with *N*-acetyl-D-phenylalanine have been investigated by using fluorine and proton magnetic resonance.[91]

An octapeptide corresponding to the sequence around the active serine of chymotrypsin has been synthesized.[92]

The previously reported[93] pH-dependence of K_m and k_{cat} for the hydrolysis of *N*-acetyl-L-tryptophan amide catalysed by α-chymotrypsin is incorrect.[94, 95] K_m for catalysis by α- and δ-chymotrypsin is dependent on an acidic group

83b J. M. Lawlor, *Chem. Commun.*, **1967**, 404.

84 For a review on "The Foundation of Enzyme Action" see H. Gutfreund and J. R. Knowles, *Essays in Biochemistry*, **3**, 25 (1967), and for one on "The Mechanism of Enzyme Action" see S. G. Waley, *Quart. Rev.* (*London*), **21**, 379 (1967).

85 Covalent labelling of active sites has been reviewed by S. T. Singer, *Adv. Protein Chem.*, **22**, 1 (1967).

86 B. W. Matthews, P. B. Sigler, R. Henderson, and D. M. Blow, *Nature*, **214**, 652 (1967).

87 Cf. *Organic Reaction Mechanisms*, **1966**, 352.

88 J. Kraut, H. T. Wright, M. Kellerman, and S. T. Freer, *Proc. Nat. Acad. Sci. U.S.*, **58**, 304 (1967).

89 B. Meloun, L. Morávek, and F. Šorm, *Coll. Czech. Chem. Commun*,, **32**, 1947 (1967).

90 D. P. Hollis, G. McDonald, and R. L. Biltonen, *Proc. Nat. Acad. Sci. U.S.*, **58**, 758 (1967).

91 T. McL. Spotswood, J. M. Evans, and J. H. Richards, *J. Am. Chem. Soc.*, **89**, 5052 (1967).

92 D. A. Laufer and E. R. Blout, *J. Am. Chem. Soc.*, **89**, 1246 (1967).

93 M. L. Bender, G. E. Clement, and F. J. Kézdy, *J. Am. Chem. Soc.*, **86**, 3680 (1964).

94 A. Himoe, P. C. Parks, and G. P. Hess, *J. Biol. Chem.*, **242**, 919 (1967).

95 M. L. Bender, M. J. Gibian, and D. J. Whelan, *Proc. Nat. Acad. Sci. U.S.*, **56**, 833 (1966).

of pK_a *ca.* 8.5, and k_{cat} is dependent on a basic group of pK_a *ca.* 6.8 but is independent of pH in the region 8—9.2.[94,95] The values of K_m for the hydrolyses of *N*-acetyl-L-tryptophan amide and *N*-acetyl-L-phenylalanine amide catalysed by α-chymotrypsin were shown[96] to be, within experimental error, equal to the independently determined values of the enzyme–substrate dissociation constant, $K_s{}^1$. It was proposed that the pH-dependence of K_m resulted from a pH-dependent equilibrium between two major conformations of the enzyme which was investigated independently by studying the dependence of optical rotation on pH and shown[97] to depend on the ionization of a single group of pK_a *ca.* 8.3; and this was identified tentatively as the NH_2-terminal isoleucyl α-amino-group.[96,98] Interestingly, the *X*-ray structure determination[86] shows this group to be turned towards the interior and ion-paired with the carboxylate group of Asp-194, and it is tempting to speculate that it is this interaction which holds the enzyme in its active conformation; deprotonation of the amino-group would then allow the chymotrypsin molecule to revert to an inactive conformation resembling that of chymotrypsinogen.[96,99,100] Consistently with this picture, binding of a competitive inhibitor *N*-acetyl-D-tryptophan to α-chymotrypsin depends on the ionization of a group on the free enzyme with apparent pK_a 9.3 at 5° (equivalent to 8.7 at 25°).[101] Also the pH-dependence of proton uptake on binding of benzyl alcohol, a competitive inhibitor, to chymotrypsin is a sigmoid curve with apparent pK_a *ca.* 8.8, there being an absorption of about one proton per mole at pH 10 and of no proton at pH 7.[102] The rate of reaction of α-chymotrypsin with L-1-chloro-4-phenyl-3-(toluene-*p*-sulphonamido)butan-2-one also depends on a group of pK_a 8.9.[103]

The acylation of α-chymotrypsin by *p*-nitrophenyl acetate at pH 7 and 25° in propan-2-ol–water and dioxan–water mixtures follows second-order kinetics, not Michaelis–Menten kinetics as previously reported.[104]

The relative rates of deacylation of acylchymotrypsins, normalized for the electronic effects of the side chains (Table 1), were interpreted in terms of a three-locus model for the active site in which the acyl group from a specific substrate (e.g., an *N*-acyl-L-phenylalanyl derivative) is fixed (*a*) by a covalent bond to Ser-195 (A), (*b*) by a hydrophobic interaction between the side chain and its locus at the active site (B), and (*c*) by hydrogen bond(s) of the peptide

96 K. G. Brandt, A. Himoe, and G. P. Hess, *J. Biol. Chem.*, **242**, 3963, 3973 (1967).
97 H. L. Oppenheimer, B. Labouesse, and G. P. Hess, *J. Biol. Chem.*, **241**, 2720 (1966).
98 C. Ghelis, J. Labouesse, and B. Labouesse, *Biochem. Biophys. Res. Commun.*, **29**, 101 (1967).
99 See also W. O. McClure and G. M. Edelman, *Biochemistry*, **6**, 559, 567 (1967).
100 See also B. H. Havsteen, *J. Biol. Chem.*, **242**, 769 (1967).
101 C. H. Johnston and J. R. Knowles, *Biochem. J.*, **103**, 428 (1967).
102 M. L. Bender and F. C. Wedler, *J. Am. Chem. Soc.*, **89**, 3052 (1967).
103 F. J. Kézdy, A. Thomson, and M. L. Bender, *J. Am. Chem. Soc.*, **89**, 1004 (1967).
104 L. Faller and J. M. Sturtevant, *J. Biol. Chem.*, **241**, 4825 (1967); see, however, ref. 95 and H. Kaplan and K. J. Laidler, *Can. J. Chem.*, **45**, 547, 559 (1967).

link in the acylamino-group (C) (see **32**). If (*b*) or (*c*) or both are absent, as in the *N*-acetylglycyl, 3-phenylpropionyl, or acetyl enzyme, respectively, rotation of the —CH_2—CO—O— group is possible and other conformations besides that most favourable for deacylation become populated. The only

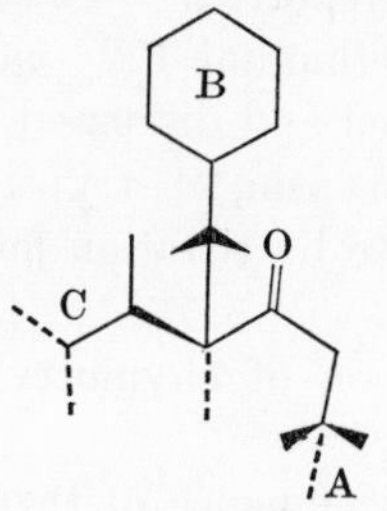

(32)

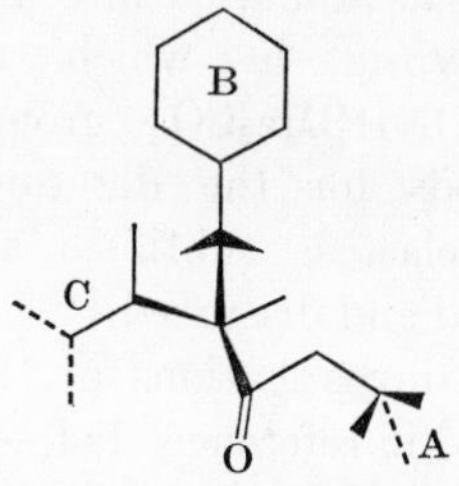

(33)

difference in the binding of the D-acyl enzymes (see **33**) is that the serine ester carbonyl group points the opposite way and is presumably placed unfavourably with respect to the catalytic groups of the enzyme.[105]

Table 1. Normalized deacylation rate constants of a series of acyl-α-chymotrypsins.

Acyl group	k_3^{norm}	Acyl group	k_3^{norm}
N-Acetyl-L-Trp	560	*N*-Acetyl-D-Leu	0.166
N-Acetyl-L-Phe	296	*N*-Acetyl-D-Phe	0.047
N-Acetyl-L-Leu	24.1	*N*-Acetyl-D-Trp	0.0302
N-Acetylglycyl	1.0	Acetyl	0.185

The apparent pK_a^r's for the hydrolysis of a series of acylchymotrypsins vary from 6.70 for 3,5-dinitrobenzoylchymotrypsin to 7.68 for [3-(3-indolyl)-acryloyl]chymotrypsin, possibly as a result of hydrogen bonding of the imidazole group of the enzyme to the oxygen of the acyl group (see **34**).[106]

The UV spectra and ORD curves of monoarylacryloyl-α-chymotrypsins suggest that the ester linkage has an *s-cis*-conformation (**35**) in contrast to small-molecule model compounds which are probably *s-trans* as (**36**).[107]

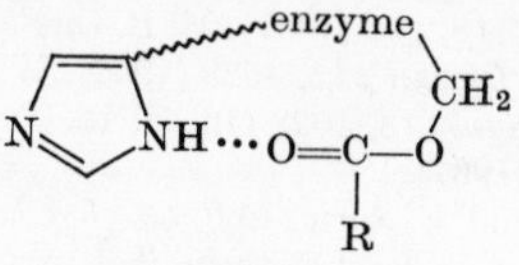

(34)

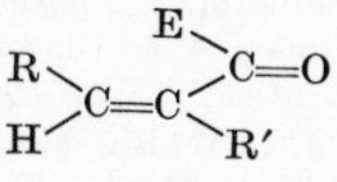

s-cis

(35)

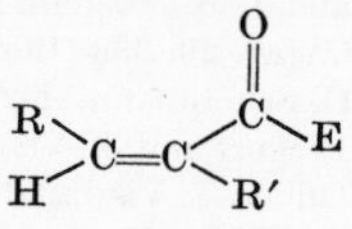

s-trans

(36)

[105] D. W. Ingles and J. R. Knowles, *Biochem. J.*, **104**, 369 (1967).
[106] S. A. Bernhard, E. Hershberger, and J. Keizer, *Biochemistry*, **5**, 4120 (1966).
[107] E. Charney and S. A. Bernhard, *J. Am. Chem. Soc.*, **89**, 2726 (1967).

Chymotrypsin in which methionine-192 is alkylated by α-bromo-4-nitroacetophenone shows a charge-transfer band (λ_{max} 350 mμ; ε 7.55×10^3 l. mole^{-1} cm^{-1}) which probably involves an indole residue acting as a donor.[108] Spectroscopic properties of chymotrypsin with methionine-192 alkylated by 2-(bromoacetamido)-4-nitrophenol have also been reported.[109] Catalysis by chymotrypsin in which the sulphur of methionine-192 carries a —$CH_2CONHCMe_2CO_2^-$ group has been investigated and discussed.[110]

Methods for the determination of α-chymotrypsin,[111] trypsin,[111,112] papain, elastase, subtilisin, and acetylcholinesterase by titration have been described and discussed.[113]

Other investigations of the mechanism of action of chymotrypsin are described in references 114—116.

Details of the determination of the amino-acid sequence of trypsin and trypsinogen have been published.[117] The conversion of trypsinogen into trypsin has been investigated.[118] Additional evidence that the imidazole group of His-46 is the catalytically active imidazole of trypsin has been published.[119] Other investigations of the mechanism of action of trypsin are reported in references 120 and 121.

The rate of hydrolysis of peptides by papain is affected by structural and configurational changes in amino-acid residues of the peptide remote from the bond being cleaved. It was suggested that there is a large active site with

108 D. S. Sigman and E. R. Blout, *J. Am. Chem. Soc.*, **89**, 1747 (1967).

109 M. B. Hille and D. E. Koshland, *J. Am. Chem. Soc.*, **89**, 5945 (1967).

110 F. J. Kézdy, J. Feder, and M. L. Bender, *J. Am. Chem. Soc.*, **89**, 1009 (1967).

111 See also D. T. Elmore and J. J. Smyth, *Biochem. J.*, **103**, 36P, 37P (1967).

112 See also T. Case and E. Shaw, *Biochem. Biophys. Res. Commun.*, **29**, 508 (1967).

113 M. L. Bender, M. L. Begué-Cantón, R. L. Blakeley, L. J. Brubacher, J. Feder, C. R. Gunter, F. J. Kézdy, J. B. Killheffer, T. H. Marshall, C. G. Miller, R. W. Roeske, and J. K. Stoops, *J. Am. Chem. Soc.*, **88**, 5890 (1966); **89**, 2241 (1967).

114 M. S. Silver and T. Sone, *J. Am. Chem. Soc.*, **89**, 357 (1967); S. G. Cohen and R. M. Schultz, *Proc. Nat. Acad. Sci. U.S.*, **57**, 247 (1967); W. B. Lawson, *J. Biol. Chem.*, **242**, 3397 (1967).

115 J. de Jersey and B. Zerner, *Biochem. Biophys. Res. Commun.*, **28**, 173 (1967); K. Brocklehurst and K. Williamson, *ibid.*, **26**, 175 (1967); K. Brocklehurst and K. Williamson, *Chem. Commun.*, **1967**, 666.

116 M. J. Hawkins, J. R. Knowles, L. Wilson, and D. Witcher, *Biochem. J.*, **104**, 762 (1967); J. Keizer and S. A. Bernhard, *Biochemistry*, **5**, 4127 (1966); J. R. Rapp, C. Niemann, and G. F. Hein, *ibid.*, p. 4100; K. Morimoto and G. Kegeles, *ibid.*, **6**, 3007 (1967); T. H. Fife and J. B. Milstein, *ibid.*, p. 2901; B. Halpern, J. Ricks, and J. W. Westley, *Australian J. Chem.*, **20**, 389 (1967); H. J. Schramm, *Z. Physiol. Chem.*, **348**, 232 (1967); E. B. Ong and G. Schöllmann, *ibid.*, **344**, 13 (1966); A. N. Glazer, *J. Biol. Chem.*, **242**, 4528 (1967); W. S. Rickert and T. Viswanatha, *Biochem. Biophys. Res. Commun.*, **28**, 1028 (1967); R. W. A. Oliver, T. Viswanatha, and W. J. D. Whish, *ibid.*, **27**, 107 (1967).

117 O. Mikeš, H. G. Müller, V. Holeyšovský, V. Tomášek, and F. Šorm, *Coll. Czech. Chem. Commun.*, **32**, 620 (1967); O. Mikeš, V. Tomášek, V. Holeyšovský, and F. Šorm, *ibid.*, p. 655.

118 T. M. Radhakrishnan, K. A. Walsh, and H. Neurath, *J. Am. Chem. Soc.*, **89**, 3059 (1967).

119 E. Shaw and S. Springhom, *Biochem. Biophys. Res. Commun.*, **27**, 391 (1967); see *Organic Reaction Mechanisms*, **1965**, 274.

120 D. T. Elmore, D. V. Roberts, and J. J. Smyth, *Biochem. J.*, **102**, 728 (1967).

121 S. T. Scrimger and T. Hofmann, *J. Biol. Chem.*, **242**, 2528 (1967).

seven subsites, each binding one amino-acid residue of the peptide.[122] Other investigations of the mechanism of action of papain are reported in references 123—126 and of ficin in reference 127.

Kinetic analysis of the bromelin-catalysed hydrolysis of α-*N*-benzoyl-L-arginine ethyl ester and α-*N*-benzoyl-L-arginine amide on the basis of the kinetic scheme:

$$\begin{array}{ccccccc} E + S & \xrightarrow{k_s} & ES & \xrightarrow{k_2} & ES^1 & \xrightarrow{k_3} & E + P_2 \\ & & & & + & & \\ & & & & P_1 & & \end{array}$$

leads to the conclusion that the supposed common deacylation rate constant, k_3, is 190 times greater for the hydrolysis of the ester than of the amide. Possibly ester and amide are bound by the enzyme in different ways or the ammonia released in hydrolysis of the amide inhibits deacylation.[128]

The sequence about the active cysteine of stem bromelin is Cys-Gly-Ala-Cys*-Try.[129]

The plots of k_{cat} and k_{cat}/K_m for the hydrolysis of *N*-acetyl-L-phenylalanyl-L-3,5-dibromotyrosine catalysed by pepsin are bell-shaped with maxima at pH 2.1 and 1.9, respectively. The plot of K_m against pH has a broad minimum between pH 1 and 2. It was proposed that the enzyme binds only substrate with the carboxyl un-ionized and that catalysis is provided by two carboxyl groups with pK_a 0.75 and 2.67 for the unbound and 0.89 and 3.44 for the bound enzyme.[130] Other investigations of pepsin are reported in reference 131.

An electron-density map of carboxypeptidase A at 2.8-Å resolution has been reported but it has not yet been fitted to the amino-acid sequence.[132] Other investigations of carboxypeptidase are reported in reference 133.

[122] I. Schechter and A. Berger, *Biochem. Biophys. Res. Commun.*, **27**, 157 (1967); see also L. J. Brubacher and M. L. Bender, *ibid.*, p. 176.

[123] A. W. Lake and G. Lowe, *Biochem. J.*, **101**, 402 (1966).

[124] L. J. Brubacher and M. L. Bender, *J. Am. Chem. Soc.*, **88**, 5871 (1966); M. L. Bender and L. J. Brubacher, *ibid.*, p. 5880.

[125] K. Morihara, *J. Biochem.* (*Tokyo*), **62**, 250 (1967).

[126] W. Cohen and P. H. Petra, *Biochemistry*, **6**, 1047 (1967); A. C. Henry and J. F. Kirsch, *ibid.*, p. 3536.

[127] M. J. Stein and I. R. Liener, *Biochem. Biophys. Res. Commun.*, **26**, 376 (1967).

[128] K. Brocklehurst, E. M. Crook, and C. W. Wharton, *Chem. Commun.*, **1967**, 1185.

[129] L.-P. Chao and I. E. Liener, *Biochem. Biophys. Res. Commun.*, **27**, 100 (1967).

[130] E. Zeffren and E. T. Kaiser, *J. Am. Chem. Soc.*, **89**, 4204 (1967).

[131] K. Inouye and J. S. Fruton, *J. Am. Chem. Soc.*, **89**, 187 (1967); *Biochemistry*, **6**, 1765 (1967); W. T. Jackson, M. Schlamowitz, and A. Shaw, *ibid.*, **5**, 4105 (1966); B. Jirgensons, *J. Am. Chem. Soc.*, **89**, 5979 (1967).

[132] M. L. Ludwig, J. A. Hartsuck, T. A. Steitz, H. Muirhead, J. C. Coppola, G. N. Reeke, and W. N. Lipscomb, *Proc. Nat. Acad. Sci. U.S.*, **57**, 511 (1967).

[133] O. A. Roholt and D. Pressman, *Proc. Nat. Acad. Sci. U.S.*, **58**, 280 (1967); F. A. Quiocho, W. H. Bishop, and F. M. Richards, *ibid.*, **57**, 525 (1967); S. Awazu, F. W. Carson, P. L. Hall, and E. T. Kaiser, *J. Am. Chem. Soc.*, **89**, 3627 (1967); I. Schechter and A. Berger, *Biochemistry*, **5**, 3362 (1966); J. F. Riordan, M. Sokolovsky, and B. L. Vallee, *ibid.*, **6**, 3609 (1967).

The complete amino-acid sequence of subtilisins BPN and Carlsberg,[134] and a partial sequence of elastase,[135] have been reported.

Investigations have also been reported on the following enzymes: ε-peptidase,[136] elastase,[137] subtilisin,[138] acetylcholinesterase,[139] a lytic protease from *Sorangium* sp.,[140] and carbonic anhydrase.[141]

Other reactions. The carboxyl- and hydroxyl-protons of salicylic acid are exchanged in methanol at –80° to –95° in independent parallel reactions. The rate for exchange of the carboxyl-protons is greater than that for benzoic acid, suggesting that salicylic acid is not intramolecularly hydrogen-bonded as this should lead to a reduced rate.[142]

The rates of hydrolysis of *tert*-butyl acetate via $A_{\mathrm{Alk}}1$ and $A_{\mathrm{Ac}}2$ mechanisms depend quite differently on the DMSO concentration in DMSO–H_2O mixtures. In water the predominant mechanism is $A_{\mathrm{Alk}}1$, but in the mixture with mole-fraction of DMSO 0.601 the $A_{\mathrm{Alk}}2$ mechanism predominates.[143]

The entropies of activation for the hydrolysis of *tert*-butyl and methyl trifluoroacetate are +14.8 and –32.3 e.u., respectively, and the solvent isotope effects, $k_{\mathrm{H_2O}}/k_{\mathrm{D_2O}}$, are 3.49 and 1.21, respectively, consistently with the former compound's reacting by a unimolecular and the latter by a bimolecular mechanism.[144] Other investigations of the hydrolyses of alkyl trihalogenoacetates are reported in references 145 and 146.

134 E. L. Smith, F. S. Markland, C. B. Kasper, R. J. DeLange, M. Landon, and W. H. Evans, *J. Biol. Chem.*, **241**, 5974 (1966).
135 J. R. Brown, D. L. Kauffman, and B. S. Hartley, *Biochem. J.*, **103**, 497 (1967).
136 J. D. Padayatty and H. Van Kley, *Biochemistry*, **5**, 1394 (1966).
137 A. Gertler and T. Hofmann, *J. Biol. Chem.*, **242**, 2522 (1967).
138 A. N. Glazer, *J. Biol. Chem.*, **242**, 433 (1967); K. E. Neet and D. E. Koshland, *Proc. Nat. Acad. Sci. U.S.*, **56**, 1606 (1966); L. Polgar and M. L. Bender, *Biochemistry*, **6**, 610 (1967).
139 H. Bockendahl, T.-M. Müller, and H. Verfürth, *Z. Physiol. Chem.*, **348**, 167 (1967); R. M. Krupka, *Biochemistry*, **6**, 1183 (1967); N. Engelhard, K. Prchal, and M. Nenner, *Angew. Chem. Intern. Ed. Engl.*, **6**, 615 (1967); H. P. Metzger and I. B. Wilson, *Biochem. Biophys. Res. Commun.*, **28**, 263 (1967).
140 L. B. Smillie and D. R. Whitaker, *J. Am. Chem. Soc.*, **89**, 3350 (1967); H. Kaplan and D. R. Whitaker, *ibid.*, p. 3352.
141 A. Nilsson and S. Lindskog, *European J. Biochem.*, **2**, 309 (1967); A. Thorslund and S. Lindskog, *ibid.*, **3**, 117 (1967); Y. Pocker and J. Meany, *J. Am. Chem. Soc.*, **89**, 631 (1967); Y. Pocker and J. Meany, *Biochemistry*, **6**, 239 (1967); Y. Pocker and J. T. Stone, *ibid.*, p. 668; A. S. Mildvan and M. C. Scrutton, *ibid.*, p. 2978; J. A. Verpoorte, S. Mehta, and J. T. Edsall, *J. Biol. Chem.*, **242**, 4221 (1967); P. L. Whitney, G. Fölsch, P. O. Nyman, and B. G. Malmström, *ibid.*, p. 4206; P. L. Whitney, P. O. Nyman, and B. G. Malmström, *ibid.*, p. 4213; M. E. Riepe and J. H. Wang, *J. Am. Chem. Soc.*, **89**, 4229 (1967).
142 M. S. Puar and E. Grunwald, *J. Am. Chem. Soc.*, **89**, 4403 (1967).
143 B. G. Cox and P. T. McTigue, *Australian J. Chem.*, **20**, 1815 (1967).
144 J. G. Martin and J. M. W. Scott, *Chem. Ind.* (*London*), **1967**, 665; see also S. V. Anantakrishnan and R. V. Venkataratnam, *Indian J. Chem.*, **4**, 379 (1966).
145 I. Talvik and M. Heinloo, *Reakts. Sposobnost Org. Soedin, Tartu. Gos. Univ.*, **3**, 244 (1966); *Chem. Abs.*, **67**, 63443a (1967).
146 I. Talvik, *Reakts. Sposobnost Org. Soedin, Tartu. Gos. Univ.*, **3**, 233 (1967); *Chem. Abs.*, **67**, 72991e (1967).

The kinetics of reactions in water–DMSO–mixtures have been reviewed.[147] Solvent effects on the following reactions have also been investigated: acid-catalysed hydrolysis of butyl acetate in aqueous dioxan,[148] of *p*-nitrophenyl acetate in aqueous acetone,[149] and of benzyl acetate in aqueous glycol and glycerol;[150] the alkaline hydrolysis of ethyl benzoate in aqueous sulpholan,[151] dioxan,[152] and ethanol,[152] of benzoate esters in aqueous DMSO,[153a] and of ethyl acetate in aqueous glycol and glycerol.[153b]

The rates of the alkaline hydrolysis of aryl benzoates are correlated best by a four-parameter equation of the Yukewa–Tsuno type.[154]

The hydrolysis of glycine ethyl ester in acidic solution is slower than that of ethyl acetate, and not faster as previous results suggested.[155] The hydrolysis of esters of amino-acids and sugars have been investigated.[156]

Acetyl fluoride has been isolated as an intermediate in the fluoride ion-catalysed hydrolysis of acetic anhydride.[157]

Rates and equilibrium constants for the formation of succinic anhydride in aqueous solution have been measured by trapping the anhydride with aniline. As expected, methyl substitution increases both the rate and the equilibrium constants for anhydride formation.[158] Hydrolysis of pyridine-2-carboxylic anhydride[159] and propionic anhydride,[160] alcoholysis of acetic anhydrides[161] and of ketens,[162] aminolysis of acetic,[163] propionic,[164] and isobutyric anhydride,[164] and the reaction of acetic anhydride with *tert*-butyl peroxide[165] have been investigated.

[147] E. Tommila, *Suomen Kem., A*, **40**, 3 (1967).

[148] O. Landauer and C. Mateescu, *Bul. Inst. Politeh. Gheorghiu Gheorghiu-Dej, Bucuresti*, **28**, 39 (1966); *Chem. Abs.*, **66**, 1961n (1967).

[149] J. Martinmaa and E. Tommila, *Suomen Kem., B*, **40**, 222 (1967).

[150] H. Sadek and F. Y. Khalil, *Suomen Kem., B*, **40**, 59 (1967).

[151] E. Tommila and J. Martinmaa, *Suomen Kem., B*, **40**, 216 (1967).

[152] E. Tommila and M.-L. Savolainen, *Suomen Kem., B*, **40**, 212 (1967).

[153a] D. D. Roberts, *J. Org. Chem.*, **31**, 4037 (1967).

[153b] H. Sadek, M. S. A. Elamayem, and A. S. El Kholy, *Suomen Kem., B*, **40**, 111 (1967).

[154] A. A. Humffray and J. J. Ryan, *J. Chem. Soc., B*, **1967**, 468.

[155] R. W. Hay and L. Main, *Australian J. Chem.*, **20**, 1757 (1967).

[156] N. K. Kochetkov, V. A. Derevitskaya, and L. M. Likhosherstov, *Izv. Akad. Nauk SSSR, Ser. Khim.*, **1967**, 367; *Chem. Abs.*, **67**, 32918e (1967); J. Yoshimura, M. Funabashi, S. Ishige, and T. Sato, *Carbohydrate Res.*, **3**, 214 (1966).

[157] C. A. Bunton and J. H. Fendler, *J. Org. Chem.*, **32**, 1547 (1967).

[158] T. Higuchi, L. Eberson, and J. D. McRae, *J. Am. Chem. Soc.*, **89**, 3001 (1967).

[159] H. J. Nestler and J. K. Seydel, *Chem. Ber.*, **100**, 1983 (1967).

[160] F. P. Cavasino and S. D'Alessandro, *Atti Accad. Sci. Letters Arti Palermo, P. I*, **25**, 119 (1964–5); *Chem. Abs.*, **65**, 19949 (1966).

[161] Y. N. Skoulikidis and D. Zorbalas, *Kolloid Z. Z. Polym.*, **212**, 46 (1966); J. Koskikallio and K. Koivula, *Suomen Kem., B*, **40**, 138 (1967).

[162] A. Tille and H. Pracejus, *Chem. Ber.*, **100**, 196 (1967).

[163] T. Jasinski and Z. Szponar, *Roczniki Chem.*, **41**, 1099 (1967).

[164] P. J. Lillford and D. P. N. Satchell, *J. Chem. Soc., B*, **1967**, 360.

[165] V. L. Antonovskii and O. K. Lyashenko, *Kinetika i Kataliz*, **7**, 767 (1966); *Chem. Abs.*, **65**, 18438 (1966).

2-Naphthol reacts with diacetyl sulphide at least twice as fast as with acetic anhydride.[166] Benzoylation of aromatic amines by thiobenzoic anhydride has been investigated.[167]

The volumes of activation for the ethanolysis of benzoyl and anisoyl chloride in ethanol at 0° are –29.1 and –20.1 ml mole^{-1}, respectively. This difference would result if the transition state of the former were S_N2-like and that of the latter S_N1-like.[168] The volumes of activation for the hydrolyses of benzoyl chlorides in water–tetrahydrofuran mixtures,[169] and $\Delta S^{\ddagger}$ and $\Delta C_p^{\ddagger}$ for the hydrolyses of a series of alkyl chloroformates and dimethylcarbamoyl chloride,[170] have been determined.

The relative rates of hydrolysis of 8-methyl- and 8-[$^2H_3$]methyl-1-naphthoyl chloride are 1.029, smaller than the value calculated for a steric isotope effect by Bartell's method (see also p. 63).[171]

Acylation of phenols with acetyl chloride is slightly faster than acylation of the analogous thiophenols.[172] Alcoholysis[173] and aminolysis[174] of acid chlorides have been studied.

The volume of activation for the reaction of sodium methoxide and methyl benzoate to yield dimethyl ether and sodium benzoate is very strongly negative, –32 cm^3 mole^{-1}.[175]

Esterification of cyanoacetic acid and monochloroacetic acid,[176] and of 4-*tert*-butylcyclohexane- and *trans*-decalin-carboxylic acids,[177] and the reaction of the latter with diphenyldiazomethane[178] have been studied.

The reactions of acids with diphenyldiazomethane have been reviewed.[179]

The reactions of *N*-substituted ureas with alcohols involve dissociation of

166 J. Hipkin and D. P. N. Satchell, *J. Chem. Soc., B*, **1967**, 365.

167 L. M. Litvinenko and N. M. Oleinik, *Zh. Organ. Khim.*, **2**, 1671 (1966); *Chem. Abs.*, **66**, 75459c (1967).

168 A. Kivinen and A. Viitala, *Suomen Kem., B*, **40**, 19 (1967).

169 H. Heydtmann and H. Stieger, *Ber. Bunsenges. Phys. Chem.*, **70**, 1095 (1966).

170 A. Queen, *Can. J. Chem.*, **45**, 1619 (1967).

171 G. J. Karabatsos, G. C. Sonnichsen, C. G. Papioannou, S. E. Scheppele, and R. L. Shone, *J. Am. Chem. Soc.*, **89**, 463 (1967).

172 J. Hipkin and D. P. N. Satchell, *J. Chem. Soc., B*, **1967**, 367.

173 E. Y. Bekhli, O. V. Nesterov, and S. G. Entelis, *J. Polymer Sci., Part C*, No. 16, 209 (1966); A. F. Popov, N. M. Oleinik, and L. M. Litvinenko, *Reakts. Sposobnost Org. Soedin, Tartu. Gos. Univ.*, **3**, 11 (1966); *Chem. Abs.*, **67**, 43080b (1967); L. M. Litvinenko and R. S. Popova, *Zh. Org. Khim.*, **3**, 718 (1967); *Chem. Abs.*, **67**, 43082d (1967).

174 D. N. Kevill and F. D. Foss, *Tetrahedron Letters*, **1967**, 2837; D. N. Kevill and W. F. K. Wang, *Chem. Commun.*, **1967**, 1178; E. K. Euranto and R. S. Leimu, *Acta Chem. Scand.*, **20**, 2028 (1966); F. Akiyama, K. Sugino, and N. Tokura, *Bull. Chem. Soc., Japan*, **40**, 359 (1967).

175 R. Hakala and J. Koskikallio, *Suomen Kem., B*, **40**, 172 (1967).

176 M. Yonezawa, S. Suzuki, H. Ito, and K. Ito, *Yuki Gosei Kagaku Kyokai Shi.*, **24**, 580 (1966); *Chem. Abs.*, **65**, 18443 (1966).

177 N. B. Chapman, A. Ehsan, J. Shorter, and K. J. Toyne, *J. Chem. Soc., B*, **1967**, 570.

178 N. B. Chapman, A. Ehsan, J. Shorter, and K. J. Toyne, *J. Chem. Soc., B*, **1967**, 256.

179 R. M. O'Ferrall, *Adv. Phys. Org. Chem.*, **5**, 331 (1967).

the former into isocyanate and amine and subsequent reaction of the isocyanate with the alcohol.[180] The reactions of isocyanates with nucleophiles have been investigated.[181]

The hydrolyses of the following compounds have also been investigated: *S*-alkyl esters of substituted thiobenzoic acids,[182] *S*-2-(octanamido)ethyl thiopyruvate,[183] santonin and related lactones,[184] methyl 4′-substituted biphenyl-2-carboxylates,[185] N^4-acylsulphanilamides,[186] *para*-substituted benzoylglycolic acids,[187] bornyl and isobornyl acetate,[188] ethyl and methyl formate,[189] diethyl malate,[190] glutarate,[191] adipate,[191] azelate,[191] and sebacate,[191] mono- and di-ethyl malonate and succinate,[192] vinyl esters,[193] ethyl glucuronate,[194] *tert*-butyl perbenzoate,[195] perylene-3,4,9,10-tetracarboxylic anhydride,[196] acetylhydrazine,[197] *o*-nitroacetanilide,[198] carbamates,[199] 6-aminothiouracils (to 6-aminouracils),[200] xanthates,[201] and sydnones.[202]

180 W. Van Pee and J.-C. Jungers, *Bull. Soc. Chim. France*, **1967**, 158; E. de Cooman and I. de Aguirre, *ibid.*, p. 165; H. Van Landeghem and I. de Aguirre, *ibid.*, p. 172; W. D'Olieslager and I. de Aguirre, *ibid.*, p. 179.

181 S. G. Entelis and O. V. Nesterov, *Russian Chem. Rev.*, **35**, 917 (1966); B. E. Myuller, N. V. Panova, K. V. Nel'son, N. V. Kozlova, and N. P. Apukhtina, *Zh. Fiz. Khim.*, **40**, 2567 (1966); *Chem. Abs.*, **66**, 37283h (1967); O. V. Nesterov, V. B. Zabrodin, Y. N. Chirkov, and S. G. Entelis, *Kinet. Katal.*, **7**, 805 (1966); *Chem. Abs.*, **66**, 37000p; R. P. Tiger and S. G. Entelis, *Kinet. Katal.*, **8**, 54 (1967); *Chem. Abs.*, **67**, 21224z (1967).

182 G. Losse, M. Mayer, and K. Kuntze, *Z. Chem.*, **7**, 104 (1967).

183 J. Knappe and U. Herzog-Wiegand, *Ann. Chem.*, **701**, 217 (1967).

184 A. J. N. Bolt, M. S. Carson, W. Cocker, and T. B. H. McMurry, *J. Chem. Soc., B*, **1967**, 165.

185 T. Drapala, *Roczniki Chem.*, **41**, 229 (1967).

186 F. Muzalewski, *Roczniki Chem.*, **41**, 1902 (1967).

187 C. Concilio and A. Bongini, *Ann. Chim. (Rome)*, **57**, 292 (1967).

188 E. A. Kalinovskaya and G. A. Rudakov, *Zh. Org. Chem.*, **3**, 307 (1967); *Chem. Abs.*, **66**, 115038y (1967); G. A. Rudakov and A. A. Vereshchagina, *Zh. Org. Khim.*, **3**, 311 (1967); *Chem. Abs.*, **66**, 115039z (1967).

189 M. Zaheeruddin and H. A. Kazmi, *Pakistan J. Sci. Ind. Res.*, **9**, 206 (1966); *Chem. Abs.*, **66**, 104464h (1967).

190 W. J. Svirbely and F. A. Kundell, *J. Am. Chem. Soc.*, **89**, 5354 (1967).

191 S. V. Anantakrishnan and R. V. Venkatarathman, *Proc. Indian Acad. Sci., Sect. A*, **65**, 188 (1967); **66**, 40 (1967).

192 N. V. Sapozhnikova, N. I. Darienko, and Z. G. Maizel, *Izv. Vysshikh Uchebn. Zavedenii. Khim. i Khim. Tekhnol.*, **9**, 410 (1966); *Chem. Abs.*, **66**, 28146p (1967).

193 S. I. Sakyzh-Zade and A. S. Rzaeva, *Dokl. Akad. Nauk Azerb. SSR*, **22**, 28 (1966); *Chem. Abs.*, **66**, 37274f (1967).

194 T. Yamana, Y. Mizukami, and F. Ichimura, *Chem. Pharm. Bull. (Tokyo)*, **14**, 479 (1966).

195 V. L. Antonovskii, M. M. Buzlanova, and Z. S. Frolova, *Kinetika i Kataliz*, **8**, 671 (1967); *Chem. Abs.* **67**, 90190x (1967).

196 W. Wojtkiewicz, Z. Jankowski, and J. Szadowski, *Przemysl Chem.*, **45**, 561 (1966); *Chem. Abs.* **66**. 104462f (1967).

197 J. le Coarer, A. Bahsoun, and A. Broche, *Bull. Soc. Chim. France*, **1967**, 440.

198 M. I. Vinnik, I. M. Medvetskaya, L. R. Andreeva, and A. E. Tiger, *Zh. Fiz. Khim.*, **41**, 252 (1967).

199 R. F. Hudson, R. J. G. Searle, and A. Mancuso, *Helv. Chim. Acta*, **1967**, 997.

200 T.-F. Chin, W.-H. Wu, and J. L. Lach, *J. Pharm. Sci.*, **56**, 562 (1967).

201 Z. Cichowski, *Polimery*, **11**, 326 (1966); *Chem. Abs.*, **67**, 11088u (1967).

202 E. R. Garrett and P. J. Mehta, *J. Pharm. Sci.*, **56**, 1468 (1967).

The hydrolysis of esters in the presence of ion-exchange resins has been studied.[203] Other reactions studied include: acid-catalysed hydrolysis and oxygen-exchange of acetic acid,[204] reaction of esters with phosphorus pentachloride and thionyl chloride,[205] conversion of isocyanates into carbodiimides with isopropyl methyl phosphofluoridate as catalyst,[206] rearrangement of 1-(3-hydroxypropionyl)-3-methylmorpholine-2,5-dione to a cyclic tridepsipeptide,[207] reaction of aldonic lactones with arylhydrazines,[208] aminolysis of xanthates,[209] decomposition of monoperoxyphthalic acid,[210] reaction of benzoyl peroxide with benzylamine,[211] cleavage of the imidazole ring of benzimidazole,[212] and the ring-opening of 5-amino-3-aryl-1,2,3-oxadiazole in HCl.[213]

There have been many investigations of decarboxylation.[214]

Non-Carboxylic Acids

Phosphorus-containing acids.[215] The hydrolyses of a series of monoaryl phosphates follow a rate law of the form

$$\text{Rate} = k_1[\text{Monoanion}] + k_2[\text{Dianion}].$$

203 M. B. Ordyan, Y. T. Edius, L. A. Sarkisyan, and A. E. Akopyan, *Arm. Khim. Zh.*, **19**, 632 (1966); *Chem. Abs.*, **66**, 54705q (1967); R. Tartarelli, G. Nencetti, M. Baccaredda, and S. Cagianelli, *Ann. Chim. (Rome)*, **56**, 1108 (1966); I. Sakurada, T. Ono, Y. Ohmura, and Y. Sakuvada, *Kobunshi Kagaku*, **24**, 87 (1967); *Chem. Abs.*, **67**, 32098n (1967).

204 C. O'Connor and T. A. Turney, *J. Chem. Soc. B*, **1966**, 1211.

205 M. Green and D. M. Thorp, *J. Chem. Soc., B*, **1967**, 1067.

206 J. O. Appleman and V. J. DeCarlo, *J. Org. Chem.*, **32**, 1505 (1967).

207 R. Kazmierczek and G. Kupryszewski, *Roczniki Chem.*, **41**, 103 (1967).

208 H.-H. Stroh and D. Henning, *Chem. Ber.*, **100**, 388 (1967).

209 H. Yoshida, S. Inokawa, and T. Ogata, *Nippon Kagaku, Zasshi*, **87**, 1212 (1966); *Chem. Abs.*, **66**, 85206g (1967).

210 R. E. Ball, J. O. Edwards, M. L. Haggett, and P. Jones, *J. Am. Chem. Soc.*, **89**, 2331 (1967).

211 N. M. Beileryan, F. O. Karapetyan, and O. A. Chaltykyan, *Arm. Khim. Zh.*, **19**, 828 (1967); *Chem. Abs.*, **67**, 43136z (1967).

212 J. Jasinka, *Roczniki Chem.*, **40**, 1429 (1966).

213 L. E. Kholodov and V. G. Yashunskii, *Zh. Obshch. Khim.*, **37**, 670 (1967); *Chem. Abs.*, **67**, 53352f (1967).

214 M. Caplow and M. Yager, *J. Am. Chem. Soc.*, **89**, 4513 (1967); J. V. Rund and K. G. Claus, *ibid.*, p. 2256; L. W. Clark, *J. Phys. Chem.*, **71**, 2597 (1967); D. B. Bigley and J. C. Thurman, *J. Chem. Soc., B*, **1967**, 941; *Tetrahedron Letters*, **1967**, 2377; D. B. Bigley and R. W. May, *J. Chem. Soc., B*, **1967**, 557; A. Hosaka, A. Sugimori, T. Genka, and G. Tsuchihashi, *Bull. Chem. Soc. Japan*, **40**, 1799 (1967); A. Hosaka, A. Sugimori, and G. Tsuchihashi, *ibid.*, p. 1803; S. Gerchakov and H. P. Schultz, *J. Org. Chem.*, **32**, 1656 (1967); D. S. Noyce and E. H. Banitt, *ibid.*, **31**, 4043 (1966); J. J. Tufariello and W. J. Kissel, *Tetrahedron Letters*, **1966**, 6145; T. Koenig, *ibid.*, **1967**, 2751; C. S. Tsai, *Can. J. Chem.*, **45**, 873 (1967); F. Wiloth and E. Schindler, *Chem. Ber.*, **100**, 2373 (1967); J. E. Rassing and V. S. Andersen, *Dansk Tidsskr. Farm.*, **40**, 257 (1966); *Chem. Abs.*, **66**, 45907e (1967); R. W. Hay and M. A. Bond, *Australian J. Chem.*, **20**, 1823 (1967); R. W. Hay and K. N. Leong, *Chem. Commun.*, **1967**, 800.

215 See A. J. Kirby and S. G. Warren, "Organic Chemistry of Phosphorus", Elsevier, Amsterdam, 1967.

Both k_1 and k_2 are increased by electron-withdrawing substituents in the aryl ring, but the effect on k_2 is much greater than on k_1, leading to an inversion in their relative magnitudes. Thus k_1 for *p*-nitrophenyl phosphate is about 70 times greater than k_2, whereas for 2,4-dinitrophenyl phosphate it is about 35 times smaller and for 2-chloro-4-nitrophenyl phosphate $k_1 \approx k_2$.[216]

The k_1 term in the rate law was considered to arise from a specific acid-catalysed hydrolysis of the dianion (eqn. 13). There is a linear correlation

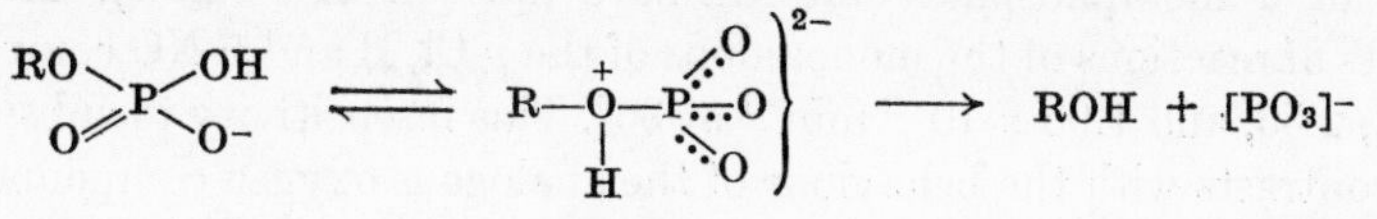

... (13)

(37)

between $\log k_1$ for a series of alkyl and aryl phosphates and the pK_a of the leaving group when these are relatively strongly basic, but the values of $\log k_1$ for aryl phosphates with strongly electron-withdrawing groups fall off the line. It was suggested that, with these compounds, species (**37**) decomposed rapidly at a rate approximating to that at which it is formed from the mono-anion, so that the latter was partly rate-determining; the change in solvent isotope effect k_1^{H}/k_1^{D} from 0.87 at 100° for methyl phosphate to 1.45 at 39° for 2,4-dinitrophenyl phosphate is consistent with this interpretation.

The k_2 term was considered to result from a unimolecular decomposition of the dianion (eqn. 14) which is consistent with its high sensitivity to substituents in the aryl ring.[216]

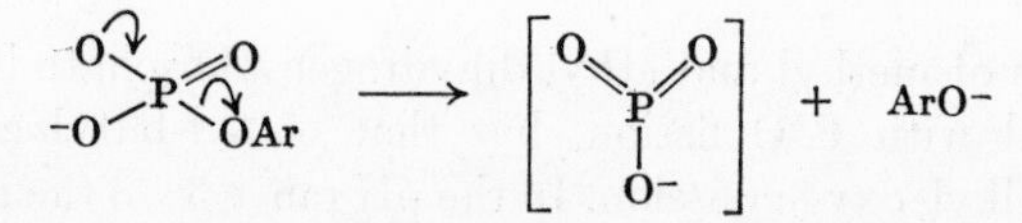

... (14)

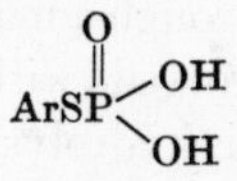

(38)

Hydrolyses of *S*-aryl phosphorothioates (**38**) have also been investigated.[217] The rates of hydrolysis of the monoanions of the phenyl and the *p*-chlorophenyl compound are faster than those of the dianions, and bell-shaped pH–rate

[216] A. J. Kirby and A. G. Varvoglis, *J. Am. Chem. Soc.*, **89**, 415 (1967); see also C. A. Bunton, E. J. Fendler, and J. H. Fendler, *ibid.*, p. 1221; C. A. Bunton, E. J. Fendler, E. Humeres, and K.-U. Yang, *J. Org. Chem.*, **32**, 2806 (1967).

[217] S. Milstien and T. H. Fife, *J. Am. Chem. Soc.*, **89**, 5820 (1967).

profiles are obtained. On the other hand, the dianion of the *p*-nitrophenyl compound reacts 16.2 times faster than the monoanion. Mechanisms similar to those of equations (13) and (14) were proposed. The ratio of methyl phosphate to inorganic phosphate formed from *p*-nitrophenyl phosphorothioates in methanol–water mixtures is 2—3 times greater than the ratio of the mole fractions of methanol to water in the solvent, but from the phenyl and the *p*-chlorophenyl compound it is only slightly greater. It was concluded that product-composition studies cannot be employed as evidence for the intervention of a metaphosphate intermediate (cf. ref. 226 below). The rate constants of reactions of the monoanions of the *p*-Cl, H and *p*-NO compounds are 2.12, 2.55, and 1.95×10^{-2} min^{-1} at 35°. This insensitivity to substituent effects contrasts with the behaviour of the analogous oxygen compound with which the *p*-nitrophenyl compound reacts more than ten times faster than the phenyl compound, and it was suggested that protonation of the leaving group is more important than bond-breaking in the transition state. The rates of hydrolysis of all three phosphorothioates decreased slightly on going from 1.0M- to 6.0M-HCl, and so there is no acid-catalysed hydrolysis of the neutral esters.[217]

The hydrolysis of 3,5-dinitrophenyl phosphate is highly light-sensitive.[218] The dianion is stable for a month in the dark at 39° but has a half-life of five minutes on irradiation with a 125-watt medium-pressure mercury lamp. The hydrolysis of the monoanion is also light-sensitive. In aqueous methanol some methyl phosphate is formed but no 3,5-dinitroanisole, which shows that P–O bond cleavage occurs. These high rates would be expected from a phosphate of a phenol with pK_a 3, and it was suggested that the reactive species is the excited singlet which should be 6 p*K* units more acidic than the ground state.[218]

The hydrolysis of methyl and ethyl dihydrogen phosphate in strong acid solutions proceed with P–O fission, but that of *tert*-butyl and isopropyl phosphate with alkyl–oxygen fission. In the pH range 3—8 the rate of hydrolysis of the isopropyl ester passes through a maximum like those of the methyl and the ethyl ester, and reaction proceeds with P–O fission, but the *tert*-butyl ester shows no such maximum and is hydrolysed with alkyl–oxygen fission.[219]

Substituent effects on the alkaline hydrolysis of aryl diethyl phosphates have been determined.[220]

The hydrolysis of acetyl phenyl phosphate catalysed by Ca^{2+} and Mg^{2+} probably does not involve the initial formation of a chelate.[221]

The hydrolysis of oligophosphates, catalysed by ethylenediamine–

218 A. J. Kirby and A. G. Varvoglis, *Chem. Commun.*, **1967**, 405, 406.
219 L. Kugel and M. Halmann, *J. Org. Chem.*, **32**, 642 (1967).
220 C. van Hooidonk and L. Ginjaar, *Rec. Trav. Chim.*, **86**, 449 (1967).
221 C. H. Oestreich and M. M. Jones, *Biochemistry*, **6**, 1515 (1967).

copper(II) and poly-L-lysine–copper(II) complexes,[222] and the metal ion-catalysed elimination of carbamoyl phosphate dianion,[223] have been studied.

The hydrolysis of phosphate ester (**39**) is buffer-catalysed in imidazole and morpholine buffers.[224] In morpholine buffer the reaction is apparently general acid-catalysed at high ratios of base:(base hydrochloride), but apparently general base-catalysed at low buffer ratios. This dependence can be expressed by the equation:

$$k_{cat} = (43.3 \times 10^{-5})\, a_H/(a_H + 2.26 \times 10^{-9}),$$

and the mechanism of equation (15) was proposed. Kinetic analysis of this leads to an equation of the correct form

$$k_{cat} = k_1 k_3 K_1 a_H/(K_a k_2 + k_3 a_H) K_a$$

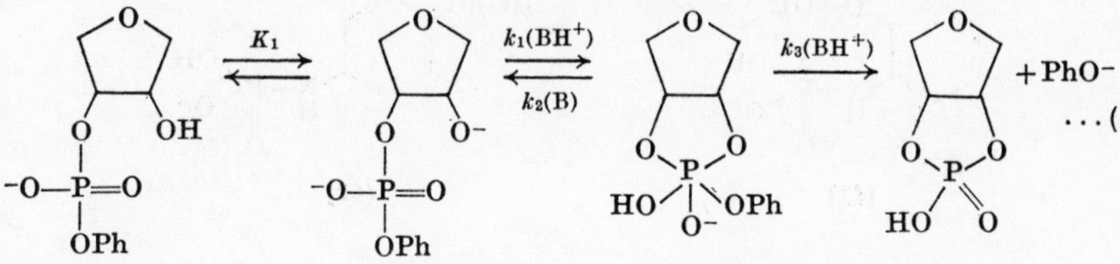

...(15)

(**39**)

Phosphate ester (**40**) is hydrolysed very rapidly in the pH range 2—4, as shown in equation (16) with a rate 2×10^7 times greater than that of the hydrolysis of ethyl *p*-nitrophenylphosphonate to give ethanol. Possible

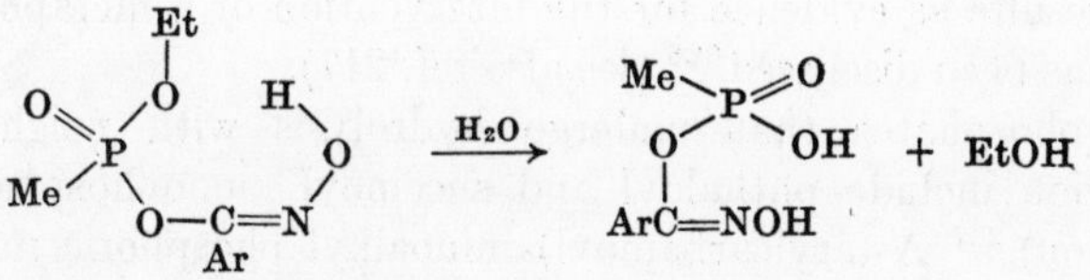

...(16)

(**40**)

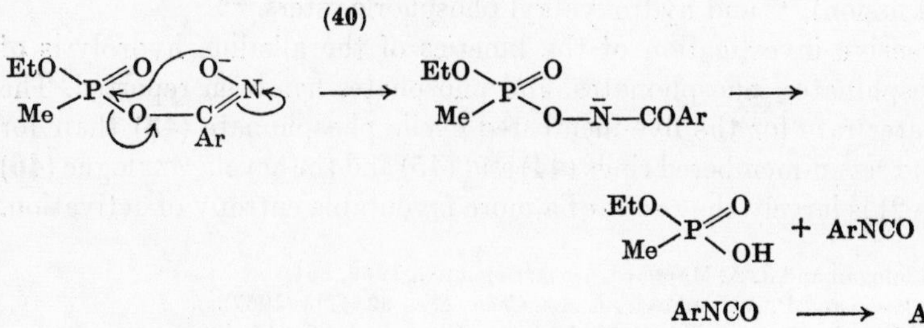

...(17)

[222] C. H. Oestreich and M. M. Jones, *Biochemistry*, **5**, 3151 (1966).
[223] Y. Moriguchi, *Bull. Chem. Soc. Japan*, **39**, 2656 (1966).
[224] D. G. Oakenfull, D. I. Richardson, and D. A. Usher, *J. Am. Chem. Soc.*, **89**, 5491 (1967).

mechanisms involve intramolecular nucleophilic or general acid-catalysis. At pH 5 the product is aniline, which may be formed by nucleophilic attack on phosphorus, as shown in equation (17).[225]

The hydrolyses of *o*- and *p*-carboxyphenyl phosphoramidate (**41**) and (**42**) follow rate laws of the form:

$$v = k_{H^+}[a_H][HO_2C\text{–R–}NHPO_3H_2] + k_1[HO_2C\text{–R–}NHPO_3H_2] + k_2[HO_2C\text{–R–}NHPO_3H^-] + k_3[^-O_2C\text{–R–}NHPO_3H^-].$$

k_{H^+}, k_1, and k_2 are 2—3-fold greater for the *ortho*- than for the *para*-compound, but k_3 is about twice as large for the *para*- as for the *ortho*-compound.

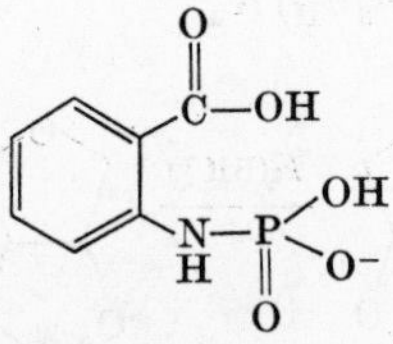

(**41**)

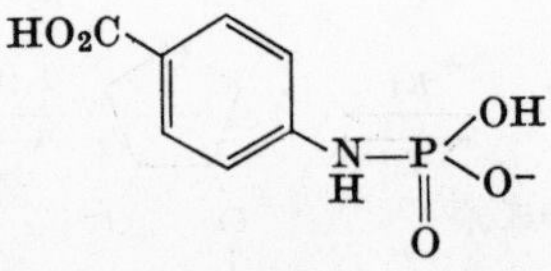

(**42**)

Mechanisms involving intramolecular catalysis were considered to explain the larger values of k_1 and k_2 for the *ortho*-compound, but electrostatic or steric effects offer alternative explanations. The mole fraction of methyl phosphate produced in methanol–water mixtures is larger than the mole fraction of methanol in the solvent, is independent of the state of ionization of the substrate, and is not affected by the *ortho*-carboxyl group. The relevance of these results as evidence for the intervention of a metaphosphate intermediate has been discussed[226] (see also ref. 217).

Other phosphates that undergo hydrolysis with neighbouring-group participation include phthaloyl and succinoyl monophosphate (with C–O bond fission),[227] *N*-(arylcarbamoyl)aminoalkyl phosphoric monesters (with C–O bond fission),[228] and hydroxyalkyl phosphoric esters.[229]

An extensive investigation of the kinetics of the alkaline hydrolysis of cyclic phosphinates, phosphonates, and phosphates has been reported. The much greater rate for the five-membered cyclic phosphonate (**43**) than for the six- and seven-membered rings (**44**) and (**45**) and the acyclic analogue (**46**) (see Table 2) is largely the result of a more favourable entropy of activation.

225 J. I. G. Cadogan and J. A. Maynard, *Chem. Commun.*, **1966**, 854.
226 S. J. Benkovic and P. A. Benkovic, *J. Am. Chem. Soc.*, **89**, 4714 (1967).
227 T. Higuchi, G. L. Flynn, and A. C. Shah, *J. Am. Chem. Soc.*, **89**, 616 (1967).
228 E. Cherbuliez, B. Baehler, O. Espejo, E. Frankenfeld, S. Jaccard, and J. Rabinowitz, *Helv. Chim. Acta*, **50**, 979 (1967).
229 J. Songstad, *Acta Chem. Scand.*, **21**, 1681 (1967).

This suggests that factors other than release of strain may be important in contributing to the enhanced rate.[230]

Table 2. Alkaline hydrolysis of some phosphoric esters.

	43	**44**	**45**	**46**
10^4k (l. mole^{-1} sec^{-1}) at 50°	5.4×10^5	26.2	1.09	1.72
E (kcal mole^{-1})	11.7	13.7	14.0	14.0
$\Delta S^{\ddagger}$	−16.6	−30.3	−35.7	−34.2

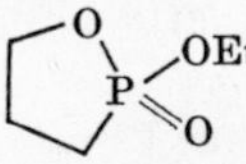

(43)

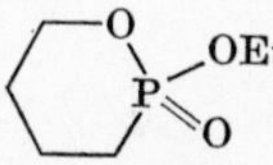

(44)

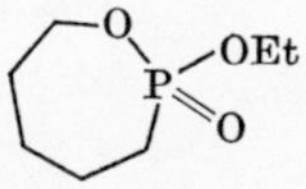

(45)

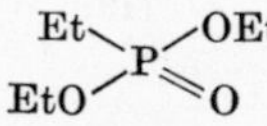

(46)

On the other hand, the much greater rates of the acid and the alkaline hydrolysis of glycerol 1,2-cyclic phosphate than of dimethyl phosphate are the result of more favourable energies of activation.[231]

As reported last year,[232] the rates of hydrolysis of five-membered cyclic phosphinates are not appreciably faster than those of their acyclic analogues. This was attributed to one of the ring carbon atoms having to take up an apical position in the pentaco-ordinated intermediate, which was considered to be unfavourable; but it was pointed out that an enhanced rate might be found with a cyclic ester in which release of strain on going to the transition state was sufficiently great to outweigh this unfavourable factor. This possibility has beeen realized by showing that the highly strained ester groups at position 7 of bicyclic esters (**47**) and (**48**) are hydrolysed 10^4—10^5 times faster than their monocyclic analogues.[233]

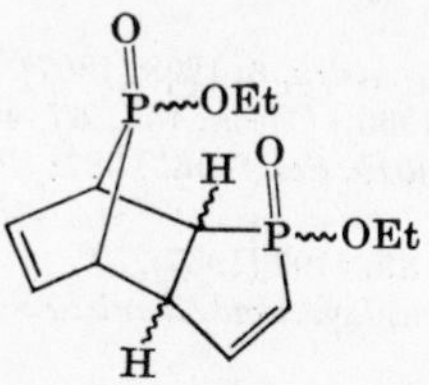

(47)

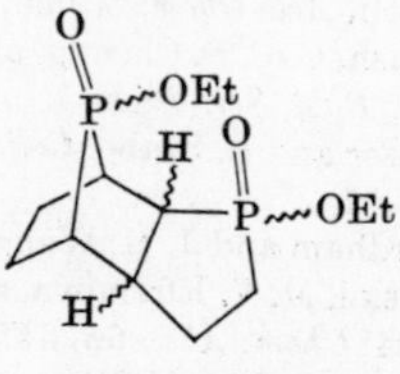

(48)

[230] G. Aksnes and K. Bergesen, *Acta Chem. Scand.*, **20**, 2508 (1966).
[231] L. Kugel and M. Halmann, *J. Am. Chem. Soc.*, **89**, 4125 (1967).
[232] *Organic Reaction Mechanisms*, **1966**, 362; see also ref. 230.
[233] R. Kluger, F. Kerst, D. G. Lee, E. A. Dennis, and F. H. Westheimer, *J. Am. Chem. Soc.*, **89**, 3918, 3919 (1967); see also D. G. Gorenstein and F. H. Westheimer, *ibid.*, p. 2762; *Proc. Nat. Acad. Sci. U.S.*, **58**, 1747 (1967).

The kinetics of hydrolysis of a large number of 1,3,2-dioxaphospholans (**49**) and 1,2,3-dioxaphospholan-2-ones and -2-thiones (**50**) in 50% aqueous sodium hydroxide–dioxan have been studied. The relative rates decrease as R^5 is varied in the order SEt > OEt > NMe and as methyl substituents are introduced into positions 4 and 5. The six-membered cyclic esters (**51**) were hydrolysed at similar rates to the analogous acyclic compounds and much more slowly than analogous five-membered cyclic esters.[234]

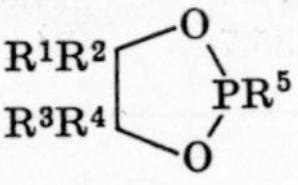

(49)

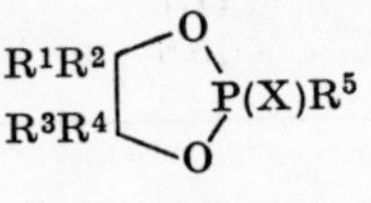

(50)

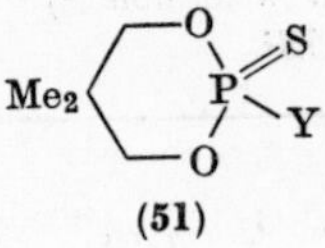

(51)

The hydrolysis of some other cyclic phosphorus esters are discussed in references 235—237.

An investigation of ^{18}O-exchange accompanying the isomerization of the glycerol monophosphates has shown that this reaction proceeds partly through a cyclic phosphate ester and partly by a direct displacement.[238]

The following reactions have also been investigated: hydrolysis of esters of bis(chloromethyl)phosphinic acid, $(ClCH_2)_2P(O)OR$,[239] acid hydrolysis of a pyrophosphoric amide,[240] transesterification of phosphinate esters, RPH(O)OR′, with glycols,[241] and chlorination of quinquevalent phosphorus esters;[242] and the hydrolyses of phosphorylcholine and 3,3-dimethylbutyl phosphate have been compared.[243]

The kinetics of hydrolysis of polyphosphates have been studied,[244] and the free energy of the hydrolysis of pyrophosphate has been determined.[245]

The three-dimensional structure of ribonuclease has been determined from

234 R. S. Edmundson and A. J. Lambie, *J. Chem. Soc., B*, **1967**, 577.

235 K. Bergesen, *Acta Chem. Scand.*, **21**, 1587 (1967).

236 H. I. Abrash, C.-C. S. Cheung, and J. C. Davis, *Biochemistry*, **6**, 1298 (1967).

237 C. Hetzer, *Publ. Sci. Tech. Min. Air. Fr.* No. 153 (1966); *Chem. Abs.*, **67**, 44053g (1967); C. D. Hetzer and L. Szabo, *Colloq. Nat. Centre Nat. Rech. Sci.*, **1965**, 327; *Chem. Abs.*, **67**, 73829v (1967).

238 W. D. Fordham and J. H. Wang, *J. Am. Chem. Soc.*, **89**, 4197 (1967).

239 V. E. Bel'skii, M. V. Efremova, and I. M. Shermergorn, *Izv. Akad. Nauk SSSR, Ser. Khim.*, **1966**, 1654; *Chem. Abs.*, **66**, 64761u (1967).

240 V. E. Bel'skii and M. V. Efremova, *Izv. Akad. Nauk SSSR, Ser. Khim.*, **1967**, 249; *Chem. Abs.*, **67**, 32092f (1967).

241 A. N. Pudovik and M. A. Pudovik, *Zh. Obshch. Khim.*, **36**, 1658 (1966); *Chem. Abs.*, **66**, 54715t (1967).

242 M. Green and D. M. Thorp, *J. Chem. Soc., A*, **1967**, 731.

243 J. Attias and C. Marmignon, *Compt. Rend., Ser. C.*, **264**, 606 (1967).

244 E. J. Griffith and R. L. Buxton, *J. Am. Chem. Soc.*, **89**, 2884 (1967).

245 C. Wu, R. J. Witonsky, P. George, and R. J. Rutman, *J. Am. Chem. Soc.*, **89**, 1987 (1967).

an electron-density map at 2-Å resolution,[246] and of ribonuclease-S from a map at 3.5-Å resolution.[247]

The action of ribonuclease with cytidine 2′,3′-cyclic phosphate and cytidyl-3′,5′-cytidine has been investigated by a stopped-flow temperature-jump technique;[248] and the mechanism of action of ribonuclease has been discussed.[249] There have been investigations of the hydrogen ion titration curve,[250] of denaturation,[251] and of chemical modification[252] of ribonuclease. The CD,[253] ORD,[254] and NMR[255] spectra of ribonuclease have been discussed.

Sulphur-containing acids. The rate of hydrolysis of *p*-nitrophenyl sulphate is independent of pH in the region 7—12 and is proportional to [$^{-}$OH] at higher pH's. The pH-independent reaction could be a unimolecular elimination of SO_3 (eqn. 18) or a bimolecular reaction, and possibly the latter is the correct description as the entropy of activation is −18 e.u. and the reaction is slightly accelerated by F^- and SO_3^{2-}. The rates of aminolysis of *p*-nitrophenyl sulphate

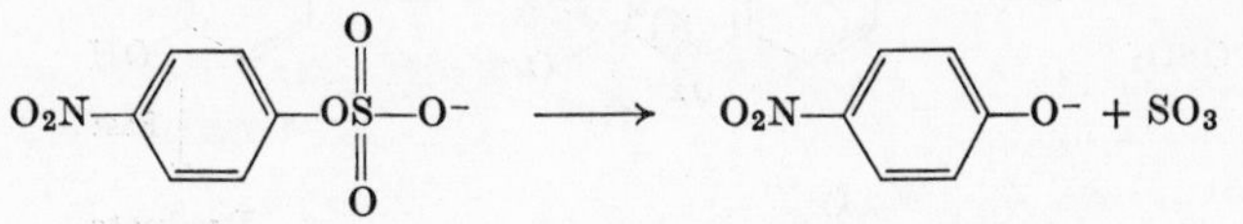 . . . (18)

depend only weakly on the basicity of the amine ($\beta = 0.20$), and the relative rate decreases in the order tertiary > secondary > primary. Hydroxylamine and hydrazine are only two to four times more reactive than predicted from their pK_a's and so the α-effect (see p. 335) is not very important here. These reactions, therefore, resemble very closely those of *p*-nitrophenyl phosphate with amines for which $\beta = 0.13$.[256]

The pH–rate profile for hydrolysis of *o*-carboxyphenyl sulphate is sigmoid

246 G. Kartha, J. Bello, and D. Harker, *Nature*, **213**, 862 (1967); see also H. P. Avey, M. O. Boles, C. H. Carlisle, S. A. Evans, S. J. Morris, R. A. Palmer, B. A. Woolhouse, and S. Shall, *ibid.*, p. 557; G. Kartha, *ibid.*, **214**, 234 (1967).

247 H. W. Wyckoff, K. D. Hardman, N. M. Allewell, T. Inagami, D. Tsernoglou, L. N. Johnson, and F. M. Richards, *J. Biol. Chem.*, **242**, 3749 (1967); H. W. Wyckoff, K. D. Hardman, N. M. Allewell, T. Inagami, L. N. Johnson, and F. M. Richards, *ibid.*, p. 3984.

248 J. E. Erman and G. G. Hammes, *J. Am. Chem. Soc.*, **88**, 5607, 5614 (1966).

249 E. N. Ramsden and K. J. Laidler, *Can. J. Chem.*, **44**, 2597 (1966); A. P. Mathias and B. R. Rabin, *Biochem. J.*, **113**, 62P (1967); K. Brocklehurst, E. M. Crook, and C. W. Wharton, *Chem. Commun.*, **1967**, 63.

250 Y. Nozaki and C. Tanford, *J. Am. Chem. Soc.*, **89**, 742 (1967).

251 J. F. Brandts and L. Hunt, *J. Am. Chem. Soc.*, **89**, 4826 (1967).

252 M. E. Friedman, H. A. Scheraga, and R. F. Goldberger, *Biochemistry*, **5**, 3770 (1966); M. Wilchek, A. Frensdorff, and M. Sela, *ibid.*, **6**, 247 (1967); H. Marzotto, P. Jajetta, and E. Scoffone, *Biochem. Biophys. Res. Commun.*, **26**, 517 (1967).

253 B. Jirgensons, *J. Am. Chem. Soc.*, **89**, 5979 (1967).

254 N. S. Simmons and A. N. Glazer, *J. Am. Chem. Soc.*, **89**, 5040 (1967).

255 C. C. McDonald and W. D. Phillips, *J. Am. Chem. Soc.*, **89**, 6332 (1967).

256 S. J. Benkovic and P. A. Benkovic, *J. Am. Chem. Soc.*, **88**, 5504 (1966).

in the pH region 2—5.5 with rate proportional to [$^{-}O_3S \cdot O \cdot C_6H_4 \cdot CO_2H$]. At pH 3 the rate is about 200 times greater than that for the hydrolysis of *p*-carboxyphenyl sulphate. Intramolecular catalysis is clearly operative and mechanisms involving general acid (eqn. 19) and nucleophilic catalysis (eqns. 20 and 21) were suggested as the most likely.[257]

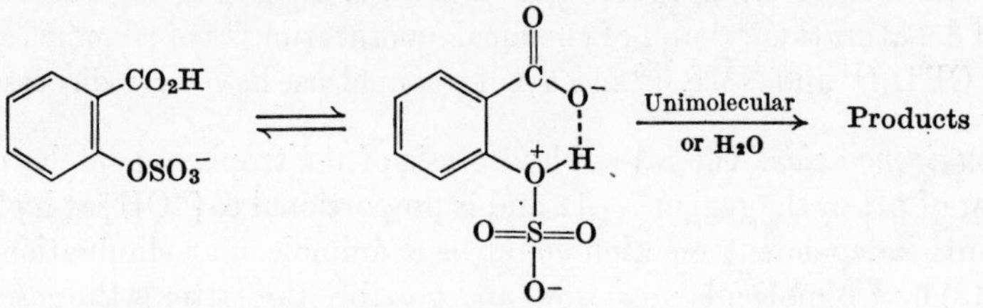

...(19)

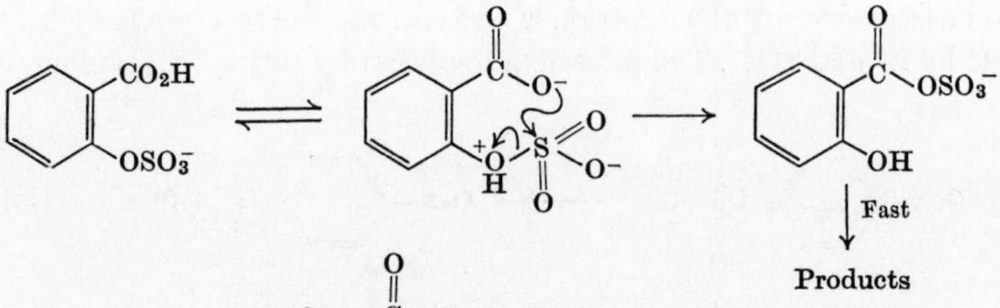

...(20)

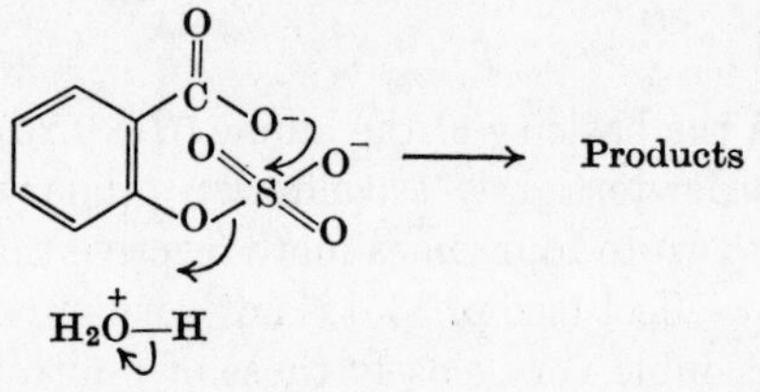

...(21)

The hydrolysis of 8-quinolyl sulphate is strongly catalysed by cupric ions, presumably via complex (**52**).[258]

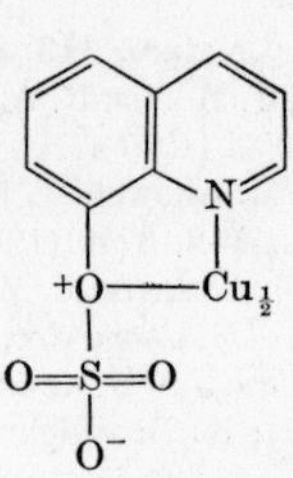

(52)

A bimolecular mechanism has been proposed for the hydrolysis of sulphonic anhydrides in aqueous acetone on the grounds that the rate is strongly increased by hydroxyl ions.[259]

[257] S. J. Benkovic, *J. Am. Chem. Soc.*, **88**, 5511 (1966).
[258] R. W. Hay and J. A. G. Edmonds, *Chem. Commun.*, **1967**, 969.
[259] N. H. Christensen, *Acta Chem. Scand.*, **20**, 1955 (1966); **21**, 899 (1967).

The rate of hydrolysis of the five-membered cyclic sultone (**54**), which cannot react via a sulfene intermediate, is similar[260] to that of the *o*-hydroxytoluene-α-sulphonic acid sultone investigated previously.[261] This supports the view that five-membered cyclic sultones are especially susceptible to nucleophilic attack at sulphur. The six-membered sultone (**55**) reacts much more slowly, but about 10 times faster than phenyl toluene-α-sulphonate.

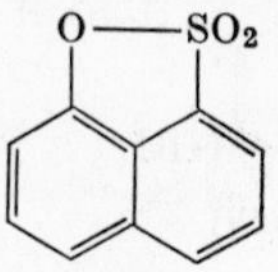

(**54**)

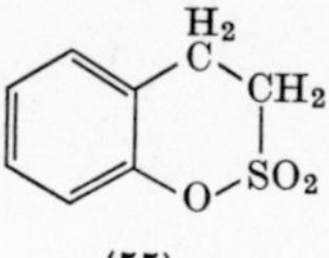

(**55**)

About 3 kcal $mole^{-1}$ of the 5 kcal $mole^{-1}$ more favourable free energy of activation for the hydrolysis of ethylene sulphite, compared with that of dimethyl sulphite, is the result of a more positive entropy of activation. It was suggested that in the transition state for hydrolysis of dimethyl sulphite the molecule is held rigidly and that the molecular motions of the alkyl groups are suppressed, but that in ethylene sulphite the molecule is already rigid so that far less entropy loss occurs when the transition state is formed (cf. ref. 230, p. 363).[262]

Intermediate (**56**) has been detected by NMR spectroscopy in the transesterification of ethylene glycol by dimethyl sulphite (eqn. 22).[263]

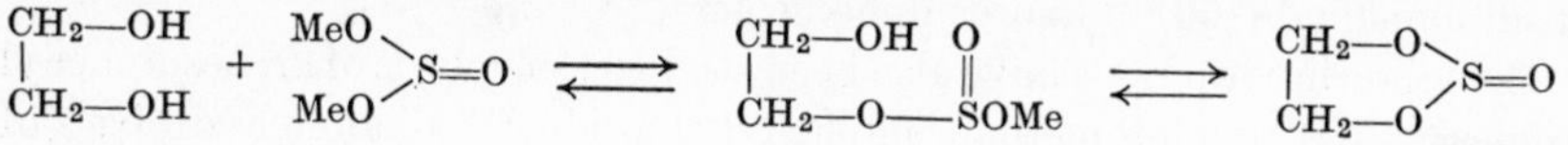

(**56**) . . . (22)

The hydrolysis of sultone (**57**) is catalysed by chymotrypsin. When (**57**) is added to a solution of chymotrypsin at pH 7—8 there is a burst in the absorbance at 400 mμ accompanying formation of a sulphonyl-enzyme (**58**), and this is followed by a slow rise in absorbance to that of acid (**59**). The ready hydrolysis of (**58**) contrasts with the stability of α-(toluene-*p*-sulphonyl)chymotrypsin and suggests that the phenolic hydroxyl group is participating. The pH–rate profile for hydrolysis is bell-shaped with pK_a (app.) 6.47 and 7.95. The former is presumably that of the imidazole group of the enzyme but the latter differs from the spectroscopically determined pK_a of the phenolic group of (**58**), which is 7.2. Therefore at present the mechanism is uncertain.[264]

[260] E. T. Kaiser, K. Kudo, and O. R. Zaborsky, *J. Am. Chem. Soc.*, **89**, 1393 (1967).
[261] See *Organic Reaction Mechanisms*, **1966**, 363.
[262] P. A. Bristow and J. G. Tillett, *Chem. Commun.*, **1967**, 1010.
[263] P. A. Bristow and J. G. Tillett, *Tetrahedron Letters*, **1967**, 901.
[264] J. H. Heidema and E. T. Kaiser, *J. Am. Chem. Soc.*, **89**, 460 (1967).

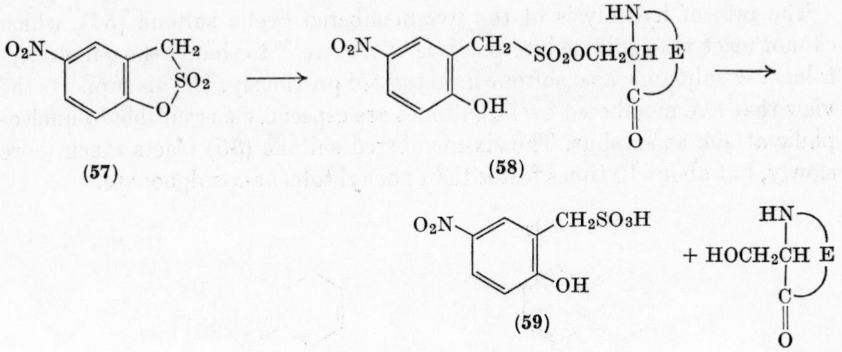

(57) (58) (59)

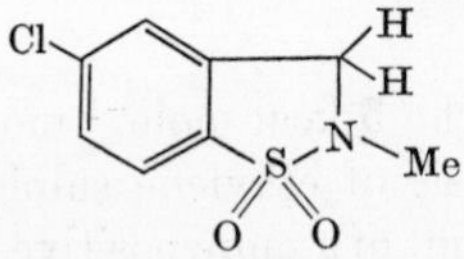

(60)

The pepsin-catalysed hydrolysis of methyl, phenyl, and *p*-bromophenyl sulphite has been investigated.[265]

Oxidation of hydroquinone monosulphate with $NaIO_4$ in methanol, but not in water, results in sulphonation of the solvent, presumably owing to formation of SO_3.[266]

NMR spectroscopic studies indicate that nitrogen is the site of protonation of sulphonamide (**60**) in fluorosulphonic acid.[267]

The following reactions have also been studied: solvolysis of arenesulphenyl halides[268] and trichloromethanesulphenyl chloride;[269] alkaline hydrolysis of esters of arenesulphenic acids[270] and trichloromethanesulphenic acid;[271] solvolysis of toluene-*p*-sulphinates in 1-methylpyrrolidone;[272] hydrolysis of sulphinyl-sulphones in aqueous dioxan;[273] the reaction of aryl thiolsulphinates with aromatic sulphinic acids (see p. 91);[274] reactions of sulphenes;[275]

265 T. W. Reid and D. Fahrney, *J. Am. Chem. Soc.*, **89**, 3941 (1967).

266 S. W. Weidman, D. F. Mayers, O. R. Zaborsky, and E. T. Kaiser, *J. Am. Chem. Soc.*, **89**, 4555 (1967).

267 F. M. Menger and L. Mandell, *J. Am. Chem. Soc.*, **89**, 4424 (1967); see also R. G. Laughlin, *ibid.*, p. 4268.

268 L. Di Nunno, G. Modena, and G. Scorrano, *Ric. Sci.*, **36**, 825 (1966); *Chem. Abs.*, **66**, 64768b (1967).

269 J. Horák, *Coll. Czech. Chem. Commun.*, **32**, 868 (1967); see also J. Horák and K. Krajkářová, *ibid.*, p. 1619.

270 C. Brown and D. R. Hogg, *Chem. Commun.*, **1967**, 38.

271 J. Horák and K. Krajkářová, *Coll. Czech. Chem. Commun.*, **32**, 2134 (1967).

272 J. W. Wilt, R. G. Stein, and W. J. Wagner, *J. Org. Chem.*, **32**, 2097 (1967).

273 J. L. Kice and G. Guaraldi, *J. Am. Chem. Soc.*, **89**, 4113 (1967).

274 J. L. Kice, C. G. Venier, and L. Heasley, *J. Am. Chem. Soc.*, **89**, 3557, 4817 (1967).

275 L. A. Paquette and M. Rosen, *J. Am. Chem. Soc.*, **89**, 4102 (1967).

^{18}O-exchange between sodium thiosulphate and water;[276] reactions of arenesulphonyl halides with amines;[277] methanolysis of thionyl chloride and methyl chlorosulphite;[278] and the reaction of *N*-sulphinylaniline with ethanol.[279]

Nitrogen-containing acids. The following reactions have been investigated: *O*-nitration of benzyl and thienyl alcohols,[280] and of methanol;[281] deuteration of nitroguanidines;[282] acid hydrolysis of nitrate esters of polyols;[283] hydrolysis of *N*-nitrosoacetanilides;[284] conversion of *anti*-diazoates into diazonium ions;[285] dissociation of diazosulphonates;[286] and diazotization of 2- and 4-aminopyridines.[287]

276 W. A. Pryor and U. Tonellato, *J. Am. Chem. Soc.*, **89**, 3379 (1967).
277 L. M. Litvinenko and V. A. Savelova, *Zh. Obshch. Khim.*, **36**, 1524 (1966); *Chem. Abs.*, **66**, 54734y (1967); L. M. Litvinenko and A. F. Popov, *Zh. Obshch. Khim.*, **36**, 1517 (1966); *Chem. Abs.*, **66**, 54733x (1967); L. M. Litvinenko, A. F. Popov, and V. A. Savelova, *Ukr. Khim. Zh.*, **33**, 57 (1966); *Chem. Abs.*, **66**, 94610m (1967); L. M. Litvinenko, A. F. Popov, and A. M. Borovenski, *Reakts. Sposobnost Org. Soedin Tartu. Gos. Univ.*, **3**, 93 (1966); *Chem. Abs.*, **66**, 115190s (1967); L. M. Litvinenko, A. F. Popov, and D. I. Sorokina, *Reakts. Spisobnost Org. Soedin Tartu. Gos. Univ.*, **3**, 211 (1966).
278 H. Minato and A. Fujii, *Bull. Chem. Soc. Japan*, **40**, 202 (1967).
279 L. Senatore and L. Jannelli, *Ann. Chim. (Rome)*, **57**, 308 (1967).
280 B. Östman, *Acta Chem. Scand.*, **21**, 1257 (1967).
281 A. P. Genich and L. V. Kustova, *Izv. Akad. Nauk SSSR, Ser. Khim.*, **1967**, 752; *Chem. Abs.*, **67**, 53258e (1967).
282 J. C. Lockhart, *J. Chem. Soc., B*, **1966**, 1174.
283 B. S. Svetlov and V. P. Shelaputina, *Zh. Fiz. Khim.*, **40**, 2889 (1966); *Chem. Abs.*, **66**, 85196d (1967).
284 B. A. Porai-Koshits and V. V. Shaburov, *Zh. Organ. Khim.*, **2**, 1666 (1966); *Chem. Abs.*, **66**, 64777d (1967).
285 E. S. Lewis and M. P. Hanson, *J. Am. Chem. Soc.*, **89**, 6268 (1967).
286 L. K. H. van Beek, J. Helfferich, H. Jonker, and T. P. G. W. Thijssens, *Rec. Trav. Chim.*, **86**, 405 (1967).
287 E. Kalatzis, *J. Chem. Soc., B*, **1967**, 273, 277.

CHAPTER 13

Photochemistry

This topic has maintained its rapid rate of growth, and more books devoted to the subject have appeared,[1] including new collections of reviews by specialists in particular areas of photochemistry.[2] Other reviews of both general[3] and specialist (carbonyl compounds,[4a] aromatic compounds,[4b] quinones,[4c] cycloadditions[5]) interest have also become available, and Mauser has derived a series of general mathematical expressions relating photochemical mechanisms and quantum-yield data.[6]

Some sub-headings have been included for the first time as a rough guide to the contents of this chapter.

Carbonyl Compounds

The representation of $n\pi^*$-excited states has been examined by Taylor[7] who has suggested that promotion of an electron to an excited π-orbital should be emphasized as in (**1**) and (**3**). These structures show both the electron-deficient nature of oxygen, and the presence of unpaired electrons. Illustrated is the now familiar "type A" rearrangement of diphenylcyclohexadienone (**2**). Whilst this kind of nomenclature is clearly helpful, it can also be most misleading, as, for example, when migration of an aryl group occurs instead of "type A" rearrangement [examples are discussed below, e.g., (**6a**), and in previous volumes]. The aryl group which migrates preferentially is the one best able to accommodate an odd electron and not (as the above scheme might seem to imply) a positive charge.†

† "Taylor notation" has been rejected by Zimmerman on a number of counts; *Chem. Commun.*, **1968**, 174.

[1] D. C. Neckars, "Mechanistic Organic Photochemistry", Reinhold, New York, 1967.

[2] "Organic Photochemistry", ed. O. L. Chapman, Vol. 1, Marcel Dekker, Inc., New York, 1967; aspects of organic photochemistry are also dealt with by P. G. Ashmore, F. S. Dainton, and T. M. Sugden in "Photochemistry and Reaction Kinetics", Cambridge University Press, London, 1967.

[3] N. J. Turro, *Chem. Eng. News*, May 8th 1967, p. 84; H. E. Zimmerman, *Science*, **153**, 837 (1966).

[4a] K. Tokumaru, A. Sugimori, T. A. Kiyama, and T. Nakata, *Yuki Gosei Kagaku Kyokai Shi*, **24**, 1183 (1966); A. S. Davies and R. B. Cundall, *Progr. Reaction Kinetics*, **4**, 149 (1967).

[4b] E. Havinga, *Chimia*, **21**, 413 (1967).

[4c] J. M. Bruce, *Quart. Rev.*, **21**, 405 (1967); see also O. L. Chapman *et al.*, *Record Chem. Progr.*, **28**, 167 (1967).

[5] R. Steinmetz, *Fortschr. Chem. Forsch.*, **7**, 445 (1967).

[6] H. Mauser, *Z. Naturforsch.* **22**b, 367, 371, 465 (1967).

[7] G. A. Taylor, *Chem. Commun.*, **1967**, 896.

Zimmerman's group has published a substantial new paper[8] on the electronic details of the rearrangement of (**2**) to (**4**), and discuss at length the timing of electron demotion. Although there is evidence which suggests that demotion occurs from rearranged (or partially rearranged) dienone $n\pi^*$-triplets to give the bicyclic zwitterion as shown below in "Taylor notation",

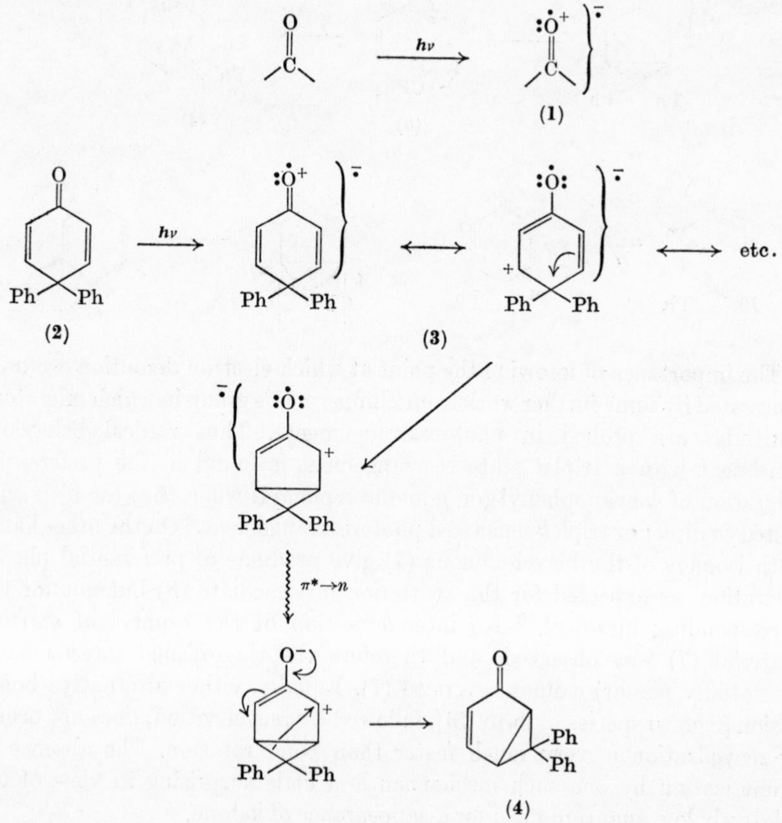

we feel that this remains the one slightly weak link in the overall mechanism; for example, there still seems only slender evidence to contradict the sequence overleaf involving demotion in formation of biradical (**5**). This would circumvent arguments which do disallow triplet rearrangement directly to the $n\pi^*$-triplet of (**4**) followed by electron demotion.

It was also emphasized that the high quantum yield (0.85) means that the supposed zwitterion cannot readily revert to starting dienone. This was rationalized by a molecular-orbital argument. We would point out that

[8] H. E. Zimmerman and J. S. Swenton, *J. Am. Chem. Soc.*, **89**, 906 (1967).

dienone formation would constitute an example of the more general electrocyclic ring-opening illustrated, for which conrotatory opening is preferred. However, in the present case this is clearly prohibited by common bonding to $C_{(4)}$ (dotted lines).

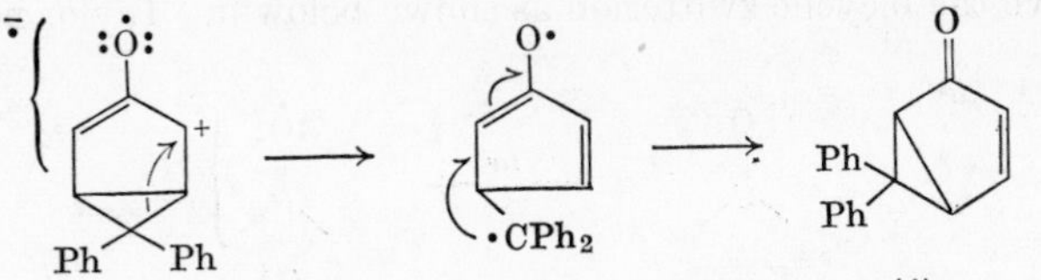

(5) (4)

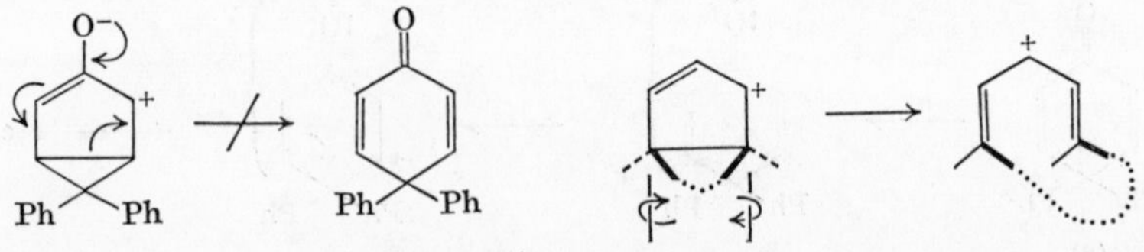

The importance of knowing the point at which electron demotion occurs is illustrated by some further work from Zimmerman's group in which migratory aptitudes are probed in photorearrangements. Thus radical behaviour consistent with a triplet state rearrangement is found in the preferential migration of *p*-cyanophenyl (or *p*-methoxyphenyl) when (**6a**) (or **b**) is subjected to direct or triplet-sensitized photorearrangement.[9] On the other hand, both isomers of the bicyclic enone (**7**) give products of preferential phenyl migration, as expected for the zwitterion intermediate (**8**) but not for the corresponding biradical.[10] No interconversion of the isomers of starting material (**7**) was observed, and therefore the ring-opened intermediates (essentially planar) cannot revert to (**7**). Equally, either alternative bond-fission [e.g., to species akin to (**5**)], followed by recyclization, does not occur, or recyclization is very much faster than bond rotation. The absence of isomerization by one such mechanism is a little surprising in view of the relatively low quantum yield for disappearance of ketone.

Zimmerman *et al.* have also examined the photochemistry of the hydrocarbons (**9**)[11] and (**10**).[12] Here, of course, only $\pi\pi^*$-excitation is possible, and both compounds on direct irradiation react as shown. The triene was stable to attempted photosensitized rearrangement (though the sensitizer triplets were quenched) but the diene gave polymer. Therefore the reactions shown probably involve singlet excited states.

9 H. E. Zimmerman, R. D. Rieke, and J. R. Scheffer, *J. Am. Chem. Soc.*, **89**, 2033 (1967).
10 H. E. Zimmerman and J. O. Grunewald, *J. Am. Chem. Soc.*, **89**, 3354, 5163 (1967).
11 H. E. Zimmerman and G. E. Samuelson, *J. Am. Chem. Soc.*, **89**, 5971 (1967).
12 H. E. Zimmerman, P. Hackett, D. F. Juers, and B. Schröder, *J. Am. Chem. Soc.*, **89**, 5973 (1967).

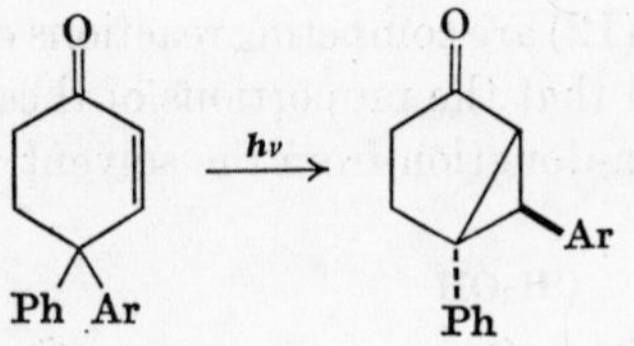

(6) **a** Ar = *p*-CNC_6H_4–
b Ar = *p*-$MeOC_6H_4$–

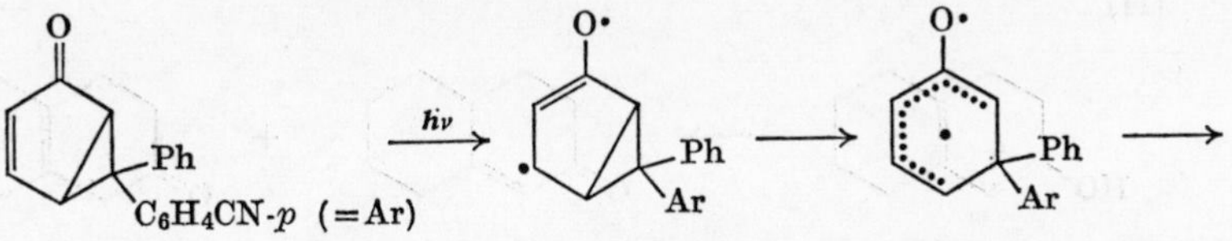

(7)

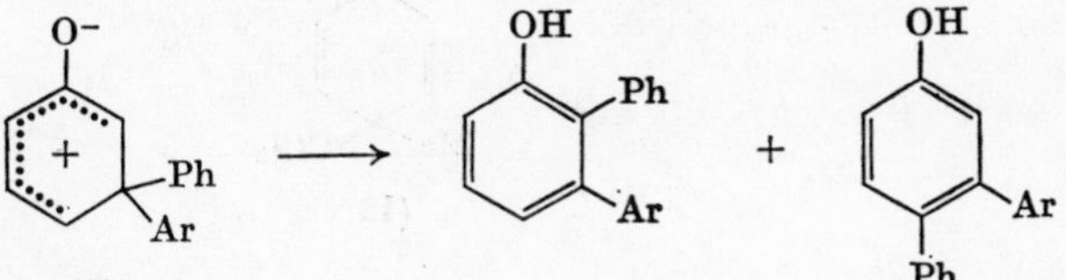

(8)

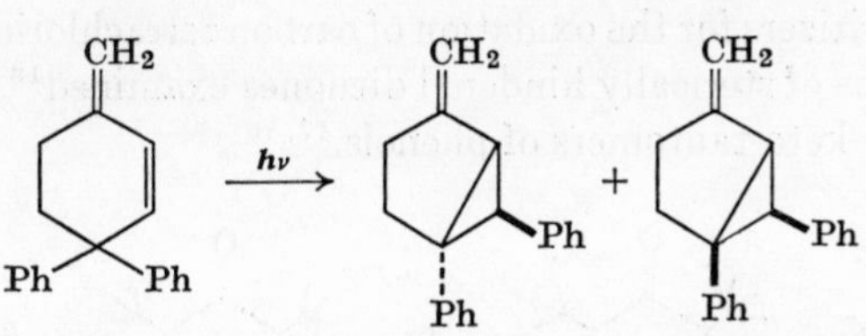

(9)

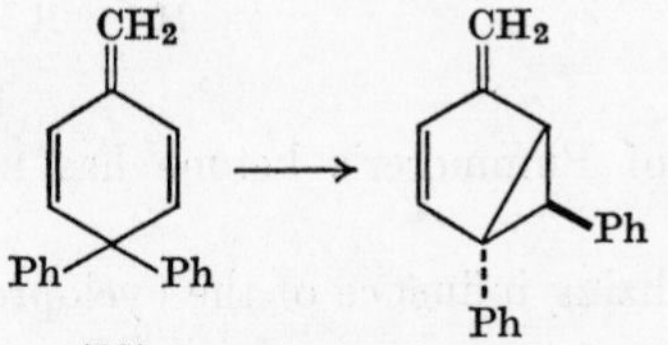

(10)

Other radical-type reactions of dienone excited states have been observed by Schuster and his co-workers. These include the fragmentation reactions of the alcohol (**11**), which were observed in hydrogen-donor solvents such as toluene.[13] In *tert*-butyl alcohol only type A rearrangement occurs. These workers have also presented evidence that the radical fragmentation and

[13] D. I. Schuster and D. F. Brizzolara, *Chem. Commun.*, **1967**, 1158.

type A photorearrangement of (**12**) are competing reactions open to a common ($n\pi^*$-triplet) excited state, and that the proportions of these processes again depend on the ease of hydrogen-donation from the solvent.[14]

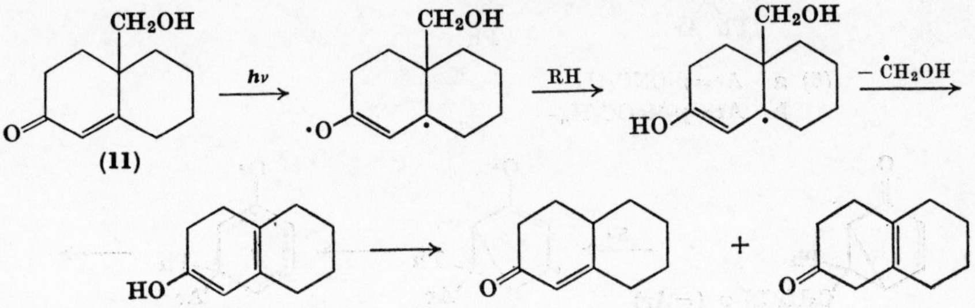

(**11**)

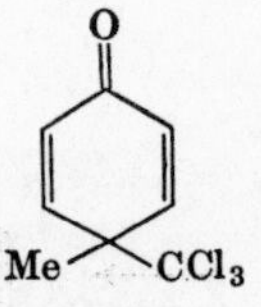

(**12**)

The general observation has also been made that α,β-unsaturated ketones act as photosensitizers for the oxidation of carbon tetrachloride to phosgene.[15]

Photoreactions of sterically hindered dienones examined[16–18] include some involving stable keto-tautomers of phenols.[17, 18]

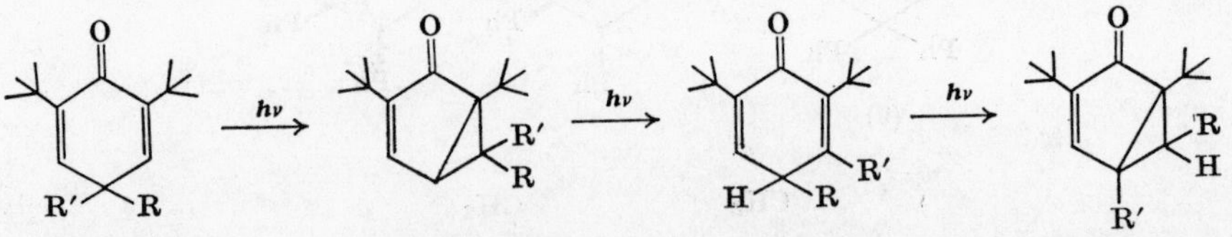

The rearrangement of Pummerer's ketone has now been discussed in detail.[19]

The interesting stabilizing influence of the cyclopropane ring in (**13**) and

14 D. J. Patel and D. I. Schuster, *J. Am. Chem. Soc.*, **89**, 184 (1967).
15 H. D. Mettee, *Can. J. Chem.*, **45**, 339 (1967).
16 B. Miller and H. Margulies, *J. Am. Chem. Soc.*, **89**, 1678 (1967); see also *Organic Reaction Mechanisms*, **1965**, 286; B. Miller, *J. Am. Chem. Soc.*, **89**, 1685 (1967); see also *Organic Reaction Mechanisms*, **1966**, 373; T. Matsuura and K. Ogura, *J. Am. Chem. Soc.*, **89**, 3850 (1967).
17 B. Miller, *J. Am. Chem. Soc.*, **89**, 1690 (1967).
18 T. Matsuura and K. Ogura, *J. Am. Chem. Soc.*, **89**, 3846 (1967).
19 T. Matsuura and K. Ogura, *Bull. Chem. Soc. Japan*, **40**, 945 (1967); see *Organic Reaction Mechanisms*, **1966**, 371.

(**14**) has been noted. Of these, all but the β-methano-dienone are photostable at 254 mμ. The last compound rearranges in the normal type A fashion [to give (**15**)].[20] Hart and his colleagues have examined the reactions of a series

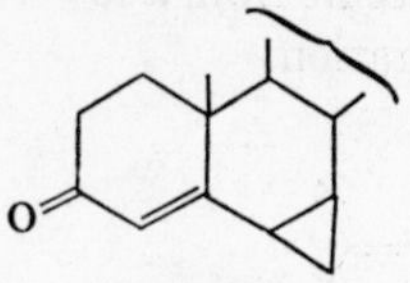

(**13**)

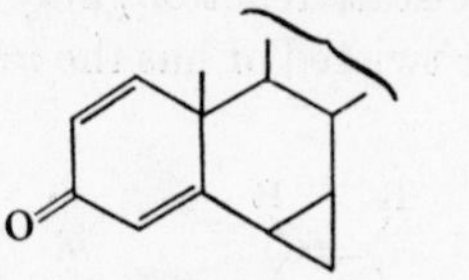

(**14**)

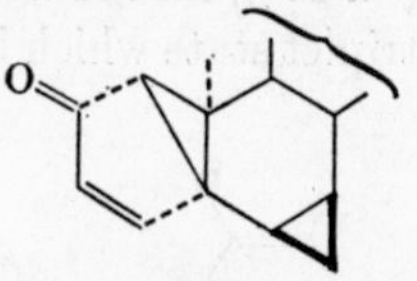

(**15**)

of methylated cyclohexa-2,4-dienones in an attempt to delineate factors influencing the competing reaction pathways encountered.[21]

Suggested[22] mechanisms are illustrated which accommodate the divergent behaviour of the isomeric hydroxy-dienones (**16**) and (**17**).

The photochemistry of dibenzoylethylenes to give phenoxy-esters has been

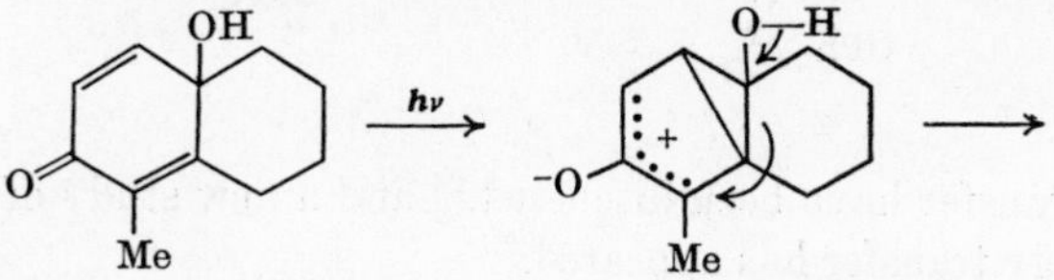

(**16**)

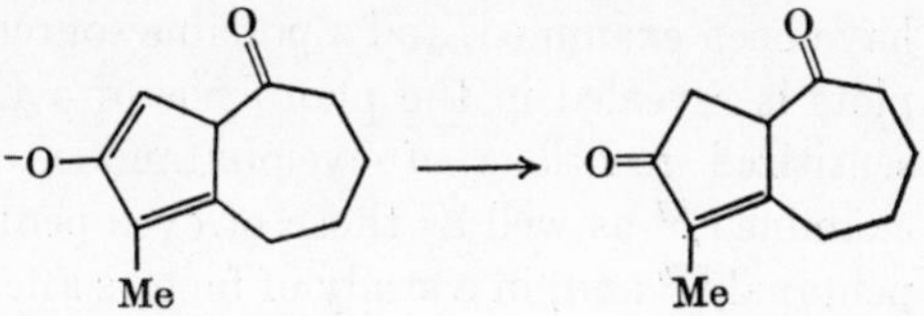

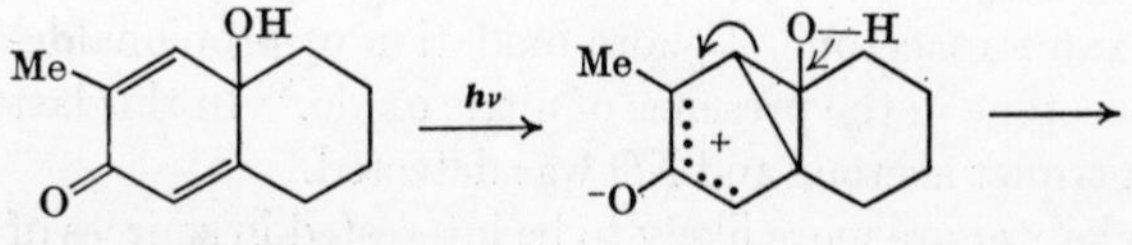

(**17**)

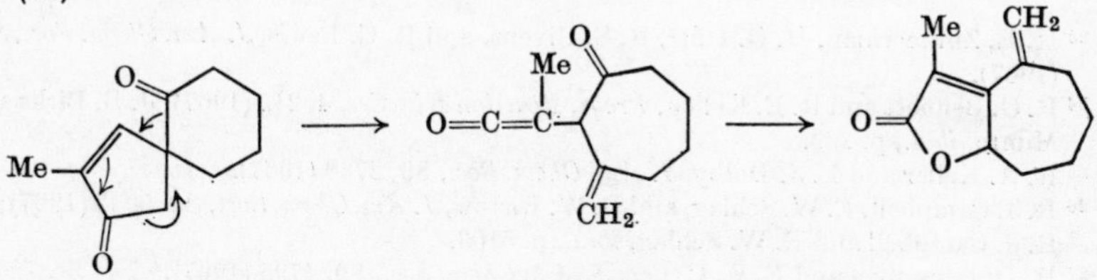

[20] J. Pfister, H. Wehrli, and K. Schaffner, *Helv. Chim. Acta*, **50**, 166 (1967).

[21] P. M. Collins and H. Hart, *J. Chem. Soc., C*, **1967**, 895, 1197; H. Hart and R. K. Murray, *J. Org. Chem.*, **32**, 2448 (1967); H. Hart and D. W. Swatton, *J. Am. Chem. Soc.*, **89**, 1874 (1967).

[22] G. F. Burkinshaw, B. R. Davis, and P. D. Woodgate, *Chem. Commun.*, **1967**, 607.

discussed in detail.[23] The results are consistent with the mechanism shown, in which it is the *cis*-isomer that reacts, predominantly in its singlet $n\pi^*$-excited state. The low efficiency of a comparable process from the triplet state (*ca.* 10% that of the singlet-state reaction) may possibly result from a lowest triplet state which is either twisted or has the $\pi\pi^*$ configuration.

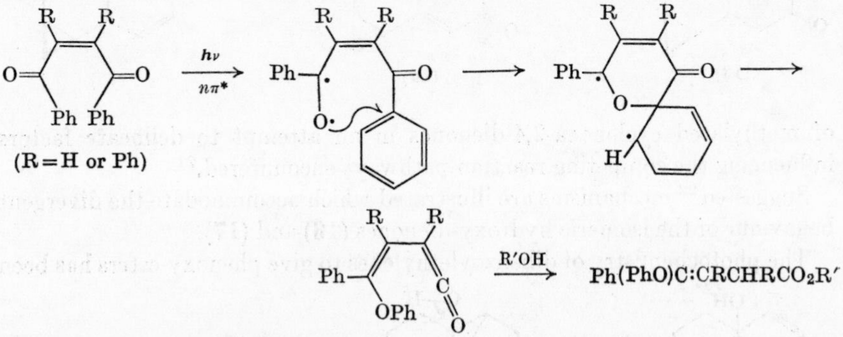

Mechanisms of energy transfer have been discussed,[24] and a new study of intramolecular triplet-energy transfer has appeared.[25]

Interpretations of non-linear Stern–Volmer plots for gas-phase quenching have been examined, and a possible source of error in interpretation of linear plots is revealed in the photolysis of cyclobutanone.[26] The benzene-photo-sensitized reactions of cyclobutanone in the gas-phase have also been examined,[27] as well as those of cyclopentanone,[28] pent-4-enal,[29a] 3-methyl-pentanal,[29b] and, in a study of factors affecting the proportions of type I and type II cleavage, a series of alkyl propyl ketones.[29c] Other gas-phase studies relate to photolysis of hexafluoroacetone,[30] photo-oxidation of propionaldehyde,[31] and photolysis of acetone in the presence of nitric oxide.[32] In this last study, a complex between triplet acetone and NO was detected.

Readers of the present chapter are more likely to be interested in sources of

23 H. E. Zimmerman, H. G. Dürr, R. S. Givens, and R. G. Lewis, *J. Am. Chem. Soc.*, **89**, 1863 (1967).
24 R. G. Bennett and R. E. Kellog, *Progr. Reaction Kinetics*, **4**, 215 (1967); J. B. Birks and I. H. Munro, *ibid.*, p. 239.
25 R. A. Keller and L. J. Dolby, *J. Am. Chem. Soc.*, **89**, 2768 (1967).
26 R. J. Campbell, E. W. Schlag, and B. W. Ristow, *J. Am. Chem. Soc.*, **89**, 5098 (1967); see also R. J. Campbell and E. W. Schlag, *ibid.*, p. 5103.
27 H. O. Denschlag and E. K. C. Lee, *J. Am. Chem. Soc.*, **89**, 4795 (1967).
28 E. K. C. Lee, *J. Phys. Chem.*, **71**, 2804 (1967).
29a E. K. C. Lee and N. W. Lee, *J. Phys. Chem.*, **71**, 1167 (1967).
29b R. E. Rebbert and P. Ausloos, *J. Am. Chem. Soc.*, **89**, 1573 (1967).
29c C. H. Nicol and J. G. Calvert, *J. Am. Chem. Soc.*, **89**, 1790 (1967).
30 L. M. Quick and E. Whittle, *Can. J. Chem.*, **45**, 1902 (1967).
31 A. P. Altshuller, I. R. Cohen, and T. C. Purcell, *Can. J. Chem.*, **44**, 2973 (1966).
32 J. M. Edwards and M. I. Christie, *Chem. Commun.*, **1967**, 789.

error in Stern–Volmer plots in solution photochemistry. Wagner[33] has pointed out that the normal mathematical analysis breaks down at high quencher concentrations such as may be employed to estimate the lifetime of short-lived triplets. This is because quenching is commonly diffusion-controlled, and, at high concentrations of quencher, a reactant molecule may undergo photoexcitation with a quencher molecule as nearest neighbour. The modified analysis required in these circumstances was presented.

Heterogeneous energy-transfer has been achieved by using polyacrylophenone as a triplet sensitizer, e.g., for the geometrical isomerization of piperylene in isopentane (in which the polymer is insoluble).[34]

The reactivity of ketone triplets in homolytic hydrogen abstraction has been examined. Cyclopentanone triplets react significantly faster than those of cyclohexanone or benzophenone, possibly because of relief of angle strain in formation of the transition state for hydrogen abstraction.[35] A steric effect has been demonstrated for reactions of a series of secondary alcohols with benzophenone triplets; for example, abstraction from methylneopentylcarbinol occurs at only one-fifth of the rate of abstraction from propan-2-ol.[36] The ease of photoreduction of substituted anthraquinones has been correlated with the effect of the substituent on the molecular energy levels.[37]

New studies of the photoreduction of benzophenone in amines,[38] and of substituted benzophenones in propan-2-ol,[39] have been reported. *p*-Dimethylaminobenzophenone readily gives a pinacol on irradiation in propan-2-ol–HCl, the lowest triplet being $n\pi^*$. However, in neutral solution the amine is not protonated, the lowest triplet is then of the charge-transfer type, and the compound is photostable.[39b] On the other hand, *p*-aminobenzophenone is reduced in tertiary amine solvents with high quantum efficiency. This suggests that hydrogen abstraction is by initial electron-transfer from an amine to the charge-transfer triplet, and proton transfer follows.[39a]

A coloured by-product of the photoreduction of benzophenone was last year[40] reported as being of the charge-transfer type. It has now been identified as the "*para*-coupled" isomer of benzpinacol (**18**).[41]

[33] P. J. Wagner, *J. Am. Chem. Soc.*, **89**, 5715 (1967).
[34] P. A. Leermakers and F. C. James, *J. Org. Chem.*, **32**, 2898 (1967).
[35] R. Simonaitis, G. W. Cowell, and J. N. Pitts, *Tetrahedron Letters*, **1967**, 3751; see also J. C. W. Chien, *J. Am. Chem. Soc.*, **89**, 1275, (1967).
[36] D. E. Pearson and M. Y. Moss, *Tetrahedron Letters*, **1967**, 3791.
[37] H. H. Dearman and A. Chan, *J. Chem. Phys.*, **44**, 416 (1966).
[38] S. G. Cohen and R. J. Baumgarten, *J. Am. Chem. Soc.*, **89**, 3471 (1967); M. Fischer, *Chem. Ber.*, **100**, 3599 (1967).
[39] (*a*) S. G. Cohen and J. I. Cohen, *J. Am. Chem. Soc.*, **89**, 164 (1967); (*b*) S. G. Cohen and M. N. Siddiqui, *ibid.*, p. 5409; (*c*) S. G. Cohen, R. Thomas, and M. N. Siddiqui, *ibid.*, p. 5845.
[40] See *Organic Reaction Mechanisms*, **1966**, 374.
[41] G. O. Schenk, M. Cziesla, K. Eppinger, G. Matthias, and M. Pape, *Tetrahedron Letters*, **1967**, 193; H. Mauser, B. Nickel, U. Sproesser, and V. Bihl, *Z. Naturforsch.*, **22**b, 903 (1967); see also H. Mauser and V. Bihl, *ibid.*, p. 1077; see also H. L. J. Bäckström and R. J. V. Niklasson, *Acta Chem. Scand.*, **20**, 2617 (1966).

Some complex and interesting new reactions have been encountered in the photo-oxidation of various phenols by aryl ketones.[42] For example, in methanol containing a trace of acid, (**19**) is formed from 2,6-di-*tert*-butylphenol and benzophenone. In the majority of these reactions the key step is identified as oxidation of phenol to a phenoxy-radical by ketone triplets.

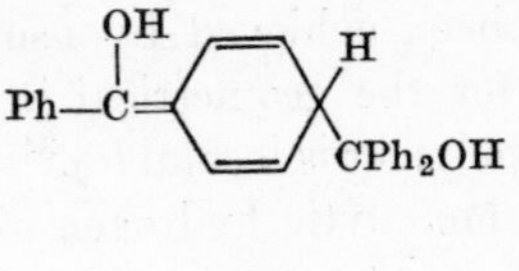

(**18**)

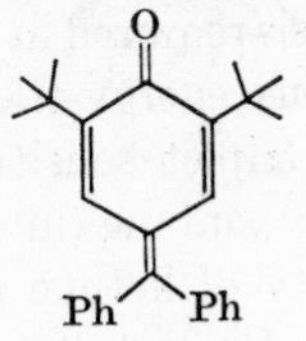

(**19**)

Quantum yields for photoreduction of a series of alkyl- and alkoxy-substituted acetophenones are in accord with calculations and spectroscopic measurements, in that these substituents stabilize the $\pi\pi^*$-excited triplet with respect to the $n\pi^*$-triplet, and it is the latter which is reactive in the photoreduction.[43]

It was proposed last year[44a] that $n\pi^*$-singlet and -triplet states should show comparable chemical reactivity, and that general failure to observe reactions of the singlet could be associated with its brief lifetime. This idea has now been scrutinized by Wagner,[44b] who has measured the reactivity of $n\pi^*$-excited acetone towards reduction by tributylstannane and has estimated that the singlet state is less reactive than the triplet by a factor of at least 10^3. However, Turro *et al.*[45] find that $n\pi^*$-singlet acetone does react both rapidly and stereospecifically with the electrophilic olefin, fumaronitrile, to form an oxetan. This contrasts with the usually non-stereospecific nature of oxetan formation which involves a 1,4-biradical intermediate in which free rotation may occur before ring-closure.[45] There have been numerous other reports of oxetan formation. Keten imines,[46] allenes,[47] and furan[48] have been employed as the olefin component of the reaction, and novel carbonyl components

[42] H.-D. Becker, *J. Org. Chem.*, **32**, 2115, 2124, 2131, 2136, 2140 (1967).
[43] N. C. Yang, D. S. McClure, S. L. Murov, J. J. Houser, and R. Dusenbery, *J. Am. Chem. Soc.*, **89**, 5466 (1967).
[44a] *Organic Reaction Mechanisms*, **1966**, 375 (Footnote, and ref. 26).
[44b] P. J. Wagner, *J. Am. Chem. Soc.*, **89**, 2503 (1967).
[45] N. J. Turro, P. Wriede, J. C. Dalton, D. Arnold, and A. Glick, *J. Am. Chem. Soc.*, **89**, 3950 (1967).
[46] L. A. Singer and G. A. Davis, *J. Am. Chem. Soc.*, **89**, 158, 598, 941 (1967).
[47] H. Hogeveen and P. J. Smit, *Rec. Trav. Chim.*, **85**, 1188 (1966); H. Gotthardt, R. Steinmetz, and G. S. Hammond, *Chem. Commun.*, **1967**, 480.
[48] M. Ogata, H. Watanabe, and H. Kanō, *Tetrahedron Letters*, **1967**, 533; G. R. Evanega and E. B. Whipple, *ibid.*, p. 2163; C. Rivas and E. Payo, *J. Org. Chem.*, **32**, 2918 (1967).

include acetyl cyanide,[49] diethyl oxomalonate,[50] ethyl cyanoformate,[51] and vinylene carbonate (1,3-dioxol-2-one) (**20**).[52] Photolysis of 9-anthraldehyde with tetramethylethylene gives the expected oxetan, but affords an example of non-linear Stern–Volmer behaviour when di-*tert*-butyl nitroxide is used as

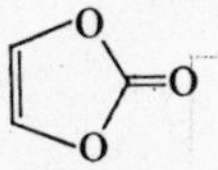

(**20**)

quencher.[53] One possible explanation is that here also, some oxetan is produced by reaction of the aldehyde singlet, and that the nitroxide not only quenches the triplets, but also promotes intersystem crossing of the excited singlets. Oxetan formation is also observed as one of many reaction paths in new studies of the photochemistry of quinones.[54]

Photolysis of thiobenzophenone in cyclohexene does not lead to a thietan, but to the dithiadecalin shown;[55a] thietan formation has, however, been claimed in the reaction with α-phellandrene.[55b]

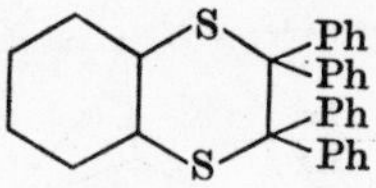

There have been numerous reports of studies of the photochemistry of cyclic ketones with an element of ring strain. In protic solvents cyclobutanones give, *inter alia*, tetrahydrofuran derivatives,[56a] and with methan[^{2}H]ol the product is that expected from the oxocarbene intermediate (**21**).[56b]

A similar carbene intermediate may be involved in the conversion of fenchone into the diol (**22**).[57]

49 Y. Shigemitsu, Y. Odaira, and S. Tsutsumi, *Tetrahedron Letters*, **1967**, 55.
50 M. Hara, Y. Odaira, and S. Tsutsumi, *Tetrahedron Letters*, **1967**, 2981.
51 Y. Odaira, T. Shimodaira, and S. Tsutsumi, *Chem. Commun.*, **1967**, 757.
52 S. Farid, D. Hess, and C. H. Krauch, *Chem. Ber.*, **100**, 3266 (1967).
53 N. C. Yang, R. Loeschen, and D. Mitchell, *J. Am. Chem. Soc.*, **89**, 5465 (1967).
54 J. M. Bruce, D. Creed, and J. N. Ellis, *J. Chem. Soc., C*, **1967**, 1486; J. A. Barltrop and B. Hesp, *ibid.*, p. 1625; S. P. Pappas and B. C. Pappas, *Tetrahedron Letters*, **1967**, 1597; J. Petránek, O. Ryba, and D. Doskočilová, *Coll. Czech. Chem. Commun.*, **32**, 2140 (1967).
55a G. Tsuchihashi, M. Yamauchi, and M. Fukuyama, *Tetrahedron Letters*, **1967**, 1971; see also H. C. Heller, *J. Am. Chem. Soc.*, **89**, 4288 (1967).
55b Y. Omote, M. Yoshioka, K. Yamada, and N. Sugiyama, *J. Org. Chem.*, **32**, 3676 (1967).
56a H. V. Hostettler, *Helv. Chim. Acta*, **49**, 2417 (1966).
56b N. J. Turro and R. M. Southam, *Tetrahedron Letters*, **1967**, 545.
57 P. Yates and A. G. Fallis, *Tetrahedron Letters*, **1967**, 4621.

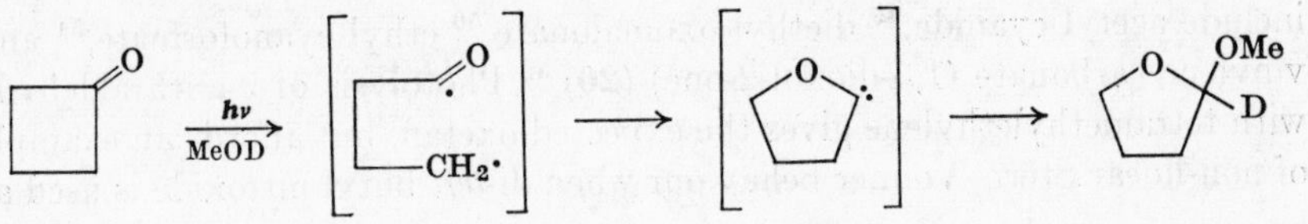

(21)

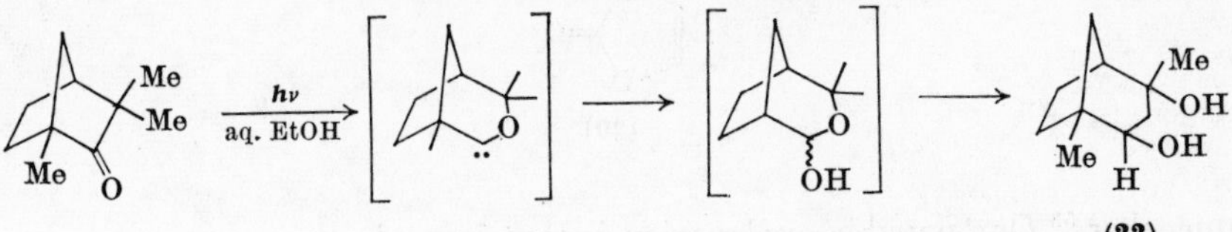

(22)

Mercury-sensitized photo-decarbonylations of norcamphor[58a] and of the bicyclo[2.1.1]hexanone (**23**)[58b] proceed as shown, the latter affording a route to bicyclo[1.1.1]pentane.

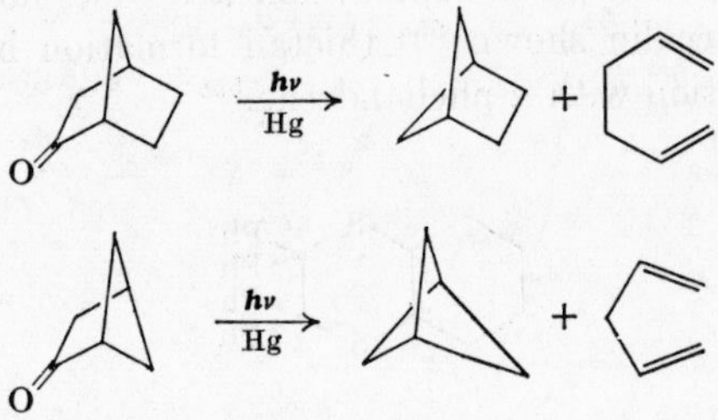

(23)

From the relative ease of photocleavage of a series of strained cycloalkenones, Matsui[59] has suggested that two mechanisms are operative. When a β-hydrogen atom is orientated suitably for effective overlap with the excited carbonyl group (as in **24**), then rapid photoreaction is observed, and it is proposed that hydrogen transfer and carbon–carbon cleavage are concerted.

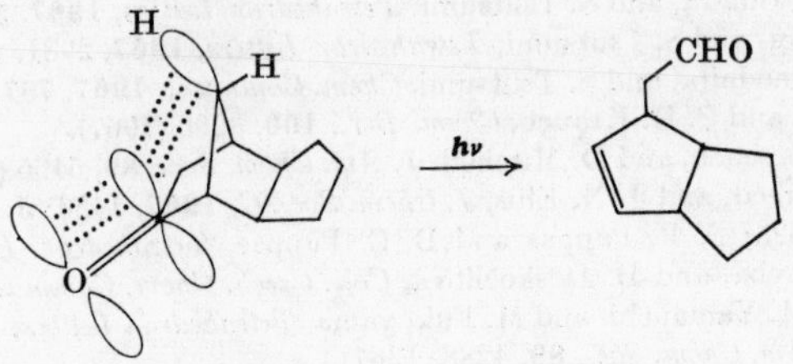

(24)

[58a] J. E. Baldwin and J. E. Gano, *Tetrahedron Letters*, **1967**, 2099.
[58b] J. Meinwald, W. Szkrybalo, and D. R. Dimmel, *Tetrahedron Letters*, **1967**, 731.
[59] T. Matsui, *Tetrahedron Letters*, **1967**, 3761.

When this orientation is absent, then hydrogen transfer in a biradical is considered to lead to a slower photoreaction, presumably because of competing ring closure back to starting ketone. A concerted transfer, though this time not to carbonyl carbon, may also be responsible for the observed stereospecificity of transfer of the *exo*-hydrogen in the photoreaction of carvonecamphor (**25**) in methanol.[60] The oxime of camphor itself undergoes N–O

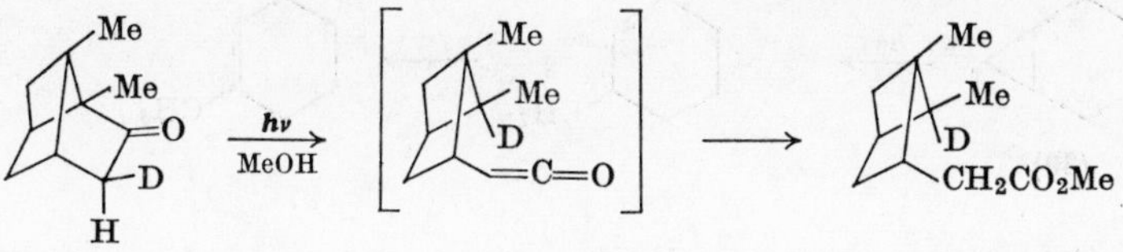

(**25**)

bond homolysis, then ring opening, as shown.[61] Its behaviour on pyrolysis is similar, unlike that of the parent ketone for which pyrolysis breaks the 1,7-bond, but photolysis cleaves the 1,2-bond. Radical (**26**) was likened to the ketone excited state.

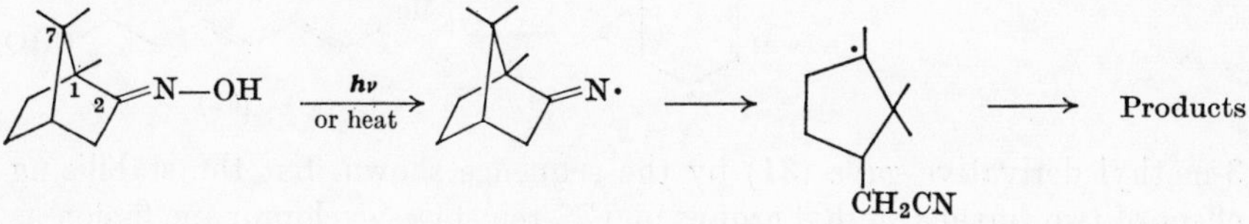

(**26**)

Substantial similarities have been found between the gas-phase photochemistry of a series of unsaturated and cyclopropyl-substituted ketones.[62] For example, both (**27**) and (**28**) decarbonylate with high quantum yield. The

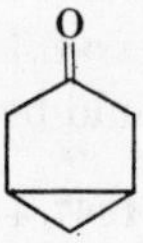

(**27**)

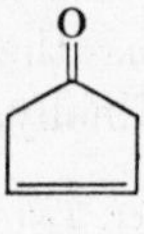

(**28**)

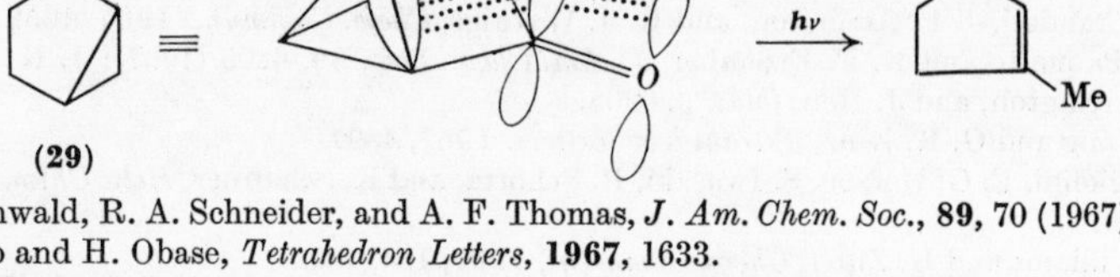

(**29**)

[60] J. Meinwald, R. A. Schneider, and A. F. Thomas, *J. Am. Chem. Soc.*, **89**, 70 (1967).
[61] T. Sato and H. Obase, *Tetrahedron Letters*, **1967**, 1633.
[62] L. D. Hess and J. N. Pitts, *J. Am. Chem. Soc.*, **89**, 1973 (1967); L. D. Hess, J. L. Jacobson, K. Schaffner, and J. N. Pitts, *ibid.*, p. 3684.

preference for opening of an external rather than an internal double bond in (**29**) is probably associated with the more favourable overlap, as shown. Similar considerations relate to the solution photochemistry of the bicycloheptanone (**30**).[63] In this case, the effect of substituents was also probed. The

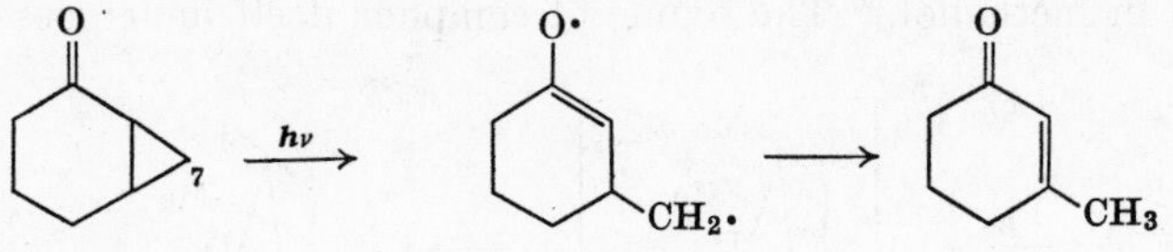

(**30**)

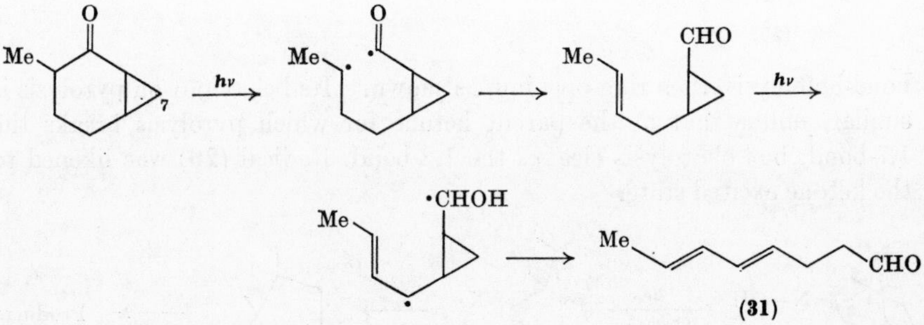

(**31**)

3-methyl derivative gave (**31**) by the sequence shown, but the stabilizing effect of two further methyl groups on $C_{(7)}$ reinstates cyclopropane fission as a major reaction pathway. Other examples of photolysis of cyclopropyl ketones to unsaturated isomers have also been noted.[64]

Miscellaneous rearrangements of unsaturated carbonyl compounds are illustrated with appropriate references (65—72) in the formulae on p. 383. Tentative mechanisms are also indicated. Schneider and Meinwald,[72] and Yang and Thap,[73] also discuss the failure of *s-cis*-α,β-unsaturated ketones to undergo photoenolization. Finally, mention should be made of papers dealing

63 W. G. Dauben and G. W. Shaffer, *Tetrahedron Letters*, **1967**, 4415.
64 B. Beugelmans, *Bull. Soc. Chim. France*, **1967**, 244; R. E. K. Winter and R. F. Lindauer, *Tetrahedron Letters*, **1967**, 2345.
65 R. G. Carlson and J. H. Bateman, *Tetrahedron Letters*, **1967**, 4151.
66 D. E. Bays and R. C. Cookson, *J. Chem. Soc., B*, **1967**, 226.
67 W. F. Erman and H. C. Kretschmar, *J. Am. Chem. Soc.*, **89**, 3842 (1967).
68a K. J. Crandall, J. P. Arrington, and R. J. Watkins, *Chem. Commun.*, **1967**, 1052.
68b L. A. Paquette and R. F. Eizember, *J. Am. Chem. Soc.*, **89**, 6205 (1967); J. K. Crandall, J. P. Arrington, and J. Hen, *ibid.*, p. 6208.
69 N. C. Yang and G. R. Lenz, *Tetrahedron Letters*, **1967**, 4897.
70 E. Baggiolini, E. G. Herzog, S. Iwasaki, R. Schorta, and K. Schaffner, *Helv. Chim. Acta*, **50**, 297 (1967).
71 J. R. Williams and H. Ziffer, *Chem. Commun.*, **1967**, 194, 469.
72 R. A. Schneider and J. Meinwald, *J. Am. Chem. Soc.*, **89**, 2023 (1967).
73 N. C. Yang and D.-M. Thap, *J. Org. Chem.*, **32**, 2462 (1967).

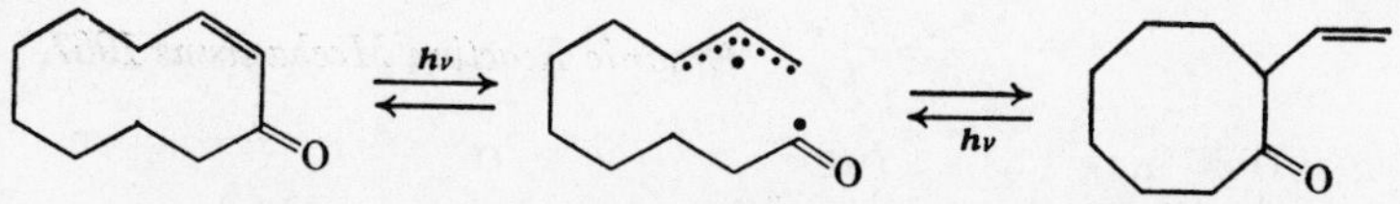

(ref. 65)

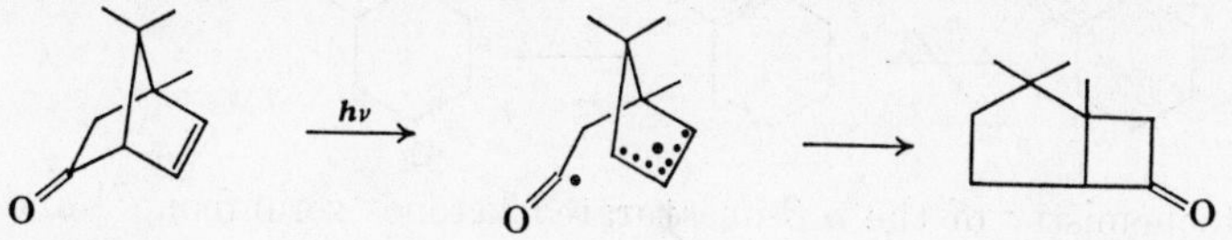

(ref. 66)

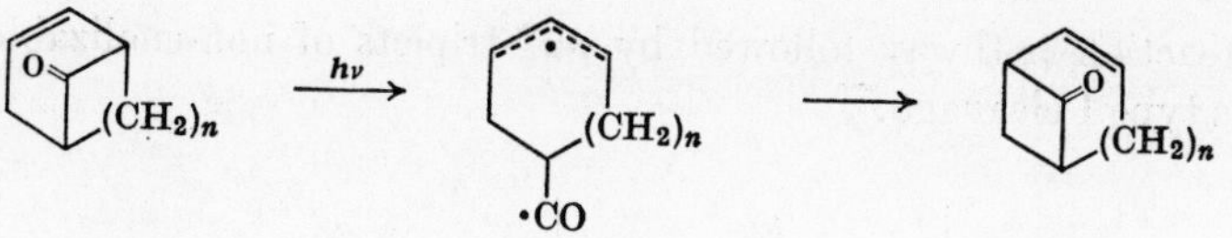

(ref. 67)

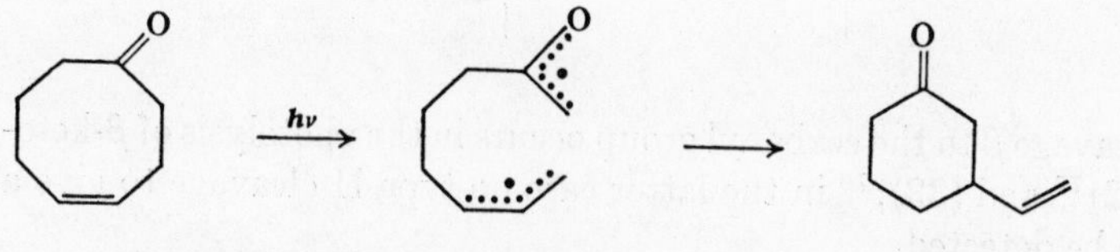

(ref. 68a)

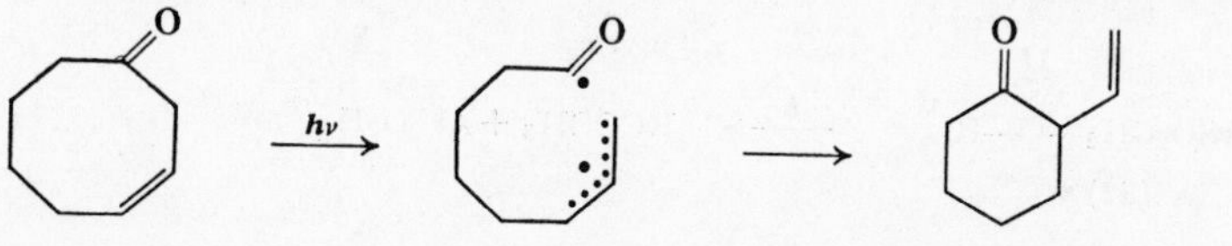

(ref. 68b)

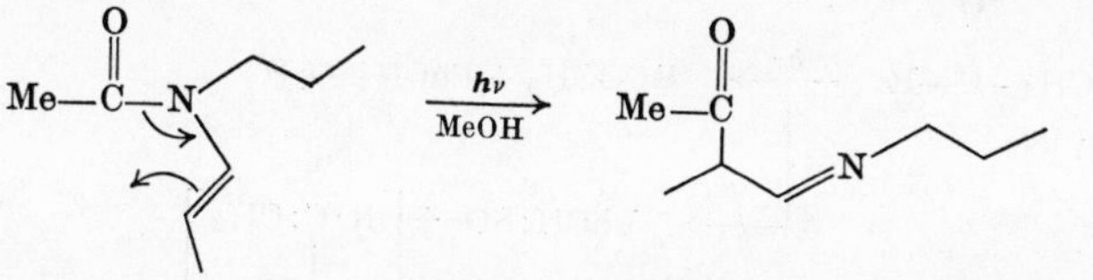

(ref. 69)

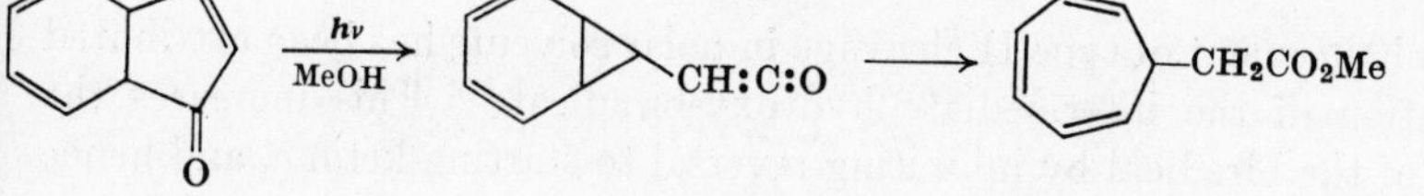

(ref. 70)

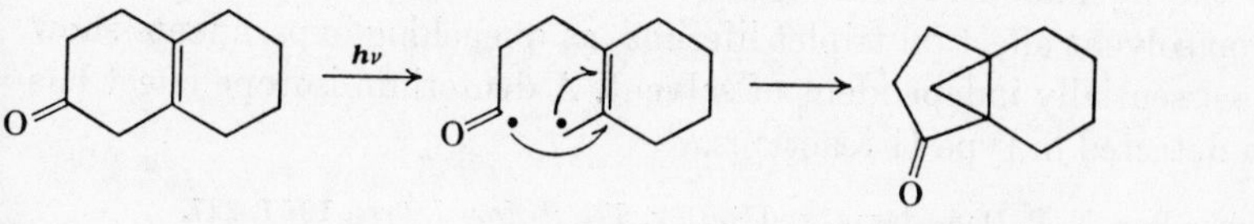

(ref. 71)

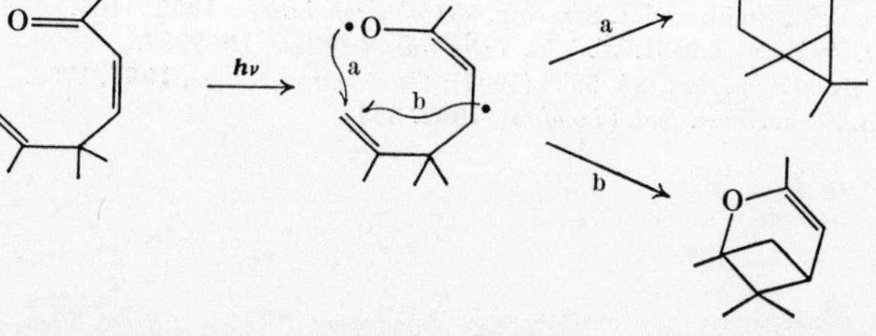

(ref. 72)

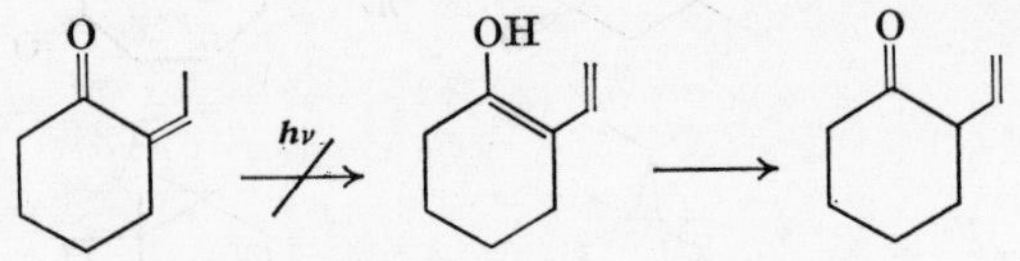

with the photochemistry of the α,β-unsaturated ketones zerumbone[74] and verbenone.[75]

The major reaction pathway followed by $n\pi^*$-triplets of non-enolizable 1,3-diketones is type I cleavage:[76]

$$\text{—CO—}\overset{|}{\underset{|}{\text{C}}}\text{—CO—} \xrightarrow{h\nu} \text{—CO}\cdot + \cdot\overset{|}{\underset{|}{\text{C}}}\text{—CO—}$$

Radical cleavage β to the carbonyl group occurs in the photolysis of β-keto-sulphones (**32**)[77] and (**33**);[78] in the latter case no type II cleavage to give a sulfene could be detected.

$$ArSO_2CH_2\text{—}\overset{O}{\overset{\|}{C}}\text{—}R \xrightarrow{h\nu} RCOCH_3 + ArSO_2H$$

(**32**)

$$PhCH_2SO_2CH_2\text{—}\overset{O}{\overset{\|}{C}}\text{—}R \xrightarrow{h\nu} RCOCH_3 + PhCH_2SO_2H$$

(**33**)

$$PhCH_2SO_2CH_2\text{—}\overset{O}{\overset{\|}{C}}\text{—}R \not\xrightarrow{h\nu} PhCH{:}SO_2 + \left[\begin{array}{c} R\cdot C{:}CH_2 \\ | \\ OH \end{array}\right]$$

The enhancement of type II cleavage in polar solvents has been attributed to solvation of the intermediate hydroxy-biradical.[79a] This increases the lifetime of the biradical by inhibiting reversal to starting ketone, and hence increases the likelihood of cleavage. This is illustrated for valerophenone. There is no solvent effect on triplet lifetime, as quenching experiments show this to be essentially independent of solvent. A deuterium isotope effect has also been detected in type II reactions.[79b]

[74] H. N. Subba Rao, N. P. Damodaran, and S. Dev, *Tetrahedron Letters*, **1967**, 227.
[75] W. F. Erman, *J. Am. Chem. Soc.*, **89**, 3828 (1967).
[76] H. Nozaki, Z. Yamaguti, T. Okada, R. Noyori, and M. Kawanisi, *Tetrahedron*, **23**, 3993 (1967).
[77] A. M. van Leusen, P. M. Smid, and J. Strating, *Tetrahedron Letters*, **1967**, 1165.
[78] C. L. McIntosh, P. de Mayo, and R. W. Yip, *Tetrahedron Letters*, **1967**, 37.
[79a] P. J. Wagner, *J. Am. Chem. Soc.*, **89**, 5898 (1967); *Tetrahedron Letters*, **1967**, 1753.
[79b] C. Djerassi and B. Zeek, *Chem. Ind. (London)*, **1967**, 358.

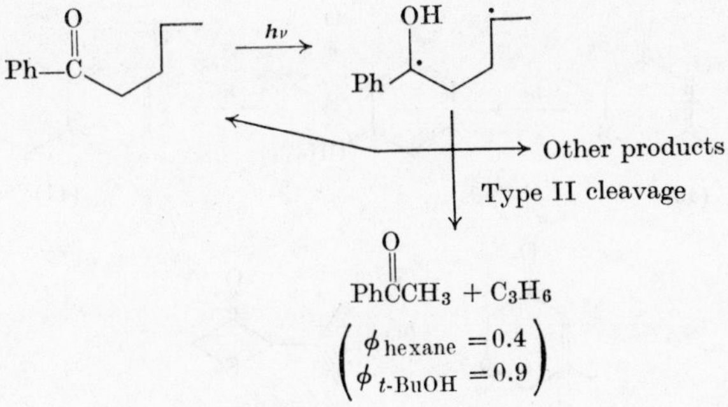

The diketone (**34**), on photolysis in methanol, gives tetramethylindanone and the homolytic methanol adduct (**35**).[80] Surprisingly, a significant product

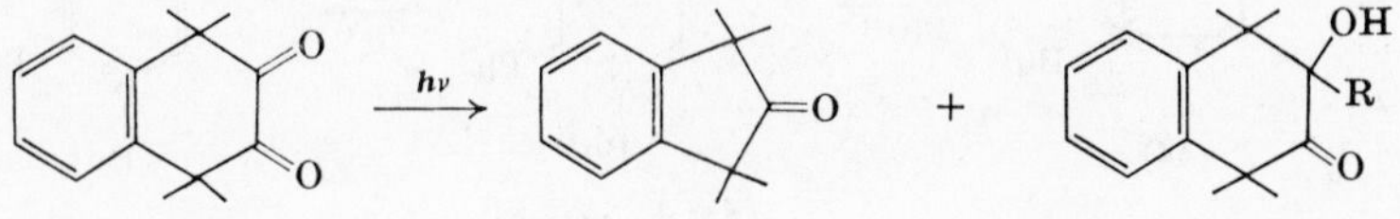

(**34**)

(**35**) $R = CH_2OH$

(**36**) $R = H$

in benzene is (**36**). A marked solvent polarity effect is observed in the proportions of (**38**) and (**39**) obtained from the (direct or sensitized) photolysis of (**37**). In non-polar solvents the product is almost exclusively (**38**).[81]

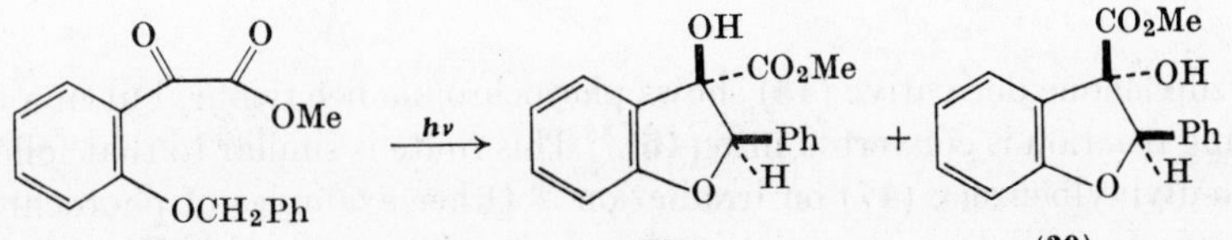

(**37**) (**38**) (**39**)

Work on the photochemistry of keto-sulphides includes observation of the rearrangement (**40**) → (**41**). The proposed two-step sequence shown is supported by isolation of (**43**) from (**42**).[82] The transformation of keto-amine (**44**) appears to involve a novel migration to carbonyl-carbon. Ionic and radical intermediates were discussed.[83]

80 G. E. Gream, J. C. Paice, and C. C. R. Ramsay, *Australian J. Chem.*, **20**, 1671 (1967).
81 S. P. Pappas, B. C. Pappas, and J. E. Blackwell, *J. Org. Chem.*, **32**, 3066 (1967).
82 W. C. Lumma and G. A. Berchtold, *J. Am. Chem. Soc.*, **89**, 2761 (1967); P. Y. Johnson and G. A. Berchtold, *ibid.*, p. 2761.
83 A. Padwa and L. Hamilton, *J. Am. Chem. Soc.*, **89**, 3077 (1967).

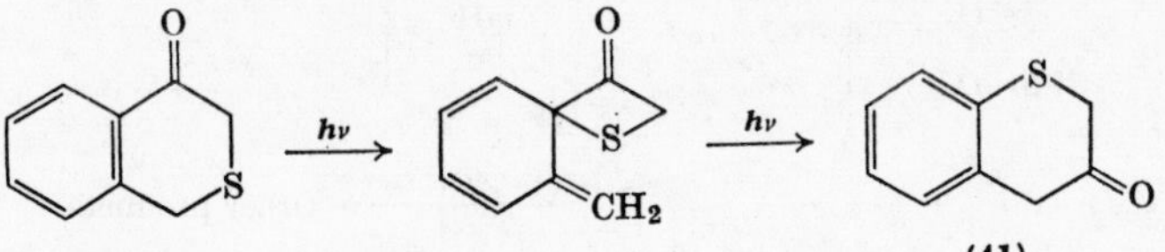

(40) **(41)**

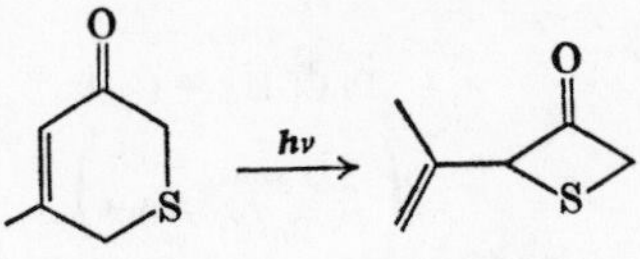

(42) **(43)**

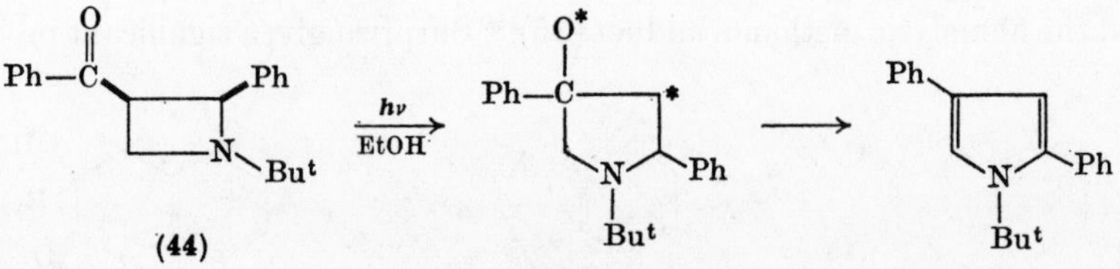

(44)

(* = +, −, or ·)

A type of photo-Tishchenko reaction is observed on photolysis of *o*-phthalaldehyde in carbon tetrachloride; phthalide is obtained.[84] The related

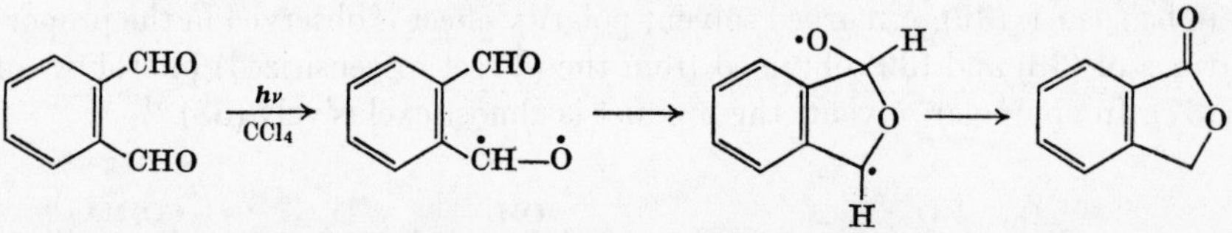

benzophenone derivative (**45**) shows photochromic behaviour, but in a competing reaction is converted into (**46**).[85] This route is similar to that followed by *o*-divinylbenzene (**47**) on irradiation.[86] Other examples of photochromic behaviour, including photoenolization, have also appeared.[87] The hindered ketone (**48**) does not photoenolize, but instead forms a cyclobutenol,[88] and

[84] J. Kagan, *Tetrahedron Letters*, **1966**, 6097.
[85] K. R. Huffman and E. F. Ullman, *J. Am. Chem. Soc.*, **89**, 5629 (1967).
[86] M. Pomerantz, *J. Am. Chem. Soc.*, **89**, 694 (1967); J. Meinwald and P. H. Mazzocchi, *ibid.*, p. 696.
[87] E. Fischer, *Fortschr. Chem. Forsch.*, **7**, 605 (1967); E. F. Ullman and W. A. Henderson, *J. Am. Chem. Soc.*, **89**, 4390 (1967); R. S. Becker and W. F. Richey, *ibid.*, 1298; G. Kortum and K.-W. Koch, *Chem. Ber.*, **100**, 1515 (1967); J. R. Huber, U. Wild, and H. H. Günthard, *Helv. Chim. Acta*, **50**, 589, 841 (1967); D. Leupold, H. Kobischke, and U. Geske, *Tetrahedron Letters*, **1967**, 3287; G. Löber, *Z. Phys. Chem.* (*Frankfurt*), **54**, 73 (1967); J. Lemaire, *J. Phys. Chem.*, **71**, 2653 (1967); M. Ottolenghi and D. S. McClure, *J. Chem. Phys.*, **46**, 4613, 4620 (1967); T. R. Evans, A. F. Toth, and P. A. Leermakers, *J. Am. Chem. Soc.*, **89**, 5060 (1967); see also reference 139.
[88] T. Matsuura and Y. Kitaura, *Tetrahedron Letters*, **1967**, 3309.

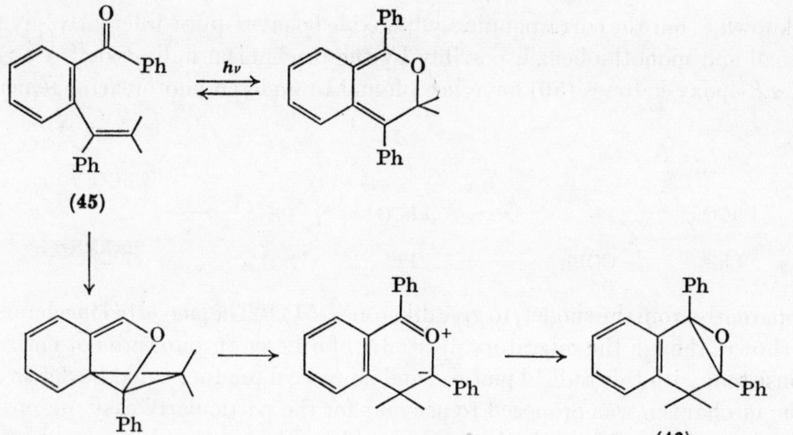

(45) (46)

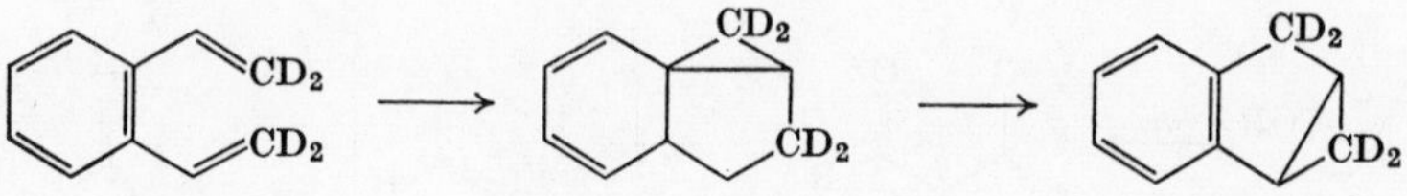

(47)

the interesting observation has also been made that the hindered phenol (**49**) undergoes ready photochemical de-*tert*-butylation.[89] The corresponding methoxy-compound is photostable.

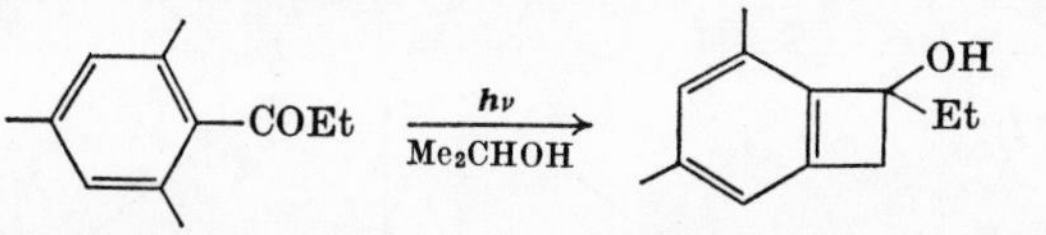

(**48**)

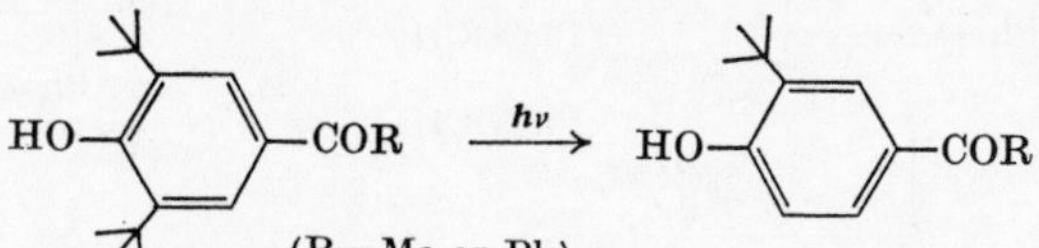

(R = Me or Ph)

(**49**)

Compounds with Three-membered Rings or Olefinic Unsaturation

The photodeamination of certain substituted aziridines[90] has now been reported in detail.[91] Extrusion of sulphur from dibenzoylstilbene episulphide

[89] T. Matsuura and Y. Kitaura, *Tetrahedron Letters*, **1967**, 3311.
[90] See *Organic Reaction Mechanisms*, **1965**, 296.
[91] A. Padwa and L. Hamilton, *J. Am. Chem. Soc.*, **89**, 102 (1967).

is known,[90] but the corresponding sulphoxide behaves quite differently, giving benzil and monothiobenzil, possibly by the mechanism indicated.[92] A series of α,β-epoxy-ketones (**50**) have been found to undergo photorearrangement,

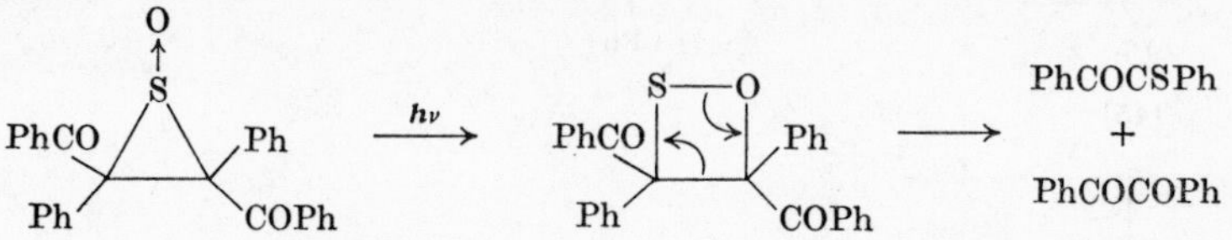

apparently from the singlet, to give diketones (**51**).[93] The suggested mechanism is shown, though the migratory aptitudes of different groups are not entirely consistent with this radical picture, and no radical product could be detected. The mechanism was proposed to account for the particularly easy migration of β-benzyl and β-benzyhydryl groups. Photolysis of the β,γ-epoxy-ketone

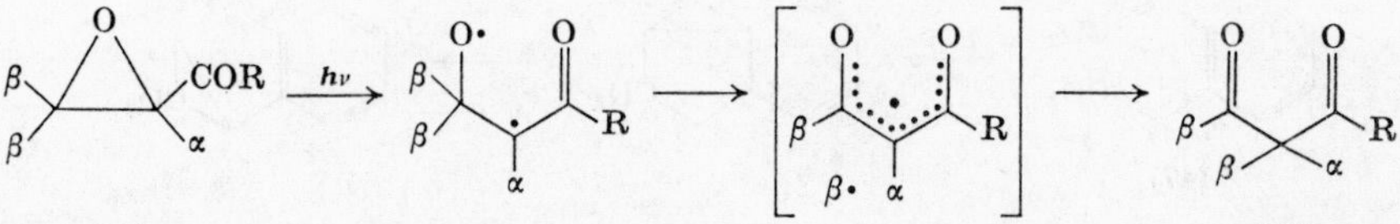

(**50**) (**51**)

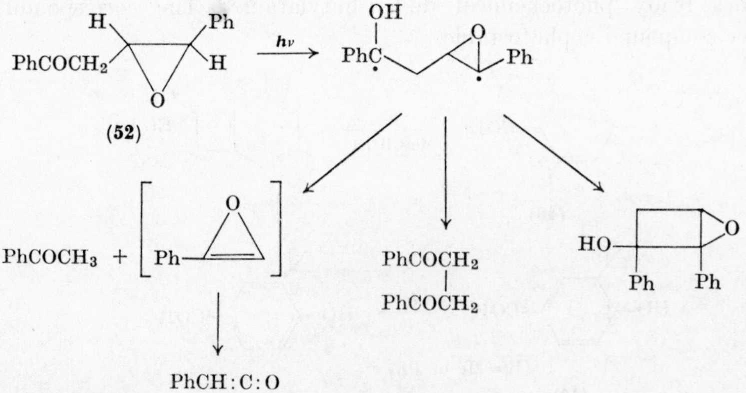

(**52**)

(**52**) gives the products shown[94] via the $n\pi^*$-triplet; and photolysis of styrene oxide in neutral ethanol gives β-ethoxyphenethyl alcohol. In basic solution, the α-phenyl derivative is produced.[95]

Photolysis of the diene (**53**) gives a vinylcyclopropane (**54**) in a singlet-state

[92] D. C. Dittmer, G. C. Levy, and G. E. Kuhlmann, *J. Am. Chem. Soc.*, **89**, 2793 (1967).
[93] C. S. Markos and W. Reusch, *J. Am. Chem. Soc.*, **89**, 3363 (1967).
[94] A. Padwa, D. Crumrine, R. Hartman, and R. Layton, *J. Am. Chem. Soc.*, **89**, 4435 (1967).
[95] K. Tokumaru, *Bull. Chem. Soc. Japan*, **40**, 242 (1967).

process. Interaction between the diene and the aromatic ring is evident from the spectrum of (**53**).[96] Further photolysis of (**54**) gives new products. Isomeric biradicals are proposed as intermediates in the two reactions. In each case, the failure to detect cyclopentenes is attributed to a *trans*-configuration of the allylic radical. Phenyl migration and ring-closure to a cyclopropane have also been observed on irradiation of latifolin.[97]

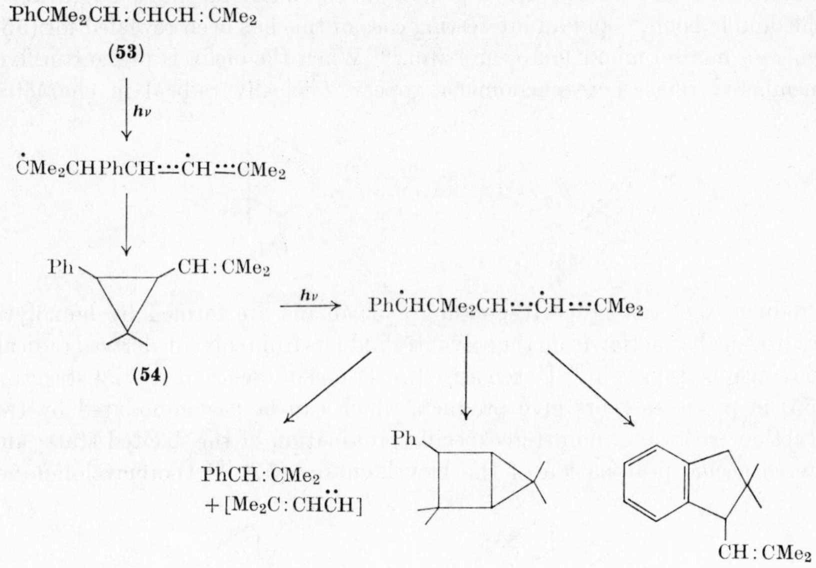

Just as styrene oxide adds alcohols,[95] so photolysis of phenylcyclopropane in protic solvents leads to heterolytic addition as shown.[98] This is suggestive of an ionic opening of the cyclopropane ring and appears to involve a singlet

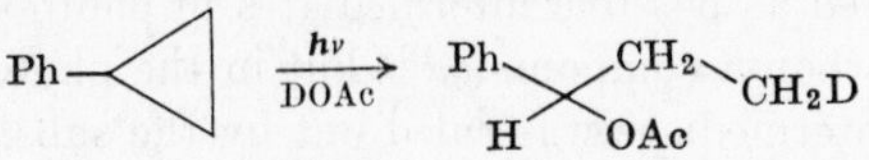

excited state. It is reminiscent of the photoaddition of alcohols to cyclohexenes,[99] though that reaction is benzene-sensitized. A new example of that type of reaction, which may involve a transient *trans*-cyclohexene, is shown for (**55**);[100] it is accompanied by a photochemical allylic rearrangement. The

[96] H. Kristinsson and G. S. Hammond, *J. Am. Chem. Soc.*, **89**, 5968, 5970 (1967).
[97] D. Kumari and S. K. Mukerjee, *Tetrahedron Letters*, **1967**, 4169.
[98] C. S. Irving, R. C. Petterson, I. Sarkar, H. Kristinsson, C. S. Aaron, G. W. Griffin, and G. J. Boudreaux, *J. Am. Chem. Soc.*, **88**, 5675 (1966).
[99] See *Organic Reaction Mechanisms*, **1966**, 392, and P. J. Kropp and H. J. Krauss, *J. Am. Chem. Soc.*, **89**, 5199 (1967).
[100] J. Pusset and R. Beugelmans, *Tetrahedron Letters*, **1967**, 3249.

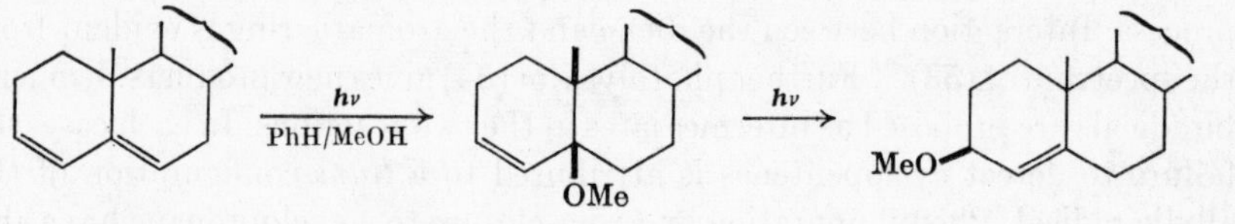

(55)

major reaction in the absence of primary or secondary alcohol is migration of the double bond,[99] and an interesting case of this has been revealed for (**56**), where a methyl migration occurs also.[101] When the olefin is present in five-membered rings, new phenomena appear, typically radical in character.

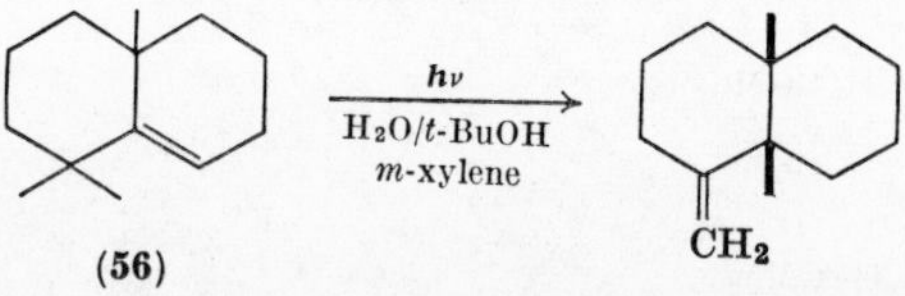

(56)

Products such as the corresponding cycloalkane are formed by homolytic hydrogen abstraction from the solvent. Products from solvent-derived radicals may also be isolated.[102] Unsensitized reactions of dienes of partial structure (**55**) in protic solvents give products which can be accommodated by two reaction pathways: non-stereospecific protonation of the excited state, and stereospecific protonation of the bicyclobutane (**57**).[103] Oxabicyclobutanes

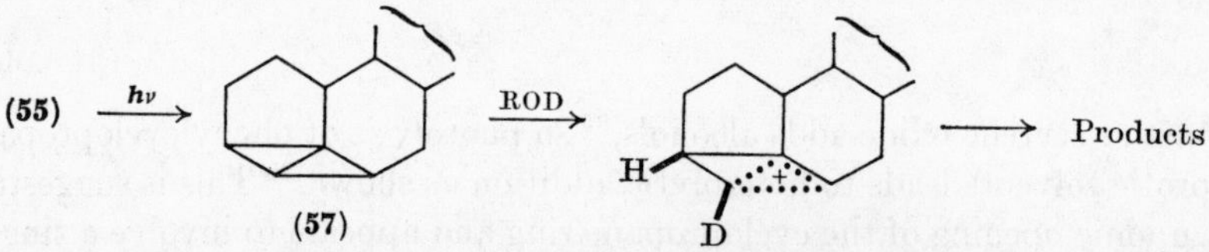

(57)

have been considered as possible intermediates in photo-additions to cyclic α,β-unsaturated carbonyl compounds,[104] but in the photorearrangement of (**58**) this type of intermediate was ruled out by the substitution pattern of the product.[105] An alternative rationalization of addition of ROH to cyclohexenes conjugated to carbonyl either in or out of the ring would involve a *trans*-cyclohexene derivative.[104, 106, 107] Interestingly, a typically radical reaction is found in the case of the five-membered ring compound (**59**).[108]

[101] J. A. Marshall and A. R. Hochstetler, *Chem. Commun.*, **1967**, 732.
[102] P. J. Kropp, *J. Am. Chem. Soc.*, **89**, 3650 (1967).
[103] G. Bauslaugh, G. Just, and E. Lee-Ruff, *Can. J. Chem.*, **44**, 2837 (1966).
[104] B. J. Ramey and P. D. Gardner, *J. Am. Chem. Soc.*, **89**, 3949 (1967).
[105] O. L. Chapman and W. R. Adams, *J. Am. Chem. Soc.*, **89**, 4243 (1967).
[106] P. J. Kropp and H. J. Krauss, *J. Org. Chem.*, **32**, 3222 (1967).
[107] P. de Mayo and J. S. Wasson, *Chem. Commun.*, **1967**, 970.
[108] P. Bladon and I. A. Williams, *J. Chem. Soc., C*, **1967**, 2032.

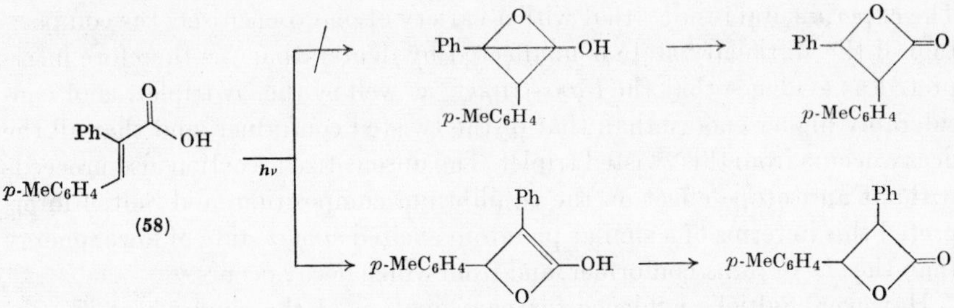

(58)

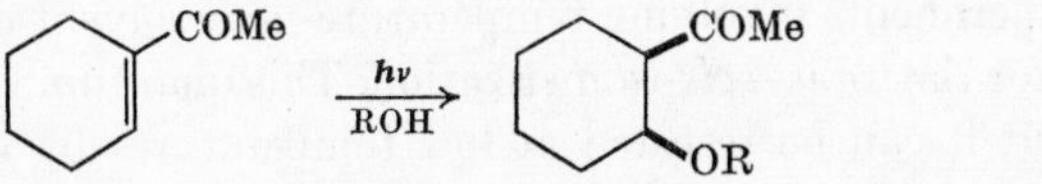

(ref. 104)

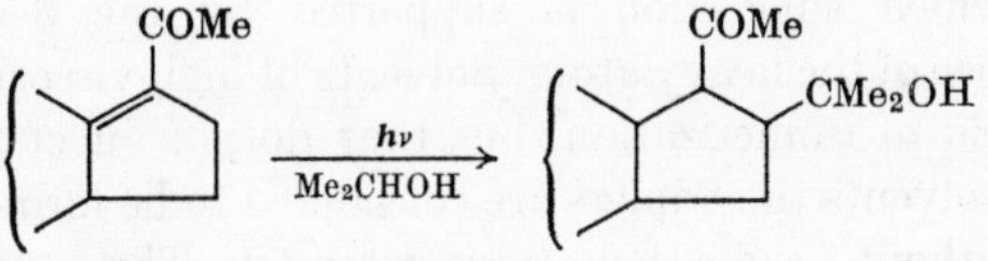

(ref. 108)

(59)

Intramolecular photoaddition is observed on irradiation of *o*-allylphenols. The proportions of five- and six-membered cyclic ethers that are formed have been found to be comparable to those from the acid-catalysed reaction.[109]

Some interesting new reports relate to the geometrical photoisomerization of stilbene. Self-consistent field M.O. calculations[110] amply confirm the idea of a "phantom triplet", i.e., a non-spectroscopic triplet (with twisted geometry) of lower energy than that of the triplet corresponding to the isomer from which it was derived. Saltiel has presented further experimental evidence bearing on this.[111] In reliance on the theoretical prediction that the extent to which deuteration will reduce the rate of radiationless decay of a triplet state to the ground-state depends on the energy separation between the two states, coupled with the supposedly similar energies of *trans*- and phantom triplets but higher energy of the *cis*-triplet, it was predicted that the photostationary state of perdeuteriostilbene would contain more *cis*-isomer than the perhydro-compound does. This is because decay could occur from both *trans*- and phantom triplets, but in the latter case there is only a very small energy drop to the highly energetic twisted conformer of the ground state. For this component of the radiationless decay, deuteration should have only a small effect.

109 G. Fráter and H. Schmid, *Helv. Chim. Acta,* **50**, 255 (1967); W. M. Horspool and P. L. Pauson, *Chem. Commun.*, **1967**, 195.
110 P. Borrell and H. H. Greenwood, *Proc. Roy. Soc.* (*London*), *A*, **298**, 453 (1967).
111 J. Saltiel, *J. Am. Chem. Soc.*, **89**, 1036 (1967).

The experimental result – that with a variety of photosensitisers the composition of the stationary state is unaffected by deuteration - is therefore interpreted as evidence that the *trans*-triplet, as well as the *cis*-triplet, is of considerably higher energy than that of the twisted conformer, and that all the decay occurs from this twisted triplet. The unsensitized reaction also proceeds without an isotope effect on the equilibrium composition, and Saltiel interpreted this in terms of a similar phantom excited *singlet* state of lower energy than the *cis*- or *trans*-conformer, and from which decay occurs.

However, Saltiel's evidence for participation of the singlet state in unsensitized photoisomerization[112] has been challenged[113] by the interpretation of a series of experiments involving temperature- and solvent-dependence of quantum yield for the *trans*→*cis*-isomerization. This quantum yield falls with temperature, but it can be restored at low temperatures by a heavy-atom solvent. This is considered to help overcome the small energy barrier for intersystem crossing, which suggestion is supported by the decreased fluorescence in the presence of the heavy atom. Solvents of high viscosity also reduce the quantum yield of isomerization, but they do not affect that of fluorescence. In viscous solvents the triplets are considered to be formed, but many of them decay without losing their *trans*-geometry. Thus, even the direct photoisomerization would appear to involve the phantom triplet.

Anomalous results have been obtained in stilbene isomerization photosensitized by some charged heterocycles (pyrylium and pyridinium salts of triplet energy *ca.* 50—60 kcal $mole^{-1}$).[114] The photostationary-state concentration of *trans*-isomer was greatly in excess of that predicted by the results of Hammond and Herkstroeter.[112] No explanation was offered.

Wagner[115] has found that the geometry of the most stable conformation of biphenyl is likewise different from that of its first excited triplet state.

The photoisomerization of some hydroxycinnamic acids[116] and the sensitized isomerization of azomethane[117] have been examined, and the interesting result has been obtained that a magnetic field catalyses the iodine-photosensitized isomerization of butenes.[118a] This results from magnetically induced predissociation of excited iodine molecules, to give atomic iodine. Other photoisomerizations of azo-compounds[118b] and also of nitroso-dimers[118c] have also been described.

112 See *Organic Reaction Mechanisms*, **1966**, 379.
113 K. A. Muszkat, D. Gegiou, and E. Fischer, *J. Am. Chem. Soc.*, **89**, 4814 (1967).
114 R. Searle, J. L. R. Williams, D. E. DeMeyer, and J. C. Doty, *Chem. Commun.*, **1967**, 1165.
115 P. J. Wagner, *J. Am. Chem. Soc.*, **89**, 2820 (1967).
116 G. Kahnt, *Phytochemistry*, **6**, 755 (1967).
117 P. S. Engel, *J. Am. Chem. Soc.*, **89**, 5731 (1967).
118a W. E. Falconer and E. Wasserman, *J. Chem. Phys.*, **45**, 1843 (1966).
118b H. van Zwet and E. C. Kooyman, *Rec. Trav. Chim.*, **86**, 993 (1967).
118c A. Mackor, T. A. J. W. Wajer, and T. J. de Boer, *Tetrahedron Letters*, **1967**, 2757; see also A. Bluhm and J. Weinstein, *Nature*, **215**, 1478 (1967).

New examples of photocyclization of stilbenes to phenanthrenes and of related photocyclizations continue to appear.[119] Irradiation of solutions of the cyclopentene (**60**) in the absence of an oxidizing agent gives substantial concentrations of the coloured ring-closed isomer (**61**).[120] This constitutes a

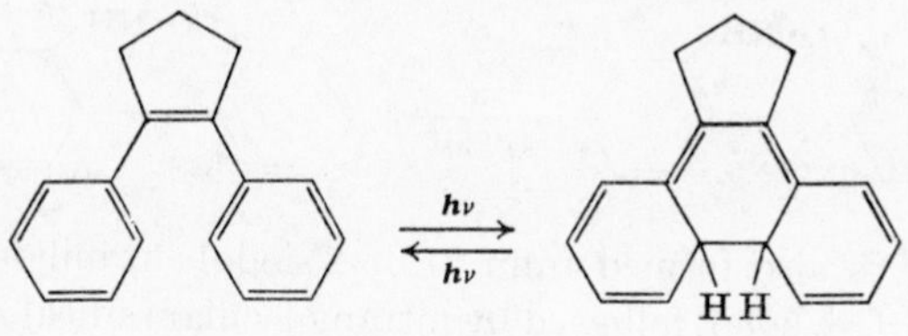

(**60**) (**61**)

particularly simple example of photochromic behaviour. Neither the forward nor the back reaction could be photosensitized.

Certain α,α'-dimethylstilbenes have been found to undergo competing oxidative cleavage to acetophenones when oxygen is present to oxidize the dihydrophenanthrene. However, this competition may be effectively suppressed by copper ions.[121]

The irradiation of tetracyclone (**62**) in oxygen-free propan-2-ol gives phenanthrene derivatives (**63**) and (**64**) with simultaneous reduction elsewhere in the molecule.[122] No oxidizing agent was added to photocyclization reactions of diphenylamine or diphenyl ether, though the carbazole and dibenzofuran formed were accompanied by decomposition products which

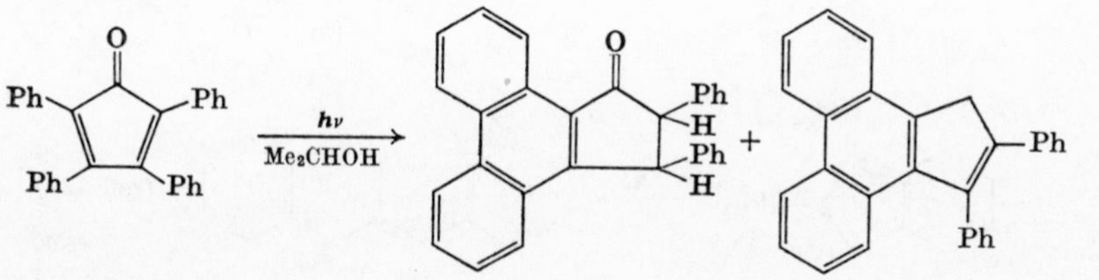

(**62**) (**63**) (**64**)

[119] M. Scholz, M. Mühlstädt, and F. Dietz, *Tetrahedron Letters*, **1967**, 665, 743; M. Scholz, F. Dietz, and M. Mühlstädt, *Z. Chem.*, **7**, 329 (1967); R. M. Kellogg, M. B. Groen, and H. Wynberg, *J. Org. Chem.*, **32**, 3093 (1967); C. E. Loader and C. J. Timmons, *J. Chem. Soc.*, *C*, **1967**, 1343; W. Carruthers, *ibid.*, p. 1525; E. J. Levi and M. Orchin, *J. Org. Chem.*, **31**, 4302 (1966); C. P. Joshua and G. E. Lewis, *Australian J. Chem.*, **20**, 929 (1967); G. E. Lewis and J. A. Reiss, *ibid.*, p. 1451; N. C. Jamieson and G. E. Lewis, *ibid.*, p. 321.

[120] K. A. Muszkat and E. Fischer, *J. Chem. Soc.*, *B*, **1967**, 662; a somewhat similar case is reported by H. Blaschke and V. Boekelheide, *J. Am. Chem. Soc.*, **89**, 2747 (1967).

[121] D. J. Collins and J. J. Hobbs, *Australian J. Chem.*, **20**, 1905 (1967).

[122] N. Toshima and I. Moritani, *Bull. Chem. Soc. Japan*, **40**, 1495 (1967); *Tetrahedron Letters*, **1967**, 357; I. Moritani, N. Toshima, S. Nakagawa, and M. Yakushiji, *Bull. Chem. Soc. Japan*, **40**, 2129 (1967).

presumably incorporated the lost hydrogen atoms. Diphenylmethane and diphenyl sulphide were inert.[123]

A further novel type of extension of the stilbene cyclization is to benzanilide, which on irradiation in benzene containing iodine gives phenanthridone.[124]

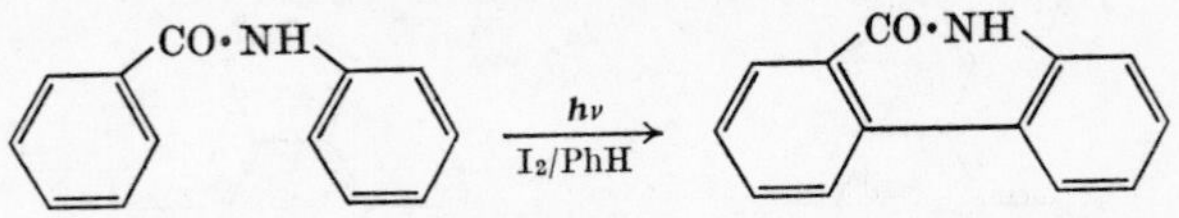

Phenanthridone is also formed from 2- or 2′-iodobenzanilide, probably by photolysis of the C–I bond, followed by intramolecular radical substitution.[124] Some other photocyclizations related to the above reactions are illustrated below, with references.

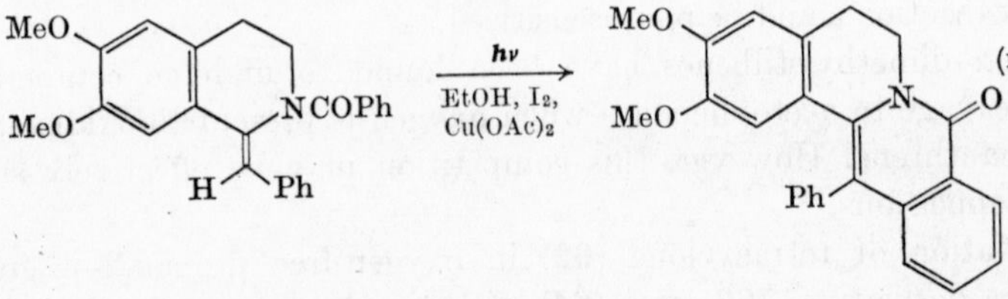

(ref. 125)

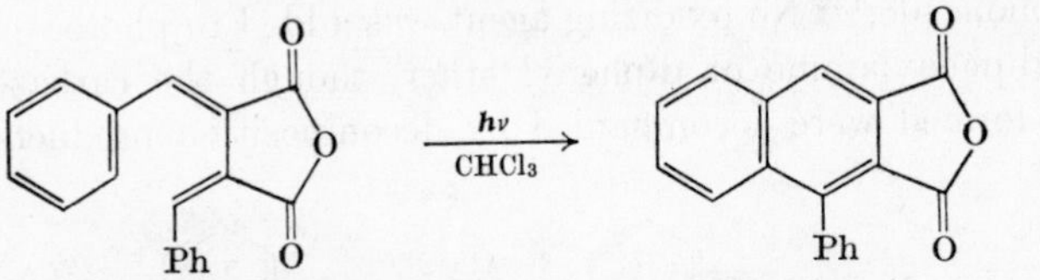

(ref. 126)

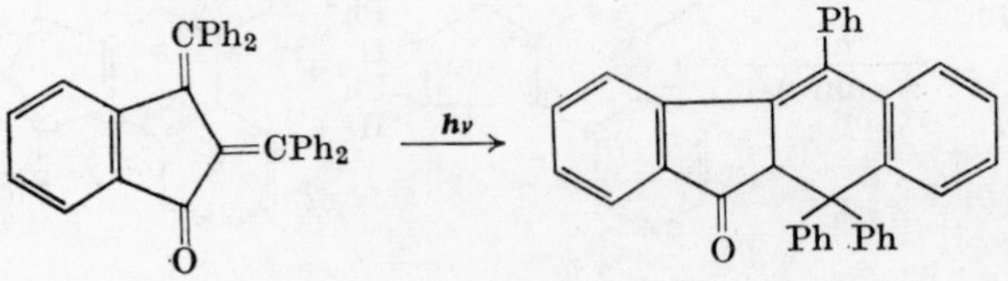

(ref. 127)

The photocyclization of the anilide (**65**) to give (**66**) under non-oxidizing conditions presumably involves hydrogen transfer in an initially formed intermediate (**67**). Deuterium-labelling experiments[128] showed this to be by a combination of 1,3-shifts; 5 → 3, and 5 → N, followed by N → 3.

123 H. Stegemeyer, *Naturwissenschaften*, **53**, 582 (1966).
124 B. S. Thyagarajan, N. Kharasch, H. B. Lewis, and W. Wolf, *Chem. Commun.*, **1967**, 614.
125 M. P. Cava and S. C. Havlicek, *Tetrahedron Letters*, **1967**, 2625.
126 B. Weinstein and D. N. Brattesani, *Chem. Ind.* (*London*), **1967**, 1292.
127 H. G. Heller, D. Auld, and K. Salisbury, *J. Chem. Soc.*, *C*, **1967**, 2457.
128 P. G. Cleveland and O. L. Chapman, *Chem. Commun.*, **1967**, 1064.

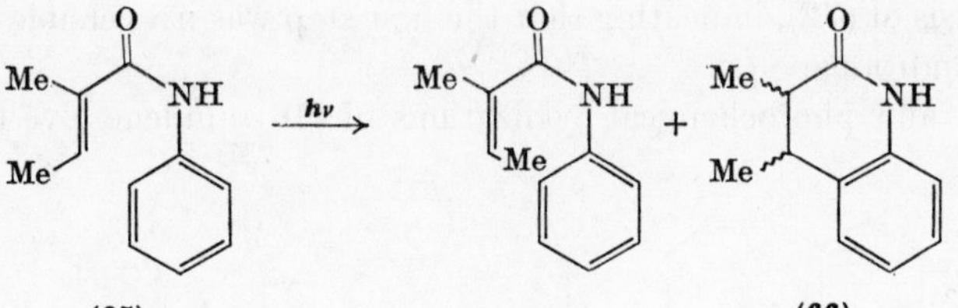

(65) (66)

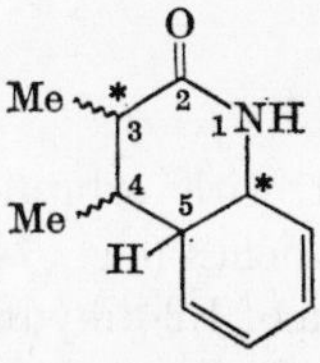

(67) (* = +, −, or ·)

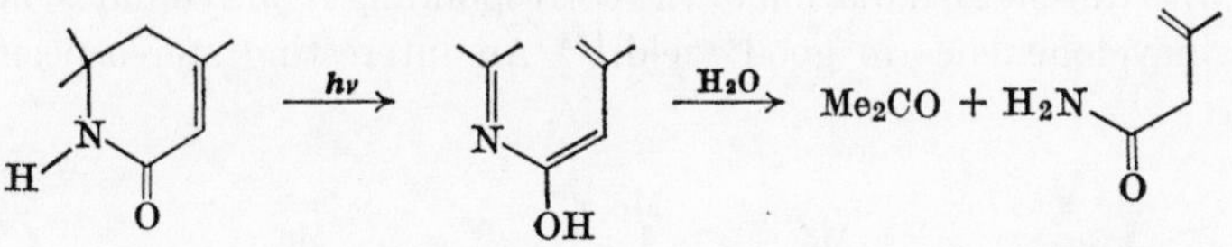

(68) (69) (70)

Photolysis of lactam (**68**) in water gives acetone and the amide (**70**), probably by hydrolysis of the ring-opened amide tautomer (**69**).[129] Flash-photolysis studies have shown that cyclo-octatriene undergoes photolysis to *cis,cis*-octatetraene, but that this product spontaneously recyclizes.[130] The photochemistry of the diazacyclohexadiene (**71**) has been found to give an imidazole derivative, probably after initial ring opening.[131] Deuterium was incorporated in the *N*-methyl group as predicted by this mechanism, and also at $C_{(2)}$ by exchange. There was no racemization of recovered starting material

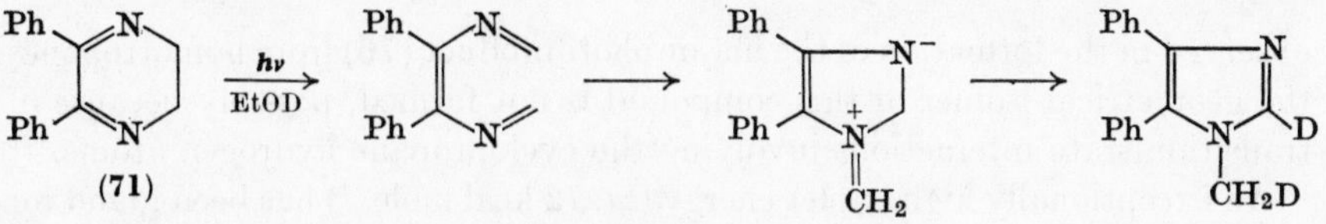

(71)

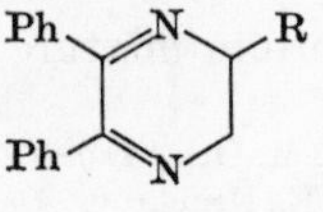

(72)

[129] E. Cavalieri and D. Gravel, *Tetrahedron Letters*, **1967**, 3973.
[130] T. D. Goldfarb and L. Lindqvist, *J. Am. Chem. Soc.*, **89**, 4588 (1967).
[131] P. Beak and J. L. Miesel, *J. Am. Chem. Soc.*, **89**, 2375 (1967).

on photolysis of (**72**), indicating that the first step was irreversible under the reaction conditions.

Thermal and photochemical cyclizations of [16]annulene give isomers of

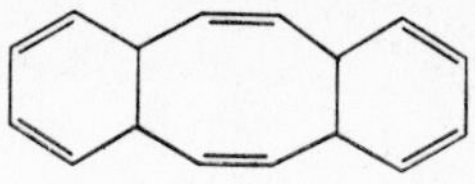

(**73**)

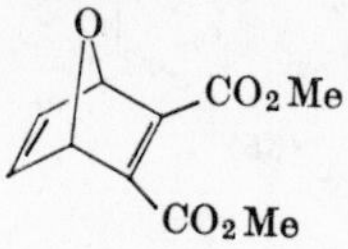

(**74**)

(**73**), with geometry consistent with orbital-symmetry considerations.[132] Photolysis of 7-oxabicycloheptadienes (e.g., **74**) has been found to produce oxaquadricyclanes.[133] Irradiation of 1,2-dihydrophthalic anhydride is known to afford a bicyclohexene derivative, but the only organic product isolated from (**75**) is durene: irradiation of the corresponding *N*-alkylimides, however, affords bicyclohexenes in good yield.[134] An interesting stereospecificity is

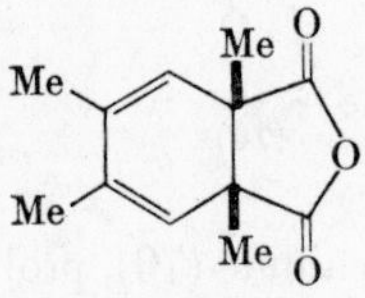

(**75**)

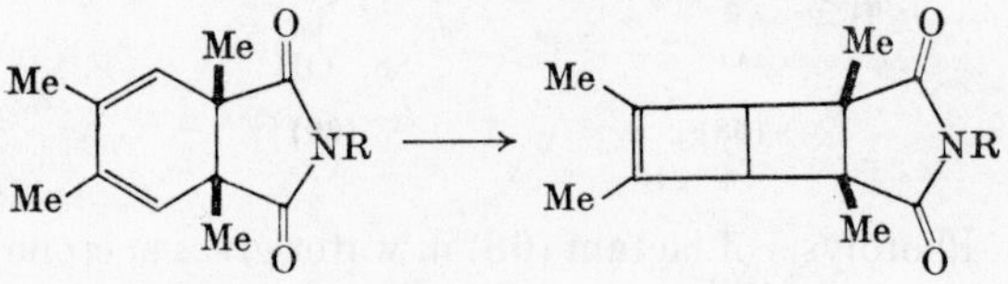

(*cis* + *trans*)

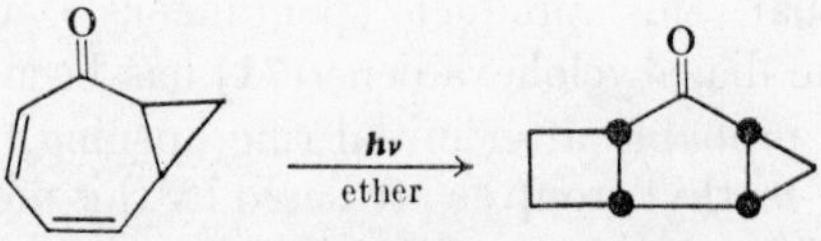

(**76**)

observed in the formation of the major photoproduct (**76**) from homotropone: the geometrical isomer of this compound is not formed, possibly because of transition-state interactions involving the cyclopropane hydrogen atoms.[135]

An exceptionally high triplet energy (*ca.* 72 kcal mole^{-1}) has been found for *cis,cis*-cyclo-octa-1,3-diene. Only with sensitizers of triplet energy greater than this is a constant equilibrium mixture of *cis,cis*- and *cis,trans*-isomers

[132] G. Schröder, W. Martin, and J. F. M. Oth, *Angew. Chem., Intern. Ed. Engl.*, **6**, 870 (1967).
[133] H. Prinzbach, M. Arguelles, and E. Druckrey, *Angew. Chem.*, **78**, 1057 (1966); E. Payo, L. Cortés, J. Mantecón, C. Rivas, and G. de Pinto, *Tetrahedron Letters*, **1967**, 2415, 4296; P. Deslongchamps and J. Kallos, *Can. J. Chem.*, **45**, 2235 (1967).
[134] J. B. Bremner and R. N. Warrener, *Chem. Commun.*, **1967**, 926.
[135] L. A. Paquette and O. Cox, *J. Am. Chem. Soc.*, **89**, 1969, 5633 (1967).

obtained. This was attributed to a non-planar diene structure. At 80° the *cis,trans*-isomer undergoes thermal conrotatory cyclization to (**77**);[136] these results together account for the previously reported sensitized cyclization of

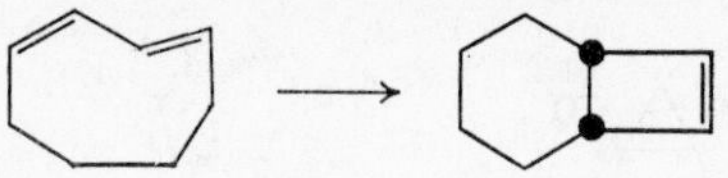

(**77**)

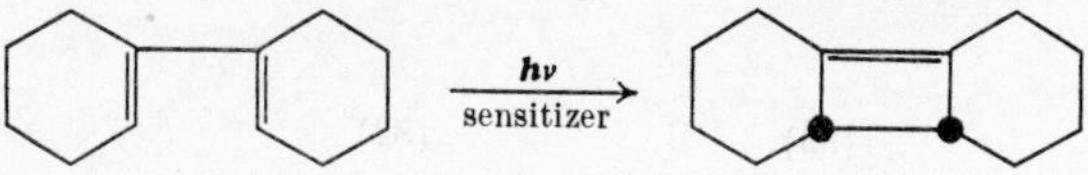

(**78**) (**79**)

the *cis,cis*-isomer. They also suggest an explanation of the anomalous behaviour of bicyclohexenyl (**78**) reported last year.[137] Perhaps photosensitized cyclization to (**79**) involves initial formation of a *trans*-cyclohexene (see above) which undergoes rapid thermal cyclization with the second (*cis*) cyclohexene unit.[136]

Photocyclization of (**80**) gives (**81**), a potential precursor of a very interesting bicyclic 10π-compound. Dehydrogenation could not be effected, and photocyclization of a monochloro-derivative took a completely different course.[138]

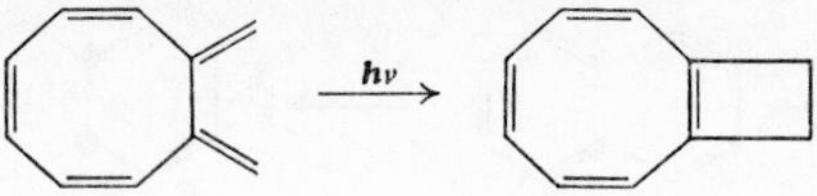

(**80**) (**81**)

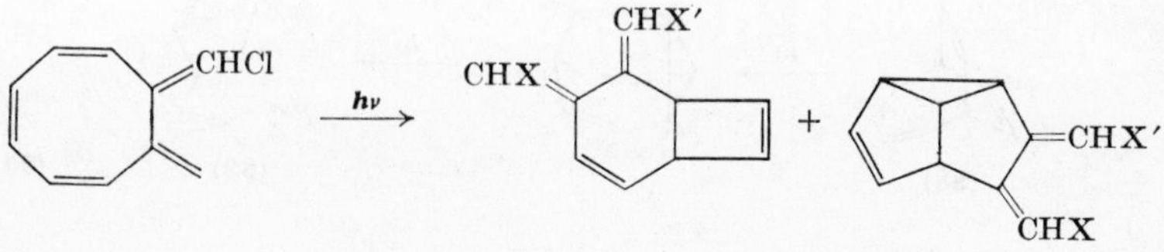

(X and X′ = H and Cl)

Other examples of photocyclization to the bicyclo[3.1.0]hexene ring system have been noted this year from conjugated[139] and non-conjugated[140] cyclohexadienes, both without the intermediacy of ring-opened products or

[136] R. S. H. Liu, *J. Am. Chem. Soc.*, **89**, 112 (1967).
[137] See *Organic Reaction Mechanisms*, **1966**, 384.
[138] J. A. Elix, M. V. Sargent, and F. Sondheimer, *J. Am. Chem. Soc.*, **89**, 180, 5081 (1967).
[139] K. R. Huffman, M. Loy, W. A. Henderson, and E. F. Ullman, *Tetrahedron Letters*, **1967**, 931; E. F. Ullman, W. A. Henderson, and K. R. Huffman, *ibid.*, p. 935.
[140] E. Druckrey, M. Arguelles, and H. Prinzbach, *Chimia*, **20**, 432 (1966); J. P. N. Brewer and H. Heaney, *Chem. Commun.*, **1967**, 811.

isomeric dienes, and from hexatrienes.[141, 142] The formation of (**83**), labelled as shown, from (**82**) rules out participation of intermediates such as (**84**) or (**85**) which have a plane of symmetry.[142]

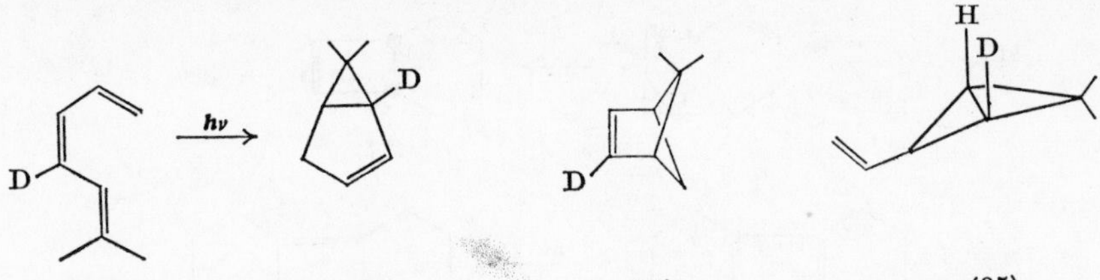

(**82**) (**83**) (**84**) (**85**)

Evidence has been obtained for the formation of cyclodecapentaene by low-temperature photolytic conrotatory opening of (**86**). Thermal recyclization gives the isomer (**87**).[143] It has also been shown that (**88**) is an intermediate in the photoisomerization of (**87**) to bullvalene,[144] and, in their turn, two new $C_{10}H_{10}$ isomers (**89**) and (**90**) have been identified after photolysis of bullvalene.[145] The former is a product of a vinylcyclopropane-to-cyclopentene rearrangement which is symmetry-allowed in the excited state.

Vinylcyclopropane-to-cyclopentene rearrangement is presumably a triplet-state process in the photosensitized rearrangement of the carene-4α-methanol

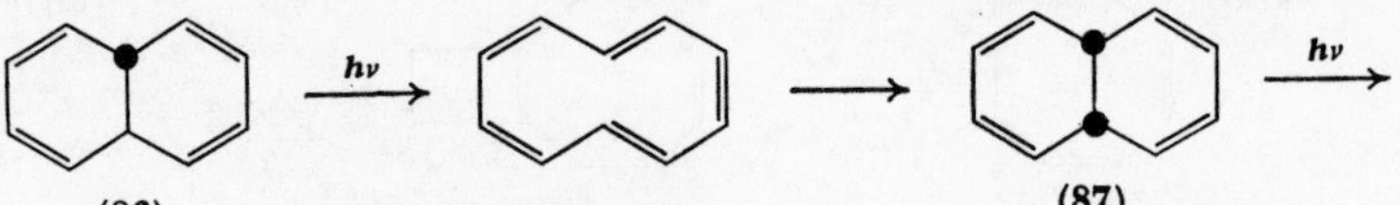

(**86**) (**87**)

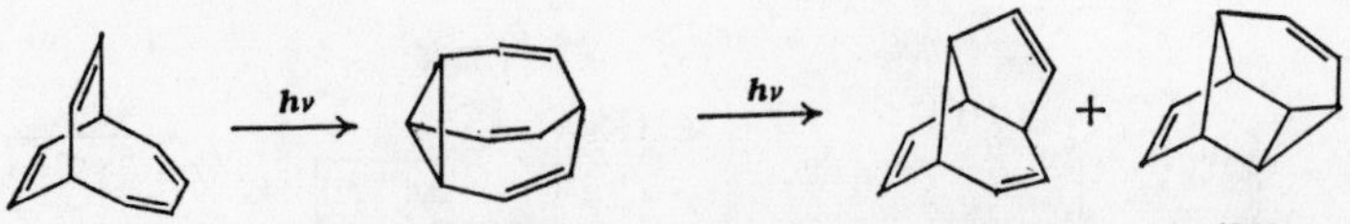

(**88**) (**89**) (**90**)

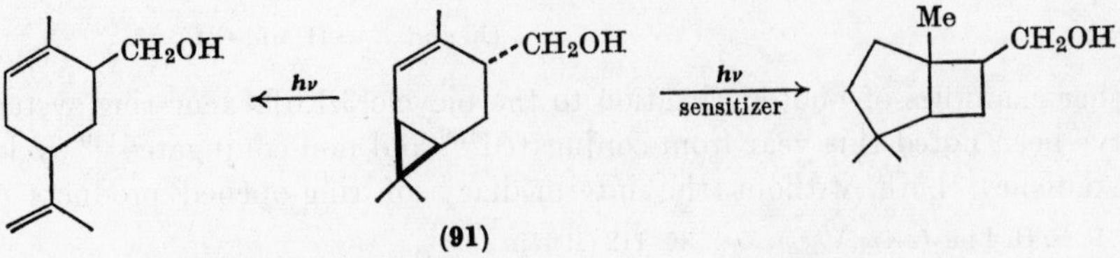

(**91**)

141 R. J. Theis and R. E. Dessy, *J. Org. Chem.*, **31**, 4248 (1966); W. G. Dauben and J. H. Smith, *ibid.*, **32**, 3244 (1967).
142 J. Meinwald and P. H. Mazzocchi, *J. Am. Chem. Soc.*, **89**, 1755 (1967).
143 E. E. van Tamelen and T. L. Burkoth, *J. Am. Chem. Soc.*, **89**, 151 (1967).
144 W. von E. Doering and J. W. Rosenthal, *Tetrahedron Letters*, **1967**, 349.
145 M. Jones, *J. Am. Chem. Soc.*, **89**, 4236 (1967).

(**91**).[146a] At the long-wavelength absorption limit the same product is also formed by direct irradiation; however, with light of shorter wavelength an external cyclopropane bond is cleaved.

It could not be decided whether the rearrangement of (**92**) involved migration of nitrogen from $C_{(1)}$ to $C_{(5)}$ or to $C_{(7)}$, both processes being symmetry-allowed in the excited state.[146b]

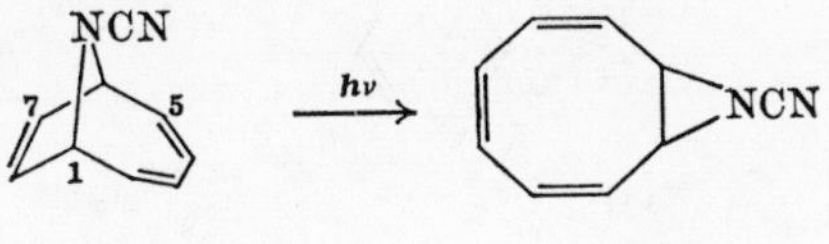

(**92**)

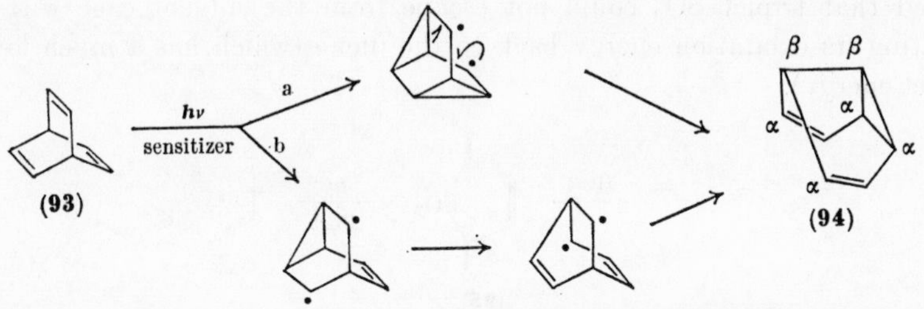

Zimmerman and his co-workers[147] have examined the sensitized isomerization of bicyclo[2.2.2]octatriene to semibullvalene [(**93**) → (**94**)]. The fate of the bridgehead protons of (**93**) were examined by deuterium-labelling; route (*a*) predicts that these two hydrogen atoms will occupy α-positions in the product; route (*b*) predicts one α and one β. The experimental result of $1\frac{1}{2}\alpha$ and $\frac{1}{2}\beta$ was consistent with route (*b*) with complete symmetrization of the allylic radical before coupling. This result was completely consistent with molecular-orbital calculations of excited-state potential energy versus reaction co-ordinate curves for the alternative mechanisms.

The photoproduct from 2,2′-bis(phenylethynyl)biphenyl does not contain a small ring structure but is in reality the phenyldibenzanthracene (**94a**).[148]

Intramolecular cyclization has been observed on irradiation of a palladium–cyclo-octene complex,[149] and thermal and photosensitized extrusion of SO_2 from (**95**) give olefins with stereochemistry in accord with orbital-symmetry considerations. The benzene-sensitized decomposition involves the triplet sulphone, and it appears that intersystem crossing must be concerted with the

[146a] P. J. Kropp, *J. Am. Chem. Soc.*, **89**, 1126 (1967).
[146b] A. G. Anastassiou and R. P. Cellura, *Chem. Commun.*, **1967**, 762.
[147] H. E. Zimmerman, R. W. Binkley, R. S. Givens, and M. A. Sherwin, *J. Am. Chem. Soc.*, **89**, 3932 (1967).
[148] E. H. White and A. A. F. Sieber, *Tetrahedron Letters*, **1967**, 2713.
[149] C. B. Anderson and B. J. Burreson, *Chem. Ind.* (*London*), **1967**, 620.

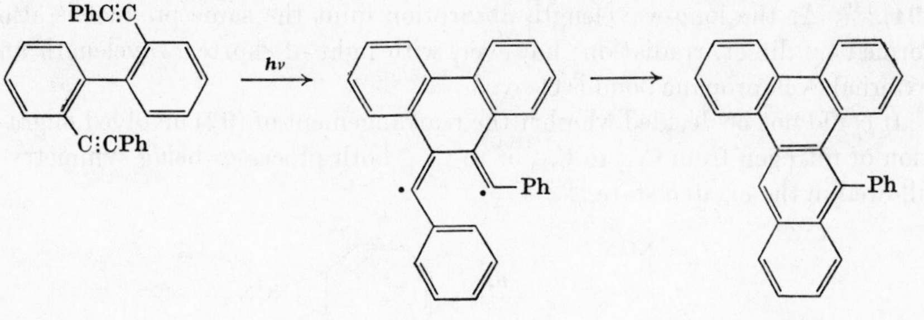

(**94a**)

decomposition to avoid further isomerization of triplet diene.[150] It was argued that triplet SO_2 could not escape from the solvent cage without donating its excitation energy back to the diene (which has a much lower triplet energy).

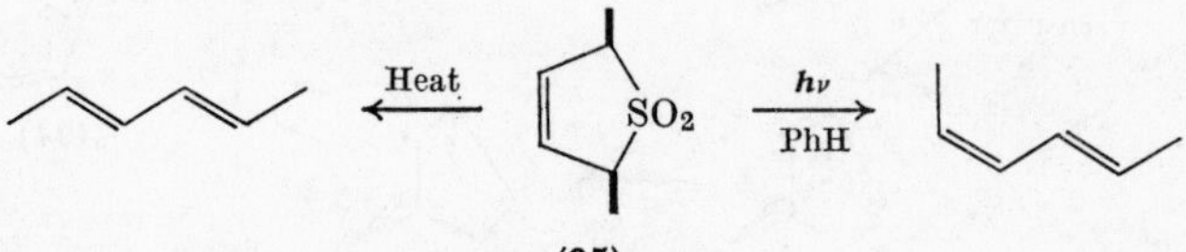

(**95**)

Photosensitized intramolecular addition of a diene to a mono-olefin has been examined by Liu and Hammond[151] who find, for example, that both the *cis*- and the *trans*-triene (**96**) give the same mixture of bicyclic products. This indicates a common (biradical) intermediate in which rotation is faster than ring-closure. It is also interesting that the products that were found could

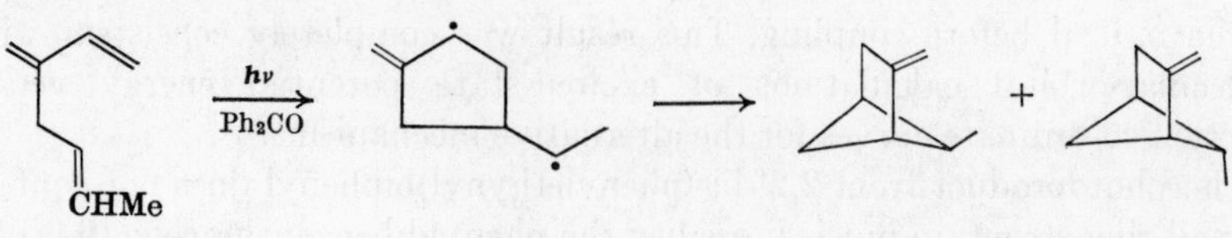

(**96**)

have originated only from a biradical with a five-membered ring. Also examined were the mercury-sensitized reactions of non-conjugated dienes, and here again preference for a five-membered-ring biradical intermediate plays a dominant role in the formation of bicyclic products.[152]

Some interesting new results have been accumulated with cycloheptatrienes

[150] J. Saltiel and L. Metts, *J. Am. Chem. Soc.*, **89**, 2232 (1967).

[151] R. S. H. Liu and G. S. Hammond, *J. Am. Chem. Soc.*, **89**, 4936 (1967).

[152] R. Srinivasan and F. I. Sonntag, *J. Am. Chem. Soc.*, **89**, 407 (1967); J. Meinwald and G. W. Smith, *ibid.*, p. 4923; R. Srinivasan and K. H. Carlough, *ibid.*, p. 4932.

and tropone derivatives. For example, the photocyclization of (**97**) to a bicyclo[3.2.0]heptadiene is accompanied by a selective [1.7]sigmatropic methyl shift exclusively in the direction of (**98**).[153] A further selective isomerization to (**99**) also occurs. The specificity in the direction of these rearrangements was attributed to the charge distribution in the excited-state transition states. For the isomerizations observed negative charge is developed adjacent to the isolated methyl group. For [1.7]sigmatropic shifts in the opposite direction, negative charge is developed at this position. The directing effect of the methyl group is thus explained.[153]

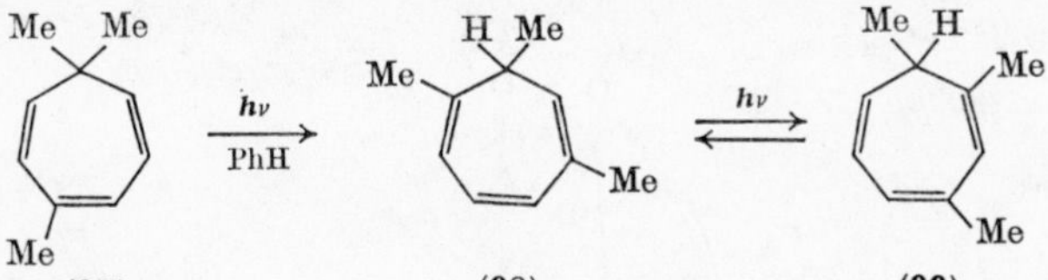

(**97**) (**98**) (**99**)

Substituent effects on the direction of cyclizations of related compounds to bicyclo[3.2.0]heptadienes have also received attention.[154]

Purpurogallin tetramethyl ether (**100**) is isomerized to (**101**) on irradiation in aqueous ethanol.[155] No reaction occurs in the absence of water, though

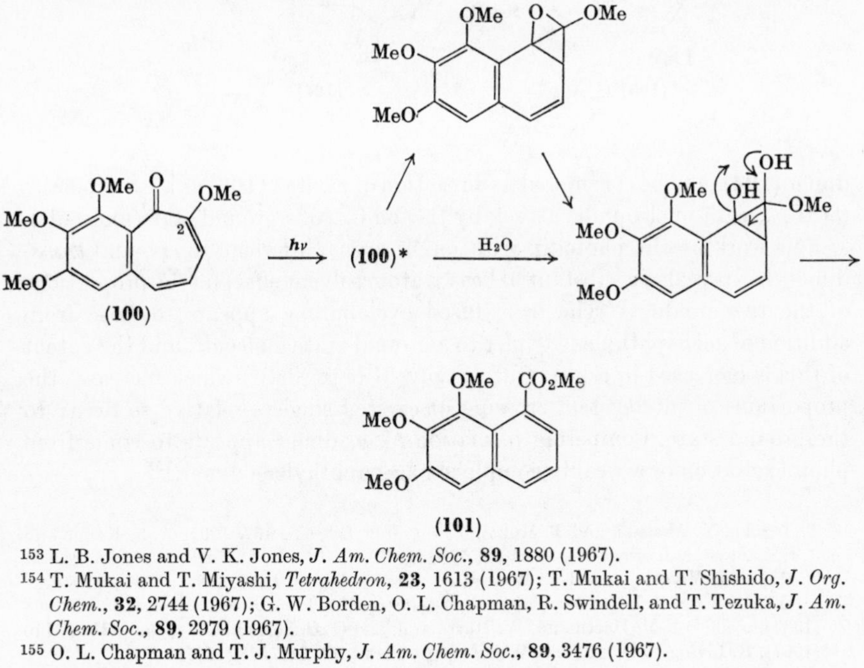

[153] L. B. Jones and V. K. Jones, *J. Am. Chem. Soc.*, **89**, 1880 (1967).
[154] T. Mukai and T. Miyashi, *Tetrahedron*, **23**, 1613 (1967); T. Mukai and T. Shishido, *J. Org. Chem.*, **32**, 2744 (1967); G. W. Borden, O. L. Chapman, R. Swindell, and T. Tezuka, *J. Am. Chem. Soc.*, **89**, 2979 (1967).
[155] O. L. Chapman and T. J. Murphy, *J. Am. Chem. Soc.*, **89**, 3476 (1967).

when water is replaced by D_2O there is no incorporation of deuterium. From these results, and the fact that ^{14}C-labelling shows that the carboxyl-carbon originates as $C_{(2)}$ in (**100**), the mechanisms shown were considered. Rapid exchange largely frustrated ^{18}O-labelling experiments.

The major photodimer (**102**) of tropone, which is formed in neutral solution, is the product of a $(6 + 4)\pi$-cycloaddition. As a concerted process this is a symmetry-forbidden reaction of photoexcited tropone. The fact that this dimerization may also be photosensitized suggests a two-step triplet process, with a biradical intermediate.[156] It has been suggested that the formation of

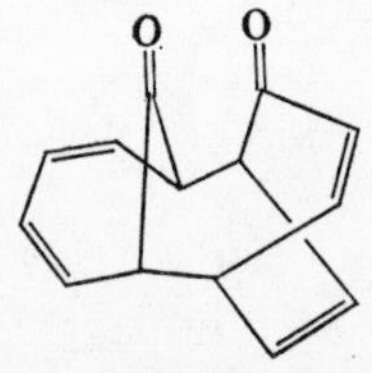

(**102**)

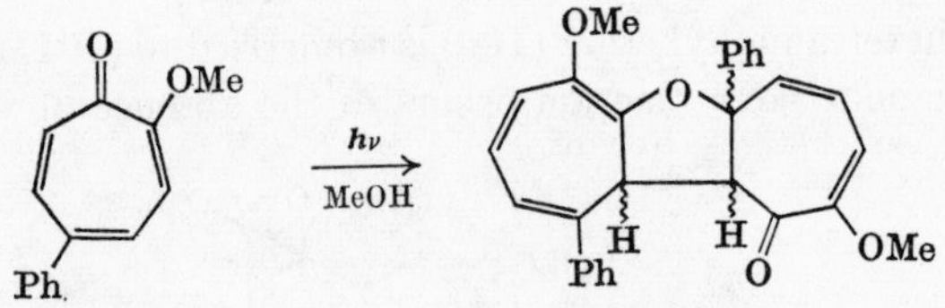

(**103**) (**104**)

dimer (**104**) may be considered as a reaction of excited (**103**) as a 1,8-dipole[157] ($O^+, C^-_{(7)}$, with nucleophilic attack by $C^-_{(7)}$ on $C_{(7)}$ of a ground-state molecule).

New work on the photodimerization of acenaphthylene to *cis*- and *trans*-dimers has revealed a substantial heavy-atom solvent effect on the proportions of the two products. The *trans*-fused cyclobutane appears to arise from addition of acenaphthylene triplet to a ground-state molecule, and the extent of this is increased in a heavy-atom solvent (e.g., PrBr) which increases the proportions of intersystem crossing of excited singlets relative to decay to the ground state. Competing formation of *cis*-dimer appears to come from photoexcitation of a weakly complexed acenaphthylene dimer.[158]

156 T. Tezuka, Y. Akasaki, and T. Mukai, *Tetrahedron Letters*, **1967**, 1397; A. S. Kende and J. E. Lancaster, *J. Am. Chem. Soc.*, **89**, 5283 (1967).
157 T. Mukai, T. Miyashi, and M. C. Woods, *Tetrahedron Letters*, **1967**, 433.
158 D. O. Cowan and R. L. Drisko, *Tetrahedron Letters*, **1967**, 1255; *J. Am. Chem. Soc.*, **89**, 3068 (1967); see also I.-M. Hartmann, W. Hartmann, and G. O. Schenk, *Chem. Ber.*, **100**, 3146 (1967); R. Livingstone and K. S. Wei, *J. Phys. Chem.*, **71**, 541, 548 (1967).

Octamethylcyclobutane is formed when tetramethylethylene is irradiated in quartz.[159] The reaction could not be sensitized. Acetone-sensitized dimerization of (**105**) gave cyclobutane dimers, but no (**106**).[160]

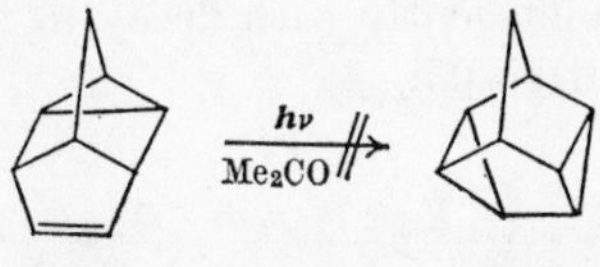

(105) (106)

Three photodimers have been obtained from 1,2,3-triphenylcyclopropene (**107**). These have been identified as (**108**), and its *cis*-isomer, from direct irradiation,[161] whilst sensitized dimerization has been found to give (**108**) and also (**109**).[162] Products (**108**) and (**109**) were considered to arise from diastereomeric biradicals (**108A**) and (**109A**); the preferred reactions of which are ring closure and hydrogen transfer, respectively. Product composition is

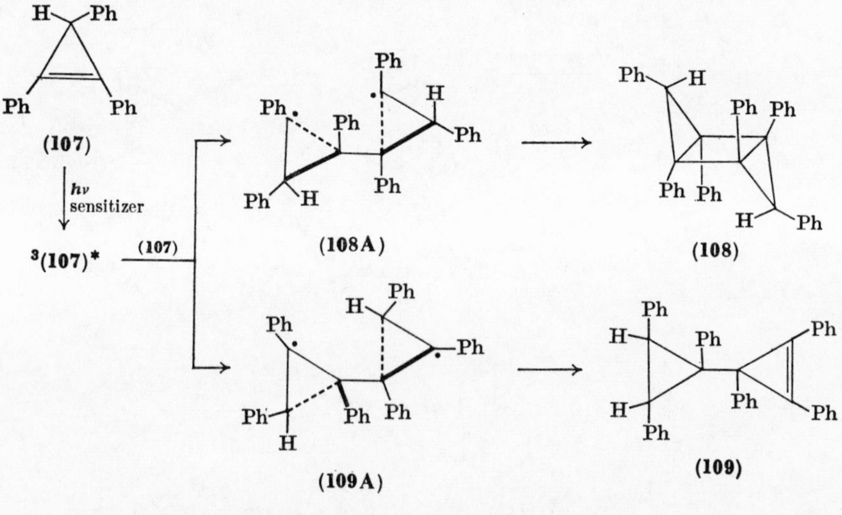

already decided at the biradical stage, as there is no deuterium isotope effect on the proportions of (**108**) and (**109**).

Several papers report the factors affecting head-to-head versus head-to-tail formation of cyclobutane in the photodimerization of various α,β-unsaturated

159 D. R. Arnold and V. Y. Abraitys, *Chem. Commun.*, **1967**, 1053.
160 H.-D. Scharf and G. Weisgerber, *Tetrahedron Letters*, **1967**, 1567.
161 H. Dürr, *Tetrahedron Letters*, **1967**, 1649.
162 C. Deboer and R. Breslow, *Tetrahedron Letters*, **1967**, 1033, 3656.

carbonyl compounds.[163] The cycloaddition of cyclopentenone to cyclohexene can be photosensitized, but only by sensitizers of triplet energy greater than 73 kcal mole^{-1}, which is considerably higher than T_1 of cyclopentenone.[164] Apparently reaction of a higher excited triplet, or collapse to a reactive species other than T_1, competes efficiently with decay to T_1, as the quantum yield for cycloaddition [→ (**110**)] is 0.5.

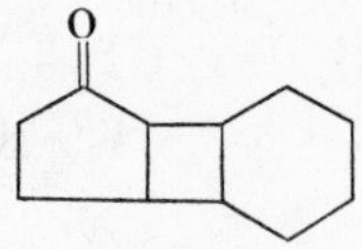

(**110**)

The formation of cyclobutanes from maleic and fumaric esters with norbornene,[165] and from cyclohexenone and norbornadiene, has been discussed.[166] In the latter instance, products of alkylation of the enone are also formed [e.g., (**111**) and (**112**)].

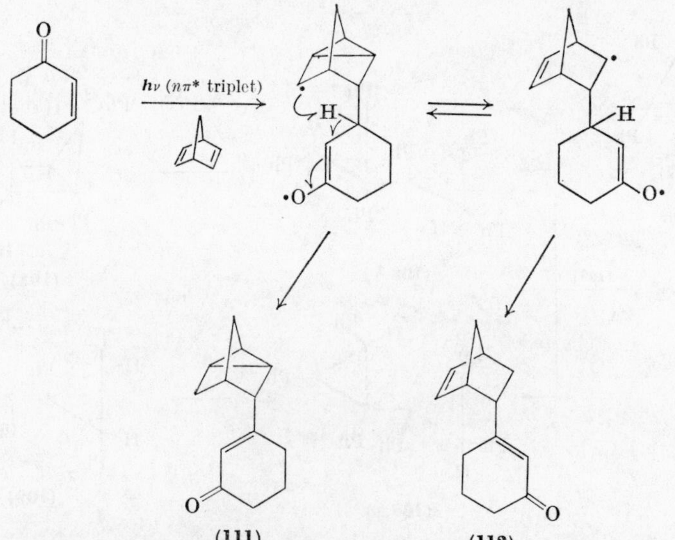

(**111**) (**112**)

The competition between formation of cyclobutanes and that of cyclohexenes as a function of sensitizer triplet energy has already been reported in

[163] J. L. Ruhlen and P. A. Leermakers, *J. Am. Chem. Soc.*, **88**, 5671 (1966); **89**, 4944 (1967); E. Y. Y. Lam, D. Valentine, and G. S. Hammond, *ibid.*, pp. 3482, 4817 (1967); H. Morrison, H. Curtis, and T. McDowell, *ibid.*, **88**, 5415 (1966).

[164] P. de Mayo, J.-P. Pete, and M. Tchir, *J. Am. Chem. Soc.*, **89**, 5712 (1967).

[165] R. L. Cargill and M. R. Willcott, *J. Org. Chem.*, **31**, 3938 (1966).

[166] J. J. McCullough and J. M. Kelly, *J. Am. Chem. Soc.*, **88**, 5935 (1966).

respect of the photosensitized dimerization of butadiene and of isoprene.[167] Similar considerations apply to the cross-dimerization of butadiene and $CH_2{=}C(CN)OAc$.[168]

Sensitized and unsensitized cycloadditions of indene to acrylonitrile have also been studied.[169]

The photodimer of *p*-benzoquinone which was reported to have the cage structure (**113**), has now been identified as (**114**).[170]

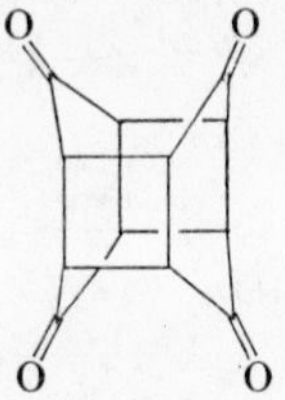

(113)

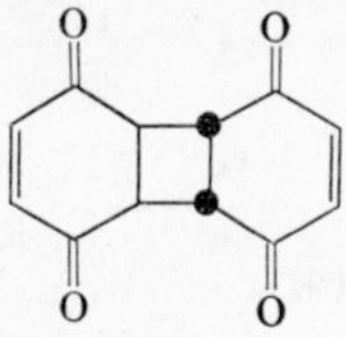

(114)

The solid-state photochemistry of several quinones and other unsaturated carbonyl compounds has been examined in relation to the crystal structure of the starting compounds.[171]

The expected photoproduct of (**115**), namely, the cyclobutane formed by intramolecular cycloaddition, was found to be accompanied by an isomer to which structure (**116**) was assigned.[172] It was suggested that energy-transfer from excited enone to the olefin initiated the formation of (**116**). This would

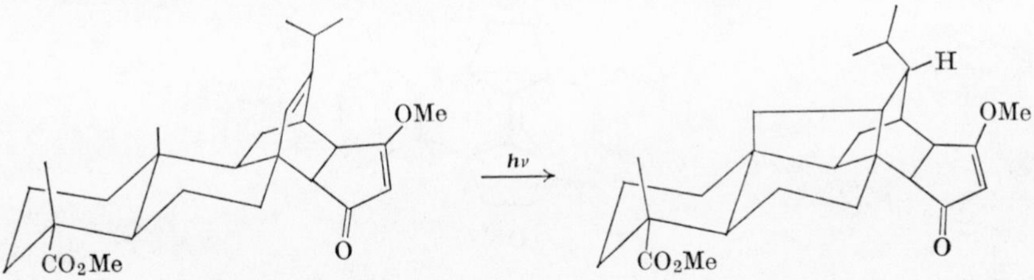

(115) **(116)**

be followed by hydrogen-abstraction from the methyl group, intramolecular hydrogen-transfer across the bridge, and finally formation of the new carbon–carbon bond. No deuterium was incorporated into (**116**) when MeOD was the

[167] See *Organic Reaction Mechanisms*, **1965**, 293.

[168] W. L. Dilling and J. C. Little, *J. Am. Chem. Soc.*, **89**, 2741 (1967); W. L. Dilling, *ibid.*, p. 2742.

[169] J. J. McCullough and C. W. Huang, *Chem. Commun.*, **1967**, 815.

[170] E. H. Gold and D. Ginsburg, *J. Chem. Soc.*, *C*, **1967**, 15.

[171] D. Rabinovich and G. M. J. Schmidt, *J. Chem. Soc.*, *B*, **1967**, 144; B. Lahev and G. M. J. Schmidt, *ibid.*, pp. 239, 312.

[172] W. Herz and M. G. Nair, *J. Am. Chem. Soc.*, **89**, 5474 (1967).

solvent. Other new examples of photoinduced hydrogen transfer and intramolecular C–C bridging have also been noted.[173]

Cookson *et al.* have continued to probe the photochemical reorganization of bisallyl compounds.[174] With the deuterated compound (**117**) they were able to show that only one allyl group is inverted in the photochemical process.[175] However, heating the products gave the familiar Cope rearrangement, both allyl groups becoming inverted. The results are consistent with a symmetry-allowed [1,3]sigmatropic rearrangement of electronically excited

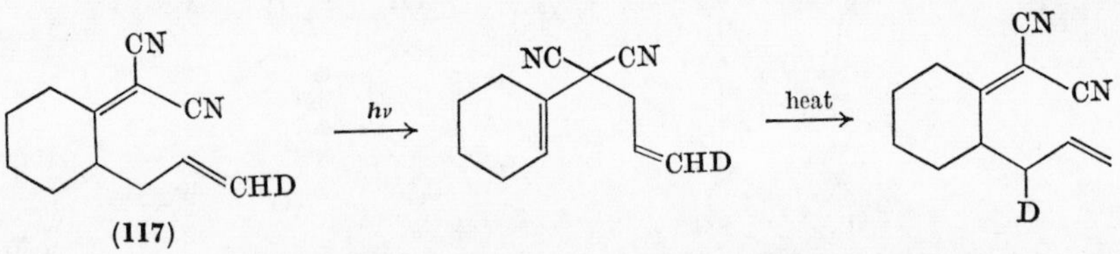

(**117**)

(**117**), which is a singlet-state process. In contrast, photolysis of benzyl and allyl benzoate in trimethylamine appeared to involve homolysis of ester triplets:

$$(ROCOPh)^* \rightarrow R\cdot + PhCO_2\cdot$$

and product formation by typically radical reactions.[176]

Aromatic Compounds

Wasserman and Keehn[177] have now succeeded in obtaining the cage dimer "dibenzoequinene" (**118**) from irradiation of [2.2]paracyclonaphthane.

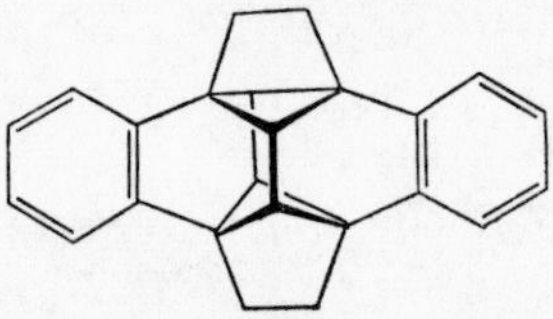

(**118**)

Various studies of anthracene photochemistry[178] range from the effects of substituents and of crystal defects on photodimerization to the effect of temperature on intersystem crossing.

[173] J. D. Rosen, *Chem. Commun.*, **1967**, 189, 364; H.-D. Scharf, *Tetrahedron*, **23**, 3057 (1967).
[174] See *Organic Reaction Mechanisms*, **1965**, 294.
[175] R. F. C. Brown, R. C. Cookson, and J. Hudec, *Chem. Commun.*, **1967**, 823.
[176] R. C. Cookson, J. Hudec, and N. A. Mirza, *Chem. Commun.*, **1967**, 824.
[177] H. H. Wasserman and P. M. Keehn, *J. Am. Chem. Soc.*, **89**, 2770 (1967).
[178] J. M. Thomas and J. O. Williams, *Chem. Commun.*, **1967**, 432; H. Bouas-Laurent and C. Leibovici, *Bull. Soc. Chim. France*, **1967**, 1847; A. Bernas, D. Leonardi, and M. Renaud, *Photochem. Photobiol.*, **5**, 721 (1966); R. Livingston and K. S. Wei, *J. Am. Chem. Soc.*, **89**, 3098 (1967); J. Adolph and D. F. Williams, *J. Chem. Phys.*, **46**, 4248 (1967); see also C. A. Parker and T. A. Joyce, *Chem. Commun.*, **1967**, 744.

The product of 185 mμ irradiation of benzene vapour has now been identified as fulvene, and not benzvalene.[179] However, benzvalene has in fact been isolated after irradiation of liquid benzene at 254 mμ,[180] and it is an intermediate in the formation of fulvene under these conditions.

Whilst Haller[181] has been able to correlate the quantum yield of formation of Dewar hexafluorobenzene from the benzenoid isomer at different irradiating wavelengths with molecular-orbital calculations on the excited states involved, Ward[182] has argued that the positional isomerization of *o*-xylene is best interpreted as a reaction of vibrationally excited ground-state molecules, and the case for a *cis,trans,trans*(or Möbius)-isomer of benzene as a key intermediate in isomerizations of benzene has been put by Farenhorst.[183] This highly energetic species is related to benzvalene and Dewar benzene by thermally allowed conrotatory electrocyclic transformations.

Photoisomerization of hexamethyl-Dewar-benzene has been found to afford hexamethylprismane (**119**) as well as the aromatic isomer.[184]

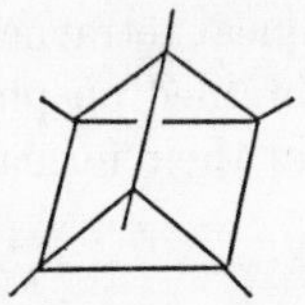

(119)

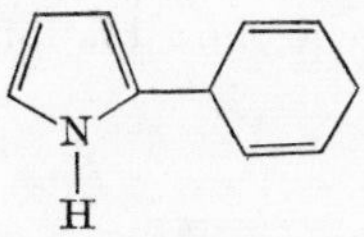

(120)

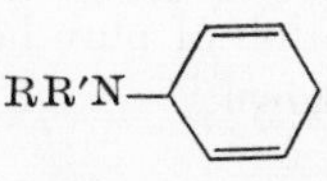

(121)

Bryce-Smith and his co-workers have reported photoaddition of pyrrole (→ **120**)[185] and primary and secondary amines (→ **121**)[186] to benzene and have discussed the mechanism for 1,3-addition of ROH reported last year.[187] New information is also available on the mechanism and geometry of photoaddition of maleic anhydride to benzene[188] and to alkylbenzenes.[189] Surprisingly all attempts to intercept the intermediate 1:1 adduct (**122**) with an

179 H. R. Ward, J. S. Wishnok, and P. D. Sherman, *J. Am. Chem. Soc.*, **89**, 162 (1967); L. Kaplan and K. E. Wilzbach, *ibid.*, p. 1030; see *Organic Reaction Mechansims*, **1966**, 380.
180 K. E. Wilzbach, J. S. Ritscher, and L. Kaplan, *J. Am. Chem. Soc.*, **89**, 1031 (1967).
181 I. Haller, *J. Chem. Phys.*, **47**, 1117 (1967).
182 H. R. Ward, *J. Am. Chem. Soc.*, **89**, 2367 (1967).
183 E. Farenhorst, *Tetrahedron Letters*, **1966**, 6465.
184 D. M. Lemal and J. P. Lokensgard, *J. Am. Chem. Soc.*, **88**, 5934 (1966).
185 M. Bellas, D. Bryce-Smith, and A. Gilbert, *Chem. Commun.*, **1967**, 263.
186 M. Bellas, D. Bryce-Smith, and A. Gilbert, *Chem. Commun.*, **1967**, 862.
187 D. Bryce-Smith, A. Gilbert, and H. C. Longuet-Higgins, *Chem. Commun.*, **1967**, 240; see *Organic Reaction Mechanisms*, **1966**, 380.
188 D. Bryce-Smith, B. Vickery, and G. I. Fray, *J. Chem. Soc., C*, **1967**, 390; W. M. Hardham and G. S. Hammond, *J. Am. Chem. Soc.*, **89**, 3200 (1967).
189 J. S. Bradshaw, *J. Org. Chem.*, **31**, 3974 (1966).

alternative dienophile were unsuccessful. Irradiation of solutions of hexamethylbenzene and maleic anhydride at the charge transfer band gave a new type of adduct (**123**).[190]

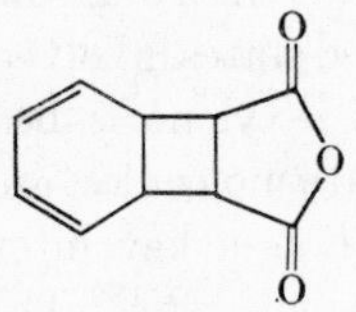

(122)

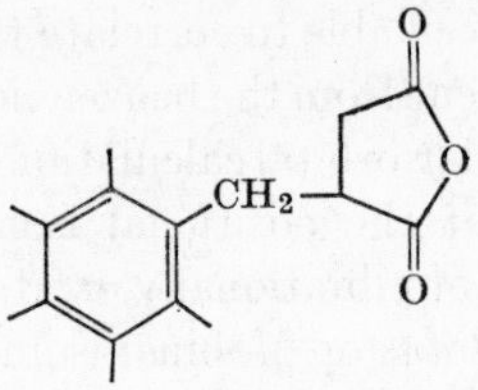

(123)

(124)

The formation of photoproducts from benzene and butadiene has been attributed to the further reactions of an initial *trans*-adduct (**124**),[191] and spectroscopic evidence has been presented suggesting that photolysis of benzene (at 254 mμ) in the presence of chloro-olefins gives linear tetraenes.[192] The unsaturated aldehydes (**125**) and (**126**) have been obtained on photo-oxidation of pure liquid benzene. Possible intermediates in their formation are shown.[193]

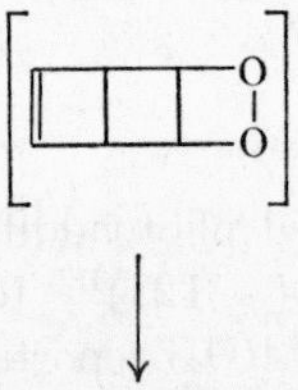

HCO·(CH:CH)$_2$·CHO

(125)

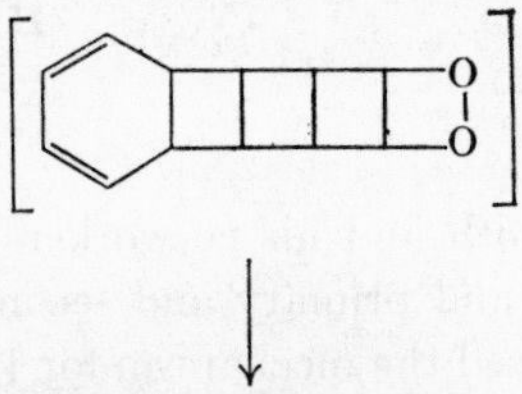

HCO·(CH:CH)$_5$·CHO

(126)

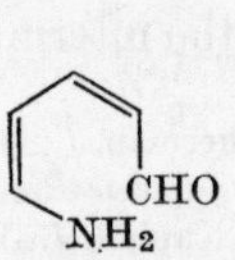

(127)

The photo-oxidation of aqueous pyridine to derivatives of glutacondialdehyde has been examined. In the absence of oxygen a photoproduct is obtained

[190] Z. Raciszewski, *J. Chem. Soc., B*, **1966**, 1147.
[191] K. Kraft and G. Koltzenburg, *Tetrahedron Letters*, **1967**, 4357, 4723.
[192] N. C. Perrins and J. P. Simons, *Chem. Commun.*, **1967**, 999.
[193] K. Wei, J.-C. Mani, and J. N. Pitts, *J. Am. Chem. Soc.*, **89**, 4225 (1967).

which reverts to pyridine in the presence of base. Photoaddition of water to the 1,2-bond was suggested, followed by ring-opening to (**127**);[194] on treatment with a base this would revert to pyridine.

An interesting approach to differentiating benzvalene and Dewar-benzene pathways for the positional photoisomerization of substituted benzenes would be to examine the photoisomerization of a compound with the substitution pattern $C_6A_2B_2C_2$. Only with substitution of this complexity do both mechanisms give primary photoisomers whose substitution pattern uniquely defines their mode of formation. This test has now been applied by using dimethylpyrazines.[195a] The methylation pattern of the pyrimidines obtained on irradiation was consistent only with the benzvalene route. A dividend resulting from the choice of heterocyclic compounds with which to carry out this experiment was the conclusion (from a study of the wavelength-dependence of the rearrangement) that it was indeed the $\pi\pi^*$-singlet which was responsible for the benzvalene isomerization.[195b]

An intermolecular mechanism has been proposed for the photoequilibration of 2- and 4-picoline; lutidines are formed as by-products, and added pyridine is methylated.[195c] Possibly related to this is the photoalkylation of benzopyridines by carboxylic acids in benzene;[196] for example, equimolar quinoline and acetic acid give 2- and 4-methylquinoline; it was suggested that the photoreaction probably involves quinolinium acetate.

Irradiation of acridine in a variety of solvents gives diacridan by initial hydrogen-abstraction from the solvent by triplet (probably $n\pi^*$) acridine.[197] Hydrogen-abstraction (to give $Me\dot{C}HOH$) has been observed[198] when naphthalene or phenanthrene is irradiated in an ethanol glass at 77°K; the rate of radical production is proportional to the square of the radiation intensity, probably implicating a two-quantum process.

Wynberg *et al.*[199] have now set out in detail their results and mechanistic conclusions on the photoisomerizations of arylthiophens.

The mercury-sensitized photochemistry of furan vapour has been discussed by Srinivasan.[200] Decarbonylation is a major reaction, and the resulting

194 J. Joussot-Dubien and J. Houdard, *Tetrahedron Letters*, **1967**, 4389.
195a F. Lahmani and N. Ivanoff, *Tetrahedron Letters*, **1967**, 3913.
195b See *Organic Reaction Mechanisms*, **1966**, 382.
195c O. S. Pascual and L. O. Tuazon, *Philippines Nucl. J.*, **1**, 49 (1966); *Chem. Abs.*, **66**, 115127 (1967).
196 H. Nozaki, M. Katô, R. Noyori, and M. Kawanisi, *Tetrahedron Letters*, **1967**, 4259.
197 A. Kellmann, *J. Chim. Phys.*, **63**, 936 (1966).
198 B. N. Shelimov, V. G. Vinogradova, V. I. Mal'tsev, and N. V. Fok, *Dokl. Akad. Nauk SSSR*, **172**, 655 (1967); *Chem. Abs.*, **66**, 94523k (1967).
199 H. Wynberg, H. van Driel, R. M. Kellogg, and J. Buter, *J. Am. Chem. Soc.*, **89**, 3487 (1967); R. M. Kellogg and H. Wynberg, *ibid.*, p. 3495; H. Wynberg, G. E. Beekhuis, H. van Driel, and R. M. Kellogg, *ibid.*, p. 3498; H. Wynberg, R. M. Kellogg, H. van Driel, and G. E. Beekhuis, *ibid.*, p. 3501; see also *Organic Reaction Mechanisms*, **1965**, 291; **1966**, 382.
200 R. Srinivasan, *J. Am. Chem. Soc.*, **89**, 1758, 4812 (1967).

cyclopropene may be trapped by excited-state furan to give (**128**). A similar product (**129**) is obtained in the presence of cyclopentene, but the yield of the decarbonylation process is unaffected, leading to the suggestion that different excited states may be involved in the two reaction modes.

The benzvalene analogue (**131**) may be an intermediate in the photolysis of (**130**), to the (minor) products (**132**) and (**133**).[201]

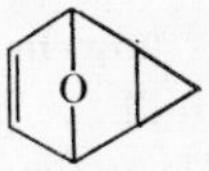

(**128**)

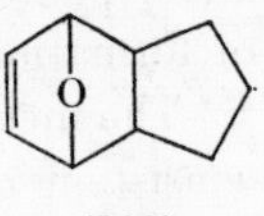

(**129**)

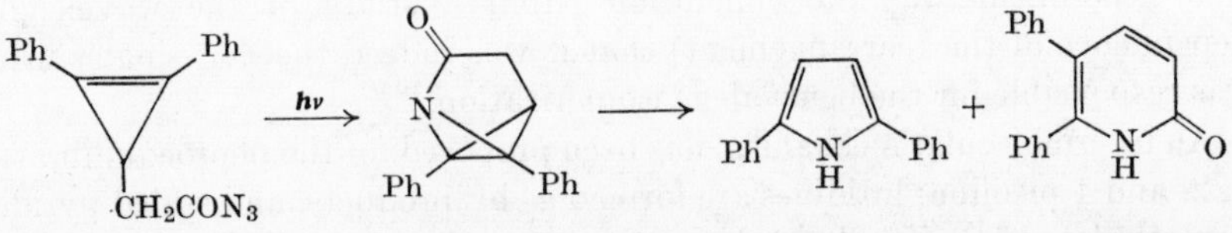

(**130**) (**131**) (**132**) (**133**)

In new studies of the photochemistry of heterocyclic *N*-oxides, several previously assigned oxaziridine structures have been revised, the photoproducts now being identified as oxazepines, e.g., (**134**) → (**135**).[202, 203] However, most of the available data on these and other products can be interpreted best in terms of initial oxaziridine formation with subsequent reactions governed by environment (substitution pattern, solvent, etc.). From the abundance of new results,[202–205] two further examples (for **136** and **137**) are illustrated with tentative mechanistic rationalizations.[203, 205]

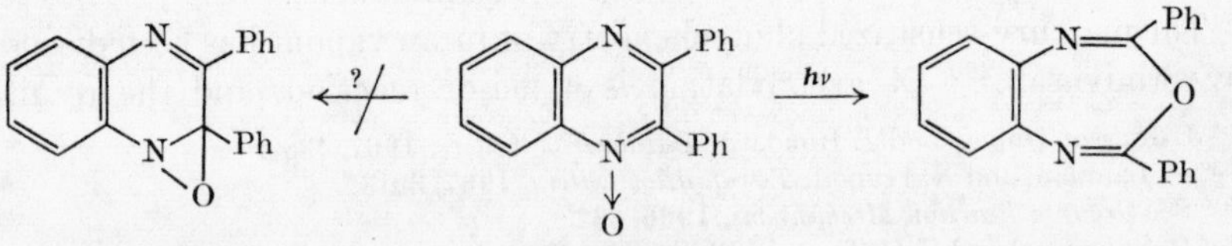

(not isolable) (**134**) (**135**)

[201] N. C. Castellucci, M. Kato, H. Zenda, and S. Masamune, *Chem. Commun.*, **1967**, 473.
[202] O. Buchardt, *Tetrahedron Letters*, **1966**, 6221; O. Buchardt, C. Lohse, A. M. Duffield, and C. Djerassi, *ibid.*, **1967**, 2741; O. Buchardt and J. Feeney, *Acta Chem. Scand.*, **21**, 1399 (1967).
[203] C. Kaneko, S. Yamada, I. Yokoe, and M. Ishikawa, *Tetrahedron Letters*, **1967**, 1873.
[204] M. J. Haddadin and C. H. Issidorides, *Tetrahedron Letters*, **1967**, 753; O. Buchardt, J. Becher, and C. Lohse, *Acta Chem. Scand.*, **20**, 2467 (1966).
[205] N. Ikekawa and Y. Honma, *Tetrahedron Letters*, **1967**, 1197.

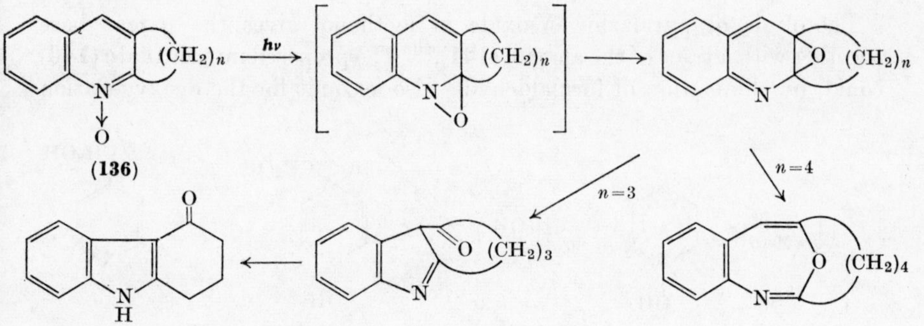

(136)

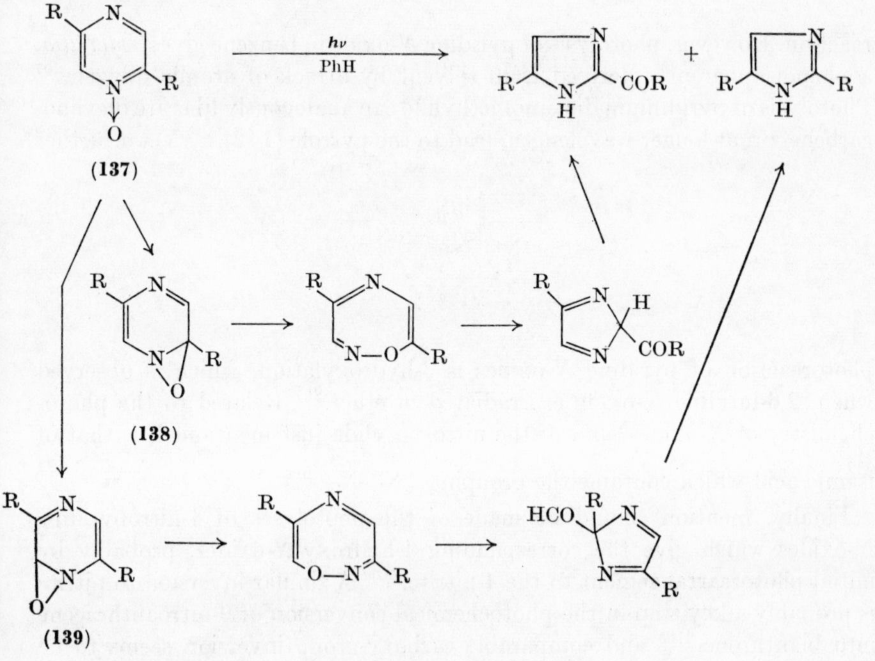

(137) (138) (139)

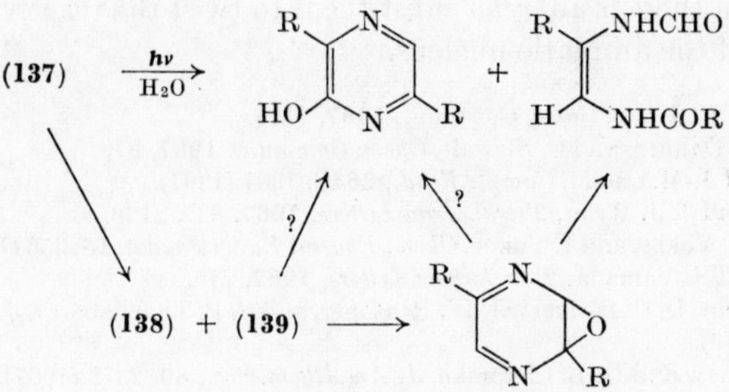

Photolysis of pyridazine *N*-oxide in methanol gives the parent base together with a trace of the alcohol (**141**).[206] The proposed intermediate (**140**) could, by elimination of formaldehyde, also account for the deoxygenation

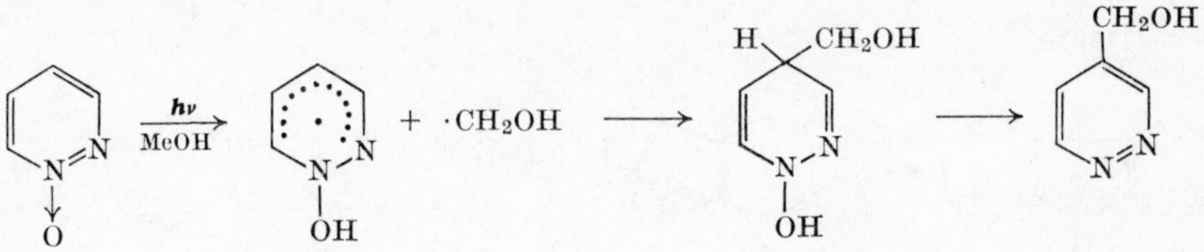

(**140**) (**141**)

reaction. However, photolysis of pyridine *N*-oxide in benzene gives, *inter alia*, a phenol apparently derived from solvent by attack of atomic oxygen.[207] Photolysis of pyridinium dicyanomethylide can analogously liberate dicyanocarbene, or, at longer wavelength, lead to the pyrrole (**142**).[208] Yet a further

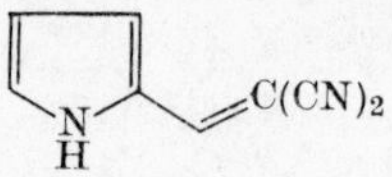

(**142**)

photoreaction of pyridine *N*-oxides is 3-hydroxylation, which is observed when, 2,6-lutidine *N*-oxide is irradiated in ether.[207] Related to the photochemistry of *N*-oxides, and of the nitrogen ylide just mentioned, is that of isamic acid which contains the grouping $>\overset{+}{N}-\overset{-}{N}<$.[209]

Finally, mention should be made of the photolyses of 4-nitropyridine *N*-oxides which give the corresponding 4-hydroxy-*N*-oxides, probably by initial photorearrangement to the 4-nitrite.[210a] A similar inversion to nitrite is probably a key step in the photochemical conversion of 9-nitroanthracene into bianthrone,[210b] and comparable carboxy-group inversion seems to be involved in the photolysis of sodium 9-anthroate.[210c] These reactions are favoured when there is an *ortho*-substituent to twist the reactive function out of the plane of the aromatic nucleus.

[206] M. Ogata and K. Kanō, *Chem. Commun.*, **1967**, 1176.
[207] J. Streith, B. Danner, and C. Sigwalt, *Chem. Commun.*, **1967**, 979.
[208] J. Streith and J.-M. Cassal, *Compt. Rend.*, **264**C, 1307 (1967).
[209] P. de Mayo and J. J. Ryan, *Tetrahedron Letters*, **1967**, 827, 2128.
[210a] C. Kaneko, S. Yokoe, and I. Yokoe, *Chem. Pharm. Bull.* (*Tokyo*), **15**, 356 (1967); C. Kaneko, I. Yokoe, and S. Yamada, *Tetrahedron Letters*, **1967**, 775.
[210b] O. L. Chapman, D. C. Heckert, J. W. Reasoner, and S. P. Thackaberry, *J. Am. Chem. Soc.*, **88**, 5550 (1966).
[210c] A. W. Bradshaw and O. L. Chapman, *J. Am. Chem. Soc.*, **89**, 2372 (1967).

New examples of the photochemical Fries rearrangement have been examined.[211] No carbon isotope effect was detected when the photolysis of *p*-methoxyphenyl[*carbonyl*-^{14}C]acetate was compared with that of the unlabelled compound.[212]

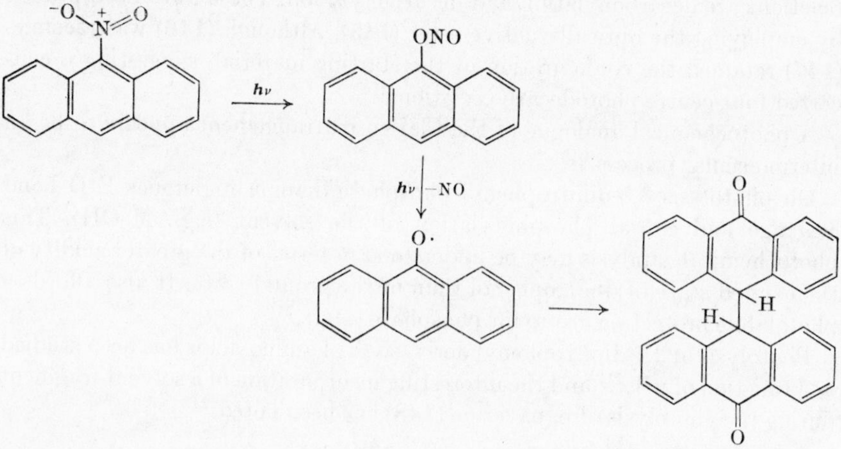

It has been found that the quantum yield of the photo-Fries rearrangement product (**144**) from *p*-tolyl acetate is independent of solvent viscosity, suggesting a tight transition state (**143**) for the rearrangement. The quantum

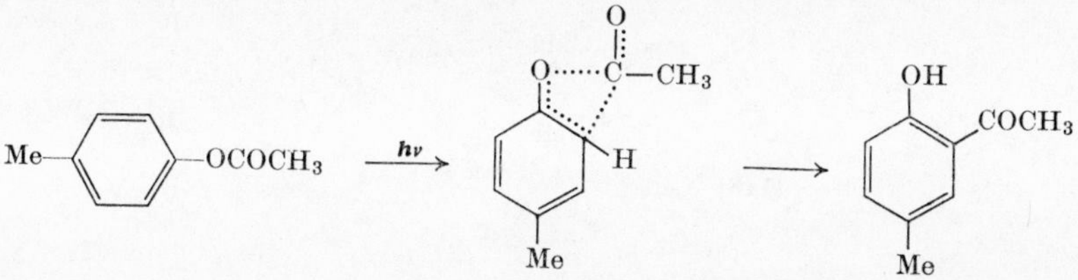

(**143**) (**144**)

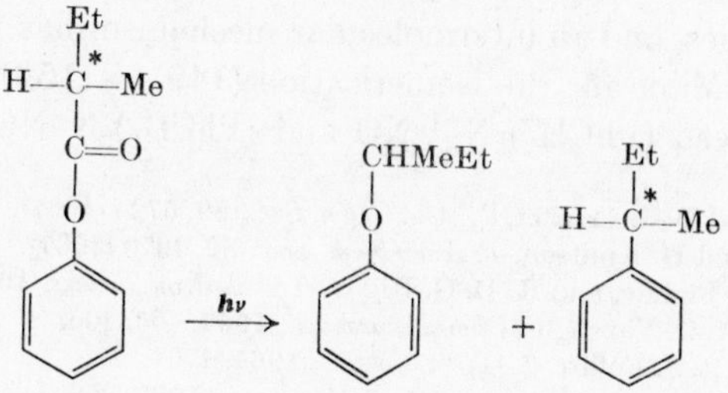

(**145**) (**146**) (**147**)

[211] H. Obara and H. Takahashi, *Bull. Chem. Soc. Japan*, **40**, 1012 (1967); D. V. Rao and V. Lamberti, *J. Org. Chem.*, **32**, 2896 (1967).

[212] L. Schutte and E. Havinga, *Tetrahedron*, **23**, 2281 (1967).

yield of *p*-cresol from the same reaction falls rapidly with increasing viscosity, presumably because of increased radical cage recombination. In support of the latter suggestion was the observation that the yield of cresol was paralleled by the yield of acetone when propan-2-ol was used as solvent.[213] Other side reactions are decarbonylation and decarboxylation. These have been probed by employing the optically active ester (**145**). Although (**146**) was racemic, (**147**) retained the configuration of the starting material, suggesting a concerted four-centre photodecarboxylation.[214]

A photochemical analogue of the Claisen rearrangement appears to be an intermolecular process.[215]

On photolysis 3,5-dinitrophenyl phosphate dianion undergoes P–O bond cleavage and causes phosphorylation of the solvent (e.g., MeOH). This photochemical catalysis may be understood in terms of the greater acidity of the excited state of dinitrophenol than of the ground state. It also affords a photolabile protecting group for phosphate esters.[216]

Photolysis of 2,4-dinitrophenyl derivatives of amino-acids has been studied as a function of pH,[217] and the interesting incorporation of a solvent fragment during the photolysis of papaverine (**148**) has been noted.[218]

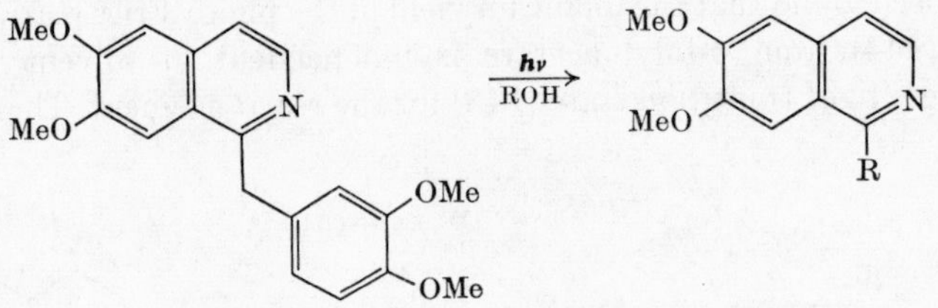

(**148**)

Miscellaneous

Chow's group[219] have extended their study of the acid-catalysed photolysis of *N*-nitroso-amines, and an intermolecular mechanism has been established by Axenrod and Milne for the isomerization (**149**) → (**150**) by a cross-over experiment between $(PhCH_2)_2N{-}^{15}NO$ and $(PhCD_2)_2N{-}NO$.[220] In compar-

213 M. R. Sandner and D. J. Trecker, *J. Am. Chem. Soc.*, **89**, 5725 (1967).
214 R. A. Finnegan and D. Knutson, *J. Am. Chem. Soc.*, **89**, 1970 (1967).
215 D. P. Kelly, J. T. Pinhey, and R. D. G. Rigby, *Tetrahedron Letters*, **1966**, 5953.
216 A. J. Kirby and A. G. Varvoglis, *Chem. Commun.*, **1967**, 405, 406.
217 D. J. Neadle and R. J. Pollitt, *J. Chem. Soc., C*, **1967**, 1764.
218 F. R. Stermitz, R. Pua, and H. Vyas, *Chem. Commun.*, **1967**, 326.
219 Y. L. Chow, *Can. J. Chem.*, **45**, 53 (1967); Y. L. Chow and A. C. H. Lee, *ibid.*, p. 311; Y. L. Chow, C. J. Colón, and S. C. Chen, *J. Org. Chem.*, **32**, 2109 (1967); Y. L. Chow and C. J. Colón, *Can. J. Chem.*, **45**, 2559 (1967).
220 T. Axenrod and G. W. A. Milne, *Tetrahedron Letters*, **1967**, 4443.

able rearrangements, Chow identified hyponitrous acid (the dimer of nitroxyl) as a by-product.

N-Nitroso-amides, unlike the nitroso-amines, are not stable to photolysis under neutral conditions. The N–N bond is broken, and several reports of

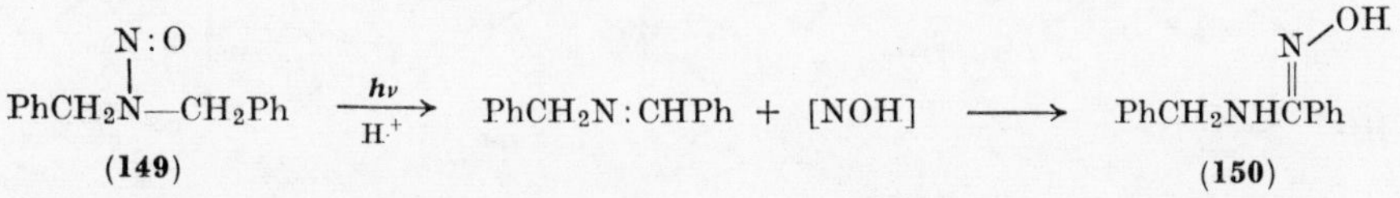

intramolecular hydrogen-abstraction by the resulting amido-radical have appeared, e.g., the transannular reaction of (**151**).[221a] In open-chain amides, intramolecular abstraction occurs only from an alkyl chain on nitrogen and not from one on carbonyl.[221b]. This may be due to an unfavourable effect of the sp^2 carbonyl on the geometry of the cyclic transition state for hydrogen abstraction in the latter case.

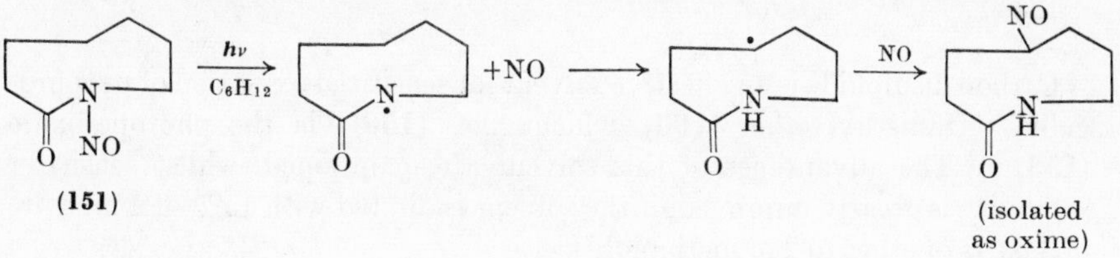

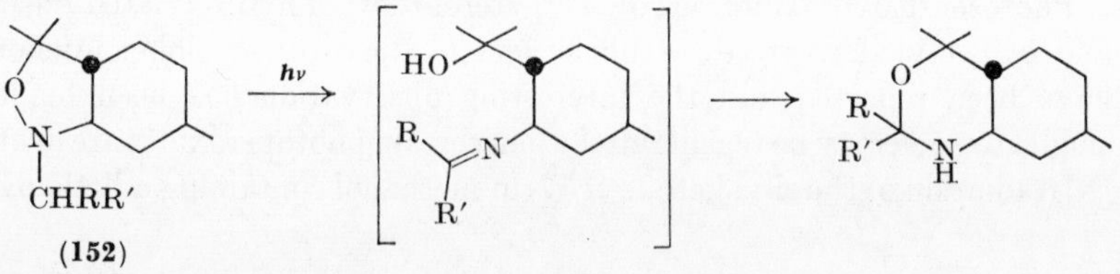

An N–O bond cleavage and subsequent hydrogen transfer and recyclization constitute the probable pathway for photorearrangement of (**152**).[222]

A particularly interesting result in the area of photosensitized oxidation is the observation that steroidal Δ^4-3β-ols give different products according to the sensitizer energy. The formation of epoxy-ketones (**153**) with low-energy triplet sensitizers was considered to be a reaction of $O_2(^1\Delta_g)$; that of enones (**154**), with sensitizers of higher triplet energy, was a reaction of $O_2(^1\Sigma^+_g)$.

[221a] O. E. Edwards and R. S. Rosich, *Can. J. Chem.*, **45**, 1287 (1967).

[221b] Y. L. Chow and A. C. H. Lee, *Chem. Ind.* (*London*), **1967**, 827; L. P. Kuhn, G. G. Kleinspehn, and A. C. Duckworth, *J. Am. Chem. Soc.*, **89**, 3858 (1967).

[222] N. A. LeBel, T. A. Lajiness, and D. B. Ledlie, *J. Am. Chem. Soc.*, **89**, 3076 (1967).

With high-energy sensitizers, but low initial olefin concentration, the proportion of (**154**) was reduced, presumably because of internal conversion into the lowest oxygen singlet before reaction occurs.[223]

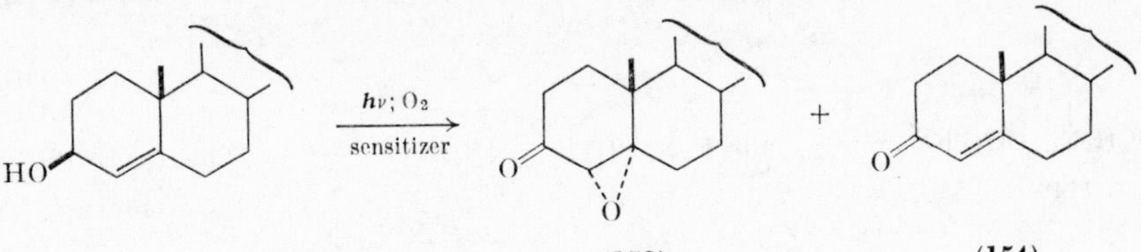

(**153**) (**154**)

(**100**) 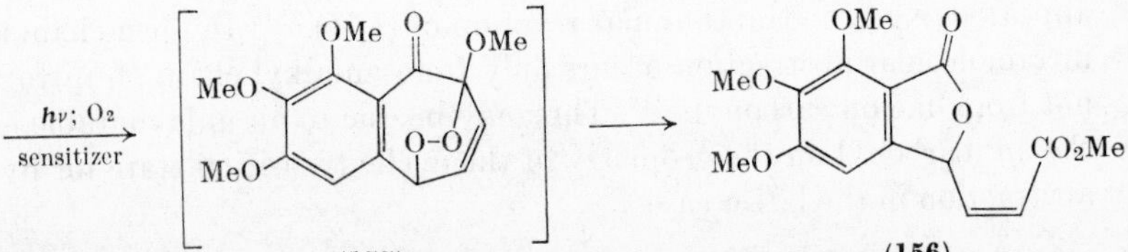

(**155**) (**156**)

Carbon disulphide is an effective solvent for sensitized oxidation of purpurogallin tetramethyl ether (**100**), which gives (**156**) via the photoperoxide (**155**).[224] The advantages of this solvent are maintained, whilst sensitizer solubility is greatly improved, if the solvent is diluted with 15% of a mixture of 3 parts of ether to 2 of methanol.[225]

Photosensitized oxidations of $\Delta^{8(9)}$-steroids,[226] of hydroxylated purines,[227] and of acetyl derivatives of filicic acid (a dihydroxycyclohexadienone),[228] have been reported and the interesting observation has been made that singlet oxygen may be regenerated when certain photoperoxides are heated.[229]

Irradiation of the silyl ketone (**157**) in methanol containing a little pyridine

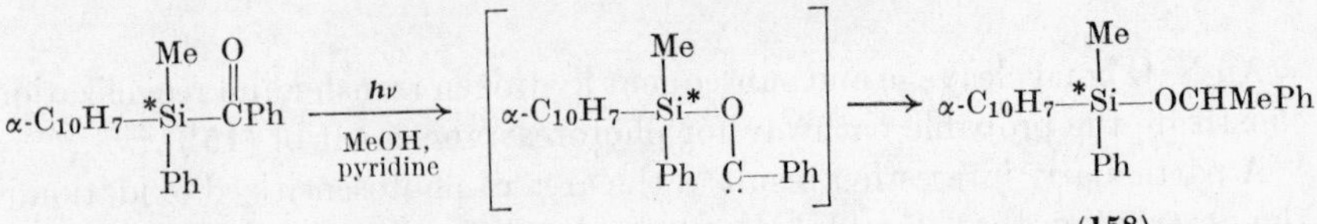

(**157**) (**158**)

[223] D. R. Kearns, R. A. Hollins, A. U. Khan, R. W. Chambers, and P. Radlick, *J. Am. Chem. Soc.*, **89**, 5455 (1967); D. R. Kearns, R. A. Hollins, A. U. Khan, and P. Radlick, *ibid.*, p. 5456.

[224] E. J. Forbes and J. Griffiths, *J. Chem. Soc., C*, **1967**, 601.

[225] E. J. Forbes and J. Griffiths, *Chem. Commun.*, **1967**, 427.

[226] J. E. Fox, A. I. Scott, and D. W. Young, *Chem. Commun.*, **1967**, 1105.

[227] T. Matsuura and I. Saito, *Chem. Commun.*, **1967**, 693.

[228] R. H. Young and H. Hart, *Chem. Commun.*, **1967**, 827, 828.

[229] H. H. Wasserman and J. R. Scheffer, *J. Am. Chem. Soc.*, **89**, 3073 (1967).

gives (**158**) with 90% retention of configuration at asymmetric silicon. Concerted rearrangement and addition of solvent to the resulting carbene was proposed.[230] The role of base is uncertain but crucial, for in its absence the major products arise from Si–C bond fission.

The photolysis of potassium tetraphenylborate in water in the absence of oxygen gives 2,5-dihydrobiphenyl and Ph_2BOK as major products. These products, together with isomeric dienes formed as by-products, can be accommodated by one of several mechanisms leading to the key intermediate (**159**).[231] Similar dienes are also found on photolysis of triphenylboron in co-ordinating solvents (MeOH, piperidine), suggesting tetrahedral boron as a prerequisite for this type of reaction.[232]

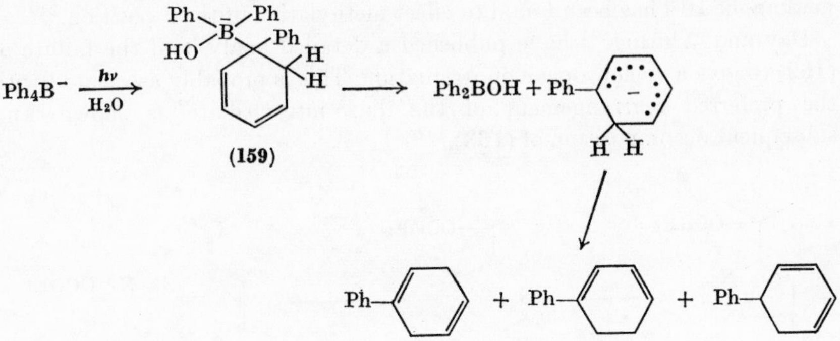

(**159**)

A heavy-atom effect has now been reported in the photolysis of triphenylmethyl halides,[233] and at 77°K photolysis of triphenylmethanethiol[234a] and other thiols[234b] involves the S–H bond.

Irradiation of sodium cyclopentadienide in THF–Bu^tOH gives a mixture of the isomeric 3,3′-bicyclopentenyls in low yield. In THF–Bu^tOD, $C_5H_5^-$ gives $C_{10}D_{14}$.[235]

In alcoholic solvents the displacement reaction (**160**) → (**161**) depends on the nature of anion Y^- and substituent X.[236] With X = H and Y = I, a relatively high-energy charge-transfer state crosses to a dissociative state,

[230] A. G. Brook and J. M. Duff, *J. Am. Chem. Soc.*, **89**, 454, 5314 (1967).
[231] J. L. R. Williams, J. C. Doty, P. J. Grisdale, T. H. Regan, and D. G. Borden, *Chem. Commun.*, **1967**, 109; J. L. R. Williams, J. C. Doty, P. J. Grisdale, R. Searle, T. H. Regan, G. P. Happ, and D. P. Maier, *J. Am. Chem. Soc.*, **89**, 5153 (1967).
[232] J. L. R. Williams, P. J. Grisdale, and J. C. Doty, *J. Am. Chem. Soc.*, **89**, 4538 (1967).
[233] H. G. Lewis and E. D. Owen, *J. Chem. Soc., B*, **1967**, 422.
[234a] J. K. S. Wan, *Chem. Commun.*, **1967**, 429.
[234b] D. H. Volman, J. Wolstenholme, and S. G. Hadley, *J. Phys. Chem.*, **71**, 1798 (1967).
[235] E. E. van Tamelen, J. I. Brauman, and L. Ellis, *J. Am. Chem. Soc.*, **89**, 5073 (1957).
[236] T. D. Walsh and R. C. Long, *J. Am. Chem. Soc.*, **89**, 3943 (1967).

but with Y = BF_4 the absence of charge-transfer precludes reaction. Finally, with X = CN and Y = I the charge-transfer state is so stable that again no reaction occurs. The complex behaviour displayed in photochemical halogen exchange by aryl halides has also been rationalized.[237]

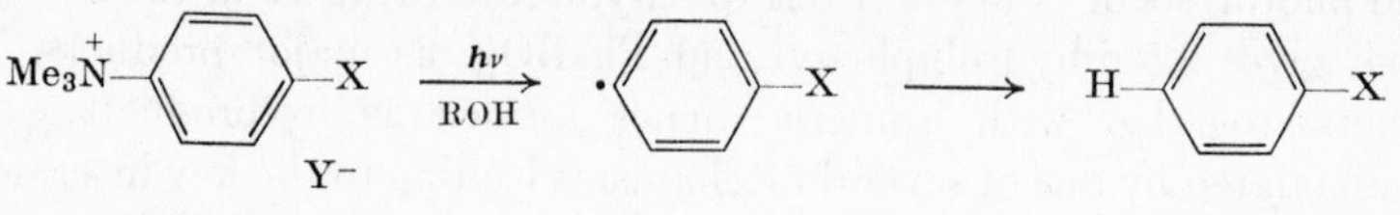

(160) **(161)**

Products of photolysis of "amidopyrene"[238] and "antipyrene"[239] have been identified, and irradiation of some hydroxylated pyrimidines in methanolic HCl has been found to effect methylation of the 2-position.[240]

Day and Whiting[241] have published a detailed analysis of the failure of (**162**) to give a cyclopropene on irradiation. This is probably associated with the preferred rearrangement of the diazo-intermediate as shown, and subsequent decomposition of (**163**).[241]

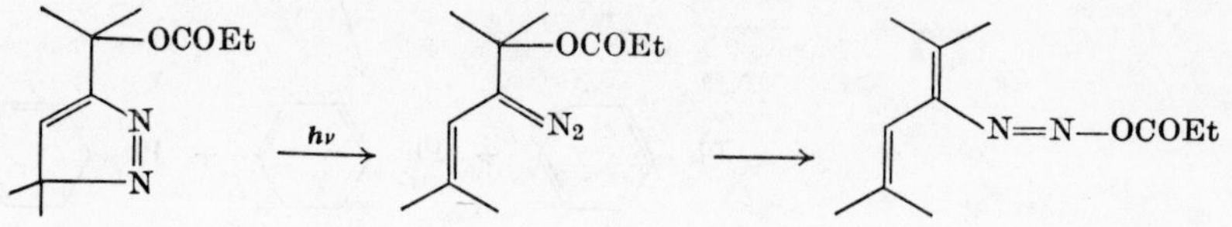

(162) **(163)**

New publications also describe: the trapping of cyclobutadiene from photolysis of tricarbonylcyclobutadieneiron;[242] further reductions of aromatic molecules by irradiation in the presence of BH_4^- or SO_2^{2-};[243] further photoreactions of *N*-chloroacetyl-amino-acids,[244] and the photofragmentation of 1,3,2-dioxaphosph(v)oles.[245]

Irradiation of 2,4,6-tri-*tert*-butylphenoxide in glass at 77°K causes development of an ESR signal, presumably by photoionization to the corresponding

[237] J. T. Echols, V. T.-C. Chuang, C. S. Parrish, J. E. Rose, and B. Milligan, *J. Am. Chem. Soc.*, **89**, 4081 (1967).
[238] J. Reisch and A. Fitzek, *Tetrahedron Letters*, **1967**, 4513.
[239] S. N. Eğe, *Chem. Commun.*, **1967**, 488.
[240] M. Ochiai and K. Morita, *Tetrahedron Letters*, **1967**, 2349.
[241] A. C. Day and M. C. Whiting, *J. Chem. Soc., B*, **1967**, 991.
[242] W. J. R. Tyerman, M. Kato, P. Kebarle, S. Masamune, O. P. Strausz, and H. E. Gunning, *Chem. Commun.*, **1967**, 497.
[243] J. A. Waters and B. Witkop, *J. Am. Chem. Soc.*, **89**, 1022 (1967).
[244] O. Yonemitsu, B. Witkop, and I. L. Karle, *J. Am. Chem. Soc.*, **89**, 1039 (1967).
[245] W. G. Bentrude, *Chem. Commun.*, **1967**, 174.

radical and an electron.[246] Luminescence which is observed on warming was attributed to the sequence:

$$Ar\cdot + e \rightarrow (Ar^-)^* \xrightarrow{-h\nu} Ar^-$$

A number of other publications have dealt with aspects of chemiluminescence.[247]

Energy distribution in the fragments from gas-phase photolysis of 2,3-diazabicyclo[2.2.1]hept-2-ene has been studied,[248] as have the properties of the excited states of this and related azo-compounds.[249] Solvent effects on the absorption and fluorescence spectra of aromatic diazines show that the $n\pi^*$-singlet is not a hydrogen-bond acceptor.[250] Isotope effects have been detected in the fluorescence spectra of compounds (e.g., β-naphthol) with a proton-donor group, and this has been associated with an isotope effect on the rate of proton transfer during the lifetime of the excited state involved.[251]

Mention may also be made of a further extension of the technique of phosphorescence-excitation spectroscopy;[252] of the gas-phase photolyses of methyl formate[253] and nitromethane;[254] of the flash photolyses of methyl iodide (involving reactions of excited iodine atoms)[255] and of ethylene;[256] of the vacuum-ultraviolet photolyses of propane[257] and cyclopentane;[258] and of new data on the mercury-sensitized trimerization of acetylene;[259] also of the quenching of excited mercury atoms by gaseous paraffin hydro-

246 H. Hogeveen and H. R. Gersmann, *Rec. Trav. Chim.*, **85**, 1230 (1966).

247 D. L. Maricle and A. Maurer, *J. Am. Chem. Soc.*, **89**, 188 (1967); A. Zweig, A. K. Hoffmann, D. L. Maricle, and A. H. Maurer, *Chem. Commun.*, **1967**, 106; S. F. Mason and D. R. Roberts, *ibid.*, p. 476; E. H. White, M. M. Bursey, D. F. Roswell, and J. H. M. Hill, *J. Org. Chem.*, **32**, 1198 (1967); Y. Omote, T. Miyake, S. Ohmori, and N. Sugiyama, *Bull. Chem. Soc. Japan*, **40**, 899 (1967); Y. Omote, S. Ohmori, N. Sugiyama, *ibid.*, p. 1693; E. H. White and D. F. Roswell, *J. Am. Chem. Soc.*, **89**, 3944 (1967); R. F. Vassilev, *Progr. Reaction Kinetics*, **4**, 305 (1967); N. Sugiyama and M. Akutagawa, *Bull. Chem. Soc. Japan*, **40**, 240 (1967); A. Zweig, G. Metzler, A. Maurer, and B. G. Roberts, *J. Am. Chem. Soc.*, **89**, 4091 (1967); A. A. Vichutinskii, A. F. Guk, V. F. Tsepalov, and V. Y. Shlyapintokh, *Izv. Akad. Nauk SSSR, Ser. Khim.*, **1966**, 1672; A. Weller and K. Zachariasse, *J. Chem. Phys.*, **46**, 4984 (1967); A. Zweig, A. H. Maurer, and B. G. Roberts, *J. Org. Chem.*, **32**, 1322 (1967); E. H. White and K. Matsuo, *J. Org. Chem.*, **32**, 1921 (1967).

248 T. F. Thomas, C. I. Sutin, and C. Steel, *J. Am. Chem. Soc.*, **89**, 5107 (1967).

249 S. D. Andrews and A. C. Day, *Chem. Commun.*, **1967**, 477; C. Steel and T. F. Thomas, *ibid.*, **1966**, 900.

250 H. Baba, L. Goodman, and P. C. Valenti, *J. Am. Chem. Soc.*, **88**, 5410 (1966).

251 L. Stryer, *J. Am. Chem. Soc.*, **88**, 5708 (1966).

252 A. P. Marchetti and D. R. Kearns, *J. Am. Chem. Soc.*, **89**, 768 (1967); see also *Organic Reaction Mechanisms*, **1966**, 369.

253 M. J. Yee Quee and J. C. J. Thynne, *Trans. Faraday Soc.*, **63**, 1656 (1967).

254 I. M. Napier and R. G. W. Norrish, *Proc. Roy. Soc.* (*London*), *A*, **299**, 317, 337 (1967).

255 R. T. Meyer, *J. Chem. Phys.*, **46**, 4146 (1967).

256 R. A. Back and D. W. L. Griffiths, *J. Chem. Phys.*, **46**, 4839 (1967).

257 R. E. Rebbert and P. Ausloos, *J. Chem. Phys.*, **46**, 4333 (1967).

258 R. D. Doepker, S. G. Lias, and P. Ausloos, *J. Chem. Phys.*, **46**, 4340 (1967).

259 D. J. LeRoy, *J. Chem. Phys.*, **45**, 3482 (1966); S. Shida, M. Tsukada, and T. Oka, *ibid.*, p. 3483.

carbons,[260] and of the photolysis of methylacetylene;[261] of the photoreduction of a conjugated diacetylene to an enyne in pentane solution;[262] and new aspects of the photochemistry of diphenylacetylene.[263]

260 K. Yang, *J. Am. Chem. Soc.*, **89**, 5344 (1967).
261 A. Galli, P. Harteck, and R. R. Reeves, *J. Phys. Chem.*, **71**, 2719 (1967).
262 D. A. Ben-Efraim, *Tetrahedron Letters*, **1967**, 957.
263 R. C. Henson, J. L. W. Jones, and E. D. Owen, *J. Chem. Soc., A*, **1967**, 116; R. C. Henson, and E. D. Owen, *Chem. Commun.*, **1967**, 153.

CHAPTER 14

Oxidations and Reductions

Ozonolysis

There have been several further investigations of the steric and electronic effects of substituents on a double bond on the course of ozonolysis reactions. Details[1] of Murray, Youssefyeh, and Story's work on the influence of olefin stereochemistry and steric factors on the ozonide *cis*:*trans* ratio, and of the mechanistic scheme required to explain them[2] have appeared. Further evidence for the intermediacy of a Criegee zwitterion (carbonyl oxide, $R_2\overset{+}{C}\text{—}O\text{—}\overset{-}{O} \leftrightarrow R_2C\text{=}\overset{+}{O}\text{—}\overset{-}{O}$) in the ozonolysis of tetramethylethylene is provided by the isolation of hydroxyacetone formed, it is believed, from the zwitterion (**1**) through the vinylic hydroperoxide (**2**), and by the isolation of the peroxy hydroperoxide (**3**) formed by 1,3-dipolar addition of (**1**) to the double bond of (**2**) (see p. 422).[3]

A systematic study of the competition between complete cleavage (ozonolysis) and partial cleavage (to epoxides and their rearrangement products) in the treatment of 1,1-disubstituted ethylenes with ozone shows that as the size of the substituents increases the ratio of partial to complete cleavage also increases. This is explained on the basis of the mechanism shown. The initial π-complex (**4**) collapses by a 1,3-dipolar cycloaddition to the primary ozonide (**5**); but if this process is sterically hindered the π-complex collapses to the σ-complex (**6**) which loses oxygen to give the epoxide and other partial cleavage products.[4] The mechanism of Murray *et al.*[1] was thought to be less important here than when there are bulky substituents on both olefinic carbons.[4]

The proportions of the two modes of cleavage of the primary ozonides (**7**) from styrenes have been measured; as would be expected, stabilization of positive charge on the benzylic carbon favours that in which this carbon becomes part of the zwitterion.[5] However, the direction of opening of the primary ozonides from propenylbenzene and 2-methylpropenylbenzene in methanol (which traps the zwitterion as isolable α-methoxy hydroperoxide)

[1] R. W. Murray, R. D. Youssefyeh, and P. R. Story, *J. Am. Chem. Soc.*, **89**, 2429 (1967).
[2] *Organic Reaction Mechanisms*, **1966**, 399—401.
[3] P. R. Story and J. R. Burgess, *J. Am. Chem. Soc.*, **89**, 5726 (1967).
[4] P. S. Bailey and A. G. Lane, *J. Am. Chem. Soc.*, **89**, 4473 (1967).
[5] S. Fliszár and J. Renard, *Can. J. Chem.*, **45**, 533 (1967); S. Fliszár, *Tetrahedron Letters*, **1966**, 6083; see also S. D. Razumovsky and Y. N. Yuriev, *ibid.*, **1967**, 3939.

was the opposite to that expected on this basis since the formation of benzaldehyde and the aliphatic zwitterion was greatly favoured. It was suggested that the direction of cleavage is controlled by polarization due to the combined inductive effects of the phenyl and the methyl group.[6] The products of ozonolysis of indene in ethanol show that the primary ozonide cleaves about equally to give

$$o\text{-OCH}\cdot C_6H_4\cdot CH_2\overset{+}{C}H\text{—O—}\overset{-}{O} \quad \text{and} \quad o\text{-OCH}\cdot CH_2\cdot C_6H_4\cdot\overset{+}{C}H\text{—O—}\overset{-}{O}.^7$$

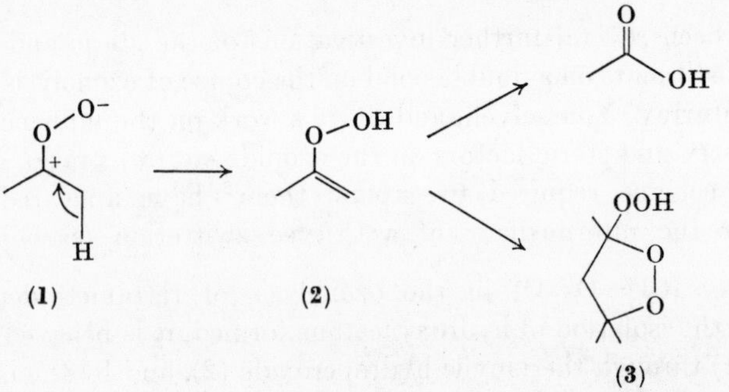

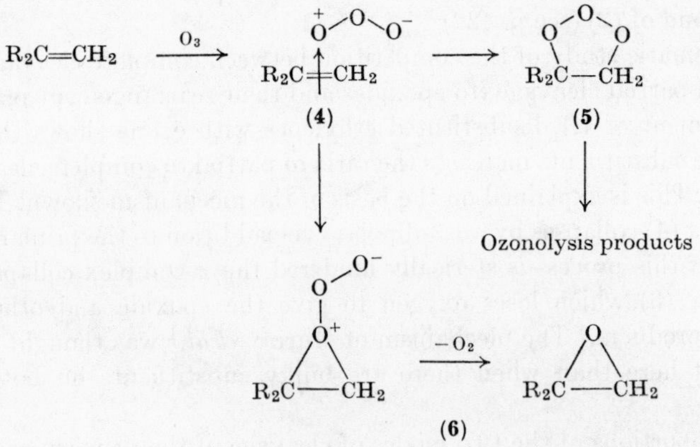

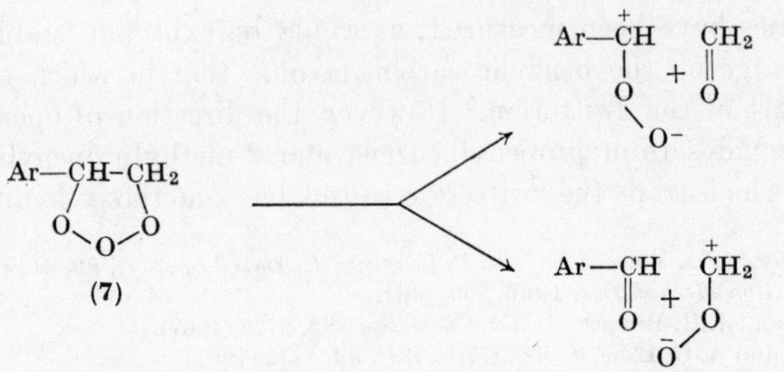

[6] W. P. Keaveney, M. G. Berger, and J. J. Pappas, *J. Org. Chem.*, **32**, 1537 (1967).
[7] S. Fliszár, C. Belżecki, and J. B. Chylińska, *Can. J. Chem.*, **45**, 221 (1967).

The beneficial effect of pyridine in producing carbonyl compounds, rather than ozonides and peroxides, in ozonolyses has been ascribed to reduction of the zwitterion with the formation of pyridine *N*-oxide. This is now rendered untenable by the demonstration that only a little pyridine is consumed in the reaction, pyridine *N*-oxide is not formed, and the mixture remains peroxidic. A new explanation of the role of pyridine is required.[8]

The formation and structure of primary ozonides and their conversion into the rearranged ozonides have been observed directly by PMR spectroscopy.[9] Some decomposition of ozonides during gas–liquid chromatography, a technique used recently for their quantitative analysis, has been demonstrated.[10]

Other reactions studied include: the ozonolysis of benzene and its methyl derivatives[11] and the solvolysis of various ozonides[12] in formic and acetic acids, the formation and structure of the oligomer formed in ozonolysis of the but-2-enes, pent-2-enes, and hex-3-enes,[13] the ozonolysis of 1,4-diphenylbut-2-ene[14] and of 1-alkylcycloalkenes,[15] and the selective reduction of ozonides by boranes.[16]

Ozonolysis reactions[17] and cyclic peroxides, including ozonides,[18] have been reviewed.

Oxidations by Metallic Ions

Results for the chromic acid oxidation of propan-2-ol and its tri- and hexafluoro-derivatives in fairly concentrated sulphuric acid fit the Westheimer mechanism of rapid, pre-equilibrium formation of the chromate ester followed by its slow cleavage at the $C_{(\alpha)}$–H bond. However, in more concentrated acid, formation of the ester appears to become rate-determining since the primary deuterium isotope effect decreases gradually from 10.5 in 57.6% sulphuric acid to 1.3 in 95.3% sulphuric acid.[19] Further support for the Westheimer mechanism comes from the spectrophotometric observation of a series of consecutive reactions and the ESR spectrum of Cr(v) in the chromic acid oxidation of propan-2-ol.[20] The chromic acid oxidation of allyl alcohols has been shown to proceed by the same mechanism; however, equatorial alcohols are oxidized faster than axial alcohols and this is explained by the better

[8] K. Griesbaum, *Chem. Commun.*, **1966**, 920.
[9] L. J. Durham and F. L. Greenwood, *Chem. Commun.*, **1967**, 843.
[10] H. Rubinstein, *J. Org. Chem.*, **32**, 3236 (1967).
[11] E. Bernatek, E. Karlsen, and T. Ledaal, *Acta Chem. Scand.*, **21**, 1229 (1967).
[12] E. Bernatek, H. Hagen, and T. Ledaal, *Acta Chem. Scand.*, **21**, 1555 (1967).
[13] F. L. Greenwood and H. Rubinstein, *J. Org. Chem.*, **32**, 3369 (1967).
[14] S. Fliszár and J. B. Chylińska, *Can. J. Chem.*, **45**, 29 (1967).
[15] D. G. M. Diaper, *Can. J. Chem.*, **44**, 2819 (1966).
[16] D. G. M. Diaper and W. M. J. Strachan, *Can. J. Chem.*, **45**, 33 (1967).
[17] R. W. Murray, *Trans. N.Y. Acad. Sci.*, **29**, 854 (1967).
[18] M. Schulz and K. Kirschke, *Adv. Heterocyclic Chem.*, **8**, 191 (1967).
[19] D. G. Lee and R. Stewart, *J. Org. Chem.*, **32**, 2868 (1967).
[20] K. B. Wiberg and H. Schafer, *J. Am. Chem. Soc.*, **89**, 455 (1967).

overlap of the electrons from the departing axial hydrogen with the allylic double bond. The enhanced oxidation rate of these unsaturated alcohols indicates some contribution from the α,β-unsaturated ketone delocalization in the transition state.[21] The lack of reactivity of the tricyclic alcohol (**8**) towards chromic acid oxidation was thought to be due to steric strain in the formation of the ketone. However, this is unlikely since it has now been shown

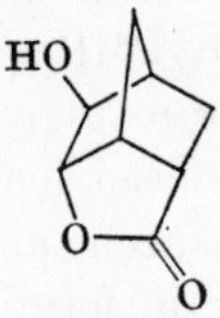

(8)

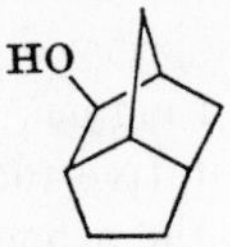

(9)

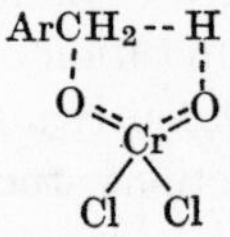

(10)

that the structurally similar 2-*exo*-brendanol (**9**) is oxidized at about the same rate as norbornan-*exo*-2-ol, and the rate retardation in the lactone (**8**) was ascribed to dipolar repulsion between the carbonyl functions in the transition state.[22] The effects of *meta*- and *para*-substituents on the rate of oxidation of toluenes with chromyl chloride in carbon disulphide are consistent with the cyclic transition state (**10**).[23a] The reactions of chromyl chloride with phenol and derivatives were also studied.[23b] Other chromate oxidations investigated include those of ethanol,[24] cyclic alcohols,[25] *cis*-caranols,[26] *endo*-5,6-trimethylenenorbornyl alcohols,[27] and secondary-tertiary vicinal glycols.[28]

In the oxidation of alkenes by aqueous potassium permanganate the initial reaction was of the first order in alkene and in permanganate, but independent of base concentration, although the products varied with pH. Substituted cinnamic acids all reacted at essentially the same rate as cinnamic acid; in this the permanganate oxidation showed strong similarities with 1,3-dipolar cycloadditions.[29] The oxidation of formate and deuterioformate ions,[30] and of alcohols in the presence of boric acid,[31] by potassium permanganate have been investigated.

From a kinetic study of the oxidation of *p*-methoxytoluene to *p*-methoxy-

[21] S. H. Burstein and H. J. Ringold, *J. Am. Chem. Soc.*, **89**, 4722 (1967).
[22] A. K. Awasthy, J. Roček, and R. M. Moriarty, *J. Am. Chem. Soc.*, **89**, 5400 (1967).
[23a] H. C. Duffin and R. B. Tucker, *Tetrahedron*, **23**, 2803 (1967).
[23b] J. A. Strickson and C. A. Brooks, *Tetrahedron*, **23**, 2817 (1967).
[24] S. Bretsznajder and R. Marcinkowski, *Bull. Acad. Polon. Sci., Ser. Sci. Chim.*, **14**, 865 (1966).
[25] G. Srinivasan and N. Venkatasubramanian, *Proc. Indian Acad. Sci., Sect. A*, **65**, 30 (1967).
[26] W. Cocker, A. C. Pratt, and P. V. R. Shannon, *Tetrahedron Letters*, **1967**, 3919.
[27] I. Rothberg and R. V. Russo, *J. Org. Chem.*, **32**, 2003 (1967).
[28] B. H. Walker, *J. Org. Chem.*, **32**, 1098 (1967).
[29] K. B. Wiberg and R. D. Geer, *J. Am. Chem. Soc.*, **88**, 5827 (1966).
[30] R. P. Bell and D. P. Onwood, *J. Chem. Soc., B*, **1967**, 150.
[31] B. V. Tronov, M. A. Sherova, and N. G. Babynina, *Uch. Zap. Khim. Fak. Kirg. Univ. Sb. Statei Molodykh Uch.*, **1965**, 34; *Chem. Abs.*, **66**, 64833u (1967).

benzyl acetate with manganic acetate in acetic acid an electron-transfer mechanism was proposed (see also p. 276). The initial reversible electron-transfer giving a radical ion is followed by slow proton loss to the benzyl radical which is rapidly oxidized to the carbonium ion:

$$ArCH_3 + Mn(OAc)_3 \rightleftharpoons [ArCH_3^+, AcO^-] + Mn(OAc)_2$$

$$[ArCH_3^+, AcO^-] \xrightarrow{\text{slow}} ArCH_2\cdot + AcOH$$

$$ArCH_2\cdot + Mn(\text{III}) \xrightarrow{\text{fast}} ArCH_2^+ + Mn(\text{II})$$

$$ArCH_2^+ + AcOH \xrightarrow{\text{fast}} ArCH_2OAc + H^+$$

Since the rate is insensitive to added sodium acetate, initial reaction between the substrate and undissociated manganic acetate to give an intimate ion-pair was proposed; proton-transfer then occurs within this ion-pair.[32] A similar mechanism probably holds for the same oxidation of various other alkoxy and dialkylamino aromatic compounds.[33] The oxidation of toluenes by Ce(IV)[34] and by V(V)[35] has also been reported.

The kinetics of the oxidation of benzyl ethers, di-isopropyl and di-(2-chloroethyl) ethers by cobaltic perchlorate in aqueous methyl cyanide have been fully investigated and explained on the basis of initial α-hydrogen abstraction. Somewhat surprisingly diphenyl ether is oxidized faster than benzyl ethers, presumably by π-electron abstraction; the rapid oxidation of biphenyl and polycyclic hydrocarbons is explained similarly.[36]

Other reactions investigated include the oxidation by Ce(IV) of ethylene glycol,[37] tartaric acid,[38] and cyclic alcohols,[39] by V(V) of glycols[40] and carbonyl compounds (induced by light),[41] and by alkaline ferricyanide of disaccharides.[42] Several reports on the oxidation of olefins and alcohols by palladium salts have appeared.[43] Rhodium(III) chloro-complexes catalyse the dehydrogenation of propan-2-ol to acetone.[44]

32 P. J. Andrulis, M. J. S. Dewar, R. Dietz, and R. L. Hunt, *J. Am. Chem. Soc.*, **88**, 5473 (1966).
33 T. Aratani and M. J. S. Dewar, *J. Am. Chem. Soc.*, **88**, 5479 (1966); P. J. Andrulis and M. J. S. Dewar, *ibid.*, p. 5483.
34 P. S. R. Murti and S. C. Pati, *Chem. Ind.* (*London*), **1967**, 702.
35 P. S. R. Murti and S. C. Pati, *Chem. Ind.* (*London*), **1966**, 1722.
36 T. A. Cooper and W. A. Waters, *J. Chem. Soc., B*, **1967**, 455, 464, 687.
37 P. G. Sant, V. M. Bhale, and W. V. Bhagwat, *Indian J. Chem.*, **4**, 469 (1966).
38 S. M. Ali and A. Aziz, *Pakistan J. Sci. Ind. Res.*, **9**, 113 (1966).
39 H. L. Hintz and D. C. Johnson, *J. Org. Chem.*, **32**, 556 (1967).
40 S. Senent and M. G. Mayo, *Anales Real Soc. Espan. Fis. Quim., Ser. B*, **62**, 1301 (1966); *Chem. Abs.*, **67**, 53310r (1967).
41 K. S. Panwar and J. N. Guar, *Talanta*, **14**, 127 (1967).
42 R. K. Srivastava, N. Nath, and M. P. Singh, *Tetrahedron*, **23**, 1189 (1967).
43 H. Okada and H. Hashimoto, *Kogyo Kagaku Zasshi*, **69**, 2137 (1966); *Chem. Abs.*, **66**, 85242r (1967); I. I. Moiseev, A. P. Belov, V. A. Igoshin, and Y. K. Syrkin, *Dokl. Akad. Nauk SSSR*, **173**, 863 (1967); *Chem. Abs.*, **67**, 53332z (1967); C. F. Kohll and R. van Helden, *Rec. Trav. Chim.*, **86**, 193 (1967); P. M. Henry, *J. Org. Chem.*, **32**, 2575 (1967); W. G. Lloyd, *ibid.*, p. 2816; A. Aguiló, *Adv. Organometal. Chem.*, **5**, 321 (1967).
44 H. B. Charman, *J. Chem. Soc., B*, **1967**, 629.

Other Oxidations

Details[45] of Torssell's work[46] on the mechanism of dimethyl sulphoxide (DMSO) oxidation of halides (Kornblum reaction) and of alcohols (Moffatt reaction) have appeared. The former reaction involves the alkoxydimethylsulphonium salt which decomposes either through the ylide and a cyclic mechanism,[46] or by a 1,2-elimination if the α-proton is sufficiently activated. The latter reaction requires a more complex mechanism, but that proposed by Torssell is impossible as written,[45, 46] since the carbonyl-oxygen atom is required to expand its octet; when rewritten, this again appears to involve the dimethylsulphonium salt. Further evidence for the alkoxydimethylsulphonium ion intermediate in the oxidation of cholestanol to cholestanone by DMSO with dicyclohexylcarbodi-imide, with acetic anhydride, or with collidine has been provided by tritium-labelling[47] although there are indications that free alkoxydimethylsulphonium ions are not intermediates in some sterol oxidations.[48] Details of the oxidation of alcohols by DMSO and acetic anhydride have appeared; the probable reaction mechanism is nucleophilic attack by the alcohol on the acetoxysulphonium ions.[49] Other oxidations reported include those of isocyanides,[50] of 2,6-disubstituted phenols,[51] of thio- and seleno-phosphoric acids,[52] and of alkyl benzyl sulphides to benzaldehydes by DMSO and benzoyl chloride.[53]

A reaction constant ($\rho = -2.29$ at 25°) has been measured for the first time for a hydride-transfer oxidation, that of (eleven) benzyl alcohols to benzaldehydes by bromine in aqueous acetic acid. The oxidation is general base-catalysed and benzyl alcohol is oxidized 4.2 times faster than its α,α-dideuterio-derivative at 25°. These results agree with simultaneous removal of the hydroxylic proton by base and an α-hydrogen as hydride.[54] Aqueous bromine is effective for the oxidative cleavage of aliphatic ethers at 25°; primary alkyl groups are converted into carboxylic acids and secondary alkyl groups into ketones. The mechanism proposed for this oxidation and for that of alcohols by bromine was, somewhat surprisingly, the simultaneous loss of the α-hydrogen as a proton (*not* as hydride) and an electron pair from the ether oxygen.[55] The large accelerating effects of alkyl substituents in the oxidation

45 K. Torssell, *Acta Chem. Scand.*, **21**, 1 (1967).
46 *Organic Reaction Mechanisms*, **1966**, 406.
47 F. W. Sweat and W. W. Epstein, *J. Org. Chem.*, **32**, 835 (1967).
48 S. M. Ifzal and D. A. Wilson, *Tetrahedron Letters*, **1967**, 1577.
49 J. D. Albright and L. Goldman, *J. Am. Chem. Soc.*, **89**, 2416 (1967).
50 D. Martin and A. Weise, *Angew. Chem. Intern. Ed. Engl.*, **6**, 168 (1967); see also D. Martin, A. Weise, and H.-J. Niclas, *ibid.*, p. 318.
51 M. G. Burdon and J. G. Moffatt, *J. Am. Chem. Soc.*, **89**, 4725 (1967).
52 M. Mikolajczyk, *Chem. Ind.* (*London*), **1966**, 2059.
53 R. Oda and Y. Hayashi, *Tetrahedron Letters*, **1967**, 3141.
54 P. Aukett and I. R. L. Barker, *Chem. Ind.* (*London*), **1967**, 193.
55 N. C. Deno and N. H. Potter, *J. Am. Chem. Soc.*, **89**, 3550, 3555 (1967).

of alcohols by bromine, permanaganate, and Hg(II), and the ready oxidative fission of di-isopropyl ether support a hydride-transfer mechanism (eqn. 1). In fact, bromine and Hg(II) oxidize an isopropyl ether faster than a primary alcohol, as was strikingly demonstrated by showing that hexane-1,6-diol was

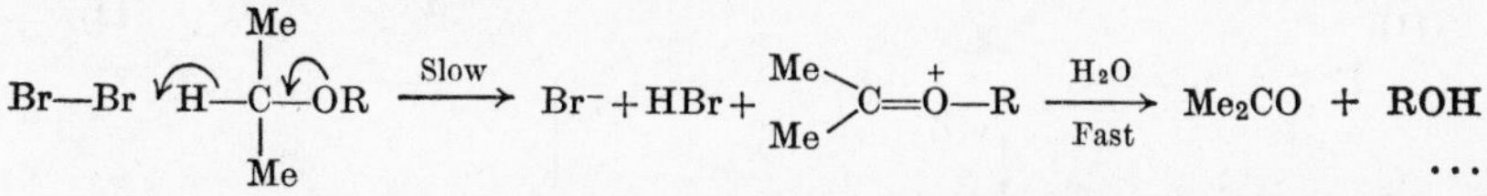

$$\text{Br—Br} \quad \overset{\text{Me}}{\underset{\text{Me}}{\text{H—C—OR}}} \quad \xrightarrow{\text{Slow}} \quad \text{Br}^- + \text{HBr} + \overset{\text{Me}}{\underset{\text{Me}}{>}}\text{C}{=}\overset{+}{\text{O}}\text{—R} \quad \xrightarrow[\text{Fast}]{\text{H}_2\text{O}} \quad \text{Me}_2\text{CO} + \text{ROH} \qquad \dots (1)$$

the main product of oxidation of its monoisopropyl ether.[56] The effect of structural variation on the oxidation of alcohols by bromine in aqueous acetic acid,[57] and the participation of the imidoyl phosphate intermediate in the oxidation of 2-methylnaphthalene-1,4-diol diphosphate by bromine in dimethylformamide,[58] have been described.

The oxidation of propan-2-ol by *N*-bromosuccinimide (NBS) in aqueous acetic acid has been shown to consist of two separate reactions, both of the first order in each reactant, a relatively slow reaction for about the first 20% followed by a faster reaction. Onset of the latter roughly coincides with the solution's becoming yellow, owing to the liberation of bromine formed from NBS and bromide ions. The rate of the second part of the reaction is the same as that obtained with bromine as oxidant; this reaction only is observed if potassium bromide is added initially to the NBS reaction. In the presence of mercuric acetate, which removes free bromide ions, the formation of molecular bromine is prevented and the second, faster reaction is not observed. Oxidation by NBS, in the presence of mercuric acetate, is between 70 and 100 times slower than by bromine, possibly because of a less favourable cyclic transition state with the former. It was also noticed that the bromine reaction supervened later in the reaction as the proportion of acetic acid in the solvent increased, as expected, since the concentration of bromide ions will decrease with the dielectric constant.[59] The oxidation of tribenzylamine by NBS has also been studied.[60]

In contrast with the iodoform reaction, the mechanism for enolizable non-methyl ketones has not been much studied. Freiberg[61] has now shown that, in the oxidation of 3β-hydroxyandrost-5-en-17-one (partial structure 11) with iodine in methanolic methoxide, oxygen is involved in the novel mechanism given. Treatment of (11) with 1 and 2 equivalents of iodine under nitrogen gave the mono- and di-iodo (12) derivatives which were fairly stable. However,

[56] R. M. Barter and J. S. Littler, *J. Chem. Soc., B*, **1967**, 205.

[57] V. Thiagarajan and N. Venkatasubramanian, *Current Sci. (India)*, **36**, 10 (1967); *Chem. Abs.*, 64837y (1967).

[58] J. S. Cohen and A. Lapidot, *J. Chem. Soc., C*, **1967**, 1210.

[59] N. Venkatasubramanian and V. Thiagarajan, *Tetrahedron Letters*, **1967**, 3349.

[60] M. Zador, *Can. J. Chem.*, **44**, 2031 (1966).

[61] L. A. Freiberg, *J. Am. Chem. Soc.*, **89**, 5297 (1967).

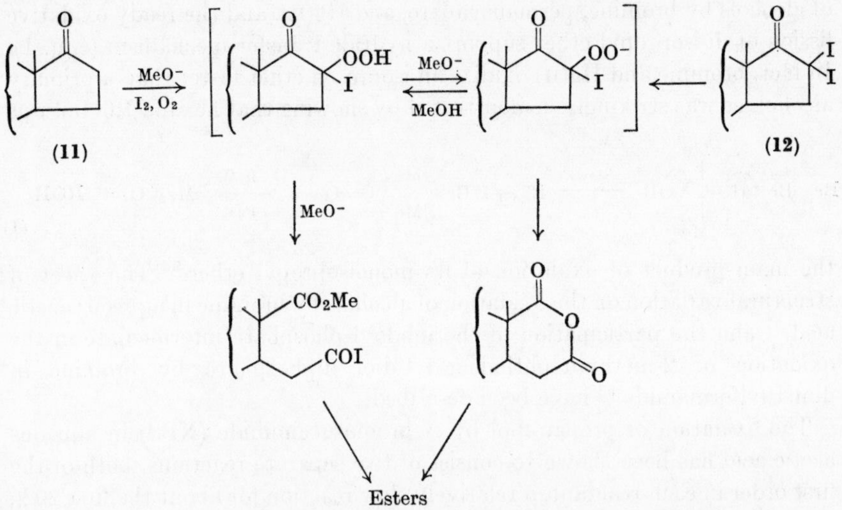

(11) (12)

$$\left.\begin{array}{l} R_2S + I_2 \rightleftharpoons R_2\overset{+}{S}I + I^- \\ R_2\overset{+}{S}I + OH^- \xrightarrow{\text{Slow}} R_2\overset{+}{S}{-}O^- + H^+ + I^- \end{array}\right\} \quad \ldots (2)$$

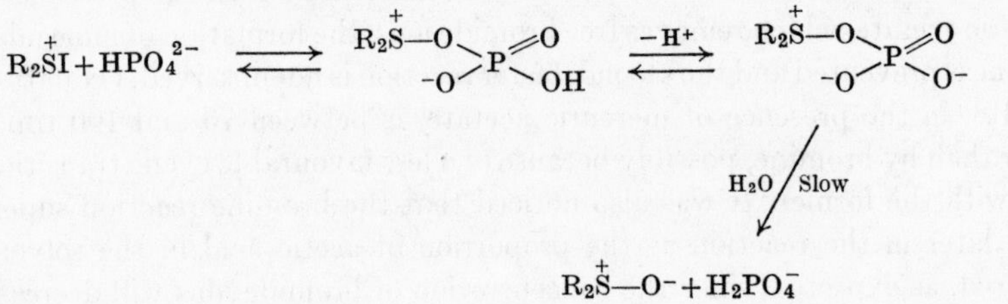

. . . (3)

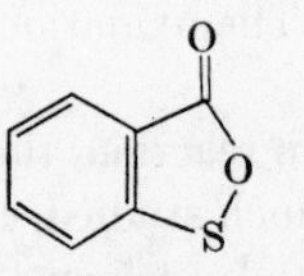

(13)

rapid and complete oxidation of (**11**) occurred in an air-saturated solution with only 1.25 equivalents of iodine, and (**12**) was rapidly oxidized on introduction of air, 1 equivalent of oxygen being consumed per mole of (**12**). The anhydride could be isolated from methyl cyanide solution. Partitioning between the two routes depended only on the concentration of base employed.[61]

The rate of oxidation of tetrahydrothiophen with iodine in aqueous solution is a complex function of the concentration of the sulphide, iodine, iodide, and any nucleophile present. The rate is greatly increased by nucleophiles, e.g., 10^4-fold by $0.1\text{M-}HPO_4^{2-}$. On the basis of the rate law the mechanism of equation (2) was proposed for the reaction in the absence, and of equation (3) in the presence, of phosphate.[62] As reported last year[63] this oxidation is also strongly catalysed by phthalate buffer; phthalic anhydride was formed during the reaction and an intermediate incorporating phthalate ions was invoked. Further evidence for incorporation of the catalyst has now been provided since when benzyl methyl sulphide is oxidized by iodine in (+)-2-methyl-2-phenylsuccinate buffer the benzyl methyl sulphoxide formed is optically active, being 6.4% richer in one enantiomer.[64] Thiols are oxidized to disulphides by aqueous potassium tri-iodide unless there is a β-carboxyl group present; in the latter case further oxidation occurs (to sulphinic and sulphonic acids) also, without going through the disulphide. *o*-Mercaptobenzoic acid, for example, gave *o*-sulphobenzoic acid. It was suggested that a sulphenyl iodide was formed first in all cases and was attacked either intermolecularly by thiol to give disulphide, or intramolecularly by carboxyl to give a cyclic intermediate (e.g., **13**) which was hydrolysed and oxidized further.[65]

The mechanism of periodate oxidation of organic compounds has been discussed.[66] Details have been published[67] of Weidman and Kaiser's measurements of the periodate oxidation of catechol[68] which have been extended to cover pH 0—10. An intermediate, formed in a second-order reaction and decomposed to products in a first-order reaction, was clearly involved, but the obvious cyclic structures (no intermediate was formed in the similar oxidation of quinol) previously considered are in doubt since the intermediate has a higher extinction coefficient than the product, *o*-benzoquinone, at 390 mμ.[67] The periodate oxidation of polycyclic hydrocarbons,[69] 2-aminoethanols,[70] glyoxal, pyruvaldehyde, and biacetyl,[71] and the overoxidation of carbohydrates[72] have also been studied.

A comparison has been made of the cyclizations which result when *ortho*-substituted anilines are oxidized with phenyl iodosoacetate and when the corresponding *ortho*-substituted aryl azides are pyrolysed, and neighbouring-group participation mechanisms are proposed.[73] The oxidative dealkylation

[62] T. Higuchi and K.-H. Gensch, *J. Am. Chem. Soc.*, **88**, 5486 (1966).
[63] *Organic Reaction Mechanisms*, **1966**, 411.
[64] T. Higuchi, I. H. Pitman, and K.-H. Gensch, *J. Am. Chem. Soc.*, **88**, 5676 (1966).
[65] J. P. Danehy and M. Y. Oester, *J. Org. Chem.*, **32**, 1491 (1967).
[66] B. Sklarz, *Quart. Rev. (London)*, **21**, 3 (1967).
[67] S. W. Weidman and E. T. Kaiser, *J. Am. Chem. Soc.*, **88**, 5820 (1966).
[68] *Organic Reaction Mechanisms*, **1965**, 306.
[69] A. J. Fatiadi, *Chem. Commun.*, **1967**, 1087.
[70] G. Dahlgren and E. M. Rand, *J. Phys. Chem.*, **71**, 1955 (1967).
[71] G. Dahlgren and K. L. Reed, *J. Am. Chem. Soc.*, **89**, 1380 (1967).
[72] B. G. Hudson and R. Barker, *J. Org. Chem.*, **32**, 2101 (1967).
[73] L. K. Dyall and J. E. Kemp, *Australian J. Chem.*, **20**, 1625 (1967).

of aliphatic amines with chlorine dioxide,[74] and the vanadium-catalysed oxidation of *p*-phenetidine with chlorate ions,[75] have been reported.

The peroxidation of various olefins, amines, sulphur and phosphorus compounds, ketones and other organic compounds,[76] the reaction of nitrogen-containing compounds with molecular oxygen[77] as well as metal-catalysed autoxidation,[78] have been reviewed. It has been shown that in the liquid-phase autoxidation of hydrocarbons, primary, secondary, and tertiary peroxy-radicals are not about equally reactive, as often assumed, but that reactivity decreases in the order: primary > secondary > tertiary, roughly in the order 5:2:1, for attack of the same hydrocarbon. The differences are probably steric in origin. It was also shown that the relative reactivities of hydrocarbons are virtually independent of the nature of the attacking peroxy-radical.[79] The absolute rate constant for the combination of two hydroperoxy-radicals in non-polar solvents (the main chain-terminating process in the autoxidation of olefins) is very much greater than previously reported values for aqueous media. In the autoxidation of a large number of hydrocarbons at 30° it was found that in general the rate constants for hydrogen-abstraction increase in the order primary < secondary < tertiary; and for compounds losing a secondary hydrogen the rate constants increase in the order unactivated < acyclic activated by one π-system < cyclic activated by one π-system < acyclic activated by two π-systems < cyclic activated by two π-systems.[80] The rate of oxidation of triphenylmethane by oxygen in DMSO–*tert*-butyl alcohol containing potassium *tert*-butoxide is equal to the rate of ionization to the carbanion, which is of the first order each in hydrocarbon and in base. The subsequent reaction between the carbanion and oxygen (which is faster than between the triphenylmethyl radical and oxygen) is of a free-radical or electron-transfer type.[81] The same basic mechanism applies in the similar oxidation of *p*-nitrotoluene and its derivatives. Substituents that decrease the rate of ionization decrease the rate of oxidation; substituents such as α-cyano or 2,6-dinitro stabilize the *p*-nitrobenzyl anions so much that they are oxidized very slowly.[82]

The oxidation of α,β- and β,γ-unsaturated aldehydes and ketones by

[74] D. H. Rosenblatt, L. A. Hull, D. C. De Luca, G. T. Davis, R. C. Weglein, and H. K. R. Williams, *J. Am. Chem. Soc.*, **89**, 1158 (1967); L. A. Hull, G. T. Davis, D. H. Rosenblatt, H. K. R. Williams, and R. C. Weglein, *ibid.*, p. 1163.
[75] P. R. Bonschev and B. G. Ieliazokova, *Mikrochim. Acta*, **1967**, 125.
[76] J. B. Lee and B. C. Uff, *Quart. Rev.* (*London*), **21**, 429 (1967).
[77] E. Höft and H. Schultze, *Z. Chem.*, **7**, 137 (1967).
[78] S. Fallab, *Angew. Chem. Intern. Ed. Engl.*, **6**, 496 (1967).
[79] B. S. Middleton and K. U. Ingold, *Can. J. Chem.*, **45**, 191 (1967).
[80] J. A. Howard and K. U. Ingold, *Can. J. Chem.*, **45**, 785, 793 (1967).
[81] G. A. Russell and A. G. Bemis, *J. Am. Chem. Soc.*, **88**, 5491 (1966).
[82] G. A. Russell, A. J. Moye, E. G. Janzen, S. Mak, and E. R. Talaty, *J. Org. Chem.*, **32**, 137 (1967).

oxygen in methanol containing a base and a cupric–pyridine complex as catalyst involves rate-determining proton-abstraction to give the dienolate anion, oxidation of this to a dienyloxy-radical, and oxygenation of the radical to a peroxy-radical which is reduced by the cuprous complex to the hydroperoxide anion. In the absence of oxygen the dienyloxy-radical dimerizes.[83] The rates of oxidation of ascorbic acid by oxygen in the absence and in the presence of cupric and ferric ions have been measured. In absence of a catalyst molecular oxygen reacts directly with ascorbate anion; for the catalysed reactions an intermediate with molecular oxygen as a ligand in a metal–ascorbate complex was proposed. Cu(II) was more effective than Fe(III) for oxidation of the anion, whilst the reverse held for oxidation of the neutral molecule, and this was attributed to the relative tendencies of the two ions to form complexes with the two chelating species.[84] In their autoxidation, 1-alkylpyrroles react with oxygen in a free-radical peroxy addition process typical of conjugated dienes.[85]

Oxidation of optically active α-methylbenzylboronic acid (**14**) in benzene with oxygen gave racemic product (**15**) and the reaction is inhibited by radical scavengers. It was concluded, therefore, that the autoxidation is a radical chain process and not the polar cyclic process often assumed.[86] It has

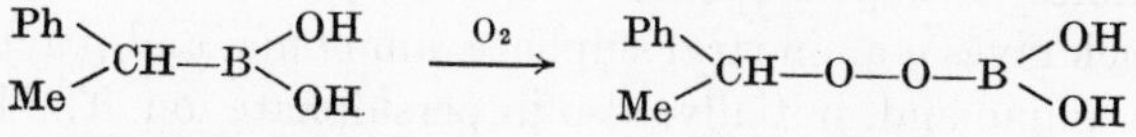

(**14**) (**15**)

also been shown that air-oxidation of trialkylboranes involves an intermolecular reaction of an alkylboron compound with an alkylperoxyboron compound, rather than the previously assumed intramolecular reaction.[87]

Many other autoxidation reactions have been studied, including those of cumene,[88] alkylchlorobenzenes,[89] dihydroxybenzenes,[90] 3,6-dimethylbenzene-1,2,4-triol,[91] xanthene,[92] acetophenones,[93] cyclohexene in the presence of benzaldehyde,[94] di-isopinocampheylbutylboranes,[95] *N*-alkylamides,[96] 5-mer-

[83] H. C. Volger and W. Brackman, *Rec. Trav. Chim.*, **85**, 817 (1966).
[84] M. M. T. Khan and A. E. Martell, *J. Am. Chem. Soc.*, **89**, 4176 (1967).
[85] E. B. Smith and H. B. Jensen, *J. Org. Chem.*, **32**, 3330 (1967).
[86] A. G. Davies and B. P. Roberts, *J. Chem. Soc., B*, **1967**, 17.
[87] S. B. Mirviss, *J. Org. Chem.*, **32**, 1713 (1967).
[88] D. G. Hendry, *J. Am. Chem. Soc.*, **89**, 5433 (1967).
[89] Z. N. Moleva and V. V. Voronenkov, *Zh. Org. Khim.*, **3**, 78 (1967); *Chem. Abs.*, **66**, 94507h (1967).
[90] H. Musso and H. Döpp, *Chem. Ber.*, **100**, 3627 (1967).
[91] J. F. Corbett, *J. Chem. Soc., C*, **1967**, 611.
[92] A. P. Ter Borg, H. R. Gersmann, and A. F. Bickel, *Rec. Trav. Chim.*, **85**, 899 (1966).
[93] H. J. Den Hertog and E. C. Kooijman, *J. Catal.*, **6**, 347, 357 (1966).
[94] T. Ikawa, T. Fukushima, M. Muto, and T. Yanagihara, *Can. J. Chem.*, **44**, 1817 (1966).
[95] P. G. Allies and P. B. Brindley, *Chem. Ind. (London)*, **1967**, 319.
[96] B. F. Sagar, *J. Chem. Soc., B*, **1967**, 428, 1047.

capto-uracil and -deoxyuridine,[97] the triene side chain in ebelin lactone,[98] and the cobalt-catalysed autoxidation of propene and acetaldehyde,[99] cyclohexene,[100] toluene,[101] and mesitylene.[102]

Other oxidations with oxygen that have been discussed include the competitive oxidation of acetylene and methylacetylene,[103] the oxidation of 1-naphthylamine catalysed by copper stearate,[104] the autoxidation of cumene in the presence of substituted copper phthalocyanines and related complexes,[105] the inhibition by nitroxides and hydroxylamines of the autoxidation of styrene,[106] the photosensitized oxidation of hydroxylated purines,[107] the photochemical oxidation of benzyl alcohols to benzaldehydes by oxygen and DMSO,[108] and a mathematical model for the liquid-phase oxidation of cyclohexanol.[109]

The alkaline hydrogen peroxide oxidation of benzyl methyl ketones was rationalized on the basis of rate-determining nucleophilic addition of hydroperoxide anion to the enol double bond;[110] the ferric ion-catalysed oxidation of ethanol by hydrogen peroxide has also been studied.[111] The rearranged products from the oxidation of 1,2,3,4-tetramethylnaphthalene with trifluoroperoxyacetic acid and boron trifluoride were all explained by initial electrophilic hydroxylation of the α- and the β-positions.[112] The reaction between aromatic amines and persulphate ions in aqueous base (Boyland–Sims reaction) which gives *o*-aminoaryl sulphate, ammonia, and polymer, is of the first order in amine and, initially, also in persulphate ion. The lack of effect of radical traps, the effect of substituents on the rate, and the Arrhenius parameters accord with nucleophilic displacement by the nitrogen on peroxide oxygen to give an arylhydroxylamine *O*-sulphonate, which rearranges to the *o*-sulphate or is oxidized by persulphate to ammonia or polymer.[113] The

97 T. I. Kalman and T. J. Bardos, *J. Am. Chem. Soc.*, **89**, 1171 (1967).
98 R. A. Eade, J. Ellis, J. J. H. Simes, and J. S. Shannon, *Chem. Commun.*, **1967**, 60.
99 E. A. Blyumberg and T. V. Filippova, *Neftekhimiya*, **6**, 863 (1966); *Chem. Abs.*, **66**, 94503d (1967).
100 M. Prévost-Gangneux, G. Clément, and J. C. Balaceanu, *Bull. Soc. Chim. France*, **1966**, 2085, 2905.
101 T. Morimoto and Y. Ogata, *J. Chem. Soc., B*, **1967**, 62.
102 S. A. Samodumov and K. I. Matkovskii, *Katal. Katal. Akad. Nauk Ukr. SSR, Respub. Mezhvedom. Sb.* No. 2, 71 (1966); *Chem. Abs.*, **66**, 64823r (1967).
103 J. M. Hay and D. Lyon, *J. Chem. Soc., B*, **1967**, 970.
104 L. N. Denisova and E. T. Denisov, *Izv. Akad. Nauk SSSR, Ser. Khim.*, **1966**, 2220; *Chem. Abs.*, **66**, 85240p (1967).
105 H. Kropf and H. Hoffmann, *Tetrahedron Letters*, **1967**, 659.
106 I. T. Brownlie and K. U. Ingold, *Can. J. Chem.*, **45**, 2427 (1967).
107 T. Matsuura and I. Saito, *Chem. Commun.*, **1967**, 693.
108 T. Sato, H. Inoue, and K. Hata, *Bull. Chem. Soc. Japan*, **40**, 1502 (1967).
109 V. V. Karitonov, *Zh. Fiz. Khim.*, **40**, 2699 (1966); *Chem. Abs.*, **66**, 85235r (1967).
110 D. D. Jones and D. C. Johnson, *J. Org. Chem.*, **32**, 1403 (1967).
111 C. Heitler, D. B. Scaife, and B. W. Thompson, *J. Chem. Soc., A*, **1967**, 1409.
112 H. Hart and R. K. Murray, *J. Org. Chem.*, **32**, 2448 (1967).
113 E. J. Behrman, *J. Am. Chem. Soc.*, **89**, 2424 (1967).

oxidation of primary alcohols by peroxydisulphate ions has been investigated.[114]

There have been several further investigations of lead tetra-acetate (LTA) oxidations this year. Oxidation of the olefins (**16**; Ar = phenyl and *p*-methoxyphenyl) with LTA gave the ketones (**17**). This double rearrangement, demonstrated for the first time, provides good support for the Criegee mechanism shown (Scheme 1). A carbonium ion is formed by initial electrophilic addition of LTA to the double bond and rearranges; the carbon–lead bond of the

$$
\underset{(\mathbf{16})}{\overset{\overset{\displaystyle Ar}{|}}{Ph_2C-CH=CH_2}} \xrightarrow{Pb(OAc)_4} \underset{}{\overset{\overset{\displaystyle Ar}{|}}{Ph_2C-\overset{+}{C}H-CH_2Pb(OAc)_3}} \longrightarrow \underset{}{\overset{\overset{\displaystyle Ar}{|}}{Ph_2\overset{+}{C}-CH-CH_2Pb(OAc)_3}}
$$

$$
\underset{(\mathbf{17})}{\overset{\overset{\displaystyle Ph}{|}}{PhCOCHCH_2Ar}} \longleftarrow \underset{\overset{\displaystyle |}{Ph}}{\overset{\overset{\displaystyle AcO\ \ Ar}{|\ \ \ |}}{Ph-C-CH-CH_2-Pb(OAc)_2}} \overset{}{\underset{OAc}{}}
$$

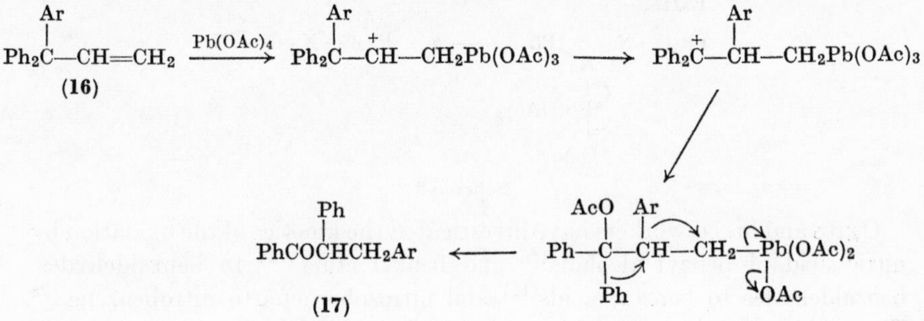

Scheme 1

intermediate is cleaved heterolytically to give another carbonium ion which again rearranges. Prévost oxidation (iodine and silver acetate) followed the same course [Scheme 1; I^+ for $Pb^+(OAc)_3$].[115] The complex oxidation of styrene by LTA has been carefully investigated and the various products have been shown to arise by heterolytic or radical-chain reactions or by both.[116] From the preliminary results of LTA oxidation of propanol and butanol in various solvents it appears that exchange of acetate for alcohol ligand is not rate-determining, that the lability of LTA towards alcohols in co-ordinating solvents is determined by the strength of co-ordination of solvent to Pb(IV), that the conversion of butanol into butyraldehyde is not base-catalysed, and that separation of the α-hydrogen is involved in the rate-determining step.[117] Benzophenone phenylhydrazone is converted by LTA in alcohols into a mixture of the azoacetate and, concurrently, an azoether as shown (Scheme 2). The effect of substituents in both rings suggests that the rate-determining step is nucleophilic displacement of an acetate on lead by the saturated nitrogen, followed by (probably intramolecular) attack by

114 L. R. Subbaraman and M. Santappa, *Proc. Indian Acad. Sci., Sect. A*, **64**, 345 (1966); *Chem. Abs.*, **66**, 115100n (1967).
115 R. O. C. Norman and C. B. Thomas, *J. Chem. Soc., B*, **1967**, 604.
116 R. O. C. Norman and C. B. Thomas, *J. Chem. Soc., B*, **1967**, 771.
117 R. Partch and J. Monthony, *Tetrahedron Letters*, **1967**, 4427.

acetate or by the solvent on the ketonic carbon.[118] The oxidation by LTA of the following have also been studied: norcamphene and 2-methylnorbornene,[119] longifolols,[120] thiocarbonates,[121] primary and secondary amines,[122] oximes,[123] azines,[124a] and benzoylhydrazones.[124b]

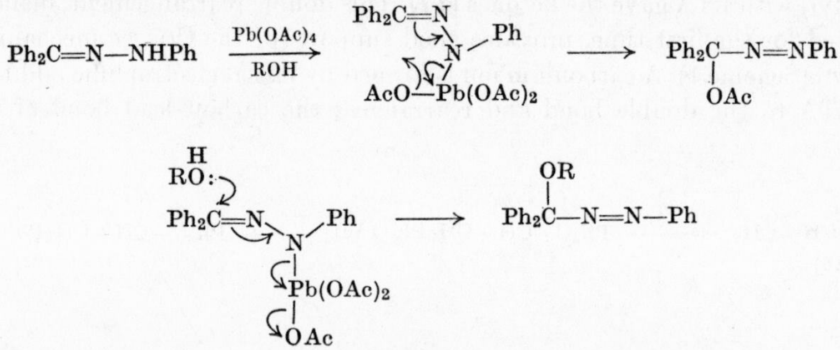

Scheme 2

Ogata and his co-workers have investigated the kinetics of the oxidation by nitric acid of benzyl alcohols[125] and benzyl ethers[126] to benzaldehydes, benzaldehydes to benzoic acids,[127] and nitrosobenzene to nitrobenzene.[128] The oxidation of chloroacetaldehyde with nitric acid has also been studied.[129]

The mechanisms of several enzymic oxidations have been investigated further.[130]

The radiolytic oxidation of purines[131] and some electrochemical oxidations[132] have been reported.

118 M. J. Harrison, R. O. C. Norman, and W. A. F. Gladstone, *J. Chem. Soc., C*, **1967**, 735.
119 W. F. Erman, *J. Org. Chem.*, **32**, 765 (1967).
120 J. Lhomme and G. Ourisson, *Chem. Commun.*, **1967**, 436.
121 T. J. Adley, A. K. M. Anisuzzaman, and L. N. Owen, *J. Chem. Soc., C*, **1967**, 807.
122 A. Stojiljković, A. Andrejević, and M. L. Mihailović, *Tetrahedron*, **23**, 721 (1967).
123 H. Kropf and R. Lambeck, *Ann. Chem.*, **700**, 1, 18 (1966).
124a B. T. Gillis and M. P. LaMontagne, *J. Org. Chem.*, **32**, 3318 (1967).
124b R. W. Hoffmann and H. J. Luthardt, *Tetrahedron Letters*, **1967**, 3501.
125 Y. Ogata, Y. Sawaki, F. Matsunaga, and H. Tezuka, *Tetrahedron*, **22**, 2655 (1966).
126 Y. Ogata and Y. Sawaki, *J. Am. Chem. Soc.*, **88**, 5832 (1966).
127 Y. Ogata and H. Tezuka, *J. Am. Chem. Soc.*, **89**, 5428 (1967).
128 Y. Ogata, H. Tezuka, and Y. Sawaki, *Tetrahedron*, **23**. 1007 (1967).
129 B. G. Yasnitskii and A. P. Zaitsev, *Zh. Prikl. Khim.*, **40**, 694 (1967).
130 B. C. Saunders and J. Wodak, *Tetrahedron*, **23**, 473 (1967); B. C. Saunders and B. P. Stark, *ibid.*, p. 1867; K. Bailey, B. R. Brown, and B. Chalmers, *Chem. Commun.*, **1967**, 618; K. Bailey and B. R. Brown, *ibid.*, p. 408; K. Brocklehurst and K. Williamson, *Biochem. Biophys. Res. Commun.*, **26**, 175 (1967); T. C. Bruice, B. Holmquist, and T. P. Stein, *J. Am. Chem. Soc.*, **89**, 4221 (1967); F. S. Brown and L. P. Hager, *ibid.*, p. 719; O. Gawron, A. J. Glaid, K. P. Mahajan, G. Kananen, and M. Limetti, *Biochem. Biophys. Res. Commun.*, **25**, 518 (1966).
131 J. Holian and W. M. Garrison, *Chem. Commun.*, **1967**, 676.
132 W. J. Koehl, *J. Org. Chem.*, **32**, 614 (1967); J.-P. Billon, G. Cauquis, J. Raison, and Y. Thibaud, *Bull. Soc. Chim. France*, **1967**, 199; B. Mooney and H. I. Stonehill, *J. Chem. Soc., A*, **1967**, 1.

The elimination mechanism of the dehydrogenation of acenaphthene by quinone closely resembles, in reverse, that of the addition of hydrogen halides. The large isotope effect, the lack of 1,2-shifts, and the predominantly *cis*-stereochemistry are explained by the formation, and partial collapse, of a classical carbonium ion pair by removal of hydride hydrogen. The amount of *cis*-elimination decreases as the solvent polarity increases (see p. 123).[133]

Other topics studied include the kinetics of quinone–hydroquinone redox reaction,[134] oxidative decarboxylation of acids by pyridine *N*-oxide,[135] oxidative phosphorylation,[136] oxidation of thiols,[137] and of catecholamines,[138] intramolecular oxidation by the nitro-group,[139] the mechanism of the Guerbet reaction (oxidation–reduction of alcohols),[140] the methanolysis of the intermediate phosphonium ions in the oxidation of trivalent phosphorus compounds by disulphides,[141] and a new solvent system for photo-oxidations.[142]

Oxidation–reduction in non-aqueous solvents[143] and heterogeneous oxidations[144] have been reviewed.

Reductions

There have been few detailed investigations of reduction mechanisms this year though many reductions, and especially their stereochemistry, have been rationalized mechanistically.

para-Substituted azobenzene-*p*-sulphonates are reduced in aqueous solution by a soluble quinol (9,10-dihydroxyanthracene-2-sulphonate, H_2Q) to the anilines when the substituents are electron-releasing, and to the hydrazobenzene when they are electron-withdrawing. Variation of the observed second-order rate coefficients with acidity in the region $H_0 = 0.0$ to pH $= 7.5$ indicates a rate law, $k_2 = k_0 + k_H a_{H^+} + k_B/a_{H^+}$. The acid-catalysed reaction is between the monoprotonated azo-compound and H_2Q, the pH-independent reaction is between the unprotonated azo-compound and H_2Q, and the base-catalysed reaction is between unprotonated azo-compound and HQ^-. A general acid-catalysed process is observed with electron-withdrawing substituted compounds and its extent increases with the electron-

133 B. M. Trost, *J. Am. Chem. Soc.*, **89**, 1847 (1967).
134 S. Carter, J. N. Murrell, E. J. Rosch, N. Trinajstić, and P. A. H. Wyatt, *J. Chem. Soc., B*, **1967**, 477.
135 T. Cohen, I. H. Song, J. H. Fager, and G. L. Deets, *J. Am. Chem. Soc.*, **89**, 4968 (1967).
136 C. D. Snyder and H. Rapoport, *J. Am. Chem. Soc.*, **89**, 1269 (1967).
137 T. J. Wallace and A. Schriesheim, *J. Appl. Chem.*, **17**, 48 (1967).
138 M. D. Hawley, S. V. Tatawawadi, S. Piekarski, and R. N. Adams, *J. Am. Chem. Soc.*, **89**, 447 (1967).
139 A. E. Luetzow and J. R. Vercellotti, *J. Chem. Soc., C*, **1967**, 1750.
140 S. Veibel and J. I. Nielsen, *Tetrahedron*, **23**, 1723 (1967).
141 R. S. Davidson, *J. Chem. Soc., C*, **1967**, 2131.
142 E. J. Forbes and J. Griffiths, *Chem. Commun.*, **1967**, 427.
143 B. Kratochvil, *Record Chem. Progr.*, **27**, 253 (1966).
144 W. F. Pickering, *Rev. Pure Appl. Chem.*, **16**, 185 (1966).

withdrawing power of the substituent. The constants k_0, k_H, and k_B do not vary greatly with structure and give non-linear free-energy relationships. A kinetic isotope effect, $k^H{}_0/k^D{}_0 = 2.7$, was demonstrated for the pH-independent reduction of one substrate. These and related facts suggest that the more electrophilic azo-compounds are reduced by electron-transfer from H_2Q with concerted proton-transfer from external acid; and the less electrophilic, more basic, azo-compounds are reduced through a 4-centre transition state in which quinol hydrogen is transferred to the basic β-nitrogen simultaneously with electron-transfer to the α-nitrogen (e.g., **18**).[145]

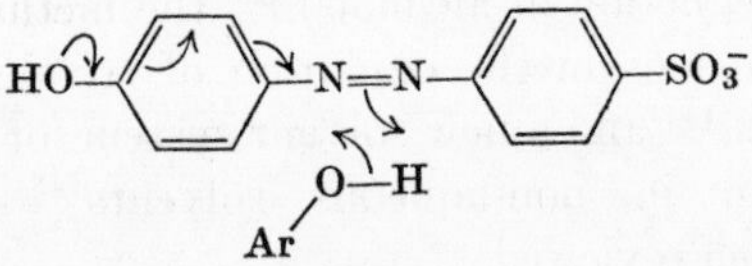

(18)

The reduction of alkoxysulphonium salts to the parent sulphides with sodium borohydride appears to proceed by nucleophilic displacement of methoxide by hydride on sulphur.[146] Reduction of aliphatic ketones with an ethanol-modified lithium aluminium hydride–sugar complex yields alcohols of up to 70% optical purity having the *R*-configuration.[147] Stereoselectivity resulting from neighbouring-group participation has been demonstrated in the reduction of a sugar with lithium aluminium hydride,[148] and of propiophenone derivatives with sodium borohydride.[149] Other complex metal hydride reductions reported include those of cyclohexenones,[150] thiazolium salts,[151] cyclic immonium salts,[152] δ-enol-lactones,[153] skeletal rearrangements during reduction of bicyclo[2.2.2]octenyl toluene-*p*-sulphonates,[154] carbon–carbon bond cleavage in the reduction of an isoquinolinium salt,[155] and asymmetric induction in the reduction of substituted acetophenonechromium tricarbonyls.[156] Stereoselective reduction of ketones with aluminium

145 R. L. Reeves and R. W. Andrus, *J. Am. Chem. Soc.*, **89**, 1715 (1967).
146 C. R. Johnson and W. G. Phillips, *J. Org. Chem.*, **32**, 3233 (1967).
147 S. R. Landor, B. J. Miller, and A. R. Tatchell, *J. Chem. Soc., C*, **1967**, 197; O. Červinka and A. Fábryová, *Tetrahedron Letters*, **1967**, 1179.
148 E. J. Hedgley, O. Mérész, and W. G. Overend, *J. Chem. Soc., C*, **1967**, 888.
149 S.-I. Yamada and K. Koga, *Tetrahedron Letters*, **1967**, 1711.
150 S. R. Landor and J. P. Regan, *J. Chem. Soc., C*, **1967**, 1159.
151 G. M. Clarke and P. Sykes, *J. Chem. Soc., C*, **1967**, 1411.
152 E. Toromanoff, *Bull. Soc. Chim. France*, **1966**, 3357.
153 J. Martin, W. Parker, B. Shroot, and T. Stewart, *J. Chem. Soc., C*, **1967**, 101.
154 R. A. Appleton, J. C. Fairlie, and R. McCrindle, *Chem. Commun.*, **1967**, 690.
155 J. L. Neumeyer, M. McCarthy, and K. K. Weinhardt, *Tetrahedron Letters*, **1967**, 1095.
156 J. Tirouflet and J. Besancon, *Tetrahedron Letters*, **1967**, 4221.

hydride[157] and the hydrogenolysis of cyclopropyl ketones with diborane and boron trifluoride[158] have also been reported.

Alkali-metal reductions in ammonia or amines have been further investigated.[159]

Electrochemical reductions of α-keto-sulphonium and -ammonium salts, α-formyl quaternary ammonium salts, and other α-substituted aldehydes,[160] of tris-(*p*-nitrophenyl) phosphate,[161] of nitroalkanes,[162] and of porphyrins and chlorins[163] have been reported. Baizer and his co-workers have continued their study of electrolytic reductive coupling reactions.[164]

Boiling ethylene glycol containing potassium hydroxide reduces ketones to the secondary alcohol, sometimes very effectively; equilibrium transfer of a hydride ion from ethylene glycol monoanion to the ketone was proposed.[165] The stereochemistry of the reduction of cyclohexane-1,2-dione by aluminium isopropoxide has been described.[166] Reduction of sulphoxides by iodide ions in acid solution is facilitated by a neighbouring carboxylic acid group; initial cyclodehydration between these groups in the conjugated acid was proposed.[167]

Other investigations were into the stereochemistry of reduction of 3-methyl- and 3-phenyl-1-tetralone with various reagents,[168] the stereospecificity in reductions (and oxidations) involving $C_{(17)}$ of lupanine,[169] dibenzenechromium as a catalyst in the reduction of alkyl halides to hydrocarbons,[170] the reductive cyclization of keto-esters,[171] chromous ion-reduction of prop-2-yn-1-ol,[172]

157 D. C. Ayres and R. Sawdaye, *J. Chem. Soc., B*, **1967**, 581.

158 E. Breuer, *Tetrahedron Letters*, **1967**, 1849.

159 W. Hückel, *Fortschr. Chem. Forsch.*, **6**, 197 (1966); S. K. Pradhan, G. Subrahmanyam, and H. J. Ringold, *J. Org. Chem.*, **32**, 3004 (1967); L. H. Slaugh and J. H. Raley, *ibid.*, p. 2861; U. R. Ghatak, J. Chakravarty, A. K. Banerjee, and N. R. Chatterjee, *Chem. Commun.*, **1967**, 217; P. Markov, D. Lasarov, and C. Ivanov, *Ann. Chem.*, **704**, 126 (1967).

160 J.-M. Savéant, *Bull. Soc. Chim. France*, **1967**, 471, 481, 486, 493.

161 K. S. V. Santhanam, L. O. Wheeler, and A. J. Bard, *J. Am. Chem. Soc.*, **89**, 3386 (1967).

162 P. E. Iversen and H. Lund, *Tetrahedron Letters*, **1967**, 4027.

163 H. H. Inhoffen, P. Jäger, R. Mählhop, and C.-D. Mengler, *Ann. Chem.*, **704**, 188 (1967).

164 J. H. Wagenknecht and M. M. Baizer, *J. Org. Chem.*, **31**, 3885 (1966); J. D. Anderson, M. M. Baizer, and J. P. Petrovich, *ibid.*, p. 3890; J. P. Petrovich, J. D. Anderson, and M. M. Baizer, *ibid.*, p. 3897.

165 D. C. Kleinfelter, *J. Org. Chem.*, **32**, 841 (1967).

166 C. H. Snyder, *J. Org. Chem.*, **31**, 4220 (1966).

167 S. Allenmark, *Arkiv Kemi*, **26**, 37 (1967); see also S. Allenmark and H. Johnsson, *Acta Chem. Scand.*, **21**, 1672 (1967).

168 K. Hanaya, *Nippon Kagaku Zasshi*, **87**, 991, 995 (1966); *Chem. Abs.*, **65**, 18473, 18474 (1966).

169 M. Wiewiorowski, O. E. Edwards, and M. D. Bratek-Wiewiorowska, *Can. J. Chem.*, **45**, 1447 (1967).

170 D. D. Mozzhukhin, B. G. Gribov, G. A. Tychin, A. S. Strizhkova, and M. L. Khidekel, *Izv. Akad. Nauk SSSR, Ser. Khim.*, **1967**, 175; *Chem. Abs.*, **66**, 115105t (1967).

171 C. D. Gutsche, I. Y. C. Tao, and J. Kozma, *J. Org. Chem.*, **32**, 1782 (1967).

172 O. N. Efimov, A. Y. Gerchikov, and A. E. Shilov, *Teor. i Eksperim. Khim. Akad. Nauk Ukr. SSR*, **2**, 424 (1966); *Chem. Abs.*, **65**, 18451 (1966).

the reduction of azo- and azoxy-benzene with sodium sulphides,[173] the reduction of benzil with hydrogen sulphide,[174] and the highly specific reduction of certain reactive bromides when heated with benzoin.[175]

Hydrogenations

Di-imide reduction of several 7-substituted norbornadienes gave, unexpectedly, the *anti*-7-substituted norbornenes, rather than the *syn*-isomers, and dideuteriodi-imide gave the *anti*-norbornene with both deuteriums *exo* (eqn. 4). The 99:1 predominance of hydrogenation on the *syn*-side indicates that the transition state for this is favoured by about 4.5 kcal mole^{-1}. Stabilization of the *exo-syn*-transition state by electron-donation from the 7-substituent to the partially positive di-imide provides a possible explanation.[176] *cis*-9,10-Dihydronaphthalene hydrogenates cyclohexenes, on heating, stereospecifically *cis*; a cyclic mechanism similar to that for di-imide seems probable.[177]

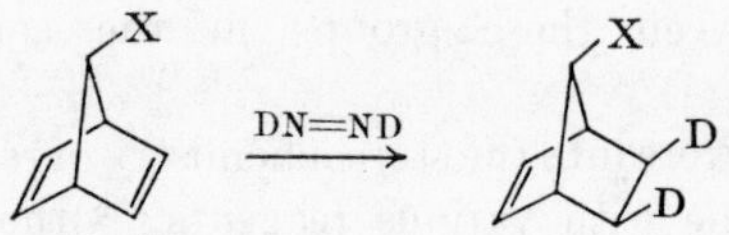

. . . (4)

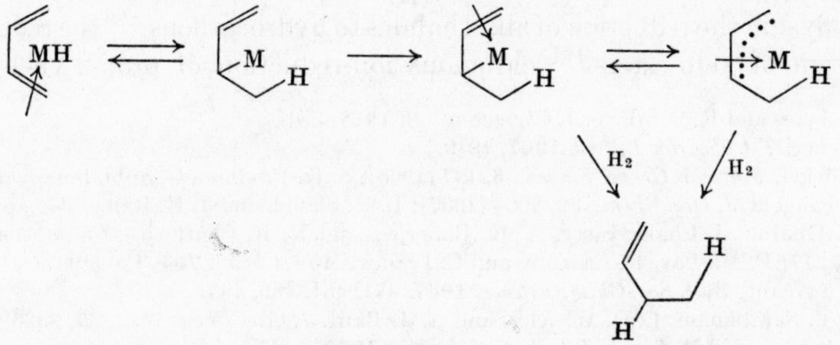

Scheme 3

Homogeneous catalytic hydrogenation has been studied further. Complexes of the type, $MX_2(QPh_n)_2$, where M = Pt or Pd, X = halide, and Q = P or As ($n = 3$) and S or Se ($n = 2$), catalyse the hydrogenation of polyolefins, in the presence of stannous chloride, to monoenes. The double bonds first migrate into conjugation and it is the conjugated diene that is specifically hydrogenated. From a detailed study of the effect of variation in olefin,

[173] S. Hashimoto and J. Sunamoto, *Yuki Gosei Kagaku Kyokai Shi*, **24**, 1225, 1231 (1966); *Chem. Abs.*, **66**, 64835w, 64836x (1967).
[174] M. Scheithauer and R. Mayer, *Chem. Ber.*, **100**, 1402 (1967).
[175] M. Michman, Y. Halpern, and S. Patai, *J. Chem. Soc., B*, **1967**, 93.
[176] W. C. Baird, B. Franzus, and J. H. Surridge, *J. Am. Chem. Soc.*, **89**, 140 (1967).
[177] W. von E. Doering and J. W. Rosenthal, *J. Am. Chem. Soc.*, **89**, 4534 (1967).

solvent, and catalyst and from isolation of products and intermediates, the π- $\rightarrow$ σ-complex mechanism (Scheme 3) was proposed (only the essential hydride bond in the M–Sn complex is shown).[178] Pent-1-ene is also isomerized in the presence of hydrogen and a Pt(II)–Sn(II) complex at room temperature, to give pent-2-enes with the *trans*-isomer predominating.[179]

Chlorohydridotris(triphenylphosphine)ruthenium(II) is a highly selective catalyst for the homogeneous hydrogenation of terminal olefins; since it catalyses hydrogen exchange with internal, as well as terminal, olefins the lack of hydrogenation of the former was attributed to steric hindrance in the hydrogenolysis of the Ru–C bond in the alkyl intermediate.[180] In the homogeneous deuteration of ethylene catalysed by *trans*-chlorocarbonylbis-(triphenylphosphine)iridium(I), extensive H–D exchange was observed in the product and in both recovered reactants; this suggests that both reactants are activated reversibly on the same metal atom at which the hydrogen transfer to the double bond occurs.[181] Hydrogenation catalysed by chlorotris(triphenylphosphine)rhodium is not affected by organic sulphides; it is retarded but not inhibited by thiophenol.[182] The homogeneous hydrogenation of butadiene catalysed by a cyanocobaltate complex in aqueous solution,[183] and of olefins in the presence of triethylaluminium and the acetylacetonates of chromium, iron, cobalt, and nickel,[184] and the effect of pH on the rate of hydrogenation in buffer solutions[185] have been investigated.

During the deuteration of cyclohexene over platinum or palladium catalysts hydrogen is exchanged for deuterium not only at the allylic position, readily explained by greater release of *peri*-strain on hydrogenation than when for this a new, non-classical surface intermediate (or transition state) in which an intramolecular hydrogen shift can occur (eqn. 5) was proposed.[186] In the catalytic hydrogenation of 1-alkyl- and 1,4-dialkyl-naphthalenes, methyl groups are rate-retarding, as expected, but *tert*-butyl groups are rate-enhancing and cause the substituted ring to be hydrogenated preferentially. This was

178 H. A. Tayim and J. C. Bailar, *J. Am. Chem. Soc.*, **89**, 4330 (1967); J. C. Bailar and H. Itatani, *ibid.*, p. 1592; H. Itatani and J. C. Bailar, *ibid.*, p. 1600; E. N. Frankel, E. A. Emken, H. Itatani, and J. C. Bailar, *J. Org. Chem.*, **32**, 1447 (1967); H. Itatani and J. C. Bailar, *J. Amer. Oil Chemists' Soc.*, **44**, 147 (1967); *Chem. Abs.*, **67**, 2663a (1967).

179 G. C. Bond and M. Hellier, *J. Catal.*, **7**, 217 (1967).

180 P. S. Hallman, D. Evans, J. A. Osborn, and G. Wilkinson, *Chem. Commun.*, **1967**, 305.

181 G. G. Eberhardt and L. Vaska, *J. Catal.*, **8**, 183 (1967).

182 A. J. Birch and K. A. M. Walker, *Tetrahedron Letters*, **1967**, 1935.

183 T. Suzuki and T. Kwan, *Nippon Kagaku Zasshi*, **87**, 926 (1966); *Chem. Abs.*, **66**, 64817s (1967).

184 I. V. Kalechits, V. G. Lipovich, and F. K. Schmidt, *Neftekhimiya*, **6**, 813 (1966); *Chem. Abs.*, **66**, 94632v (1967).

185 A. M. Sokol'skaya, S. M. Reshetnikov, K. K. Kuzembaev, S. A. Ryabinina, and E. N. Bakhanova, *Tr. Inst. Khim. Nauk, Akad. Nauk Kaz. SSR*, **14**, 57 (1966); *Chem. Abs.*, **66**, 85212f (1967).

186 G. V. Smith and J. R. Swoap, *J. Org. Chem.*, **31**, 3904 (1966).

explained by greater release of *peri*-strain on hydrogenation than when the substituted ring becomes saturated.[187] The hydrogenation of 2-cyclopentylidenecyclopentanol (**19**) and 2-isopropylidenecyclopentanol over various metal catalysts was highly stereoselective, giving the *trans*-products; the allylic alcohol is presumably adsorbed preferentially with its hydroxyl group directed towards the catalyst surface from which the hydrogen is delivered.[188]

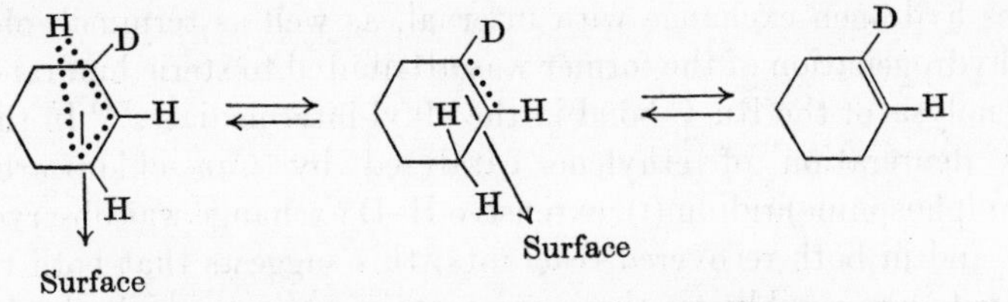

. . . (5)

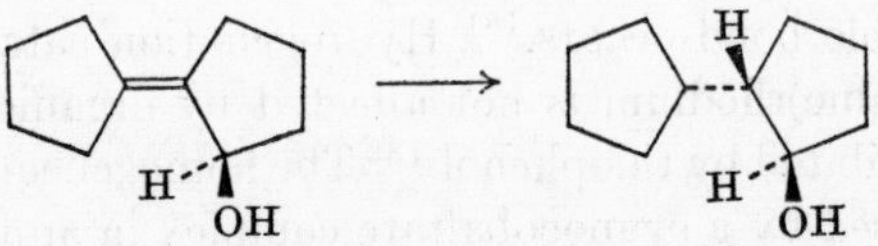

(**19**)

The rate-determining step in the hydrogenation of propene on nickel, palladium, and platinum appears to be addition of hydrogen to the half-hydrogenated species, C_3H_7.[189] The hydrogenation of ethylene over copper,[190] copper–nickel and nickel–gold alloys,[191] and over a platinum–tin chloride catalyst[192] has been studied. Other reports include the hydrogenation of hex-1-ene and cyclohexene,[193] and of methylacetylene,[194] the stereochemistry of the hydrogenation of allenes,[195] new platinum-metal catalysts,[196] modification of nickel catalysts by cadmium,[197] the hydrogenation of cyclohexene on iron films,[198] of acetophenone on nickel, platinum, and palladium,[199] of

187 H. van Bekkum, T. J. Nieuwstad, J. van Barneveld, and B. M. Wepster, *Tetrahedron Letters*, **1967**, 2269.
188 T. J. Howard and B. Morley, *Chem. Ind.* (*London*), **1967**, 73.
189 K. Hirota and Y. Hironaka, *Bull. Chem. Soc. Japan*, **39**, 2638 (1966).
190 D. T. Messervy and K. E. Hayes, *Can. J. Chem.*, **45**, 629 (1967).
191 J. S. Campbell and P. H. Emett, *J. Catal.*, **7**, 252 (1967).
192 A. P. Khrushch, L. A. Tokina, and A. E. Shilov, *Kinetika i Kataliz*, **7**, 901 (1966).
193 A. M. Sokol'skaya and S. A. Ryabinina, *Vestn. Akad. Nauk Kaz. SSR*, **22**, 52 (1967); *Chem. Abs.*, **66**, 115090j (1967).
194 R. S. Mann and K. C. Khulbe, *Can. J. Chem.*, **45**, 2755 (1967).
195 L. Crombie, P. A. Jenkins, D. A. Mitchard, and J. C. Williams, *Tetrahedron Letters*, **1967**, 4297; L. Crombie and P. A. Jenkins, *Chem. Commun.*, **1967**, 870.
196 H. C. Brown and C. A. Brown, *Tetrahedron*, Suppl. 8, Part 1, 149 (1966).
197 L. K. Freidlin, N. V. Borunova, I. E. Neimark, and L. I. Gvinter, *Izv. Akad. Nauk SSSR, Ser. Khim.*, **1966**, 1328; *Chem. Abs.*, **66**, 54783p (1967).
198 J. Erkelens, *J. Catal.*, **8**, 212 (1967).
199 N. S. Barinov, D. V. Mushenko, E. G. Lebedeva, and A. A. Balandin, *Dokl. Akad. Nauk SSSR*, **172**, 1109 (1967); *Chem. Abs.*, **66**, 115091k (1967).

α,β-unsaturated ketones,[200] of 2-acetoxy-1-tetralone and 3-acetoxy-4-chromanone,[201] and of *p*-dimethylaminobenzenes.[202]

A general, qualitative mechanism for catalytic hydrogenolysis of benzylic compounds has been proposed. Hydrogenolysis of *cis*- and *trans*-isomers of various 1-substituted 1-*tert*-butyl-4-phenylcyclohexanes can result in stereospecific retention or inversion of configuration or can give the same mixture of products. The last, stereoconvergent, hydrogenolyses proceed through the monoadsorbed substrate as a common intermediate.[203] The hydrogenolysis of the corresponding homobenzylic compounds appears to follow a fundamentally similar mechanism.[204] Hydrogenolysis of benzyl alcohol derivatives at palladium occurred with inversion, and a metal–benzyl complex intermediate was proposed.[205] Hydrogenolysis of benzhydrol to diphenylmethane, under hydroformylation conditions with octacarbonyldicobalt as catalyst proceeds by the following mechanism:

$$H_2 + Co_2(CO)_8 \rightleftarrows 2HCo(CO)_4$$

$$HCo(CO)_4 + Ph_2CHOH \rightleftarrows Ph_2CHOH_2^+\ Co(CO)_4^- \xrightarrow{\text{Slow}}$$

$$H_2O + Ph_2CHCo(CO)_4 \xrightarrow{H_2} Ph_2CH_2 + HCo(CO)_4$$

Carbonylhydridocobalt and benzhydrol give the oxonium salt which loses water in the rate-determining step, to form a complex which is hydrogenated to product.[206] Hydrogenolysis of *cis*- and *trans*-1,2-dimethylcyclobutane has also been investigated.[207]

[200] R. L. Augustine, D. Migliorini, R. Foscante, and C. Sodano, *Amer. Chem. Soc., Div. Petrol. Chem. Reprints*, **11**, A53 (1966).
[201] K. Hanaya, *Bull. Chem. Soc. Japan*, **40**, 1884 (1967).
[202] A. V. Finkel'shtein, V. L. Pogrebnaya, and V. P. Kumarev, *Zh. Fiz. Khim.*, **41**, 666 (1967); *Chem. Abs.*, **67**, 2547r (1967).
[203] E. W. Garbisch, L. Schreader, and J. J. Frankel, *J. Am. Chem. Soc.*, **89**, 4233 (1967).
[204] E. W. Garbisch, *Chem. Commun.*, **1967**, 806.
[205] A. M. Khan, F. J. McQuillin, and I. Jardine, *J. Chem. Soc., C*, **1967**, 136.
[206] Y. C. Fu, H. Greenfield, S. J. Metlin, and I. Wender, *J. Org. Chem.*, **32**, 2837 (1967).
[207] G. Maire and F.-G. Gault, *Bull. Soc. Chim. France*, **1967**, 894.

Author Index 1967

Cumulative Subject Index 1965 to 1967

Errata for Organic Reaction Mechanisms, 1965

P. 21, line 5: Volumes of activation should be -14.3, -17.7 and -17.8 cm^3 $mole^{-1}$.
P. 180: The terminal methyl group of the formula after formula (**41**) should be replaced by a methoxyl group.

Errata for Organic Reaction Mechanisms, 1966

P. 7, line 5: *Insert* with sulphuric acid *after* diol (**13**).
P. 7, 5 lines from the bottom:
For norbonyl *read* norbornyl.
P. 32: Three hydrogen atoms are missing from formula (**140**). Formulae (**141**) and (**142**) should carry positive charges.
P. 64: The formula after formula (**53**) should have an extra carbon atom.
P. 83, 2 lines from the bottom:
For k_{Br}/k_{OTs} read k_{OTs}/k_{Br}.
P. 114, line 4: *For* 6-*O*-unsubstituted *read* 6-*O*-substituted.
P. 115, line 3: *For* benzyloxy *read* benzoyloxy.
P. 137, line 7: *For* poxidation *read* Epoxidation.
P. 155: The formula between formulae (**78**) and (**79**) should be a diradical.
P. 218, line 2: *For* sopropylidene *read* isopropylidene.
P. 290: Key reactions missing from reaction scheme (8) include

$$PhCCl_2Li \xrightarrow{LiI} PhCClILi \xrightarrow{MeLi} PhCClLiMe$$

A modified mechanism is given in this volume p. 290.
P. 291, formula (**40**): Insert a lone pair of electrons on the bivalent carbon atom.
P. 320, line 2: *For* direct expulsion of ^-OH *read* direct expulsion of ^-OH.[59]
P. 374: The top right-hand formula is wrong and should be replaced by the two intermediates:

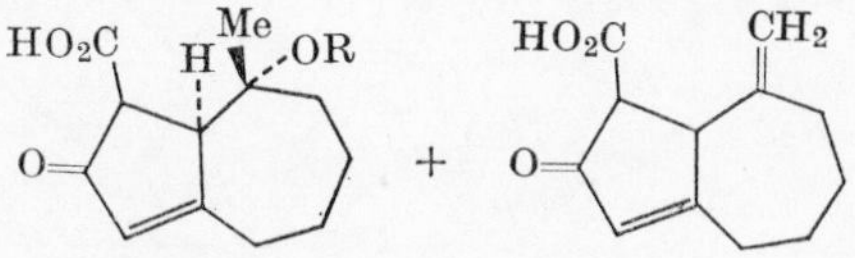

P. 423: The page reference 39 under Bell, R. P., should be 93.
P. 438: *For* Hoffman, R. W., read Hoffmann, R. W.
P. 473: *Add* **66,** 43 *to the entry* "Dicarbonium ions"
P. 477: *For* Osazene *read* Osazone.